Lecture Notes in Civil Engineering

Volume 87

Lecture Notes in Civil Engineering (LNCE) publishes the latest developments in Civil Engineering - quickly, informally and in top quality. Though original research reported in proceedings and post-proceedings represents the core of LNCE, edited volumes of exceptionally high quality and interest may also be considered for publication. Volumes published in LNCE embrace all aspects and subfields of, as well as new challenges in, Civil Engineering. Topics in the series include:

- Construction and Structural Mechanics
- Building Materials
- Concrete, Steel and Timber Structures
- Geotechnical Engineering
- Earthquake Engineering
- Coastal Engineering
- Ocean and Offshore Engineering; Ships and Floating Structures
- Hydraulics, Hydrology and Water Resources Engineering
- Environmental Engineering and Sustainability
- Structural Health and Monitoring
- Surveying and Geographical Information Systems
- Indoor Environments
- Transportation and Traffic
- Risk Analysis
- Safety and Security

To submit a proposal or request further information, please contact the appropriate Springer Editor:

- Mr. Pierpaolo Riva at pierpaolo.riva@springer.com (Europe and Americas);
- Ms. Swati Meherishi at swati.meherishi@springer.com (Asia - except China, and Australia, New Zealand);
- Dr. Mengchu Huang at mengchu.huang@springer.com (China).

All books in the series now indexed by Scopus and EI Compendex database!

More information about this series at http://www.springer.com/series/15087

Laxmikant Madanmanohar Gupta ·
Maya Rajnarayan Ray ·
Pawan Kumar Labhasetwar
Editors

Advances in Civil Engineering and Infrastructural Development

Select Proceedings of ICRACEID 2019

Springer

Editors
Laxmikant Madanmanohar Gupta
Department of Civil Engineering
Visvesvaraya National Institute
of Technology
Nagpur, Maharashtra, India

Maya Rajnarayan Ray
Department of Civil Engineering
G. H. Raisoni College of Engineering
Nagpur, Maharashtra, India

Pawan Kumar Labhasetwar
Water Technology & Management
Division (WTMD)
National Environmental Engineering
Research Institute (NEERI) (Council
of Scientific and Industrial Research)
Nagpur, Maharashtra, India

ISSN 2366-2557 ISSN 2366-2565 (electronic)
Lecture Notes in Civil Engineering
ISBN 978-981-15-6462-8 ISBN 978-981-15-6463-5 (eBook)
https://doi.org/10.1007/978-981-15-6463-5

This Springer imprint is published by the registered company Springer Nature Singapore Pte Ltd.
The registered company address is: 152 Beach Road, #21-01/04 Gateway East, Singapore 189721, Singapore

Preface

We are delighted to be the editor(s) of the proceedings of the "**International Conference on Recent Advancements in Civil Engineering and Infrastructural Developments (ICRACEID 2019)**," which is being published by Springer Nature. The **ICRACEID 19** is a major civil engineering conference organized with the objective of bringing together academia, industry personals and engineering graduates to analyze scientific studies focusing on infrastructure technologies, environmental sustainability, smart transportation, smart materials and sustainable environmental management. The **ICRACEID 19** was organized at G. H. Raisoni College of Engineering (GHRCE), Nagpur, during July 5–6, 2019.

The **ICRACEID 19** was planned to organize and integrate various research and innovation activities for a sustainable infrastructure development. This conference was aimed to provide a platform for the discussion and analysis of various aspects of sustainable infrastructure development, environmental issues related to the development and the optimal solutions, sustainable materials and transportation systems in the civil engineering and various aspects of the smart technology for the smart cities. These key topics of central discussion brought effective participation from the academic Alma mater, industry professionals, governmental policy makers, as well as NGOs and community service providers. It is seen that due to the growth of urban areas, urban population has increased exponentially, and to meet these increasing needs of development, an integrated approach of sustainable development is required. The **ICRACEID 19** has provided a common platform to academia, industry practitioners, policy makers, engineering graduates to share their innovation, research and ideas with a special focus on the recent trends and advances in infrastructure development, smart technologies for sustainable development. The purpose of this special issue is to identify innovative technologies for infrastructure development, environmental management, sustainable materials and transportations technologies and polices for sustainable infrastructural development for the society. The peer-reviewed research papers of the **ICRACEID 19** have been divided into four parts: (i) Structural Engineering, (ii) Environmental and Water Resources Engineering, (iii) Transportation and Geotechnical Engineering and (iv) Smart Materials and Construction Management with focus on sustainable infrastructure development. This book series is going to provide a basis for academia, industry practitioners, engineering graduates and policy makers to analyze scientific studies focusing on sustainable infrastructure development and environment.

As editors of this book series of Springer's proceedings of ICRACEID 19, we wish to express our sincere appreciation to the authors, plenary speakers, participants, international advisory committee, national and international reviewers, technical committee members as well as conference organizing committee members of the G. H. Raisoni College of Engineering, Nagpur (India), who have actively contributed in the success of the ICRACEID 19 international conference and publication of this book.

Our sincere gratitude to Shri. Sunil Gyanchandji Raisoni, Chairperson of the Raisoni Group of Institutions; Dr. Preeti Bajaj, Director, G. H. Raisoni College of Engineering (GHRCE), Nagpur; Dr. P. Y. Pawade and Dr. B. V. Khode, Program Chairs, and Prof. S. S. Sanghai for handling the Easy Chair conference management system of the ICRACEID-19 for successfully organizing this conference.

Nagpur, India

Dr. Laxmikant Madanmanohar Gupta
Dr. Pawan Kumar Labhasetwar
Dr. Maya Rajnarayan Ray

Contents

About the Editors

Dr. Laxmikant Madanmanohar Gupta is working as professor (Higher Administrative Grade, HAG) in the department of Applied Mechanics and former Dean (Planning and Finance) Visvesvaraya National Institute of Technology (VNIT), Nagpur. He obtained his B.E. (Civil Engineering) from Government College of Engineering Amravati, M.Tech. (Structural Engineering) from Visvesvaraya Regional College of Engineering (VRCE), Nagpur and Ph.D. (Civil Engineering) from Nagpur University. Major area of his research is structural engineering. He has published almost 50 papers in national and international journals, whereas almost 100 papers in the national and international conferences. Dr. Gupta is also having the membership of several professional bodies of national and international repute. In addition to this he has organized many short term courses for the quality improvement. Moreover he has completed several R & D and consultancy projects successfully. He has several awards to his credit given by VRCE Nagpur in 1995 and PWD (Government of Maharashtra) for some of his distinguish works.

Dr. Maya Rajnarayan Ray is currently a professor at the department of civil engineering, G.H. Raisoni college of Engineering, Nagpur. She obtained B.E. and M.E. (Civil Engineering) from the M.S. University of Baroda, Vadodra and Ph.D. (Civil Engineering) from the Indian Institute of Technology, Guwahati. Major area of her research interest includes, hydraulic and water resources engineering, hydraulic structures, computational hydraulics, free surface flow, water quality monitoring, ground water hydrology, operation of reservoir, study of downstream impacts and stochastic hydrology. She has published more than a dozen papers in the respected international journals and proceedings. She has also completed several consultancy projects successfully. Dr. Maya is a member of several professional bodies of national and international repute.

Dr. Pawan Kumar Labhasetwar, is Chief Scientist and Head in Water Technology & Management Division (WTMD), National Environmental Engineering Research Institute (NEERI), Nagpur, A CSIR laboratory of Gov. of India.

He obtained B. E. (Civil Engineering) and M. E. (Env. Engineering) from Government Engineering College Jabalpur and Ph. D. (Env. Engineering) from NEERI, Nagpur University. His area of research interest includes water resource management, water quality monitoring, water safety plan, environmental impact assessment, grey water treatment reuse, environmental sanitation. He has published more than 40 papers in national and international journals, where as he has completed more than 300 R & D and consultancy projects successfully, funded by government of India and different bodies. He has many awards and achievements to his credit like 'Jal Nirmalata' Award of Indian Water Works Association 2016-17, best paper award, Nagpur 'Gaurav Sanman', Oklahoma University Water Prize, project innovation award, DST-Lockheed Martin Innovation Growth award from FICCI, NEERI foundation day award, Commonwealth Science Council fellowship etc.

Structure Engineering

Analysis of Tall Building Using IS 16700-2017 and ASCE 7-10

Deepika Nair, S. P. Raut, and S. V. Denge

Abstract With the increase in population, the lack of space in cities becomes the main issue. Due to this limitation in space, the constructions of tall or high-rise buildings have become the necessity of time. The design criteria for high-rise buildings are very complex in comparison with low- and mid-rise buildings. When a tall building is considered, it is exposed to static as well as dynamic loading. The life of a tall building is affected by various parameters such as bending, shear, torsion and drifts, and these effects have to be considered while analyzing and designing it. Each country develops its own codes and standards for analysis and design. Till date, the high-rise constructions in India are carried out according to various Indian standards and codal provisions. Recently, a new code IS 16700-2017 "Criteria for Structural Safety of Tall Concrete Buildings" specifically for tall buildings has been introduced in India. This code focuses on various aspects of tall buildings. In the present work, a G + 20 reinforced concrete tall building of height 63 m is considered. The seismic and wind analysis of the structure is carried out using ETABS 2016 version 16.2.1 software. The results are obtained using Indian and American standards. The analysis result shows that American standard gives higher values than that of Indian standard.

Keywords Tall building · Static loading · Dynamic loading · Bending · Shear · Torsion · ETABS

D. Nair (✉) · S. P. Raut
Department of Civil Engineering, Yeshwantrao Chavan College of Engineering, Nagpur, India
e-mail: nairdeepika20@gmail.com

S. P. Raut
e-mail: sprce@rediffmail.com

S. V. Denge
DGM-Design and Engineering, JSW Severfield Structures Ltd., Mumbai, India
e-mail: sanjaydenge@gmail.com

L. M. Gupta et al. (eds.), *Advances in Civil Engineering and Infrastructural Development*, Lecture Notes in Civil Engineering 87,
https://doi.org/10.1007/978-981-15-6463-5_1

1 Introduction

At present, there is rapid emerging trend of construction of tall buildings. In last few decades, these constructions of tall buildings have been increased [8]. With design point of view, it is much more complex than any low- or mid-rise buildings, but with the increase in new technologies and design tools that mainly focus on safety, economy and stable construction this issue can be dealt with [25]. It necessitates the development of various methods and tools by which the analysis and design procedure of tall building could be made easier [17].

Huang et al. [8] studied the incremental dynamic analysis of super-tall building of height 700 m. The analysis helped to evaluate the structural collapse of the building under seismic loads along with regular behavior of the structure. Lee and Tovar [17] focused on the placement of outriggers in tall buildings by using topology optimization. Similarly, Khanorkar et al. [15] focused on the behavior of tall building using belt truss and outrigger systems to control the deflection. Xu et al. [26] studied the effect of wind loads on super-tall building of height 303 m by performing full-scale field measurement. Liang et al. [20] studied the effect of torsional dynamic wind loads on the rectangular tall buildings. Li et al. [19] described the wind-induced responses and dynamic behavior of Guangdong International Building located in China with a height of approximately 200 m. Zhang and Li [27] studied the wind effects on a super-tall building of height 600 m by wind tunnel test and field measurement. Li [18] modeled the tall buildings with shear wall having a narrow rectangular plane configuration. Lu et al. [22] studied a tall building with 60 stories and the podium structure surrounding it having 10 stories. Esteghamati et al. [6] illustrated the effect of design drift limit on the seismic performance of RC high-rise buildings. The design drift limit affects the seismic performance of the building. Alavi et al. [2] studied the design of high-rise building subjected to flexural vibration at a preferred natural frequency. Mwafy et al. [24] studied the effect of seismic performance on economy of high-rise structure when designed for varying concrete strength ranging from 45 MPa to 110 MPa. This study showed that with the increase in concrete strength, the cost of steel is reduced up to 37%, whereas unit cost of concrete increases with increasing strength. Kwon and Kareem [16] described the comparative study of major international wind codes and standards for effects of wind on tall building. Karadi and Chandra [14] had done a comparative study on analysis of tall buildings using the Indian standards IS 456:2000 [11] and IS 1893 (Part-1):2002 [12] and Eurocodes like EC2 and EC8. Binzet et al. [4] studied the analysis and design of a 43-storied 195-m-high tall building with Istanbul seismic design code for tall building. Mahmoud and Abdallah [23] carried out the study of seismic performance of a 14-story existing building with shear wall. It showed that the dynamic response spectrum method (RSM) was more preferred than equivalent static method (ESM), although ESM was used for setting the scale factor. Fu et al. [7] studied the analysis of wind-induced inter-story drift and equivalent static wind load of tall building 432 m high. The inter-story drift calculated from equivalent static method was consistent with that of spectrum decomposition method. Lu et al. [21] carried out a comparative study on the seismic design of tall building using Chinese seismic design code and US seismic design code. Jameel et al. [13] investigated the effect of walls, slabs and wall openings on structural strengthening of tall buildings. The frame + wall + slab model gave the most economic design.

The present work is the analysis of a G + 20 reinforced concrete tall building of height 63 m. The structure is modeled using a 3D structural software ETABS 2016 version 16.2.1. The structure is subjected to lateral forces. The main lateral forces are seismic and wind forces. The analysis is carried out as per Indian standard IS 16700-2017 [10] and American standard ASCE 7-10 [3] using ETABS software [5].

2 Methodology

Initially, a symmetrical model of G + 20 stories is selected. It is modeled in ETABS software. Loads are applied to MODEL-A and MODEL-B as per Indian and American standards, respectively. First stage of analysis is static analysis which is carried out for both seismic and wind loadings. It is observed from the analysis that the earthquake is predominant. Since earthquake is predominant, for further analysis only seismic force is considered. The second stage of analysis is dynamic seismic analysis which is carried out for MODEL-A and MODEL-B as per IS 1893-2016 (Part I) and ASCE 7-10, respectively, in ETABS. The load combinations are decided as per IS 456-2000 and IS 16700-2017 for MODEL-A and as per ACI 318-14 [1] for MODEL-B.

2.1 Method of Analysis

In the present study, the structure is analyzed for lateral loads using equivalent static method, response spectrum analysis (linear dynamic analysis) and static wind analysis using ETABS software.

Seismic Analysis A structure vibrates in response to the occurrence of earthquake. The seismic forces can be resolved into two horizontal directions (x and y directions) and one in vertical direction (z direction). The ground motion due to earthquake causes the structure to vibrate in all the three directions. The vibration occurs predominantly in horizontal direction. Vertical vibration should also be considered depending on the recommendations of the codes. In the present study, static and dynamic seismic analysis is considered as per the recommendations of IS 1893-2016 (Part I), IS 16700-2017 and ASCE 7-10.

Equivalent Static Analysis Method (ESAM) Equivalent static analysis method is based on the assumption of static cantilever beam. In this method, story force is generated according to the height at which the story is located from the seismic base. As the story height increases, more loads are developed by that story. The amount of base shear that is considered for designing the building is based on the approximate time period of the building, zone where the building is situated, site-specific ground acceleration, type of lateral load resisting system, etc.

Response Spectrum Analysis (RSA) Response spectrum analysis is dynamic analysis of a structure. It is referred as dynamic analysis because for different building frequencies the mode shapes and modal mass participation of the structure are considered. During an earthquake, the response of the building is a combination of different natural frequencies of the building. These frequencies of the structure are known as eigenvalues, and the shape each mode generates is the eigenvector. It provides more realistic dynamic response of the building.

Wind Analysis When a building is subjected to wind loads, both positive and negative pressures occur simultaneously. Exposed basic wind speed, building height, topography, internal and external pressures, building shapes, etc., are some of the primary factors that must be considered for wind analysis. Tall or slender buildings respond statically or dynamically to the wind effects which must be checked as per the recommended codal provisions. In the present study, static wind analysis is performed as per the codal provisions of IS 875-2015 (Part III) [9], IS 16700-2017 and ASCE 7-10.

3 Model Description

Figure 1a–c shows the plan, elevation and 3D view of the building, respectively, modeled in ETABS. Two reinforced concrete buildings are modeled: according to Indian standards—MODEL A and according to American standards—MODEL B. Table 1 shows the common parameters of the building.

3.1 Loads

The loads acting on the buildings are:

- Dead load
- Live load
- Seismic load
- Wind load.

The loads are applied to MODEL-A and MODEL-B as per Indian and American standards, respectively.

The seismic analysis for MODEL-A is performed as per IS 1893-2016 (Part I). Table 2 shows the data assumed for static analysis. The seismic analysis for MODEL-B is performed as per ASCE 7-10. Some values are taken from NEHRP seismic provisions prepared by FEMA. Table 3 shows the seismic data for MODEL-B.

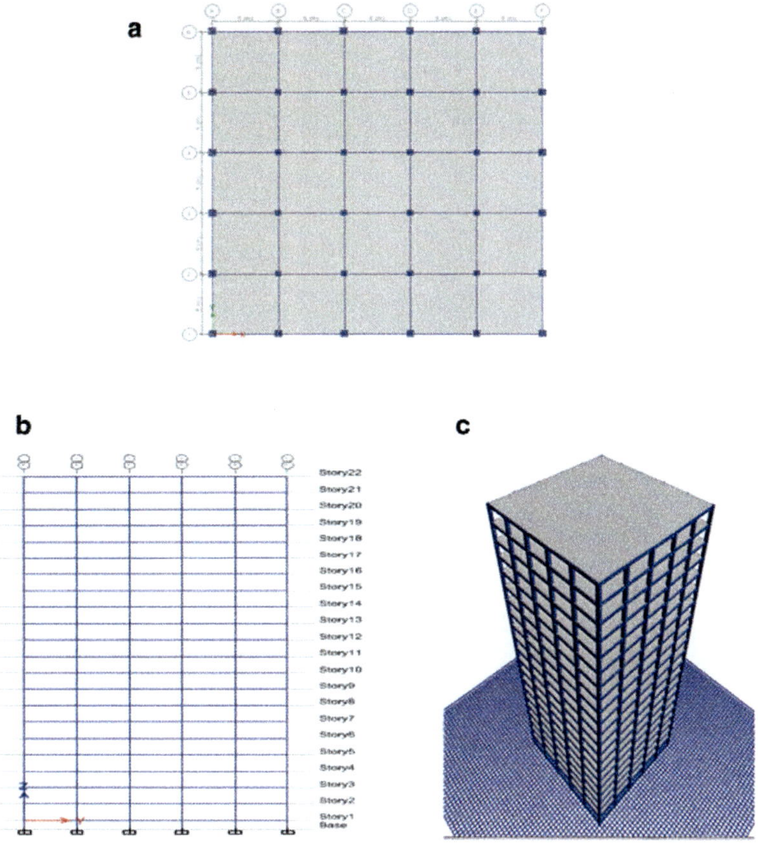

Fig. 1 **a** Plan view of G + 20 tall building. **b** Elevation of G + 20 tall building. **c** 3D view of G + 20 tall building

The static wind analysis is performed for MODEL-A as per IS 875-2015 (Part III). Table 4 shows the wind data for MODEL-A. The static wind analysis for MODEL-B is performed as per ASCE 7-10. Table 5 shows the data assumed for static wind analysis of MODEL-B.

3.2 Load Combinations

The load combinations have been applied for the two models according to the respective codes. For MODEL-A, load combinations are considered as per IS 456-2000 and IS 16700-2017, and for MODEL-B load combinations are considered as per ACI 318-14. Table 6 shows load combinations for MODEL-A and MODEL-B.

Table 1 Description of building

S. no.	Parameters	Details/value
1	Plan dimensions	30 m × 30 m
2	Stories	G + 20
3	Height of building	63 m
4	Story height	3 m
5	Grade of concrete	M 50
6	Grade of steel	Fe 500
7	Frame type	SMRF
8	Outer wall	230 mm
9	Inner wall	115 mm
10	Slab thickness	150 mm
11	Exterior column	0.7 m × 0.7 m
12	Interior column	0.6 m × 0.6 m
13	Beam	0.3 m × 0.6 m

Table 2 Seismic data for MODEL-A

IS 1893-2016 (Part I)	
Location	Mumbai
Seismic zone	III
Seismic intensity, z	0.16
Importance factor, I	1
Response reduction factor, R	5
Type of soil	Medium

Table 3 Seismic data for MODEL-B

ASCE 7-10	
Location	Seattle
Seismic zone	III
Site class	C
Risk category	III
Seismic importance factor, I_e	1.25
Response modification factor, R	8
Spectral response acceleration at short period, S_s	1.289
Spectral response acceleration at 1-s period, S_1	0.498
Long period and transition period	6 s

Table 4 Wind data for MODEL-A

IS 875-2015 (Part III)	
Location	Mumbai
Basic wind speed	44 m/s
Probability factor, k_1	1
Topography factor, k_3	1
Terrain category	4
Wind directionality factor, k_d	0.9
Area averaging factor, k_a	0.8
Combination factor	1

Table 5 Wind data for MODEL-B

ASCE 7-10	
Location	Seattle
Basic wind speed	58 m/s
Risk category	C
Surface roughness	B
Exposure category	B
Wind directionality factor, k_d	0.85
Topography factor, k_{zt}	1
Gust factor, G	0.85

Table 6 Load combinations for MODEL-A and MODEL-B

As per IS 456-2000 and IS 16700-2017	As per ACI 318-14
1.5(DL + LL)	1.4DL
1.5(DL ± EQ)	1.2DL + 1.6LL + 0.5RL
1.5 (DL ± W)	1.2DL + 1.6RL + 1.0LL
1.2(DL + LL ± EQ)	1.2DL + 1.6RL + 0.5W
1.2 (DL + LL ± W)	1.2DL ± 1.0 W + 1.0LL + 0.5RL
0.9DL ± 1.5EQ	0.9DL ± 1.0EQ
0.9DL ± 1.5W	0.9DL ± 1.0W

DL dead load, *LL* live load, *RL* roof live load, *EQ* earthquake load, *W* wind load

4 Results and Discussion

The results of the analysis for MODEL-A and MODEL-B are shown as follows:

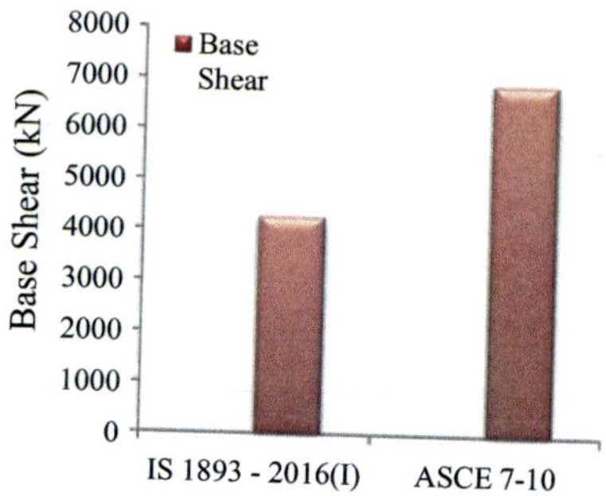

Fig. 2 Variations in base shear for MODEL-A and MODEL-B

- Base Shear

Base shear is an estimate of the maximum expected lateral force that occurs due to seismic ground motion at the base of structure. The base shear as shown in Fig. 2 for MODEL-B is more than MODEL-A. It is because of the variations in zonal parameters of both the locations.

- Overturning Moment

Figure 3 shows the variations in overturning moments for both the models. The story overturning moment varies inversely with the story height. The overturning moment for MODEL-A is less than MODEL-B.

- Story Displacement

Story displacement is the total displacement of *i*th story with respect to the base. Figure 4 shows the variations in story displacement. From Fig. 4, it can be seen that the story displacement increases with increasing height.

- Story Drift

Figure 5 shows the variations in story drift. The maximum story drift for MODEL-A and MODEL-B occurs between story 3 and story 5. The story drift for both the

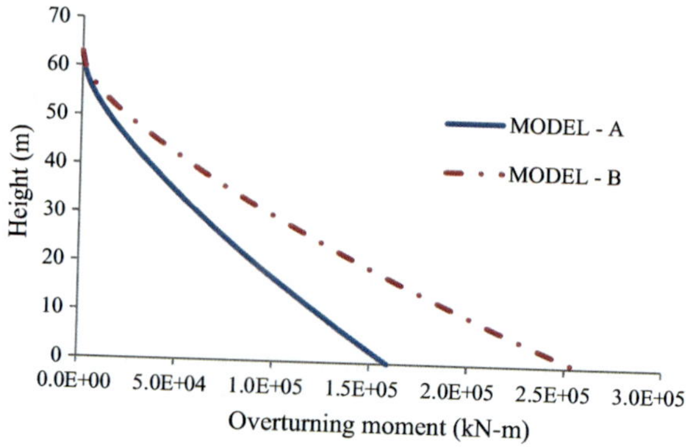

Fig. 3 Variations in overturning moment for MODEL-A and MODEL-B

Fig. 4 Variations in story displacement for MODEL-A and MODEL-B

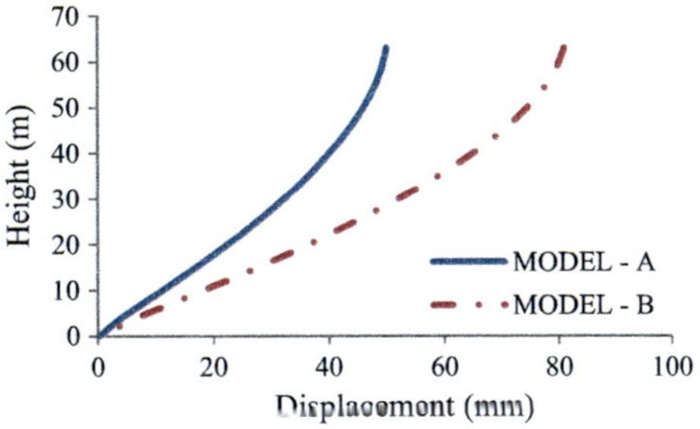

Fig. 5 Variations in story drift for MODEL-A and MODEL-B

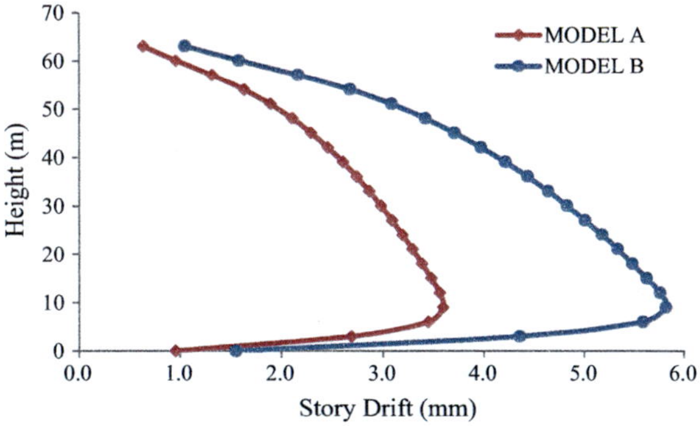

models increases with story height up to 9 m where it reaches to a maximum value and then starts decreasing. The story drift for MODEL-B is more than MODEL-A.

- Story Stiffness

Figure 6 shows the variations in story stiffness for MODEL-A and MODEL-B, respectively.

5 Conclusion

The analysis of the two tall buildings as per Indian and American standards reflects that the base shear for MODEL-B is higher than MODEL-A by 38.32%. The overturning moment for MODEL-A is less than MODEL-B. It varies by 38.14%. Similarly, the top displacement for MODEL-B is more than MODEL-A by 38.14% as higher lateral forces attract higher displacement. The story drift for MODEL-B is higher than MODEL-A by 38.18%. The story drift for both the models is within the permissible limits as specified by the respective codes. As per IS 16700-2017 and ASCE 7-10, the stiffness for any story shall not be less than 70% of that of the story above. The story stiffness for MODEL-A and MODEL-B follows the codal specifications.

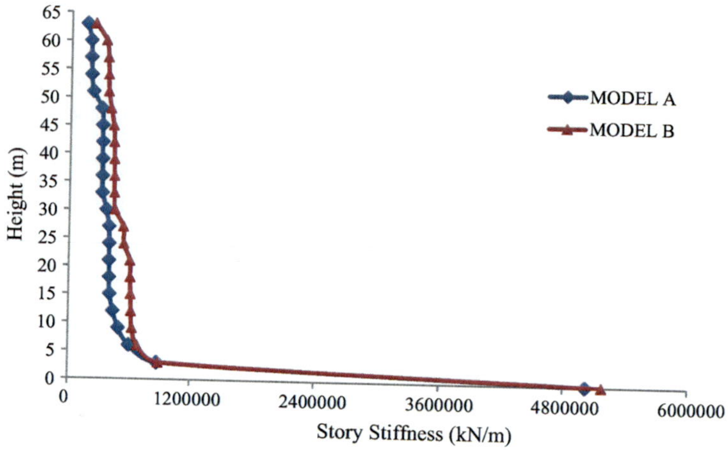

Fig. 6 Story stiffness for MODEL-A and MODEL-B

References

1. ACI 318(2014) Building code requirements for structural concrete. American Concrete Institute
2. Alavi A, Rahgozar P, Rahgozar R (2018) Minimum-weight design of high-rise structures subjected to flexural vibration at a desired natural frequency. Struct Design of Tall Spec Build 277:e1515. https://doi.org/10.1002/tal.1515
3. American Society of Civil Engineers Standard ASCE/SEI 7(2010) Minimum design loads for buildings and other structures. American Society of Civil Engineers., Reston, Virginia, USA
4. Binzet S, Tuzun C, Erdik M (2014) Performance based design of a high rise building based on Istanbul tall building seismic code. In: Second European conference on earthquake engineering and seismology
5. Computers and Structures (2016) CSI analysis reference manual for ETABS, Berkeley, California, USA
6. Esteghamati MZ, Banazadeh M, Huang Q (2017) The effect of design drift limit on the seismic performance of RC dual high-rise buildings. Struct Des Tall Spec Build 27:e1464. https://doi.org/10.1002/tal.1464
7. Fu JY, Zheng QX, Wu JR, Huang YQ (2016) Wind-induced inter-story drift analysis and equivalent static wind load for multiple targets of tall buildings. Struct Des Tall Spec Build 25:297–321
8. Huang T, Ren X, Li J (2017) Incremental dynamic analysis of seismic collapse of super-tall building structures. Struct Des Tall Spec Build 26:e1370. https://doi.org/10.1002/tal.1370
9. IS: 875(2015): Part III Design loads (Other than earthquake) for buildings and structures—code of practice. Bureau of Indian Standards, New Delhi
10. IS 16700(2017) Criteria for structural safety for tall concrete buildings. Bureau of Indian Standards, New Delhi
11. IS 456(2000) Code of practice—plain and reinforced concrete, (Fourth Revision). Bureau of Indian Standards, New Delhi
12. IS 1893(2016): Part I Criteria for earthquake resistant design of structures (General provisions and buildings). Bureau of Indian Standards, New Delhi
13. Jameel M, Islam ABMS, Hussain RR, Khaleel M, Zaheer MM (2012) Optimum structural modelling for tall buildings. Struct Des Tall Spec Build 22:1173–1185. https://doi.org/10.1002/tal.1004

14. Karadi AS, Chandra BSS (2017) Analysis and comparison of tall building using Indian and Euro code of standards. Int Res J Eng Tech 04(08):383–389
15. Khanorkar AA, Sukhdeve S, Denge SV, Raut SP (2016) Investigation into behavior of tall building using belt truss and outrigger systems to control deflection. J Struct Eng 1(2):1–13
16. Kwon DK, Kareem A (2013) Comparative study of major international wind codes and standards for wind effects on tall buildings. Eng Struct 51:23–35
17. Lee S, Tovar A (2014) Outrigger placement in tall buildings using topology optimization. Eng Struct 74:122–129
18. Li QS (2001) Stability of tall buildings with shear-wall structures. Eng Struct 23:1177–1185
19. Li QS, Wua JR, Liang SG, Xiao YQ, Wong CK (2004) Full-scale measurements and numerical evaluation of wind-induced vibration of a 63-story reinforced concrete tall building. Eng struct 26:1779–1794
20. Liang S, Li QS, Liu S, Zhang L, Gu M (2004) Torsional dynamic wind loads on rectangular tall buildings. Engineering Structures 26(1):129–137
21. Lu X, Xinzheng Lu, Li M, Guan H, Ye L (2015) A comparative case study on seismic design of tall RC frame-coretube structures in China and USA. Struct Des Tall Spec Build 24:687–702. https://doi.org/10.1002/tal.1206
22. Lu X, Gong Z, Weng D, Ren, X (2007) The application of a new structural control concept for tall building with large podium structure. Engineering Structures 29(8):1833–1844
23. Mahmoud S, Abdallah W (2014) Response analysis of multi-story RC buildings under equivalent static and dynamic loads according to Egyptian code. Int J Civ Struct Eng Res 2(1):79–88
24. Mwafy A, Hussain N, El-Sawy K (2014) Seismic performance and cost-effectiveness of high-rise buildings with increasing concrete strength. Struct Des Tall Spec Build 24:257–279. https://doi.org/10.1002/tal.1165
25. Taranath BS (2010) Reinforced concrete design of tall building. CRC Press
26. Xu A, Xie ZN, Fu JU, Wu JR, Tuan A (2014) Evaluation of wind loads on super-tall buildings from field-measured wind-induced acceleration response. Struct Des Tall Spec Build 23:641–663
27. Zhang JW, Li QS (2017) Wind tunnel test and field measurement study of wind effects on a 600-m-high super-tall building. Struct Des Tall Spec Build 26:e1385. https://doi.org/10.1002/tal.1385

Comparative Study of Prediction of 28 Days Strength Using Fuzzy Logic and Model Tree

Vardhan Nagarkar, P. S. Kulkarni, and S. N. Londhe

Abstract Concrete is a construction material which is used in *n* construction processes as a major stress resistance due to its strength characteristics. Designing a concrete mix is a tough task which includes right proportion of ingredients to be mixed in particular order and in particular environment with the objective to produce concrete mix with specified strength, durability, workability, and as economical as possible. Structural engineering field is full of nonlinear problems. This paper speaks about one of the basic nonlinear problems which is a strength prediction. We have tried to convert the nonlinear problem of strength into a linear problem by using model tree analysis and predicted the strength of various mix proportions using fuzzy logic. The objective of this research work is to study fuzzy logic tool and model tree regression analysis processes for prediction of concrete compressive strength, respectively, and its result comparison. Results of this study states that model tree regression analysis works more efficiently than a fuzzy logic.

Keywords Fuzzy logic · Model tree · Concrete strength · Concrete mix

1 Introduction

Concrete is most important material in construction industry. The property of concrete depends on various factors such as mix proportion, manufacturing method, and environmental conditions, are few of the issues that determine concrete strength. There are unit several nonlinear issues in construction industry; concrete compressive strength is one in each of them. It is terribly troublesome to predict the concrete compressive strength because of its nonlinear behavior, to work out whether or not specific needs are met or not that we have go to check the concrete.

V. Nagarkar (✉) · P. S. Kulkarni · S. N. Londhe
Department of Civil Engineering, Vishwakarma Institute of Information Technology, Pune 411048, India
e-mail: vardhanvn@gmail.com

L. M. Gupta et al. (eds.), *Advances in Civil Engineering and Infrastructural Development*, Lecture Notes in Civil Engineering 87,
https://doi.org/10.1007/978-981-15-6463-5_2

The dataset of concrete compressive strength has been collected from various construction sites. This concrete dataset contains 241 instances that contain eight attributes.

2 Objectives

The objective of this paper is to find out the reliable method among fuzzy logic and model tree regression analysis for predicting a concrete strength using a given dataset. Firstly, we need to learn concepts of both techniques which are being used for prediction. There is not any mathematical tool for strength calculation, so we have solved a nonlinear problem using a fuzzy logic and a conversion of this problem into linear one and again solving it using a model tree regression analysis.

After that we have compared the results and found out the correlation and error measures between actual values and predicted values.

3 Methodology

Some of structural engineering problems are nonlinear, which can be solved by experimental work. There is not any mathematical way to get the result of these problems. The potential of fuzzy logic concept lies in its working process and its empirical parameters to achieve research objectives. According to domain knowledge and experience, we have found out the rules for fuzzy logic model. This fuzzy logic model is based on nonlinearity of problems. We need to convert these nonlinear problems into linear once here model tree comes into picture. Model tree gives us the linear equations for strength calculation using mix proportion dataset.

This paper talks about the comparisons between the fuzzy logic results and model tree results for strength prediction algorithm. We have used MATLAB for a fuzzy logic coding and the WEKA, open-source portable software for model tree regression analysis. In a parametric study, the data consists of 241 values with eight different parameters. The instances in the dataset are pertaining to the attributes to represent the ingredient of concrete. In prediction of concrete compressive strength, composition consist of cement, fly ash, fine aggregate, 10 mm course aggregate, 20 mm course aggregate, water, and admixture (Table 1).

Table 1 Concrete compressive strength real-time dataset

Parameter	Range	Average	Std. dev.
Cement	130–460	302.38	73.12
Fly ash	0–120	25.38	31.05
Fine agg	398–1011	837.47	97.57
10 mm	0–958	390.67	143.93
20 mm	482–1242	790.45	135.25
Water	127–202	167.12	15.46
Admixture	0–5.2	2.68	1.31
Strength	12–65	34.2	10.63

4 Modeling Techniques

4.1 Fuzzy Logic

Fuzzy logic concept is a subpart of Boolean algebra. It has been invented to give solution of partial truth values. Dr. Lotfi Zadeh introduced fuzzy logic in 1960s. During his research work, he dealt with lot of criticism, but nowadays, there are many applications which works on fuzzy logic (Fig. 1).

Fuzzifier contains membership functions. This membership function ranges from 0 to 1 that is from no membership to high membership for each of the input. Triangular membership function is used for this study as a input and output as well. We have used seven input parameters which are cement, fly ash, fine aggregate, 10 mm coarse aggregate, 20 mm coarse aggregate, water, and admixture (Fig. 2).

In this work, Mamdani inference system is used for analysis (Fig. 3).

Where A and B are fuzzy sets in the antecedent, while $z = f(A, \text{and } B)$ is a crisp function in the consequent. Usually, $f(A, B)$ is a function in input variable A and B which is in the form of polynomial function, but if it describes output of model within fuzzy region specified by rule, then it can be any function.

Defuzzification is a method of extracting some values from a fuzzy set as a representative value. Basically in simple language, defuzzification is done to convert computer-aided output to human-understandable output. In this study, we have used

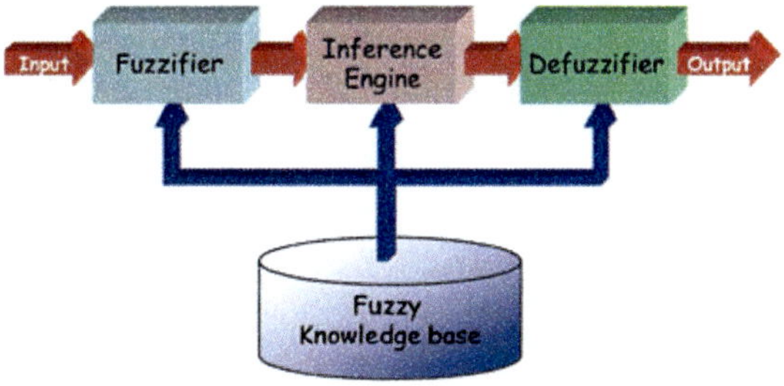

Fig. 1 Fuzzy logic system

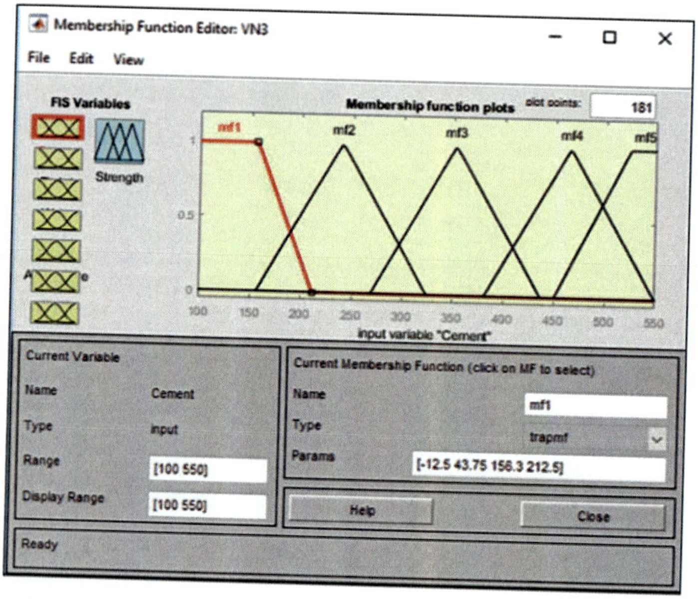

Fig. 2 Membership function

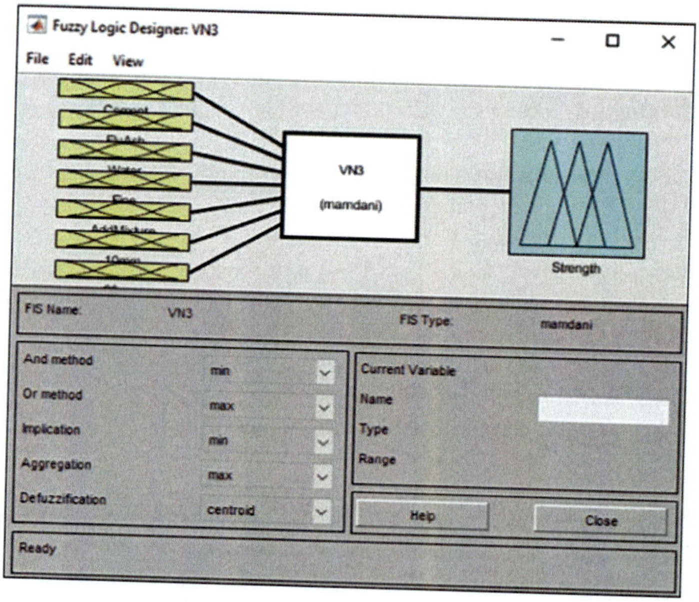

Fig. 3 Fuzzy inference system

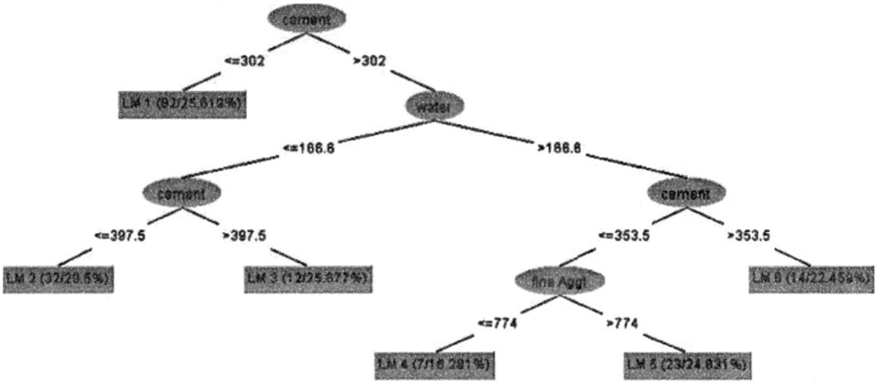

Fig. 4 Model tree

a centroid defuzzification method which a very basic one and default method if we go through inbuilt MATLAB application.

4.2 *Model Tree*

M5P model tree was introduced by Quinlan in 1992. The tree-based model works on divide and conquers method to form a relation between independent and dependent variables. It divides a dataset in different nodes (leaf), which sets a relation between variables. The tree models can be also used for qualitative and quantitative data (Fig. 4).

Model tree is simply a regression analysis tool. Depending on the database, we have it to divide our data into various crystals for specific ranges, and it gives us linear equations comparing input and output for given problem. We can predict our output using these equations.

5 Experimental Design

Parametric study is conducted using machine learning tool with default parameters. The data mining method used to build the model is classification. The WEKA, open-source, portable, GUI based workbench is used for model tree regression analysis. Another tool used was MATLAB for fuzzy logic modeling and coding. The instances in the dataset are pertaining to the attributes to represent the ingredient of concrete. In prediction of concrete compressive strength, composition consists of cement, fly ash, fine aggregate, 10 mm course aggregate, 20 mm course aggregate, water, and admixture.

Triangular function is used for input and output as well. In the current work, Mamdani inference system is used in which almost 50 rules are designed according to a database and experience. The rules are in the form of "If and then" that is if cement is mf1, water is mf2, fine aggregate is mf1, 10 mm aggregate is mf1, 20 mm aggregate is mf1, fly ash is mf1, admixture is mf1, then strength is mf1. The rules are defined according to ranges of membership functions for input and output which are defined using domain knowledge and experience. All these processes are done on MATLAB 2018.

In model tree analysis, the data is divided according to standard deviation, and regression analysis is performed on the data in which the equation for each node is defined and predicted the output according to that equation. This whole process is done on a WEKA software which is a open-source portable software.

6 Results and Discussion

Strength is the main component we need to find out to know the acceptability of concrete mix. Our conventional method takes 28 days to find the strength, so in the current work of the fuzzy logic code, you can get the strength within no time. Fuzzy logic model is prepared using domain knowledge and experience, and there is not any mathematical base to it.

To get the mathematical base to our nonlinear problem, we have used model tree regression analysis which converts our problem into linear equations and gives output according to these equations

To examine how close the predicted values to the actual compressive strength, three indices, correlation coefficient R, root mean square error RMSE, and mean absolute error MAE were calculated to evaluate the performance of the algorithms based on mix proportions.

$$R = \frac{\sum_{i=1}^{n}\left(P_i - \overline{P}\right)\left(A_i - \overline{A}\right)}{\sqrt{\sum_{i=1}^{n}\left(P_i - \overline{P}\right)^2 \sum_{i=1}^{n}\left(A_i - \overline{A}\right)^2}} \tag{1}$$

Equation 1: Correlation coefficient

$$\text{RMSE} = \sqrt{\frac{\sum_{i=1}^{n}(P_i - A_i)^2}{n}} \tag{2}$$

Equation 2: Root mean square error

$$\text{MAE} = \frac{\sum_{i=1}^{n}|P_i - A_i|}{n} \tag{3}$$

Table 2 Relations between outputs

	Fuzzy logic	Model tree
Correlation coefficient	0.90	0.94
Root mean square error	6.58	3.27
Mean absolute error	5.23	2.61

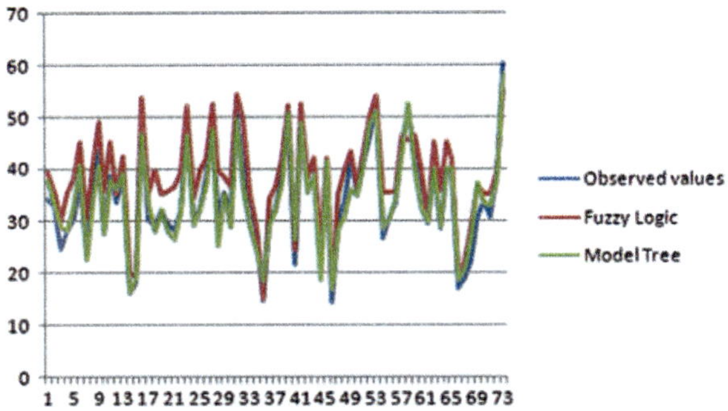

Fig. 5 Relation between outputs

Equation 3: Mean absolute error

Where A_i is the compressive strength of concrete mixtures, P_i is the predicted value, n is the total number of data points in validation, A is the mean value of dataset, and P is the mean value of predicted results.

Figure 5 shows the comparative results for both models in which we can see that results of model tree are much more precise than a fuzzy logic model, and error measures are also less than a fuzzy logic.

Figure 5 shows the relation between observed values, predicted values by fuzzy logic, and predicted values by model tree. From graph, we can clearly conclude that model tree works more efficiently than a fuzzy logic model.

7 Conclusion

In this paper, two prediction algorithms are applied on the task of classifying concrete compressive strength dataset, and the most accurate learning method is evaluated. With high correlation coefficient, lesser RMSE and MAE, the model tree is working well for this dataset in comparison with fuzzy logic. Model tree is totally a mathematical approach to solve a problem while fuzzy logic is more of a experience and logical approach, so it may affect a results of fuzzy logic.

References

1. Lotfi A, Andersen HC, Tsoi AC (1996) Matrix formulation of fuzzy rule-based systems. IEEE Trans Syst Man Cybern Part B Cybern 26:332–340
2. Nataraja MC, Jayaram MA, Ravikumar CN (2006) A fuzzy-neuro model for normal concrete mix design. Eng Lett 13(2):98–107
3. Alavi SA, Naderpour H (2015) Application of fuzzy logic in reinforced concrete structures. In: Forth international conference in civil, environmental & structural engineering, Scotland, Sept 2015
4. Diab AM, Elyamany HE, Abd Elmoaty AEM, Shalan AH (2015) Comparison between neural network and fuzzy logic on assessment of long term concrete compressive strength and expansion due to sulfate attack. Int J Res Appl Sci Eng Technol 3(9):175–192
5. Garrido A. A brief history of fuzzy logic. Faculty of Sciences, UNED, Madrid, Spain
6. Vakhshouri B, Nejadi S (2017) Prediction of compressive strength of self-compacting concrete by ANFIS models. Neurocomputing 280:13–22
7. Tayfur G, Erdem TK, Kırca Ö (2013) Strength prediction of high-strength concrete by fuzzy logic and artificial neural networks. ASCE
8. Topcu IB, Sandemir M (2008) Prediction of compressive strength of concrete containing fly ash using artificial neural networks and fuzzy logic. Comput Mater Sci 41:305–311
9. Deepa C, Sathiya Kumari K, Pream Sudha V (2010) Prediction of the compressive strength of high performance concrete mix using tree based modeling. Int J Comput Appl 6(5):18–24

Development of Cementless Recycled Concrete Aggregates Paver Blocks Using Molten Plastic Waste as Binder

Dhanesh V. Shirsat, P. S. Kulkarni, S. N. Londhe, and Shirish Phadtare

Abstract Amount of demolished concrete waste and plastic waste being accumulated in twenty-first centuries have been created big challenges for their disposal. At present, nearly 530 million of tons debris is generated in the country annually. Due to our society has become dependent on usage of plastic packaging, approximately 5.6 million ton of plastic products are consumed per annum. The degradation rate of waste is also a very slow process. Using recycled concrete aggregates and waste plastic in development of paver blocks is an interesting possibility for economy on wastage disposals. The objective of the research is to study the characteristics of developed paver block. This paper presents a recent study on properties of paver block prepared by recycled concrete aggregates by using molten plastic waste as binder. For the laboratory samples, the compressive strength, split tensile strength, abrasion resistance, water absorption and heat resistance parameters were measured.

Keywords Recycled concrete aggregates · Plastic waste · Low-density polyethylene · Paver blocks

1 Introduction

Concrete is widely used in all types of construction. There is a vast increment in construction industry as compared to past many decades. This increment is due to rapid growth in population, thus ultimately results in depletion of various construction materials due to uncontrollable use of construction materials. There has been introduction of various techniques, various materials and various substitutes of materials in early research. India is developing country. Due to its developing nature, there

D. V. Shirsat (✉) · P. S. Kulkarni · S. N. Londhe
Department of Civil Engineering, Vishwakarma Institute of Information Technology, Pune 411048, India
e-mail: dhanesh.shirsat@gmail.com

S. Phadtare
Rudra Environmental Solutions India Ltd., Pune, India

L. M. Gupta et al. (eds.), *Advances in Civil Engineering and Infrastructural Development*, Lecture Notes in Civil Engineering 87,
https://doi.org/10.1007/978-981-15-6463-5_3

are lots of new constructions and demolitions of old structure done every day. Therefore, due to rapid growth in construction industry, a large quantity of construction and demolition waste (C&D) is generated. It ranges approximately 530 million tons annually. Our society has become dependent on usage of plastic packaging. Approximately 5.6 million tons of plastic product has consumed annually. Due to liberal use of plastic metered, huge quantity of plastic waste is generated every day. It is worth mentioning that plastic waste will never degrade and remain on earth for years causing several types of pollution. Therefore, the waste disposal has become critical environmental problem in the country. The possibility of recycling waste from construction industry is thus of growing significance [1].

Paver block is adaptable, functional and cost-effective and requires literally no maintenance if precisely manufactured and laid. Due to its durability, pleasing surface made them irresistible for many industrial, commercial and residential use such as parking areas, pedestrian walks, etc. [2]. Paver blocks is made from concrete consisting cement, fine aggregate, coarse aggregate (<10 mm), water, and therefore, performance is depends on material used in it. In households, use of plastic is increasing day by day. If we use this plastic in our construction industry, it will help to reuse of waste plastic. So plastic is non-biodegradable; it has longer life and serviceability. We can use waste plastic in molten state as replacement of cement paste [3] and aggregates from construction industry waste, by recycling the concrete aggregates [4] in paver blocks. Recycle concrete aggregate (RCA) is defined as the aggregate which has been used in previous construction extracted from demolished concrete structures and used as replacement of natural aggregate because of depletion of natural resources.

Therefore, use of waste plastic and RCA materials in paver blocks is aimed at reducing natural resources and leading better durability and economy [5]. It will definitely help in defending environment effects and provide new way for safe disposal of waste plastic and construction and demolition waste.

2 Materials

2.1 *Waste Plastic*

By definition, plastic is used to made different shape of plastic things such as plastic bags, food bags, drinking containers, and after using such things, they are become waste material. Gathering of such plastic waste can results into dangerous to human beings, animals and plants [6]. Therefore, waste plastic needs proper disposal as well as use this waste plastic in recycle form. If they are not recycled, then these things become pollutant to the environment. For this research work, we collected low-density polyethylene (LPDE) waste plastic in shredded form from Pune Municipal Corporation Plastic Waste Processing Unit, Hadapsar, Pune. Plastic are classified according to their chemical sources as shown below (Fig. 1):

Fig. 1 Waste plastic. *Source* Packagingdigest.com

Table 1 Properties of waste plastic [9]

S. No.	Experiments	Results
1	Melting point	150 °C
2	Thermal coefficient of expansion	$100–200 \times 10^{-6}$
3	Density	910–950 kg/m^3
4	Tensile strength	20–30 N/mm^2

- Low-density polyethylene (LPDE)
- High-density polyethylene (HDPE)
- Polypropylene terephthalate (PET)
- Polypropylene (PP)
- Polystyrene (PS)
- Polyvinyl chloride (PVC).

The basic properties of waste plastic are provided in Table 1.

2.2 *Recycled Concrete Aggregates*

Recycled concrete aggregate (RCA) is generally produced by two-stage crushing of demolished concrete, and screening and removal of contaminants such as reinforcement. For recycled concrete aggregate, we collected tested cubes from Durocrete, Pune. As the quantity of material requirement is less, therefore, there was not any scope of having crusher to crush the cubes. Hence, we had adopted manual crushing method. After that, this material underwent for further process of sieving to produced fine aggregate and coarse aggregate according to the particle size requirement of Indian standards IS 383:1970 [7]. In this research, we use RCA < 4.75 mm, i.e. fine aggregates. The basic properties of RCA are provided in Fig. 2 and Table 2.

Fig. 2 Recycled concrete aggregates

Table 2 Properties of recycled concrete aggregate (<4.75 mm)

S. No.	Experiments	Results
1	Specific gravity	2.41
2	Water absorption	1.79
3	Fineness modulus	3.63
4	Silt content	4%
5	Bulk density	1400 kg/m^3

Table 3 Mix ratios by weight for the mixture

Trial	Mixing ratio	Waste plastic	RCA
1	1:3	1	3
2	1:2	1	2
3	1:1	1	1

3 Mixing Ratios

Three series of mixes were prepared, and identical materials were employed in each series. Each series contains 100% of recycled concrete aggregate to serve. The mixing ratios adopted in this research are 1:3, 1:2 and 1:1. The mixing ratios for the mixes are shown in Table 3.

4 Mixing Procedure

Initially required recycled concrete aggregates for mixing ratio are measured and placed in oven at temperature of 110° for 2 h. After that required waste plastic are divided into two equal parts, and one part is heated in GI sheet container at temperature of above 150 °C. After that plastic starts melting, add another part of waste plastic in it and mixed well and keep heating continued. As a result of heating plastic waste, it melts. Due to heating, its viscosity decreases with increase in temperature. After that waste plastic becomes semi-fluid. Remove the recycled concrete aggregate from oven and add it in right proportion in molten state of plastic and mix well. After that, continue heating mixture. Again, due to heating, viscosity of mixture decreases due to increase in temperature. Heat the mixture till it becomes semi-fluid and becomes workable as concrete. At the time of heating, stir the mixture continuously till it becomes united. After certain time of heating, the mixture becomes workable and can be seen by naked eyes.

5 Fabrication of Paving Blocks

Paver blocks were fabricated in galvanized iron sheet moulds having internal dimension of 200 mm long, 100 mm wide and 80 mm thick. The mixture was poured into the moulds in three layers of each of having equal thickness and compacted by single tamping of 25 strokes of tamping rod for each layer. After that, compression force is applied manually to compact mixture in the moulds. Excess mixture was then removed by trowel. After 24 h of casting, paver blocks were demoulded and were stored in dry place, free from direct sunlight until the day of testing (Fig. 3).

Fig. 3 Fabrication of paver blocks

6 Test Method

A lot of tests were carried out as per IS 15658:2006 to determine the compression strength, water absorption, split tensile strength, abrasion test and heat resistance test [8].

First of all, the one-day compressive strength was determined using compressive testing machine (CTM) having maximum capacity of 2000 KN. Paving blocks are placed in centre of CTM, and load was applied. Compressive strength is calculated by dividing failure load by loading area. Then, split tensile strength is calculated as per IS 15658:2006. Test is conducted on longest length of the section. Thereafter, specimen was concentrically packed with two steel pieces—one at top and one at bottom. Then, load was applied concentrically with the loading machine. After that load is terminated when specimen split into two pieces. And load is noted down. After that splitting, tensile strength is calculated. Then, after abrasion resistance test carried out as per IS 15658:2006.

7 Results and Discussion

In this paper, each presented result is an average of three measurements. The test results are summarized in Tables 4, 5 and 6. And each test is carried out confirming IS 15658:2006.

Table 4 Comparison of compression strength of plastic RCA paver blocks possessing various ratios

Mix designation	Plastic: RCA ratio	Compressive strength (N/mm^2)
M1	1:3	4.72
M2	1:2	5.49
M3	1:1	9.47

Table 5 Test results of paver blocks having mix ratio 1:1

Mix	Tensile splitting strength (MPa)	Abrasion resistance (mm)	Water absorption (%)
M3	1.82	0.92	0.04

Table 6 Heat resistance results

Specimen No.	Temperature (°C)	Remarks
1	50	No change
	100	Starts melting
	150	Melts

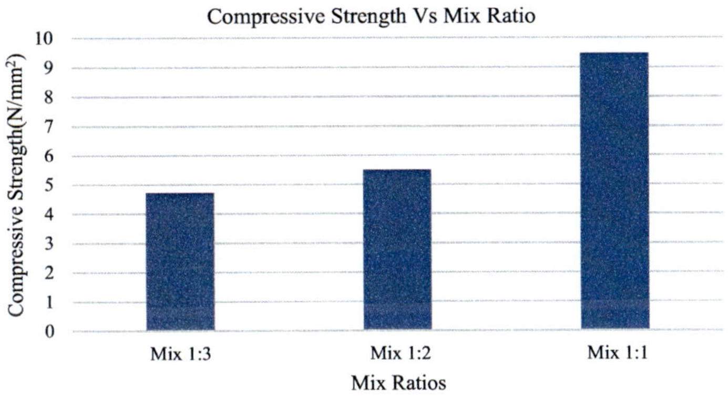

Fig. 4 Relationship between compressive strength and mixing ratios

Compression test was carried out confirming IS 15658:2006 to obtain compressive strength. The cubes were tested using compression testing machine (CTM) of capacity 2000 KN.

From Fig. 4, it shows that mix proportion ratio 1:1 (plastic: RCA) shows better result as compared to other mix proportion, although compressive strength of paver block is significantly less. Such pavers can be used in pedestrian purpose. That is why we conducted following test confirming IS 15658-2006 on mix proportion 1:1 such as water absorption, split tensile strength and abrasion resistance test and heat resistance test. The results are shown in Table 5.

As stated, above paver blocks having mix proportion 1:1 give satisfactory results. Abrasion resistance of developed paver blocks is less than 1, which means it is able to maintain its original structure and appearance, and having good resistance of wear and tear. From Table 5, it is found that water absorption of paver block containing plastic and recycled concrete aggregates is less, which is within the maximum value specified by the code.

Paver block is made from waste plastic which is required to measure its heat resistance [9]. Hence, specimen is placed in oven for 2 h. Results are shown in Table 6. Results also show that there is no any kind of change in shape of developed paver block at 50 °C. So, it is good heat resistance, and it can be withstand in atmospheric temperature.

8 Observations

In concrete paver block, fly ash is used as a substitute material for cement. The effect of fly ash on the performance of paver block has been reported in many researches. Fly ash is pozzolanic material contains silica and if reacts with calcium hydroxide form hydrate silicate which develops strength and lime to fills void. A certain amount of fly ash has been used in production of paver block satisfy the targeted value. Due to its usage, durability and workability of concrete are increased and also reduce permeability. This study aimed to developed a technique for production of paver blocks by using waste plastic and recycled concrete aggregates. Laboratory trials were carried on three series of mixes. At the time of trials, there are certain things which we are observe. On the first trial, while mixing we added whole required plastic in container; due to that plastic does not melt properly. For the second trial, we divided required plastic in two equal parts. When first part of waste plastic starts melting, add second part of waste plastic and stir continuously and continued heating. Due to this molten state of waste, plastic becomes more workable and consume less time for melting as compared to first trial. Same procedure is followed by the time of mixing of RCA in mixture. Due to dividing material into two equal parts, workability of mixture is increased. The result of the study reveals that there is increase in strength of paver blocks as mixing ratios decease. Mix ratio 1:1 gives maximum strength as compared to others. In mix ratio 1:3, there is three times of RCA used as compared to waste plastic. In RCA, it found that old mortar is stick to aggregate which is more porous. The presence of it on aggregate increases the absorption capacity and decreases specific gravity. Due to high absorption capacity, amount of molten waste plastic content which are required to cover the surface of RCA is not satisfied with this mix ratios 1:3 and 1:2 just because of less amount of plastic content as compared to RCA. Due to this, RCA surfaces are not completely cover. Just because plastic content is less bonding between RCA are not done properly results in high void ratio gives low compressive strength. On other hand, 1:1 mix ratio gives higher strength due to high content of plastic which are required to cover the surface of RCA and also fill the void between the mixtures and due to this it becomes more workable as compared to first one.

9 Conclusions

An experimental investigation on use of waste plastic recycled concrete aggregate paver blocks has been carried out and present in this paper. The study says that

- New efficient manufacturing method for suitable alternative for cement and natural resources.
- By using waste plastic and recycled concrete aggregates in production of paver block has beneficial way of disposal of waste plastic and construction and demolished waste.
- There was on an average 36% decrease in density of paver block.
- By using molten plastic and RCA in production of paver block reduces weight up to 28% as compared to conventional one.
- Water absorption of developed paver block is negligible as compared to concrete paver block.

- Compressive strength is less as compared to the conventional one though it can be used in gardens and pedestrian purpose.
- It can be used in compound walls.

References

1. Poon CS, Kou SC, Lam L (2002) Use of recycled aggregates in molded concrete bricks and blocks. Constr Build Mater 16:281–289
2. Kalingarani K, Harikrishna Devudu P, Jegan Ram M (2010) Development of paver blocks from industrial wastes. IOSR J Mech Civ Eng 12–17. e-ISSN: 2278-1684, p-ISSN: 2320-334X
3. Nivetha C, Rubiya M, Shobana S (2016) Production of plastic paver block from the solid waste. ARPN J Eng Appl Sci 11(2):1078–1079. ISSN: 1819-6608
4. Poon CS, Chan D (2006) Paving blocks made with recycled concrete aggregate and crush clay bricks. Constr Build Mater 20:569–577
5. Mohan DMS, Vignesh J, Jayyappan P, Suresh C (2018) Utilization of plastic bags in pavement blocks. Int J Pure Appl Math 119(15):1407–1414
6. Dinesh S, Kirubakaran K, Dinesh A (2016) Utilization of waste plastic in manufacturing of bricks and paver blocks. Int J Appl Eng Res 11(3):364–368. ISSN 0973-4562
7. IS 383-1970. Specification for coarse and fine aggregates from natural source of concrete. Bureau of Indian Standard
8. IS 15658-2006. Specification for precast concrete blocks for paving
9. Shanmugavalli B, Gowtham K, Nalwin PJ (2017) Reuse of plastic waste in paver blocks. Int J Eng Res Technol (IJERT) 6(2):313–315. ISSN: 2278-0181

Seismic Retrofitting of Indian RC Buildings Using Shear Walls

Saif Usmani, Kuldeep R. Dabhekar, Isha Khedikar, and Nimita R. Gautam

Abstract Shear walls are often used to increment stiffness in building frames. The increase in stiffness improves the performance of building and makes it less vulnerable to the lateral forces it is subjected to in an event of earthquake. Moreover, the overall safety of building is increased due to increased occupational safety [1]. This study tries to study analytically the effects of retrofitting of RC buildings using structural walls or shear walls.

Keywords Pushover curve · Target displacement · Plastic hinges

1 Introduction

The buildings nowadays are designed using the earthquake as a design consideration; however, this was not the practice always. Older buildings are under-designed with respect to earthquake forces. This paper discusses at length the performance of a typical under-designed frame, and this is followed by the methodology and details about modelling. Thereafter, the characteristics of the retrofitted model are studied, and the pushover analysis results are studied.

1.1 Related Work

Various studies have been forwarded to explain the relation between different performance objectives and the retrofitting strategies suitable to achieve desired performance. Some studies emphasize on energy dissipation devices to modify the drift

S. Usmani (✉) · K. R. Dabhekar · I. Khedikar
Department of Civil Engineering, G. H. Raisoni College of Engineering, CRPF Gate, No. 3, Hingna Road, Digdoh Hills, Nagpur 440016, India
e-mail: saifu527@gmail.com

N. R. Gautam
Department of Civil Engineering, PBCOE, Nagpur, India

L. M. Gupta et al. (eds.), *Advances in Civil Engineering and Infrastructural Development*, Lecture Notes in Civil Engineering 87,
https://doi.org/10.1007/978-981-15-6463-5_4

profile of building; they study the effect of different strategies on the behaviour of building (6) [2–4]. Some of them investigate the use of memory alloys or shear linkages in braces for dissipation of energy (7) [1]. While the effectiveness of these methods cannot be overlooked, they may lead to significant increase in costs and functionality which might not always be desired from a retrofitting project. Other alternatives have also been discussed which study effects on experimental models, and the scope of such studies is somewhat limited as all scenarios and building types cannot be effectively modelled. Analytical studies however focus on more realistic and performance-oriented goals.

1.2 Methodology

The paper focusses on analytically comparing the two models, so as to evaluate performance parameters affected by the use of retrofitting strategy. The main objective of the study is to study the increase in capacity and the change in target drift due to jacketing of elements. The models are evaluated using a computer-based tool, and the results are compared analytically. The modelling work is carried out on SAP2000, and a nonlinear pushover curve is plotted for the yield displacement of h/20 as per ASCE 41-17. The structural walls are inserted in the exterior periphery of building. The walls are modelled as nonlinear shear walls.

1.3 Modelling

The paper focusses on analytically comparing the bare frame model with the model after addition of shear walls, and the study is performed using SAP2000 software package. The area members including slabs are modelled as shells, and the shells are modelled as membrane in case of slabs. Constraints are modelled into joints in the form of diaphragms, and these diaphragms are rigid and are auto-generated for each individual storey. The bare frame model is modelled using M20 concrete and Fe415 grade of steel. The model is analysed using nonlinear static pushover analysis [5–7].

2 Analysis

See Fig. 1.

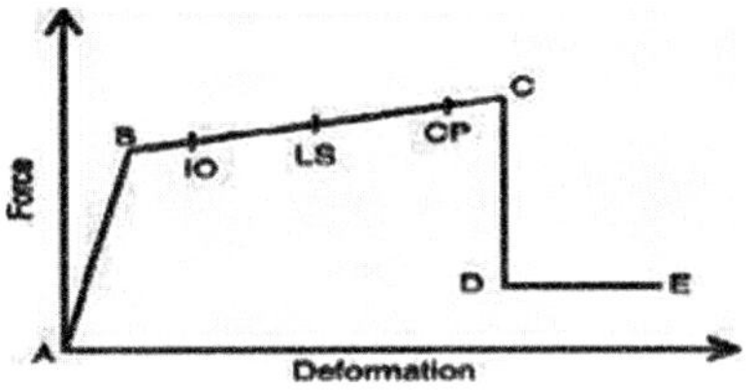

Fig. 1 Typical pushover curve

3 Nonlinear Static Pushover Analysis

Pushover analysis is a nonlinear static procedure (NSP) where the lateral force is incrementally increased in a predefined pattern. It gives better insight as to which member in the structure acts as a weak link, and the trends in formation of hinges help in identifying weak members(5). Pushover analysis helps in estimating base shear and displacement capacity of structure along with sequential formation of hinges in the hinges under analysis. The result is displayed in the form of *pushover curve* which is essentially a curve of base shear versus roof displacement. This curve can also be used as capacity curve of the structure. To incorporate the effect of inelastic behaviour of structural elements, lumped plasticity models have been used and assigned in the form of moment rotation capacity (i.e. plastic hinge) to individual element. In the present study, default properties of plastic hinges in SAP2000 as per FEMA 356 have been used (Fig. 2).

The pushover analysis procedure is validated through the study of Inel and Ozmen [8], and a frame was defined having plan dimension of 11.2 m and a storey height of 2.8 m. The material used was a grade of concrete having 16 Mpa as compressive strength, and reinforcement was selected to have a yield strength of 220 Mpa. The model was inserted with uncoupled moment (M3) hinges for beams and coupled axial force and biaxial moment hinges (P-M2-M3) as per FEMA 356-2000.

The pushover curve for bare and retrofitted curve is thoroughly studied and the following parameters are calculated,

Target displacement,

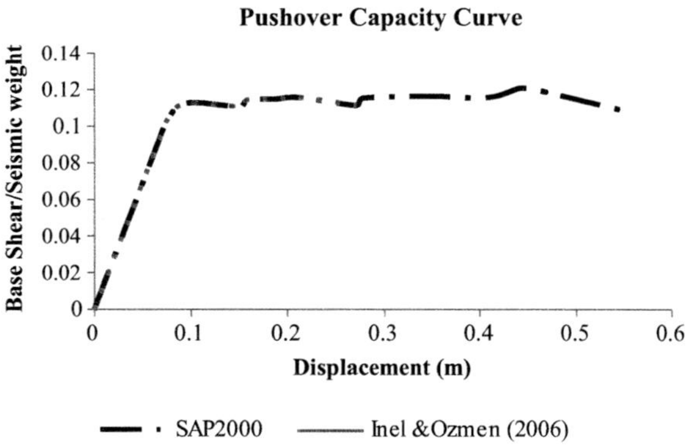

Fig. 2 Validation curve

As per 7.4.3.3.2 from ASCE/SEI 41-17, target displacement can be calculated as follows,

$\delta t = C_0 C_1 C_2 \,.\, \mathrm{Sa} \,.\, (\mathrm{Te}^2/4\pi^2) \,.\, g$

δt = Target displacement

C_0 = Modification factor to relate the spectral displacement model to an equivalent single-degree-of-freedom system (SDOF) to the multiple-degree-of-freedom (MDOF) system.

C_1 = Modification factor to relate expected maximum inelastic displacement to displacements calculated for linear elastic response. For periods less than 0.2 s, C_1 need not be taken, for periods greater than 1.0 s $C_1 = 1.0$.

C_2 = Modification factor to represent the effect of pinched hysteresis shape, cyclic stiffness degradation, and strength deterioration on the maximum displacement response.

Sa = Response spectrum acceleration at the effective fundamental period and damping ratio of the building under consideration.

Te = Effective fundamental period of the building in the direction under consideration.

4 Calculation of Capacity from Pushover Curve

The target displacement when plotted onto the pushover curve gives capacity of the building under analysis, as per ASCE 41-17. The capacity gives a vague idea regarding the performance of building in the predominant mode or first mode. This mode has the maximum period and is used primarily for pushover analysis [9, 10].

The modelling work is done on SAP2000 package; the model has a plan dimension of 12 × 14 m with three bays on either side of the structure. The building is a six-storeyed building with a storey height of 3 m; the building is situated in seismic zone V. The soil condition is hard rock and soil, and the model is subjected to lateral loads in the form of earthquake loads which are self-generated by the software. The load cases are defined as per IS 1893-Part-1-2016. The models are evaluated based on their target displacement, performance point and hinge formation characteristics.

First mode is predominantly used to perform pushover analysis on the building. Bare frame and retrofitted frame are inserted with auto hinges compliant with FEMA 356-2000, and uncoupled moment hinges (M3) are used for beams and coupled force-biaxial bending and moment hinges (P-M2-M3) [11, 12].

Response spectrum load case is defined in accordance with Table 7 of IS 1893 (Part 1)-2016. The mass source is defined to include dead load, 25 per cent of live load as well as other participating load combinations. The scale factor is calculated as follows,

$$\text{Scale factor} = (Z/2) \times (I/R) \times g$$

where

Z = Zone factor
R = Response reduction factor
I = Importance factor.
g = Acceleration due to gravity.

The model has similar beams on each floor; the beam dimensions are 230 × 450 mm. Building is designed for gravity as well as earthquake loads; the requirement for reinforcement is checked, and this helps in identifying critical members (Figs. 3 and 4).

The members shown in red have failed due to greater than maximum requirement of reinforcement; the columns failed are identified as GC-6, GC-7 at the ground storey and FC- 6, FC-7 at first storey (Fig. 5).

The members shown in red have failed due to greater than maximum requirement of reinforcement; the columns failed are identified from left to right as GC-10, GC-11 at ground storey and FC-10, FC-11 at first storey.

Name/designation of member	Location	Percentage of reinforcement required (%)
GC-6,GC-7	Ground storey	>6
FC-6,FC-7	First floor	>6

(continued)

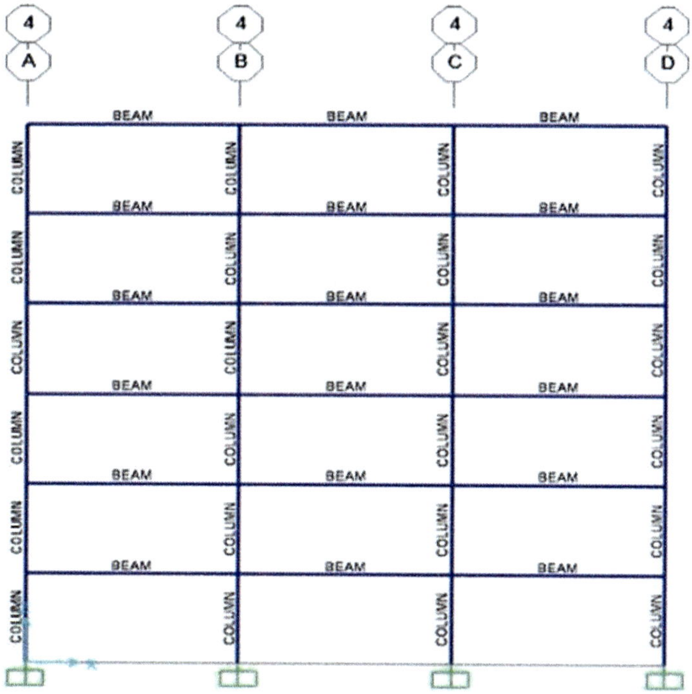

Fig. 3 Typical frame in the *X-Z* direction

(continued)

Name/designation of member	Location	Percentage of reinforcement required (%)
GC-10,GC-11	Ground storey	>6
FC-10,FC-11	First floor	>6

The identified critical members fail in design itself due to greater than maximum reinforcement requirement. It should be noted that C stands for column in this naming system. The columns in other storeys are also beyond the practical limit of reinforcement provision. Flexural members are marginally safe, although ultimate safety is dependent on the strength of compression members [4, 6, 12].

Retrofitting is done using shear walls on the periphery of building, and the walls have a thickness of 0.3 m. They are modelled as shell elements with nonlinearity in the direction *Y-Y*, and the walls have two layers of reinforcement. The pushover analysis is initiated using constraints on each storey in the form of non-rigid diaphragm. Displacement is in the form of monitored displacement, and the direction is the same as predominant mode (Fig. 6) [10, 13].

5 Results

The bare frame is evaluated for performance variables, and the ultimate performance level for a particular zone is calculated. Designs of members help in gaining insight about the critical members, and flexural members do not fail under any lateral load combination. However, the columns/compression members fail due to significantly higher reinforcement requirement (Fig. 7).

The bare frame exhibits a target displacement of 0.654 m, and the performance level as per the hinges formed in the building is limited safety (LS).

Step	Type of hinge	Member
3	B-IO	Beams except fifth and sixth storey
4	IO-LS	Beams at third storey
5	C-D	Base column
13	C-D	All columns
27	Beyond C	Total collapse

The shear walls fail prior to the yield displacement; however, they still satisfy the target displacement criteria with a target displacement of 0.079 m. The building exhibits a performance level of immediate occupancy (IO) (Fig. 8).

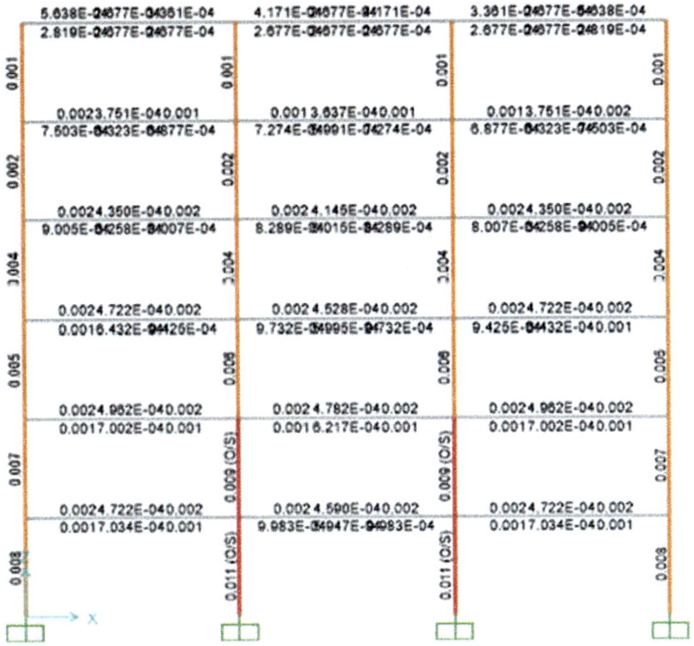

Fig. 4 Frame after analysis and design

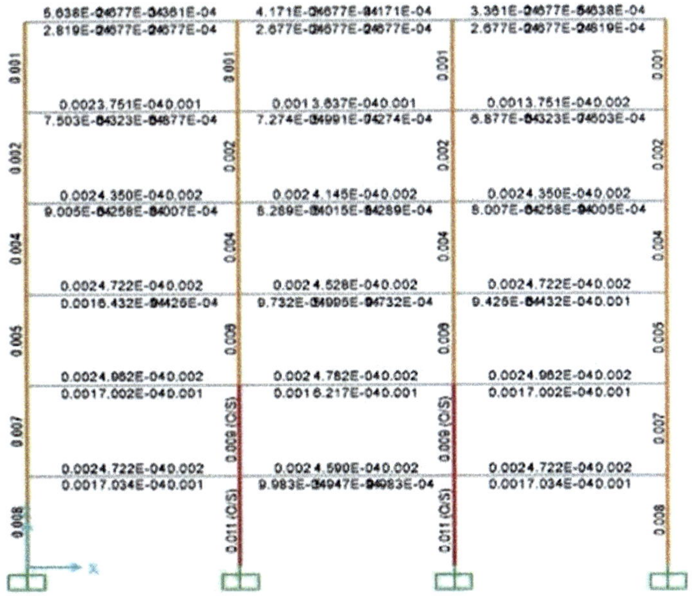

Fig. 5 Frame along *X-Y* direction post analysis

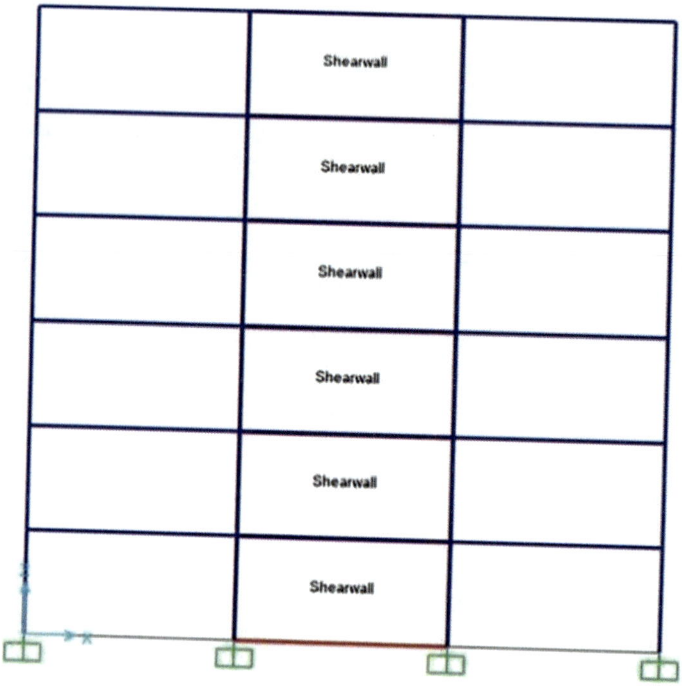

Fig. 6 Frame with added shear walls

Fig. 7 Pushover curve for bare frame

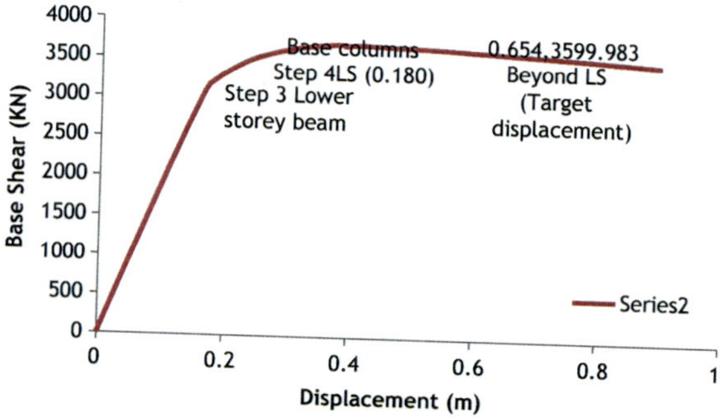

6 Conclusion

A substantial increase in the performance level of the building is observed indicated by the decrease in target displacement and performance level.

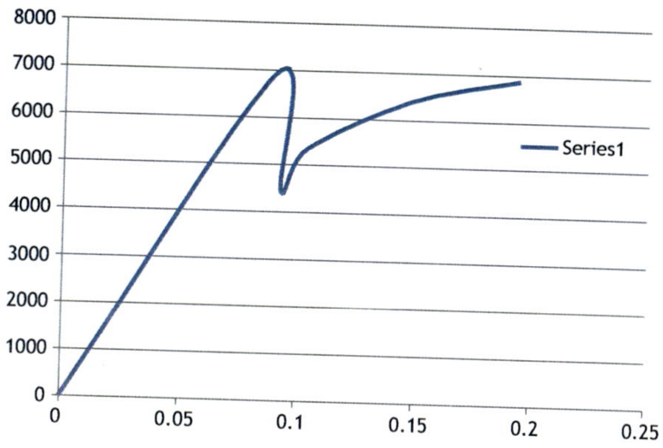

Fig. 8 Pushover curve for retrofitted frame

References

1. Rodrigues H, Furtado A, Vila-Pouca N, Varum H, Barbosa AR (2018) Int J Civ Eng
2. Impollonia N, Palmeri A (2017) Seismic performance of buildings retrofitted with nonlinear viscous dampers and adjacent reaction towers
3. Valente M, Milani G (2018) Alternative retrofitting strategies to prevent the failure of an under-designed reinforced concrete frame. Eng Fail Anal 89:271–289
4. Analytical study of moment-resisting frames retrofitted with shear slotted bolted connection
5. Durucan C, Dicleli M (2010) Analytical study on seismic retrofitting of reinforced concrete buildings using steel braces with shear link. J Eng Struct
6. Thermou GE, Pantazopolou SJ, Elnashai AS (2007) Design methodology for seismic upgrading of substandard reinforced concrete structures. J Earthq Eng
7. Thermou GE, Pantazopoulou SJ, Elnashai AS (2012) Global interventions for seismic upgrading of substandard RC buildings
8. Inel M, Ozmen HB (2006) Effects of plastic hinge properties in nonlinear analysis of reinforced concrete buildings. Eng Struct 28(11):1494–1502
9. Mazza F, Mazza M, Vulcano A (2018) Base-isolation systems for the seismic retrofitting of RC framed buildings with soft-storey subjected to near-fault earthquakes. Soil Dyn Earthq Eng 109:209–221
10. Ribakov Y, Halperin I, Pushkar S (2015) Seismic resistance and sustainable performance of retrofitted buildings by Adding Stiff Diaphragm Seismic Isolation
11. Chey M-H, Geoffrey Chase J, Mander JB, Carr AJ (2013) Innovative seismic retrofitting strategy of added stories isolation system
12. Mazza F, Mazza M, Vulcano A (2015) Displacement-based seismic design of hysteretic damped braces for retrofitting in elevation irregular r.c. framed structures. Soil Dyn Earthq Eng 69:115–124
13. Moon KH, Han SW, Lee CS (2017) Seismic retrofit design method using friction damping systems for old low- and mid-rise regular reinforced concrete buildings. Eng Struct 146:105–117

Utilization of Cupola Slag as a Sustainable Construction Material

S. S. Meshram and S. P. Raut

Abstract The present paper reviews the experimental characterization of cupola slag as a pozzolanic binder. Chemical composition of cupola slag was investigated using XRF scan. From the chemical composition, it can be indicated that the constituents of cupola slag is comparable to the other cementitious materials. Its use in production of building blocks is advantageous. The present paper explores various optimum percentage replacement level of cupola slag in the concrete and mortar.

Keywords Cupola slag (CS) · Characterization · X-ray fluorescence (XRF)

1 Introduction

Industrialization in creating nations has come about in consequent accumulation of unmanageable industrial waste. Different forms of industrial wastes generated in India are shown in Fig. 1. Pollution emerging from such wastes may be cause of concern for numerous creating countries, and huge land area is required to dump this industrial waste. Reusing such wastes into innovative material development can be a reasonable arrangement not as it were to the contamination issue, but moreover to the issue of the landfills. Due to continuously growing population, infrastructural needs are exponentially increasing, and in turn, demand for the construction materials is also increasing. This demand is continuously rising with the increasing need of houses both in rural as well as urban areas. Worldwide, researchers have made significant efforts to develop sustainable materials from agricultural and industrial waste materials as a central raw material.

S. S. Meshram (✉) · S. P. Raut
Department of Civil Engineering, Yeshwantrao Chavan College of Engineering, Nagpur, India
e-mail: sangitameshram3@gmail.com

S. P. Raut
e-mail: sprce@rediffmail.com

L. M. Gupta et al. (eds.), *Advances in Civil Engineering and Infrastructural Development*, Lecture Notes in Civil Engineering 87,
https://doi.org/10.1007/978-981-15-6463-5_5

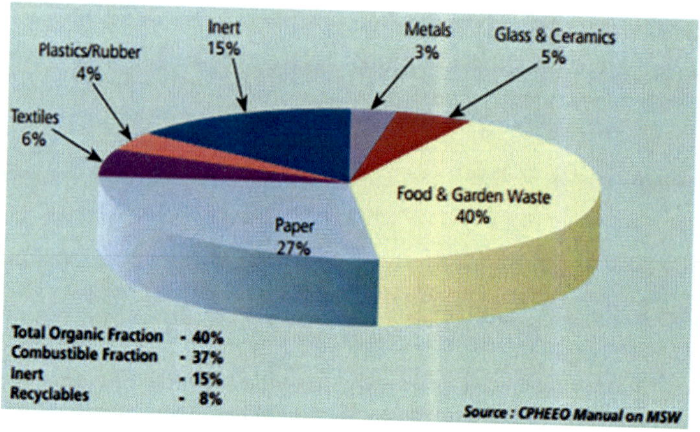

Fig. 1 Industrial waste generated in India

There is an increase in the demand of construction materials due to population growth, urbanization and industrial development. Since the huge request has been put on construction sector, and particularly during last 10–15 years due to the expanding inhabitants, demand–supply management of these materials is difficult. Subsequently, to fulfil the growing request, researchers are endeavouring to plan and create sustainable substitute solutions for the construction materials. Raise in attractiveness of using natural friendly, low-fetched and low-weight building materials within the construction sector has brought almost to explore how this may be accomplished by profiting to the environment as well as keeping the material prerequisites confirmed within the benchmarks [1, 2].

1.1 Cupola Slag

In India, 6000 cupola furnaces are introduced. Due to that larger amount of waste is generated in the form of cupola slag which is presently going to land filling only, polluting environment.

Cupola slag is a waste material of cast iron manufacturing which is formed in cupola furnaces during the separation of the molten steel from impurities. The slag is in the form of molten liquid melt and may be a composite arrangement of oxides and silicates which solidifies after cooling. Essentially, in integrated iron plants, steel and iron are now prepared using an electric arc furnace process. The open hearth furnace process is not utilized. About 5–7% of waste is generated in cupola furnaces while production of cast iron [3]. Industry produces 50–3000 Tonnes of C.I. depending upon furnace size and requirements. For each generation of 1 Ton of molten metal, 50 kg of slag is created roughly, i.e., 5% of the molten metal produced in the cupola furnace. In common, the cupola works for the period of 7–8 h/heat. Depending on the size of cupola furnace, about 20–25 T of molten metal/heat is produced. The sum of slag produced is around 1000–1200 kg/heat [4].

Cupola furnace is cylindrical shaped melting device which is used in steel industries for melting of cast iron ranging from 0.5 to 4 m in diameter. Bottom of furnace is having a door which can swing in and out. Top of the furnace is kept open. Air vent is arranged to supply the air in furnace. Shells of furnace are made up of steel and refractory bricks. There is one slag hole from which slag comes out at higher temperature with low viscosity that solidifies in black colour lumps upon cooling. Lumps size varied from 100 to 450 mm. Cupola slag (CS) is a thick solid material, and its colour changes from grey to black. A physical property of CS to be used as compared with natural coarse aggregates and IS 383-1970 requirements for material to be used as aggregates [5].

Portland cement is manufactured by blending sources of CaO, SiO_2, Al_2O_3 and Fe_2O_3 in fixed proportion in a kiln. Natural rock such as shale, dolomite and limestone contain these ingredients. In Portland cement, cupola furnace slag can be a substitute for calcium oxide, alumina and iron. In the manufacture of Portland cement, cupola furnace slag is not frequently used as a raw material because it is not available in commercial quantities. From these furnaces, cupola slag comes out about as by-product and is generally packed away and disposed [6].

2 Chemical Composition

The original samples of cupola slag are shown in Fig. 2. X–ray florescence (XRF) scan is done at Indian Bureau of Mines (IBM), Nagpur (As per Annexure 1). Chemical composition of different materials is appeared in Table 1. Physical properties of different materials are given in Table 2. The percentage of different chemical constituents of CS is comparable to the other cementitous materials; therefore, the use of cupola slag in manufacturing building blocks will be advantageous.

Fig. 2 Cupola slag

Table 1 Chemical composition of various materials

Chemical composition	PPC (Portland pozzolana cement) [7]	Fly ash [8]	Rice husk ash [7]	GGBS (Ground granulated blast furnace slag) [8]	Cupola slag (%) [XRF scan]
Lime, CaO	64.12	2.5	0.74	42.4	32.89
Silica, SiO_2	22.28	52.5	83.74	32.3	43.89
Alumina, AL_2O_3	4.72	28.2	0.29	13.3	13.28
Iron oxide, Fe_2O_3	2.75	10.5	0.67	0.3	1.59
Magnesia, MgO	1.23	1.6	0.86	6.4	2.77
Sulphur trioxide, SO_3	1.97	0.2	0.87	2.1	0.73
Sodium Oxide, NA_2O	0.28	0.04	0.091	–	0.16
Potassium oxide, K_2O	0.76	0.90	2.84	–	0.71

Table 2 Physical properties of various materials

Materials	Density (kg/m^3)	Specific gravity
PPC (Portland pozzolana cement)	1000–1500	2.9
Fly ash	1120–1500	2.96
Rice husk ash	340–400	2.19
GGBS (Ground granulated blast furnace slag)	1200–1300	2.85
Cupola slag	1200–1300	2.93

Vishwash K. Mistry replaced the cupola slag aggregates to natural aggregates with a varying percentage from 0 to 100 at a interval of 10% for M20 grade concrete. 0.5 water cement was used with a accelerated curing condition for almost three days at a temperature of 85 °C. Compressive strength of concrete was calculated. For ACC curing, getting strength drop for initial and final replacement, but in between replacement percentages, strength was increased in between 50 and 80% replacement. For ACC curing, 60% replacement gives higher strength. For normal curing, required characteristic mean strength for M20 grade concrete was also obtained at 80% replacement. So, it can be considered as optimum percentage of replacement [3].

Kiran Bhagat found that the physical properties of cupola slag and foundry sand are similar to those of coarse aggregate and fine aggregate respectively. Both materials are environmental building material. The problem of the disposal and maintenance problem of land filling is reduced. Cupola slag and foundry sand were varying from 0 to 100 percentage replacement with coarse and fine aggregates but at 60 percentage replacement the result were meets the standard requirements [5].

D. A. Aderibigbe revealed that chemical constituents of cupola slag exhibit pozzolanic properties. In the preparation of cement mortar, Portland cement was replaced by cupola slag up to 20% without sacrificing considerable strength (e.g., a 13.5% reduction in strength). There was considerable improvement in physical properties by calcining the slag at 700 °C for 5 h. These results suggested that the cupola slag may be used where cement is costly, as it is a waste product [9].

Researchers have utilized cupola slag as a partial replacement of cement and aggregates to develop sustainable concrete and mortar, but none of the researchers have used it to develop the sustainable building blocks. There is a lot of scope for curtailing the environmental pollution caused due to land filling of cupola slag as it has a potential of utilizing as a sustainable construction material.

3 Conclusion

- The utilization of cupola slag as an additive material in construction industry by means of reusing of waste, curtailing the environmental pollution caused due to land filling of cupola slag thus provides environmental and economic benefits.
- It has been observed that the chemical and physical properties of cupola slag is comparable to the other cementitious materials, and it can be used in the production of sustainable building blocks.
- The results clarify that the density of cupola slag is less as compared to normal bricks could be effectively used as an alternative in the lightweight brick production.

Annexure 1

GOVERNMENT OF INDIA
MINISTRY OF MINES
INDIAN BUREAU OF MINES
MODERN MINERAL PROCESSING
LABORATORY AND PILOT PLANT

No. 402(43)/MPD/Chem/2018-19/CB/ Nagpur, the 4th October.2018

To:
Prof. S.S.Meshram,
Project Guide.
Yeshwantrao Chavan College of Engg.,
Hingna road,Wanadongri,
Nagpur-441110 (M.S)

Sub : XRF scan of cupola slag sample of Mr. Taher Shakir, Final year Student. B.E,Civil Engineering, Yeshwantrao Chavan College of Engg., Nagpur

Ref: Your office letter no. nil dt. 27.09..2018

Sir,

With reference to the above pleas, find below the XRF scan of cupola slag sample of Mr. Taher Shakir, Final year Student, B.E.Civil Engineering, Yeshwantrao Chavan College of Engg., Nagpur.

Na2O (%)	MgO (%)	Al2O3 (%)	SiO2 (%)	P2O5 (%)	SO3 (%)	K2O (%)	SrO (%)	ZrO2 (%)	BaO (%)
0.16	2.77	13.28	43.89	0.05	0.73	0.71	0.07	0.04	0.11

Fe2O3 (%)	Rb2O (%)	MnO2 (%)	Cr2O3 (%)	CaO (%)	TiO2 (%)	Cl (%)	NiO %	LOI (%)
1.59	0.005	1.89	0.02	32.89	1.21	0.009	0.007	0.29

Thanking you.

Yours faithfully,

(Dr.A.N. Verma)
Senior Chemist & I/C. Chem.Lab.
FOR Chief Ore Dressing Officer

Plot No. L-8, MIDC, Hingna Road, Nagpur-4400016. Ph. 07104-234438 / 235282/235543 ext.210
Fax No.07104-235542, E-mail : indian75@bsnl.in, ibmhngn@bsnl.in

References

1. Raut SP, Ralegaonkar RV, Mandavgane SA (2011) Development of sustainable construction material using industrial and agricultural solid waste: a review of waste-create bricks. Constr Build Mater 25:4037–4042
2. Raut SP, Sedmake R, Dhunde S, Ralegaonkar RV, Mandavgane SA (2011) Reuse of recycle paper mill waste in energy absorbing light weight bricks. Constr Build Mater
3. Mistry VK, Patel BR, Varia DJ (2016) Suitability of concrete using cupola slag as replacement of coarse aggregate. Int J Sci Eng Res 7(2)
4. Chandiya KV, Karthika P (2017) An experimental study on strength and durability of concrete with partial replacement of coarse aggregate with cupola slag. Indo-Iran J Sci Res (IIJSR) 1(1):1–11
5. Bhagat K, Pokar NR, Rabadiya HM (2017) Suitability of concrete by using cupola slag as partial replacement of coarse aggregate and foundry sand as partial replacement of fine aggregate: a critical review. JETIR 4(3)

6. Arum C, Mark GO (2014) Partial replacement of Portland cement by granulated cupola slag—sustainable option for concrete of low permeability. Civ Environ Res 6(3)
7. Balapour M, Ramezanianpour AA, Hajibandeh E (2017) An investigation on mechanical and durability properties of mortars containing nano and micro RHA. Constr Build Mater 132:470–477
8. Gambhir ML (2017) Concrete technology, 5th edn. McGraw Hill Education
9. Aderibigbe DA, Ojobo AE (1982) Properties of cupola slag as a pozzolana and its effects on partial replacement of cement in a mortar. Conserv Recycl 5(4):203–208

Preparation of Flexural Design Charts Using IS 13920:2016

G. K. Koshti and R. K. Ingle

Abstract The code IS 13920:2016 *"Ductile design and detailing of reinforce concrete structures subjected to seismic forces—code of practice"* is used in the earthquake-resistant design of reinforced concrete buildings. The formulations and provisions given in IS 13920:2016 are different than the provisions given in IS 456:2000. The structural practice handbook, SP 16:1980, has tables and charts that help structural designers to rapidly design the simple sections according to IS 456:1978. Even though the SP 16:1980 is based on IS 456:1978, it continues to be used for IS 456:2000, without revision as there have been no major changes in Sect. 5 on which the design aid is based. While designing with IS 13920:2016, SP 16:1980 is used instead of separate design charts even though the design methodology is different, cumbersome, and iterative. The present study is aimed to propose a new design aid useful to the design of RC, sections according to the provisions of IS 13920:2016 and to provide ease to structural design. This study is validated along with detailed explanation using examples.

Keywords Ductile design · Design charts · Flexural capacity · Design aids

1 Introduction

The design of flexural members according to provisions of IS 13920:2016 such as member sizes and minimum reinforcement may not reflect their identities in IS 456:2000. Eventually, charts for these provisions are not available in SP 16:1980 containing this provisions. To design a beam for moments, it is required to refer doubly reinforced tables given in SP 16:1980 as Table 51, 52, 54, 55, 56 for corresponding Mu/bd^2 to get tension and compression reinforcement. These percentage

G. K. Koshti (✉)
VNIT Nagpur, Nagpur, India
e-mail: koshtigourav690@gmail.com

R. K. Ingle
Department of Applied Mechanics, VNIT Nagpur, Nagpur, India
e-mail: rkingle@rediffmail.com

L. M. Gupta et al. (eds.), *Advances in Civil Engineering and Infrastructural Development*, Lecture Notes in Civil Engineering 87,
https://doi.org/10.1007/978-981-15-6463-5_6

reinforcements from the above tables are only for balanced sections and not for under-reinforced sections (SP 16:1980). Most preferable sections in practice are under-reinforced sections but not balanced sections which are ductile and give warning before failure [1]. There is no proper control on the under-reinforced sections as per Indian code IS 456:2000 and its special extension SP 16:1980. This lack leads to unpredictable ductility to the beams designed to IS code.

1.1 Need of Design Aid for Earthquake-Resistant Design

In earthquake-resistant design of reinforced concrete beam, the flexural reinforcement provided should sustain the reversal moments, and it should give ductile behavior [2]. Ductility performance or the reinforced concrete beam is increased by adding compression reinforcement along with increasing grade of concrete, decrease in tension reinforcement, decreasing grade of steel, providing special confining reinforcement in the form of stirrups [3]. By increasing the compression reinforcement exactly to half of the tensile reinforcement, beam can be become a under-reinforced section which gives ductile warning as per clause 6.2.3 of IS 13920:2016.

In order to get the tension reinforcement exactly half the compression reinforcement, designer required to do manual calculations. These limitations using SP 16:1980 lead to have a separate design chart for flexure design which can give tension and compression reinforcement half the tension reinforcement in a single reference. Similarly for the development, length provision for the flexural reinforcement in IS 13920:2016 is different than IS 456:2000 [4]. Design tables in SP 16:1980 may not satisfy the ductile requirement.

1.2 Changes Required in SP 16:1980 to Make It Useful for ERD

After obtaining the top face and bottom reinforcement from flexural design capacity, moments are required to calculate for shear design according to clause 6.3.3 of IS 13920:2016. In order to get the capacity moments for sagging and hogging, designer should depend on SP 16:1980. If the percentage of top face reinforcement or bottom face reinforcement is falling in doubly reinforced tables, corresponding moment is taken as moment capacity of considered reinforcement percentage [5]. If not, it is required to refer the singly reinforced table. If the considered face reinforcement percentages are falling singly as well as doubly reinforced tables two times referring and iteration make cumber some. This may lead to incorrect values of sagging and hogging capacities.

Similarly in the joint design calculations, the safety of joint is mainly governed by shear and the flexure check. While calculating column shear in joint shear strength check, hogging moment and sagging moment capacities, i.e., M_s, M_h, M_{c_T} and M_{c_B} for top face and bottom face reinforcement are required due to formation of plastic hinges in beams [6]. Double referring iterative calculations makes joint shear strength calculations cumbersome as similar as shear strength. In column shear strength calculations also, same problem will be faced [7]. Except member size restrictions, confining spacing requirements for column the charts given in SP 16:1980 are sufficient enough to use in earthquake-resistant design.

2 Preparation of Design Charts

The provisions given in IS 13920:2016 are sufficient to provide the ductile behavior to reinforced concrete structures. But still, there are much improvements that are required to include when it is compared with other international codes [8]. To calculate flexural capacity of a section to design for a shear, much labor work can be reduced by referring only one chart for top steel and bottom steel. Moreover, this readily available charts will be helpful to designers for a rapid design. This can be modified to design for shear produced due to earthquake, considering the effects of confinement, and more accurate results can be obtained.

2.1 Different Cases Considered for the Study

Depending on the philosophies and clauses given in different standard codes, different cases are considered for this study as given in Table 1.

Above design and flexural capacity calculations procedure is discussed for a considered section. To simplify this, spreadsheet is prepared, and step by step method is followed. As per flowchart shown in Figs. 1 and 2, one of the major parameter which plays an important role while calculating the actual flexural capacity is confinement of the concrete due to hoop reinforcement. Methods are discussed by Kent et al., Park et al. [9], and Mander et al. [10]. Inducing this effect, more accurate results could be obtained.

Table 1 Different cases considered for the study

Case I	Case II
As per SP 16:1978,	As per IS 13920:2016; clause 6.2.3,
$p_t = p_{t,\lim} + p_{t,2}$	$p_t = p_{t,\lim} + p_{t,2}$
$p_{sc} = p_{t,2}$	$p_{sc} = 0.5\ p_t$

Fig. 1 Flowchart for flexural design using SP 16:1980 and IS 13920:2016 approach

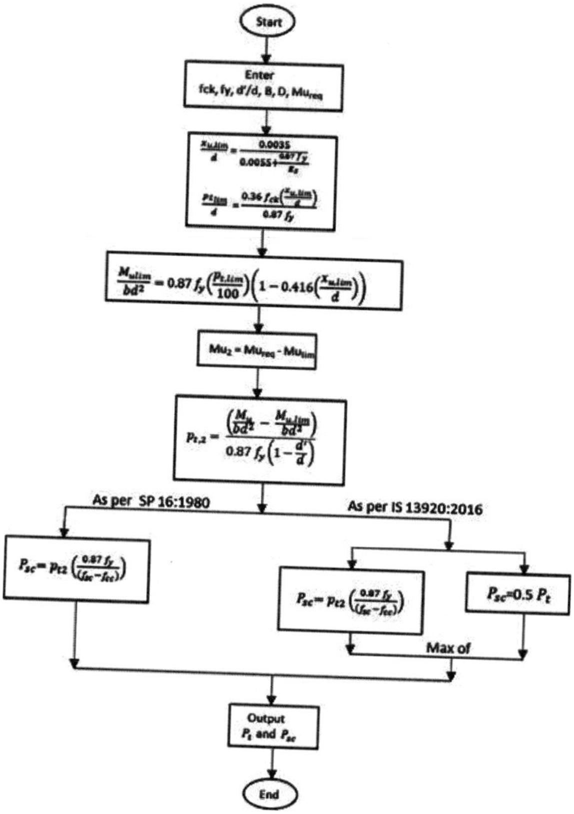

3 Flexural Design Charts

As discussed in Sect. 2, charts are prepared in spreadsheets using visual basic program, and sample of this is given in Table 2. To explain the procedure to prepare charts, the combination of $f_{ck} = 25$ MPa and $f_y = 415$ MPa is considered in the present study.

Similarly, the flexural design charts can be prepared considering M30 and M35 grade of concrete, different grades of steel, and different levels of strain at compression steel (d'/d) viz 0.125, 0.150, and 0.175. The procedure to explaining the use of these charts is discussed in the next section.

4 Use of Prepared Flexural Charts

The procedure to use the charts considering different cases is explained below with the help of example.

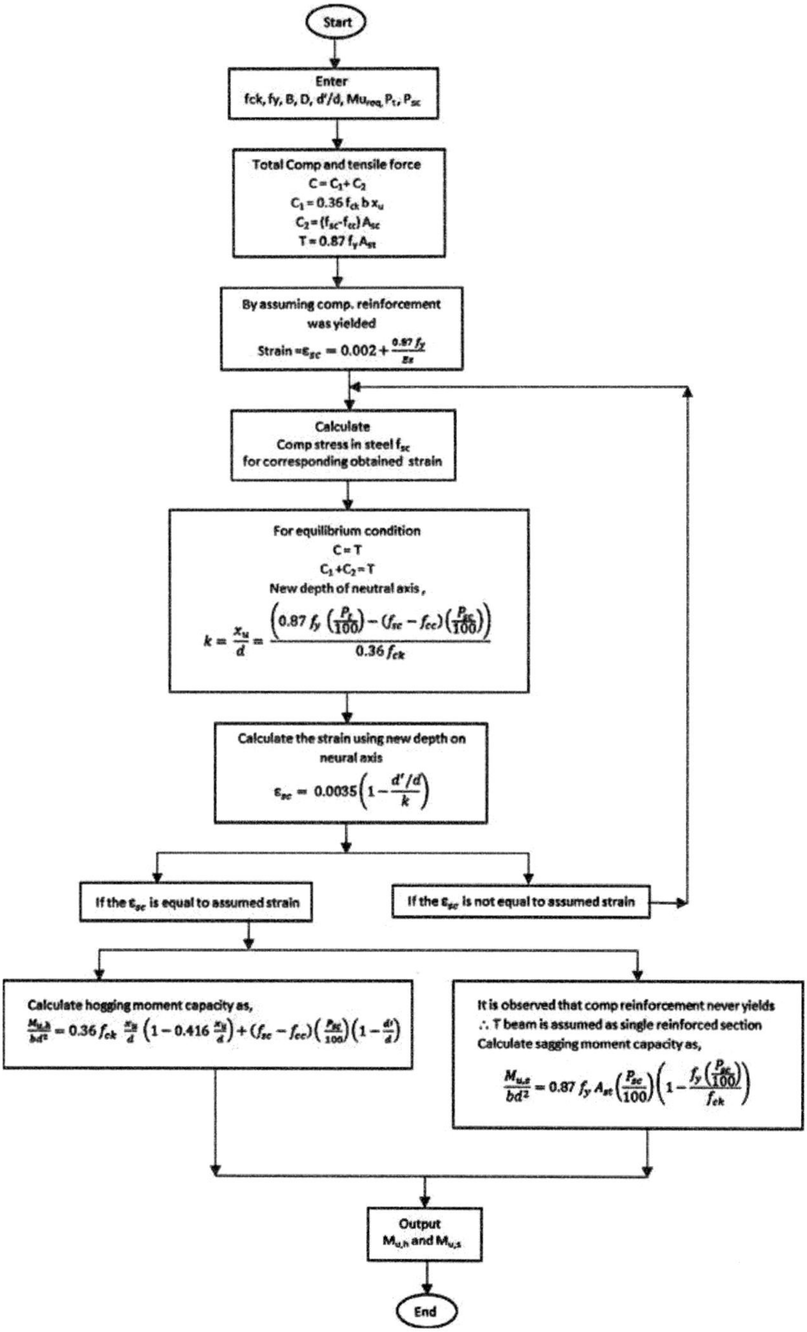

Fig. 2 Flowchart for flexural capacity calculation for shear design

Table 2 Flexural design charts

$d'/d = 0.05$			$d'/d = 0.075$			$d'/d = 0.1$		
Mu/bd^2	% Ast	% Asc = Ast/2	Mu/bd^2	% Ast	% Asc = Ast/2	Mu/bd^2	% Ast	% Asc = Ast/2
3.46	1.202	0.601	3.46	1.202	0.601	3.46	1.1822	0.591
3.5	1.214	0.607	3.5	1.214	0.607	3.5	1.1942	0.597
3.6	1.243	0.621	3.6	1.244	0.622	3.6	1.2241	0.612
3.7	1.272	0.636	3.7	1.274	0.637	3.7	1.2541	0.627
3.8	1.301	0.651	3.8	1.304	0.652	3.8	1.2840	0.642
3.9	1.330	0.665	3.9	1.334	0.667	3.9	1.3140	0.657
4.0	1.360	0.680	4.0	1.364	0.682	4.0	1.3439	0.672
4.1	1.389	0.694	4.1	1.394	0.697	4.1	1.3739	0.687
4.2	1.418	0.709	4.2	1.424	0.712	4.2	1.4038	0.702
4.3	1.447	0.724	4.3	1.454	0.727	4.3	1.4337	0.717
4.4	1.476	0.738	4.4	1.484	0.742	4.4	1.4637	0.732
4.5	1.505	0.753	4.5	1.514	0.757	4.5	1.4936	0.747
4.6	1.535	0.767	4.6	1.544	0.772	4.6	1.5136	0.757
4.7	1.564	0.782	4.7	1.574	0.787	4.7	1.5435	0.772
4.8	1.593	0.796	4.8	1.603	0.802	4.8	1.5735	0.787
4.9	1.622	0.811	4.9	1.633	0.817	4.9	1.6034	0.802
5.0	1.651	0.826	5.0	1.663	0.832	5.0	1.6333	0.817
5.1	1.680	0.840	5.1	1.693	0.847	5.1	1.6633	0.832
5.2	1.709	0.855	5.2	1.723	0.862	5.2	1.6932	0.847
5.3	1.739	0.869	5.3	1.753	0.877	5.3	1.7232	0.862
5.4	1.768	0.884	5.4	1.783	0.892	5.4	1.7531	0.877
5.5	1.797	0.898	5.5	1.813	0.907	5.5	1.7931	0.897
5.6	1.826	0.913	5.6	1.843	0.922	5.6	1.8230	0.912
5.7	1.855	0.928	5.7	1.873	0.936	5.7	1.8529	0.926
5.8	1.884	0.942	5.8	1.903	0.951	5.8	1.8829	0.941
5.9	1.914	0.957	5.9	1.933	0.966	5.9	1.9128	0.956
6.0	1.943	0.971	6.0	1.963	0.981	6.0	1.9428	0.971
6.1	1.972	0.986	6.1	1.993	0.996	6.1	1.9727	0.986
6.2	2.001	1.001	6.2	2.023	1.011	6.2	2.0027	1.001
6.3	2.030	1.015	6.3	2.053	1.026	6.3	2.0326	1.016
6.4	2.059	1.030	6.4	2.083	1.041	6.4	2.0625	1.031
6.5	2.088	1.044	6.5	2.112	1.056	6.5	2.0925	1.046

(continued)

Table 2 (continued)

$d'/d = 0.05$			$d'/d = 0.075$			$d'/d = 0.1$		
Mu/bd^2	% Ast	% Asc = Ast/2	Mu/bd^2	% Ast	% Asc = Ast/2	Mu/bd^2	% Ast	% Asc = Ast/2
6.6	2.118	1.059	6.6	2.142	1.071	6.6	2.1224	1.061
6.7	2.147	1.073	6.7	2.172	1.086	6.7	2.1474	1.074
6.8	2.176	1.088	6.8	2.202	1.101	6.8	2.1773	1.089
6.9	2.205	1.103	6.9	2.232	1.116	6.9	2.2073	1.104
7.0	2.234	1.117	7.0	2.262	1.131	7.0	2.2372	1.119
7.1	2.263	1.132	7.1	2.292	1.150	7.1	2.2671	1.134
7.2	2.293	1.147	7.2	2.322	1.182	7.2	2.2971	1.149
7.3	2.322	1.178	7.3	2.352	1.213	7.3	2.3270	1.164
7.4	2.351	1.208	7.4	2.382	1.245	7.4	2.3570	1.178
7.5	2.380	1.239	7.5	2.412	1.276	7.5	2.3869	1.193
7.6	2.409	1.270	7.6	2.442	1.308	7.6	2.4169	1.208
7.7	2.438	1.300	7.7	2.472	1.339	7.7	2.4468	1.223
7.8	2.467	1.331	7.8	2.502	1.371			
7.9	2.497	1.362						
8.0	2.526	1.392						

4.1 Design Example Using Prepared Charts

Given data: $M_{u,h} = -369$ kN m, $M_{u,s} = 320$ kN m. $B = 300, D = 600, f_y = 415$ MPa, $f_{ck} = 25$ MPa, cover for moderate exposure conditions = 30 mm, assuming 20 mm diameter bars in two layers. Calculate % reinforcement required for this (Source: Ingle and Jain [11].

$\therefore$ Effective depth = 600 – 30 – 8 – 20 – 10 = 532 mm

$$\frac{M_{u,h}}{bd^2} = \frac{-369 * 10^6}{300 * 532^2} = 4.35, \frac{M_{u,s}}{bd^2} = \frac{320 * 10^6}{300 * 532^2} = 3.77, \frac{d'}{d} = \frac{68}{532} = 0.1278$$

Referring to Table 51 of SP 16 and Table 2 of current study, for $f_{ck} = 25$ and Fe 415 for $d'/d = 0.1$. The comparative study of the output obtained considering above different cases discussed in Sect. 2 is given in Table 3.

4.2 Flexural Capacity Calculation Using Prepared Charts

In earthquake action, due to lateral vibrations, shear gets developed in structural members as well as at joints. So by knowing the capacity at a section, shear can be calculated. This is explained with the help of following example.

Table 3 Comparative study of design parameters obtained from different cases

		Case I (SP 16)	Case II (IS 13920:2016)
(i)	Top steel (P_t) for hogging action, $\frac{M_{u,h}}{bd^2} = 4.35$	1.476	1.326
(ii)	Bottom steel (P_{sc}) for hogging action	0.291	0.738
(iii)	Bottom steel P_t for saging action, $\frac{M_{u,s}}{bd^2} = 3.77$	1.300	1.260
(iv)	Top steel (P_{sc}) for sagging action	0.102	0.650
(v)	Top steel = max of (i) and (iv)	1.476	1.326
(vi)	Bottom steel = max of (ii) and (iii)	1.300	1.260

Table 4 Reinforcement details

Longitudinal reinforcement			
Beam	Left end	Center	Right end
Top reinforcement	3–16 Φ straight + 5–20 Φ + 1–16 Φ extra Steel provided = 2374 mm^2 i.e. 1.487%	3–16 Φ straight Steel provided = 603 mm^2 i.e. 0.378%	3–16 Φ straight + 5–20 Φ + 1–16 Φ extra Steel provided = 2374 mm^2 i.e. 1.487%
Bottom reinforcement	3–16 Φ straight + 3 20 Φ extra Steel provided = 1545 mm^2 i.e. 0.97%	3–16 Φ straight Steel provided = 603 mm^2 i.e. 0.378%	3–16 Φ straight + (2–16 Φ + 1–20 Φ) extra Steel provided = 1319 mm^2 i.e. 0.99%

Source Ingle and Jain [11]

Example Given Data: $f_{ck} = 25$ MPa, $f_y = 415$ MPa, $d'/d = 0.1$.

Summary of flexural reinforcement is as per Table 4.

The additional shear due to formation of plastic hinges at both the ends of beams is evaluated as per clause 6.3.3 of IS 13920:2016. Here, $M_{u,\lim}^{As}$, $M_{u,\lim}^{Bh}$ are the sagging and hogging moments at both the ends of the beam which are to be calculated on the basis of actual area of steel provided in the section.

At joint *A*, the beam is provided with $p_t = 1.487$ at top and $p_{sc} = 0.97$ at bottom, and at joint *B*, it is provided with $p_t = 1.487$ at top and $p_{sc} = 0.99$ at bottom.

Assume, shear at the ends of beam, $V_{uA}^{D+L} = V_{uB}^{D+L} = \frac{1.2\times(DL+LL)}{2} = 84$ kN, length = 5 m now, following Table 5 is prepared for flexural capacity by referring Table 51 of SP 16:1980 and Table 2 of current study.

Using equations given in IS 13920:2016 to calculate shear for beam section, flexural capacities obtained in Table 5, and shear is calculated for both the cases, and the results are given in Table 6.

Table 5 Calculation of flexural capacity using charts

	$p_t = 1.487$		$p_t = 0.97$		$p_t = 0.99$	
	SP 16	IS 13920:2016	SP 16	IS 13920:2016	SP 16	IS 13920:2016
$\frac{M_u}{bd^2}$	4.476	4.643	2.297	3.128	2.862	3.189
MR (kN m)	380	394	237	265	234	270

Table 6 Comparison of shear developed due to earthquake action

Sway direction	Right		Left	
End shear	V_{uA} (kN)	V_{uB} (kN)	V_{uA} (kN)	V_{uB} (kN)
Case I (SP 16)	89	257	257	89
Case II (IS 13920)	101	269	270	103

5 Observation and Discussions

i. It is observed that (Table 3), top steel required is less when we used the charts prepared using IS 13920:2016 compared with SP 16. Hence, section becomes under-reinforced leading to more ductility.
ii. The charts prepared using IS 13920:1993 provide ease to structural design as well as to calculate flexural capacity of a beam section easily compared to SP 16:1980 charts.
iii. From Table 5, it can be observed that more shear is developed in case II [3] compared to case I (SP 16). This is because of being more compression reinforcement is considered in case II which gives more moment of resistance.
iv. From Table 6, it can be seen that the flexural capacities required in earthquake analysis are increased, and this increased capacities need to consider for shear design as these indications are close to actual behavior of section.
v. As the increased capacities will give increased shear force this will give a safe design for shear avoiding brittle failure.

References

1. Jain SK (2016) Earthquake safety in India: achievements, challenges and opportunities. Bull Earthquake Eng 14(5):1337–1436
2. Murty CVR (2010) How do beam—column joints in RC buildings resist Earthquake? IITK-BMPTC Earthquake Tip 20
3. IS 13920:2016 (2016) Indian standard code of practice on ductile design and detailing of reinforced concrete structures subjected to seismic forces. Bureau of Indian Standards, New Delhi, India
4. IS 456:2000 (2000) Indian standard code of practice on plain and reinforced concrete. Bureau of Indian Standards, New Delhi, India

5. Priestley MJN, Paulay T (1992) Seismic design of reinforced concrete and masonry buildings, 3rd edn. Wiley
6. IS 1893:2016 (Part 1) (2016) Indian standard criteria for earthquake resistant design of structures (Part 1—General Provisions and buildings). Bureau of Indian Standards, New Delhi, India
7. Jain SK, Murty CVR (2014) Proposed draft provisions and commentary on ductile detailing of RC structures subjected to seismic forces. Final report of earthquake code IITK GSDMA EQ16-V3
8. Khose VN, Singh Y, Lang D (2010) Limitations of Indian seismic design codes for RC buildings. In: Kumar A, Sharma M (eds) 14 symposium on earthquake engineering, pp 1416–1423
9. Kent DC, Park R (1971) Flexural members with confined concrete. J Struct Div 97(7):1969–1990
10. Mander JB, Priestley MJ, Park R (1988) Theoretical stress-strain model for confined concrete. J Struct Eng 114(8):1804–1826
11. Ingle RK, Jain SK (2004) Explanatory examples on ductile detailing of RC building. IITK GSDMA EQ22. www.iitk.ac.in/nicee/IITK-GSDMA/EQ22.pdf

Seismic Response Reduction of RC Frame Staging in Elevated Water Tank

Sakshi A. Manchalwar and V. Verghese

Abstract Elevated water tank with frame staging is mostly used in India and due to its requirement after earthquake, it is considered as an important structure. Thus, it is required to control the seismic response of tank staging to minimize damages. In this work, the effectiveness of lead rubber bearing (LRB) and X-plate damper (XPD) in seismic response control of staging is investigated. Initially, two types of frame staging are considered for the same capacity of the tank and from nonlinear static analysis, best-suited staging is identified. After this, LRB and XPD are installed separately and to check the effectiveness of these devices, nonlinear time history analysis has been performed for four different earthquake records. It is observed that due to these control devices, seismic responses are significantly reduced.

Keywords Elevated water tank · Lead rubber bearing · Time history analysis · Pushover analysis · Frame staging

1 Introduction

Planning structures to react flexibly when they are exposed to quakes are uneconomical, particularly in high seismic peril locales. Along these lines, seismic plan particulars grant structures to yield under the plan premise tremor. For whatever length of time that sufficient specifying is given, structures can be intended for power levels much lower than would be required for the flexible reaction.

Traditional seismic tremor safe structures depend on inelastic conduct in assigned segments of the structures. For exceptional minute casings, where high malleability requests are normal, inelastic conduct or harm ought to be continued in the bars from the section faces and in the bar segment associations. Vitality scattering (inelastic

S. A. Manchalwar (✉)
Priyadarshini J. L. College of Engineering, Nagpur, Maharashtra 440009, India
e-mail: manchalwarsakshi@gmail.com

V. Verghese
Karmavir Dadasaheb Kannamwar College of Engineering, Nagpur, India

L. M. Gupta et al. (eds.), *Advances in Civil Engineering and Infrastructural Development*, Lecture Notes in Civil Engineering 87,
https://doi.org/10.1007/978-981-15-6463-5_7

reaction) is a consequence of basic harm (yielding), which is now and then troublesome and costly to fix. Furthermore, if the area of the vitality dissemination in the structure is not anticipated precisely, and if the harm happens in the gravity load framework, the structures may fall.

Structures ought to have the option to keep up a few cycles of inelastic distortion without a noteworthy loss of solidarity during quakes. A sensible measure of firmness misfortune can be obliged during the inelastic distortion, and the presentation of the structure can be assessed by looking into the vitality dispersal at each cycle of loading.

An assortment of strategies is accessible to manage immense tremor powers. Among these strategies, confining the structure from the beginning, inactive vitality dissemination frameworks, and controlling the inelastic reaction of the structure are the best and usually utilized.

Seismic isolation for tank is effective in decreasing seismic effect of elevated water tanks [1]. Panchal and Jangid [2] investigated seismic effect of liquid tanks using variable friction isolator and results show this isolation is very effective in decreasing seismic effect. Shekari et al. [3] investigated isolated tank for long period earthquakes which shows significant reduction in response parameters as compared to fixed base tank. Friction pendulum isolator is more efficient for slender tank [4] and isolator parameters also affect the response of tank. Soni et al [5] investigated the effectiveness of double variable frequency pendulum isolator for liquid storage slender and broad tanks. By using three types of isolation systems, Seleemah and Sharkawy [6] investigated the seismic responses of base-isolated slender and broad cylindrical liquid storage ground tanks. Panchal and Jangid [7], the seismic effect of liquid storage steel tanks was studied using Variable frequency pendulum (VFP) Isolator.

The present work is distributed into two phases: in the first phase, seismic performance evaluation of two different tank staging is done for same tank capacity and compared the nonlinear static performance. In second phase, the seismic response of staging for nonlinear dynamic analysis is the study. This nonlinear time history analysis is performed for two real ground motions in SAP [8]. From previous study, it is observed that passive dampers are not studied for tanks. Thus, base isolation (laminated rubber bearing) and passive device (metallic dampers) are used, respectively. These protective systems significantly reduced the seismic response of structure which makes safe response of conditions of staging.

2 X-Plate Damper (XPD)

An XPD may be a device that is capable of sustaining several cycles of stable yielding deformation leading to a high level of energy dissipation or damping. Its energy dissipation depends totally on relative displacement at intervals on the device and not on the relative velocities. It consists of an assembly that holds either single or multiple elements of 'X'-form plates, the amount of plates depends on the requisite of the system to dissipate the external input unstable energy. The employment of such

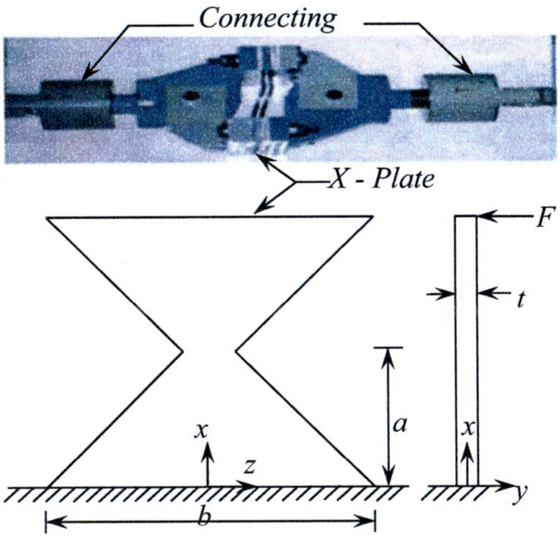

Fig. 1 Typical XPD with holding devices [17]

device was first planned by Kelly et al. [9], they played out a progression of trial tests on structures and his work was reached out by Skinner et al. [10] and Tyler [11]. An X-plate damper device has been studied via experiments by Bergman and Goel [12] and Whittaker [13, 14], subsequently employed in seismic retrofitting projects discussed by Matinez-Romero [15] and Perry et al. [16]. A typical XPD with device employed in the current work as shown in Fig. 1.

Using beam theory, the properties of XPD are expressed in Eqs. (1–3),

$$F_y = \frac{\sigma_y b t^2}{6a} n \tag{1}$$

$$q = \frac{2\sigma_y a^2}{Et} \tag{2}$$

$$K_d = \frac{F_y}{q}$$

$$K_d = \frac{Ebt^3}{12a^3} n \tag{3}$$

3 Lead Rubber Bearing Isolation

The LRB framework was invented in New Zealand in 1975 and created as an elastomeric-based framework. Afterward, it was generally utilized in created nations, for example, Japan and the USA. The working rule of this framework is like the laminated elastic bearing framework. The frameworks contrast in light of the fact that around

Fig. 2 Lead rubber bearing isolator

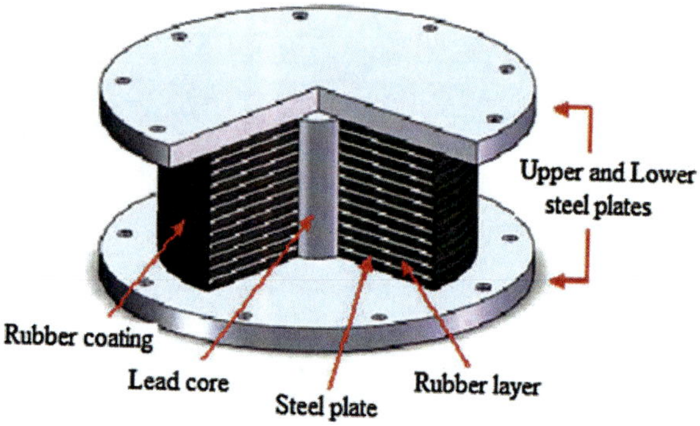

and hollow lead center is put amidst the LRB framework to give extra unbending nature to the framework (Fig. 2). The lead center builds the vitality retention limit of the framework; in this way, the parallel solidness is possibly expanded against solid ground movements. Subsequently, the principle point of the lead expansion is to increment both the solidness at moderately low even power levels and the vitality dispersal limit.

The attributes of the lead material have been considered in the generation of LRB frameworks. By and large, lead has a low yield moment that shear stress achieve 10 MPa, and it shows elasto-plastic conduct. Lead is additionally impervious to rehashed stacks and can restore itself after some time following deformation, Komur [18].

4 Details of Tank Model

In this work, elevated tank of capacity 250 m^3 with two different staging is considered. The configuration of frame staging is shown in Fig. 3. The tank frame staging is considered a special moment resistance frame with response reduction factor $R = 4$. Seismic forces are calculated according to IS 1893 Part 2 [19] and designed. Concrete

Fig. 3 Plan and elevation of staging

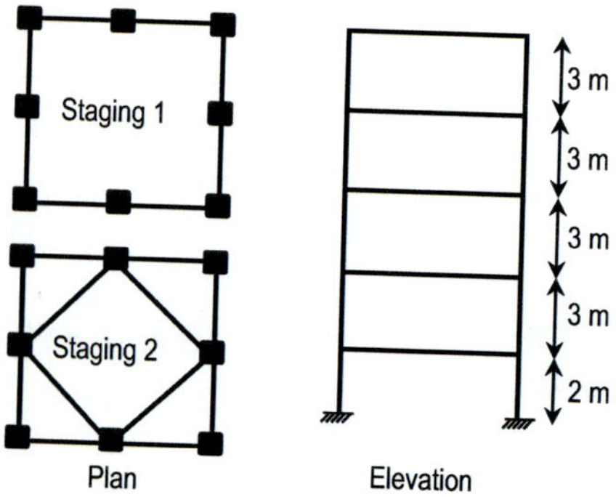

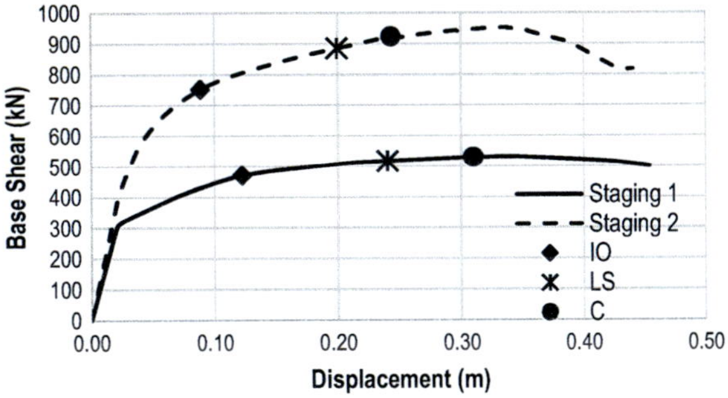

Fig. 4 Pushover curve for two different staging patterns

grade of M30 and steel grade of Fe415 is used in the design. Initially, the size of a column, top beam, and brace based on IS 1893 Part 2 [19] 450 × 450 mm, 300 × 600 mm, and 300 × 450 mm is used, respectively. All the sections are designed as per IS 456 [20].

Figure 4 shows a capacity curve which gives a comparison between two types of staging used for the same capacity of the water tank. Staging 2 having additional diagonal bracing in a plan which improves performance of staging as compared to staging 1. Performance of structure has been classified into different levels, viz. Immediate Occupancy (IO), Life Safety (LS), and Collapse (C). Immediate Occupancy implies the post-earthquake damage state in which without a doubt, very constrained basic damage has happened. Life Safety Performance Level methods the post-earthquake damage state in which critical damage to the structure has happened; however, some edge against either fractional or all out auxiliary breakdown remains. Breakdown Performance Level methods the structure is very nearly encountering an incomplete or complete collapse. To study the effectiveness of base isolation and damper, staging 2 is considered.

4.1 Ground Motion Data

For the numerical analysis, four earthquake records are listed in Table 1 which were taken from ATC-40. These time histories have been scaled to spectral compatible

Table 1 Earthquake used in the study

EQ ID	Earthquake			PGA (g)
	M	Year	Name	
120522	7.1	1999	Hector Mine	0.34
120612	6.5	1979	Imperial Valley	0.35
120812	7.5	1999	Kocaeli, Turkey	0.36
121211	6.5	1987	Superstition Hills	0.36

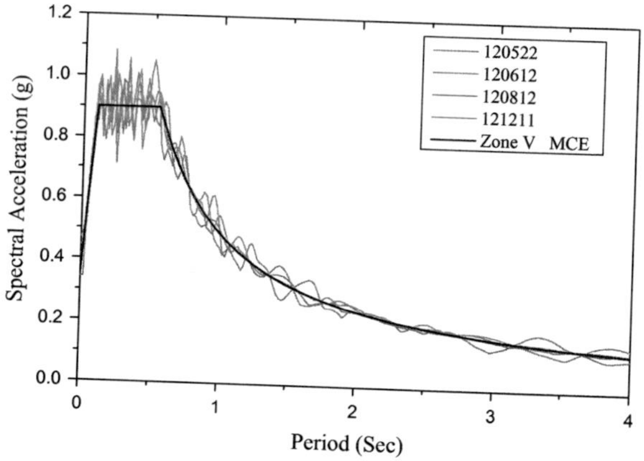

Fig. 5 Compatible response spectra for four earthquakes

time history data with peak ground acceleration (PGA) of 0.36 g with the help of a wavelet transform-based software WAVEGEN [21]. Figure 5 shows the target and the matched response spectra.

5 Dynamics Response with Isolator

Lead rubber bearing is installed at the base of staging 2. The seismic response of staging with and without LRB is shown in Fig. 6 for four different earthquakes. It is observed that base shear is significantly reduced in the case of LRB as compared to fix base structure. Base shear of the structure is depending on mass and fundament period of the structure. Fundament period of the structure and base shear is inversely proportional. The fundament period of the structure with LRB is increased by 2 to 3 times period of the fixed base structure thus reduction of base shear is observed. Because of the reduction in forces, damages in tank staging and wall will minimize.

6 Dynamic Response with Metallic Damper

Figure 7 shows displacement control of staging with damper as compared to without damper case for four earthquakes. It is observed that top staging displacement reduces approximately 60–70% in all earthquakes. Elastic drift limit of the elevated tank is restricted to 0.4 present of the height of staging as per IS 1893 Part 1 [22]. It is found that elastic drift limit has been checked in both the cases and found to be in case with damper, it is in the prescribed limit.

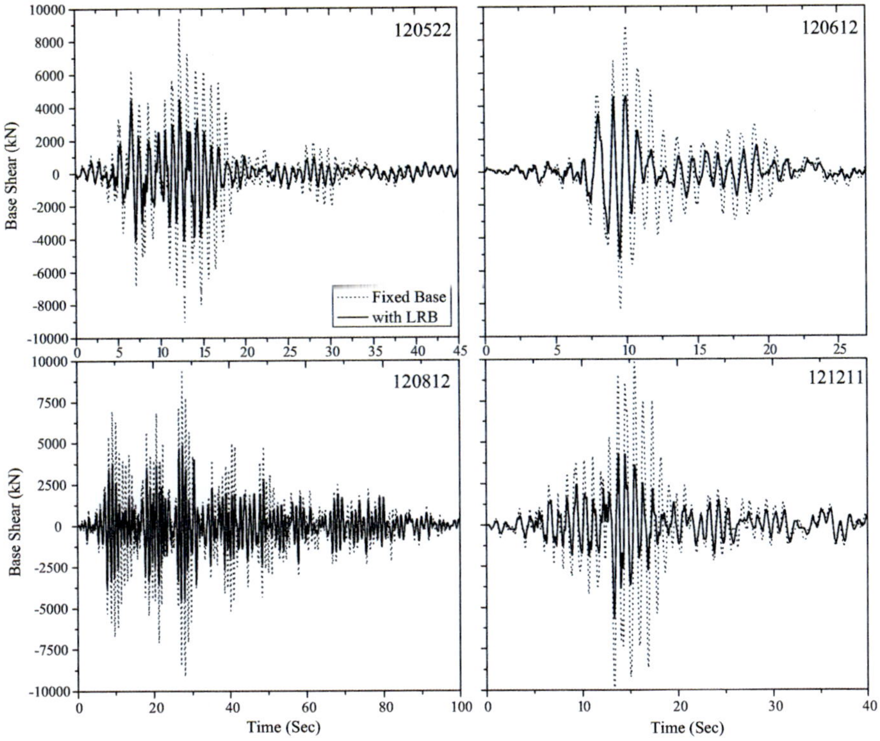

Fig. 6 Base shear comparison for with and without LRB

7 Conclusions

Most of the failures of water tanks were due to failure in staging and it was observed in past earthquakes. Seismic isolation and passive energy dissipation systems are very effective in reducing response of structure. The main objective of study is to investigate different control techniques for frame staging in elevated water tank. All analyses were performed in SAP [8] and based on the study of XPD and LRB, following conclusions are made:

1. Proper bracing system can increase capacity of tank staging.
2. Lead rubber bearing is very effective in reducing forces on structure.
3. Marginal reduction in displacement is observed due to the use of metallic dampers.

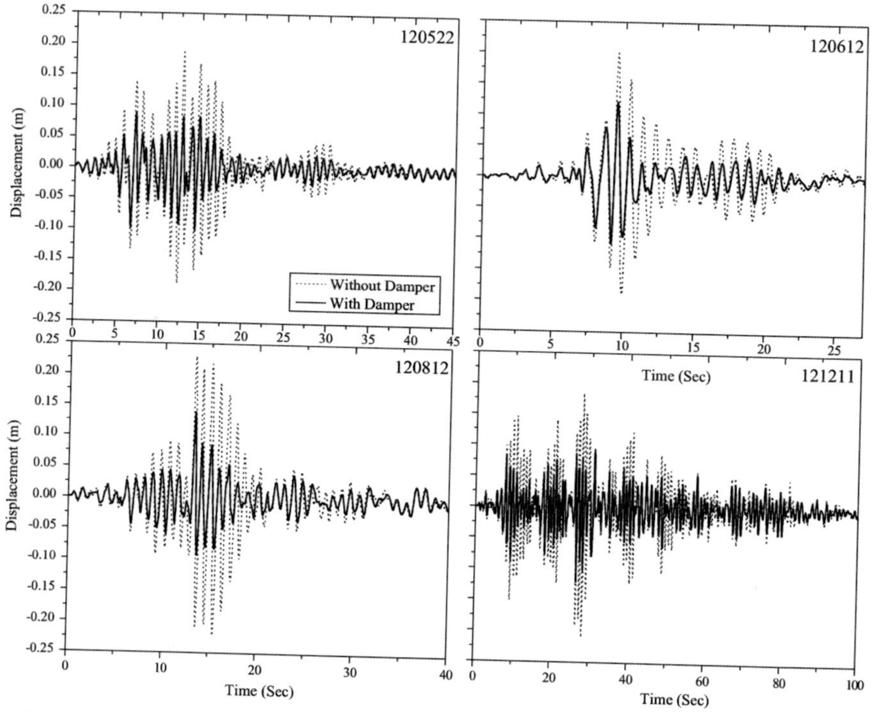

Fig. 7 Top displacement comparison for with and without damper

References

1. Shenton HW, Hampton FP (1999) Seismic response of isolated elevated water tanks. J Struct Eng 125(9):965–976
2. Panchal VR, Jangid RS (2008) Variable friction pendulum system for seismic isolation of liquid storage tanks. Nuclear Eng Des 238:1304–1315
3. Shekari MR, Khaji N, Ahmadi MT (2010) On the seismic behavior of cylindrical base-isolated liquid storage tanks excited by long-period ground motions. Soil Dyn Earthquake Eng 30:968–980
4. Abal E, Uckan E (2010) Parametric analysis of liquid storage tanks base isolated by curved surface sliding bearings. Soil Dyn Earthquake Eng 30:21–31
5. Soni DP, Mistry BB, Panchal VR (2011) Double variable frequency pendulum isolator for seismic isolation of liquid storage tanks. Nuclear Eng Des 241:700–713
6. Seleemah AA, Sharkawy ME (2011) Seismic response of base isolated liquid storage ground tanks. Ain Shams Eng J 2:33–42
7. Panchal VR, Jangid RS (2011) Seismic response of liquid storage steel tanks with variable frequency pendulum isolator. KSCE J Civil Eng 15(6):1041–1055
8. SAP (2000) Integrated software for structural analysis and design, technical reference manual. Computers and Structures, Inc
9. Kelly JM, Skinner RI, Heine AJ (1972) Mechanisms of energy absorption in special devices for use in earthquake resistant structures. Bull NZ Soc Earthquake Eng 5(3):63–88
10. Skinner RI, Kelly JM, Heine AJ (1974) Hysteretic dampers for earthquake resistant structures. Earthquake Eng Struct Dyn 3(3):287–296

11. Tyler RG (1978) Tapered steel energy dissipators for earthquake resistant structures. Bull NZ Nat Soc Earthquake Eng 11(4):282–294
12. Bergman S, Goel S (1987) Evaluation of cyclic testing of steel-plate devices for added damping and stiffness. Report UMCE 87–10, Department of Civil Engineering, University of Michigan
13. Whittaker AS, Bertero VV, Alonso LJ, Thompson CL (1989) Earthquake simulator testing of steel plate added damping and stiffness elements. Report UCB/EERC-89/02, Earthquake Engineering Research Center, University of California at Berkeley
14. Whittaker AS, Bertero VV, Thompson CL, Aionso LJ (1991) Seismic testing of steel plate energy dissipation devices. Earthquake Spectra 7(4):563–606
15. Martinez-Romero E (1993) Experiences on the use of supplementary energy dissipators on building structures. Earthquake Spectra 9(3):581–626
16. Perry CL, Fierro EA, Sedarat H, Scholl RE (1993) Seismic upgrade in San Francisco using energy dissipation devices. Earthquake Spectra 9(3):559–579
17. Manchalwar AA, Bakre SV (2016) Performance of RC structures equipped with steel and aluminium X-plate dampers. J Inst Eng India Ser A 97(4):415–425
18. Komur MA (2015) Soft-story effects on the behavior of fixed-base and LRB base-isolated reinforced concrete buildings. Arab J Sci Eng 41(2):381–391
19. IS 1893 Part 2 (2014) Criteria for earthquake resistant design of structures part 2 liquid retaining tanks. Bureau of Indian Standard, New Delhi, India
20. IS 456 (2000) Plain and reinforced concrete-code of practice (fourth revision). BIS, New Delhi, India
21. Mukherjee S, Gupta VK (2002) Wavelet-based generation of spectrum-compatible time histories. Soil Dyn Earthquake Eng 22:799–804
22. IS 1893 Part 1 (2016) Criteria for earthquake resistant design of structures part 1 general provisions and buildings. Bureau of Indian Standard, New Delhi, India

Bending Analysis of Laminated Composite Thick Beam

D. H. Tupe, A. G. Dahake, and G. R. Gandhe

Abstract The aim of this study to investigate the characteristics of a laminated composite beam with simple supports. The relations are developed between solutions of traditional theories of beams with those of the superior order shear deformation theories. Shear warp theories are those in which the effect of slanting shear strain is integrated. The interactions are developed for displacement, shear, and bending analysis. The principal equation of beam can be derived using energy and variational principles. In energy methods, the principle of virtual work or their derivatives, such as the principle of minimum potential energy is used to obtain the governing differential equations. The energy method has the advantage of providing information in the form of boundary conditions. The results of displacement, shear, and bending of a laminated composite beam are given for other shear deformation theories, in order to illustrate the influence of thick beam on the still behavior of beam.

Keywords Simple support · Laminated composite · Shear deformation · Thick beam

1 Introduction

The different beam theories, succeeding from the simple traditional lamination beam theory to the primary order beam theory then advanced order beam theory. The latter two beam theories allow for the effect of transverse shear deformation which has been neglected in traditional lamination beam theory. In point of view, the mathematical similarity of the governing equations and on the basis of load equivalence, exact relationship between the bending solutions of these three beam theories is

D. H. Tupe · G. R. Gandhe
Department of Civil Engineering, Deogiri Institute of Engineering & Management Studies, Aurangabad, MS 431001, India

A. G. Dahake (✉)
Department of Civil Engineering, G H Raisoni College of Engineering & Management, Wagholi, Pune, MS 412207, India
e-mail: ajay.dahake@raisoni.net

L. M. Gupta et al. (eds.), *Advances in Civil Engineering and Infrastructural Development*, Lecture Notes in Civil Engineering 87,
https://doi.org/10.1007/978-981-15-6463-5_8

derived. Ferreira et al. [1] pooled the Carrera's incorporated research and a circular support function involvement performance to resolve the still distortions and free quaking effects of slender and deep isotropic and cross-ply-bonded plates. Hiroyuki [2] analyzed the normal occurrences and buckling effect of laminated compound round curves exposed to primary axial stress by considering the entire belongings of perpendicular shear and regular stresses and rotatory inertia. Displacement field built on advanced order shear deformation theory is executed by Kadoli et al. [3] to study the stationary effect of functionally graded metal–ceramic beams under ambient temperature. A new plane stress model of composite beams with interlayer slips is established by Wu and Chen [4] via state space method.

Patel et al. [5] used the method of initial functions (MIF) to analyze the isotropic beam with simple supports subjected to udl. The significance of this method is that no assumptions are required about the variation of stress or displacements. Amabili and Farhadi [6] studied the nonlinear-forced sensations of isotropic and laminate composite regular shape simply supported plates. Ghugal and Kulkarni [7, 8] carried out the analysis considering the temperature stresses and deflections for orthotropic, two-layer, and three-layer square cross-ply-laminated plates. Mohite and Upadhyay [9] investigated the excellence of the stresses obtained using superior order shear flexible laminated theories for a slab.

2 Methodology

The beam considered of three layers:

First layer (0° layer) considering following parts:

$$0 \le x \le L, -b/2 \le y \le b/2, -h/2 \le z \le -h/6$$

Second layer (90° layer) considering following parts:

$$0 \le x \le L, -b/2 \le y \le b/2, -h/6 \le z \le h/6$$

Third layer (0° layer) considering following parts:

$$0 \le x \le L, -b/2 \le y \le b/2, h/6 \le z \le h/2$$

where x, y, z are Cartesian directions,
L = length, b = width and h = total depth of beam.

The beam is exposed to varying force $q(x)$ per unit length (Fig. 1).

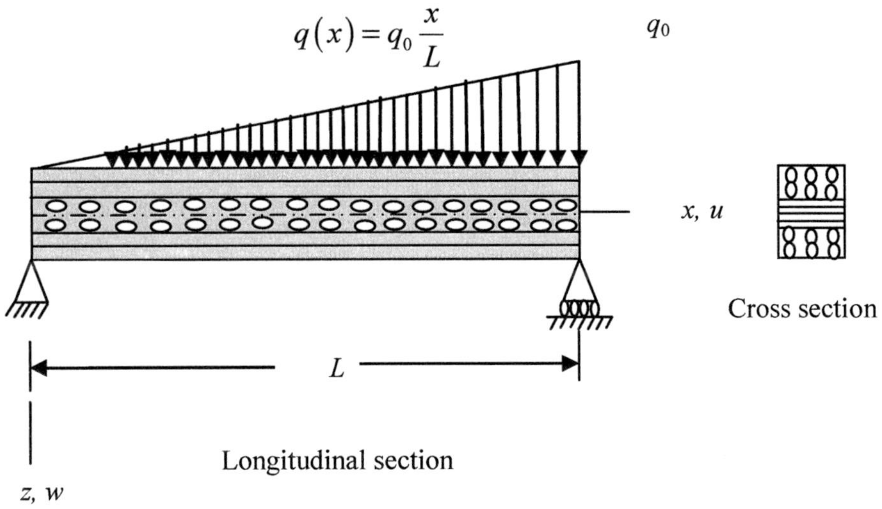

Fig. 1 Laminated beam with varying load

2.1 Displacement Field

The displacement field used in present layerwise shear deformation theory is as follows:

$$u(x, z) = u_0 - z\frac{\mathrm{d}w}{\mathrm{d}x} + z\left[1 - \frac{4}{3}\left(\frac{z}{h}\right)^2\right]\phi(x)$$

$$w(x, z) = w\langle x\rangle$$

Example A beam with simple supports three layers (0°/90°/0°) with varying load

Beam is contrasted and three equivalent thickness unidirectionally strengthened covers of graphite-epoxy and has a length to profundity proportion of four ($L/h = 4$) and ten ($L/h = 10$). The heading of graphite fibers is longitudinal in base and best (Layers 1 and 3 separately) and typical to the longitudinal pivot of the beam in center (Layer 2). The numerical values of the properties are as follows:

$$E_1 = 172.75 \times 10^3\,\text{MPa},\quad E_2 = 6.89 \times 10^3\,\text{MPa},$$
$$G_1 = 3.445 \times 10^3\,\text{MPa},\quad G_2 = 1.378 \times 10^3\,\text{MPa},$$
$$\mu_{12} = \mu_{21} = 0.25.$$

Table 1 The axial deformation ($\bar{u}$) at (x =0.25 L, z = h/2), deflection ($\bar{w}$) at (x =0.25 L, z = 0) in-plane stress ($\bar{\sigma}_x$) at (x = 0.25 L, z = h/2), max. Shear $\tau_{zx}^{\overline{CR}}$ and $\tau_{zx}^{\overline{EE}}$ (x =0, z = 0) of the simple support beam

Theory	S	$\bar{w}$	$\bar{u}$	$\bar{\sigma}x$	$\tau_{zx}^{\overline{CR}}$	$\tau_{zx}^{\overline{EE}}$
Present (HSDT)	4	2.8312	0.2379	5.2675	0.6109	2.2000
FSDT		2.8252	0.2375	5.2500	0.6108	2.0000
CLB		0.1718	0.2978	6.0000	–	2.0000
Present (HSDT)	10	2.1055	3.7171	32.9210	1.4761	5.5000
FSDT		2.1033	3.7109	32.8120	1.4758	5.0000
CLB		0.1719	4.6667	37.5000	–	5.0000

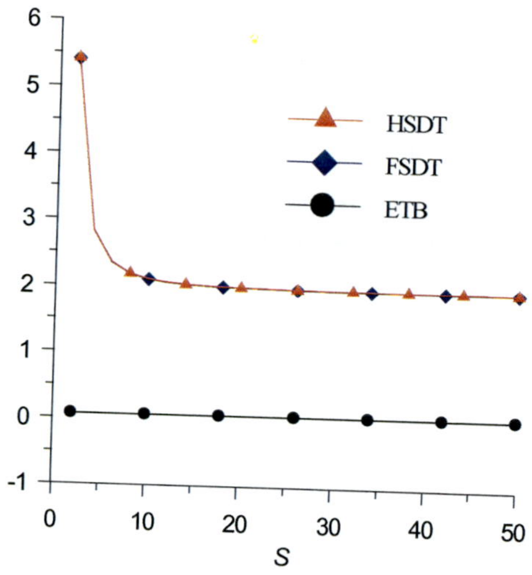

Fig. 2 Deflection $\bar{w}$ for L/h = 4

3 Results

The results are presented in Table 1 and their variations are plotted in respective figures (Figs. 2, 3, 4 and 5).

4 Conclusions

In this numerical study, higher order type of displacement function is presented. The displacement function uses higher order terms to represent the displacement across the thickness and the displacement function satisfies displacement and transverse shear stress continuity at the interface. Zero shear stress condition at top and bottom of the beam is also satisfied. Results presented for transverse displacement, in-plane

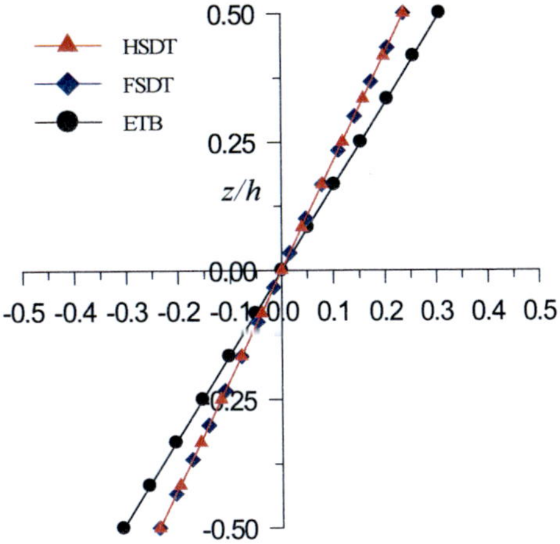

Fig. 3 Displacement $\bar{u}$ for $L/h = 4$

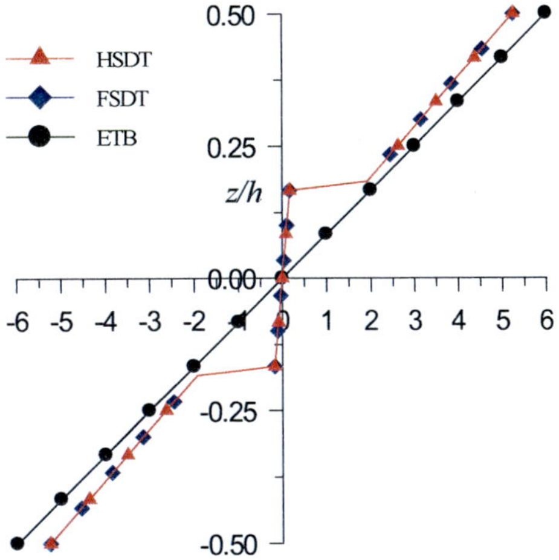

Fig. 4 Bending stresses $\overline{\sigma_x}$ for $L/h = 4$

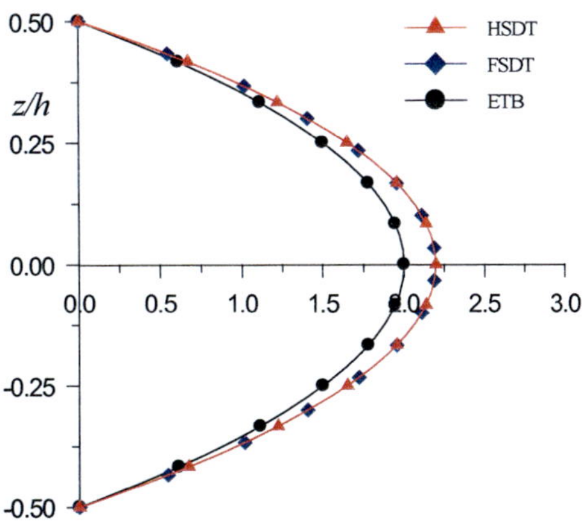

Fig. 5 Shear stresses $\overline{\tau_{zx}^{EE}}$ for $L/h = 4$

displacement, in-plane stress, and transverse shear stress are giving accurate results compare to CLB and FSDT.

References

1. Ferreira AJM, Roque CMC, Carrera E, Cinefra M (2011) Analysis of thick isotropic and cross-ply laminated plates by radial basis functions and a unified formulation. J Sound Vib 330:771–787
2. Hiroyuki M (2004) Free vibration and stability of laminated composite circular arches subjected to initial axial stress. J Sound Vib 271:651–670
3. Kadoli R, Akhtar K, Ganesan N (2008) Static analysis of functionally graded beams using higher order shear deformation theory. Applied Math Model 32:2525
4. Wu Z, Chen W (2006) Free vibration of laminated composite and sandwich plates using global-local higher-order theory. J Sound Vib 298:333–349
5. Patel R, Dubey SK, Pathak KK (2014) Analysis of isotropic beams using method of initial functions (MIF). Electron J Struct Eng 14:1–6
6. Amabili M, Farhadi S (2009) Shear deformable versus classical theories for nonlinear vibrations of rectangular isotropic and laminated composite plates. J Sound Vib 320:649–667
7. Ghugal YM, Kulkarni SK (2013) Thermal flexural analysis of cross-ply laminated plates using trigonometric shear deformation theory. Latin Am J Solids Struct 10:1001–1023
8. Ghugal YM, Kulkarni SK (2013) Flexural analysis of cross-ply laminated plates subjected to nonlinear thermal and mechanical loadings. Acta Mech 224:675–690
9. Mohite PM, Upadhyay CS (2006) Accurate computation of critical local quantities in com posite laminated plates under transverse loading. Comput Struct 82:657–675

Nonlinear Finite Element Analysis of Reinforced Concrete Haunched Beams Without Shear Reinforcement Under Static Shear Load

Hamzah Sabah Jebur

Abstract In the current paper, the verification of simply supported reinforced concrete (haunched and prismatic) beams has been done for different haunched angles. These beams have tested without shear reinforcement previously. These beams have tested by applying a concentrated static load at a mid-span. The verification has been done by comparing the results of nonlinear finite element analysis method by using ANSYS (Mechanical APDL) software with the experimental work results, where the load–deflection curves were obtained for both (FEA and experimental). The verification has done for three beams; one was as a prismatic beam and two more as a haunched beam with different haunched angles (α). The results obtained from ANSYS were in agreement with the experimental results, where the load–deflection curves of ANSYS results and experimental results were matching for all beams. The ANSYS results confirm the different behaviour of RCHBs with respect to prismatic beam which have been observed in an experimental work also. This paper shows the strength of ANSYS software in modelling and analysis.

Keywords RCHBs · Pure shear failure · Beams without shear reinforcement · Finite element analysis · ANSYS

1 Introduction

The reinforced concrete haunched beam is a beam which has varying depth along the length of beam with constant width, and it can be used as a simply supported or continuous beams in bridges and mid-rise framed buildings Fig. 1, where it is used in Latin American cities such as Mexico City in Mexico or in Ecuador [1, 2]. The purpose of using RCHBs is to ease placement of facilities (electrical cables, pipes, air conditioning, sewage, etc.) and to decrease the quantity of concrete as well as the weight of structure [1, 2, 9]. Using this beam gives more efficient lateral stiffness or moment capacity to self-weight ratio as compared with prismatic beams [2] and is

H. S. Jebur (✉)
Department of Civil Engineering, Vishwakarma Institute of Information Technology, Pune, India
e-mail: sabahjabermn@gmail.com

L. M. Gupta et al. (eds.), *Advances in Civil Engineering and Infrastructural Development*, Lecture Notes in Civil Engineering 87,
https://doi.org/10.1007/978-981-15-6463-5_9

Fig. 1 Using RCHBs in urban bridge and frame building [2, 3]

used for aesthetic reasons. However, the major disadvantage of using haunched beams is that their use usually involves higher construction prices, as special formwork and qualified construction staff area unit needed [1–3]. This type of beam is not used in many other countries because there is no enough data about RCHBs available. The first study has done for shear failure of RCHBs at the lakebed zone of Mexico City on September 19, 1985, Michoacán Earthquake [3]. To analyse this type of beam numerically, it is required using such a software which can model and analyse the varying shapes. The finite element analysis method uses the advance mathematical concept to find out the results of complex engineering problems [4]. In this method, we can model any complex engineering problem, and the results of analysis will be approximate to the experimental results [4]. The complexity of the engineering problem may be in the load applied, boundary conditions, varying shape or non-homogeneous material [4]. The nonlinear finite element analysis is widely used for analysis in structural engineering to get much clear idea about the behaviour of structures, and it shows a good agreement with experimental work. Many software are based on the concept of FEA to analyse a complex engineering problem. Some of the popular packages are STAAD-PRO, Robot Structural Analysis Professional RSA, GT-STRUDEL, NASTRAN, ABAQUS and ANSYS [4]. Handling one of this software can solve several and different problems. In this study, the strength of ANSYS software system has been shown in modelling concrete haunched beams (RCHBs) under static shear load, and it had been ready to show the behaviour of RCHB in steps from initial crack until completed failure. The comparison of ANSYS results with experimental results is shown in this study. In the current paper, the verification has been done for reinforced concrete beams without shear reinforcement to understand the behaviour of RCHB and to get idea about the capacity of this type of beams. Three beams have been presented in this paper to understand the behaviour of beams by comparing with experimental results which have been done by Nghiep [5]. The test is shown that the capacity of RCHBs is decreased by increasing the haunched angle. The shear reinforcement is providing in support region area only, whilst the longitudinal reinforcement has provided along the length of beams with enough quantity to ensure the failure will be in shear [6]. The load is applied as a point (concentrated) load at a mid-span. The material properties and dimensions have been taken from the experimental work data. The specimens were a slender (a/d > 2.5) [5, 6]. The details of reinforcement and properties of concrete were same in all beams, but the depth was varying by changing the angle α of haunch. The nonlinear finite element analysis by ANSYS is shown a good agreement with experimental work and ABAQUS software as shown in results.

2 Objectives

In this paper, the nonlinear finite element analysis method is used to analyse RCHBs without shear reinforcement by using ANSYS software by verifying a previous experimental work has been done to ensure the behaviour of this type of beams under shear failure and to see the cracks pattern by incrementing the load. Handling a software can save time and cost of studying any behaviour of any structure.

3 Description of Beams

The geometry and properties were taken from references [5, 6], where the verification is done on this experimental study.

3.1 Geometry of Beams [5]

As shown in photos and table, the beams have the same depth at mid-span, but the depth at support region is varying with changing the haunched angle (0, 3.95 and 5.91°). The dimensions of beams and reinforcement details are shown in Table 1 and Fig. 2.

Table 1 Dimensions of beams [5, 6]

Beam ID	*Ds* (mm)	*ds* (mm)	*Dm* (mm)	*dm* (mm)	*b* (mm)
1L-1	340	300	340	300	200
2L-1	240	200	340	300	200
3L-1	190	150	340	300	200

Dm: total depth at mid-span, *dm*: effective depth at mid-span, *Ds*: total depth at support region and *ds*: effective depth at support region

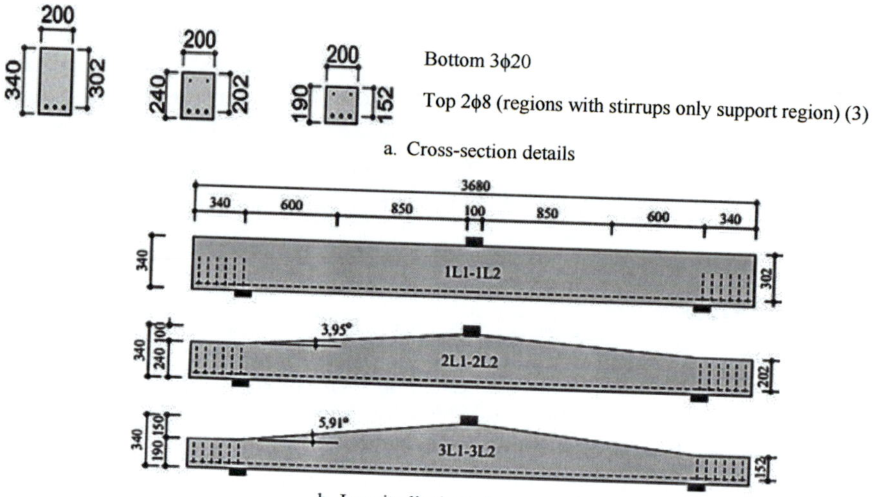

b. Longitudinal reinforcement details

Fig. 2 Geometry of beams with reinforcement details [5, 6]

3.2 *Properties of Materials [5, 6]*

The longitudinal reinforcement in bottom is used along the length of beams, but the longitudinal reinforcement at top and ties is used in support region only, where $3\phi20$ bars are used in bottom reinforcement and $2\phi8$ bars in top reinforcement with yielding strength 550 MPa, and the bar diameter of ties is $\phi8$. The compressive strength of concrete (fc') is 30 MPa. The longitudinal reinforcement bars were used much enough to ensure that shear failure will occur in all test beams.

4 ANSYS Model Progress

4.1 *Elements Type*

The element types can be selected according to ANSYS Mechanical APDL Manual as following:

(a) **Solid 65** [7]

Solid 65 is an element having eight nodes and six faces, for that, eight nodes should be available to define one Solid 65 element. This element can be used as a concrete element because it can be cracked and crashed [7, 8]. The behaviour of this element is in three directions (three degree of freedom). The element can be reinforced by

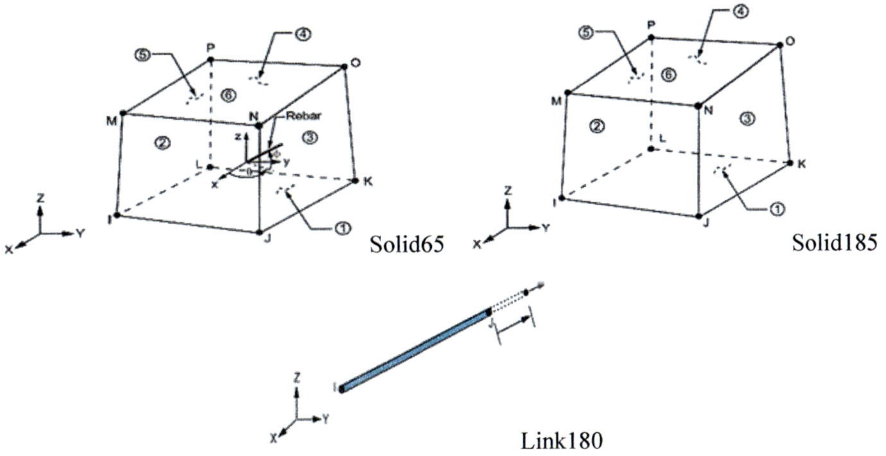

Fig. 3 Elements shape [7]

bar reinforcement also, where the bar reinforcement can be provided in the element with specific ratio.

(b) **Link 180** [7]

This element required two nodes only to define it, where it is one-dimensional element. Its behaviour is in three dimensions x, y and z (three degree of freedom). The element is a uniaxial tension–compression element, means it can carry axial tensile and compressive stresses. For that we can use this element as a reinforcement bar in modelling.

(c) **Solid 185** [7]

Solid 185 required eight nodes to define, and it has six faces similar to Solid 65, just its behaviour is different. The element deforms in three dimensions. The properties of this element as mentioned in ANSYS Manual are plasticity, stress stiffening, creep, large deflection and large strain capabilities. It is suitable to use as a bearing plate at supports and at load point to prevent the local crashing of concrete, where the used element are shown in (Fig. 3).

4.2 Geometry Model of Beams

Node to node method of modelling type has been used. The sequence of modelling defines the elements, real constant, the properties for every element and adds the

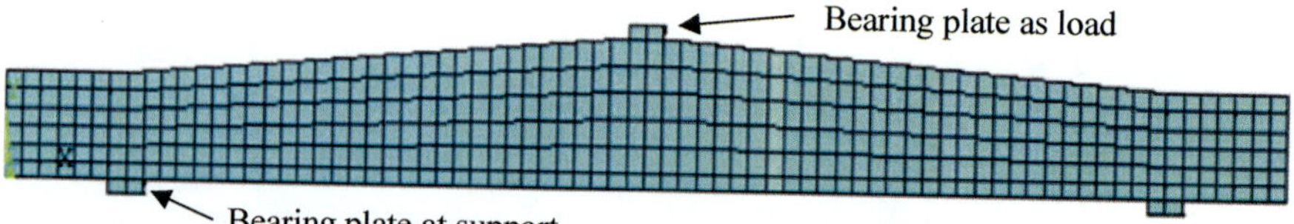

Fig. 4 Boundary condition of beam 2L1

nodes then adds the element. To define the element Solid 185 and Solid 65, there should be there eight nodes to define one element, where the link 180 requires two nodes only [7]. The meshing is defined already by defining the spacing between the nodes.

4.3 *Boundary Condition and Loads*

The boundary condition of beam was defined as a simply supported beam by defining one support from one side as a roller and from other side as a hinged support (Fig. 4). The supports are added to bearing plates not to concrete element directly. The load has applied on nodes of loading plates at mid-span of beams. The load was defined as a static load with tolerance value 0.05.

5 ANSYS Results

The results are shown that the failure occurs because of shear failure Fig. 7 and Fig. 8, and it shows the capacity of beam decreases when the haunched angle α is increased. The results are matching with the experimental results. The test results have taken from References [5, 6], where the verification has done on this experimental work (Table 2).

It can be observed that the deflection capacity is increased by increasing the angle α (Table 2). (Fig. 5) shows the maximum deflection of beam 1L2 at ultimate load.

Table 2 Results of ANSYS and test summery

Beam ID	F. Test (kN)	F. ANSYS (kN)	δ. test (mm)	δ. ANSYS (mm)
1L1	151	142.6	10.7	10.8
2L1	150	134	11.9	10.9
3L1	133	125	11.9	11.6

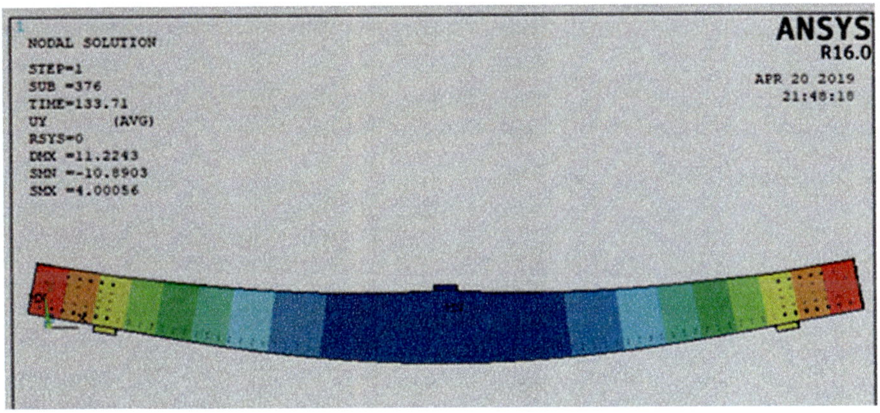

Fig. 5 Deflection of beam 2L1 as shown in ANSYS

Figure 7 shows the cracks pattern of beam 2L1 from load 30 kN (initial crack) till failure load 133 kN, where the sequence of crack pattern is shown by increasing the load every 10 kN more (30, 40, 50, …, 133).

Figure 6 shows the comparison of ANSYS results with experimental results and test results with ABAQUS results also by taking the load–deflection curve for every beam, and this comparison shows a good agreement in ANSYS and ABAQUS results with experimental work results (Fig. 6).

6 Conclusion

As shown in results, the shear capacity of beam is decreased when the depth of beam near support decreases. The angle of shear failure is about 45° similar to test results. It is worth noting that all these beams are not designed for shear, so whatever is the capacity in these beams is the capacity of concrete itself. It can be observed clearly that the ANSYS software can give a good indication about the behaviour of varying shape beams (RCHBs).

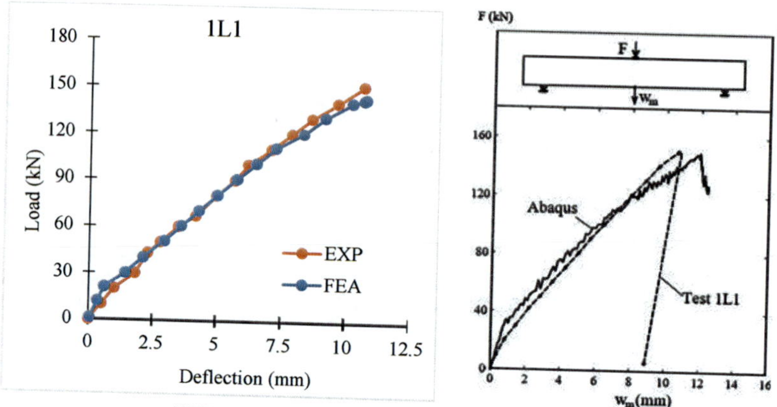

a. ANSYS relusts in current study, ABAQUS and test results of 1L1

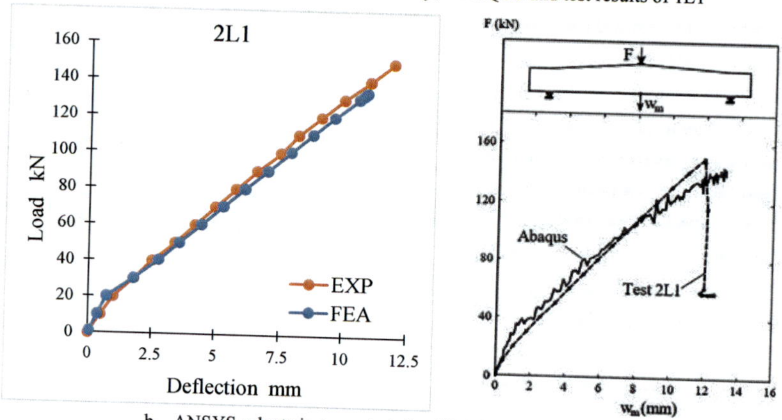

b. ANSYS relusts in current study, ABAQUS and test results of 2L1

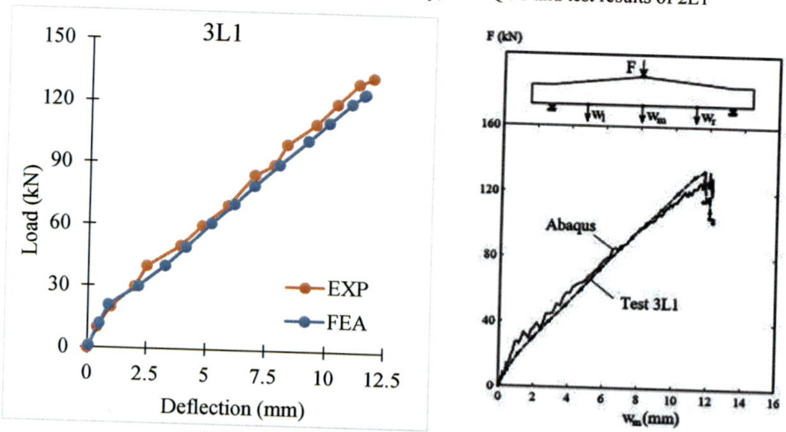

c. ANSYS relusts in current study, ABAQUS and test results of 3L1

Fig. 6 Load-Deflection curve of beams (1L1, 2L1 and 3L1), respectively, as shown in presented study by ANSYS software, and experimental and ABAQUS results as shown previously by Nghiep [6]

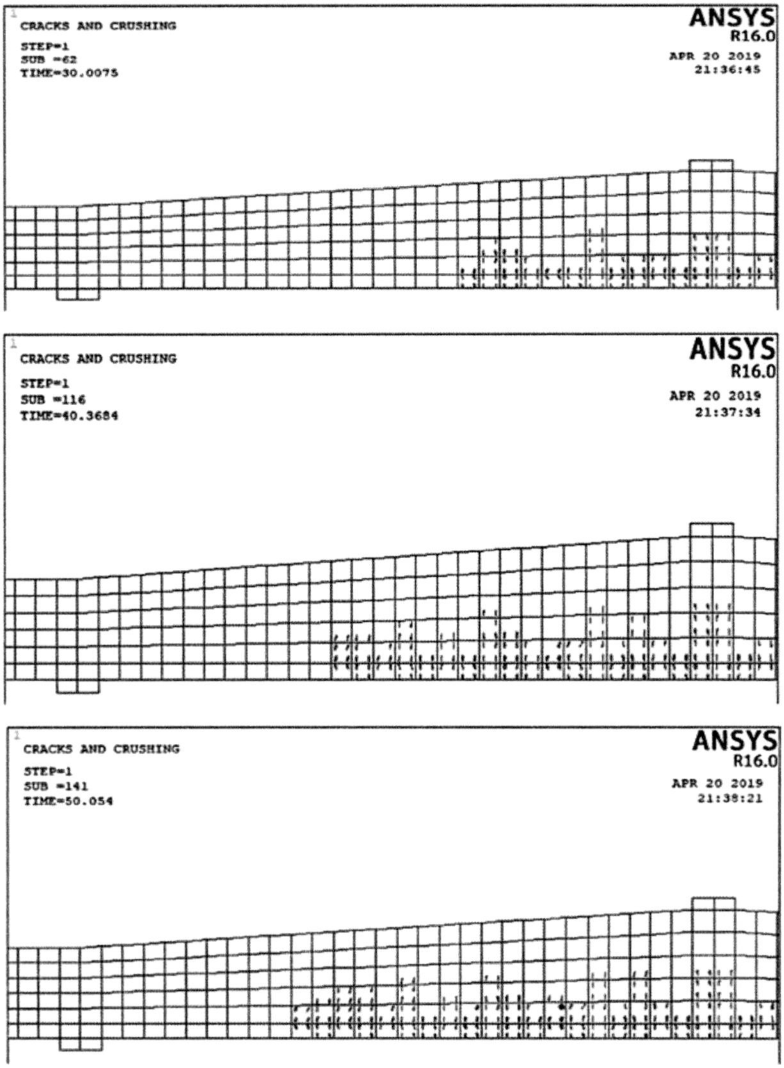

at 30kN

at 40kN

at 50kN

Fig. 7 Cracks pattern of beam 2L1 as shown in ANSYS

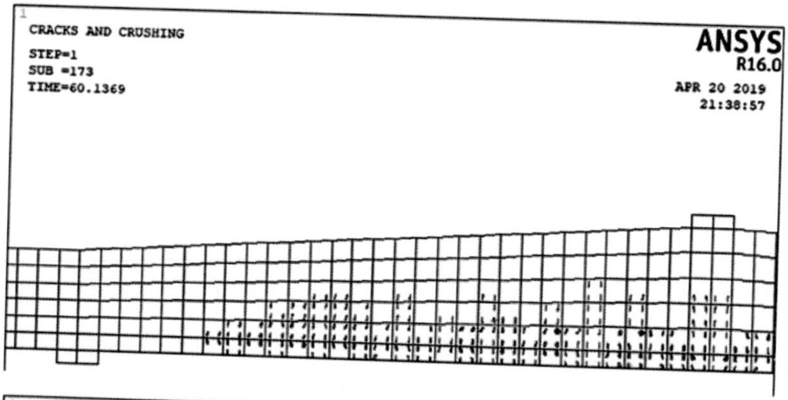

at 60kN

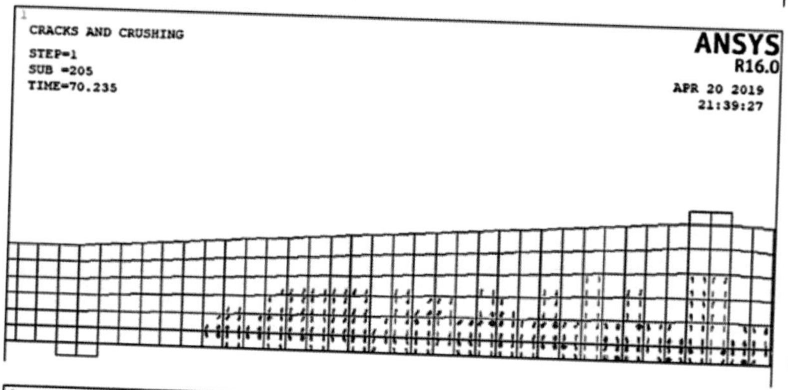

at 70kN

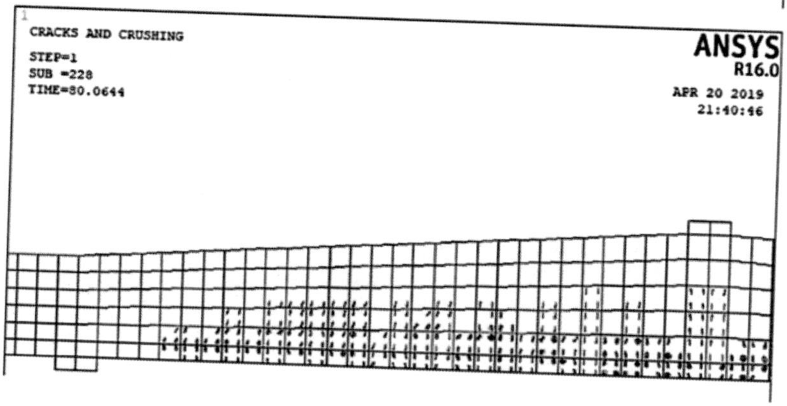

at 80kN

Fig. 7 (continued)

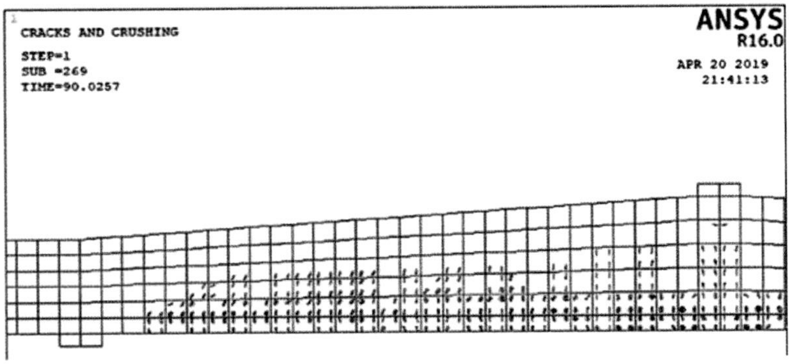

at 90kN

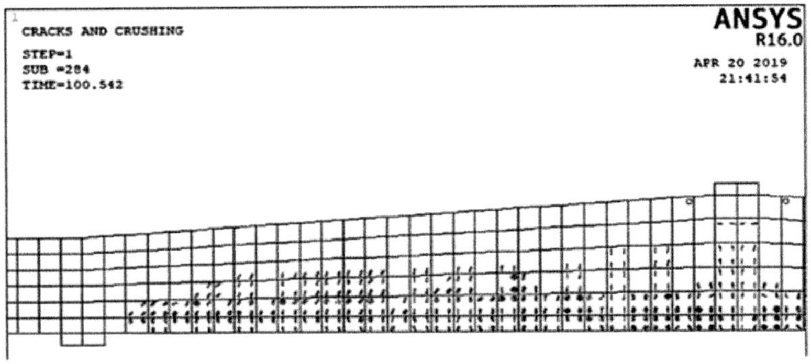

at 100kN

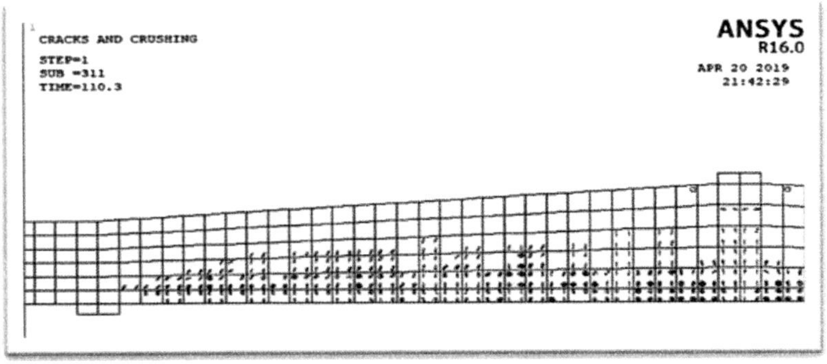

at 110kN

Fig. 7 (continued)

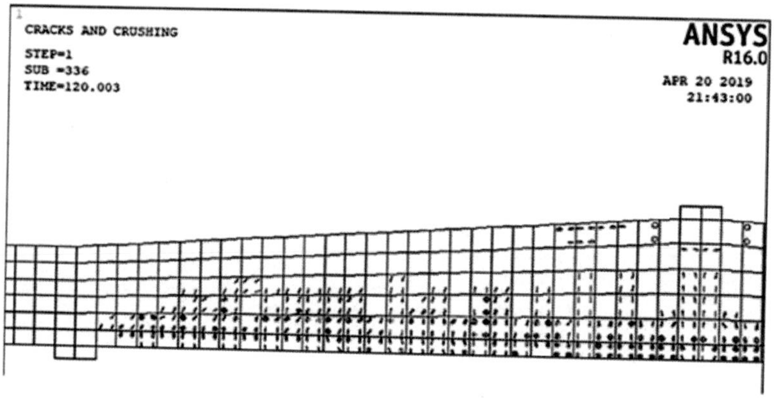

at 120kN

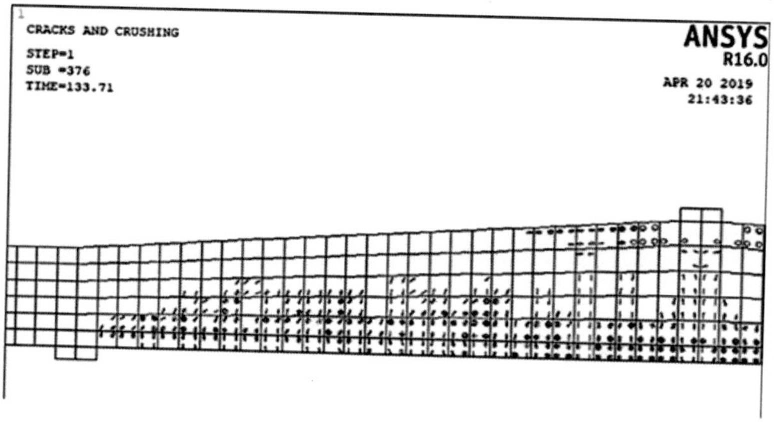

at 133kN

Fig. 7 (continued)

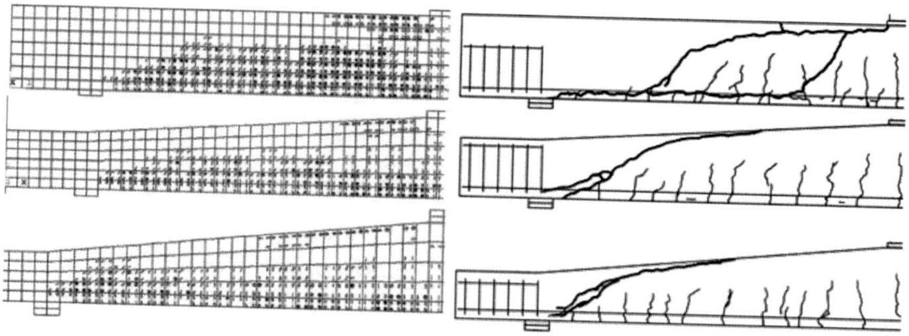

Fig. 8 Crack pattern of beams 1L1, 2L1 and 3L1, respectively, as shown in the present work by ANSYS and as shown in the previous experimental results [6]

References

1. Tena-Colunga A, Archundia-Aranda HI, González-Cuevas OM (2008) Behavior of reinforced concrete haunched beams subjected to static shear loading. 30:478–492
2. Archundia-Aranda HI, Tena-Colunga A, Grande-Vega A (2013) Behavior of reinforced concrete haunched beams subjected to cyclic shear loading. 49:27–42
3. Tena-Colunga A, Angel L, Archundia-Aranda HI (2017) Assessment of the shear strength of continuous reinforced concrete haunched beams based upon cyclic testing. J Build Eng
4. Bhavikatti SS (2005) Finite element analysis, 1st edn
5. Rombach GA, Kohl M, Nghiep VH (2011) Shear design of concrete members without shear reinforcement—a solved problem? 136–140
6. Nghiep VH (2011) Shear design for straight and hauched concrete beams without shear reinforcement
7. (2013) ANSYS mechanical APDL element reference
8. (2013) ANSYS mechanical APDL structural guide
9. Hou C, Matsumoto K, Niwa J (2015) Shear failure mechanism of reinforced concrete haunched beams. Japan Soc Civil Eng (JSCE) 3:230–245

Seismic Response of Rectangular RC Building with and Without Infill Walls Considering Soil–Structure Interaction

P. S. Bhurse, S. S. Sanghai, and N. Lalitha Kumari

Abstract In the last few years, the tall, proportioned and irregular structure exhibits more risks during earthquakes. For interior wall and exterior walls of structure, infill walls are frequently used to fill up the space between beam and column frame. In this study, the influence of the response of infill walls on the seismic performance of the reinforced concrete building prone to lateral seismic loads is investigated. The exact modeling of the structure reflects on many structural aspects like strength and stiffness of the structure to resist the lateral earthquake load initiated by the earthquake. A comparative and parametric study is carried out with the help of joint displacement, axial force, maximum bending moment, shear force, time period, etc.

Keywords Infill walls · Soil–structure interaction · Time history analysis · Equivalent diagonal strut method

1 Introduction

The masonry infill wall has the main influence on the structural reaction of reinforced concrete (RC) structures when exposed to seismic activity. Infill wall is mostly used to increase initial stiffness and strength of reinforced concrete building construction. The arrangement of infill strut panel and reinforced concrete edge is mostly used in the structure, where the section is predisposed to seismic movement. Infill wall is normally used as a partition element because of many suitable components like lighter in load, openness in structure, excellent visible view, defending fabric goods, etc.,

P. S. Bhurse (✉) · S. S. Sanghai
Department of Civil Engineering, G H Raisoni College of Engineering, Nagpur, India
e-mail: bhurse_priya.ghrcemtechstr@raisoni.net

S. S. Sanghai
e-mail: sanket.sanghai@raisoni.net

N. L. Kumari
Department of Civil Engineering, Chaitanya Bharathi College of Engineering and Technology (A), Gandipet, Hyderabad, Telangana, India
e-mail: lalitha1059@gmail.com

L. M. Gupta et al. (eds.), *Advances in Civil Engineering and Infrastructural Development*, Lecture Notes in Civil Engineering 87,
https://doi.org/10.1007/978-981-15-6463-5_10

even so in the structural investigation, solely the impact of mass is measured and its structural physical appearance such as strength and stiffness is usually neglected. Still it provides significant lateral stiffness to the bare framed structures, these existed not measured in the preceding edition of Indian standard code of seismic activity resilient project is IS 1893:2002(Part 1) [1]. But, in the new publication of Indian standard for a seismic activity resilient plan IS 1893:2016(Part 1), [2] many provisions for infill walls are specified. The condition is that unreinforced infill wall can be modeled as an equivalent sloping strut, which if not considered results in unequal structure. The ends of the equivalent strut treated as pin joint are connected to the RC frame and influence the opening on a width of the equivalent diagonal strut also stated. Infill wall is unknown, although the mutual masonry wall, brick wall, etc. For the study, we must measure the structural strength, stiffness of infill walls. As per statement 7.9 p.n. 25 IS 1893:2016(Part 1), [1]

$$W_{\mathrm{ds}} = 0.175\ \propto_h^{-0.4}\ L_{\mathrm{ds}} \tag{1}$$

where

$$\propto_h = h\left\{\sqrt[4]{\frac{E_m.t.\sin 2\theta}{4.E_f.I_c.h}}\right\} \tag{2}$$

E_m = Modulus of resistance of the material of the unreinforced brickwork infill
E_f = Modulus of resistance of the material of the RC moment resisting structure
I_C = Moment of inertia of the adjacent column
t = Width of masonry infill walls
θ = The angle of the diagonal strut with the parallel
h = Height of URM infill walls
W_{ds} = Breadth of equivalent diagonal strut
L_{ds} = Sloping distance of infill strut panel (Fig. 1). It is well known fact that,

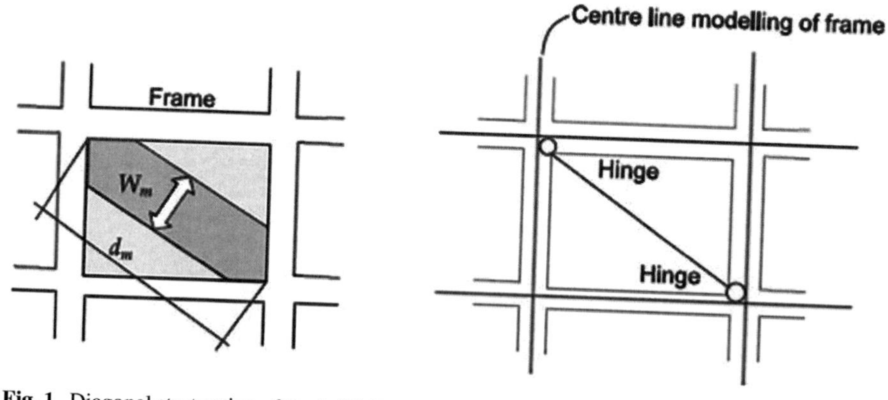

Fig. 1 Diagonal strut action of the infill [3]

characteristics of earthquake motion depends upon local site conditions. If the soil strata is very soft, then it alters the characteristics of earthquake motion. Hence, soil-structure interaction plays important role in response calculation of building. Neglecting the stiffness of infill strut and overestimating the stiffness of soil strata may lead to conservative approach in response calculation. Hence, this study is done to calculate the seismic response by considering the stiffness of infill panels for rectangular building resting on nonlinear soil-foundation system.

2 Finite Element Modeling

The structure models are modeled, while 3D structural solids with element category are allocated by FEA software, ANSYS 15. Mesh convergence study was prepared for structure and intended for soil a coarser mesh used. The ANSYS framed framework is model with 2-node beam element BEAM188, and it has six degrees of freedom at each node. Slab surface used SHELL181 also has six DOF at each node. And foundation with SOLID 186, interface with the element is CONTA174 and TARGE170; SURF154 is used for various loads and surface effect application in 3D analysis of structure. The soil is modeled with SOLID65, and Drucker–Prager model is used for nonlinear material of soil activities.

3 Material Models Used

The soil volume dimension is modeled as the solid section with dimensions as length and breadth as five times the equivalent dimension of the structure and depth of soil should be at least three times elevation of structure [4]. In this paper, soil volume is modeled by using the direct method. Dead load and live load is given as per IS875 (Part 1) and (Part 2) 1987 [5, 6], respectively. The dead load includes self-weight and wall loads (Table 1).

Table 1 Material property of concrete and soil [7, 8]

Properties	Structure	Soil		
Material	Concrete (M25)	Hard	Medium	Soft
Young's modulus E (N/m^2)	2.5×10^{10}	8.40×10^9	4.46×10^8	1.03×10^8
Poisson's ratio	0.25	0.3	0.35	0.4
Density, P (kN/m^3)	25	22	18	16
Cohesion, C (kN/m^2)	–	20	30	23
Internal friction angle (°)	–	30	0	23

4 Geometry of Building

The design is similar for G+3 and G+7 story structures shown in Figs. 2 and 3. Usual floor-to-floor altitude is taken as 3.1 m for both the story. The dimension of

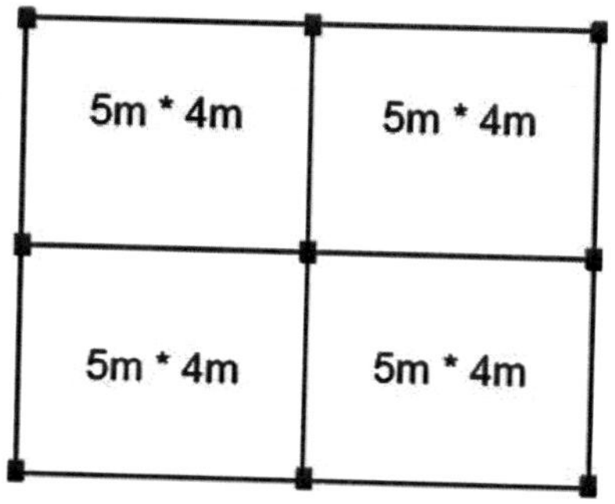

Fig. 2 Plan of structure

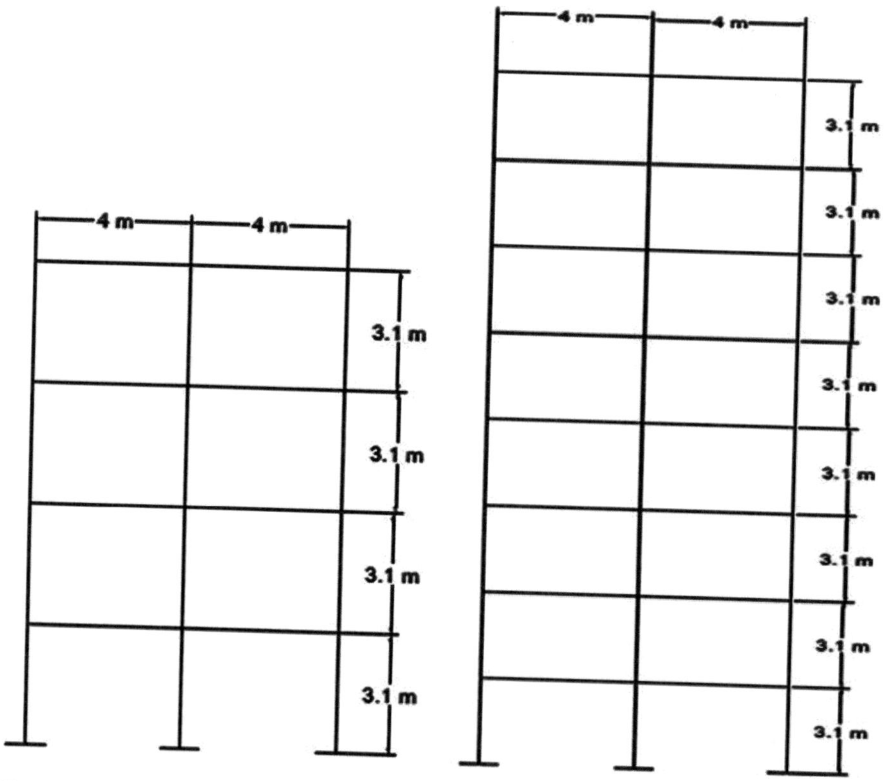

Fig. 3 Elevation of structure

Table 2 Geometric details of G+3 and G+7 story building

Parameters	Rectangle (m)
Plan dimension	10 × 8
Height of story	3.1
Depth of foundation	1.2
Size of diagonal strut	0.440 × 0.345
Thickness of slab	0.15

Table 3 Acceleration time history records

Earthquake	Peak ground acceleration (PGA)			Duration (s)
	Hard soil (g)	Stiff soil (g)	Soft soil (g)	
El-Centro	0.0922	0.1042	0.08919	40
Uttarkashi	0.3708	0.5042	0.6229	36.16
Dharamshala	1.3734	1.8639	2.29554	16.18
Kocaeli	0.1004	0.10449	0.0995	34.96
Parkfield	0.01226	0.04179	0.01226	30.33

the building is shown in Table 2. The framed structure is modeled in finite element program ANSYS 15.

The plan and elevation of all three categories of structure are given below:

Following are the acceleration time history records which are used for analysis (Table 3).

5 Demonstration with a Study on MDOF System

The finite element modeling is done for building along with the foundation system using FEM software ANSYS 15.0. The soil physical properties are applied from the material reference library in ANSYS for different linear or nonlinear soil model and structure.

In the present estimation, the analysis of loam modeled by using ANSYS 15.0.

Following models are studied.

Table 4 Results of G+3 and G+7 story structure

		G+3 square		G+7 square	
		fundamental time period (s)	Top floor displacement (mm)	fundamental time period (s)	Top floor displacement (mm)
Bare frame	Model 1:Bare frame (BF)	0.323	64.6	0.816	128.25
	Model 2:BF with SSI hard soil	0.650	82.8	1.054	146.32
	Model 3:BF with SSI stiff soil	0.857	87.80	2.133	164.6
	Model 4:BF with SSI soft soil	0.923	145.79	2.915	233.14
Strut frame	Model 5:Strut frame (SF)	0.257	0.079	0.768	39.20
	Model 6:SF with SSI hard soil	0.116	2.23	0.302	42.4
	Model 7:SF with SSI stiff soil	0.131	2.42	0.416	43
	Model 8:SF with SSI soft soil	0.340	2.78	0.431	44.4

Fig. 4 Bare frame

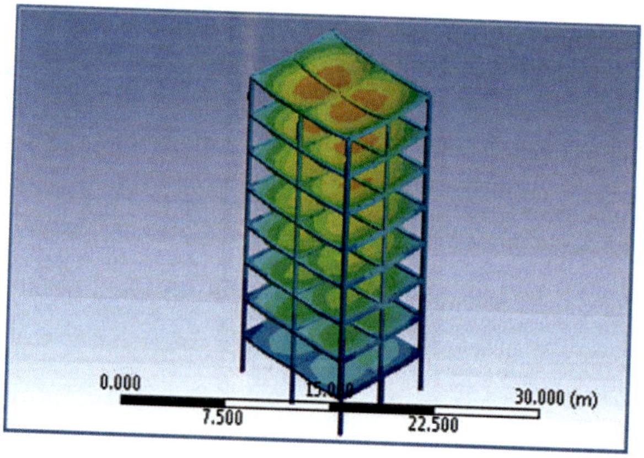

Fig. 5 Strut frame

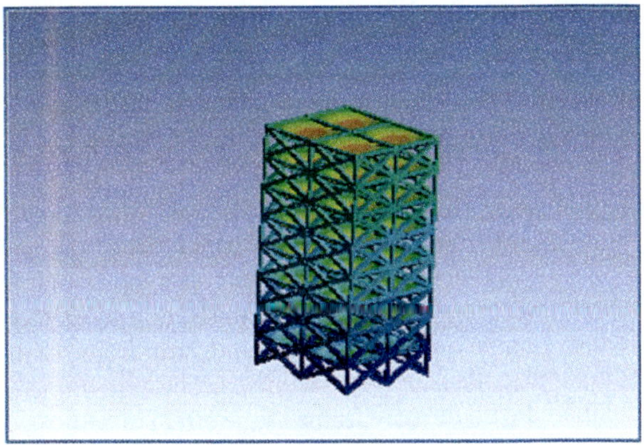

Fig. 6 Bare frame with SSI

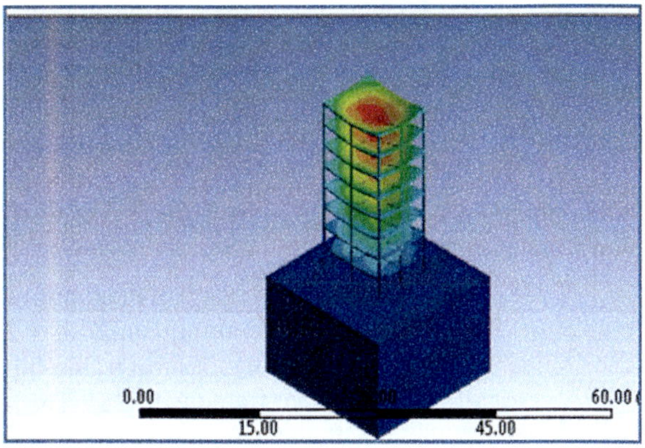

Fig. 7 Strut frame with SSI

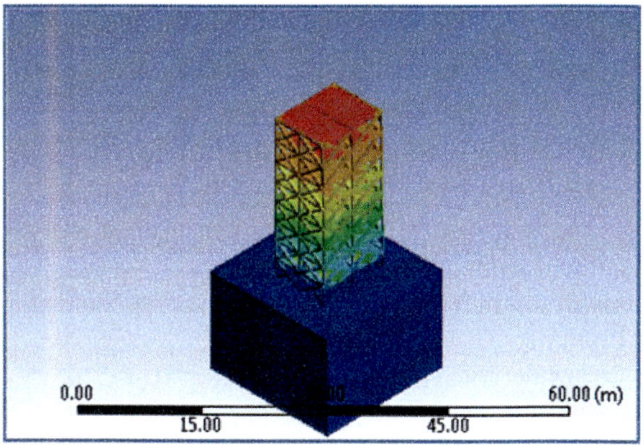

6 Results and Discussion

Displacement in X-direction is calculated for dynamic loading at every point for top of model is shown in Table 4 (Figs. 4, 5, 6 and 7).

From Table 4, it is clear that in all four types of building time period increase with consideration of the effect of SSI, while time period decreases with consideration of stiffness of infill strut panel. Since the soil is becoming softer, the time period increases consequently. And also shows that joint displacement of the structure decreases with consideration of strut frame up to 86.83% as consideration of bare frame. The maximum deformation in strut frame with SSI is decreases to 90.81% as compared to the bare frame with SSI. The bare frame with SSI increases to 30.88% as compared with the bare frame as the effect of SSI specified to the structure. However, from the results, it is clear that due to the effect of SSI, consideration of infill strut panel gives better results.

7 Maximum Axial Force for G+3 and G+7 Story Building

Maximum axial force for all models for G+3 and G+7 story rectangle building is shown in the following figures.

From Figs. 8 and 9 suggests the graph of maximum axial force in rectangle-shaped building; it is clear that the maximum quantity of axial force decreases to 76.67% due to infill strut panel. Considering the strut frame with SSI, axial force decreases to 83.92% as compared to the bare frame with SSI. The bare frame with SSI decreases to 69.14% as compared with the bare frame. The effect of SSI indicates that the axial force decreases due to the fact soil turn into softer due to expand in a time period. For high story building, SSI and infill strut play a vital role and gives better results.

8 Maximum Bending Moment for G+3 and G+7 Story Building

Maximum bending moment for G+3 and G+7 building with consideration of different soil is shown in the following figures.

From Figs. 10 and 11, it is clear that maximum bending moment changes drastically while considering infill and SSI. When only infill struts are modelled, it was observed that, bending moment decreases by 98.60% as compared to the bare frame. But when SSI effects are considered along with infill struts, bending moment decreases by 88.68%. This changes will ultimately change the design of main structural members.

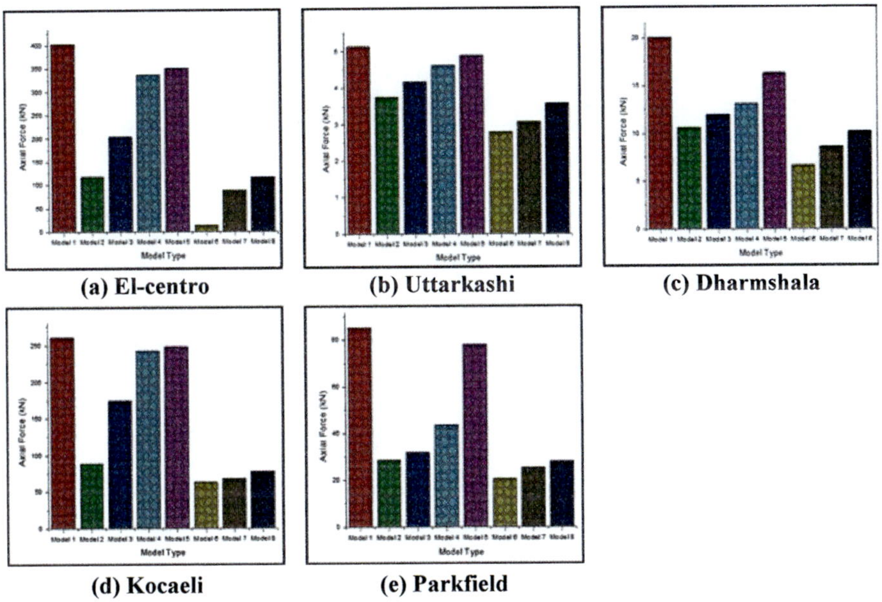

Fig. 8 Maximum axial force for G+3 story rectangle building for earthquake. **a** El-centro, **b** Uttarkashi, **c** Dharamshala, **d** Kocaeli, **e** Parkfield

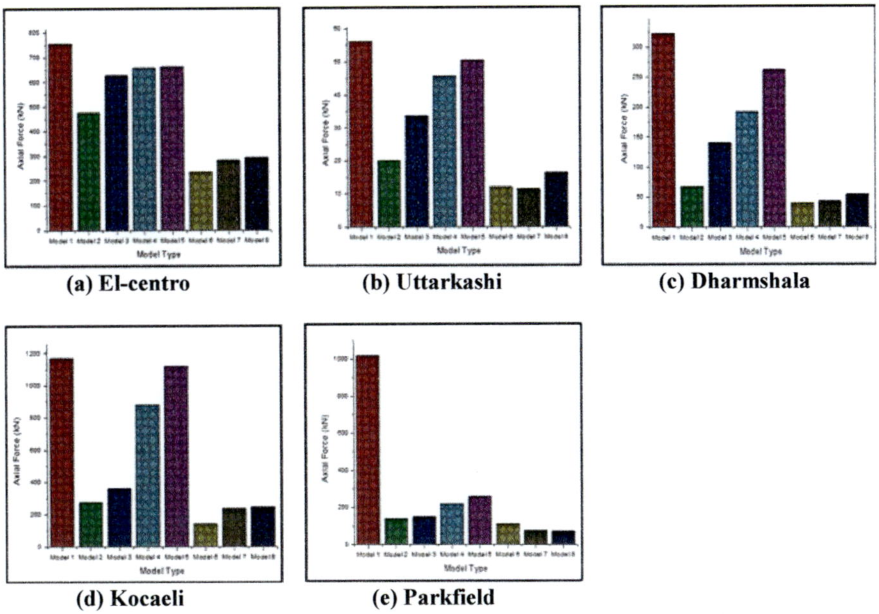

Fig. 9 Maximum axial force for G+7 story rectangle building for earthquake. **a** El-centro, **b** Uttarkashi, **c** Dharamshala, **d** Kocaeli, **e** Parkfield

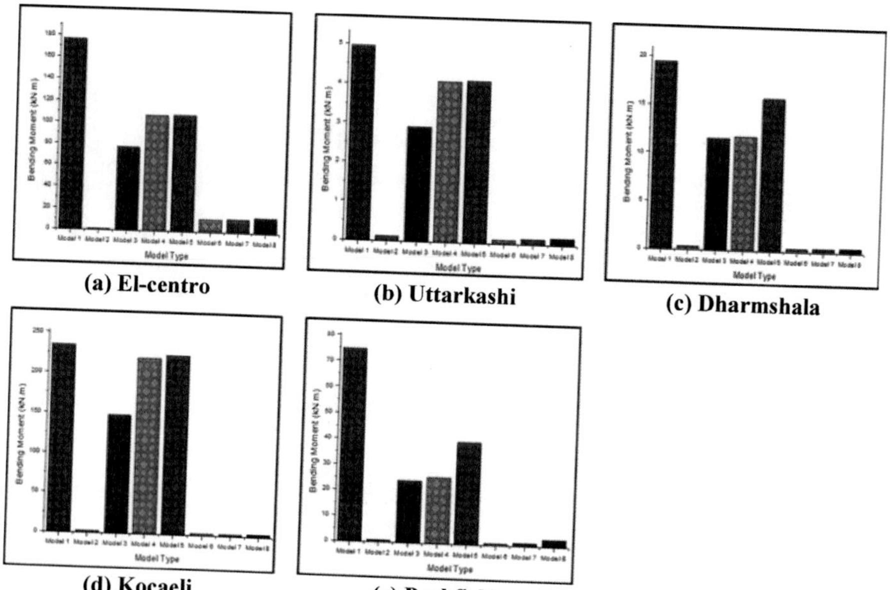

(a) El-centro (b) Uttarkashi (c) Dharmshala

(d) Kocaeli (e) Parkfield

Fig. 10 Maximum bending moment for G+3 story rectangle building for earthquake. **a** El-centro, **b** Uttarkashi, **c** Dharamshala, **d** Kocaeli, **e** Parkfield

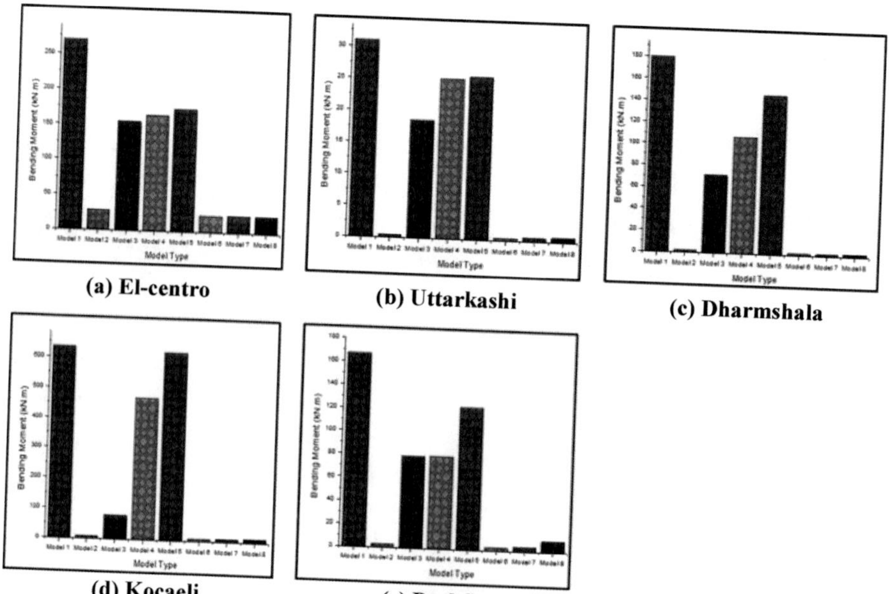

(a) El-centro (b) Uttarkashi (c) Dharmshala

(d) Kocaeli (e) Parkfield

Fig. 11 Maximum bending moment for G+7 story rectangle building for earthquake. **a** El-centro, **b** Uttarkashi, **c** Dharamshala, **d** Kocaeli, **e** Parkfield

9 Maximum Shear Force for G+3 and G+7 Story Building

The maximum shear force for G+3 and G+7 story building with consideration of soil effect is shown in figures.

From Figs. 12 and 13, it shows that the once installation of infill walls to the structure is reduced up to 98.46% as compared with the bare frame. Similarly in strut frame with SSI, the axial force the shear force is also decreasing to 87.47% as compared with a bare frame with SSI. The bare frame with SSI decreases to 87.50% as compared with the bare frame. Therefore, consideration both infill and SSI effect provide better results.

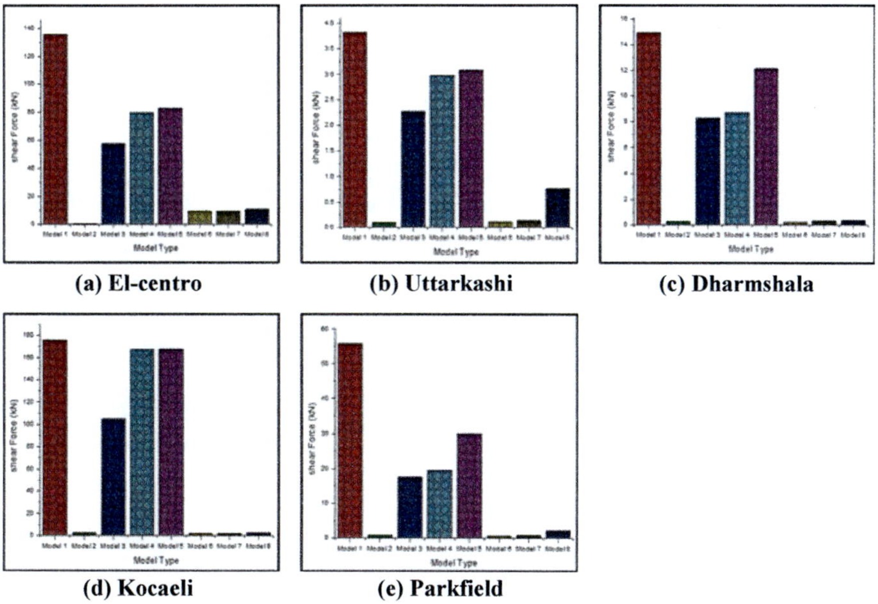

Fig. 12 Maximum shear force for G+3 story rectangle building for earthquake. **a** El-centro, **b** Uttarkashi, **c** Dharamshala, **d** Kocaeli, **e** Parkfield

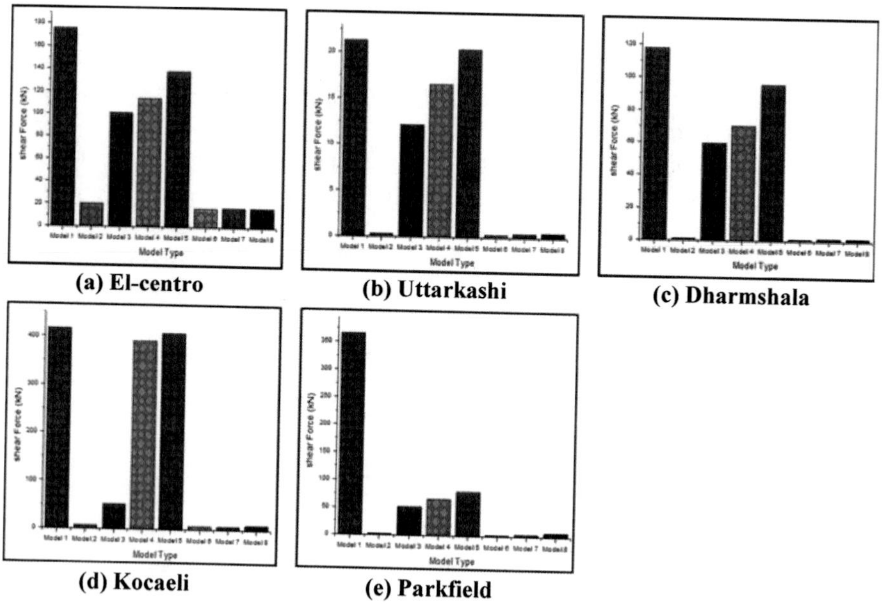

Fig. 13 Maximum shear force for G+7 story rectangle building for earthquake. **a** El-centro, **b** Uttarkashi, **c** Dharamshala, **d** Kocaeli, **e** Parkfield

10 Conclusions

In this research, the diverse effect of soil–structure interaction on infill strut panel is considered. RC structure with regarding to the loading of 3.1 m floor-to-floor height with base measurement 10 m × 8 m is examined for the impact of soil–structure interface by utilizing *Drucker–Prager* model nonlinear in ANSYS 15 with and without infill strut panel. From above results, it can be concluded that

1. Time period influence the earthquake reaction of the structure, since time period increases with consideration of SSI where as time period decreases with consideration of infill strut panel.
2. Deformation besides increases with consideration of SSI whereas reducing consideration of infill strut panel.
3. From ratio between bare frame and strut frame, it has been observed that maximum axial force, bending moment; shear force decreases after introducing infill strut panel.
4. From ratio between bare frame with SSI and strut frame with SSI, it has been observed that maximum axial force, bending moment; shear force increases due to effect of SSI. For as much as strut frame with SSI decreases the axial force, shear force; bending moment.
5. From above consequences, it is concluded that the strut frame results more as balance to a bare frame. And bare frame with SSI results more as balance to a

strut frame with SSI. So we can convey that strut frame with SSI results is much more to another model.

References

1. IS 1893[PART 1]:2016 Criteria for earthquake resistant design of the structure
2. Pulikanti S, Ramancharla PK (2014) SSI analysis of framed structure supported on pile foundations-with and without interface elements. Front Geotech Eng (FGE) 1(3)
3. Das D, Murty CVR (2004) Brick masonry infills in seismic design of RC framed buildings: part 1-cost implications. Indian Concrete J 78(7):39-44
4. Sunny NS, Mathai A (2017) Soil-structure interaction analysis of multi-storey building. Int J Sci Technol Eng
5. Standard I (1987) Code of practice for design loads (other than earthquake) for buildings and structures. Part I, Dead loads (second revision), IS-875-1987. Published by Bureau of Indian Standards, New Delhi, 110002
6. Standard I (1987) Code of practice for design loads (other than earthquake) for buildings and structures. Part II, Imposed loads (second revision), IS-875-1987. Published by Bureau of Indian Standards, New Delhi, 110002
7. Jayalekshmi BR, Chinmayi HK (2014) Effect of soil flexibility on seismic force evaluation of RC framed buildings with shear wall: a comparative study of IS 1893 and EUROCODE8. J Struct
8. Badry P (2016) Seismic soil structure interaction analysis of piled raft supported asymmetrical buildings

Identification of Enhanced Stiffness of Beam by EMI Frequency Shift Technique

Suraj Khante and Pranav Nimkar

Abstract Stiffness of the beam is the main parameter to distribute the load on the structure. As the concrete gains its strength, its stiffness also increases with it. Fundamental natural frequency of the structure is proportional to the square root of stiffness; thus, fundamental and subsequent natural frequencies increase with the increase in stiffness of the structure and with damage, the stiffness and natural frequencies decrease while mass is maintained constant. RCC beams are coated with carbon fiber-reinforced polymer (CFRP) layer, which is used to increase the flexural stiffness of beam. By using electro-mechanical impedance (EMI) technique, conductance and susceptance signatures of structure are recorded through Inductance capacitance resistance (LCR) meter by applying alternating voltage to lead zirconate titanate (PZT) patches. In the present work, the EMI conductance values of the structure are recorded against the frequency in the form of signatures. The peak in the conductance values shows the resonant frequency of the structure and indicates modal natural frequency of the structure. The increase in stiffness of the beam is monitored with EMI resonant frequency shift technique.

Keywords EMI · Resonant frequency · CFRP · PZT · Natural frequency

1 Introduction

Structural health monitoring (SHM) is a process in which certain strategies are implemented for determining the presence, location, severity of damages and the remaining life of structure after the occurrence of damage. This term is usually

S. Khante (✉) · P. Nimkar
Government College of Engineering, Amravati 444604, India
e-mail: snkhante@yahoo.com

P. Nimkar
e-mail: nimkarpranav249@gmail.com

L. M. Gupta et al. (eds.), *Advances in Civil Engineering and Infrastructural Development*, Lecture Notes in Civil Engineering 87,
https://doi.org/10.1007/978-981-15-6463-5_11

referred to aerospace, civil and mechanical engineering infrastructure. It is the continuous measurement of the loading environment and the critical responses of a structural system or its components. Health monitoring is typically used to track and evaluate the performance, symptoms of operational incidents and anomalies due to deterioration or damage as well as health during and after extreme events.

Structures are assemblies of load carrying members capable of safely transferring the superimposed loads to the foundations. They are constructed (e.g., buildings, bridges, dams, transmission towers, etc.) or manufactured (e.g., machines, trains, ships, aircraft, etc.) to serve specific functions during their design lives. Each structure forms an integral component of civil, mechanical or aerospace systems. In order to serve their designated functions, the structures must satisfy both strength and serviceability criteria throughout their stipulated design lives. However, with the passage of time, some amount of deterioration and damages are bound to occur, due to a variety of factors; such as environmental degradation, fatigue, excessive loads and natural calamities or simply due to long endurance combined with intensive usage. Even materials are not 100% immune from damages. According to Yao [1], 'damage' is defined as deficiency or deterioration in the strength of a structure, caused by external loads, environmental conditions or human errors.

2 History and Background

The use of natural frequency as a diagnostic parameter in structural assessment procedures using vibration monitoring is discussed by Salawu [2]. The approach was based on the fact that natural frequencies are sensitive indicators of structural integrity. Park et al. [3] done experimental studies on three kinds of structural members to detect the locations of cracks and loosened bolts. It was found that cracks or loosened bolts near the PZT sensors could be effectively detected by monitoring the shifts of the resonant frequencies of the impedance functions. Kim et al. [4] done experiment on pre-stressed beam and found that the resonant frequencies tend to shift left, according to the decrement of the tendon force. This indicates that the modal stiffness of interface is decreased with the reduction of the tendon force. Zou et al. [5] concluded that damage may be detected only using frequency response of the structure. The damage produces a decrease in structural stiffness, which in turn produces decrease in natural frequencies. Roh and Lee [6] proposed to analyse the changes of the impedances of the thickness modes wherein frequency range is above 1 MHz at the PZT based on its resonant frequency shifts rather than those of the lateral modes where frequency range is above 20 kHz at the PZT based on its root mean square (RMS) deviations. In order to establish trend of research and developments in the field of Structural Health Monitoring using EMI technique Na and Baek [7] presented literature review. Wang et al. [8] found that when the host structure is subjected to a tensile stress, the resonance in the admittance spectrum of PZT increases linearly and the tensile force also increases. While the host structure is subjected to a compressive load, the resonances decrease linearly as the compressive force increases. Theoretical analysis shows that while the tensile force increases the stiffness of both the host structure and the PZT material, and the compressive force decreases it, the stiffness changes cause the resonance shift in the PZT admittance spectrum. Hamid and Allan [9] done strengthening of beam by CFRP sheets. CFRP sheets are epoxy bonded to the tension face and web of concrete beams to enhance their flexural and shear strengths. The effect of CFRP sheets on strength and stiffness of the beams is evaluated for various orientations of the fibers with respect to the axis of the beam. Tests were presented by Täljsten and Blanksvärd [10] where a cement based bonding agent replaced the epoxy for retrofitting with an advantage of very good composite action. The CFRP sheets were epoxy bonded to the tension face and web of concrete beams to enhance their flexural and shear strengths by Norric et al. [11]. The effect of CFRP sheets on strength and stiffness of the beams was studied for various orientations of the fibers with respect to the axis of the beam. To monitor strength gain of precast RCC beams EMI technique was employed using reusable PZT patches by Khante and Jain [12]. Khante and Chikte [13] studied PZT based EMI technique and its utility for defect detection of concrete filled tubes.

From above study, it is evident that the natural frequency of structure is better indicator of the structural damage and it increases and decreases with the increase and decrease in stiffness of the structure. In the present work, the stiffness of the structure is estimated by using resonant frequency shift by EMI technique. The CFRP sheet is used to increase the stiffness of the beam so as to keep mass of the beam constant.

3 System Development

3.1 EMI Technique

In this technique, the PZT patches are used as piezoelectric transducers for monitoring the condition of structure. The PZT patches work as actuator and sensors simultaneously. The PZT patches are actuated by using electric signals and sensed with the help of LCR meter. The LCR meter gives the reading of admittance of the structure. The admittance of the structure has two parts, viz. conductance (G) and susceptance (B). The admittance of the structure is expressed as follows.

$$Y = G + Bj \tag{1}$$

in which conductance is the real part and susceptance is the imaginary part. Since susceptance is the imaginary part; thus, it can be ignored and only conductance is used for calculation. The RMSD index of the structure is calculated as follows.

$$M = \sqrt{\left(\frac{\sum (G_2 - G_1)^2}{\sum G_1^2}\right)} \times 100 \tag{2}$$

where

M = Damage metric (root mean square deviation

G_1 = Healthy signature

G_2 = Signature of specimen after damage.

The stiffness of the structure is the main parameter which decides the distribution of the load of the structure over the supports. So it is beneficial if the stiffness of the structure before and after damage can be calculated. The peak in the conductance signature represents the natural frequency of the structure of certain mode. This natural frequency decreases as the structure gets damaged. So the frequency shift will be measure of structural damage.

For simply supported beam, the natural frequency is related to the stiffness of the structure with following relationship.

$$f = \frac{n^2 \pi^2}{l^2} \sqrt{\frac{EI}{m}} \tag{3}$$

where

f = natural frequency in (Hz)

EI = Flexural rigidity of structure (N/m)

M = Mass per unit length of beam (kg/m)

l = Length of beam (m)

n = Mode number.

So from the above relation, it is clear that natural frequency of the structure is directly proportional to square root of stiffness of the structure. This relation can be used to find the reduced stiffness of the structure after damage. This calculated value of stiffness thus can be used to predict the load carrying capacity of structure and also to predict the future life of the structure.

3.2 Beam Specifications

The beam size is standardized as $100 \times 125 \times 1000$ mm^3. The beams are reinforced with two bars of 8 mm diameter at bottom and top. The reinforcement is provided with the clear cover of 15 mm. Bars of 6 mm diameter are used as stirrups at the spacing of 95 mm center to center. The grade of concrete used for the casting of beams is M20, and grade of steel is Fe 500. A total of three beams are casted, namely control beam, Beam 1, Beam 2. Control beam is without CFRP layer, Beam 1 is bonded with CFRP layer 1, and Beam 2 is bonded with CFRP layer 2. The PZT patches of size $10 \times 10 \times 0.2$ mm are attached to beam at one third length from left (PZT 1) and right (PZT 2) side of beam. The dimensions of beam and position of PZT patches are shown in Fig. 1, and details of reinforcement are shown in Fig. 2.

The PZT patch acts as both actuator and sensor. The PZT patches are actuated by electric voltage with the help of LCR meter. The LCR meter gives the response of the structure in the form of admittance of the electric signals. Figure 3 shows the connection of PZT patch to LCR meter.

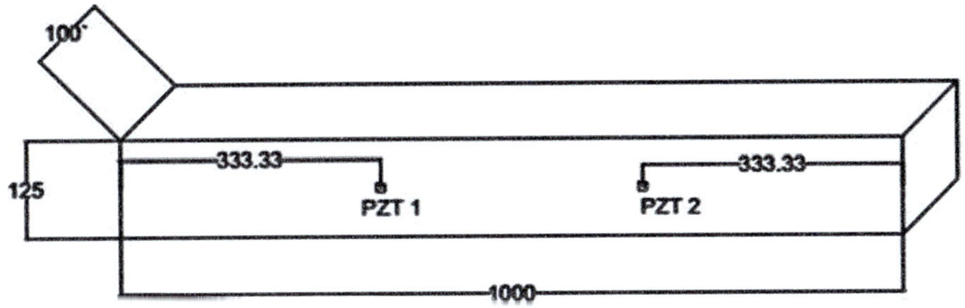

Fig. 1 Beam size in mm and position of PZT patches on beam. All dimensions in mm

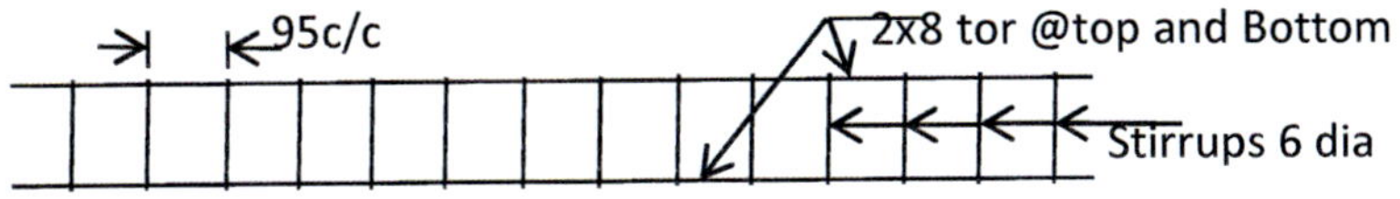

Fig. 2 Reinforcements in beam

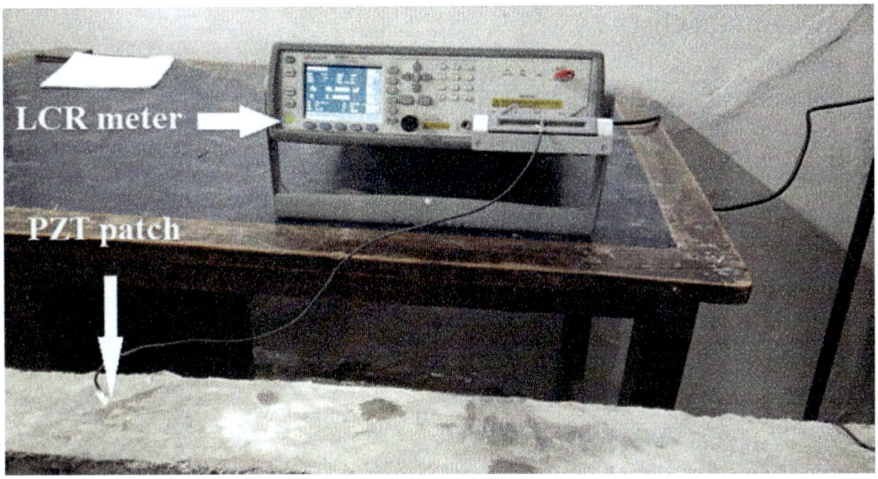

Fig. 3 LCR meter and PZT patch

4 Performance Analysis

At the beginning of the experimental work after the casting and curing of the beams for 28 days, the EMI conductance signatures are recorded for the beam with the help of LCR meter. In order to increase the stiffness of the beam, carbon fiber-reinforced polymer (CFRP) layer is attached at the bottom side of the beam by epoxy adhesive. With the addition of the CFRP layer, the flexural stiffness of beam increases. This increase in stiffness will cause the modal natural frequency of the beam to increase.

As the peak in the conductance signature represents the natural frequency of the beam of certain mode. Due to increase in natural frequency, the peaks in the conductance signature shift toward right, i.e., increase shift. In the similar manner, second CFRP layer is also bonded and conductance signatures are recorded. From these signatures, the increase shift in frequency is noted for the CFRP layer 1 and layer 2.

This increase in shift frequency is thus co-related to percentage increase in square root value of stiffness of beam. From this, the standard values of shift in frequency corresponding to percent increase in root of stiffness are recorded. These standard values will be further used for calculation of percent decrease in stiffness due to damage.

The beam is modeled in the finite element modeling software ANSYS. In ANSYS, the load verses deflection graph are plotted control beam, Beam 1 and Beam 2. The slope of load verses deflection graph gives the flexural stiffness of the beam. Flexural stiffness of the beam is calculated for the three conditions of beam, namely beam with no CFRP layer, layer 1 and layer 2. From Eq. (1), natural frequency is proportional to square root of stiffness of structure.

5 Result and Discussion

The stiffness of the beam is calculated in ANSYS for simply supported condition with concentrated load at its center. The beam is modeled in ANSYS workbench without CFRP layer, with CFRP layer 1 and CFRP layer 2. Simply supported condition is applied to the beam, and incremental concentrated load is applied to the beam. After analysis of the beam modeled in ANSYS gives the load and corresponding deformation. The load verses deformation curves are plotted and slope of this curve gives the stiffness of the beam. The results of the stiffness of the beam for beam without CFRP layer, beam with CFRP layer 1 and layer 2 are shown in Table 1. Figure 4 shows the concentrated load verses deformation curve of corresponding readings.

Table 1 Stiffness of beam

S. No.	Beam	Stiffness N/mm	Root of stiffness	Percentage increase in root of stiffness	
1	Beam without CFRP	56,363.43	237.40	6.49	Beam 1 with respect to control beam
2	Beam with CFRP 1	63,916.82	252.81	10.71	Beam 2 with respect to beam 1
3	Beam with CFRP 2	78,343.3	279.89	17.89	Beam 2 With respect to control beam

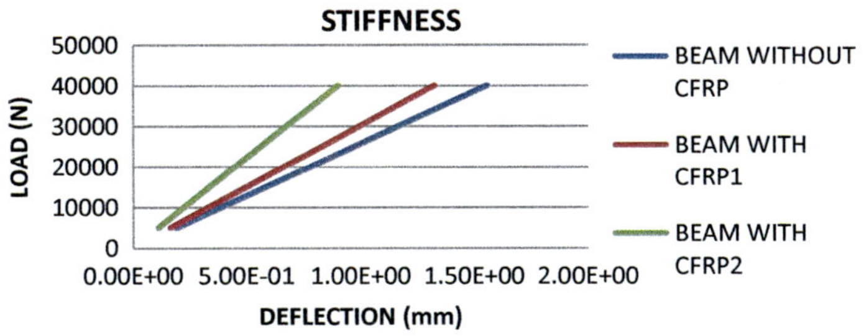

Fig. 4 Load versus deformation curve of beam

By using EMI technique, test was conducted on beam at the time of addition of CFRP layers. Figure 5a, b shows the frequency versus conductance plot of the beams. The peaks in the signature show natural frequency, which increase with the addition of CFRP layer. From these peaks in the natural frequency of beam with addition of CFRP layer, increment in corresponding natural frequency is found out. These shifts are listed in Table 2.

These shifts are due to increase in stiffness of the beam with the addition of CFRP layers. By using Eq. (1), the percentage increase in square root value of stiffness corresponding to the increase in natural frequency of beam is calculated. These shift and percentage increase of stiffness results are listed in Table 3.

By using the conductance versus frequency plot of the addition of CFRP layers, the root mean square deviation (RMSD) index is calculated for Beam 1 and Beam 2. These RMSD indices are shown in the following Figs. 6 and 7. It is observed that there is variation in the RMSD of PZT 1 and PZT 2 of beams. It is due to the variation in the bonding condition of the CFRP sheet to the beam.

6 Conclusions

The following conclusions can be drawn based on the present experiment

1. Peaks in the conductance signature of the beam indicate the resonant frequency of the beam, indicating the modal frequency.
2. With the addition of CFRP layer, the stiffness of the beam increases. This increase in stiffness results in increase in natural frequency of the beam. Thus, peaks in the conductance signature shifts toward right with addition of CFRP layers indicating increase in stiffness.
3. The percent increase in root of stiffness of the beam is co-related to the observed shift in resonant frequency peaks in EMI conductance signature of beam.
4. RMSD indices of the beam with the addition of CFRP layer to beam is monitored and found to be increased with addition of CFRP layers.
5. Conclusively, it is established that the increase in stiffness of beam can be indicated with the frequency shift technique.

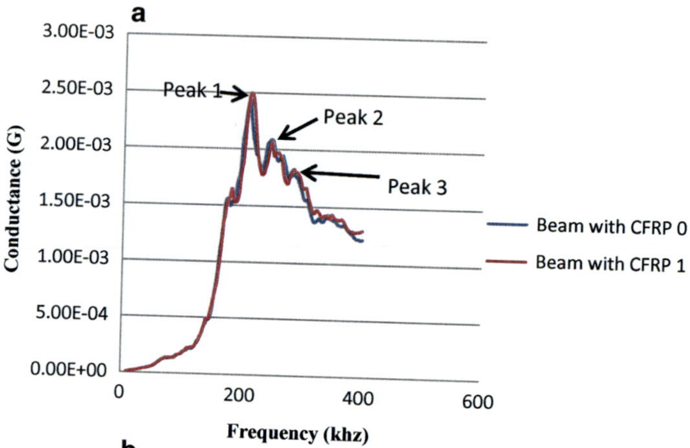

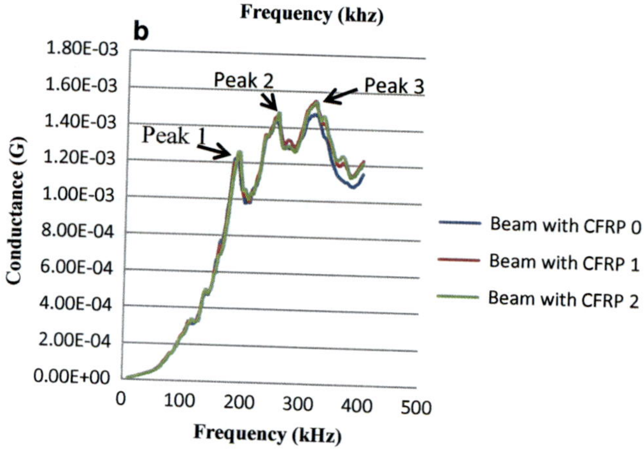

Fig. 5 **a** Conductance plot of Beam 1 with the addition of CFRP layer. **b** Conductance plot of Beam 2 with the addition of CFRP layers

Table 2 Shift of natural frequency (kHz)

S. no.	Beam number	Peak	Beam with CFRP 0	Beam with CFRP 1	Beam with CFRP 2	Shift		
						Beam CFRP 1 with respect to beam CFRP 0	Beam CFRP 2 with respect to beam CFRP 1	Beam CFRP 2 with respect to beam CFRP 0
1	1	1	208.9	210.85	–	1.95	–	–
2		2	244	245.95	–	1.95	–	–
3		3	281.05	283	–	1.95	–	–
1	2	1	187.45	189.4	191.35	1.95	1.95	3.9
2		2	251.8	253.75	255.7	1.95	1.95	3.9
3		3	316.15	318.1	320.05	1.95	1.95	3.9

Table 3 Shift results

S. no.	Comparison between	CFRP layer	Shift in natural frequency (kHz)	Percentage increase in square root value of stiffness	
1	Control beam—Beam 1	CFRP1-CFRP0	1.95	6.49	8.6
2	Beam 1—Beam 2	CFRP2-CFRP1	1.95	10.71	
3	Control beam—Beam 2	CFRP2-CFRP0	3.9	17.90	17.90

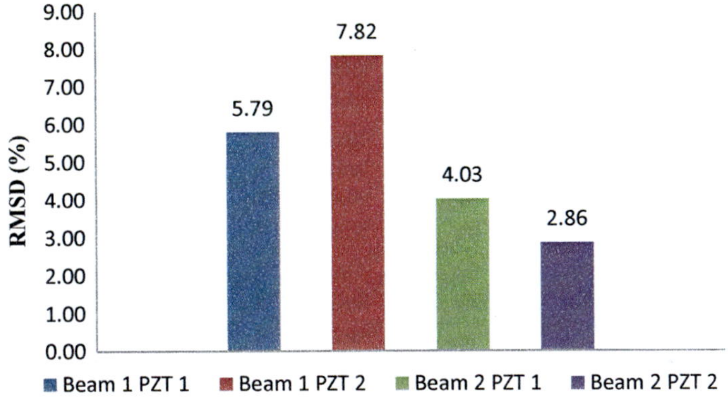

Fig. 6 RMSD of CFRP layer 1

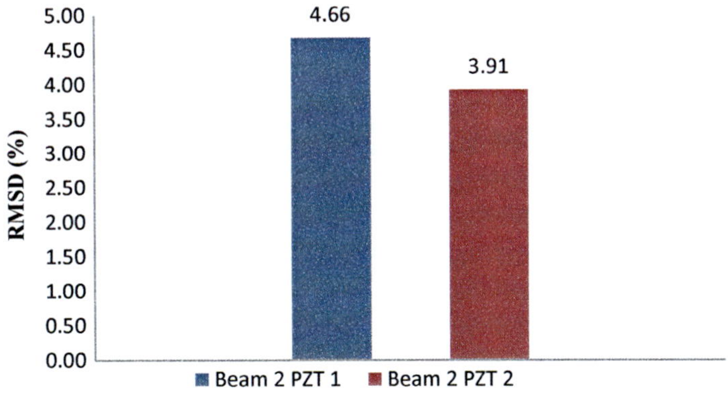

Fig. 7 RMSD of CFRP layer 2

References

1. Yao JTP (1985) Safety and reliability of existing structures. Pitman Publishing Programme, London
2. Salawu OS (1997) Detection through changes a review. Eng Struct 19(9):718–723
3. Park SH, Yi JH, Yun CB, Roh YR (2004) Impedance-based damage detection for civil infrastructures. KSCE J Civ Eng 8(4):425–433
4. Huynh TC, Lee KS, Kim JT (2015) Local dynamic characteristics of PZT impedance interface on tendon anchorage under prestress force variation. Smart Struct Syst 15(2):375–393
5. Zou Y, Tong LP, Steven GP (2000) Vibration-based model-dependent damage (delamination) identification and health monitoring for composite structures—a review. J Sound Vibr 230(2):357–378
6. Roh Y, Lee J-J (2005) Health monitoring of steel structures using impedance of thickness modes at PZT patches. Smart Struct Syst 1(4):339–353
7. Na WS, Baek J (2018) A review of the piezoelectric electromechanical impedance based structural health monitoring technique for engineering structures. Sensors 18(5):1307
8. Wang T, Wei D, Shao J, Li Y, Song G (2018) Structural stress monitoring based on piezoelectric impedance frequency shift. J Aerosp Eng 31(6):04018092
9. Hamid Rahimi, Allan Hutchinson," Concrete Beams Strengthened With Externally Bonded Frp Plates", Journal Of Composites For Construction February (2001)
10. Täljsten B, Blanksvärd T (2007) Mineral-based bonding of carbon FRP to strengthen concrete structures. J Compos Constr 11(2):120–128
11. Norris T, Saadatmanesh H, Ehsani MR (1997) Shear and flexural strengthening of R/C beams with carbon fiber sheets. J Struct Eng 123(7):903–911
12. Khante SN, Jain R (2017) Identification of strength gain of precast concrete during initial period using reusable PZT. Int J Innov Res Sci Eng Technol 6
13. Khante SN, Chikte SS (2017) Defect detection of concrete filled steel tubes with PZT based technique. Int J Innov Res Sci Eng Technol 6

Application of Maturity Meter to Estimate Location of Sensors for Plate Elements

Pranjali Hajare

Abstract The quality assurance of concrete structures mainly depends on the measure of the in-place strength of concrete. The in-place strength of concrete plays an important role in planning the activities on-site such as opening of structure of occupancy, opening of roads to traffic, construction claims, etc. The strength of concrete is a measure of temperature and time. This research work scrutinizes the application of "Maturity Method" in finding the location for placement of sensors in plate elements, specifically a squared manhole cover. A trial program is proposed to detect the distress locations in a manhole cover using maturity meter. For this study, three non-sacrificial sensors were placed in a manhole cover and hydration extent was examined. Further, the results were interpreted in excel on the basis of the correlation between the three sensors placed in the manhole cover kept at room temperature where concrete hydration was permitted.

Keywords Concrete · Hydration · Thermal behaviour · In-place strength · Plate elements · Maturity

1 Introduction

Concrete is the most important and most widely used building material in the construction industry worldwide. The material has gained its predominance mainly due to its versatile nature and decisive advantages which have proved to be a boon to the construction industry [1]. The traditional methods assuming that the cube strength represents in-place strength of concrete does not hold accurate unless the standard practices are adopted during construction. This unbalanced strength of concrete can lead to a disastrous effect on the structure.

The strength of concrete is a measure of its temperature and time. "Maturity method is a technique used to estimate the in-place strength of fresh concrete with due consideration of temperature–time effect on strength of concrete" [2]. This method

P. Hajare (✉)
Department of Civil Engineering, Vishwakarma Institutes of Information Technology, Pune, India
e-mail: pranjali.hajare@gmail.com

L. M. Gupta et al. (eds.), *Advances in Civil Engineering and Infrastructural Development*, Lecture Notes in Civil Engineering 87,
https://doi.org/10.1007/978-981-15-6463-5_12

is based on a rudimentary assumption that a given concrete mixture design poured during the interval of a project has the same compressive strength at its "maturity index." Higher the temperature, the faster is the curing of concrete. There is much saving of time and money since this method is more accurate in predicting the early age strength [3].

Several research has been carried out about this Maturity method with regards to its sensor placement which records the temperature of in-place fresh concrete. Various structural elements were understudy for the placement of sensors which included beams, columns, slabs, pavements, mass concreting structures, etc. [4]. The research being limited to the specific structural members, plate members with smaller thickness needed to considered for the research. Also, the maturity meters are used only to detect the strength of fresh concrete and not the hardened one.

2 Objectives

The objectives of this paper basically rely on the application of Maturity Method further exceeding to study the fundamental aspects of Maturity Method. Also to evaluate the criteria specified for placement of sensors in structural elements. The main objective concentrates on the experimental analysis for determination of location of sensors in plate elements.

3 Maturity Method

A systematic application of this method can result in saving time, money and workmanship. The maturity function is dependent on temperature sensitivity for initial strength development. All concrete mixes have different maturity functions. Maturity meter is much more dependable in estimating relative strength compared to the absolute strength. The specimens with higher early age temperatures result in higher initial strengths and lower long term strength provided they have equal values of maturity index [5].

In 1951, Saul suggested that maturity should be calculated with respect to datum temperature, which is the lowest temperature at which concrete starts gaining strength. Thus, maturity is computed from temperature history using the following equation,

$$M = \sum_{0}^{t} (T - T_o)\Delta t$$

where, M—maturity at age t

T—average temperature of concrete during time interval Δt

T_o—datum temperature.

The traditional method of cube/cylinder breaking test is time-consuming, material consuming, labour consuming. In order to compensate this losses, maturity Meter was developed, to know the real-time in-place strength of concrete. Even still, the traditional method is largely used in developing countries like India. The major problem found with the traditional methods of finding the compressive strength of concrete is that the strength of the cubes/cylinders cast for estimating the concrete strength and the actual strength of elements in the structure do not match with each other.

This mismatch of strength can lead to various damages to the structure. Either the structure becomes over-safe or under-safe. There is around $\pm 200\%$ variation in cube/cylinder and element strength. Hence, for this reason, maturity Meters come in practice to be used on-site, to save time and money.

The construction industry is predominantly using the traditional method of breaking cylinder test for obtaining the compressive strength of concrete. On the contrary, there are many advantages of using Maturity method compared to the breaking cylinder test in consideration of the test procedure, reliability, speed, and cost. The break test being the manual one, may have variations in testing time, whereas using the maturity meters, data is logged or retrieved by external device in real-time. Specimens used in break test may have small volumes and hence can retain less heat leading to low strength results but maturity meters can predict the actual in-place strength of concrete. Traditional method is much time-consuming since the procedure takes too long to give away the results, meters on the other side gives real-time strength results.

Casting of specimens, collecting results and repeating the process is much of time-consuming. The maturity meters save the material cost, labour cost and time by evaluating the on-site strength results. Financing cost may be increased if the project gets delayed, whereas a remarkable financial cost saving was observed when implementing the maturity meters on-site due to early cost completion of projects [4].

Figure 1 represents the temperature–time graph of a specific concrete mix. Maturity is linearly dependent on temperature is accommodated as a perspective and can be constituted by the area under the temperature curve. This area under the curve is the difference between the recorded temperature of placed concrete and the datum temperature.

Figures 2 and 3 show the maturity meter with its wired sensor used for recording the temperature of fresh concrete. This technology being the proven one in foreign countries for 50 years, the meters are manufactured by Sun Pacific company, Pune under the guidance of research experts from Vishwakarma Institutes of Information Technology, Pune, and with sponsorship of Pune Construction Engineering Research Foundation (PCERF).

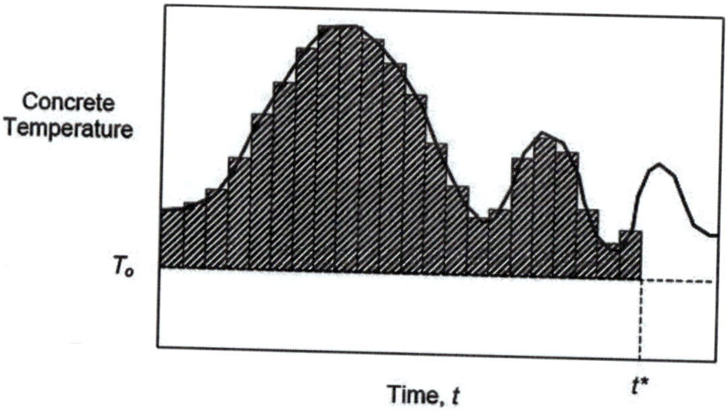

Fig. 1 Nurse—saul maturity function

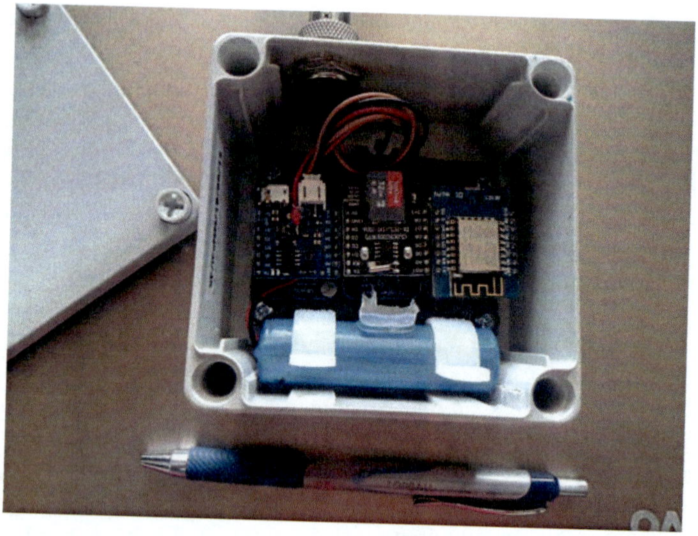

Fig. 2 Maturity meter

4 Experimental Analysis

The general procedure of Maturity method follows the following procedure—(1) determination of the concrete temperature history, (2) measurement of compressive strength development, (3) determination of the strength of concrete [6, 7].

A square manhole cover with dimensions 400 mm × 400 mm × 50 mm was cast at a precast plant. The location of the sensors depend on either of the two criteria viz. internal temperature of concrete is low or regions where highest loads are expected. The manhole cover was filled with M30 normal concrete and external vibration was provided. Three non-sacrificial (wired) sensors attached to the maturity meter were

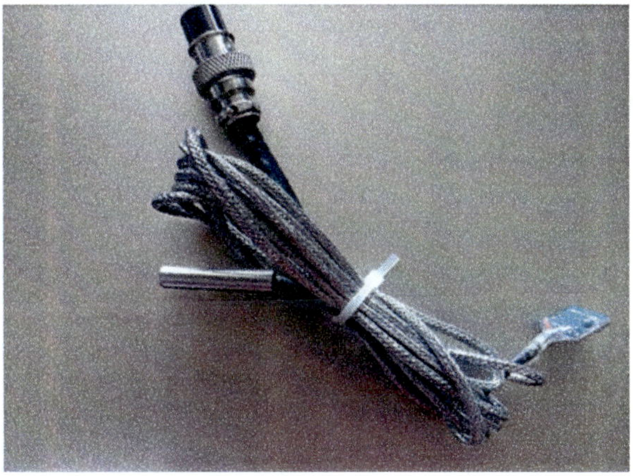

Fig. 3 Sensor of maturity meter

Fig. 4 Placement of sensors in a square manhole cover

inserted in the manhole cover. One at the bottom surface near one edge, one at the mid-depth at the center and other at the top surface on the parallel edge. The readings were recorded by the meter at a frequency period of 10 min.

The specimen was cured and readings were recorded for 14 days. The readings were then retrieved and results were interpreted using excel sheets. The sensors were then removed from the specimen by breaking it manually [8] (Fig. 4).

5 Results and Discussions

The results speculated from the experimental study for square manhole cover casted with three non-sacrificial sensors inserted in it were interpreted as follows

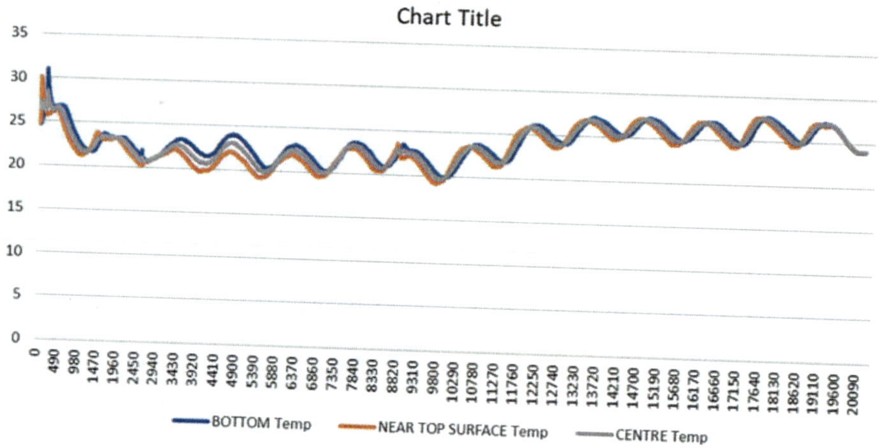

Fig. 5 Temperature versus time graph for 3 sensors

- The concrete was workable with slump of 100 mm and no segregation was observed in the manhole covers.
- The battery life of meters being 17 days, no interruption in the recording of readings was observed until 14 days.
- As we know that concrete gains its maximum strength during its first 3 days of curing, the correlation is calculated for the first 3 days only.
- The correlation coefficient calculated for all of the three sensors was as follows—(1) bottom and surface—0.8367, (2) surface and center—0.9541, (3) bottom and center—0.9622, much higher than the standard value of correlation coefficient for concrete which is 0.75.
- In order to get the precise results, percentage error values were calculated considering central sensor as the benchmark. From the central sensor, the variation in temperatures was calculated towards both top and bottom of the manhole cover.
- From Figs. 6 and 7, the graphs have predicted the variation in error as 10.72% for top-center sensors and 10.87% for center-bottom sensors.
- The % Error graphs were calculated using the difference between the top and bottom temperatures with respect to the central temperatures (Fig. 5).

6 Conclusions

The experimental study and results concluded that the correlation coefficient value obtained is superior to the standard value. Hence, it can be concluded that there exists a strong relation between the temperature and time for all the three sensors and the correlation coefficient is significant. For concrete specimens, the maximum error should be limited to 15%. Any error exceeding 15% is not contemplated for results. From Figs. 6 and 7, it can be seen that both the percentage errors are within the limiting

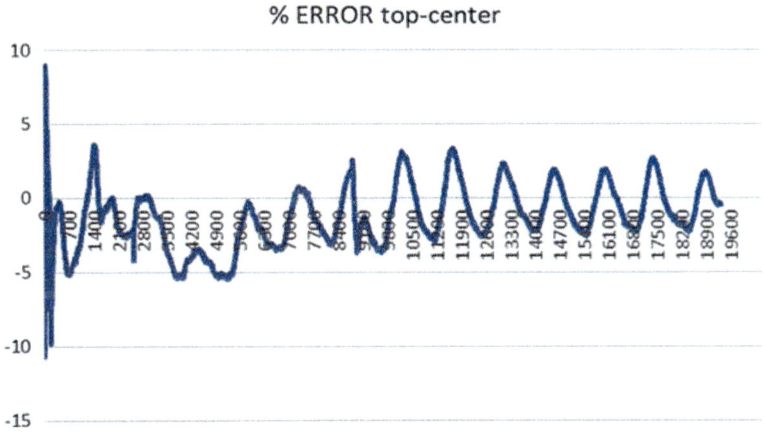

Fig. 6 % Error between top and center sensors

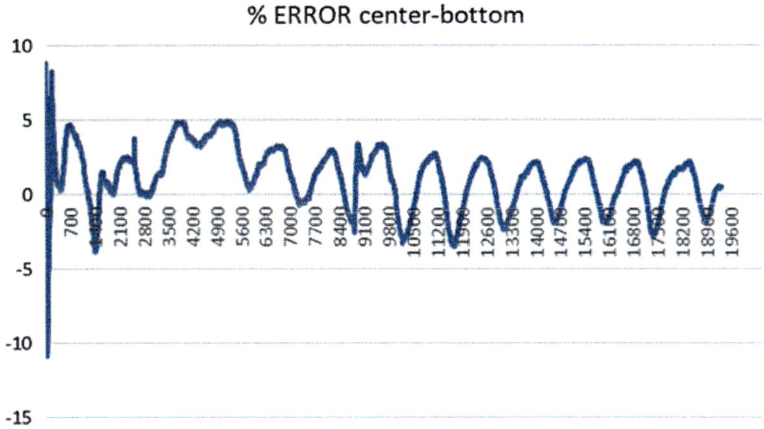

Fig. 7 % Error between center and bottom sensors

values. Hence we can conclude that there is not much variation in temperatures for both top and bottom sensors and one single sensor can be used to evaluate the strength gain of a squared manhole cover. Thus, it can be also said that the three sensor readings are strongly related to each other and do not vary much concerning the sensor placement. Consequently, it can be terminated that the placement of sensor in thin plate elements does not rely on the location of sensors and one single sensor can be replaced to represent a square manhole cover. Ideally, the location can be selected at the center of the manhole cover at mid-depth as the distress regions.

References

1. Shetty MS (2005) Concrete technology: theory and practice, 6th edn. S. Chand Publications, New Delhi
2. Irtishad A, Hesham A, Salman A (2006) Utilization of maturity meters for concrete quality assurance. Florida Department of Transportation
3. Carino NJ (2004) The maturity method. In: Malhotra VM, Carino NJ (eds) Handbook on nondestructive testing of concrete, 2nd edn. CRC Press, Boca Raton, Florida
4. De Carufel S, Andrew F, Pouria G, Aali A (2018) Concrete maturity from theory to application. Giatec Scientific Inc., Canada
5. Carino NJ, Lew HS (2001) The maturity method: from theory to application. National Institute of Standards and Technology, Gaithersburg, MD
6. Benaicha M, Burtschell Y, Alaoui AH (2016) Prediction of compressive strength at early age of concrete–application of maturity. J Build Eng 1(6):119–125
7. Yikici TA, Chen HLR, Use of maturity method to estimate compressive strength of mass concrete. West Virginia University, Civil and Environmental Engineering, Morgantown
8. ASTM C1074 (2004) Standard practice for estimating concrete strength by the maturity method. ASTM Standards, Vol 04.02, ASTM, West Conshohocken, PA

Analysis of Various Asymmetric Membrane Roofs with Opening(s)

Grace Mary Abraham and Ruksa Nizar

Abstract A roof is a part of a building which provides us protection from animals, weather, notably rain or snow and admits light if a skylight is present. The different types of roofs are gable roof, hip roof, sawtooth roof, etc., and the materials used in the roofs include grass, ceramics, Teflon, etc. Opening is the intentional introduction of ambient air into a space, and it is mainly used to control indoor air quality by diluting and displacing indoor pollutants; it can also be used for purposes of thermal comfort or dehumidification. Modern trends of roofs are now available with different modern and cheap materials with openings. This paper discusses the numerical and physical models developed for the design of different asymmetric membrane roofs and thus comparing the results of the models with those of the various models and their analysis. Generally, the shapes considered are conoid shape, saddle shape, and elliptical dome shape with one or more openings. The materials adopted for asymmetric membrane roofs are modern textile material, i.e., polytetrafluoroethylene, synthetic rubber, i.e., ethylene–propylene–diene–terpolymer and polyvinyl chloride (PVC). This initially outlines procedures for stress analysis, wind pressure analysis, and seismic analysis. Finally, construction of the actual membrane is described, and comparison is made. Determination of the mechanical properties of the fabrics used to construct the asymmetric membrane roofs which are also briefly discussed.

Keywords Roof · Conoid · Saddle · Semi elliptical · PVC · PTFE · EPDM · Stress · Deformation

G. M. Abraham · R. Nizar (✉)
Amal Jyothi College of Engineering, Koovapally P.O, Kanjirapaly, Kottayam 686518, India
e-mail: ruksanizar@gmail.com

L. M. Gupta et al. (eds.), *Advances in Civil Engineering and Infrastructural Development*, Lecture Notes in Civil Engineering 87,
https://doi.org/10.1007/978-981-15-6463-5_13

1 Introduction

Roof is a part of building envelope. It is one of the major coverings on the uppermost part of a building or shelter. It provides protection from animals and weather, rain or snow, heat, wind and sunlight. The word also denotes the framing or structure which supports the covering. Some roofs are also provided with openings [1, 2]. Opening is defined as the introduction of ambient air into any space [3]. The major categories of openings are mechanical ventilation which refers to any system that uses mechanical means, such as a fan, to introduce sub-aerial air to a space [4]. This includes positive pressure ventilation, exhausts ventilation, and balanced systems that use both supply and exhaust ventilation, natural ventilation which refer to intentionally designed passive methods of introducing sub-aerial to a space without the use of mechanical systems [5]; mixed mode ventilation (or hybrid ventilation) systems use both natural and mechanical processes, and infiltration is the uncontrolled flow of air from outdoors to indoors through fissures and leaks (unplanned openings) in a building envelope [6]. Another type of flow has been referred to as adventitious ventilation.

From the literature survey, it was understood that internal pressure responses were closely related to characteristics of opening [7], and response characteristics of the building cavity determined by opening area, building volume [8]. Many previous studies focus on wind pressures in a building with a dominant opening in the wall and only little attention has been paid to the cases of openings on the roofs [9]. Also, pressure responses of a building with various openings on a flat roof corner were not focused [10]. No studies were conducted on the basis of seismic analysis. Different zones and directions are selected for applying wind load. Architectural purposes are not focused, and application of different materials is not carried out.

This paper focuses on the following objectives:

To compare the performance of asymmetric membrane roofs with opening(s) by varying the following conditions.

(a) Shape—conoid, saddle, semi-elliptical
(b) Material—Polytetrafluoroethylene (PTFE), ethylene–propylene–diene–terpolymer, polyvinyl chloride

Scope of this paper includes

(a) The covered area of the membrane roof is 2900 m^2.
(b) The opening shape is adopted as circular.
(c) Seismic zone is adopted as zone III.

1.1 Material Properties

1.1.1 Modern Textile Material, i.e., PTFE or Polytetrafluoroethylene

Polytetrafluoroethylene (PTFE) is a synthetic fluoropolymer of tetrafluoroethylene that has numerous applications. PTFE is a fluorocarbon solid, as it is a high-molecular-weight compound consisting wholly of carbon and fluorine. PTFE has one of the lowest coefficients of friction of any solid. The density of the material is 2200 kg/m^3, Poisson's ratio is 0.46, and Young's modulus is 5.52E+08 Pa.

1.1.2 Polyvinyl Chloride (PVC)

Polyvinyl chloride is the world's third-most widely produced synthetic plastic polymer, after polyethylene and polypropylene. About 40 million tonnes are produced per year. PVC comes in two basic forms: rigid (sometimes abbreviated as RPVC) and flexible. The density is 1.35 g/cm^3, Poisson's ratio is 0.4, and Young's modulus is 0.0034 Pa.

1.1.3 Synthetic Rubber (Ethylene–Propylene–Diene–Terpolymer)

Ethylene–propylene–diene–monomer rubber (EPDM rubber), a type of synthetic rubber, is an elastomer characterized by a wide range of applications. This is an M-class rubber where the "M" in M-class refers to its classification in ASTM standard D-1418; the M-class includes rubbers having a saturated chain of the polyethylene type. The value for density is 1.1 g/cm^3, Poisson's ratio is 0.48, and Young's modulus is 6E+0.6 Pa.

2 Design and Modeling

Modeling and analysis of conoid semi-elliptical and saddle roofs have been carried out in ANSYS Workbench. The area of the roof is kept constant. The values of wind pressure and dimensions are based upon various journals. The materials considered for the analysis are modern textile membrane, polyvinyl chloride, and synthetic rubber. As per IS 875 Part III, the value of wind pressure obtained after designing is 1.64 N/mm^2 for conoid roof, 2.622 and 2.9506 N/mm^2 for semi-elliptical roof, and 2.9 N/mm^2 for saddle roof. Seismic analysis is carried on the basis of response spectrum method.

2.1 Description of the Geometry of Roof

SL NO	DESCRIPTION	CONOID	ELLIPTICAL	SADDLE
1.	Span	70m	47.6m	46m
2.	Height	20m	37.2m	35m
3.	Opening Diameter	2m	10m	-
4.	Area	2900m²	2900m²	2900m²
5.	Thickness	120mm	12mm	120mm
6.	Radius	33m	44.2m	-

3 Analysis

3.1 Wind Analysis

The wind loading is along the surface of roof, and the wind load is calculated by using code IS 875(Part 3)-1987. Basic wind speed is 39 m/s, terrain category is 2, and building class is taken as C (Fig. 1).

For the material EPDM, the maximum value for deformation obtained is 0.1201 which is at the mid of the roof which is shown in red color and the minimum value is 0. The maximum value for deformation obtained is 0.66014 cm which is at the mid of the roof which is shown in red color and the minimum value is 0 for PVC. The maximum value for deformation obtained is 3.8508 cm which is at the mid of the roof which is shown in red color and the minimum value is 0 for PTFE (Fig. 2).

The maximum value for stress obtained is 0.00128 MPa for EPDM. The maximum value for stress obtained is 1.2922 MPa for PVC. The maximum value for stress obtained is 0.1288 MPa for PTFE.

For semi-elliptical roof, the wind load obtained as per IS 875 Part 3 is 2.622 and 2.9506 MPa (Fig. 3).

The maximum value for deformation obtained is 40.765 mm for PVC. The maximum value for deformation obtained is 345.49 mm for PTFE. The maximum value for deformation obtained is 99.53 mm for EPDM (Fig. 4).

The maximum value for stress obtained is 14.71 MPa for PVC. The maximum value for stress obtained is 15.05 MPa for PTFE. The maximum value for stress obtained is 0.0263 MPa for EPDM.

For saddle roof, the wind load obtained as per IS 875 Part 3 is 2.94 MPa (Fig. 5).

The maximum value for deformation obtained is 99.44 mm for PVC. The maximum value for deformation obtained is 42.436 mm for PTFE. The maximum value for deformation obtained is 28.277 mm for EPDM (Fig. 6).

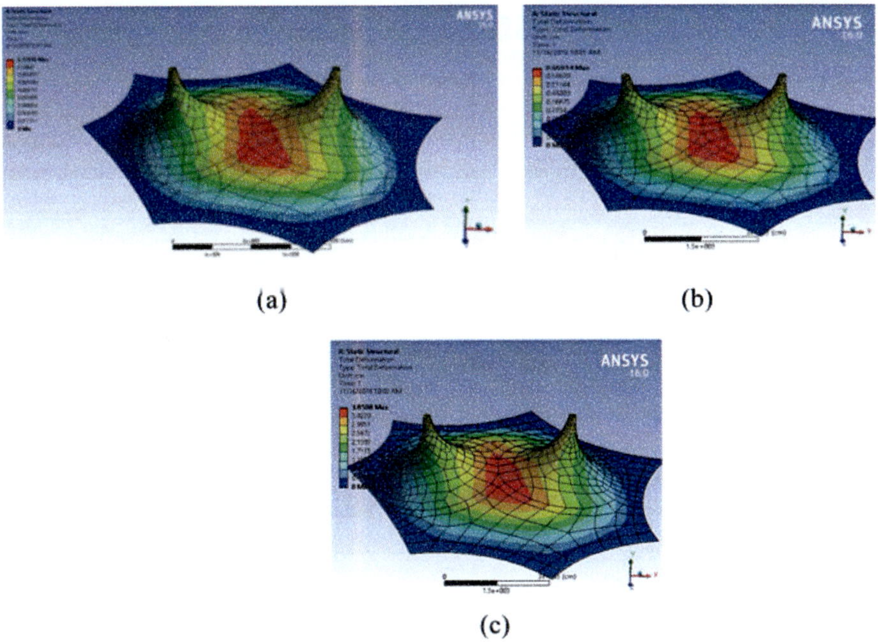

Fig. 1 Deformation diagram for conoid roof using the material **a** EPDM, **b** PVC, **c** PTFE

The maximum value for stress obtained is 20.64 MPa for PVC. The maximum value for stress obtained is 1.061 MPa for PTFE. The maximum value for stress obtained is 0.01055 MPa for EPDM.

3.2 Seismic Analysis

From IS 1893 Part I: 2002, page 16, Clause 6.4.5, the value for Sa/g is taken as 2.50 for soft soil sites. The value for frequency is obtained from the software. From frequency values, time period can be calculated. Using the values of time period, Sa/g can be obtained and thus acceleration values are calculated by multiplying Sa/g with 9810 mm/s^2.

3.2.1 Conoid Roof

Figure 7 shows the acceleration versus mode graph for conoid roof using the material PVC, PTFE, and EPDM (Fig. 8).

The maximum value for deformation obtained is 0.0791 mm for PVC. The maximum value for deformation obtained is 0.06233 mm for PTFE.

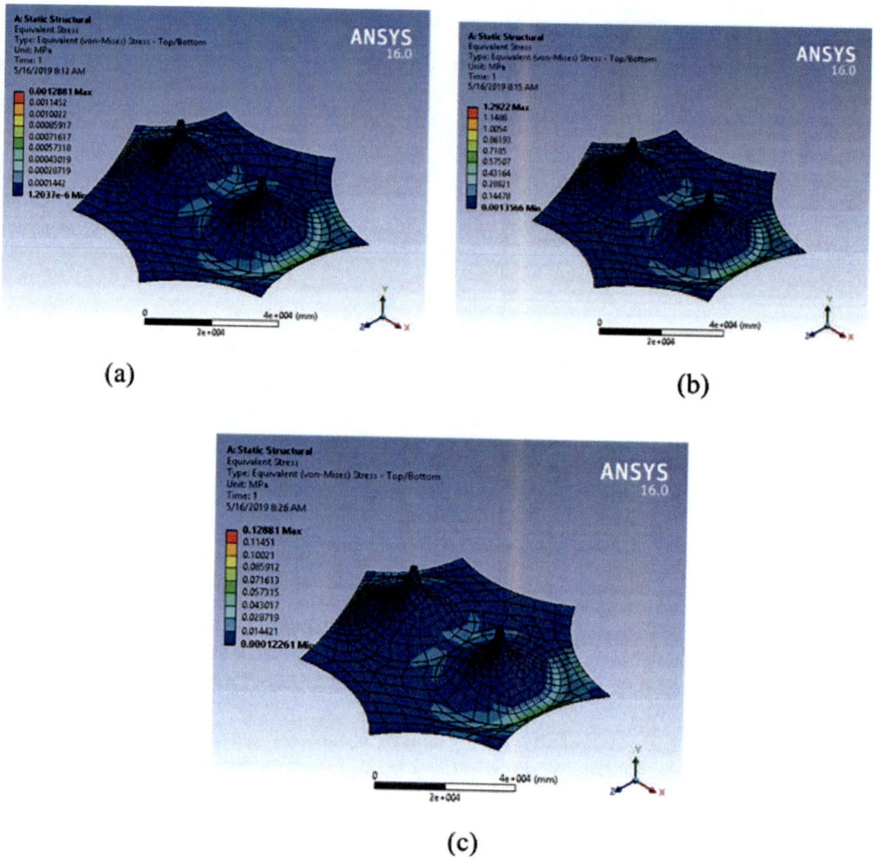

Fig. 2 Stress diagram for conoid roof using the material **a** EPDM, **b** PVC, **c** PTFE

3.2.2 Semi-elliptical Roof

The maximum value for deformation obtained is 5.2917 mm for PVC. The maximum value for deformation obtained is 78.84 mm for PTFE (Figs. 9 and 10).

3.2.3 Saddle Roof

See Figs. 11 and 12.

4 Results and Discussions

See Table 1.

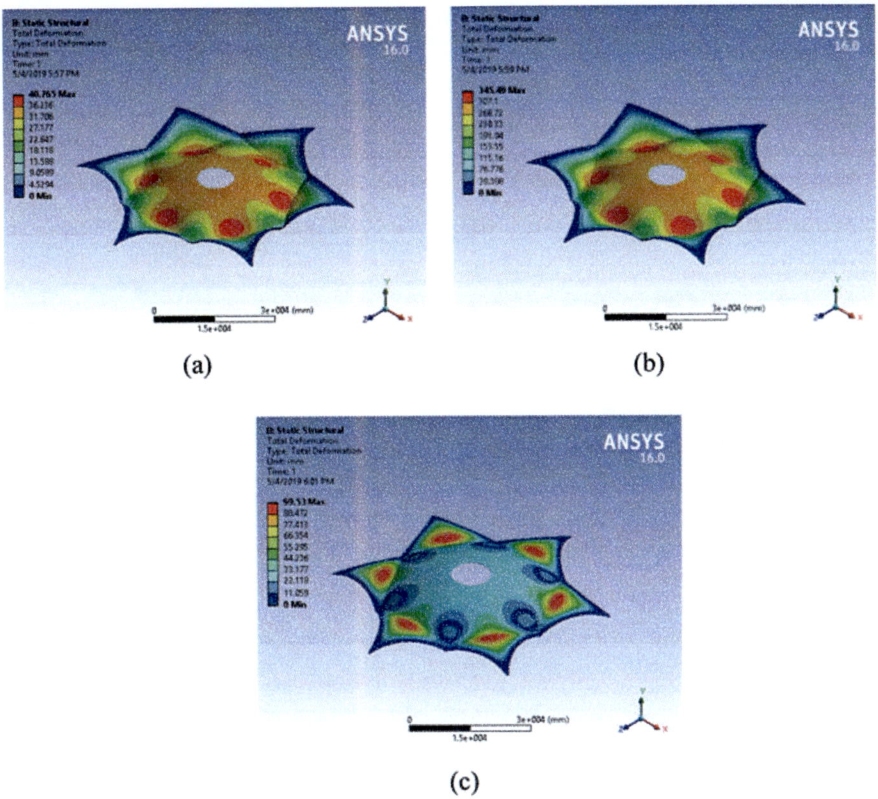

Fig. 3 Deformation diagram for semi-elliptical roof using the material **a** PVC, **b** PTFE, **c** EPDM

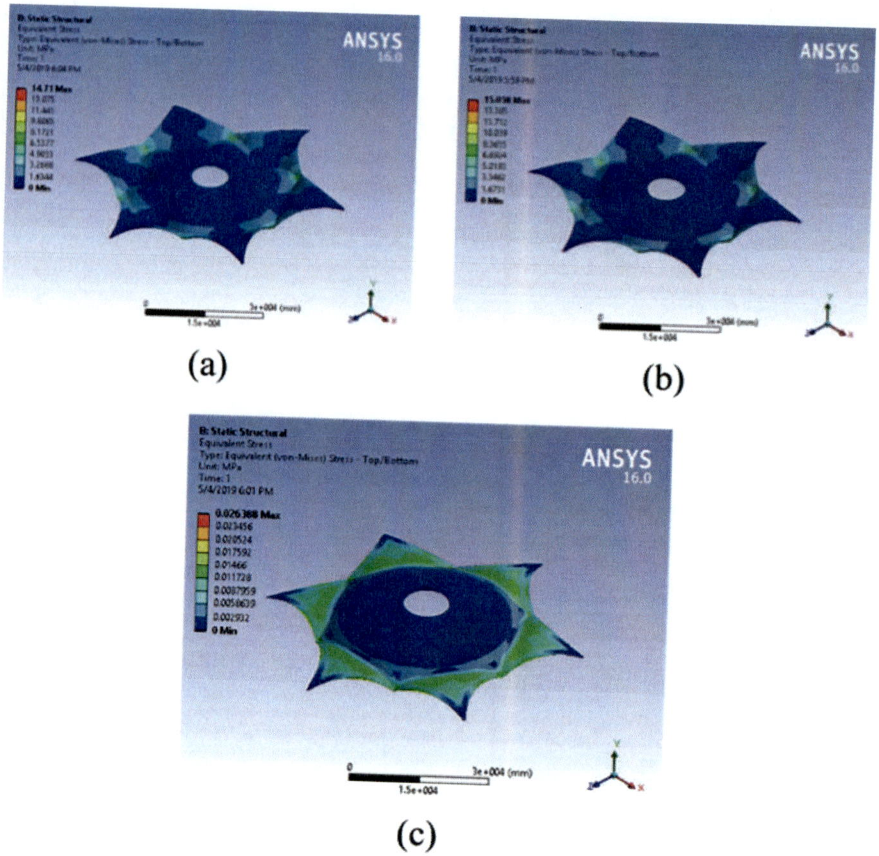

Fig. 4 Stress diagram for semi-elliptical roof using the material **a** PVC, **b** PTFE, **c** EPDM

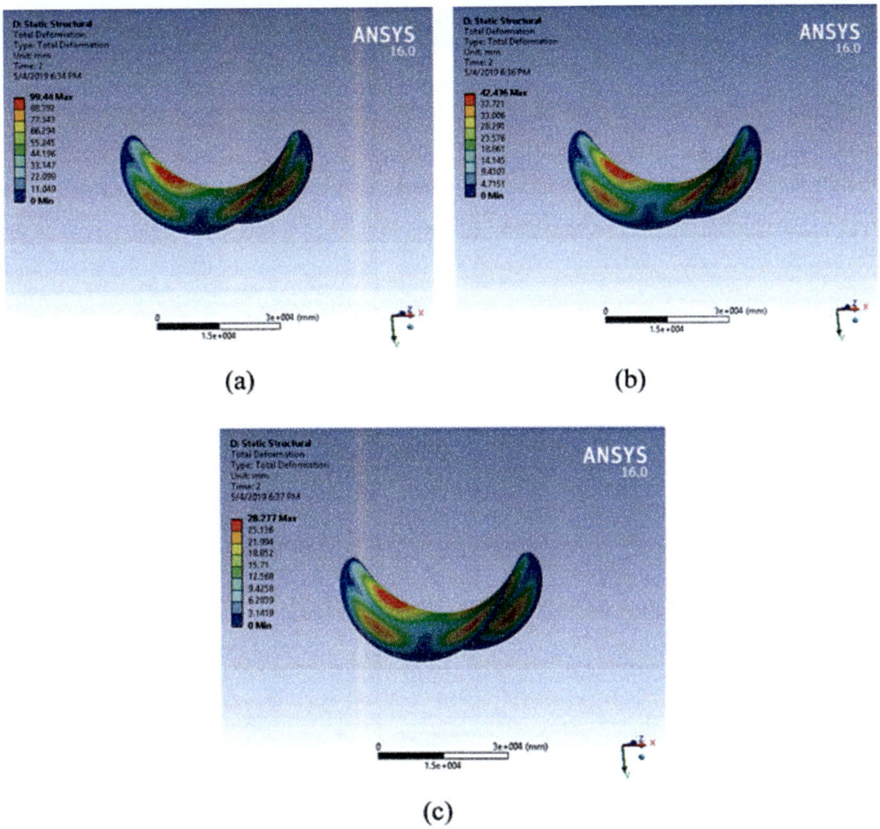

Fig. 5 Deformation diagram for saddle roof using the material **a** PVC, **b** PTFE, **c** EPDM

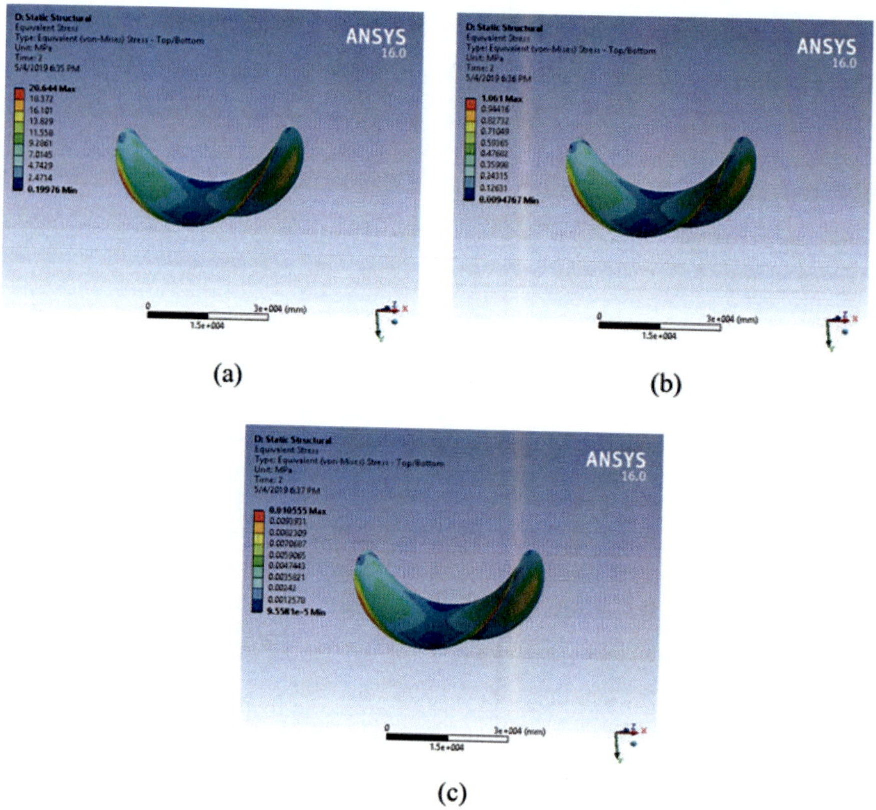

Fig. 6 Stress diagram for saddle roof using the material **a** PVC, **b** PTFE, **c** EPDM

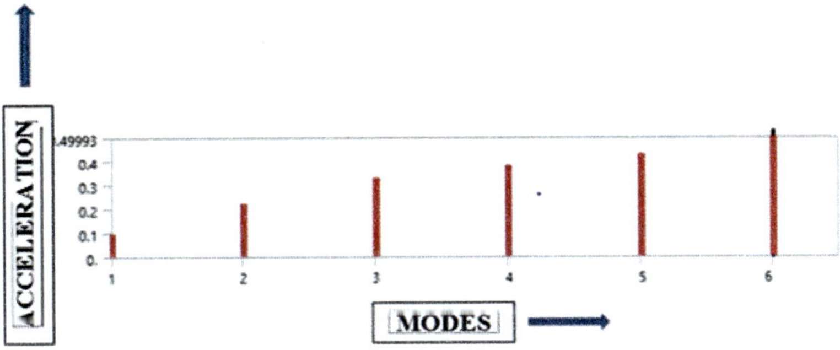

(a)

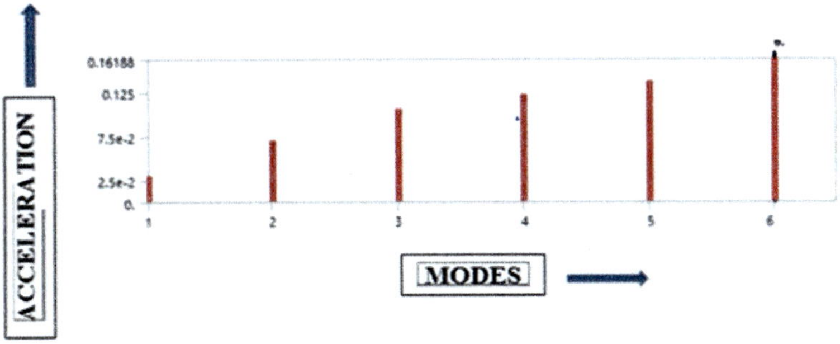

(b)

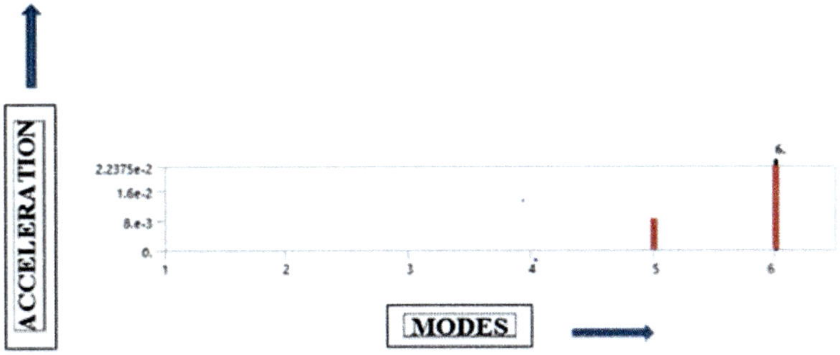

(c)

Fig. 7 Acceleration versus mode graph for conoid roof using the material **a** PVC, **b** PTFE, **c** EPDM

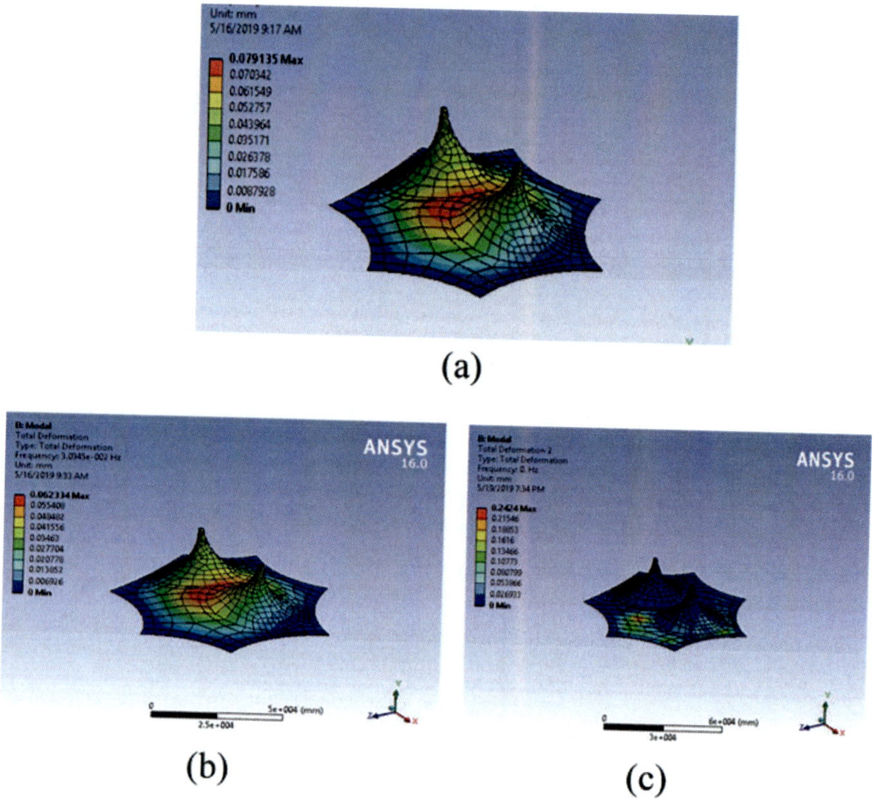

Fig. 8 Deformation diagram for conoid roof using the material **a** PVC, **b** PTFE, **c** EPDM

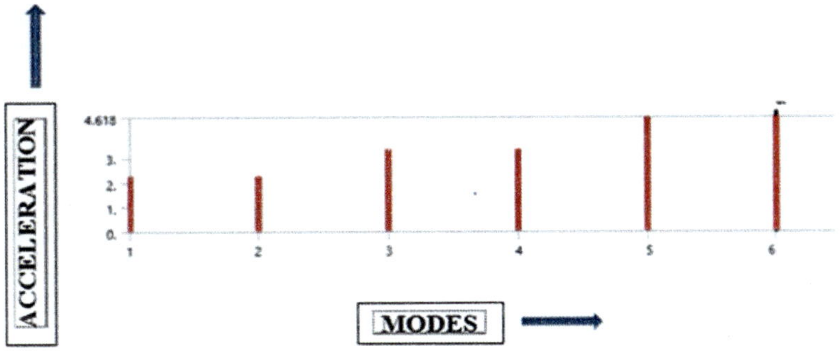

(a)

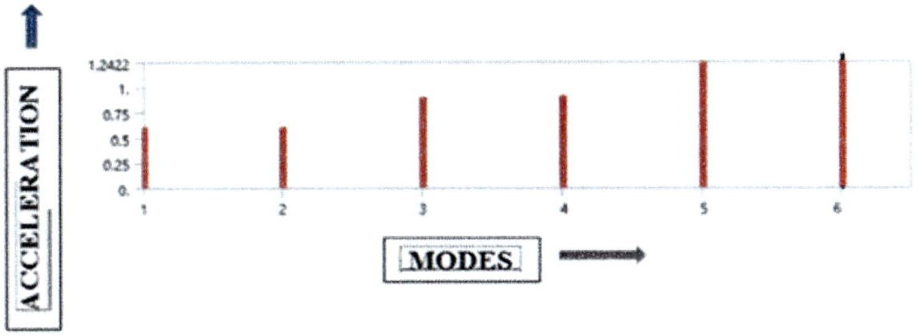

(b)

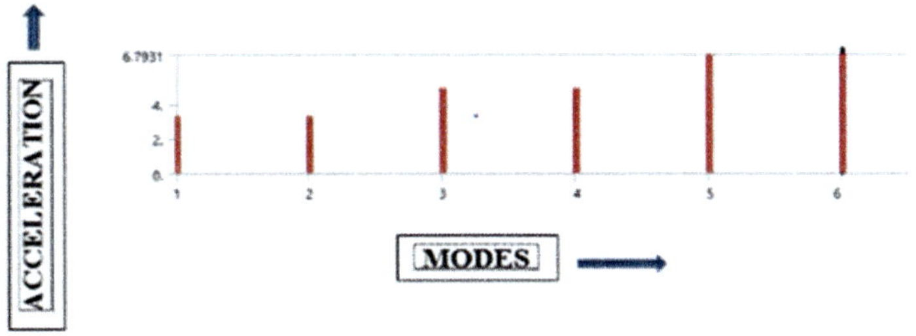

(c)

Fig. 9 Acceleration versus mode graph for semi-elliptical roof using the material **a** PVC, **b** PTFE, **c** EPDM

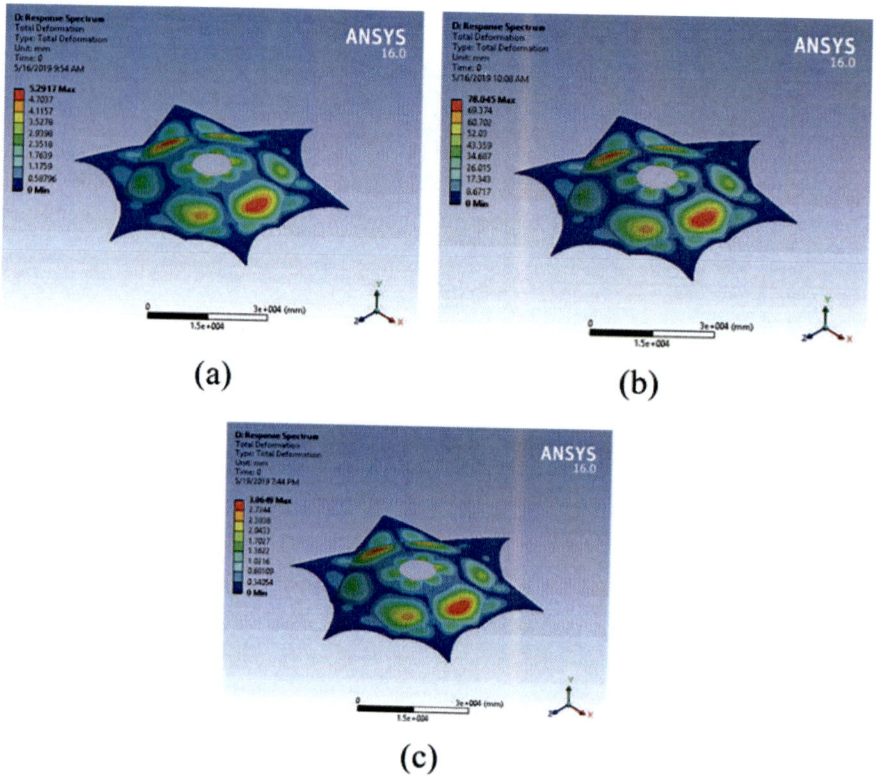

Fig. 10 Deformation diagram for semi-elliptical roof using the material **a** PVC, **b** PTFE, **c** EPDM

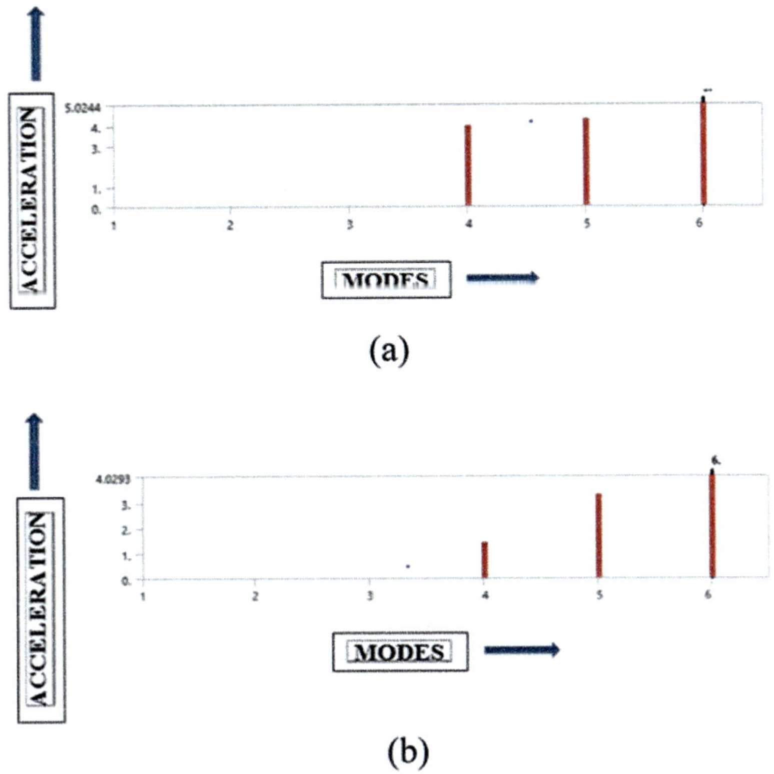

Fig. 11 Acceleration versus mode graph for saddle roof using the material **a** PVC, **b** PTFE

Fig. 12 Deformation diagram for saddle roof using the material **a** PVC, **b** PTFE

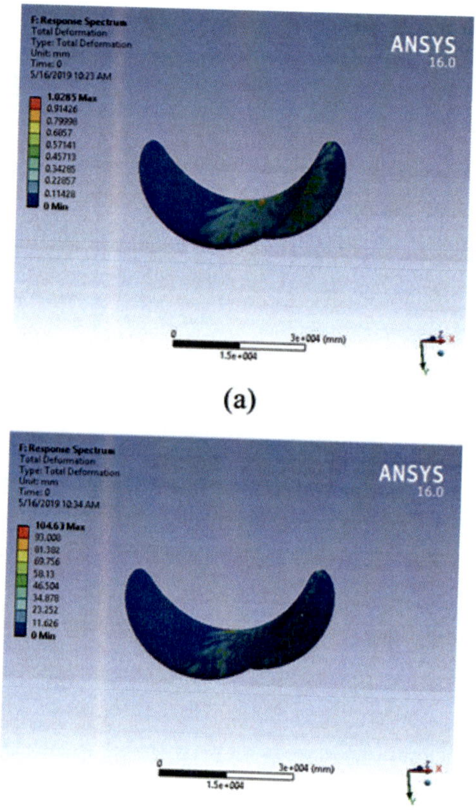

Table 1 Results and discussions of roofs for wind and seismic analysis

S. no.	Shape	Material	Deformation for wind (mm)	Deformation for seismic (mm)
1	Conoid	PVC PTFE EPDM	0.66014 3.8508 0.1201	0.0791 0.06233 0.2424
2	Semi-elliptical	PVC PTFE EPDM	40.765 345.49 99.53	5.2917 78.84 3.6
3	Saddle	PVC PTFE EPDM	99.4 42.43 28.27	1.028 104.6 –

References

1. Rizzo F, Ricciardelli F (2017) Design pressure coefficients for circular and elliptical plan structures with hyperbolic paraboloid roof. Eng Struct 139(2017):153–169
2. Rizzo F, D'Asdia P, Lazzari M (2017) Aerodynamic behavior of hyperbolic paraboloid shaped roofs: wind tunnel tests
3. Holmes JD, Ginger JD (2012) Internal pressures–The dominant windward opening case—a review. J Wind Eng Ind Aerodyn 100(2012):70–76
4. Xu H, Lou W (2016) Wind-Induced internal pressures in building with dominant opening on hemi-ellipsoidal roof. J Struct Eng 142(7):1–10. https://doi.org/10.1061/(ASCE)EM.1943-7889.0001468
5. Guhal TK, Sharma RN, Richards PJ (2014) Dynamic wind load on an internal partition wall inside a compartmentalized building with an external dominant opening. J Struct Eng https://doi.org/10.1061/(ASCE)AE.1943-5568.0000113
6. Tecle AS, Bitsuamlak GT, Chowdhury AG (2010) Opening and compartmentalization effects of internal pressure in low-rise buildings with gable and hip roofs. J Struct Eng https://doi.org/10.1061/(ASCE)AE.1943-5568.0000101
7. Wang YJ, Li QS (2015) Wind pressure characteristics of a low-rise building with various openings on a roof corner. J Wind Struct 21(1):1–23. https://doi.org/10.12989/was.2015.21.1.001
8. Guha TK, Sharma RN, Richards PJ (2015) Internal pressure in a building with a single dominant opening: an experimental and numerical case study. Wind Eng 156. https://doi.org/10.3850/978-981-07-8012-8
9. Karava P, Stathopoulos T (2009) Wind-induced internal pressures in buildings with large facade openings. In: Proceedings of the 11th americas conference on wind engineering, American association for wind engineering, Fort Collins, CO
10. Prevatt DO, Cui B, (2010) Wind tunnel studies on sawtooth and monosloped roofs. J Struct Eng 136(9):1161–1171. https://doi.org/10.1061/(ASCES)T.1943-541X.0000200
11. Pan F, Cai CS, Zhang W (2013) Wind-Induced internal pressures of buildings with multiple openings. J Eng Mech 139(3):376–385. https://doi.org/10.1061/(ASCE)EM.1943-7889.0000464
12. Chowdhury AG, Canino I, Mirmiran A, Suksawang N, Baheru T (2013) Wind-Loading effects on roof-to-wall connections of timber residential buildings. J Eng Mech 139(3):386–395. https://doi.org/10.1061/(ASCE)EM.1943-7889.0000512
13. Wang J, Cao S, Pang W, Cao J (2018) Experimental study on tornado-induced wind pressures on a cubic building with openings. J Struct Eng 144(2):04017206. https://doi.org/10.1061/(ASCE)ST.1943-541X.0001952
14. Ginger JD, Holmes JD, Kim PY (2010) Variation of internal pressure with varying sizes of dominant openings and volumes. J Struct Eng 136(10):1319–1326. https://doi.org/10.1061/ASCEST.1943-541X.0000225
15. Amin JA, Ahuja AK (2013) Effects of side ratio on wind-induced pressure distribution on rectangular buildings. J Struct 2013:12, Article ID 176739. http://doi.org/10.1155/2013/176739

Evaluation of Response Reduction Factor for RCC Moment Resisting Frame with Ductile Shear Wall

Akash Soni, Manohari P. Kulkarni, and Shardul G. Joshi

Abstract Most of the buildings are irregular and their behavior during the event of earthquake is significantly different than that of regular buildings. The actual base shear which act on the building is much higher than the designed base shear. That is why it is uneconomical to design the building for the actual base shear. The building is expected to undergo the inelastic deformations without excess damage to it. Response reduction factor (*R*) is an indication of extent to which the energy is dissipated through inelastic deformation during an event of seismic excitation. Many of the design codes recommends the different values of '*R*' accounting for their ductility, over strength and Redundancy of the structure. IS 1893 (part 1): 2016 has recommended the values of *R* based on the structural framing system. The main objective of the present work is to verify the value recommended for the dual system using Non-linear Static Pushover Analysis and compare the obtained *R* value with code specified value.

Keywords Ductility · Response reduction factor · Ductility reduction factor · Over strength · Pushover analysis

1 Introduction

Response reduction factor is mathematically expressed as the ratio of elastic base shear force to the design base shear force. Generally, *R* is expressed as the product of the factors which account for reserve strength, ductility, redundancy and added viscosity. The factor *R* reduces the elastic base shear to the design base shear and

A. Soni · M. P. Kulkarni · S. G. Joshi (✉)
Department of Civil Engineering, Vishwakarma Institute of Information Technology, Pune, India
e-mail: shardul.joshi@viit.ac.in

A. Soni
e-mail: akash.218m0094@viit.ac.in

M. P. Kulkarni
e-mail: kul.manohari@gmail.com

L. M. Gupta et al. (eds.), *Advances in Civil Engineering and Infrastructural Development*, Lecture Notes in Civil Engineering 87,
https://doi.org/10.1007/978-981-15-6463-5_14

makes the structure to behave inelastically during the seismic event and allows dissipation of energy through the inelastic deformations. Some of the building codes considers R in terms of its components. The main factors which constitutes R are Over strength factor, Ductility reduction factor and Redundancy factor. The response reduction factor can be expressed as:

$$R = R_s * R_\mu * R_r$$

where, R_s is the over strength factor, R_μ is the ductility factor and R_r is the redundancy factor.

1.1 Components of Response Reduction Factor

1.1.1 Over Strength Factor R_S

This is the ratio of the maximum base shear at yielding level to the design base shear. The first significant yield is observed beyond the design base shear in r.c. buildings. Overstrength is the difference between ultimate strength and design strength. The factors responsible for overstrength are:

(1) Code defined load factors and safety factors
(2) Contribution from non-structural components
(3) Ductility ensured through codal provisions
(4) Conservations in the design procedures

Koppas et al. [4] have estimated overstrength factors between 1.5 and 2.7. Lee et al. [6] estimated the overstrength factor between 2.3 and 2.8.

1.1.2 Ductility Reduction Factor $R\mu$

It is mathematically expressed as the ratio of elastic base shear to the base shear at yield level. It accounts for the energy dissipating capacity of properly designed and well detailed r.c. structures. Highly ductile structures perform much better during seismic events. When a structure is subjected to earthquake induced forces, it experiences inelastic deformations which is mainly dependent on displacement ductility ratio "μ". Many researchers have expressed ductility reduction factor for different values of period of vibration. Pauley and Priestley [10] expressed R_μ in terms of μ for the structures having different values of period of vibrations. Miranda and Bertero [8] have correlated to R_μ with different soil conditions.

1.1.3 Redundancy Factor R_r

The redundancy factor R_r is mathematically expressed as the ratio of the base shear at yielding level to the unfactored base shear. Redundancy means excessive i.e. more than required. If the structure is non-redundant, failure of any member is directly the failure of the building itself. In case of redundant structure, failure occur much after than the failure of a single member. It is expected that the building should have high redundancy. For multi storied building, R_r is generally taken as unity.

$$R_r = 1$$

Many researchers have worked on the evaluation of response reduction factor. Some of the research papers which are referred for the study are outlined in the subsequent paragraph.

Riddell et al. [9] studied the variation of response reduction factor with respect to the period of vibration of the buildings. Miranda [3] worked on different components of response reduction factor. Akbari et al. [7] studied the seismic behavior of steel and r.c. buildings. Tamboli et al. [5] evaluated response reduction factor for r.c. building strengthened with X bracings at different locations. Kadid and Yahiaoui [1] studied the behavior of r.c. building strengthened with different types of bracings. Brahmavrathan et al. [2] studied the variation in R factor for different heights irregular r.c. buildings. Venkatraman et al. [11] have employed nonlinear push over analysis to evaluate component wise R factor for different heights of the building.

1.2 Significance of the Study

Most of the buildings are irregular in seismic zone III which covers major portion on the seismic map of India. As mentioned in earlier paragraphs, it is not advisable to design the buildings earthquake proof but can be designed as earthquake resistant. The structure is made to behave inelastically to dissipate the energy so that the structure can be optimally designed. The response reduction factor (R) values for different structural framing systems are provided in IS 1893 (Part 1): 2016. It is believed by many researchers that these values are recommended on the basis of engineering judgement but has very little technical base. More over the value is provided for R directly and not for its components. The present work is an attempt to verify the value of R for the dual system comprising of special moment resisting frames and shear walls.

The present study calculates response reduction factor for an irregular reinforced concrete multistoried building using non linear push over analysis. Push over analysis is a nonlinear procedure which estimates capacity of the structure beyond the yielding up to the failure. The analysis predicts weak areas as it monitors progressive collapse of the building. Push over analysis is a second stage analysis as the seismic analysis using equivalent static or response spectrum procedure determines the reinforcement required. The analysis results into a push over curve which is the plot of base shear versus roof displacement. The spectral acceleration versus spectral displacement curve is obtained thereafter. This curve is known as Capacity Curve. The response spectrum plot is modified using the correlation between period of vibration and spectral displacement. This curve is known as Demand Curve. The superimposition of these two curves is known as Performance Curve.

2 Model for the Study

The model of reinforced concrete building is prepared in etabs 9.7.4 for this study. A G + 15 storied Reinforce concrete moment resisting frame with a ductile shear wall at center portion (Considered a lift pit which is surrounded by shear wall) is considered. The structure is having 15 m × 12 m plan area which is same through the elevation is considered. Bay spacing between columns along longer side is taken as 5 m c/c with 3 number of bays while along shorter direction taken as 3 m c/c. with 4 bays. The floor to floor height of building is considered as 3 m. The size of the beam is considered as 230 mm × 450 mm and that for the column is 280 mm × 280 mm which are reinforced with 0.8% steel. Figure 1 shows the details of the building considered for the study.

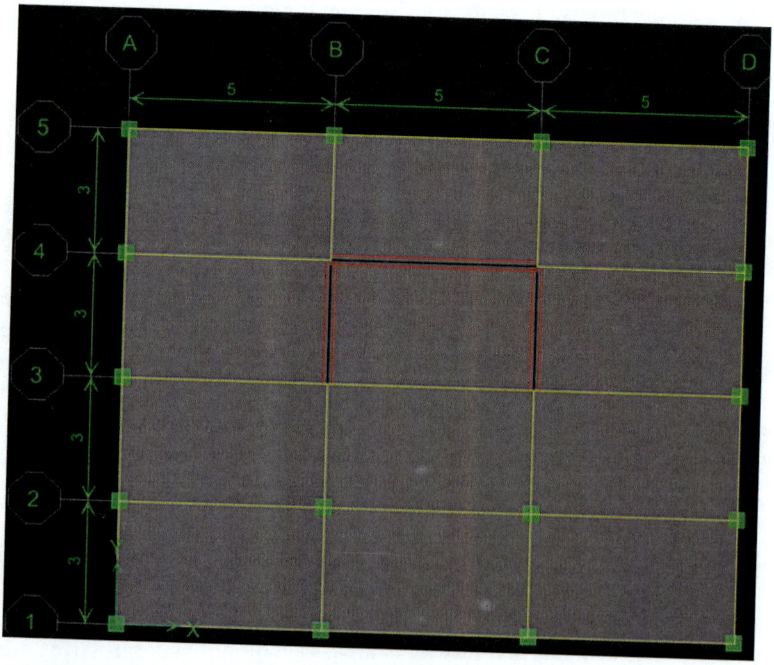

Fig. 1 Typical floor plan of the building

The seismic parameters are referred from IS1893-2016. The building is a residential building situated in zone IV. Live load of 3 KN/m^2 distributed over the slab area with floor finish load of 1 KN/m^2. DL + 0.25LL is considered as lumped on each floor for seismic weight calculations. Building is considered as general importance with $I = 1.0$. The building is special moment resisting frame (SMRF) with Ductile shear wall (dual system). Damping is considered as 5% of critical damping. The soil Type is medium soil (Soil type II). A nonlinear push over analysis is carried out using Etabs software.

3 Results and Discussions

The performance curve obtained through the analysis is shown in Fig. 2. The base shear corresponding to the performance point is obtained as 2594.55 KN and that using equivalent static method is obtained as 534.05 KN. The location of hinges developed are shown in Fig. 3.

The Response reduction factor for particular model of dual system (SMRF with Shear wall) is obtained by taking the ratio of Performance point base shear to the design base shear.

$$R = 2594.55\ \text{KN}/534.05\ \text{KN}$$

$$R = 4.86$$

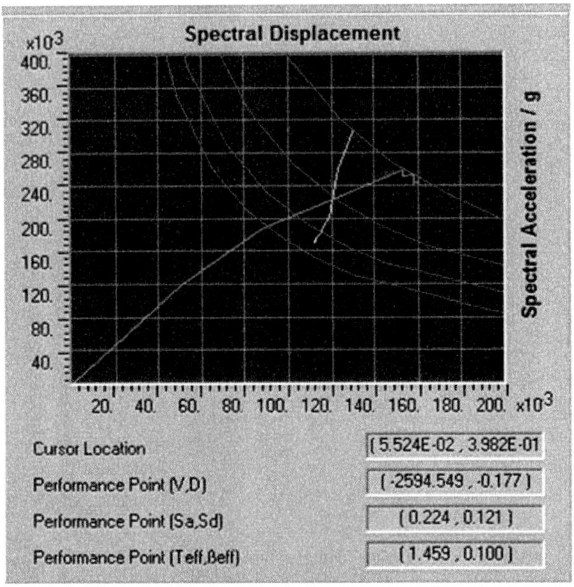

Fig. 2 Performance curve of the building

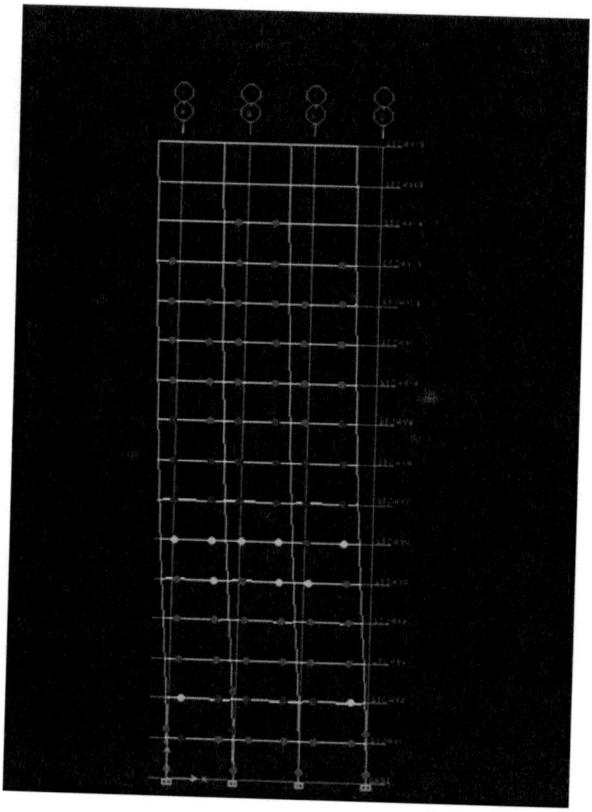

Fig. 3 Location of plastic hinges

R value is obtained by experimental analysis is 4.86 which fairly matches with the one as specified in IS code.

Value for SMRF with Ductile shear wall recommended by IS 1893 (part 1): 2016 is 5.

4 Conclusions

The performance of RCC frame with Ductile shear wall by pushover analysis is investigated by using ETABS 9 software and compare with the design values of base shear obtained by static equivalent method. Following conclusions are drawn:

(1) The performance of well detailed reinforced concrete frame building is adequate as indicated by the intersection of the demand and capacity curves and the distribution of hinges in the beams and the columns. Most of the hinges developed in the beams and few in the columns but with limited damage.

(2) The value of R obtained through the analysis fairly matches with the value recommended by Indian standard code.
(3) The actual value of R in real designs may be considered as even less than computed here, because of various reasons, such as, irregularity in dimensions leading to minor to moderate torsional effects, Lack of quality control and poor workmanship during the construction, not following the ductile detailing requirements exactly as per the guidelines, etc.

References

1. Kadid A, Yahiaoui D (2011) Seismic assessment of braced RC frames. In: The proceedings of the twelfth East Asia Pacific conference on structural engineering and construction—EASEC12, Procedia Eng 14, 2899–2905
2. Brahmavrathan D, Arunkumar C (2016) Evaluation of response reduction factor of irregular reinforced concrete framed structures. Indian J Sci Technol 9(23):1–8
3. Miranda E (1997) Strength reduction factors in performance-based design. In: EERC-CURE symposium in Honour of Vitelmo V. Bertero, Berkeley, California
4. Koppas AJ (1999) Evaluation of behavior factors on the basis of ductility and overstrength studies. Eng Struct 21:823–835
5. Tamboli K, Amin JA (2015) Evaluation of response reduction factor and ductility factor for RC braced frame. J Mater Eng Struct 2(3):120–129
6. Lee DG, Cho SH, Ko H (2015) Response modification factors for seismic design of building structures in low seismicity region. Korea Earthquake Research Centre
7. Maheri MR, Akbari R (2003) Seismic behavior factor 'R' for steel X-braced and knee-braced RC buildings. Eng Struct 25:1505–1513
8. Miranda E, Bertero VV (1994) Evaluation of strength reduction factors for earthquake-resistant design. Earthquake Spectra 10(2):357–379
9. Riddell R, Hidalgo P, Cruz E (1989) Response modification factors for earthquake resistant design of short period buildings. Earthquake Spectra 5(3):571–590
10. Pauley T, Priestley MJN (1992) Seismic design of reinforced concrete and masonry buildings. A Wiley Inter Science Publications
11. Venkat Raman RL, Ingle RK (2016) Evaluation of response reduction factor of RC framed buildings by push over analysis. Int J Adv Mech Civ Eng 3(4):100–105

Study on Properties of Lightweight Concrete Using Expanded Clay Aggregate and Its Value Engineering

Shaikh Mohd. Tazir Asif and A. S. Wayal

Abstract Comparison was made between normal concrete (NC) and lightweight concrete (LWC) in terms of fresh, hardened and durability properties of concrete like fresh and hardened density, workability, compressive and flexural strength, modulus of elasticity, rapid chloride ion penetration and water absorption test. NC of M30 grade with normal aggregate and LWC having densities 1400, 1600 and 1800 kg/m^3 from expanded clay aggregate (ECA) with complete replacement were prepared. The specimens prepared were tested on 7, 28 and 56 days. Key observations were that the density of the mixes remained constant throughout 56 days for all the mixes. Strengths of the LWC decrease with an increase in the percentage of ECA in the mix. However, LWC showed better results in durability properties than NC. Value engineering between NC and LWC showing best result among lightweight concrete was considered. Complete replacement of structural members was made by LWC and compared to get additional floors with the same load of NC for G + 5, G + 10 and G + 15 structures.

Keywords Lightweight concrete · Expanded clay aggregate · Water absorption · Value engineering

1 Introduction

A considerable amount of load coming onto a structure is from the dead load of the structure. An increase in the consumption of lightweight concrete for the past few decades due to the load reduction benefits resulting in ease of transport and placing of materials was observed. Rajprakash et al. (2017) found out the dead

S. Mohd. T. Asif
Construction Management, Civil and Environmental Engineering, Department, Veermata Jijabai Technological Institute, Matunga, Mumbai 400019, India

A. S. Wayal (✉)
Civil and Environmental Engineering, Department, Veermata Jijabai Technological Institute, Matunga, Mumbai 400019, India
e-mail: aswayal@ci.vjti.ac.in

L. M. Gupta et al. (eds.), *Advances in Civil Engineering and Infrastructural Development*, Lecture Notes in Civil Engineering 87,
https://doi.org/10.1007/978-981-15-6463-5_15

weight of the concrete considerably reduced when 20 and 40% of ECA are replaced by coarse aggregate [1]. Hanamanth et al. (2015) made an attempt to replace 100% ECA by coarse aggregate and achieved 35% of reduction in overall density [2]. Aggregate having air-dried unit weight at 28 days in the range 1440–1850 kg/m3 and compressive strength not less than 17.2 MPa are called lightweight aggregate [3, 4].

Other lightweight concrete properties like workability, less dead load, resistance to freezing and thawing, etc., are proven to be better than normal concrete. However, lightweight concrete offers less strength comparable to normal weight concrete. Moravia et al. (2010) found out that though the concrete made by ECA has a less compressive strength than the normal concrete for all the mixes [5]. The strength of lightweight can further be increased by adding mineral additives. Subasi (2009) studied the effects of fly ash usage in lightweight concrete from ECA and found out that strength of lightweight concrete can be enhanced by the addition of fly ash instead of cement [6]. Murat et al. (2015), Properties like compressive strength, Poisson's ratio, modulus of elasticity of lightweight aggregate concrete (LWAC) mixtures were conducted and found out that within the same range of compressive strength, LWAC mixtures show reduction in MOE and Poisson's ratios of both LWAC and CC mixtures [7].

In this study, an attempt is made by using expanded clay aggregate (ECA) as coarse aggregate in the concrete, where different densities (1400, 1600 and 1800 kg/m^3) of lightweight concrete mix were prepared. Fresh, hardened and durability properties of the mix like density, workability, compressive strength, chloride ion penetration, flexural strength, capillary water absorption, water absorption under submersion and low pressure, etc., were evaluated and compared against normal concrete. Also structure of G + 5, G + 10 and G + 15 will be evaluated with NC and LWC to get the additional floors for the lightweight concrete structure.

2 Experimental Data

2.1 Raw Materials

Binders used for the study purpose were Ordinary Portland Cement (OPC) (Grade 53), fly ash (FA) and micro-silica (MS). Cement used was confirming to IS: 12269 [8] having specific gravity 3.15 gm/cm^3 and Initial and Final setting time as 203 min and 248 min, respectively. The standard consistency of cement was 29%. Chemical composition of OPC is given in Table 1. Specific gravity of fly ash according to IS: 3812 [9] and micro-silica confirming to IS: 15388 [10] were 2.2 gm/cm^3 and 2.26 gm/cm^3 respectively. Crushed angular aggregates of 10 and 20 mm sizes confirming to IS: 383-1970 with specific gravity of 2.8 and 2.82 gm/cm^3 having water absorption of 1.89 and 1.87%, respectively, were utilized as coarse aggregate for normal grade concrete. Crushed sand confirming to IS 383 limits with specific gravity of

Table 1 Chemical composition of ordinary Portland cement

Properties	Values
Lime saturation factor	0.90
Al_2O_3 to Fe_2O_3 ratio	0.84
Insoluble residue, % by mass	0.51
MgO, % by mass	2.18
SO_3, % by mass	2.04
Loss of ignition, % by mass	2.52
Cl, % by mass	0.017
Na_2O, % by mass	0.045
C_3A, % by mass	3.11

2.7 gm/cm^3 and having water absorption of 3.9% was used as fine aggregate for both normal and light weight concrete. The particle distribution for both types of aggregate was in the range given as per IS 383 [11].

Superplasticizers according to IS: 9103 [12] used in the mix are sulfonated naphthalene formaldehyde- or polycarboxylate ether (PCE)-based water-reducing superplasticizing admixture, set retarding admixture and air entraining admixture were used in the study. All admixture used was chloride free and complied to IS 9103.

2.2 *Expanded Clay Aggregate (ECA)*

It is produced from sedimentary sometimes metamorphic clay having zero or very less amount of lime. The clay is initially dried, heated and then burned in a rotary kiln at a temperature of 1100–1300 °C. They are usually round or in a shape of a potato due to their circular movement inside the rotary kiln. ECA can be brown, reddish, brown–red, yellow which may be due to the difference in its chemical compositions and also in the manufacturing. Inside the ECA, various tiny holes are interconnected to each other and are formed when gases are released inside the pellets during heating is trapped inside during the cooling.

It is an inert light material and does not possess any harmful materials with pH value of nearly 7. Also, it does not get damaged in water. It is also non-biodegradable, non-combustible against any adverse conditions. It also has excellent properties like good fire resistant, excellent thermal insulation and good sound insulation. ECA has loose bulk densities from 250 to 710 kg/m^3. ECA is available in wide range of size (from 0.1 to 25 mm) and can be suitable for coarse aggregate, fine aggregate and as well as in combination of both. In this study ECA is used in place of conventional coarse aggregate for lightweight concrete. The properties and particle size distribution are shown in Table 2.

Table 2 Properties of ECA

Properties	Values
Particle size	2–8 mm
Specific gravity	0.80 gm/cm^3
Water absorption	20%
Loose bulk density	530 kg/m^3
Ph. value	8.05
Crushing strength	2.26 N/mm^2
Thermal conductivity	0.10 W/mk
Particle size distribution	
IS sieve size (mm)	Percentage passing
12.5	100
9.5	100
4.75	20.97
2.36	1.12

2.3 *Concrete Mixes*

The concrete mix for normal concrete and lightweight concrete was designed as per IS: 10262 and ACI 213 [13] guidelines, respectively. The design mix for normal and lightweight concrete is shown in Table 3.

2.4 *Preparation, Casting and Curing*

Concrete was prepared in central shaft-type laboratory mixer having a capacity of 0.50 m^3. Moisture correction for concrete was done, and additional water was added to maintain the design mix. The concrete were casted in the molds and compacted by hand compaction as per recommendations of test-specific code. The specimens after 24 h casting were demolded then cured with complete immersion in water and removed directly on the day of testing.

3 Result and Discussion

Fresh, hardened and durability properties of normal and lightweight were studied, and the following tests were conducted.

Table 3 Design mix of concrete

S. No.	Grade of concrete	Binder (Kg/m^3)			Aggregate proportions (%)				Water (Kg)	Admixture (%)
		Cement	Fly ash	Silica fume	10 mm	20 mm	Sand	ECA		
1	M30 NC	340	120	0	25	28	47	0	160	1.2
2	D1400 LWC	450	200	50	0	0	36	64	180	1.25
3	D1600 LWC	450	200	50	0	0	65	35	180	1.45
4	D1800 LWC	450	200	50	0	0	81	19	180	0.95

Table 4 Fresh density of the mix

Mix description	M30 NC	D1400 LWC	D1600 LWC	D1800 LWC
Density (Kg/m^3)	2430	1408	1558	1747

Table 5 Slump cone/flow test result

Mix description	M30 NC	D1400 LWC	D1600 LWC	D1800 LWC
Slump cone/flow value (mm)	200	500	480	440

3.1 *Fresh Properties*

Density

The fresh density of the mixes was measured as per IS 1199 and ACI 213R-87 [13] for normal and lightweight concrete, respectively, and shown in Table 4. The target density for the lightweight concrete was achieved. Also no bleeding or segregation was observed resulting in concrete easily pumpable to high floors.

Workability

For workability, the slump cone and slump flow test was done as per IS 1199 [14] and BS EN 12350–9 for normal and lightweight concrete, respectively. The result is illustrated in Table 5, and it was observed that the workability of lightweight concrete increases with an increase in proportion of ECA percentage in the mix.

3.2 *Hardened Properties*

Hardened density

The densities were measured on 7, 28 and 56 days for both normal and lightweight concrete. It was observed that the densities for all the mixes remained in the acceptable variations to the target density and provided in Table 6.

Table 6 Density of concrete at hardened state

	Densities of concrete at different days (Kg/m^3)		
Mix description	7 Days	28 Days	56 Days
M30 NC.	2410	2445	2423
D1400 LWC	1410	1427	1445
D1600 LWC	1559	1570	1581
D1800 LWC	1741	1773	1723

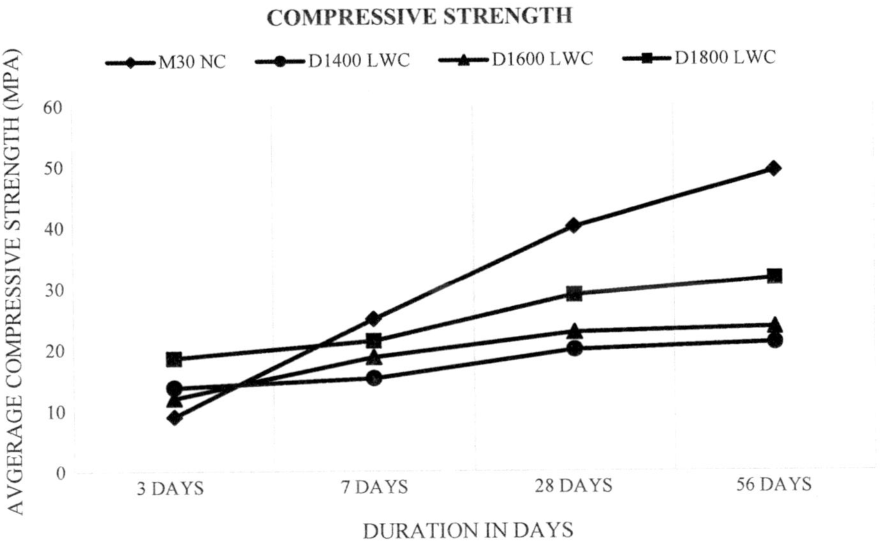

Fig. 1 Compressive strength of mixes at different age

Compressive strength

Compressive test was conducted as per IS 516 on 3, 7, 28 and 56 days. The specimen were tested with compression testing machine (CTM) having capacity of 4000 KN with an application of 0.233 Mpa/sec uniform loading until specimen failure. The compressive strength result is shown in Fig. 1. It can be seen that percentage in the reduction was 50%, 43% and 27% for D1400, D1600 and D1800, respectively, with M30 NC at 28 days. The compressive strength for lightweight concrete was less comparing it with normal concrete and was decreasing with percentage increase of ECA in the mix.

Flexural strength

Flexural test was carried as per IS 516. The test was done according to 4-point load system. The roller on the top was placed 100 mm from the center of the beam, and there was a distance of 200 mm between the rollers. The rate of loading applied was 400 kg/min until the failure of the beam, and maximum flexural strength was recorded. The result is shown in Fig. 2. The flexural strength of D1800 LWC was the highest among the lightweight concrete and also in range with strength of the mix M30 NC.

Modulus of elasticity

Modulus of elasticity (MOE) was carried as per ASTM C469. The MOE was tested on 28 days for all the mixes, and the results are provided in Fig. 3. MOE was also evaluated by the empirical formulae given below.

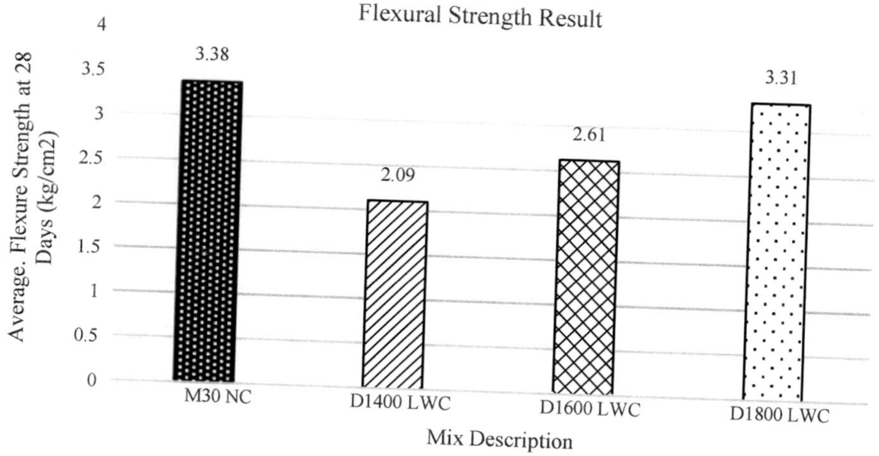

Fig. 2 Flexural strength of mixes at 28 days

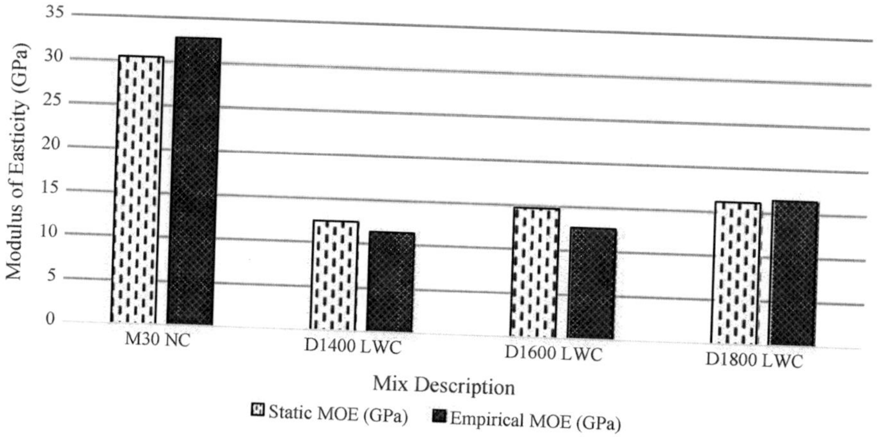

Fig. 3 Comparison of static and empirical MOE

$$E_c = 43 \times 10^{-3} \times d^{1.5} \times f_c^{0.5}$$

where E_c = modulus of elasticity, d = density of the mix in Kg/m^3 and f_c = strength in MPa at 28 days.

It was observed that the MOE values from the experiment and that obtained from empirical formulae have a slight variation. Also the MOE was directly proportional to the strength of the mixes.

3.3 Durability Properties

Rapid chloride ion penetration test

This test was performed as per ASTM C 1202 [15] in which the samples were monitored by observing the total electric charge penetrating through them. The samples were kept in test device. One test cell was filled with 3% NaCl solution, while the other test cell with 0.3 N NaOH solution. The whole assembly is then connected to a constant 60 V DC supply. The amount of charge passing through samples was noted after every 30 min for a total of 6-h duration. Then, the total charge passed in coulombs is calculated by the formula given below.

$$Q = 900\,(I_0 + 2I_{30} + 2I_{60} + \cdots + 2I_{300} + 2I_{330} + I_{360})$$

where Q = charge passed in coulomb, I_0 = current in amperes at 0 min after application of voltage, I_t = current in amperes at t mins after the application of voltage.

The following were the results obtained for the different mixes, and it is shown in the Fig. 4 given.

It was observed that total amount of electric current passed through D1800 mix was the least in all the four mixes, showing its tendency to perform better in those environments.

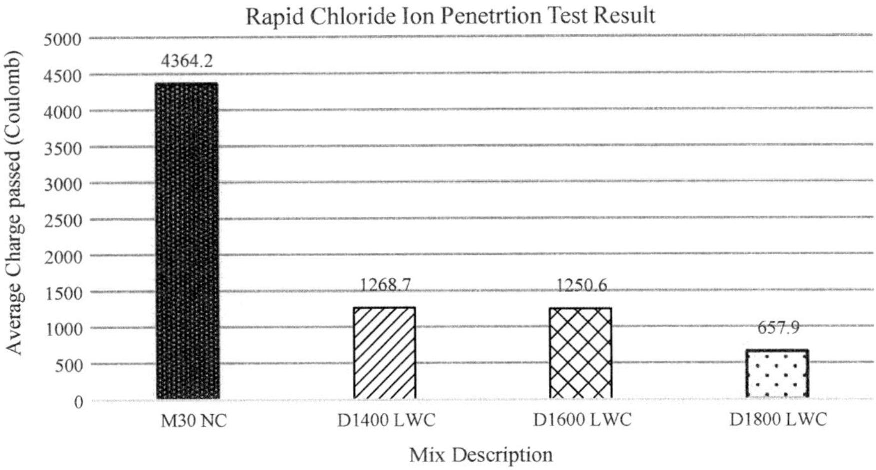

Fig. 4 Rapid chloride ion penetration test

Water absorption under submersion

Water absorption under submersion test was performed as per BS 1881: Part 122. The test was carried at 28 days after casting of the mixes. After 28 days curing, the specimen was first oven-dried to remove all the moisture and to achieve a constant weight and recorded. The sample was cooled down to bring it to the room temperature after which submerged in water for 30 min. After 30 min, the sample was again weighed to determine the water absorption of the mixture. The result is illustrated in Fig. 5.

Water absorption under low pressure

This test is used in the evaluation of concrete structures exposed to water under low pressure. This test was performed as per Rilem 11.4 test method. Rilem tube filled with water having the capacity of 5 ml is attached to a face of cube as shown in Fig. 6, and the decrease in the water level is recorded after specific time interval. Figure 7 shows the test result of percentage of water absorption under low pressure for different mixes.

Capillary water absorption

Capillary water absorption was performed as per BS-EN-480-5. The test was conducted on 28 days. The samples were submerged in water up to 1.5 cm depth, and the capillary action of water was observed on 3 h, 24 h, 48 h and 72 h by weighing in these time period. The percentage of water absorption was observed by calculating the weight difference of the samples. The result obtained is shown in Fig. 8.

The result shows that the capillary action in lightweight concrete is less in comparison to the normal concrete, and hence, the effect of water due to capillary in the lightweight concrete will be less in comparison to normal concrete.

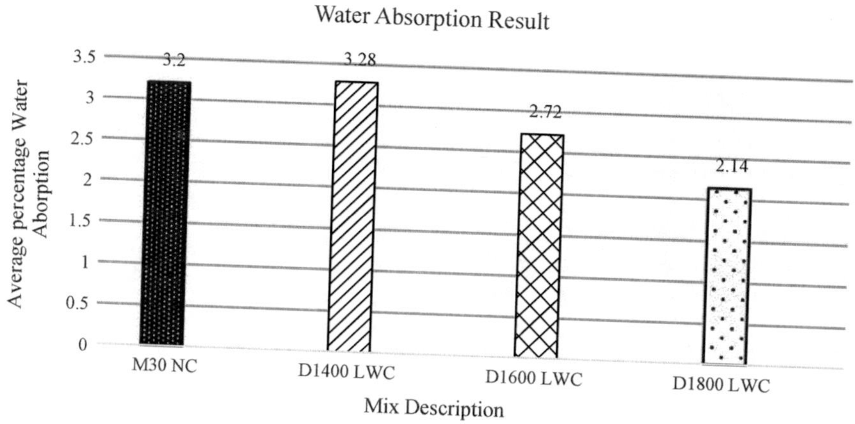

Fig. 5 Water absorption test result

Fig. 6 Experimental setup of water absorption under low pressure

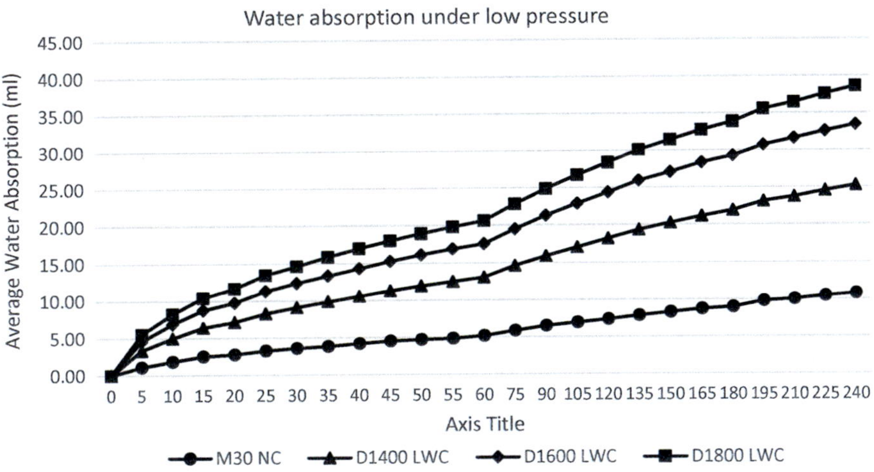

Fig. 7 Rilem water absorption test result

4 Value Engineering

Since the mix D1800 LWC performed better for all the experiments in lightweight concrete mixes, it was considered for structural design and value engineering. Structural design analysis is carried on reinforced cement concrete (RCC) residential towers of G + 5, G + 10 and G + 15 by completely replacing the members with D1800 LWC mix. Figure 9 shows the floor plan used for structural design and analysis.

Structural design and analysis were carried out in structural analysis and building design software ETABS. Only limit-state conditions of live and dead load were

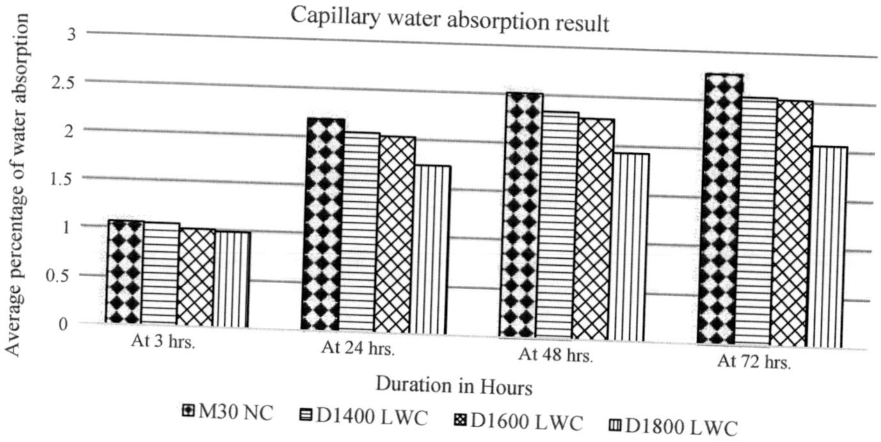

Fig. 8 Capillary water absorption test result

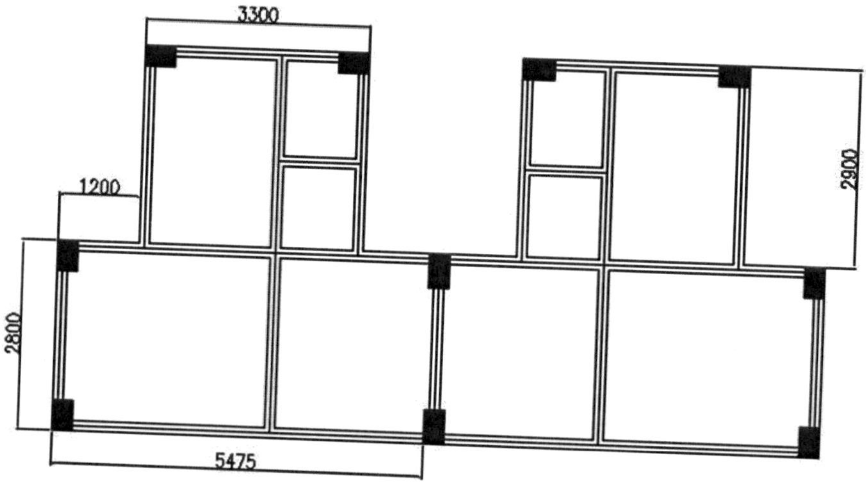

Fig. 9 Floor plan used for value engineering, all dimensions in mm

considered for the design of structures. The structures mentioned above were drafted and analyzed. The total load, total cost of steel and concrete were calculated for G + 5, G + 10 and G + 15 for both M25 NC and D1800 LWC; the complete replacement of the concrete with lightweight concrete was done. A study was done to make optimal utilization of total load of the structure of normal grade by complete of lightweight concrete for structural members. Analyses were done in order to find additional floors that can be added so as to utilize the additional load of normal concrete structure. Hence, G + 6, G + 7, G + 11, G + 12, G + 16 and G + 17 were drafted and analyzed. The results are shown (Table 7).

Table 7 Total load and total material comparison for complete structural replacement

Floors NC	Total load (KN)	Total reinforcement quantity (Kg)	Total concrete quantity (cu. m)	Floors LWC	Total load (KN)	Total reinforcement quantity (Kg)	Total concrete quantity (cu. m)
G + 5	10,445.3	11,869.32	118.62	G + 5	9355.6	11,769.01	120.79
				G + 6	10,722.5	11,927.38	142.56
				G + 7	12,099.8	12,016.64	156.73
G + 10	17,070.9	20,959.45	205.18	G + 10	15,749.8	20,707.76	208.93
				G + 11	17,095.2	21,137.6	225.21
				G + 12	18,440.7	21,670.83	246.5
G + 15	24,946.7	31,279.91	269.8	G + 15	22,422	30,321.15	273.83
				G + 16	23,767.4	30,772.42	288.9
				G + 17	24,956.2	31,357.9	303.97

Table 8 Unit cost comparison of the mixes

Mix description	M30 NC	D1400 LWC	D1600 LWC	D1800 LWC
Cost of 1 m^3 of concrete (Rs.)	6250	11,485	9330	6950
% increase in unit price from standard mix	0	83.76%	49.28%	11.2%

From Table 22 above, it can be seen that an additional floor for G + 10 and above and two additional floors for G + 15, and above structures can be implemented by completely replacing D1800 LWC by M30 NC for beams, columns and slabs. This will also result in saving the cost of reinforcement for same number of floors by lightweight concrete (Table 8).

5 Conclusion

Workability of lightweight concrete was more compared to normal concrete and directly proportional to the percentage of ECA in the mix. The compressive and flexural D1800 LWC was highest among all the lightweight concrete mixes; however, it was 27 and 2% less as compared to M30 NC mix at 28 days. The hardened densities for all the mixes remained in an acceptable range throughout the experimental work. Since modulus of elasticity is dependent on stiffness of the materials and hence it had high values for normal concrete than lightweight concrete also MOE decreases with the addition of ECA percentage in the mix.

Rapid chloride ion penetration test showed that lightweight concrete has higher tendency of resisting the chloride ion penetration than normal concrete with mix D1800 LWC having the highest tendency. In the water absorption test, i.e., under submersion and low pressure, the rate of water absorption was the least for the mix D1800 LWC, showing the property of resisting water penetration inside the concrete. Also capillary water absorption for all the lightweight concrete mixes was less in comparison to the normal concrete mixes. Mix D1800 LWC has better performance in fresh, hardened and durability state; it was chosen for comparison with mix N30 NC for value engineering. An additional floor for G + 10 and above and two additional floors for G + 15 and above structures can be implemented by completely replacing D1800 LWC by M30 NC for beams, columns and slabs.

Acknowledgements The authors sincerely acknowledge the support provided by Mr. Amith Kalathingal, Manager of Research and Development at Nuvoco Vistas Corp Ltd, and Mr. Fakhruddin Vaglawala, Sr Manager Structural Systems at Nuvoco Vistas Corp Ltd, for providing testing facility at Nuvoco Lab.

References

1. Rajprakash RN, Krishnamoorthi A (2017) Experimental study on light weight concrete using LECA. Int Chemtec Res 10(8):98–109
2. Shebannavar H, Maneeth PD, Brijbhushan S (2015) Comparative study of LECA as a complete replacement of coarse aggregate by ACI method with equivalent likeness of strength of is method. Int Res J Eng Technol (IRJET) 2(8):589–594
3. Standard specification for lightweight aggregates for structural concrete, ASTM C330, ASTM International
4. Indian Standard, Method of tests for strength of concrete, IS: 516-1959, Bureau of Indian Standards, New Delhi
5. Moravia WG, Gumieri AG, Vasconcelos WL (2010) Efficiency factor and modulus of elasticity of lightweight concrete with expanded clay aggregate. Ibracon Struct Mater J 3(2):195–204
6. Subasi S (2009) The effects of using fly ash on high strength lightweight concrete produced with expanded clay aggregate. Sci Res Essay 4(4):275–288
7. Dilli ME, Atahan HN, Şengül C (2015) A comparison of strength and elastic properties between conventional and lightweight structural concretes designed with expanded clay aggregates. Constr Build Mater 30(101):260–267
8. Indian Standard, Ordinary portland cement- grade 53- Specification, IS; 12269-1967, Bureau of Indian Standards, New Delhi
9. Indian Standard, Pulverized fuel ash-specification, Part 1 for use as pozzolana in cement, cement mortar and concrete, IS:3812 (Part1): 2003. Bureau of Indian Standards, New Delhi
10. Indian Standard, Silica fume—specification, IS: 15388-2003, Bureau of Indian Standards, New Delhi
11. Indian Standard, Specification of coarse and fine aggregate from natural sources for concrete, IS: 383-1970, Bureau of Indian Standards, New Delhi
12. Indian Standard, Concrete admixture—specification, IS: 9103-1999, Bureau of Indian Standards, New Delhi

13. Guide for structural lightweight -aggregate concrete, ACI 213R-03
14. Indian Standard, Method of sampling and analysis of concrete, IS: 1199-1959, Bureau of Indian Standards, New Delhi
15. Standard Test Method For, Electric indication of concrete's ability to resist chloride ion penetration, ASTM C 1202-05, ASTM International

Effect of Modern Chemical Admixtures on the Performance of Strength of Cement Mortar Cubes

Alima Fernandes and K. G. Guptha

Abstract Admixtures are chemical additives in concrete and play a vital role in modifying the desired properties in fresh concrete or green concrete. The evolution of these admixtures in their performance contributed a lot to the modern concrete over the conventional concrete. In the present experimental study, effect of low, medium and high end admixtures is considered to evaluate their performance on strength with their respective optimum dosages. Cement mortar cubes tested for 3 days, 7 days and 28 days with standard water–cement ratio determined from the standard consistency test of cement paste following IS procedures reveal notable improvements in the strength. This study highlights the minimum water–cement ratio required in each category of the admixture, and its effect on strength is discussed.

Keywords Admixture · Strength · Performance

1 Introduction

Admixtures are the widely used additives in modern concrete to produce high-performance concrete. The addition of these admixtures in the conventional concrete modifies the desired properties in fresh or green stage of concrete. The availability of admixtures can be categorized into many types; however, water reducers, accelerators and retarders have a wide range of applications in the construction field. The combination of more than one admixture in required dosages shall yield a high-performance concrete. Hence, laboratory experiments are carried out before the application of these admixtures in the field to determine the behavior on properties of concrete in green state.

A. Fernandes (✉) · K. G. Guptha
Department of Civil Engineering, Goa College of Engineering, Farmagudi, Goa, India
e-mail: alimaferns2204@gmail.com

K. G. Guptha
e-mail: kgg@gec.ac.in

L. M. Gupta et al. (eds.), *Advances in Civil Engineering and Infrastructural Development*, Lecture Notes in Civil Engineering 87,
https://doi.org/10.1007/978-981-15-6463-5_16

The present study includes experimental tests to observe the behavior of performance of low, medium and high end admixtures on the strength. Conplast SP 430 ES2, Conplast SP 600, Auramix 201, Auramix 300 plus, Auramix 402 and Auramix 450 are six different admixtures having different chemical compositions used in this study. The Auramix series chemical admixtures are second-generation admixtures based on polycarboxylic ethers. However, the conplast series admixtures are first-generation admixtures.

Compressive strength being the most important property of concrete makes it necessary to determine the effect of modern admixtures on strength. Studies have shown that factors affecting strength include quality of materials and the water–cement ratio. Hence, admixtures are added to produce workable mix keeping water–cement ratio as low as possible in order to achieve high-performance concrete.

Generally, test cubes of standard sizes are tested to determine the compressive strength. The present study aims at determining the strength of cement mortar cubes with the addition of optimum dosage of admixtures. The study reveals clear differentiation of strength for 3 days, 7 days and 28 days of cement mortar with the use of different admixtures enlisted above. Studies were carried out under controlled environment as per relevant IS codes.

2 Materials and Methodology

2.1 Materials

Laboratory experiments were carried out to determine the strength of cement mortar cubes having 70.6 mm size using admixtures. The materials included are cement, standard sand, water and admixture. The detailed description of these materials and the apparatus used are listed below:

Cement—Cement is manufactured from calcareous materials and is found to be the most reactive material when it is in contact with water. All the tests were carried out using OPC 43 grade UltraTech cement conforming to IS 8112: 2013 (43 grade) of 38th week. The cement was sieved through 90 μm standard sieve to have a uniform sample.

Standard sand—Standard manufactured sand corresponding to IS 650: 1991 (reaffirmed 2008) was used in all the experiments. The standard manufactured sand consists of three different grades. Equal proportions of these grades, i.e., particles smaller than 2 mm and greater than 1 mm, particles having size smaller than 1 mm and greater than 500 μm and particles having size below 500 μm and greater than 90 μm, were used.

Water—Fresh and clean water corresponding to IS 456: 2000 was used in all the experimental tests. Standard water–cement ratio was found to be 0.4 for OPC 43 grade UltraTech cement as determined from the standard consistency test of cement paste as per IS 4031 (Part 4): 1988.

Table 1 Parameters analyzed

S. No	Admixture	Minimum w/c ratio	Optimum w/c ratio	Optimum admixture dosage (in %)	Optimum flow achieved (s)
1	Conplast SP 430ES2	0.4	0.5	0.9	21.84
2	Conplast SP 600	0.4	0.5	0.6	23.34
3	Auramix 201	0.3	0.42	0.9	19.65
4	Auramix 300 plus	0.4	0.5	0.6	20.41
5	Auramix 402	0.3	0.42	0.6	27.94
6	Auramix 450	0.3	0.42	0.9	19.43

Chemical admixtures—Low, medium and high end admixtures pertaining to IS 9103 were used for the analysis of all the test cubes. The optimum dosage of all the six different chemical admixtures is given in Table 1. The chemical admixtures used in the study were Conplast SP 430 ES2 and Conplast SP 600 identified as low end admixture, Auramix 201 and Auramix 300 Plus identified as medium end admixtures and Auramix 402 and Auramix 450 identified as high end admixtures as per the data provided by Fosroc Constructive Solutions.

2.2 Apparatus

The strength test of all the analysis was carried out using different apparatus. The detailed description of the main apparatus used in the present study is discussed as follows:

Cube molds—Cube molds of size 70.6 mm × 70.6 mm × 70.6 mm of high-level accuracy with ISI certification mark along with a base plate was used for all the experiments. All the molds were applied with mold releasing oil to avoid rusting and to give a smooth and easy removal of the cubes.

Tamping rod and vibration machine—The molds were filled in three different layers. The mix was compacted by temping with a rod for 25 times per layer. The mix was vibrated on a vibration table as per procedure laid down to achieve full compaction.

2.3 Methodology

The methodology adopted for the present study was divided into two parts. In the first part or preliminary test, Marsh cone test was performed for water–cement ratios varying from 0.2 to 0.6 and admixture dosage varying from 0.3 to 1.5% by weight of cement at an interval of 0.3 to determine the minimum water–cement ratio required for the respective admixture to perform. Optimum water–cement ratio and optimum admixture dosage for corresponding flow value were also determined from this test.

In the second part, the strength of cement water cubes was tested with the optimum admixture dosage values determined from the Marsh cone test. This test included a mixture of cement and standard sand with standard water–cement ratio of 0.4 of the cement used corresponding to IS 4031 (Part 6): 1988 and optimum dosage of admixture. Figure 1 shows a picture of cubes casted. Figure 2 shows a detailed process of the present experimental study carried out. The details of the proposed methodology to determine the strength are shown in Fig. 3 for low, medium and high end admixtures.

Typical test procedure to determine the performance of admixture on strength had the following steps:

(a) Dry mix of cement and standard sand in the proportion of 1:3 was prepared.

Fig. 1 Picture of cubes casted

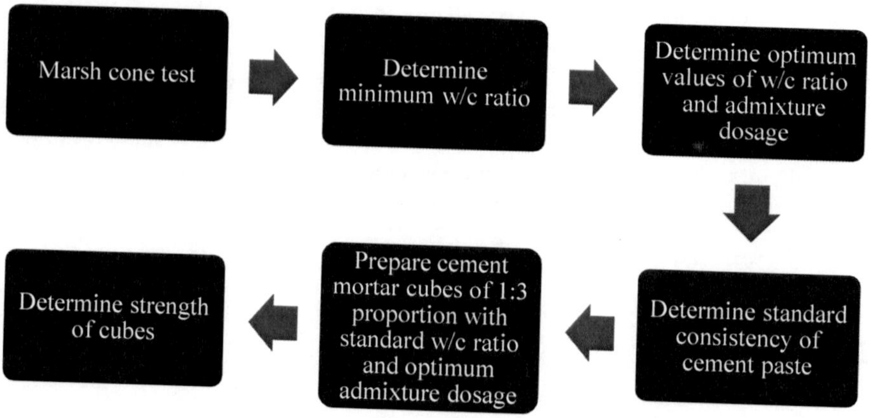

Fig. 2 Flowchart of the detailed process of experimental study

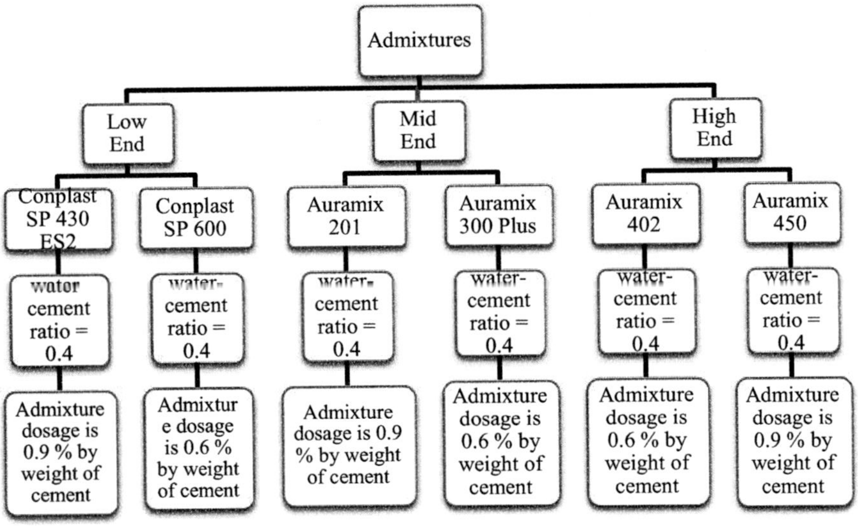

Fig. 3 Flowchart of the methodology proposed of strength test

(b) Taking water–cement ratio as 0.4 admixture was added to water.
(c) Care was taken to see that the mortar prepared was homogeneous through the process of mechanical mixing.
(d) Prepared mortar was filled in the molds in three layers, each layer being tampered 20–25 times with the tamping rod.
(e) Prepared samples of molds were vibrated as per code to achieve full compaction.
(f) Cubes were cured in freshwater curing tank.
(g) Strength of 3 days, 7 days and 28 days was recorded.
(h) Tests were also conducted for mid and high end admixtures.
(i) All readings were recorded for reference.

3 Result Interpretation

Based on the analysis performed on all the admixtures under controlled environmental conditions following IS procedures, the results are summarized and discussed in the subsequent chapter.

The minimum water–cement ratio required for the performance of admixture determined from the Marsh cone test is tabulated in Table 1. From the experiments conducted, it was observed that the optimum dosage of the admixtures used ranged from 0.6 to 0.9% by weight of cement. Hence, this study highlights the behavior of flow values corresponding to 0.6% and 0.9% of admixture dosage. Saturation point on the flow curve is the optimum admixture dosage corresponding to the admixture used. Up to the saturation point, the flow value reduces at a high rate, but beyond this point there is a nominal reduction in the flow value. The optimum values of water–cement ratio and admixture dosage are shown in Figs. 4 and 5.

Referring to Table 1, it can be seen that the optimum dosage value of Conplast SP 430 ES2, Auramix 201 and Auramix 450 is 0.9% by weight of cement used, whereas

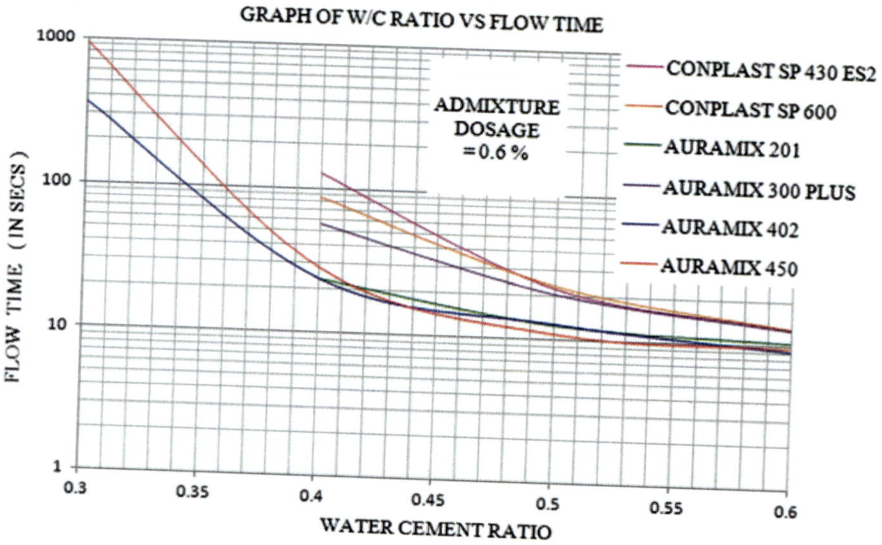

Fig. 4 Behavior of flow value for 0.6% of admixture dosage by weight of cement

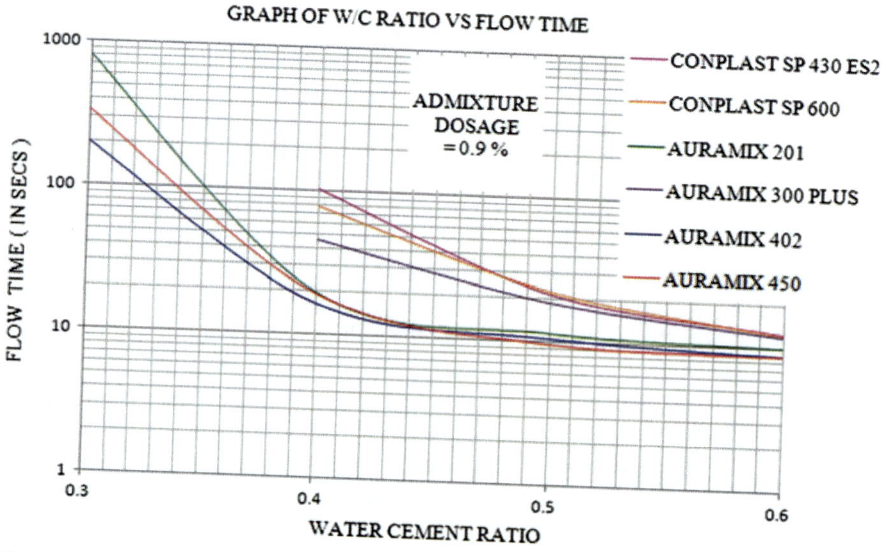

Fig. 5 Behavior of flow value for 0.9% of admixture dosage by weight of cement

Table 2 Average cube strength of cement mortars

Description	Strength in N/mm^2		
	3 days	7 days	28 days
Without admixture	28.05	37.5	55.68
Conplast SP 430ES2	27.5	37.16	55.01
Conplast SP 600	27.71	37.44	55.42
Auramix 201	27.98	37.81	55.96
Auramix 300 plus	28.13	38.02	56.27
Auramix 402	28.52	38.54	57.05
Auramix 450	29.09	39.31	58.19

Conplast SP 600, Auramix 300 Plus and Auramix 402 is 0.6% by weight of cement used.

Referring to Table 2, the behavior of strength curve observed was similar for all the mortar mix prepared. Marginal improvement in the strength of cement mortar cube in comparison with low, medium and high end admixtures for 3 days, 7 days and 28 days is observed. From Fig. 6, it is observed that Auramix 450 has the highest strength compared to the strength of cubes containing other admixtures, whereas Conplast SP 430 ES2 has the lowest strength. It is also observed that the high end admixtures show higher strength in comparison with medium and low end admixtures.

The performance of cube strength inclusive of medium end admixtures was found to be similar to the performance of cube strength without the addition of any admixtures. However, in this case, it is observed that the performance of cubes containing low end admixtures has underperformed.

Figure 7 shows an average curve of performance of admixture on strength for cement mortar cubes of the admixtures used in the present study. A logarithmic

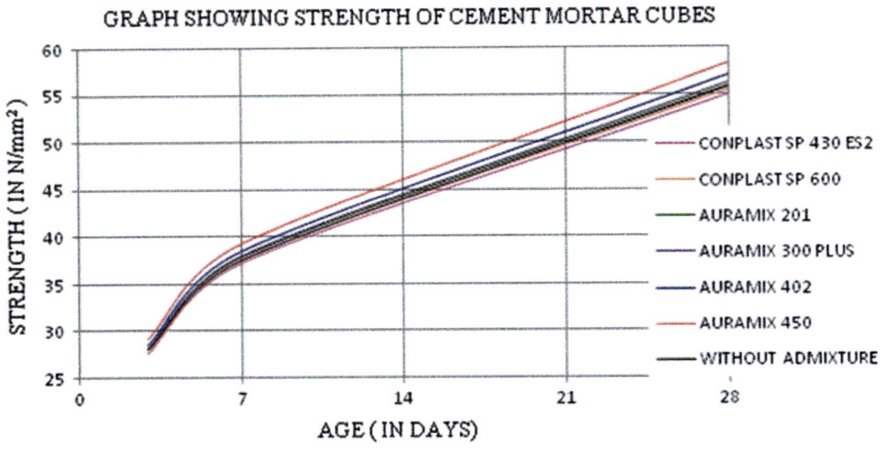

Fig. 6 Behavior of strength of different cement mortars

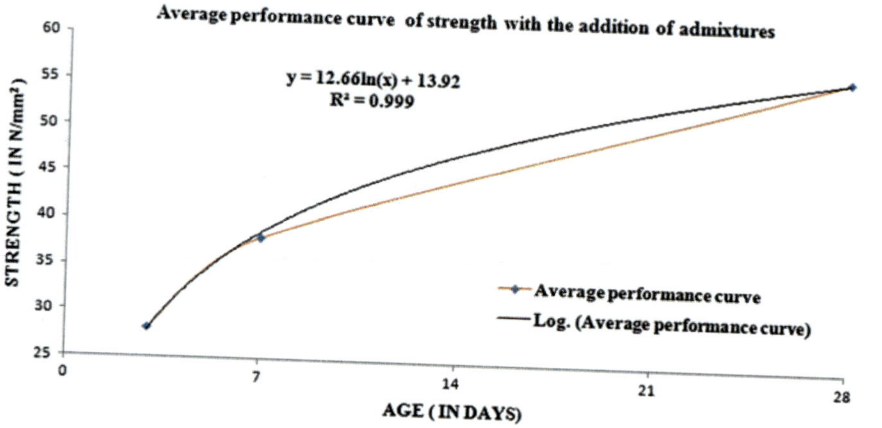

Fig. 7 Average performance curve of strength inclusive of chemical admixtures

equation is fixed having regressive coefficient as 0.999 with tolerance value of ± 5% of strength.

The expression of the average strength of cube with the addition of admixtures is found to be

$$y = 12.66\,\ln(x) + 13.92 \tag{1}$$

where

x = age of cube in days
y = strength in N/mm^2

4 Conclusions

From the above experimental study, the following conclusions are drawn:

1. The required standard water–cement ratio for OPC 43 grade UltraTech cement is 0.4.
2. From the above study, it can be concluded that the addition of admixtures does not yield high performance on the strength of mortars in comparison with the cubes tested without any addition of admixtures. However, high end admixtures show exhibited improvement as compared to medium and low end admixtures.
3. The low end admixtures have underperformed in comparison with the performance without the use of admixtures.
4. The performance of medium end admixtures is similar to that without the use of admixtures.

5. An appropriate expression was fixed to determine the strength of mortar cubes with the addition of chemical admixtures having regressive coefficient as 0.999 and $\pm 5\%$ tolerance which is given by,

$$y = 12.66 \ln(x) + 13.92$$

References

1. Alsadey S (2013) Effects of super plasticizing and retarding admixtures on properties of concrete. In: International conference on innovations in engineering and technology, pp 271–274. ICIET, Bangkok, Thailand
2. Lateef HA (2016) Studying of effect the high range, water-reducer/superplasticizer, retarding admixture on properties of concrete. Int J Sci Eng Res 219–223. ISSN 2229-5518
3. Shah DS, Shah MP, Pitroda J (2014) Chemical admixtures: a major role in modern concrete materials and technologies. In: National conference on trends and challenges of civil engineering in today's transforming world. ISBN: 978-81-929339-0-0
4. Mire A, Singh RC (2017) Study of different admixtures and their effects on concrete. Int J Res Appl Sci Eng Technol (IJRASET) 5(V):147–148. ISSN: 2321-9653
5. Khatib JM, Mohammed TU, Zhang JS, Hamada H (2010) New generation water-reducing admixture for concrete. In: Second international conference on sustainable construction materials and technologies. ISBN 978-1-4507-1490-7
6. IS 456 (2000) Plain and reinforced concrete- code of practice. (Fourth Revision)
7. IS 9103 (1999) (Reaffirmed 2004) Concrete Admixtures—Specification. (First Revision)
8. IS 4031 (Part 6) (1988) Methods of physical tests for hydraulic cement—determination of compressive strength of hydraulic cement other than masonry cement. (First Revision)
9. IS 4031 (Part 4) (1988) Methods of physical tests for hydraulic cement—determination of consistency of standard cement paste. (First Revision)

Review of Performance-Based Design of RC Shear Walls

Tarak Santosh Parab and Vikas Chodankar

Abstract Reinforced concrete shear walls are generally used as predominant in-plane lateral force-resisting members in tall buildings. National Building Code of India has clauses guiding their design process as well as there are numerous research papers detailing empirical formulations that can be used for complicated designs. Many structural design related software packages come with comprehensive shear wall modelling, analysis and designing modules. But in performance-based design philosophy it is necessary to consider not only the strength and the stiffness aspects of a shear wall but also its performance and its ultimate failure mode when subjected to strong dynamic loads. Especially RC shear walls present more challenges when it comes to accurate modelling and design because of the heterogeneous nature of concrete, variable concrete-aggregate and concrete-rebar bonding, strength degradation or pinched hysteretic behaviour, etc. This paper presents a summary of literature available of performance-based design methodologies and their applicability to the design of reinforced concrete shear walls.

Keywords Performance-based design · RC walls · Seismic design

1 Introduction

Shear walls are generally used to resist lateral force borne by tall buildings due to earthquake ground motions or wind forces along with other gravity loads. A shear wall increases the stiffness of a building in the direction parallel to its plane. This helps in reducing lateral sway of the building and increasing stability in that plane. Multiple shear walls can be 'coupled' to behave as a single wall by using coupling beams. Shear walls can be provided as perimeter wall or can form the main central core of a tall building.

When a structure is subjected to dynamic forces, lateral force-resisting systems resist the overturning effect and structural sway by means base fixity and lateral

T. S. Parab (✉) · V. Chodankar
Dept. of Civil Engineering, Goa College of Engineering, Farmagudi, Ponda, Goa, India
e-mail: tarakparab23@gmail.com

L. M. Gupta et al. (eds.), *Advances in Civil Engineering and Infrastructural Development*, Lecture Notes in Civil Engineering 87,
https://doi.org/10.1007/978-981-15-6463-5_17

stiffness. When shear walls are employed along with Special Moment Resisting Frame Systems (SMRF), the structure is designed such that 75% of the effective lateral load is borne by the shear walls while rest of the 25% is carried by frame system. Hence, it is important to model and detail shear wall appropriately so that the performance of the actual structure will be as per the predicted performance.

2 Evaluation Procedures

Performance evaluation of a structure can be performed by various methods. These procedures can be classified into the following four categories:

a. Linear Static Procedure (LSP)
b. Linear Dynamic Procedure (LDP)
c. Non-linear Static Procedure (NSP)
d. Non-linear Dynamic Procedure (NDP).

The first two design procedure employs empirical data and formulations in designing a structural member. Generally, the design is force-based with certain modifications to accommodate inelastic non-linear behaviour of the members when subjected to strong dynamic forces, system ductility, redundancy and over-strength when subjected to design forces. As per IS: 1893 (Part 1) [1], response reduction factor R is used to reduce seismic loads so as to allow for permissible displacements. This is the simplest and least computationally intensive method but the design can be more conservative and the validity of performance can be the least.

The non-linear static procedure also employs assumptions to accommodate the dynamic characteristics of the structural response. This procedure is generally only valid for the first mode dominated structures as the procedure can not accurately account for higher mode effects. NDP procedure is the most rigorous and the results can most accurately describe the performance of a real structure. As the model increases in details, the validity of the results increases. The procedure is highly accurate as it modifies stiffness of the members with each time iteration. But it is the most time consuming and computationally intensive. And as the structure incorporates more complex elements, such as RC shear walls, the designer needs to utilise rational judgement while deciding whether to simplify the model and use rigorous evaluation procedures or to rely on empirical formulations.

The analytical study conducted by [2] NDP showed the best structural performance while LSP and LDP resulted in the most conservative design. The study concluded that additional computational expense for NDP has the potential for reducing unnecessary costs for building owners on seismic improvements. The structure was within code specified permissible limits.

Ductile behaviour is necessary for the design of shear wall so as to allow for permissible deformation and redistribution of loads to adjacent elements without collapse during hazardous events. This can be achieved using capacity-based design approach. To prevent unpredictable brittle shear failure it is essential for the flexural end plastic hinge formation before shear failure. Adequate transverse reinforcements should be provided to avoid shear failure mode. It is also essential to provide confinement to concrete at the base as well as ends of a shear wall to allow permissible strain in the concrete without exceeding ultimate compressive stress.

3 Shear Wall Modelling

3.1 Factors Influencing Shear Wall Modelling

- Aspect Ratio, i.e., wall height to width ratio. A slender pier/wall will behave more like a frame element or a continuous beam in resisting lateral forces whereas wider pier will behave more like a cantilever.
- Irregularities and openings in-wall, as well as positions of the openings within the wall.
- Coupling beam characteristics and behaviour. Generally, coupling beams are considered as sacrificial so as to sustain deformations. This helps in absorbing and dissipating energy during cyclic loading and preventing major damage to shear walls.
- Finite element meshing of the panels. A simple uniform rectangular meshing scheme is adequate to capture elastic behaviour of the wall. But more comprehensive meshing techniques need to be utilised to capture precise inelastic behaviour.
- Modelled plastic hinge length.
- Strength degradation in concrete when exposed to dynamic loading as seen as 'pinching' behaviour in concrete hysteresis graph.
- Variations in cement-aggregate and concrete-rebar bonding strengths.
- Variations in concrete strength due to confinement effect.
- Eccentrically placed walls in the plan of the structure can add unforeseen out of plane stresses and rotations.

3.2 A Shear Wall Model Should be Able to Recreate Following Types of Responses

- Vertical axial and bending deformations
- Horizontal axial and bending deformations
- Shear deformations
- Rigid body displacements.

3.3 Shear Wall Element Models

Shear wall models can be broadly classified as:

i. macro models—generally uses empirical formulations for design
ii. micro models—uses finite element modelling and non-linear methodologies for design

Following are recommended types of wall element models:

a. **Two-Component Element:**

This element consists two components, an elastoplastic beam and a linear elastic beam (see Fig. 1), combined in series to represent a vertical element, such as wall or column was proposed by [3]. The model cannot accommodate strength or stiffness degradation in cyclic loading.

b. **One Component Element:**

In this model, the element consists of a single element with lumped hinges at each end (see Fig. 2). It was proposed by [4]. The non-linear deformations were assumed to be lumped at the ends of the elements and elements deform in double curvature. Different hysteretic rules were assigned to both the models to consider the degradation in cyclic loading.

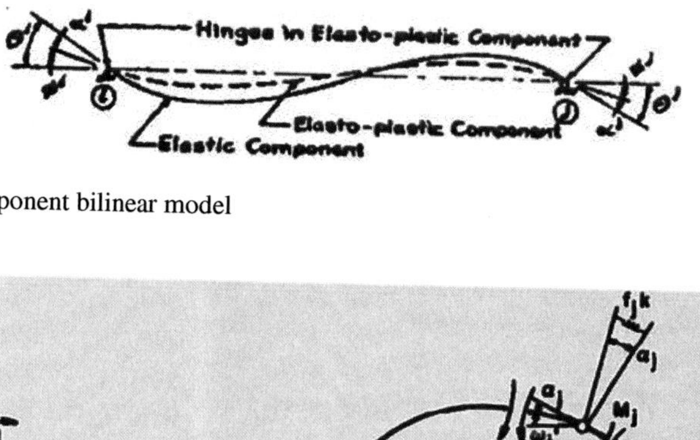

Fig. 1 Two component bilinear model

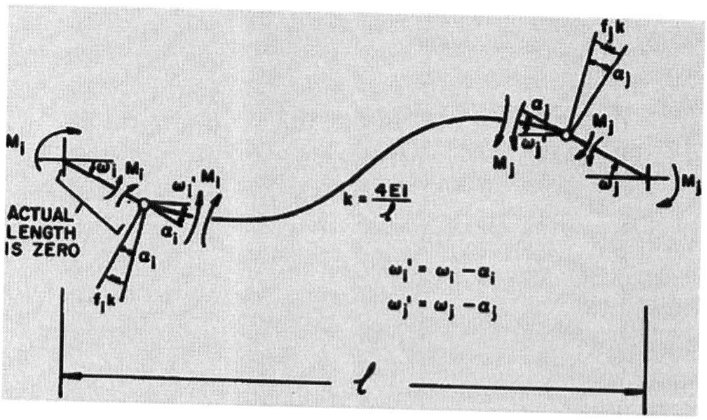

Fig. 2 One component model

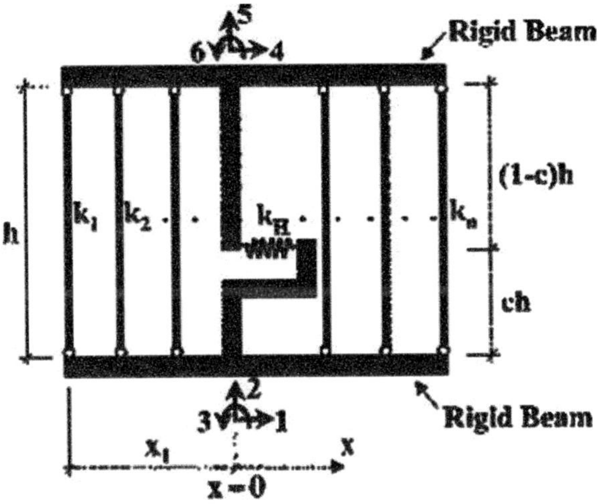

Fig. 3 Multiple vertical line element model (MVLM)

c. **Three Vertical Line Element Model (TVLM)**

Kabeyasawa et al. [5] proposed the model in which the boundary elements are modified as two vertical springs and the middle element is modelled as multi-spring element for rotational shear and axial deformations of the wall. The model then was modified by [6] to the Multiple Vertical Line Element Model (MVLM) in which multiple vertical elements were used to represent different sections of the shear wall web (see Fig. 3). The model is successful in representing important parameters of shear walls. Orakcal et al. [7] studied the behaviour of the model with respect to cyclic response of the flexural walls.

d. **Finite Element Methods**

A finite element algorithm was proposed by [8] for modelling shear walls. The mesh is made up of rectangular elements to which non-linear characteristics of concrete and steel are assigned.

e. **Multilayer Flexibility-based Finite Element**

The model was developed by [9] which used multilayer flexibility-based finite element with multilayer interfaces (see Fig. 4). The hysteretic behaviour of concrete (pinching), strength degradation, effect of slip was modelled. It was possible to incorporate confinement behaviour and inelastic shear deformation in the model. The model was verified with experiments on slender walls.

f. **Fibre Cross-Section Elements**

Fibre cross-section elements are the most recommended models for RC shear walls for most predicting non-linear responses. In these models, steel and concrete fibres

Fig. 4 Multi-layer finite element model

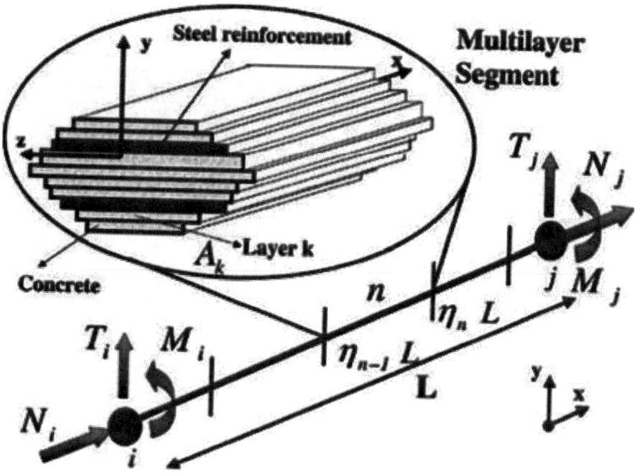

are provided running in the direction of stress to be resisted (see Fig. 5). These are now already being implemented in design software packages like SAP2000 and Perform-3D by CSI.

Basic characteristics of the fibres:

- Steel fibres are allowed to yield and may degrade under cyclic loading.
- Concrete fibres are allowed to crack and may crush.
- Concrete fibres have zero tension strength.

Fig. 5 Fibre cross-section element showing concrete and steel fibre sections arrangement

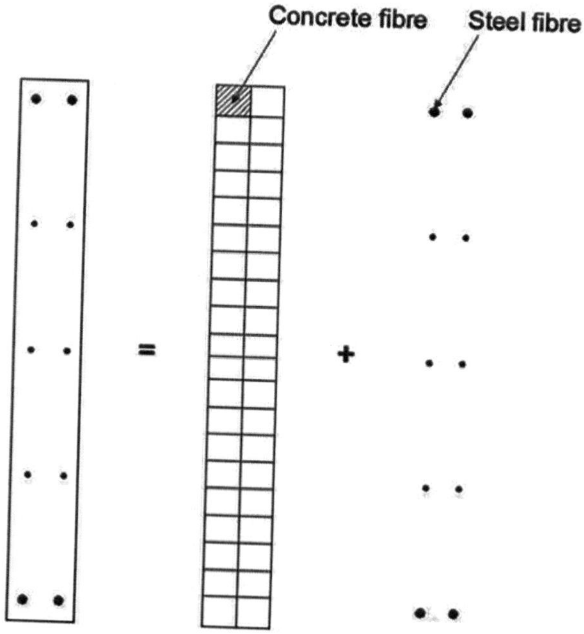

- Usually, brittle strength loss is omitted in concrete fibres.
- Only vertical axial and bending deformations affect the fibres.
- The fibre sections account for P-M interaction.

g. **Multi-layered Shell Elements**

Xinzheng [10] modelled and studied behaviour of shear wall using Multi-layered shell elements. The shear wall was modelled using OpenSees application. The element simplifies the three-dimensional wall into shell which has multiple fully-bonded layers in thickness direction (see Fig. 2). Different properties were assigned to each layer depending upon the assumed location of concrete (confined/unconfined), reinforcements and cover. The reinforcement layer is usually smeared within the concrete layer (see Figs. 6 and 7).

Similar experiment was conducted by [11]. The behaviour of a typical plan of 10 storey RC Frame with shear walls building was studied where the shear walls were modelled as a multi-layered shell elements using SAP2000. Since this type of model does not directly provide information regarding hinge formation in the elements, it was concluded that the indirect approach is necessary to be derived to interpret the results.

Fig. 6 Multi-layered shell element

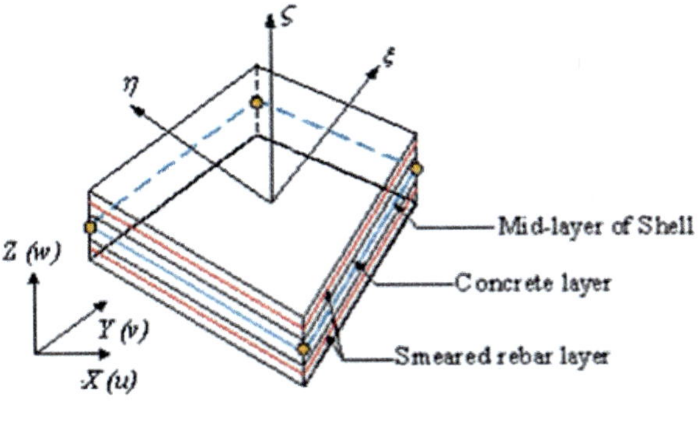

Fig. 7 Distribution of rebar layers

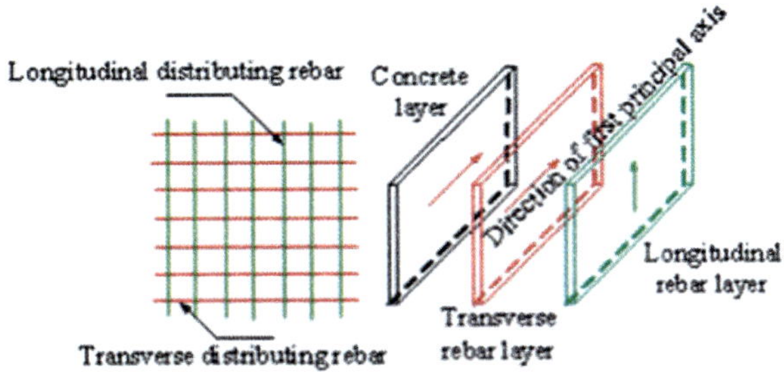

4 Verification of Design

Verification of design can be done using scaled models having similar aspect ratios and reinforcements detailing. By comparing various parameters such as elastic and plastic hinge rotations, deflections and sway, and overall stiffness degradation of the digital model with experimental data the validity of the design can be determined. It is important to focus on both local as well as overall structural behaviour as a result of the performance of the shear walls.

5 Conclusion

RC shear wall presents a challenge in regards to achieving realistic modelling due to various material and structural related complexities. Many structural designing application has incorporated modelling and designing modules based on recent developments in this field. But further studies are required for optimising the computation and enhancing reinforcement detailing as per the strength and ductility demands at different sections of the wall.

While deciding the type of element to be used it can be concluded that fibre cross-section elements can produce the most accurate results but needs rational judgement while choosing number and placement of fibres. Whereas multi-layered shell elements need further research so as to invent methodologies to simplify the modelling of the elements while incorporating numerous variables such as cracking in concrete, slip in rebar, etc.

With the refinement in modelling and detailing procedures, exponential increase in available computing power, developments in analytical and experimental research along with powerful proprietary and open-source software applications the procedure of performance-based design of RC shear walls can be easily used for engineering applications.

References

1. IS 1893 (Part-1) (2016) Criteria for earthquake resistant design of structures: general provisions and buildings, bureau of Indian standards, New Delhi
2. Hagen G (2012) Performance-based analysis of a reinforced concrete shear wall building
3. Clough RW, Benuska KL, Wilson EL (1965) Inelastic earthquake response of tall buildings. In: 3rd World conference on earthquake engineering on proceedings, New Zealand
4. Giberson F (1967) The response of nonlinear multi-story structures subjected to earthquake excitation
5. Kabeyasawa T, Shiohara H, Otani S, Aoyama H (1983) Analysis of the full-scale seven-story reinforced concrete test structure. In: 3rd Joint technical coordinating committee on proceedings, U.S. Japan cooperative earthquake research program, building research institute, Tsukuba

6. Vulcano A, Bertero V, Colotti V (1988) Analytical modeling of RC structural walls. In: 9th World conference on earthquake engineering on proceeding, Tokyo—Kyoto, Japan, vol 6, pp 41–46
7. Orakcal K, Wallace JW (2006) Flexural modeling of reinforced concrete walls-experimental verification. ACI Strcut J 103(2):196–206
8. Vecchio FJ (1999) Toward cyclic load modeling of reinforced concrete. ACI Struct J 96(2):193–202
9. Belmouden Y, Lestuzzi P (2007) Analytical model for predicting nonlinear reversed cyclic behaviour of reinforced concrete structural walls. J Eng Struct 29(7):1263–1276
10. Lu X, Xie L, Guan H, Yuli Huang, Xiao Lu (2015) a shear wall element for nonlinear seismic analysis of super-tall buildings using opensees. Finite Elem Anal Des 98:14–25
11. Debnath PP, Choudhury S (2017) Nonlinear analysis of shear wall in unified performance based seismic design of buildings. Asian J Civ Eng (BHRC) 18(4):633–642

Environmental and Water Resources Engineering

Effect of Urban Land Use on Agriculture, Forest, and River Beds: A Case Study of Dehradun City, Uttarakhand, India

Kunal Sawant, Rishi Prakash, and Nitin Mishra

Abstract Urbanization has led to a tremendous pressure on available land in Dehradun, Uttarakhand, India. Therefore, proper land planning measures are required to mitigate the effect of urbanization on agriculture, forest, and water bodies. This study illustrates the use of remote sensing and GIS to detect the change in urban sprawl of Dehradun region during the period 2013–18 and its effect on agriculture, forest, and water bodies. Landsat 8 imagery has been used in this study. Supervised classification has been adopted in Landsat 8 images of the study area. Four different land cover classes have been considered in classification stage. These are urban, forests, agriculture and vegetation, and seasonal river beds. The accuracy obtained for both the images after classification was above 85% for good change detection results. The change detection technique used is post-classification comparison method which is the matrix union method. The study has shown that the built-up area in Dehradun city has expanded from 75.07 km^2 in 2013 to 105.51 km^2 in 2018 and 0.26 km^2 of forest area, 34.40 km^2 of agricultural land, and 2.56 km^2 of water beds have been converted into built-up area in the duration of 5 years. The increase in built-up area from 2013 to 2018 is 40.54%.

Keywords Landsat 8 · Change detection · Recoding · Supervised classification

1 Introduction

The term land can be defined as a figure on the earth's surface including all physical attributes of the biosphere and the environment which focuses on the soil and rock forms, major water bodies (river, shallow lakes, swamps, marshes, and even underground water reserves), major biodiversity and their respective distribution, and major physical changes due to human activities [1]. Change detection is the process of visualizing the difference in a body or phenomenon by observing at different times. The study about change in land cover is one of the important areas to understand the

K. Sawant · R. Prakash · N. Mishra (✉)
Graphic Era Deemed to be University, Dehradun 248001, India
e-mail: nitinuag@gmail.com

L. M. Gupta et al. (eds.), *Advances in Civil Engineering and Infrastructural Development*, Lecture Notes in Civil Engineering 87,
https://doi.org/10.1007/978-981-15-6463-5_18

degree of interaction between man and ecology. To implement adequate development plans, a highly maintained and timely organized data regarding land use detection is must. The change in land use patterns affects aquatic life, water quality, and other processes that combine together to affect weather and biosphere. Uttarakhand is a state in the northern part of India formed after its division from Uttar Pradesh in 2000 [2]. Dehradun, the capital city of Uttarakhand, has witnessed a tremendous growth in construction and built-up area. Therefore, a balanced land use planning is required for the proper development of Dehradun city. The goal of my project is to identify the urban sprawl of Dehradun city over time using satellite imagery.

2 Datasets Used

2.1 *Study Area*

Dehradun extends over an area of 300 square kilometers. It lies between 77° 34′ E and 78° 18′ E longitudes and 29° 58′ north to 31° 2′ N latitudes [3]. Dehradun comprises Song River in the east, the Tons River in the west direction, and the Himalayan Mountains in the north and covered by Sal Forests in the south as shown in Fig. 1.

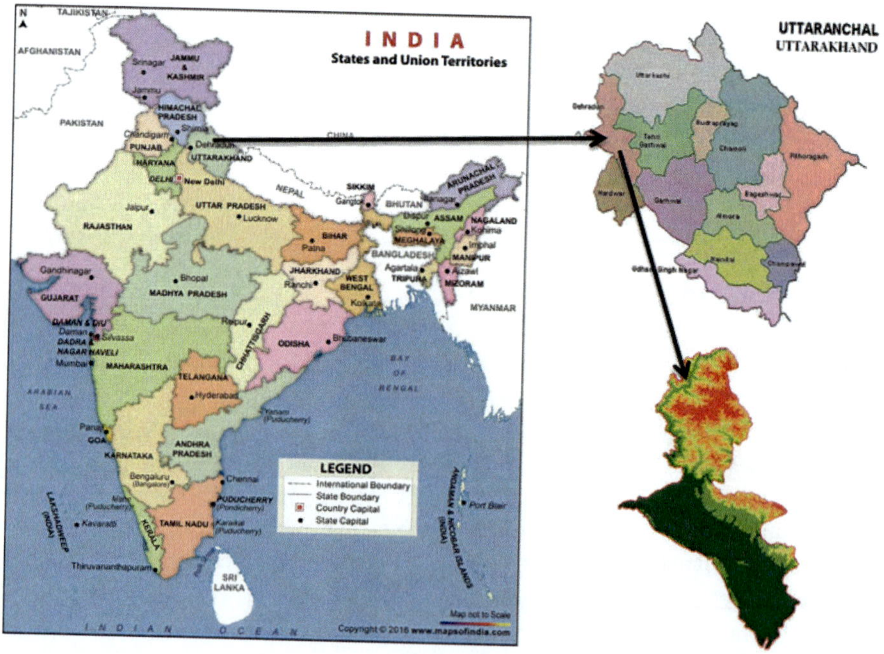

Fig. 1 Study area showing Doon Valley

2.2 *Dataset Used*

- Landsat 8 images of the year 2013 and 2018 have been used to study urban land cover changes in Dehradun city. Landsat 8 satellite was launched on February 11, 2013. It consists of 11 spectral bands. It has a spatial resolution of 30 m, thermal resolution of 100 m, and panchromatic resolution of 15 m.
- Two images of Dehradun city were downloaded from United States Geological Survey (USGS) Web site of same place but different dates of duration of 5 years. Landsat 8 image of Dehradun city of date May 29, 2013, and Landsat 8 image of Dehradun city of date May 11, 2018, were used.
- Both the images of Landsat 8 were used. We have used band combination of 2, 4, and 7 for layer stacking.

3 Methodology

Following steps have been used for urban land change detection of Dehradun, Uttarakhand (India):

Step 1: Layer stacking of two Landsat 8 images is performed of dates May 29, 2013, and May 11, 2018.
Step 2: Sub-setting of Dehradun area from both the layer stacked images.
Step 3: Supervised classification of both the subset images.
Step 4: Recoding of all urban and non-urban classes using Thematic Recode option.
Step 5: Change detection using matrix union method. Change detection matrix generation. The methodology adopted is shown in Fig. 2.

3.1 *Layer Stacking*

Landsat 8 images of 2013 and 2018 were downloaded using the earth explorer option of United States Geological Survey (USGS). The images downloaded are of dates May 29, 2013, and May 11, 2018. Once the images were downloaded, they were imported into ERDAS Imagine software. Layer stacking is a preprocessing operation which combines several bands of Landsat and provides a single image with improved quality and resolution. The band combination used is 2, 4, and 7 for layer stacking in both the images.

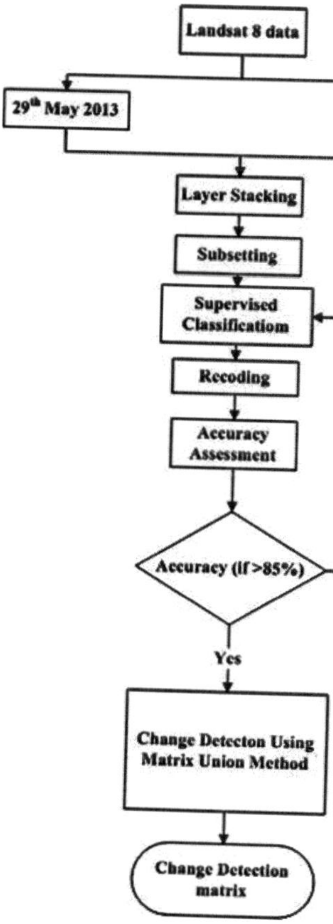

Fig. 2 Flowchart of methodology adopted

3.2 Subset of Stacked Images

After the process of layer stacking, we need to extract or cut our study area from the stacked images because our study area is a very small part of the stacked image. Therefore, we do not require the entire stacked image for further processing. Figure 3a shows the subset image of year 2013, and Fig. 3b shows the subset image of year 2018.

3.3 Supervised Classification

The classification method used for this project is supervised classification technique. In supervised classification, the user himself performs the pixel classification process

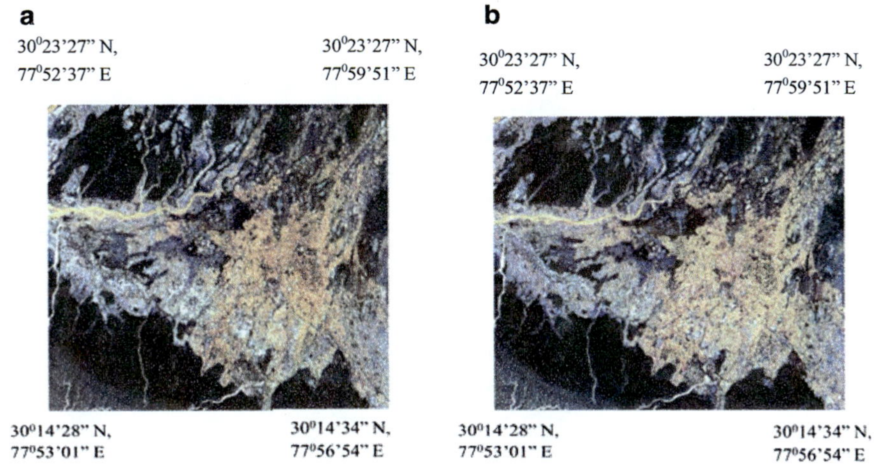

Fig. 3 **a** Subset image of year 2013 and **b** subset image of year 2018

[4]. The user creates training samples consisting of various pixels that should be associated with each class. This is done by choosing sample sites of known cover type called training areas. Signature editor method is employed to select the training samples and technique used is maximum likelihood classification. Figure 4a displays the classified image of year 2013, and Fig. 4b illustrates the classified image of year 2018. There are a total of four classes taken in classification: Red is for urban, green is for forestland, yellow is for agricultural land, and blue is for seasonal water beds. The samples taken for forest area are 25, for agriculture and vegetation is 40, 25 samples for urban area, and 15 samples for seasonal river bed. Supervised classification is different from unsupervised classification in the sense that it requires external training samples to execute the classification operation [5].

3.4 Recoding

Recoding is the process of reclassification of the classes. This is done by combining them based on similar characteristics and having a more general class. The goal of this project is to identify the change of the urban class. Recoding is done using thematic option in the raster tab and all the urban classes were recorded as 1, all the forest classes were recorded as 2, all the agriculture and vegetation classes were recorded as 3, and all water classes were recorded as 4. These recoded images will be further used as inputs in the matrix union for change detection. In case, the pixels of a particular class are showing in some other class, thematic recoding helps in this error correction process by first drawing a polygon around the boundary of the area in which the error in pixels is showing and then using the thematic recode option, we correct the pixels in that area. In this project, water beds were getting classified

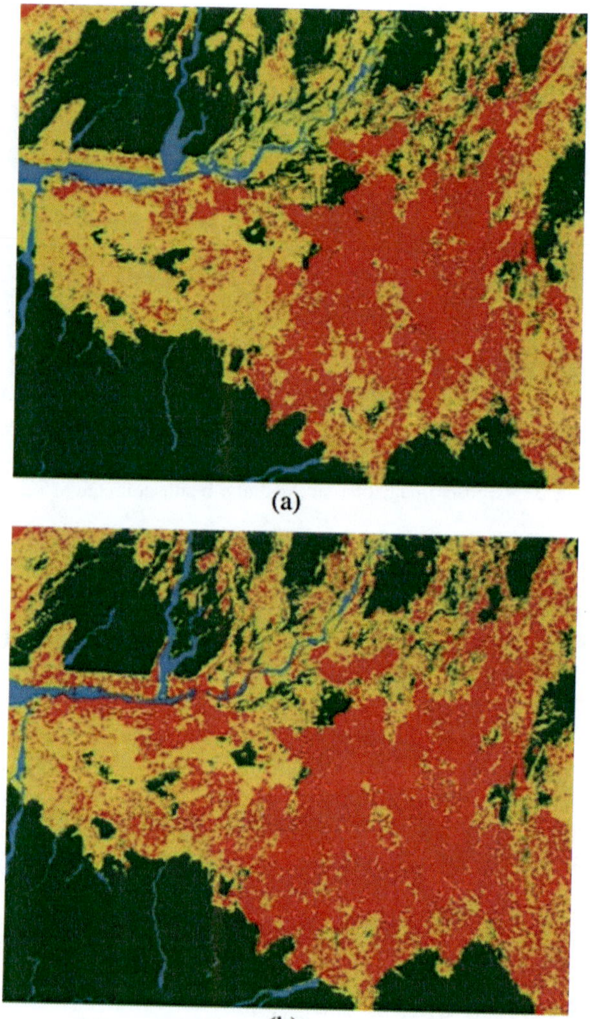

(a)

(b)

Fig. 4 **a** Classified image of year 2013 and **b** classified image of year 2018. Red is for urban, green is for forests, yellow for agriculture, and blue for river beds

under the urban region due to same reflectance of rocks and sand in both river beds and urban regions so we were able to solve this discrepancy using thematic recode option which enabled us to recode the wrong classes to correct one. Recode option is available in raster tab of ERDAS IMAGINE.

3.5 Accuracy Assessment

After performing recoding and thematic corrections, accuracy assessment is performed on both the recoded images. Accuracy assessment is performed to evaluate the accuracy of classification of our images. The accuracy of 2013 image is 88% and of 2018 image is 85% which is desirable and good for change detection.

3.6 Change Detection Using Matrix Union Method

Change detection is defined as a process used to identify the change that occurred in a specific area over a span of time. By observing the same area at different time intervals using satellites or aerial photography, the user can identify the change of land use and land cover in that area. Several other analyses can also be done with the help of change detection techniques such as change in the condition of the water or the degrading of a plant species. Since the images used for this study were already classified, the post-classification comparison method was used. This is a simple and more promising method which produces accurate results if the classification is done accurately. Since the images taken at the different times were already independently classified, these classified images can be used to produce change maps that will visualize the pixels that have changed from one class to another.

The method used for change detection is matrix union method present in thematic option of the raster tab. It takes both the recoded images as input and can be used to create a visualization map which will display the change of pixels from 2013 to 2018. We saved the matrix union file by the name 2013–18. We have to open this file in Arc GIS software to generate a change detection matrix in Excel sheet from there which will display the change in pixel count between 2013 and 2018 and also the change in area. The matrix union function will create a matrix from two input thematic rasters. To open this dialog, go to the raster tab, raster GIS group, click the drop down below thematic and select matrix union. Figure 5 shows the matrix union image 2013–18 opened in Arc GIS software. We will open the attribute table of this image in Arc GIS software from which we will generate a change detection matrix in Excel sheet denoting the change in area during the time duration of 2013–18.

3.7 Change Detection Matrix

A change detection matrix helps us to detect the change in pixel count during a given time duration from which we can compute the change in area in km^2 [6]. Table 1 shows the attribute table of matrix union file 2013–18 opened in Arc GIS from where we extract its information to generate a change detection matrix [7, 8].

Fig. 5 Matrix union image 2013–18 opened in Arc GIS software

Table 1 Attribute table of matrix union file 2013–18 in Arc GIS

OID	Value	Count	Recode2013	Reccode2018
0	0	0	0	0
1	1	75,884	1	1
2	2	42	1	2
3	3	7429	1	3
4	4	53	1	4
5	5	288	2	1
6	6	131,784	2	2
7	7	15,350	2	3
8	8	1161	2	4
9	9	38,218	3	1
10	10	2348	3	2
11	11	124,032	3	3
12	12	759	3	4
13	13	2845	4	1
14	14	156	4	2
15	15	1689	4	3
16	16	7734	4	4

Table 1 shows the change in pixel count from year 2013 to 2018 between the recoded classes: 1 (urban), 2 (forests), 3 (agriculture and vegetation), and 4 (seasonal river beds). We can use this information to generate a change detection matrix from where we can also calculate the urban land change detection in km^2. We will create the change detection matrix in Microsoft Excel 2016 for better reliability and performance. A change detection matrix represents the change in pixel count between 2013

Table 2 Change detection matrix for change in pixel count

	1	2	3	4	2018
1	**75,884**	288	38,218	2845	117,235
2	42	**131,784**	2348	156	134,330
3	7429	15,350	**124,032**	1689	148,500
4	53	1161	759	**7734**	9707
2013	83,408	148,583	165,357	12,424	**409,772**

Table 3 Change detection matrix for area count in km^2

	1	2	3	4	2018
1	**68.30**	0.26	34.40	2.56	105.51
2	0.04	**118.61**	2.11	1.14	120.90
3	6.69	13.82	**111.63**	0.52	133.65
4	0.05	1.04	0.68	**6.96**	8.74
2013	75.07	133.72	148.82	11.18	**368.79**

and 2018 of urban area and also detects how much area of forests, agriculture, and rivers have been converted into built-up or urban area during this time period and how much urban area has been converted into forests, agriculture, and water beds. From Table 1, it is seen that 288 pixels of forest lands, 38,218 pixels of agriculture and vegetation, and 2845 pixels of seasonal river beds have been converted into urban land. Table 1 also shows that 42 pixels of urban area have been converted to forests, 7429 pixels to agricultural land, 52 pixels to river bed, and 75,884 pixels remained unchanged during this time period. Table 2 shows the change detection matrix between the years 2013 and 2018 depicting the change in pixel count. Table 3 shows the change detection matrix depicting the change in area in km^2. Table 3 shows that the urban area was 75.07 km^2 in 2013 and has been increased to 105.51 km^2 in 2018 and has been increased by 30.44 km^2, and percentage increase in area is 40.54%. From the year 2013 to 2018, 0.26 km^2 of forestland, 34.40 km^2 of agriculture and vegetation, and 2.56 km^2 of seasonal river beds have been converted into urban area, and 68.30 km^2 of urban or built-up area remains unchanged during a span of 5 years. Our main focus is on the change of pixels in urban class, and hence, there is an increase of 33,827 pixels in urban class from 2013 to 2018. In order to convert pixel, count into km^2, it is necessary to know about the cell size of the classes and then multiplying them by their corresponding pixel count to compute the area.

4 Conclusion

Preprocessing operations of layer stacking and sub-setting are performed on both the Landsat 8 imagers of years 2013 and 2018. Supervised classification technique is used for land cover classification, and accuracy obtained for both the images is above 85%. After the classification process, recoding is performed such that all urban classes are recoded to 1, forest class to 2, agricultural land to 3, and seasonal river beds to 4. After that matrix union method is used for land change detection. A change detection matrix is generated which depicts the change in pixel count from which change in km^2 can be computed during a span of 5 years. The change in urban area observed in the duration of 5 years is 30.44 km^2, and percentage increase is 40.54%. From 2013 to 2018, 0.26 km^2 of forestland, 34.40 km^2 of agriculture and vegetation, and 2.56 km^2 of seasonal river beds have been converted into urban area during a span of 5 years. This study is very important for monitoring and future planning of Dehradun city and its natural resources.

Acknowledgements Authors are very thankful to Mr. Lakhwinder Singh, Ph.D Research Scholar, IIT Roorkee, for his constant help regarding various GIS topics related to image processing and classification.

References

1. Patidar S, Sankhala V (2015) Change detection of land-use and land-cover of dehradun city: a spatio-temporal analysis. Intl J Adv Remote Sensing GIS 5(1):1170–1180
2. Rawat V, Puri M (2017) Land use/land cover change study of district Dehradun, Uttarakhand using Remote sensing and GIS technologies. Intl J Adv Remote Sens GIS 6(1):2223–2233
3. Bhat PA, Shafiq MU, Mir AA, Ahmed P (2017) Urban sprawl and its impact on land use/land cover dynamics of Dehradun city, India. Int J Sustain Built Environ 6(2):513–521
4. Deb SK, Nath RK (2012) Land use/cover classification—an introduction review and comparison. Global J Res Civ Struct Eng 12(1):5–16
5. Devi MR, Baboo SS (2011) Land use and land cover classification using RGB and L bases supervised classification algorithm. Int J Comput Sci Eng Technol 2(10):167–180
6. Kuldeep T, Kamlesh K (2011) Land use/land cover change detection in Dehradun using GIS and remote sensing techniques. Int J Geomatics Geosci 2(1):34–41
7. Kumar KS, Valsala SS, Subhramanyam JV, Mallampati VM, Shaikh K, Ekkirala P (2015) Prediction of future land use/land cover changes of Vijayawada City using remote sensing and GIS. Int J Innovative Res Adv Eng 2(3):91–97
8. Selim M (2018) Change detection analysis using new nano satellite imagery. Int J Eng Adv Technol 7(3):4–10

Spatio-Temporal Trend Analysis of Rainfall for Kumaon Region of Uttarakhand

Pooja Negi, Nitin Mishra, and Amit Kumar Sharma

Abstract Environmental change is upsetting the key climatic parameters at a world-wide dimension. In any case, the progressions having limited power are not rise to for all areas particularly in India. These progressions must be measured locally to deal with the normal water assets all the more adequately. Precipitation is a standout amongst the most significant climatic parameter. The reason for this investigation is to watch the worldly fluctuation of precipitation for the time of 1901–2017 (117 years), of Kumaon area regions of Uttarakhand. The point of the examination is to decide the pattern in yearly precipitation time arrangement utilizing the M-K and Sen T-tests. The pattern size in precipitation has been evaluated by Sen's estimator strategy. Auto correlation impact is decreased prior applying the M-K test for the pattern in rainfall. On the yearly premise, examination of M-K analysis shows expanding (positive) no criticalness pattern in precipitation time arrangement over the Kumaon areas of Uttarakhand.

Keywords Mann–Kendall test · Non parametric test · Sen slope · Climate change · Rainfall trend · Trend analysis · Time series · Kumaon region

1 Introduction

Climate change is the change in earth's atmospheric system which results in new climate pattern over a critical time-frame. It is regularly used to elude anthropogenic climate change which is known as global warming. Global warming is due to human activities and is referred as the long term increase in level of surface temperature since 1900 for the most part brought about by the emissions of greenhouse gases. Climate change includes both global warming and its effects such as changing pattern of rainfall and temperature. As per IPCC Fifth Assessment Report which states that human influence (emission of green house gases such as Methane, Carbon dioxide and Nitrous oxide) has been the main cause leading to the warming since the mid

P. Negi · N. Mishra · A. K. Sharma (✉)
Graphic Era Deemed to be University, Dehradun 248001, India
e-mail: amitsharma812010@gmail.com

L. M. Gupta et al. (eds.), *Advances in Civil Engineering and Infrastructural Development*, Lecture Notes in Civil Engineering 87,
https://doi.org/10.1007/978-981-15-6463-5_19

twentieth century. During the twenty-first century, surface temperature is likely to increase a further 0.3–1.7 °C to 2.6–4.8 °C depending on the rate of emission of greenhouse gases. Effect of environmental change in future is very serious as given by IPCC reports which imply that there will be decrease in the freshwater accessibility in view of environmental change. Precipitation is fundamental component of climate [1]. Changes in precipitation can affect human lives, ecosystem, plants and animals [2].

To examine the trend of precipitation there are several non parametric and parametric tests available. Parametric tests are said to be more influential than non-parametric tests, but data series required in them should be independent and normally distributed [3]. Man–kendall test is the most used and important test used for analysis of trend of Rainfall [4].

2 Study Area

Kumaon is one of the locales and regulatory division of the Indian region of Uttarakhand, the other being Garhwal. It consists of Almora, Bageshwar, Champawat, Nainital, Pithoragarh, and Udham Singh Nagar. Its miles constrained on the north by Tibet, at the east by Nepal, on the south by means of the territory of Uttar Pradesh and on the west by Garhwal locale.

Kumaon locale consists of a massive Himalayan strip, together with submontane strips known as the Terai and the Bhabhar. The submontane traces had been as much as 1850 a nearly invulnerable woodland, supplied up to wild creatures; yet after 1850, the various clearings pulled in a wider population from the slopes, who developed the rich soil at some point of the hot and bloodless seasons, coming back to the slopes in the downpours. The remainder of Kumaon is a labyrinth of mountains, some part of the Himalayan range, some of which can be some of the loftiest acknowledged. In a tract not in extra of 225 km length and 65 km in broadness, there are greater than thirty pinnacles ascending to heights surpassing 5500 m (Fig. 1).

3 Trend Analysis

(a) **Mann–Kendall Test**: The non-parametric MK test has been ordinarily utilized for recognizing monotonic patterns arrangement of natural information, atmosphere information or hydrological information. The invalid speculation, H_0, is that the information originated from a populace with autonomous acknowledges and are indistinguishably conveyed [5]. The elective speculation, H_A, are that the information pursues a monotonic pattern. The MK test measurement are determined supportive:

Fig. 1 Study area showing 13 districts

$$S = \sum_{k=1}^{n-1} \cdot \sum_{j=k+1}^{n} sgn(Xj - Xk) \tag{1}$$

for

$$sig(x) = \begin{cases} 1 & \text{if } X > 0 \\ 0 & \text{if } X = 0 \\ -1 & \text{if } X < 0 \end{cases} \tag{2}$$

The mean of S is $E[S] = 0$ and the variance σ^2 is:

$$\sigma^2 = \frac{\left[n(n-1)(2n+5)\sum_{j=1}^{p} t(t-1)(2t+5)\right]}{18} \tag{3}$$

In which p is the variety of the tied agencies in the records set and t are the quantity of facts points inside the jth tied group. The statistic S is about regular allotted if the subsequent Z-transformation is employed:

$$Z = \begin{cases} (S-1)/\sigma & \text{if } S > 0 \\ 0 & \text{if } S = 0 \\ (S+1)/\sigma & \text{if } S > 0 \end{cases} \quad (4)$$

(b) **Sen's slope**: The test computes equally the slope (i.e., linear rate of changes) and intercept according to Sen's methods [6]. First, a set of linear slopes are premeditated:

$$T_i = \frac{(x_j - x_k)}{(j-k)} \quad \text{for } i = 1, 2, \ldots .N. \quad (5)$$

In which x_j and x_k are represented as value of data at time j and k (where $j > k$) correspondingly. The median of the N values of T_i is considered as Sen's slope estimator of which is given as:

$$Q_i = \begin{cases} T_{\frac{N+1}{2}} & N \text{ is Odd} \\ {}^{1}/_{2}\left(T_{\frac{N}{2}} + T_{\frac{N+2}{2}} \right) & N \text{ is Even} \end{cases} \quad (6)$$

If N is odd, the Sen's estimator is estimated as $Q_i = T(N + 1)/2$, and if N is even it is estimated as $Q_i = [TN/2 + T(N + 2)/2]/2$. Lastly Q_i is estimated at 100 $(1 - \alpha)$ % confidence interval by a two sided test and then by the non-parametric test Q_i, a exact slope can be derived with a positive value signifies an upward or increasing trend and a negative value signifies a downward or decreasing trend in the time series [7].

4 Results

(a) **Non-Parametric Test Results** (Fig. 2; Tables 1 and 2)

Parametric Test Result: This test gives statistics for Mann–kendall test i.e., value of 'Z' which will implies the significance of the test. It also provides Sen's slope value which tells the magnitude of the trend.

5 Conclusion

Rainfall data of 117 years from 1901–2017 is analyzed to find the rainfall trend in the Kumaon region. Due to rapid growth, changes in the pattern of rainfall will have impact on this region and people living here from study, we found that the value of Z shows an increasing trend and from this we can conclude that climate change

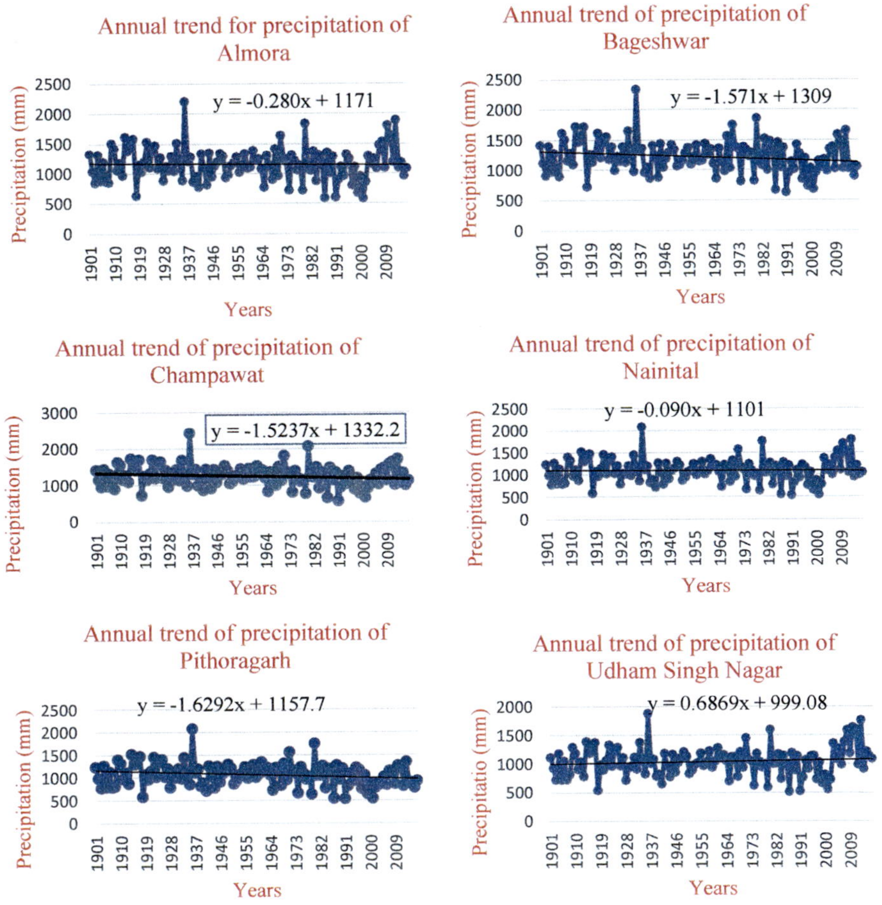

Fig. 2 Rainfall trend graph represented for Kumaon stations

Table 1 Z value and trend magnitude for precipitation

Station Name	Pre-monsoon		Monsoon		Post-monsoon	
	'Z' statistics	Sen's Slope	'Z' statistics	Sen's Slope	'Z' statistics	Sen's Slope
Almora	3.78	0.712	2.46	4.287	−1.22	−0.903
Bageshwar	3.40	0.786	2.25	3.490	−1.49	−1.162
Chamoli	−0.39	−0.377	0.28	1.493	2.38	1.765
Nainital	3.67	0.591	2.65	4.338	−1.33	−0.801
Pithoragarh	3.15	0.614	1.75	2.385	−1.68	−0.800
Udham S.Nagar	3.58	0.515	2.75	4.736	−1.67	−0.722

Table 2 Value of 'z' and magnitude of trend for precipitation

Station Name	Winter		Annual	
	'Z' statistics	Sen's Slope	'Z' statistics	Sen's Slope
Almora	1.96	0.204	2.33	4.776
Bageshwar	1.53	0.195	2.11	3.312
Champawat	1.76	0.165	2.22	3.560
Nainital	1.74	0.160	2.51	4.809
Pithoragarh	2.00	0.175	1.97	2.776
Udham S.Nagar	1.57	0.133	2.55	5.294

impact is present and there will be heavy rainfall with time. Sen's slope also shows an increase in slope magnitude. The study shows increasing trend for pre-monsoon, winter, summer while shows decreasing trend in post-monsoon for some regions.

References

1. Jain SK, Kumar V (2012) Trend analysis of rainfall and temperature data for India. pp 37–49
2. Mondal A et.al (2012) Rainfall trend analysis by Mann–Kendall test: a case study of north-eastern part of Cuttack district, Orissa. Int J Geol Earth Environ Sci 2271–2081
3. Jaswal AK, Bhan SC, Karandikar AS, Gujar MK (2015) Seasonal and annual rainfall trends in Himachal Pradesh during 1951–2005. pp 247–264
4. Ganguly A, Chaudhuri RR, Sharma P (2015) Analysis of trend of the precipitation data: a case study of Kangra district, Himachal Pradesh. Int J Res 87–95
5. Bandyopadhyay A, Pal A, Debnath S (2011) Development of an Arc GIS toolbar for trend analysis of climatic data. Int J Geol Environ Eng 5:777–784
6. Shahid S (2010) Trends in extreme rainfall events of Bangladesh. Theor Appl Climatol 489–499
7. Duhan D, Pandey A (2013) Statistical analysis of long term spatial and temporal trends of precipitation during 1901–2002 at Madhya Pradesh, India. Atmos Res 1–8

Water Body Mapping of Chennai Region Using GIS and Remote Sensing

Ashish Bhandari, Nitin Mishra, and K. K. Gupta

Abstract This venture depicts the improvement of water body mapping (WBM), utilizing a computerized calculation to process Landsat pictures from the United States Geological Survey (USGS) database. We utilized Landsat 2, 3, 5, 8 band pictures so as to depict a precise water body map, with no normal unsettling influence. Perpetual water bodies were recognized from fleeting water-shrouded regions by computing the recurrence of water body presence from covering, multi-transient, Landsat scenes. By examining the recurrence of water body presence, the WBM isolates stream channels and floodplains more obviously than past examinations. This recommends the utilization of multi-transient pictures is as significant as examination at higher goals for worldwide water body mapping. We screen the Chennai and adjacent surface water region changes from 1977 to 2016. Three scenes from the Multispectral Scanner System (MSS), three scenes from the Thematic Mapper (TM), and three scenes from the Operational Land Imager (OLI) and Thermal Infrared Sensor (TIRS) remote detecting information of Landsat satellites, the standardized distinctive water file (NDWI), standardized diverse vegetation record (NDVI), and NIR technique were utilized to quantitatively gauge the Chennai buffer surface water territory during the 1970s, 1980s, 2000s, and 2010s, separately. The examination closed the estimation of human exercises sway on surface water spatiotemporal dissemination. The point was to decide the exactness of utilizing basic computerized picture handling strategies to delineate water bodies with Landsat 2(MSS), Landsat 3(MSS), Landsat 5(MSS/TM), and Landsat 8(OLI/TIRS) information.

Keywords Water body mapping · Landsat images · NDVI · NDWI

A. Bhandari
National Institute of Hydrology, Roorkee, India

N. Mishra (✉) · K. K. Gupta
Department of Civil Engineering, Graphic Era (Deemed to be University), Dehradun 248001, India
e-mail: nitinuag@gmail.com

L. M. Gupta et al. (eds.), *Advances in Civil Engineering and Infrastructural Development*, Lecture Notes in Civil Engineering 87,
https://doi.org/10.1007/978-981-15-6463-5_20

1 Introduction

Surface water is one of the extremely valuable vital assets of human survival and social improvement. It is essential for biological systems, human, and yields [1]. Land surface water is a significant segment of the water cycle. Further more, water data extraction assumes a significant job in wetland depiction, surface water study and the executives, flood observing and flood catastrophe hazard evaluation, and condition checking [2]. Among the correlative research, surface water data extraction or the spatiotemporal change location of surface water zone is a key issue [3]. To sum things up, the exact extraction and estimation of water highlights and surface water zone change are the most significant examination objective [4]. Moreover, water highlight is a standout among the most significant articles on the earth, and its extraction is of extraordinary noteworthiness to many related looks into in remote detecting and hydrology areas [4]. Some broadly utilized information is MSS, TM, and OLI/TIRS pictures given via Landsat arrangement satellites. In the present examination, the surface water zone changes Chennai support which is evaluated dependent on time arrangement Landsat remote detecting pictures from 1977 to 2016 [5].

2 Study Area

Chennai is arranged on the north-east end of Tamil Nadu on the bank of Bay of Bengal. It is 12° 9′ to 13° 9′ of the north scope and 80° 12′ to 80° 19′ of the south longitude on a 'sandy racking breaker cleared shoreline.' It extends about 25.60 km along the Bay coast from Thiruvanmiyur in the south to Thiruvottiyur in the north and runs inland in a rough semi-roundabout design. It is encompassed on the east by the Bay of Bengal and on the staying three sides by Chengalpattu and Thiruvallur districts. Chennai is a low-lying zone, and the land surface is level. Geology of the land all through the locale renders subdivisions into characteristic areas rather troublesome. It rises scarcely as the separation from the ocean coast increments, however, the normal rise of the city is under 22′ above mean ocean level, while a large portion of the regions are exactly adrift dimension and waste in such territories remains a difficult issue.

3 Data Sources

In order to do our project, we needed the data for processing, and for further work, we download the data from USGS Earth Explorer by using the shape file of Chennai buffer, then select the year from where and till what date we want the image of the area and then select and download the image of comparatively good quality and minimum cloud cover from Landsat data [6] (Figs. 1 and 2).

Buffer Used—20 km around the district boundary given for Chennai district (Table 1).

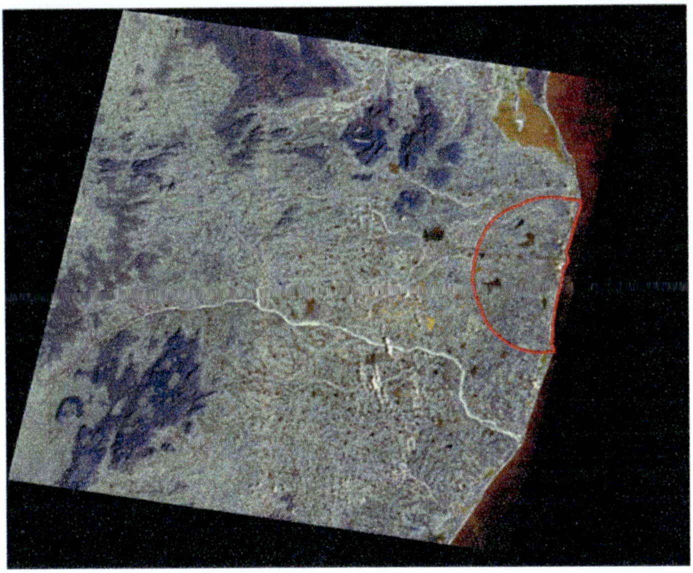

Fig. 1 Landsat image acquired from USGS

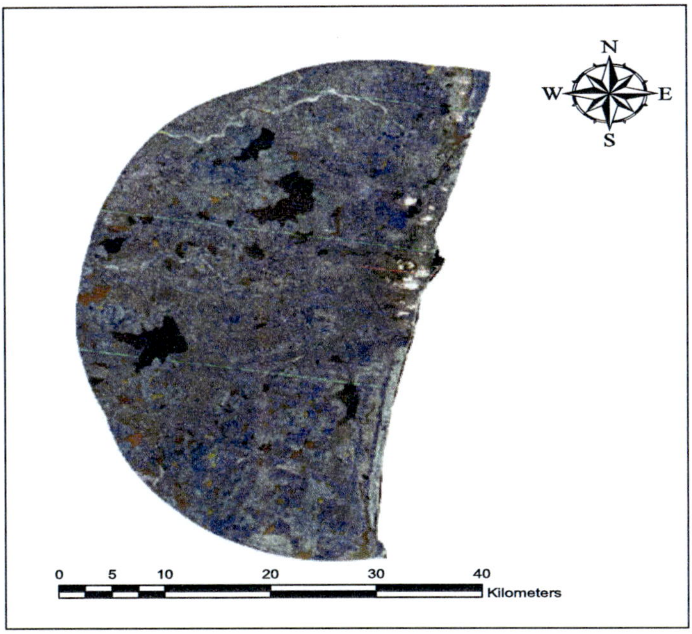

Fig. 2 Landsat image of Chennai buffer boundary

Table 1 Specifications of the multispectral scanner (MSS), thematic mapper (TM), operational land imager (OLI), and thermal infrared sensor (TIRS) data used in this study

Satellite	Sensor	Path/Row	Acquisition date	Resolution (m)	Wavelength (μm)
Landsat 2	MSS	153/51	07-02-1977	79	Band 4 (Green): 0.5–0.6
				79	Band 5 (Red): 0.6–0.7
				79	Band 6 (NIR): 0.7–0.8
				79	Band 7 (NIR): 0.8–1.1
Landsat 3	MSS	153/51	23-07-1980	79	Band 4 (Green): 0.5–0.6
				79	Band 5 (Red): 0.6–0.7
				79	Band 6 (NIR): 0.7–0.8
				79	Band 7 (NIR): 0.8–1.1
				240	Band 8 (Thermal): 10.4–12.6
Landsat 5	MSS	142/51	11-09-1989	79	Band 4 (Green): 0.5–0.6
				79	Band 5 (Red): 0.6–0.7
				79	Band 6 (NIR): 0.7–0.8
				79	Band 7 (NIR): 0.8–1.1
Landsat 5	TM	142/51	25-09-2005 02-03-2006 11-08-2006	30	Band 1 (Blue): 0.45–0.52
				30	Band 2 (Green): 0.52–0.6
				30	Band 3 (Red): 0.63–0.69
				30	Band 4 (NIR): 0.76–0.9
				30	Band 5 (NIR): 1.55–1.75
				120	Band 6 (TIR): 10.4–12.5

(continued)

Table 1 (continued)

Satellite	Sensor	Path/Row	Acquisition date	Resolution (m)	Wavelength (μm)
				30	Band 7 (MIR): 2.08–2.35
Landsat 8	OLI/TIRS	142/51	28-07-2015 11-04-2016 12-09-2016	30	Band 1 (Coastal aerosol): 0.43–0.45
				30	Band 2 (Blue):0.45–0.51
				30	Band 3 (Green): 0.53–0.59
				30	Band 4 (Red): 0.64–0.67
				30	Band 5 (NIR): 0.85–0.88
				30	Band 6 (SWIR 1): 1.57–1.65
				30	Band 7 (SWIR 1): 2.11–2.29
				15	Band 8 (Pan): 0.50–0.68
				30	Band 9 (Cirrus): 1.36–1.38
				100	Band 10 (TIRS 1): 10.60–11.19
				100	Band 11 (TIRS 2): 11.50–12.51

4 Software Used

1. ArcMap10.2
2. ERDAS IMAGINE 2014

5 Method

Single-band thickness cutting and the multispectral most extreme probability calculation were utilized to separate the satellite information [7]. The thickness cutting endeavored to figure the exactness of a basic method utilizing a solitary band to guide water bodies. The insights of every territory were broke down to figure out which band or groups best segregated water bodies from the encompassing scene. A straightforward thickness cut characterization of each band was utilized to discover which groups were most valuable for water limit outline. The subsequent water classes were then consolidated into one super 'water' class. The thickness cut arrangement of each band and the most extreme probability characterization were then contrasted and different information for discovering the adjustment in water body [8].

5.1 *Normalized Difference Water Index (NDWI)*

Numerical indicator uses the green and near-infrared bands of the electromagnetic spectrum and is adopted to study remote sensing measurements and assess whether the result being observed contains water or not [9, 10]. Normalized difference water index (NDWI) was calculated as follows:

$$\mathrm{NDWI} = \left(\frac{\mathrm{GREEN} - \mathrm{NIR}}{\mathrm{GREEN} + \mathrm{NIR}}\right)$$

5.2 *Normalized Difference Vegetation Index (NDVI)*

This is a numerical indicator that uses the red and near-infrared bands of the electromagnetic spectrum and is adopted to study remote sensing measurements and assess whether the result being observed contains green vegetation or not. Normalized difference vegetation index (NDVI) was calculated as follows:

$$\mathrm{NDVI} = \left(\frac{\mathrm{NIR} - \mathrm{RED}}{\mathrm{NIR} + \mathrm{RED}}\right)$$

The main aim is to delineate a permanent water body map [11, 12]. For that purpose, water frequency was used to distinguish permanent water bodies from areas of temporal water covered [13]. The NDVI, NDWI, and NIR of the images show the water bodies present in the different area [14]. We then compare these results with the other results of different years, and thus, we calculate the change of water body either increasing or decreasing in the Bangalore city. However, the ideal threshold of each index for separating water varies regionally and over time due to local similarities and mixing with other feature types [15–17] (Fig. 3).

5.3 *Image Threshold Segmentation*

Image division is a significant advance in separating data about water bodies from NDVI, NDWI, and NIR information. The limit division connected in NDVI and NDWI to isolate the picture into two classes: foundation highlights and water bodies. The edge esteems for NDVI, NDWI, and NIR were set to zero, yet limit esteem

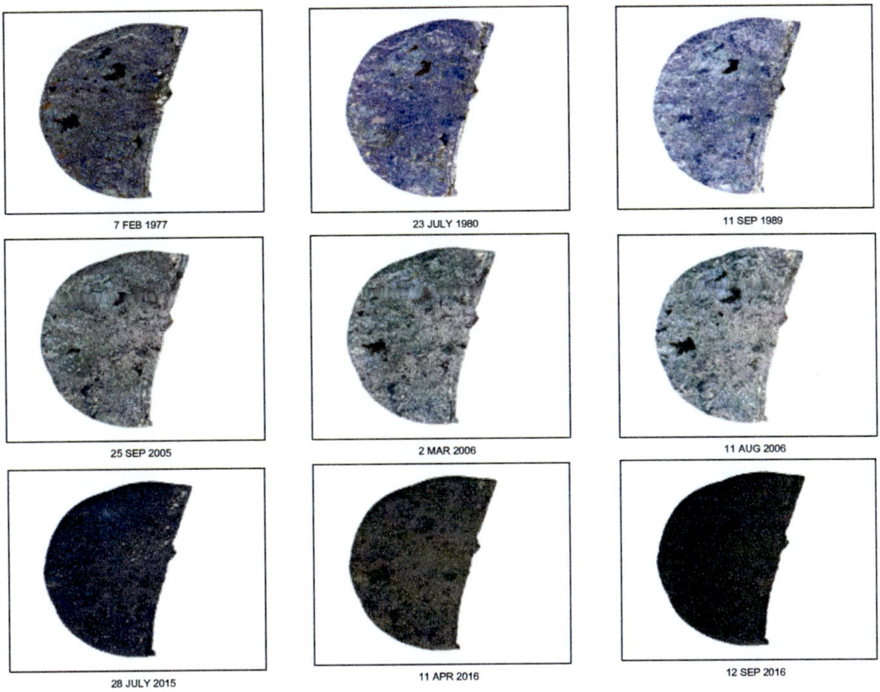

Fig. 3 Landsat images of Chennai region

change dependent on genuine condition is fundamental. Also, accomplish a legitimate outcome for the water bodies' data. Consequently, changing limits are required when various stages or diverse area of remote detecting information are utilized to distinguish water highlights data. Numerous strategies can be utilized for edge division of picture. The Otsu technique is a profitable edge strategy that used to water bodies depiction and change of water territory checking. It has been utilized to decide the different limits for isolating water highlights from the foundation include in the present investigation. The limit esteems we connected are as per the underneath criteria: Value of water pixel in normalized difference water index (NDWI) is $> = 0$ up to 1.

- Value of water pixel in normalized difference vegetation index (NDVI) is $< = 0$ or from -1 to $+1$.
- In near-infrared (NIR) band, value of water is zero (Table 2).

6 Result and Discussion

NDWI models were used to extract information of water bodies from the MSS, TM, and OLI/TIRS Landsat data. The results showing information of the water bodies'

Table 2 Threshold used for to reclassify NDVI, NDWI, and IR images

Image Date	NDWI (G)	NDVI (L)	IR(Pixel stretched value) (L)
Landsat 8			
July 28, 2015	0.05	0	7800
April 11, 2016	0.05	0	9500
September 12, 2016	0	0	700
Landsat 5			
September 11, 1989	0.15	−0.12	50
September 25, 2005	0.05	0	40
March 2, 2006	0.12	0	30
August 11, 2006	0.05	0	30
Landsat 2 and 3			
February 7, 1977(L2)	0.2	−0.05	20
July 23, 1980(L3)	0.2	0	25

L—Values lesser than, G—Values greater than

detection from images were compared with the other results of different years to extract the information about water bodies' change in the area. NDVI is extracted and then both indices and NIR data are used to water information extraction. Following images are given and showing the NDVI, NDWI, and NIR band images of the study area and showing the water body present in the area, as the resulting images are classified in only two classes so this becomes easy to find visually that which are the different water bodies present in there.

By applying threshold according to the above criteria, we concluded water body present in the area [18]. Now we merge the images we get by using raster calculator so that we can get the more accurate data for the water body, for that we merge the NDWI, NDVI, and NIR band image which we reclassified already, and the resulting image is a combination of the data of water body by different methods [19, 20]. The classes it divided at the end are shown below with pixel values where the class divided for water is '3' as seen in figure below, and it is then further calculated accurately for different aspects for further discussion (Figs. 4, 5, 6, 7, 8, 9, 10, 11; Tables 3 and 4).

0—Not classified as water by any method; **1**—Classified as water by one method; **2**—Classified as water by two methods; **3**—Classified as water by three methods (Fig. 12).

The area above is calculated using the no. of water pixels as follows:

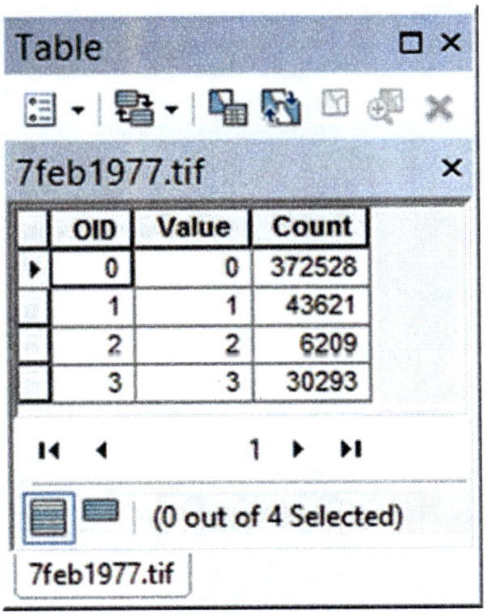

OID	Value	Count
0	0	372528
1	1	43621
2	2	6209
3	3	30293

Fig. 4 Attribute table of landsat image showing no. of pixels for different classes

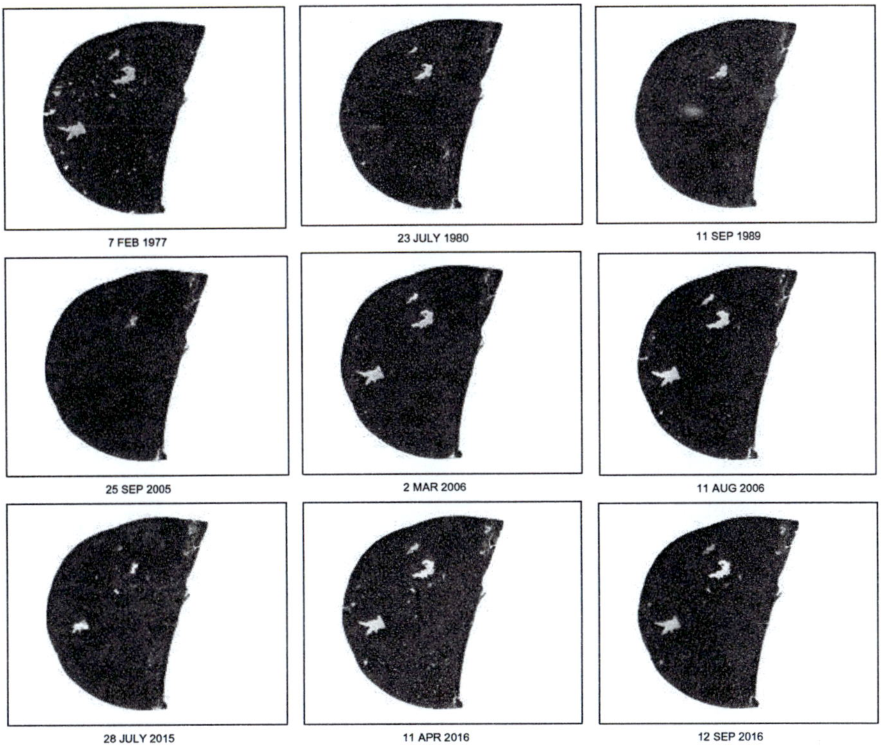

Fig. 5 Normalized difference water index (NDWI) images of Chennai region

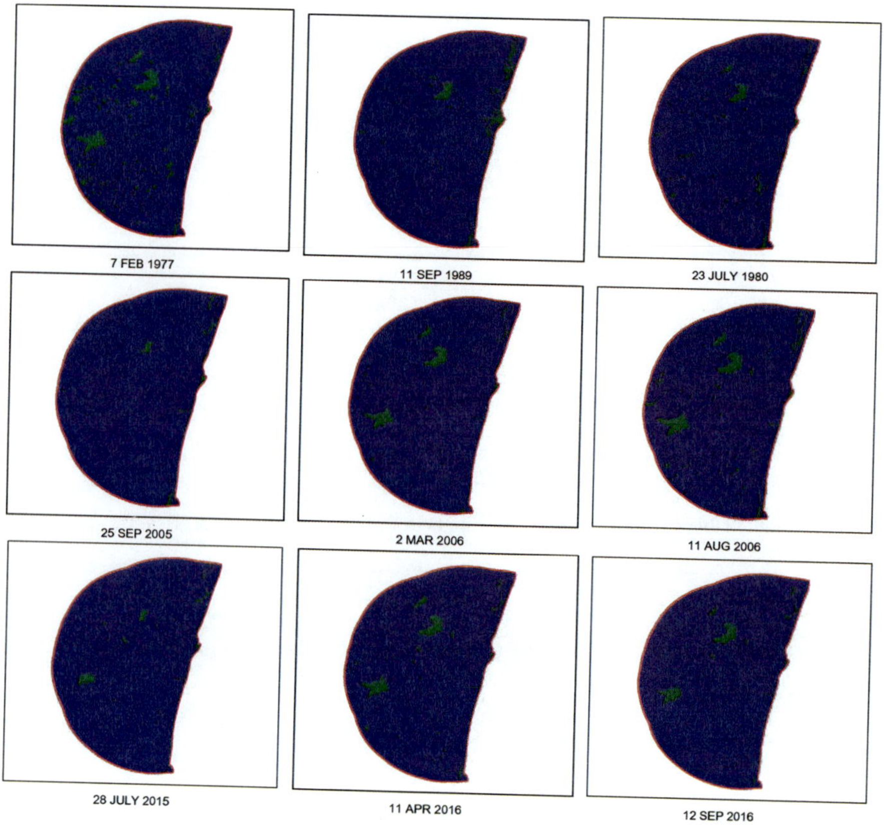

Fig. 6 Classified NDWI images of Chennai region

$$\text{Area} = (\text{Pixelisizei} \times \text{No . of water Pixels})$$

We could not identify a clear trend either increasing or decreasing (may be because of different threshold and different sensor imagery).

- Area of Chambarambakkam lake reduced from 80 to 2005 (three images), but after 2006 it is of approximately same area as of during 1977.
- India's coastline near Chennai is almost constant since 1977. (No receding or advancing of sea water could be observed).

Major water bodies are in and around Chennai Puzhal Lake and Chembarambakkam Lake, Alambadhi Lake, Cholavaram tank, Kusasthalai River and Palar River, and the Great Salt Lake.

The area of water bodies highly decreases from 109.0548 to 28.5948 km^2 in 1977–1989. Water bodies area increased in 2005, 2006, and 2007 was due to tsunami in December 2004 from 28.8045–72.7623 km^2. In the late 2015, heavy rainfall in

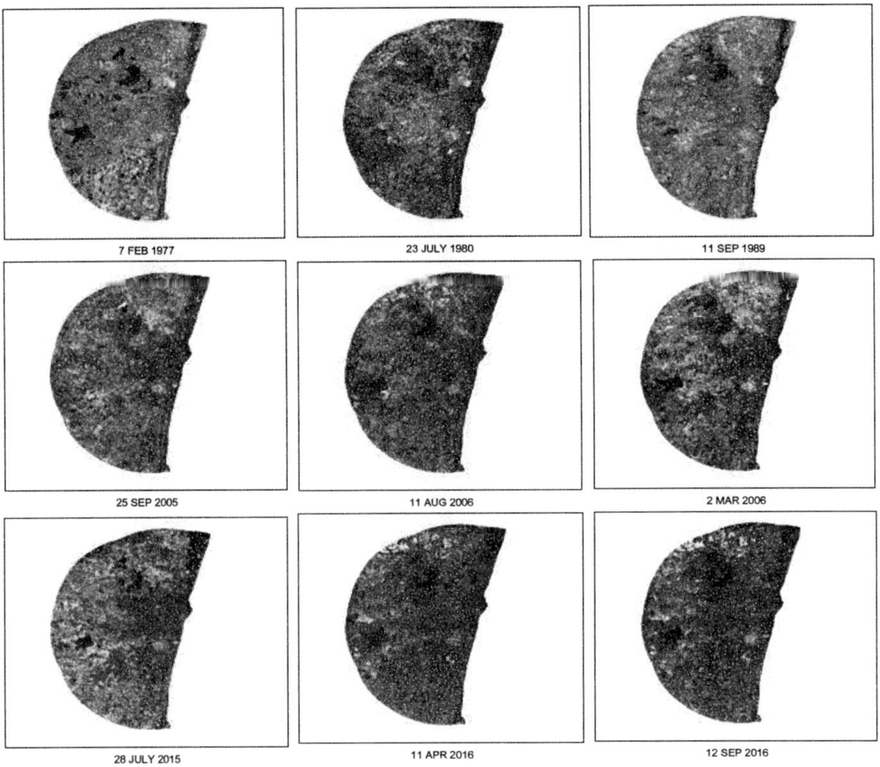

Fig. 7 Normalized difference vegetation index (NDVI) images of Chennai region

Chennai and nearby area resulted in little increasing water bodies' area in 2016 compared to July 2015.

7 Conclusion

We determined nine water data extricating information (NDWI, NDVI, and NIR pictures) in light of MSS, TM, and OLI/TIRS fleeting Landsat information of the Chennai and adjacent from 1977 to 2016. The outcomes present that 1. Area of surface water exceptionally declines from 1977 to 1989 and change of water region was, for the most part, influenced by human abuse; 2. Chennai surface water region had not moderately transformed from 1989 to 2005; 3. The surface water zone of Chennai had critical changes from 2005 to 2006, for the most part in the upstream and downstream; and 4. The principle explanations behind the expansion of the zone of surface water because of substantial flood in Chennai and close-by region in December 2015. The outcomes additionally connote that NDWI, NDVI, and NIR can separate territory of

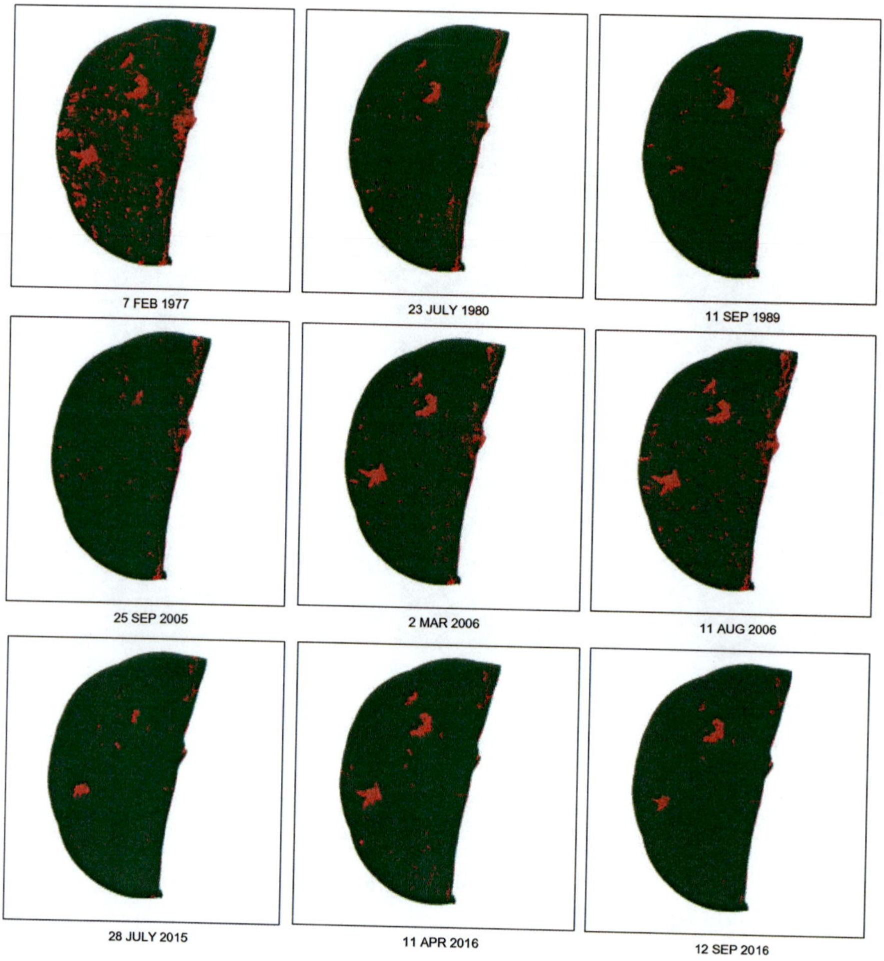

Fig. 8 Classified NDVI images of Chennai region

surface water from Landsat information and used to screen changes of dynamic of surface water and flood fiasco distinguish. At that point, diverse flood influenced zone can be marked to direct appraisal of flood hazard and harm with urban information, populace information, crop information, and other information. Engineered opening radar (SAR) used to beat little lakes and streams detection. Small lakes may assume a significant job in biogeochemical cycles, and redressed identification of those little water bodies ought to be processed. The investigation of Chennai was centered around utilizing less number of information assets and gadgets, so as to recognize the water bodies surface change. The informational indexes were prepared utilizing ERDAS IMAGINE 2014 and ArcGIS 10.2. The investigation featured the diverse water bodies present in Chennai and close-by and how they got influenced or changed when and by various elements including precipitation, vegetation, urbanization, and

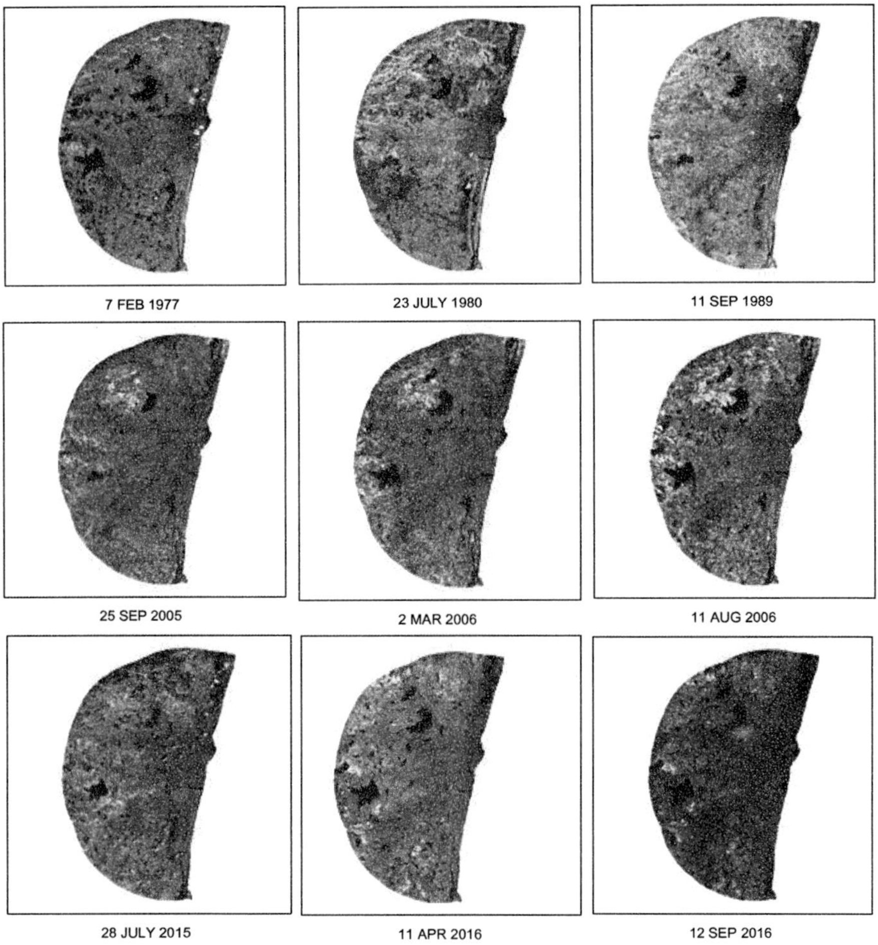

Fig. 9 Near-infrared (NIR) images of Chennai region

different elements. Portrayal, watching, and using water assets have dependably been a noteworthy worry so as to improve human foundation, from early years until today. The examination is a piece of studies gone for Chennai.

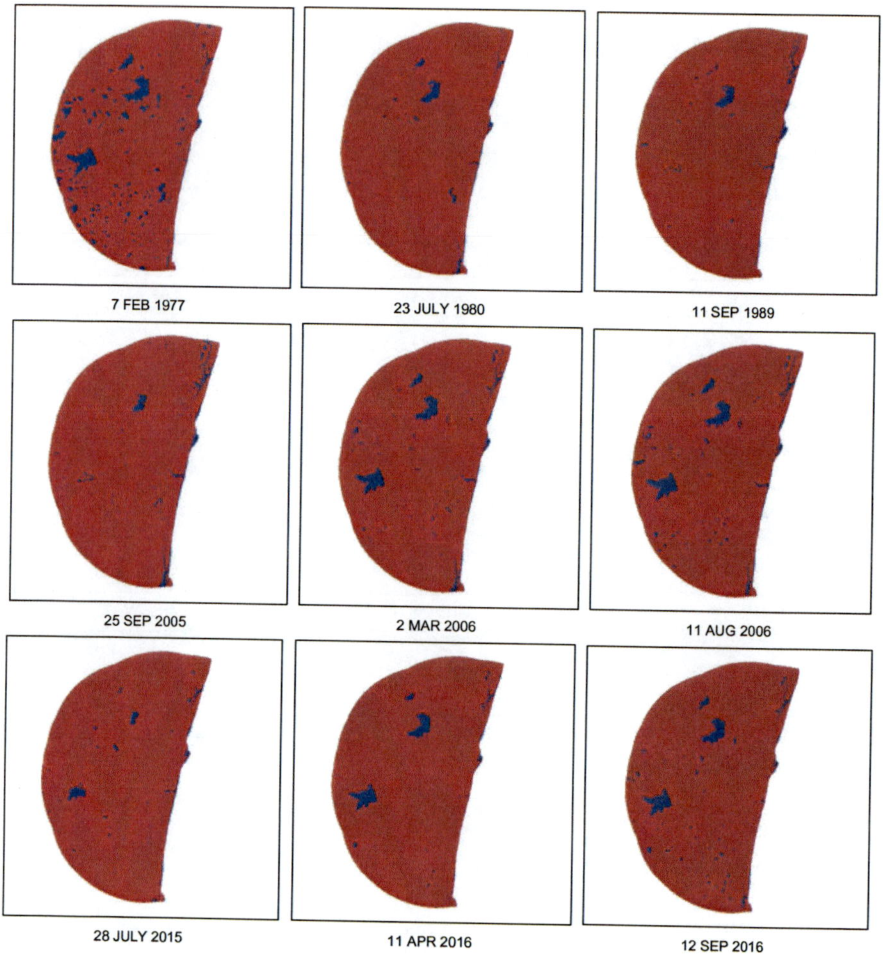

Fig. 10 Classified NIR images of Chennai region

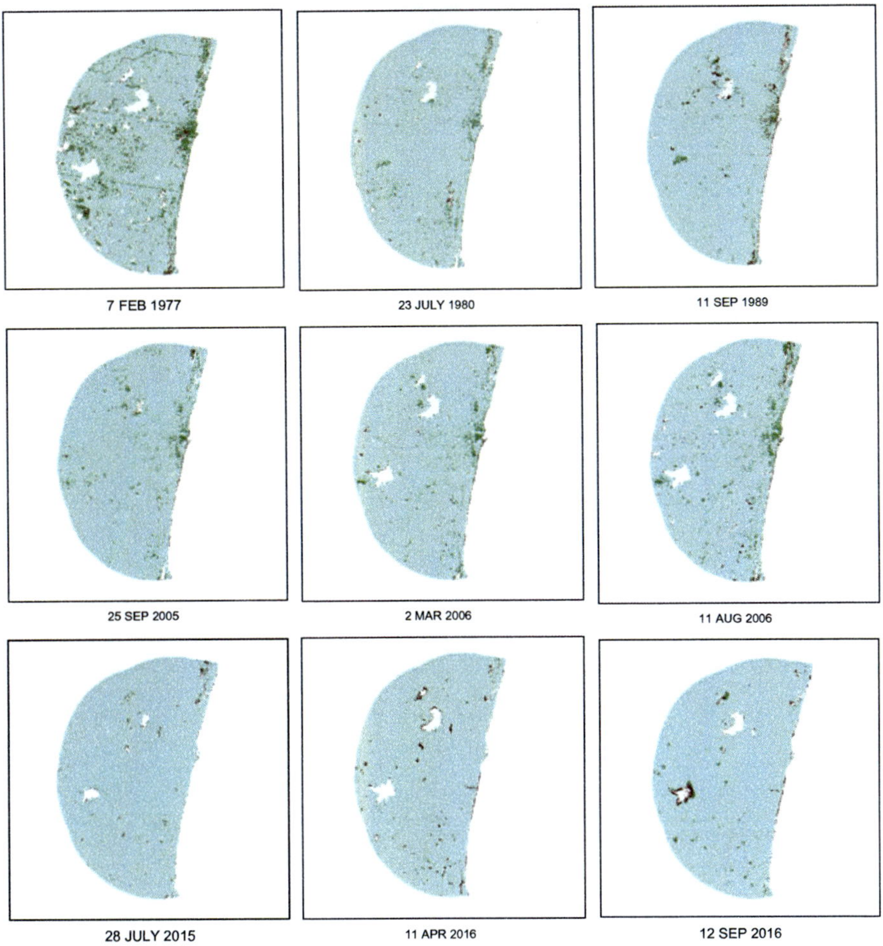

Fig. 11 Images after merging classified NDVI, NDWI, and NIR using raster calculator in ArcGIS

Table 3 No. of pixels of different classes

Image date	No of pixels of different classes			
	0	1	2	3
Landsat 8				
July 28, 2015	1,767,042	4941	10,679	27,922
April 11, 2016	1,733,664	7982	21,953	46,985
September 12, 2016	1,737,702	17,776	18,448	36,658
Landsat 5				
September 11, 1989	426,401	9914	8393	7943
September 25, 2005	1,709,435	61,101	8043	32,005
March 2, 2006	1,670,170	69,771	9335	61,308
August 11, 2006	1,624,073	83,000	22,664	80,847
Landsat 2 and 3				
February 7, 1977 (L2)	372,528	43,621	6209	30,293
July 23, 1980 (L3)	422,923	14,159	3174	12,395

Table 4 Area classified as water body

Image date	No. of water pixels	Pixel size (m)	Area (km^2)
Landsat 8			
July 28, 2015	27,922	30 × 30	25.1298
April 11, 2016	46,985	30 × 30	42.2865
September 12, 2016	36,658	30 × 30	32.9922
Landsat 5			
September 11, 1989	7943	60 × 60	28.5948
September 25, 2005	32,005	30 × 30	28.8045
March 2, 2006	61,308	30 × 30	55.1772
August 11, 2006	80,847	30 × 30	72.7623
Landsat 2 and 3			
February 7, 1977 (L2)	30,293	60 × 60	109.0548
July 23, 1980 (L3)	12,395	60 × 60	44.6220

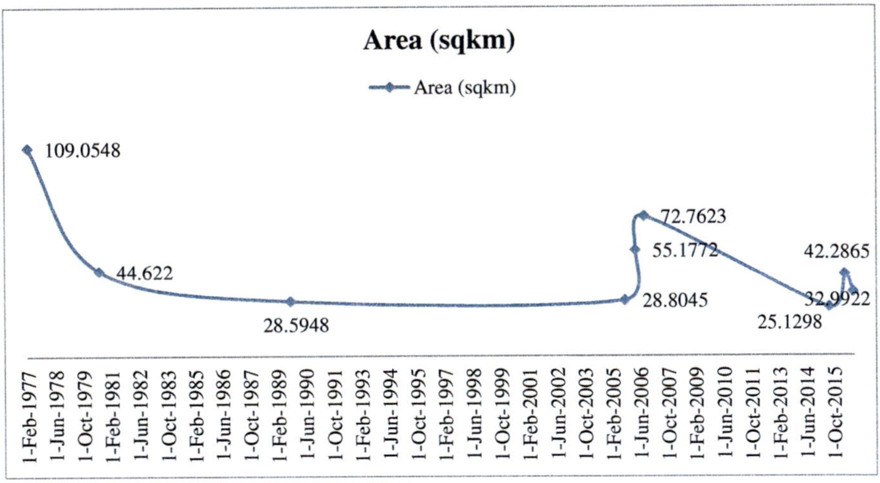

Fig. 12 Time series of water pixel area in Chennai region

References

1. Du Z, Linghu B, Ling F, Li W, Tian W, Wang H, Gui Y, Sun B, Zhang X (2012) Estimating surface water area changes using time-series landsat data in the qingjiang river basin, China. J Appl Remote Sens
2. Downing JA, Cole JJ, Duarte CA, Middelburg JJ, Melack JM, Prairie YT, Tranvik LJ (2012) Global abundance and size distribution of streams and rivers. Inland Waters 2(4):229–236
3. Feng M, Sexton JO, Channan S, Townshend JR (2015) A global, high-resolution(30 m) inland water body dataset for 2000: first results of a topographic-spectral classification algorithm. Int J Digit Earth
4. Roknia K, Ahmad AS, Hazine S (2014), Water feature extraction and change detection using multitemporal landsat imagery. Remote Sens 6:4173–4189
5. Lee KH, Lunetta RS (1995) Wetland detection methods, wetland and environment applications of GIS U.G. In: Lyon McCarthy J (eds), Lewis, Boca Raton, Florida, pp 249–284
6. Http://landsat.usgs.gov/
7. Lu S, Wu B, Yan N, Wang H (2011) Water body mapping method with HJ-1A/B satellite imagery. Int J Appl Earth Obs Geoinf 13:428
8. Li P, Jiang L, Feng Z (2014) Cross-comparison of vegetation indices derived from landsat enhanced thematic mapper plus-7 (ETM+) and landsat-8 operational land imager (OLI) sensors. Remote Sens 6, 310–329
9. Manavalan P, Sathyanath P, Rajegowda GL (1993) Digital image analysis techniques to estimate waterspread for capacity evaluations of reservoirs. Photogram Eng Remote Sens 59(9):1389–1395
10. Marangoz AMA, Karakış, His ORUC, M, McFeeters, SK (1996) The use of the normalized difference water index (NDWI) in the delineation of open water features. Int J Remote Sens 17:1425–1432
11. McFeeters SK (1996) The use of normalized difference water index (NDWI) in the delineation of open water features. Int J Remote Sens 17(7):1425–1432
12. NASA/NGA (2003) SRTM water body data product specific guidance, version 2.0. Available online
13. Morse A, Zarriello TJ, Kramber WJ (1990) Using remote sensing and Gis technology to help adjudicate Idaho water rights. Photogram Eng Remote Sens 56(3):365–370

14. Rouse JW, Haas RH, Schell JA, Deering DW (1973) Vegetation monitoring systems in the great plains with ERTS (earth resources technology satellite), Greenbelt, ON, Canada, 10–14; vol SP-351, pp 309-317
15. Oki T, Kanae S (2006) Global hydrological cycles and world water resources. Science 313:1068–1072
16. Palmer SCJ, Kutser T, Hunter PD (2015) Remote sensing of inland waters: challenges, progress and future directions. Remote Sens Environ 157:1–8
17. Ridd MK, Liu J (1998) A comparison of four algorithms for change detection in urban environment year. Remote Sens Environ 63:95–100
18. Shen L, Li C (2010) Water body extraction from landsat ETM+ imagery using adaboost algorithm. In: Conference on Geoinformatics, Beijing, China; pp 1–4
19. Sun F, Sun W, Chen J, Gong P (2012) Comparison and improvement of methods for identifying Water bodies in remotely sensed imagery. Int J Remote Sens 33:6854–6875
20. Xu H (2006) Modification of normalized difference water index (NDWI) to enhance features in remotely sensed open water imagery. Int J Remote Sens 27:3025–3030

Investigation of Hydraulic Jump Over Rough Sloping Floor in Prismatic Rectangular Channel—An Experimental Study

Murari Kumar, Nitin Mishra, and Sanjeev Kumar

Abstract Hydraulic jump is the one, which has widely been studied in hydraulic engineering. It is the most regular encountered cases of rapidly varied flow in hydraulic engineering. Very limited studies have been reported in literature on the performance of the behavior of hydraulic jump, only for smooth and rough horizontal channel beds. This study attempts to investigate the characteristics of hydraulic jumps formed on rough sloping channel beds under different flow conditions using laboratory experiments. A series of experiments were conducted in a rectangular flume having artificially roughened beds which is formed by placing concrete with different roughness and tilting flume slope. This study concerns with the investigations of effects on slopes of channel and channel bed roughness height at hydraulic characteristics. To achieve this, the bed of flume was provided the roughness with the help of different size of gravels ($d_{50} = 1$, 1.2, 1.5 and 1.7 cm) and three different channels slopes (0.0115, 0.0146, 0.0174) were used. In all 48 different runs were conducted with varying discharge and approach supercritical flow depth. The available empirical relationships for estimating the sequent depth ratio were first checked for their accuracy for new data. As the existing relationships were inadequate, new relationship are proposed, herein for parameter of the hydraulic jump. Comparison of observed values of sequent depth ratio with calculating values coming from respective proposed relationship showed satisfactory agreement.

Keywords Bed roughness height · Sequent depth ratio · Length of roller · Energy loss

M. Kumar
Department of Agriculture, Bhagalpur, Bihar, India

N. Mishra · S. Kumar (✉)
Department of Civil Engineering, Graphic Era University, Dehradun, Uttarakhand 248001, India
e-mail: er.sanjeevkr@gmail.com

L. M. Gupta et al. (eds.), *Advances in Civil Engineering and Infrastructural Development*, Lecture Notes in Civil Engineering 87,
https://doi.org/10.1007/978-981-15-6463-5_21

1 Introduction

Hydraulic jump is the process of transition of flow from supercritical to subcritical in which large scale turbulence is generated with consequent dissipation of energy [1]. The hydraulic jump is used as energy dissipater in hydraulic structures. It is one of the most frequently encountered cases of rapidly varied flow which has extensively been studied in the field of hydraulic engineering. The supercritical flow jumps up to meet its alternate depth. There is a sudden rise in the water surface and considerable loss of energy in a hydraulic jump. The energy is dissipated into heat through these eddies [1]. Schematic three-dimension view of a typical hydraulic jump in sloping channel has been shown in Fig. 1.

For channels of large slope, the weight effect of water in the jump may become so pronounced that it must be included in the analysis. The momentum formulas for jumps on horizontal floors cannot be applied straight forwardly to jumps on sloping floor. Thus, the hydraulic jump on sloping channels has widely attracted attention of designers [2], hydraulic jump occurs in channels with large bed slope is also affected by the gravitational force acting on the flow. The major attributes associated with obtaining a useful solution to the problems related to the hydraulic jump on sloping floor are [3].

1. The term $W \sin \theta$ is not defined, because the length and shape of the jump are not well defined. Here, W is the weight of the water element in the section and θ is the slope in degrees.
2. The specific weight of the fluid in the control volume changes significantly owing to air entrainment. The pressure term cannot be accurately quantified.

Whenever a hydraulic jump occurs in a channel there will be heavy amount of turbulence and considerable energy loss, thus, momentum formulas for jump on horizontal floor cannot be applied straight forwardly to jumps on sloping channel. This momentum formula needs to be derived from the second law of Newton [2, 4].

Figure 2 shows the jump occurring on a positive slope in a rectangular channel of unit width. Neglecting boundary friction and assuming hydrostatic distribution of pressure, the momentum equation between Sects. 1 and 2 can be written as [1]:

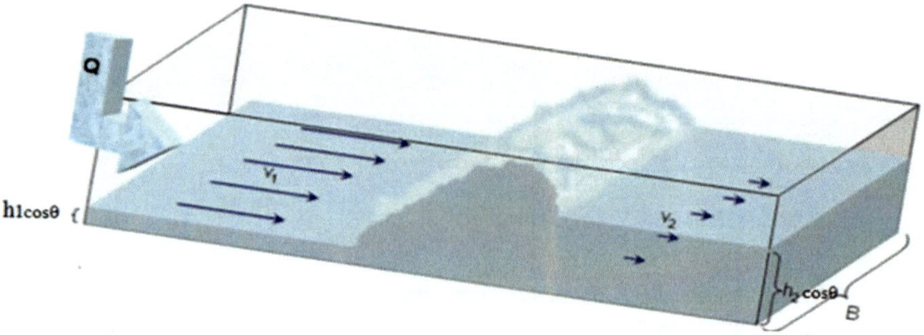

Fig. 1 Three-dimensional view of hydraulic jump on sloping floors (Einstein.atom.colostate.edu)

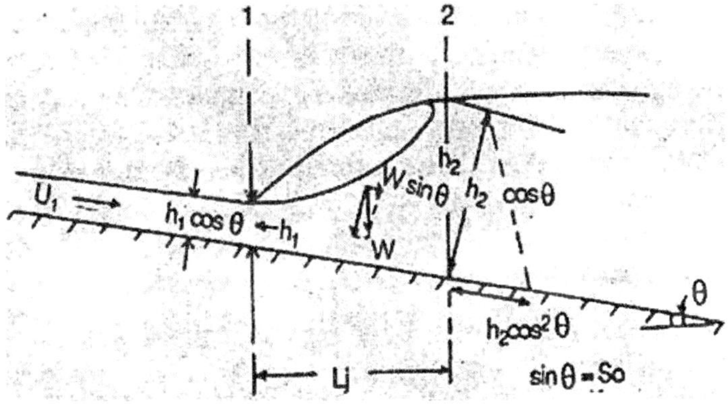

Fig. 2 Hydraulic jump on a sloping channel [1]

$$(\rho g h_1^2 \cos^3\theta)/2 - (\rho g h_2^2 \cos^3\theta)/2 + W \sin\theta = \rho g(U_2 - U_1) \tag{1}$$

where

q = discharge per unit width
h_1, h_2 = depths of water
θ = angle of bed with the horizontal.

Here,

$$U_1 = q/h_1 \cos\theta$$
$$U_2 = q/h_2 \cos\theta$$

If the weight of water is assumed equal to

$$W = K\rho g L_j \left(\frac{h_1 + h_2}{2}\right) \cos\theta$$

Equation (1) may be simplified to

$$\left(\frac{h_2}{h_1}\right)^3 - \left(2G_1^2 + 1\right)\left(\frac{h_2}{h_1}\right) + 2G_1^2 = 0 \tag{2}$$

Here,

$$G_1^2 = K_1^2 F_1^2 \tag{3}$$

$$F_1 = U_1/\sqrt{g h_1 \cos\theta} \tag{4}$$

Here, U_1 = average velocity at the section where the jump starts.

Provided the following relationship for K_1 [5]

$$K_1 = 10^{0.027\theta}$$

where θ is in degrees.

The empirical equation shown above proposed by [5] is used in channels to find G_1 in Eq. (5) and hence the conjugate depth ratio or solution of Eq. (2) is [6] (Figs. 3 and 4)

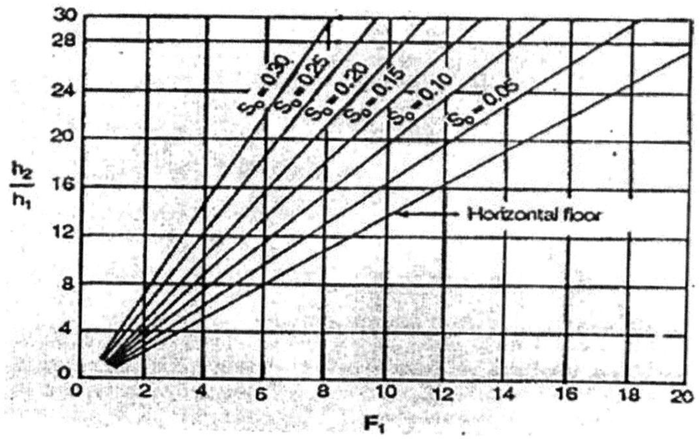

Fig. 3 Conjugate depth relation for jump on sloping floors [1]

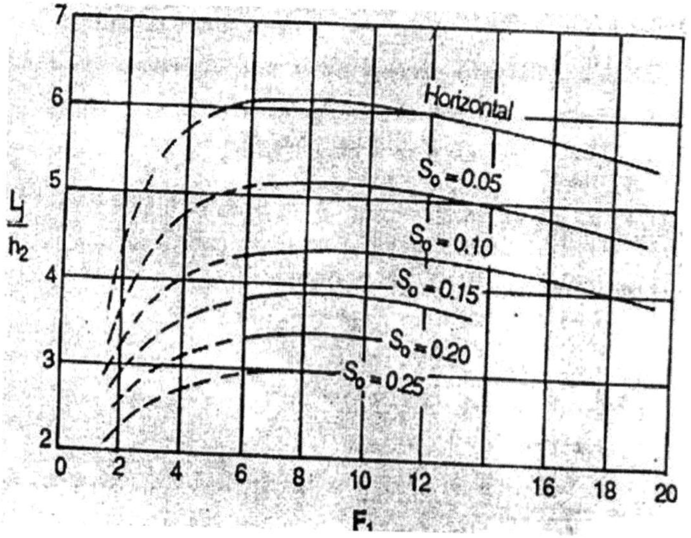

Fig. 4 Length of hydraulic jump on sloping floors [1]

$$h_2/h_1 = 0.5\left(\sqrt{1 + 8G_1^2}\right) \tag{5}$$

The main objective of this study is to investigate the hydraulic jump on slopping floors with different roughness. To achieve this objective, the experiment was carried out for both smooth and rough sloping beds. Five rough beds made up of closely packed crushed gravel particles cemented to the bottom of bed. The grain size distributions were characterized by $d_{50} = 0.002$, 0.00398, 0.0056, 0.007, and 0.011 m, d_{50} being the diameter of the bed particles for which 50% are finer.

The other objective is to investigate the relationship for hydraulic jump on rough slopping floors [7–9]

$$G_1^2 = K_m^2 F_1^2 \tag{6}$$

where K_m is the modified value of K_1.

2 Experiments

For the experiments rectangular flume having a length of 8.02 m and width of 0.6 m. Sidewalls of the flume are made of glass for facilitating easy visual access. There was a constant overhead tank used to provide discharge in the flume. The upstream supercritical flow depth was controlled by the sluice gate while the tail water depth was controlled by downstream sluice gate and a rectangular weir for the measurement of the discharge at any point of time through the flume. For preparation of rough slopping bed used different size of gravels as ($d_{50} = 0.010$, 0.012, 0.015, 0.017 m). To measure the discharge in the flume, a sharp-crested weir is used at the downstream of the flume. Discharge through the weir can be measured with Eq. (7).

$$Q = \left(\frac{2}{3}\right) C_d b \sqrt{2g} (H)^{1.5} \tag{7}$$

where

Q = discharge
C_d = coefficient of discharge = 0.611 + 0.075 (H/Z)
b = width of the weir
g = acceleration due to gravity
H = height of the water above the weir section
Z = height of the sharp crest weir.

For slope measurement with the help of water level in container and distance between container.

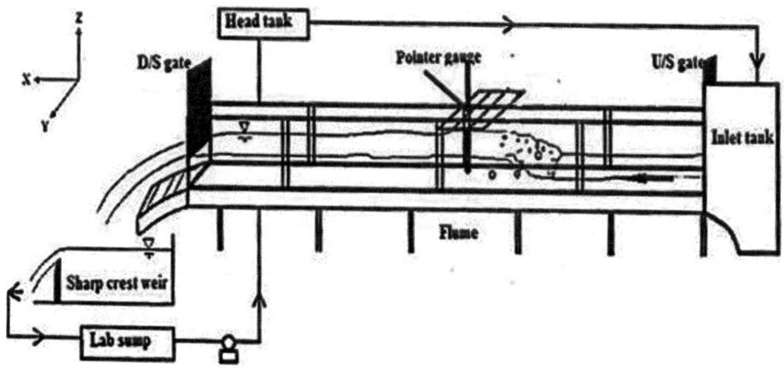

Fig. 5 Experimental set up

$$\text{Slope} = \left(\frac{h_2 - h_1}{L}\right) \tag{8}$$

h_2, h_1 = water level in the container
L = distance between the container.

2.1 Experimental Data Collection

Experiments on hydraulic jump have been carried out for various slopes and different roughness of the channel and at several values of approach flow Froude number F_1 in the hydraulics engineering laboratory. Using Manning's equation obtained value of roughness's coefficient was 0.012 (Fig. 5).

Computation and variation of "K_1"
The Ballenger's equation for the sequent depth ratio for a sloping floor as

$$h_2/h_1 = 0.5(\sqrt{1 + 8G_1^2} - 1) \tag{9}$$

Here,

$$G_1^2 = K_1^2 F_1^2 \tag{10}$$

and

$$K_1 = 10^{0.027\theta} \tag{11}$$

where θ is in degrees.

Here, K_1 is the parameter accounting for the effect of slope on conjugate depth relationship [6, 10]. Numerous run have been taken with varying slopes. The pre-jump, post jump data and roller length were collected corresponding to channel bed

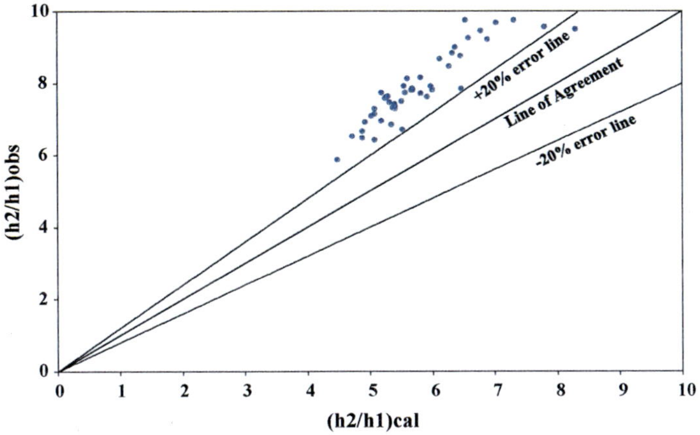

Fig. 6 Comparison of observed values of sequent depth ratio with calculated value of sequent depth ratio

slope, collection of data was carried out at slope 0.0115, 0.0146, and 0.0174. The observed value of sequent depth ratio obtained in the laboratory and then calculated with the help of Eq. (9) [6] (Fig. 6).

It is clearly seen that all the data points lie in one side of line of agreement so this comparison is not good. Based on this analysis we can say there is a need to improve this analysis. In the new relationship parameter K_1 has been replaced by a modified factor K_m. By substituting the observed values of sequent depth ratio in Eq. (9) we obtain value of G_1, and we can calculate the values of K_m by substituting observed values of G_1 and F_1 in Eq. (9) With these all values of K_m a new relationship has been formulated between K_m and bed angle (Fig. 7)

$$K_m = 1.2892\mathrm{e}^{0.04360\theta} \tag{12}$$

where θ = channel bed slope (degree).

By using Eqs. (9) and (10) we have calculated the new value of sequent depth ratio and compare these values with observed value of sequent depth ratio and it can clearly be seen from Fig. 8 that all data points lie to the line of agreement with a maximum error of ±15%.

Variation of energy loss with G_1

Energy loss for a hydraulic jump is given as [1]

$$\Delta E = h_1 \cos\theta + \frac{U_1^2}{2g} + L_j \tan\theta - h_2 \cos\theta - \frac{U_2^2}{2g} \tag{13}$$

Relative energy loss for a hydraulic jump [1]

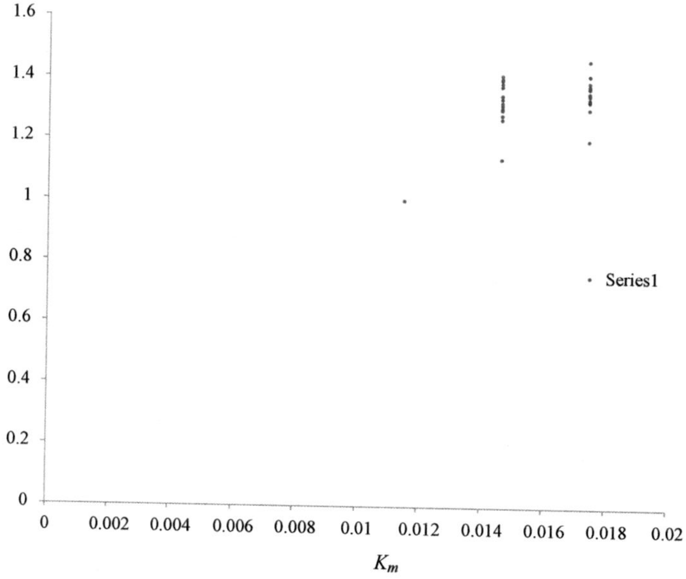

Fig. 7 Relation between K_m and channel slope (θ)

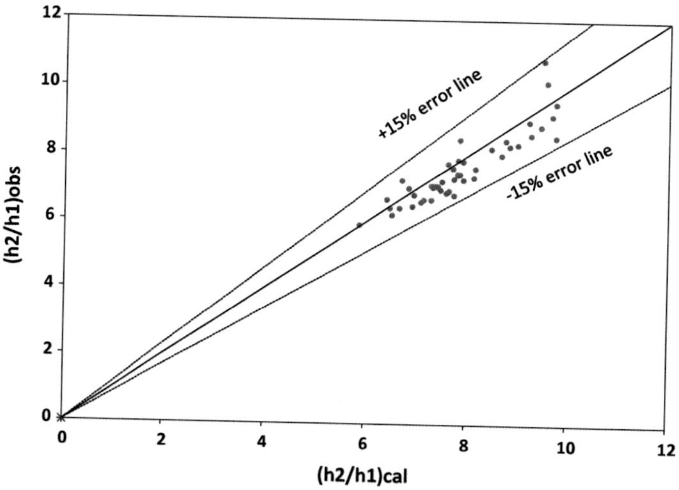

Fig. 8 Comparison between observed values of sequent depth ratio to calculated values of sequent depth ratio

$$\frac{\Delta E}{E_1} = \frac{1 - \frac{h_2}{h_1} + \frac{U_1^2 - U_2^2}{2gh_1 \cos_\theta} - \frac{L_r \tan\theta}{h_1}}{1 + \frac{U_1^2}{2gh_1 \cos\theta}} \tag{14}$$

A new relationship has been obtained between relative energy loss and G_1 [11].

$$\frac{\Delta E}{E_1} = 0.0016 G_1^{2.7631} \tag{15}$$

By using Eq. (15) we calculate the values of $(\Delta E/E_1)$ and compare these values with observed values of relative energy loss. From Fig. 9, it is clearly seen that all data points within the ±25% error (Fig. 10).

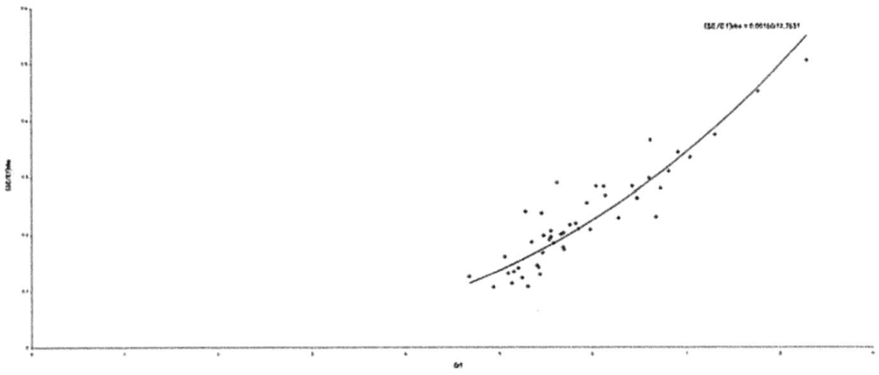

Fig. 9 Relation between relative energy loss and G_1

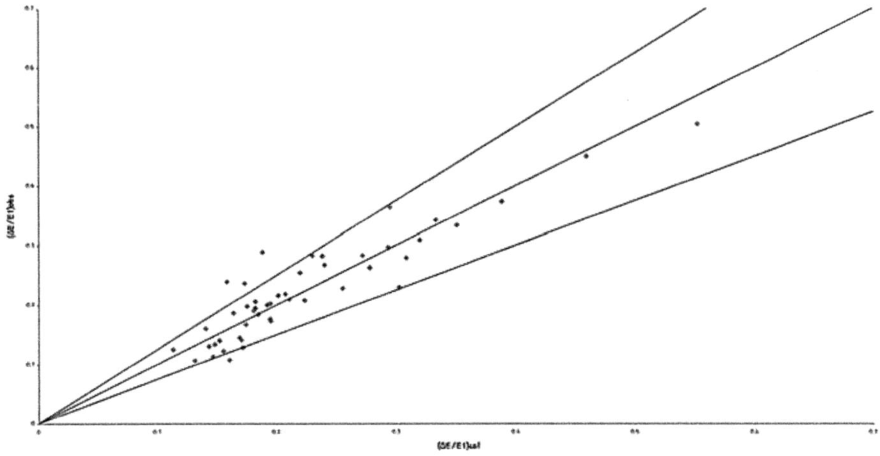

Fig. 10 Comparison between $(\Delta E/E_1)_{obs}$ and $(\Delta E/E_1)_{cal}$

3 Conclusion

A comparison of the values of the length of jump obtained in case of smooth bed relative to rough bed showed that the effect of roughness can be presented as the percentage length reduction. For this purpose, three slopes equal to 0.0115, 0.0146, 0.0174 have been taken and four different bed roughness ($k_s = 0.010, 0.012, 0.015, 0.017$ m) have been used. The different 48 runs were conducted by varying channel discharge and approach flow, Froude number. The existence relationship on sequent depth ratio for hydraulic jump on rough sloping floors is not satisfactory so new relationship has been derived.

The following broad conclusions that can be drawn from present study:

(i) A new parameter "K_m" is introduced as per Eqs. (9) and (10) for relating thru parameter F_1 and G_1
(ii) The modified parameter "K_m" is dependent upon slope of the channel ant the relationship between these variable was found to be

$$K_m = 1.2892e^{0.0436\theta}$$

where θ = slope of channel (degree).
(iii) Relative energy loss ($\Delta E/E_1$) is a function of G_1, which is a function of K_m so relative energy loss is also a function of K_m

$$\frac{\Delta E}{E_1} = 0.0016 G_1^{2.7631}$$

With this relationship we calculate the values of relative energy loss and compare it with observed values which show all the points lie within ±25% error.

References

1. Ranga Raju KG (1993). Flow through open channels, 2nd edn. McGraw Hill, New Delhi
2. Chow VT (1959) Open channel hydraulics. McGraw-Hill Book Co., Ltd., London
3. Gupta M (2010) Characteristics of hydraulic jump on a sloping floor. M.Tech thesis, Dept. of Civil Engg., IIT Roorkee
4. Bakhmeteff BA, Matzke AE (1938) The hydraulic jump in a sloping channels. Trans Am Soc Civ Eng 60(2):111–118
5. Rajaratnam N (1967) Hydraulic jumps. In: Chow VT (ed) Advance in hydro science, vol 4. Academic Press, New York, NY, pp 197–280
6. Beirami MK, Chamani MR (2006) Hydraulic jump in a sloping channels: sequent depth ratio. J Hydraulic Eng ASCE 132:1062–1067
7. Rajaratnam N (1965) The hydraulic jump as a wall jet. J Hydraulic Div ASCE Eng 91(5):107–132

8. Ali HS (1991) Effects of roughness-bed stilling basin on length of rectangular hydraulic jumps. J Hydraulic Eng ASCE 117(1):83–93
9. Carollo FG, Ferro V, Pampalone V (2007) Hydraulic jumps on rough beds. J Hydraulic Eng 133(9):989–999
10. Carollo FG, Ferro V, Pampalone V (2011) Sequent depth ratio of B-jump. J Hydraulic Eng ASCE 137:651–658
11. Carollo FG, Ferro V, Pampalone V (2009) New solution of classification hydraulic jump. J Hydraulic Eng 135(6):521–531

Groundwater Storage Analysis in Changing Land Use/Land Cover for Haridwar Districts of Upper Ganga Canal Command (1972–2011)

Nitin Mishra and Amit Kumar Sharma

Abstract Over the last century, a key change in land use has happened locally, provincially and all around. These will prolong in the future also. The affect of built-up area on groundwater has a key worry over past many years, and in exacting, to those worried in groundwater amount and subjective examinations. Augmentation in impermeable region due to urbanization results in decreased infiltration and finally affecting the groundwater storage. So, land use change has to be calculated with spatial information technologies; especially, the application of remotely sensed data and geographical information systems (GIS) has been used. The present study investigates the urban growth of Haridwar district, Uttrakhand, using satellite data for the years 1972, 1980, 1992, 2002 and 2011. Unscrupulous population increase, rise in food, silage and energy demands in combination with industrial action have led to quick modification in land use patterns. In this study, GIS tool is used to classify different land use classes from remote sensing data and groundwater potential in Haridwar district, and supervised classification technique is applied to categorize urban LULC. The study of results shows the extreme increase of urban area and reduced green cover within the city boundary limit. Due to increase of settlement area will directly impact the decrease of groundwater level. Proper planning of land use and groundwater organization is an input to socioeconomic upliftment of a region.

Keywords Urbanization · Groundwater · LULC · Change detection · Remote sensing · GIS

1 Introduction

Water is necessary for sustaining life on earth. The demand of water is increasing year to year with the ever rising population. For masses population, water is the basic requirement for drinking and food production [5]. For agriculture sector, supply of

N. Mishra (✉) · A. K. Sharma
Department of Civil Engineering, Graphic Era Deemed to be University, Dehradun, Uttarakhand 248001, India
e-mail: nitinuag@gmail.com

L. M. Gupta et al. (eds.), *Advances in Civil Engineering and Infrastructural Development*, Lecture Notes in Civil Engineering 87,
https://doi.org/10.1007/978-981-15-6463-5_22

water for irrigation becomes not easy assignment, after having restricted source and irregular distribution sample in space and time of water (cgwb.gov.in). Depending on a single sources for any purpose is much unreliable, and the idea of using water for two or more alternative sources with economic condition is drawing the attention of planners, engineers and decision makers. The prospects of involving conjunctive use of surface and groundwater will be better, provided that one may understand the hydrological balance cycle [1, 4]. The various hydrological components are predictable on a seasonal basis for monsoon, pre-monsoon and post-monsoon seasons. By understanding and monitoring the water balance system, it will help in better conjunctive use (cgwb.gov.in).

Groundwater as the main basis of drinking water all around a globe acts as a very important role in altering the ecological value of numerous regions [9]. Nonetheless, the amount and nature of groundwater because of human action are changing [6] risking reasonableness of the groundwater as a basis of drinking water and disturbing the existing resources. Indicating effect on groundwater system and analyzing a level of alter later on are thus a main scientific task [19]. LULC change is an important human-induced activity changing the groundwater structure [3]. In any case, the effect of later on LULC changes on the groundwater system has not been calculated broadly. All over, total narration of mankind, extreme use of land assets by human have resulted in a considerable change of LULC [2]. From the period of industrialization and quick population increase, land use alter phenomena have rapidly increased in most of the area. The impact of LULC changes on local water equilibrium is the important vital research in global hydrological fields, and bunch of study indicates that major LULC changes are significant components bringing about the local climate and hydrological cycle adjustments [8, 20]. Thereby, IGBP IHDP WCRP and DIVERSITAS, etc., acquire the relationship between Biosphere Aspects of the Hydrological Cycle (BAHC) and the LULC changes, in addition to its climate frangibility, as the foundation plans [7, 10].

The amount and quality of groundwater are altering due to human activity which is one of the major scientific tasks. From the era of industrialization and quick increase in population, land use change phenomenon has strongly increased in most areas. Management strategies on the development of groundwater resource, while the growing demand for water by communities and industries for example Urban/Land use, Cropping Pattern, Industrialization etc, is increased, which is to meet with groundwater.

The impact of urbanization implicates in groundwater amount and nature studies. Augmentation in urbanization decreases infiltration. Land use planning and management approaches are input for growth. It has been cited that in most regions globally, water table is depleting at the rate of 1–2 m/year (thirdworldcentre.org). The amount and nature of groundwater are changing due to human interference. Since the era of industrial development and rapid increase in population, land use change phenomena have rapidly accelerated in many areas which directly impacting the hydrology of the catchment area.

The impact of LULC changes on the local water balance is a major vibrant study in the international hydrological fields. Such work marks that the local vegetation ecosystem alters are due to LULC changes extraordinarily affect the regional hydrological cycle. About 75–80% of human requirements like urban waste disposal into water bodies cause groundwater contamination [16, 18]. Groundwater pollution has become one of the most vital problems in India. In January, 1994 CPCB, Delhi carried out the first main groundwater quality monitoring exercise [4]. Many of the serious regions have to depend on groundwater resources for many needs due to lacking of surface water. In the built-up central parts of cities, the subsoil water by now has been adulterated by industrial wastes.

Furthermore, in LULC recognized by IGBP and IHDP, one core trouble is to recognize the affect of local LULC alters on hydrological method and water resources [15]. Such effort specifies that local vegetation ecosystem alters affected by LULC changes extraordinarily influence the local hydrological cycle [20]. Therefore, the mechanism of LULC changes in the basin affecting on hydrological practice turns out to be prior fields in the development of hydrology [7]. Therefore, actual assessment of affect on groundwater system by human behavior is serious to create sensible operation program of local groundwater resources [13, 14]. Past investigations of the effect on groundwater system by human actions mainly emphasize on the aspects of groundwater utilization intensity and reasonability while ignoring the affect of land use changes on the groundwater system in the catchment. Actually, as the important part of local hydrological cycle, groundwater system has burly response to LULC changes [1, 11].

About 75–80% requirements of water for human are fulfilled by groundwater. Uncontrolled discarding of waste into water bodies, open dump areas and poorly land filing design cause groundwater contamination [16, 18]. Groundwater pollution has turned out to be the major toxicological and environmental problem in India. In January, 1994 CPCB, Delhi has undertaken the first foremost groundwater quality monitoring exercise. In December 1995, the reports published recognized 22 locations in 16 states of India as "critical" sites of groundwater pollution and also originate industrial effluents to be the primary cause of groundwater pollution [4]. Many of the critical regions have to depend on groundwater resources for different needs due to scarcity of surface water. In the industrial and urban fringe zones of cities, the subsoil water in the area has already been contaminated by industrial wastes. Industry releases high concentrations of poisonous substance. Wells in many built-up areas are infected with nitrate and detergents. High fluoride content of groundwater has harmful effects and is suspected to be a severe health risk in the neighboring region.

2 Study Area

Haridwar covers approximately 2360 km^2 of area, lies in the west side of Uttarakhand, ranges from 29.58° north and 78.13° east longitude, respectively, and a mean sea level of 249.70 mts. District came into existence on December 28, 1988, former to its inclusion in the state of Uttarakhand, previously the district was part of Saharanpur Commissionary. The district shares its boundary with Saharanpur in the west,

Dehradun in north and east, Pauri Garhwal in east, Muzzaffarnagar and Bijnor in the south. The district is administratively subdivided into three tehsils, i.e., Haridwar, Roorkee and Laksar and six blocks, i.e., Bhagwanpur, Roorkee, Narsan, Bahadrabad, Laksar and Khanpur. Haridwar is the first cities in which Ganga emerges from the mountains to touch the planes. As in line with the 2001 census, 1,444,213 is the populace of the district (Fig. 1).

The land use/cover map was generated by using satellite image of 1972, 1980, 1992, 2002 and 2011, through ERDAS IMAGINE (version 9.2), which is powerful

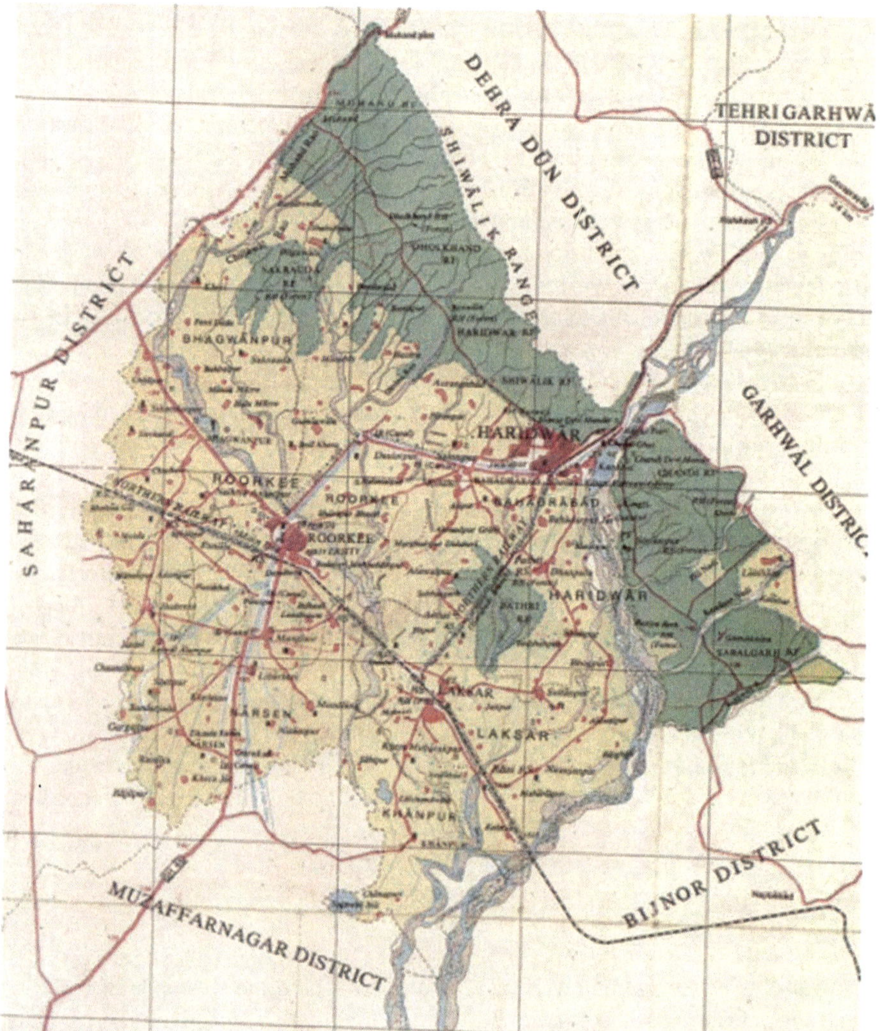

Fig. 1 Map of Haridwar district

image processing software. The reason of digital LULC classification is to have a significant information class value, which can evaluate the landscape in exact. Land cover refers to the cover containing greenery, built up, water, barren land, etc. Land cover is a basic variable that influences on and hyperlinks many parts of the human and physical environments. Land cover is extremely impacted by human made rather than natural events. Today, primarily agricultural growth, burning actions or fuel–wood utilization, deforestation, some construction works and urbanization originate land cover changes. Accordingly, such adjustments might also have terrific impacts on the catchment with the aid of various hydrological approaches which include infiltration, groundwater recharge and base drift and runoff. Land use trade is the principle situation for worldwide surroundings trade and is utilized by town and states planners to design eco-friendly and sustainable economic growth (Table 1).

3 Groundwater Analysis

Groundwater resources are facing increasing pressure as urbanization is increasing and climatic conditions are changing all over the world. The affect of LULC changes and climate changes are problematic to distinct as they partly result in similar changes in the ecosystems. These cause major risks to our most valued water resource and on ecosystems (nks-umwelt.pt-dlr.de).

Demand for groundwater is increasing, while the water resources remain almost constant. Therefore, continuous withdrawals of groundwater may result in depletion of water table. It is necessary to maintain the water level fluctuations ought to be stored inside a specific variety over the monsoon and non-monsoon seasons (www.sys.virginia.edu). Inspite of the favorable average accessibility of groundwater, while some areas in the country having lack of water. From time immemorial, groundwater has played a vital role in growing the food production in India [5, 17].

For this, 12 wells are marked in the area, and the data was acquired by Saharanpur Divisions of UP State Groundwater Department. Storage per year was estimated by using data. Both the techniques, i.e., recharge due to rainfall using rainfall infiltration factor and groundwater storage equation are utilized in this study to find out storage of Haridwar district. The recommended significance for specific yield is 16% by GEC1997. The soil properties are assumed same for whole Haridwar district for this study and in the absence of any pumping test. Study area is having area of 1928 km^2, a specific yield value of 0.16 (recommended values Geological Estimation Committee, 1997), considering the sandy loam soil alluvium.

Storage Equation

The estimation of groundwater storage is done by applying the storage equation. The specific yield values considered in the computations are to be taken preferably from field tests, in the lack of which, the encouraged values of particular yield are 16% to be taken into consideration. [12] and study area of different districts. Pre-

Table 1 LULC for Haridwar district

Class	1972		1980		1992		2002		2011	
	Area (km^2)	Area (%)	Area (km^2)	Area (%)	Area (km^2)	Area (%)	Area (km^2)	Area (%)	Area (km^2)	Area (%)
Urban	11.42	0.59	126.21	6.55	204.33	10.60	493.03	25.57	693.47	35.97
Water	51.96	2.69	49.52	2.57	45.57	2.36	18.65	0.97	9.56	0.50
Vegetation	518.85	26.91	492.48	25.54	486.58	25.24	318.21	16.50	277.70	14.40
Agriculture	1255.88	65.14	1199.47	62.21	1147.84	59.54	1062.56	55.11	935.54	48.52
Barren land	89.90	4.66	60.32	3.13	43.68	2.27	35.55	1.84	11.74	0.61

and post-monsoon data were collected from Saharanpur and Meerut Divisions of UP State Groundwater Department.

The storage equation used to find out groundwater storage given by GEC1997 is as follows:

$$S = \Delta h \times S_y \times A$$

where

$S =$ Storage,
$\Delta h =$ Well location data of pre monsoon (h_2)-well data location of post-monsoon (h_1),
$S_y =$ Specific yield.

In Haridwar district as given in Table 2 and Fig. 2, the built-up area has augmented from 11.42 km^2 in 1972 to 693.47 km^2 in 2011 which shows an increment, whereas other like water bodies are decreasing from 51.96 km^2 in 1972 to 9.56 km^2 in 2011, vegetation from 518.85 km^2 in 1972 to 277.70 km^2 in 2011, agriculture from 1255.88 km^2 in 1972 to 935.54 km^2 in 2011 and barren land from 89.90 km^2 in 1972 to 11.74 km^2 in 2011 which led to the decrease in storage from 0.629 km^3 in 1972 to 0.259 km^3 in 2011.

Table 2 Year wise change in storage with respect to land use of Haridwar

Year	Urban	Water	Vegetation	Agriculture	Barren land	Change in storage
1972	11.42	51.96	518.85	1255.88	89.90	0.629
1980	126.21	49.52	492.48	1199.47	60.32	0.562
1992	204.33	45.57	486.58	1147.84	43.68	0.336
2002	493.03	18.65	318.21	1062.56	35.55	0.287
2011	693.47	9.56	277.70	935.54	11.74	0.259

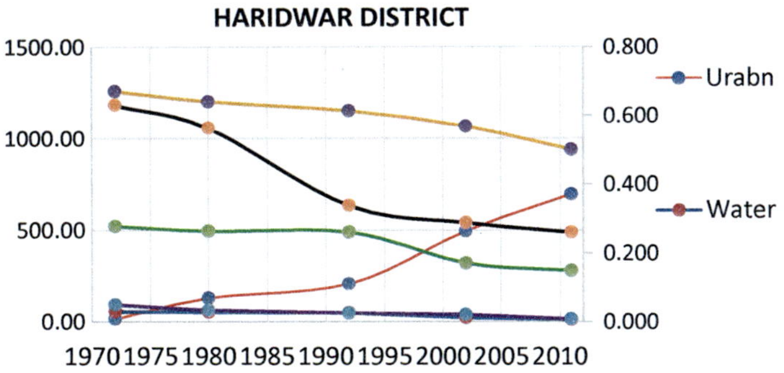

Fig. 2 Year wise change in storage with respect to land use of Haridwar

4 Conclusions

From the last few years, there has been quick increase in water utilization owning to the population blast, extraordinary upgrading in living standard and massive growth. The problem is caused not only by natural factors but also due to mismanagement and inadequate knowledge of present water resources. Therefore, an appropriate management plan can only be implemented by knowing the balance of different hydrologic components. The current study is an effort to evaluate the impact of land use change on groundwater storage in Haridwar district having an area of 1928 km^2 of Uttarakhand. It has been observed that the land use intensely affects the groundwater storage in the area. The current study deals with implementing a physical storage equation of a groundwater, remote sensing data, ERDAS IMAGINE 9.2 for Index Map registration and GIS techniques to locate the well location on Shapefile, estimating the water table contour which shows the variation of groundwater fluctuation in pre-monsoon season and post-monsoon Season.

This result shows an increase in paved area which leads to increase in runoff volume and decrease in time of concentration, which effect the decrease in percolation rate, and once the percolation rate gets affected, it is obvious the groundwater storage will decline. In addition as per decadal calculation results shows groundwater levels are going down due excessive withdrawal in Haridwar District area and also reduction in natural recharge area.

References

1. Alley WM, Reilley TE (1999), Sustainability of groundwater resources. U.S. Geol. Surv. Circutar, pp 11–86
2. Bronstert A (2004) Rainfall-runoff modeling for assessing impacts of climate and land use change. Hydrol Process 18:567–570
3. Calder IR (1993) Hydrologic effects of land use change. In: Maidment DR (ed) Handbook of hydrology. McGraw-Hill, New York, USA, pp 13.1–13.50
4. CPCB (1998) Down to earth analysis on groundwater. Report on groundwater pollution in NCT of Delhi. Central Pollution Control Board, New Delhi, India, pp 30–41
5. Singh DK, Singh AK (2002) Groundwater situation in India: problems and perspective. Int J Water Resour Dev 18(4)
6. Gehrels H, Peters NE, Hoehn E, Jensen K, Leibundgut C, Griffioen J, Webb B, Zaadnoordijk W-J (eds) (2001) Impacts of human activity on groundwater dynamics. IAHS Publ. No. 269
7. Hoff H (2002) The water challenge: Joint Water Project. Global Change Newslett 50:46–48
8. Hutjes RWA (1998) Biosphere aspects of the hydrological cycle. J Hydrol 212–213:1–21
9. IPCC: Climate Change (2001) The scientific basis. In: Houghton JT, Ding Y, Griggs DJ, Noguer M, van der Linden PJ, Dai X, Maskell K, Johnson CA (eds) Contribution of Working Group I to the third assessment report of the Intergovernmental Panel on Climate Change. Cambridge University Press, Cambridge, United Kingdom, New York, NY, USA, 881 pp
10. Lambin EF, Baulies X, Bockstael NE et al (2002) Land use and land-cover change implementation strategy. IGBP Report No. 48 and IHDP Report No. 10, Stockholm, pp 21–66
11. Mtembezeka P, Andrews AJ, Appiah SO (1997) Groundwater management in drought-prone areas of Africa. Water Resour Dev 13(2):241–261

12. Purohit R, Chatterjee R, Gupta K, Mohiddin SK, Singh N, Shekhar S (2009) Dynamic groundwater resources of National Capital Territory, Delhi: assessment, development and management options. Environ Earth Sci 59:669–686
13. Sato K, Iwasa Y (2003) Groundwater hydraulics. Springer, Tokyo
14. Schwarts WF, Hubao Z (2003) Fundamentals of groundwater. Wiley, New York
15. Serneels S (2001) Priority questions for land use/cover change research in the next couple of years. LUCC Newslett 1–9
16. Singh RB (1999) Urban impacts on groundwater quality in the Delhi region. In: Ellis JB (ed) Impacts of urban growth on surface water and groundwater quality. Proceeding symposium during IUGG 99, Birmingham, Jul 1999. IAHS Publ. no. 259, pp 227–236
17. Singh RB (2001) Impact of land use change on groundwater in the Punjab-Haryana plains, India. In: Impact of human activity on groundwater dynamics (Proceedings of a symposium held during the sixth IAHS scientific assembly at Maastricht, The Netherlands, Jul 2001). IAHS Publ. no. 269
18. Singh RB (2000) Environmental consequences of agricultural development; a case study from the green revolution state of Haryana, India. Agric Ecosyst Environ 82:97–103
19. Tang Z, Engel BA, Pijanowski BC, Lim KJ (2005) Forecasting land use change and its environmental impact at a watershed scale. J Environ Manag 76:35–45
20. Zhang L, Dawas WR, Reece PH (2001) Response of mean annual evapotranspiration to vegetation changes at catchment scale. Water Resour Res 37(3):701–708

Design of Rising Main for Amravati Water Supply Scheme

R. K. Rai and S. R. Khandeshwar

Abstract Rising main or pumping main is used to convey the water from source or reservoir to water treatment plant and further to the balancing reservoir or elevated service reservoir. If water treatment plant is located above reservoir sump level, water can be transmitted using direct pumping or combined pumping and gravity technique. This paper reviews the method for the analysis and design of combined pumping and gravity transmission system by provision of break pressure tank for Amravati water supply scheme. The water is lifted to break pressure tank by pumping main and thereafter, it is conveyed to water treatment plant using gravity main. The variable cost parameters of the transmission system such as cost of pipes, pumps and energy are studied. Diameters of both the pipes were found to achieve cost effectiveness and performance. Further, an alternate water transmission system was studied and designed. Lagrange multiplier method was used to develop mathematical model.

Keywords Break pressure tank · Lagrange multiplier method · Pumping main · Water transmission system

1 Introduction

Water transmission is one of the most important component of water supply scheme like a water treatment plant and water distribution. Water transmission system (WTS) is used to transmit water from the supply source reservoir to receiving reservoirs. The gravity system, pumping system or combined system are three water transmission systems used for water transmission. If water treatment plant (WTP) is located above reservoir sump level, water can be transmitted using direct pumping or combined pumping and gravity technique.

R. K. Rai
Department of Civil Engineering, Government College of Engineering, Amravati 444604, India

S. R. Khandeshwar (✉)
Environmental Engineering, Government College of Engineering, Amravati 444604, India
e-mail: khandeshwarswaraj@gmail.com

L. M. Gupta et al. (eds.), *Advances in Civil Engineering and Infrastructural Development*, Lecture Notes in Civil Engineering 87,
https://doi.org/10.1007/978-981-15-6463-5_23

In case of direct pumping the water is transmitted using direct single stage pumping. The combined system includes transmission by use of pumping and gravity main along with break pressure tank (BPT). BPT is provided in between the source or reservoir and WTP at the higher elevation [1]. The water is lifted to BPT using pumping main and further conveyed to WTP by gravity main.

The huge investment of cost has been made over the maintenance, construction, and operation of water supply scheme. The water transmission with pumping system demands more funds due to pumps and other accessories associated with it along with energy costs. Thus any small saving in cost for water network leads to significant markdown over the total cost of scheme. The traditional methods generally used for design and analysis are proved to be tedious and uneconomical. The prime elemental objective of this paper is analysis and design of combined pumping and gravity transmission system by provision of BPT. Optimal diameters for gravity main, pumping main and optimal cost transmission network, are determined by using the method of the Lagrange multipliers. The design equations for optimization had been developed by using Hazen-Williams equation.

2 Hydraulic Equations for Design and Analysis of WTS

2.1 Hydraulic Equation

Hydraulic equations commonly used for design and analyses of WTS are as follows:

1. Darcy-Weisbach equation; and
2. Hazen-Williams equation.

In this paper, the design and analysis was worked out by using the most popular Hazen-Williams equation. This equation is conventionally acceptable equation for design of water conveyance system as it is simple to use. Hazen-Williams equation with hydraulic mean depth, slope and velocity is given by Eq. (1):

$$V = 0.849 C_{\mathrm{H}} R^{0.63} S^{0.54} \tag{1}$$

where

C_{H} = Hazen Williams constant;
R = Hydraulic mean depth, m; and
S = Slope of the energy line.

For the design and analysis, the value for constant C_{H} was taken from Manual, 1993. The value of C_{H} for Cast Iron pipe is 130 and for PSC pipe is 140. The frictional head loss is considered as the major loss in pipe network which can be computed by Eq. (2) [2, 3]:

$$h_f = \frac{10.674 L Q^{1.852}}{C_{\mathrm{H}}^{1.852} D^{4.870}} \tag{2}$$

where

h_f = Frictional head loss, m;
D = Pipe diameter, m;
Q = Flow of water, m^3/s; and
L = Length of pipe, m.

2.2 *Optimal Diameter of Pumping Main*

Lagrange multiplier method was used to calculate the optimal diameter of pumping main. The variable cost function was added and differentiated with respect to diameter. In the pumping main, water flows in pipe by creating the pressure head difference using pumps. The optimal diameter for pumping main was calculated by using Eq. (3):

$$D = \left(\frac{51.98 \gamma Q^{2.852} K_{\mathrm{T}}}{m C_{\mathrm{H}}^{1.852} K_m} \right)^{\frac{1}{m+4.870}} \tag{3}$$

where,

K_m and m = Cost parameters for pipes
K_{T} = Cost coefficient for pumping

The value of K_{T} is calculated using Eq. (4) [4]:

$$K_{\mathrm{T}} = \frac{K_{\mathrm{P}}(1 + s_{\mathrm{b}})}{1000n} + \frac{8760 F_{\mathrm{A}} F_{\mathrm{D}} R_{\mathrm{e}}}{1000rn} \tag{4}$$

3 Cost Functions

Cost function was developed for various components of water transmission network which are variable in nature. The total cost of pumping system worked out by calculating cost function for each unit such as pipes, pumps and energy. Cost functions for different components used in water transmission system are as follows:

3.1 Cost of Pipes [4]

Pipe is the prime component of water supply scheme. Pipe involves the huge investment of cost as compare to other components. Cast iron or ductile iron pipes are most preferable in pumping system and prestressed pipes for gravity main. CI pipes can sustain the large pressure which is developed in pumping system. CI Pipes are much easier for connections. Cost of pipe is given by Eq. (5):

$$C_m = K_m L D^m \tag{5}$$

To determine the cost parameters for pipes, the cost data was taken from Current Schedule of Rates (CSR) of Maharashtra Jeevan pradhikaran (MJP) for Amravati region, 2018–19 [5, 6]. The cost of pipes for different diameters were taken from schedule of rate and it was multiplied by capitalization factor calculated using Eq. (6):

$$C_f = 1 + \frac{1-\alpha}{(1+r)^T - 1} + \frac{\beta}{r} \tag{6}$$

where,

α = components salvage fraction;
β = the component maintenance fraction;
r = rate of interest as ₹/year; and
T = components useful life, years.

The values of α, β, r and T were taken as 0.10, 0.02, 0.08 and 60 years for obtaining capitalization factor for CI pipe, which came out to be 1.259. The values of α, β, r and T were taken as 0.08, 0.015, 0.08 and 50 years for obtaining capitalization factor for PSC pipe, which came out to be 1.208. In order to determine the cost for the any diameter of CI and PSC pipe, a graph of cost of pipe in ₹ per m versus diameter of pipe in m can be plotted [6]. The values of cost parameters were plotted by obtaining the best fit of the curve and comparing it with standard equation of pipe. Capitalized Cost versus Diameter of CI Pipe and PSC for cost parameters calculation is shown in Figs. 1 and 2.

The cost parameters for CI and PSC pipes are given in Table 1.

3.2 Cost of Pumping [1]

In case of rising terrain or head difference, it becomes necessary to provide some external energy to fulfil water demand in such condition pump is used. The pump cost data was taken from CSR of MJP for Mechanical and Electrical 2018–19 for Amravati region [5]. The value of cost parameter K_P = 17,695 ₹/kW was obtained from equation of curve. Cost of pumping plant is given by Eq. (7):

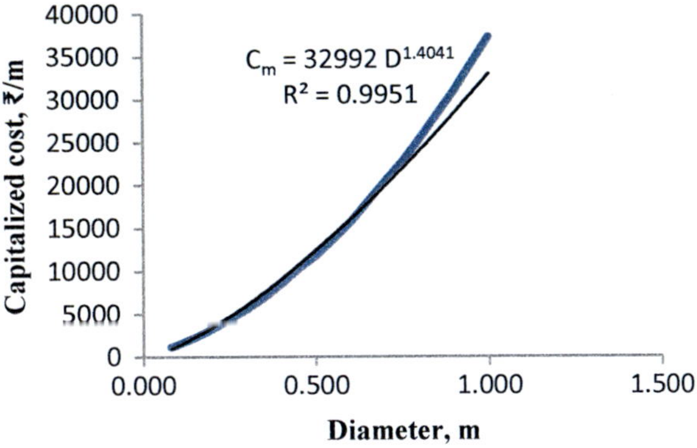

Fig. 1 Capitalized cost of CI pipe versus diameter

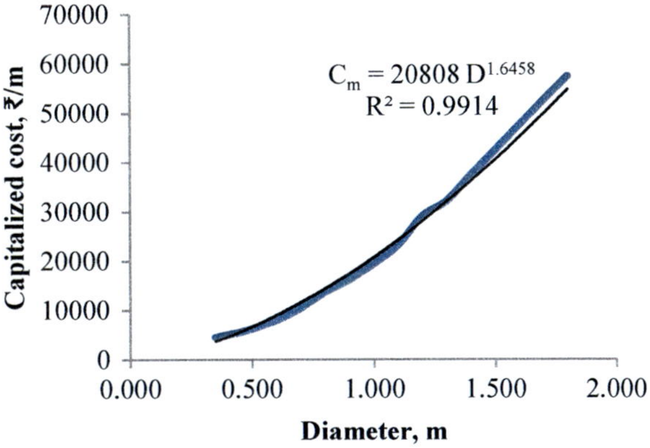

Fig. 2 Capitalized cost of PSC pipe versus diameter

Table 1 Parameters for CI and PSC pipes

Material of pipe	Cost parameter		
	K_m	m	R^2
Cast iron	32,992	1.4041	0.9951
Prestressed concrete	20,808	1.6458	0.9914

$$C_{\rm p} = K_{\rm p}\left(\frac{(1+S_{\rm b})\rho g Q h_0}{1000n}\right) \tag{7}$$

where,

$K_{\rm P}$ = Cost parameter of pump; and
P = Power of pump, kW.
η = Efficiency of pump;
ρ = Mass density of fluid; and
h_0 = Net head, m.
$S_{\rm b}$ = Standby fraction.

Where, Net Head = Static head + frictional head loss + minor head loss + residual head [7]

$$h_0 = h_s + h_f + h_m + h_r \tag{8}$$

3.3 Cost of Energy

Total cost of energy can be obtained by multiplying the power by number of hours in year i.e. 8760 h and rate of electricity R_e in ₹/kWh. The total cost of energy required annually given by Eq. (9) [4]:

$$C_e = \frac{8760 \rho g Q h_0 F_A F_D R_e}{1000 r \eta} \tag{9}$$

where,

F_A = Annual averaging factor;
F_D = Daily averaging factor for the discharge;
R_e = Electricity rate, ₹/kWh; and
r = Rate of interest.

4 Methodology

4.1 Procedure for Design, Analysis and Cost Calculation of Pumping Main

1. Find the range of set of diameters for transmission pipe on basis of minimum velocity of 0.3 m/s and maximum velocity of 3.0 m/s;
2. Calculate the value of pumping cost coefficient, K_T using Eq. (4);
3. Calculate the optimal diameter for pumping main using Eq. (3);
4. Find out cost of pipe using cost parameters using Eq. (5);
5. Calculate the head loss for optimal diameter using Eq. (2);
6. Net head calculation is done by adding frictional head loss, static head, minor loss (10% of frictional head loss) and residual head (3 m). Static head is positive is case of rising topography;
7. Find out cost of pumps and cost of energy for pipe using Eqs. (7) and (9); and
8. Finally, the total cost is calculated by adding the cost of pipe, pump and energy.

4.2 *Procedure for Design, Analysis and Cost Calculation of Gravity Main*

1. Find the range of set of diameters for transmission pipe on basis of minimum velocity of 0.3 m/s and maximum velocity of 3.0 m/s;
2. Net Head is sum of Static head, frictional head loss, minor head loss (10% of frictional head loss) and residual head (3 m). Static head is negative in case of falling topography. In case of gravity main net head is always zero or negative. For optimal condition equate the net head to zero.
3. The headloss obtained is used for optimal diameter calculation using headloss Eq. (2) and;
4. Find out cost of pipe for obtained diameter using cost parameters using Eq. (5).

5 Case Study for Amravati Water Supply Scheme

The case study of Amravati water supply scheme (AWSS) was considered for the design, analysis and cost calculation. Amravati is a city in the state of Maharashtra, India, 2nd largest and most populous city of Vidarbha after Nagpur, located at coordinates 20°55′33″ N and 77°45′53″ E. The major source of water for Amravati city is upper Wardha dam also called Simbhora dam located near Morshi. The sump level of dam is 324 m from where the water is lifted to BPT located at Ner Pinglai at elevation of 406.5 m by pumping main. The water from BPT is further transmitted to the WTP for aeration and further treatment located in Tapovan at level of 396 m by gravity main. The data for transmission system of AWSS is given in Table 2 and Fig. 3 represents its schematic diagram.

The design parameters taken in consideration for AWSS are given in Table 3.

The value for F_A and F_D was assumed as 0.75, rate of electricity, R_e as 8 ₹/kWh and rate of interest, r as 8%.

5.1 *Design and Analysis of AWSS*

The step by step procedure as discussed under heading 4 was followed for both of the pumping and gravity main. The value of K_T was obtained as 676.90. The optimal

Table 2 Data for transmission system of AWSS

Parameters	Values
Discharge (m^3/s)	1.869
Level of source (m)	324.000
BPT level (m)	406.500
WTP level (m)	396.000

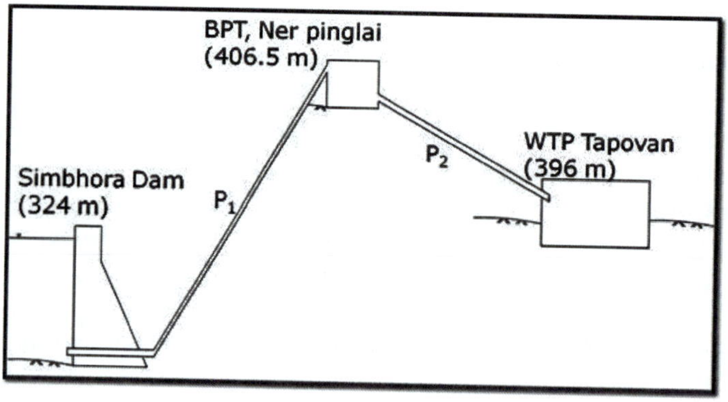

Fig. 3 Schematic diagram for transmission system of AWSS

Table 3 Parameters for design of transmission system

Parameters	Pumping main, P_1	Gravity main, P_2
Material	Cast iron	PSC
C_H	130	140
L (m)	23,000	25,000
h_s (m)	82.5	−10.5
K_m	32,992	20,808
K_p (₹/kW)	17,695	17,695
m	1.4041	1.6458

diameters were rounded off to nearest 50 mm value for its availability in market so analysis and cost calculation was done for these values. The cost of components for various range of diameter for pumping main is shown in Fig. 4 and total cost in Fig. 5.

Fig. 4 Cost parameters versus diameter

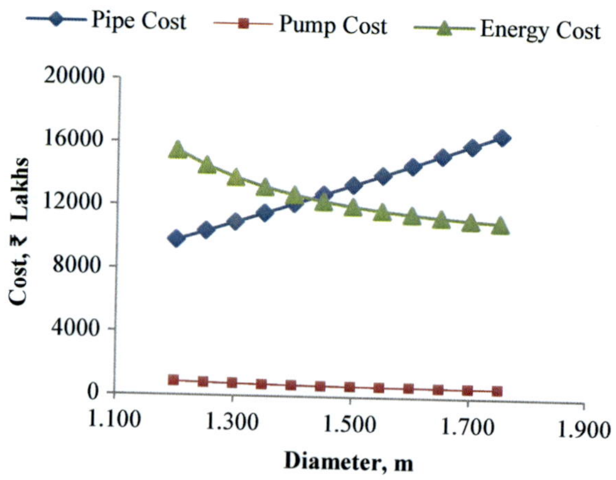

Fig. 5 Total cost versus diameter

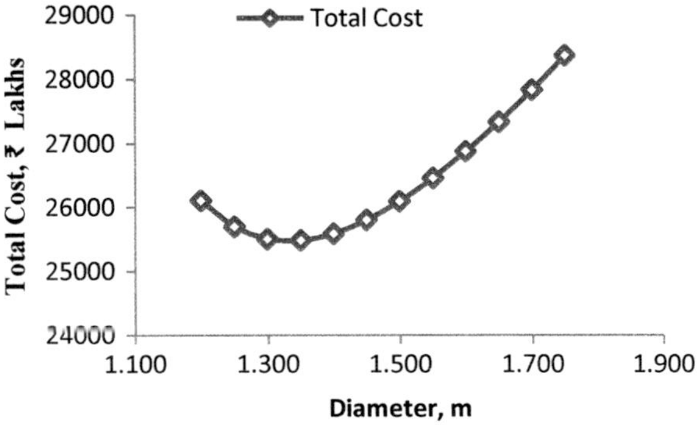

Thus, optimal diameter for pumping main for transmission of water from Simbhora dam to BPT at Ner Pinglai was 1.350 m by CI pipe with provision of pumps. The total cost of pumping main was ₹ 25,480.800 Lakhs. The optimal diameter for gravity main for transmission of water from BPT at Ner Pinglai to WTP, Tapovan was 1.700 m by PSC pipe with no provision of pumping. Thus, the total cost of gravity main was ₹ 12,452.565 Lakhs. Thus total cost of transmission system for AWSS was ₹ 37,933.365 Lakhs.

Further, alternate transmission system for AWSS was also studied. Google Earth Pro was used to study the intermediate topography of transmission system. Among many alternatives, the alternate system where BPT was located at 414.000 m was taken in consideration. The water is first pumped to BPT using pipe P_1 and further transmitted by pipe P_2 to WTP, Amravati. The length of pipe P_1 was 18,200 m and that of pipe P_2 was 26,600 m. The static head for pipe P_1 was 90 m and for pipe P_2 was -18 m. The intermediate topography using Google Earth Pro for alternative is shown in Fig. 6.

The optimal diameter for transmission of water for CI pipe from Simbhora dam to BPT was 1.300 m, with total cost as ₹ 23,160.130 Lakhs. The optimal diameter for transmission of water using PSC pipe from BPT to WTP was 1.500 m, with total cost as ₹ 10,784.055 Lakhs. Thus, total cost for transmission system for Alternative was ₹ 33,944.185 Lakhs.

6 Results

1. The total cost of transmission system for existing AWSS was ₹ 37,933.365 Lakhs;
2. The total cost of alternate transmission system taken in consideration for AWSS was ₹ 33,944.185 Lakhs.
3. The alternate WTS was found to be 10.516% more economical than existing WTS.

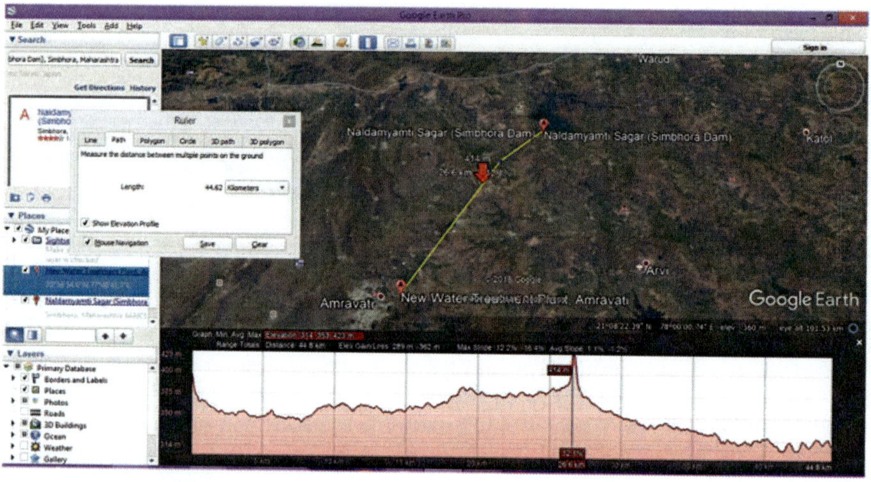

Fig. 6 Topography of alternate WTS

7 Conclusions

1. The optimal diameter gives diameter at which total cost of variable parameters like pipes, pump and energy is minimum for the given discharge, pipe material and pumps used for transmission;
2. For transmission of water by pumping main, CI pipes along with provision of high head pumps were preferred and for transmission by gravity main, PSC pipes were preferred;
3. The optimal diameter for the pumping main does not depend on elevation difference and that of gravity main depends on elevation difference;
4. Head loss, net head and power varies inversely with the increase in diameter of pipe;
5. The nature of graph of total cost of pumping main is because of its dependence on the cost of pipe, pump and energy. The cost of pump and energy varies inversely with the increase in diameter of pipe, while cost of pipe increase with increase of diameter of pipe and;
6. The total cost of complete transmission system reduce with increase in elevation of BPT.

References

1. Mahar PN, Singh RP (2014) Optimal design of pumping main considering pump characteristics. J Pipeline Syst Eng Pract ASCE 5(1):1–6
2. Swamee PK (2001) Optimal design of pumping main. J Transp Eng ASCE 127(3):264–267

3. Sodiki J, Adigio E (2014) A review on the development and application of methods for estimating head loss components in water distribution pipe work. Am J Eng Res 3(9):91–96
4. Swamee PK, Sharma AK (2008) Design of water supply pipe networks. Wiley, Hoboken, NJ
5. MJP (2018) Schedule of rates for Maharashtra Jeevan Pradhikaran works for electrical and mechanical for year 2018–19, Amravati Region
6. MJP (2018) Schedule of rates for Maharashtra Jeevan Pradhikaran works for year 2018–19, Amravati Region
7. Swamee PK (1996) Design of multistage pumping main. J Transp Eng ASCE 122(1):1–4
8. Sharp BB (1986) Economics of pumping and utilization factor. J Hydraulic Eng ASCE 111(11):1386–1395
9. Tarquin A, Dowdy J (1989) Optimal pump operation in water distribution. J Hydraulic Eng ASCE 115(2):158–168

Synthetic Stream Flow Generation of River Gomti Using ARIMA Model

Hemanshu Singh and Maya Rajnarayan Ray

Abstract Study of synthesizing of time series applies various stochastic models out of which autoregressive integrated moving average (ARIMA) model has proved to be an effective tool. Such models are useful to select the best fit model from the available past values of any time series. The objective of present study is to develop a methodology to synthetically generate the time series using ARIMA model for an Indian River Gomti at Gomti Barrage of Uttar Pradesh (U.P.), India. The approach for prediction of time series is based on the idea of predicting future values of an observed time series using a model with estimated regression parameters. ARIMA is a popularly adopted stochastic technique for the various studies in the water resources engineering where long futuristic data is necessary. The present model uses an iterative three-stage modelling approach. First stage is to identify and select model, which involves the checking of stationarity of the variables. Second stage comprises the checking of the seasonality of the dependent variables and to select the suitable model based on the plots of the autocorrelation and partial autocorrelation functions of the dependent time series data. Parameter estimation has been carried out using computation algorithms to arrive at coefficients which best fit the selected ARIMA (1, 1, 1) model. Model checking has been performed by testing whether the estimated model confirms to the least value of AIC and SBIC. By checking of the least value of the AIC and SBIC, the model is generated and hence the mathematical equation has been formulated for the generation of the forecasted stream flow for future time period from the original data of discharge collected from gauge site.

Keywords Synthetic stream flow · Autoregressive moving average · Akaike information criterion

---

H. Singh · M. R. Ray (✉)
Department of Civil Engineering, G. H. Raisoni College of Engineering, RPF Gate, No. 3, Hingna Road, Digdoh Hills, Nagpur, Maharashtra 440016, India
e-mail: mayasubodh@gmail.com

H. Singh
e-mail: hemanshusingh22@gmail.com

L. M. Gupta et al. (eds.), *Advances in Civil Engineering and Infrastructural Development*, Lecture Notes in Civil Engineering 87,
https://doi.org/10.1007/978-981-15-6463-5_24

1 Introduction

The Box–Jenkins methodology uses an iterative approach; one of such methods is autoregressive integrated moving average (ARIMA). The ARIMA model is a linear mathematical model which is able to represent stationary as well as non-stationary time series. Time step legs presented in the equation of this method are termed as autoregressive model, and model to predict is called moving average model. An ARIMA (p, d, q) model represents autoregressive terms as p, non-seasonal differences as d and q as the forecast errors in particular number of lagging.

The Box–Jenkins method of modelling is used for enabling a tentative model, to find model parameters and to perform checking. Autoregressive terms predict the value of series from its previous lag. The term moving average predicts the value using previous random errors. After arriving to a satisfactory model, one can forecast future time series.

The process followed to build an ARIMA model is model identification with the help of ACF and PCF functions, which finally helps to know the pattern of the data and its stationarity. Second step is to estimate the coefficients using least square method and maximum likelihood approach. The third step is diagnostic checking using residual of statistics, ACF and PCF parameters. The last step is forecasting of time series using statistics and confidence interval.

2 Literature Review

Sarma and Ahmed [1] presented the capability of ARIMA model through their studies to synthesize stream flow of Pagladia River, Assam, India. Box and Jenkins [2] presented several models which included ARIMA model as well. This model is useful when the data series is stationary which means mean, variance and ACF remain constant. This model finds application in the case where data series shows strong trend characteristics, and it is also applicable to seasonal and non-seasonal data series. Jianxin et al. [3] used the ARIMA model to analyse the GPS data for the period 1996 April to 1998 March. The quality of data was improved after removal of outlier, discontinuities and seasonal changes. Ong et al. [4] had used ARIMA as a method to analyse stationary univariate time series data. Mohammadi et al. [5] had studied on the Amir Kabir Reservoir in Tehran for the prediction of spring inflow; in their study, they have presented a comparison of ANN, ARIMA and regression analysis between hydro-climatologic data and inflow. Wurbs [6] presented a capability of stochastic model for synthetic generation of time series to conduct a study on reservoir operation. Stream flow, i.e. monthly minimum daily discharge forecasting for three gauge stations using stochastic approaches like ARIMA and Thomas–Fiering, has been shown by Yurekli and Kurunc. He found that the performance of ARIMA model was slightly better over the Thomas–Fiering model. Kim and Valdes [7] revealed the application of synthetic time series to assess the performance of water resources project. He presented that parametric models assume the variables follow the Gaussian characteristics. Kumar et al. [8] had demonstrated the capability of ANN technique for the river flow forecasting using feed forward recurrent neural network algorithms. Herman et al. [9] presented a study on the application of synthetic stream flow for the synthesis of draught for the stockholders. Ghimire [10] presented an over view of the application of ARIMA model for the rivers of US. Rana et al. [11] shown the application of ARIMA model for the forecasting of stream flow of Doyian station, China. Wang and Salas [12] demonstrated the forecasting of monthly stream flow using ARMA technique for the Colorado state.

3 Analysis of Data

It includes the collection of data of past years' time series of the site to be analysed. In the present study, ten years' data had been used for the analysis. The data consists of the monthly and annually minimum and maximum discharge and their respective mean discharge. Historical stream flow data of **river Gomti at Gomti Barrage site, of Lucknow, U.P. India**. The data is arranged and analysed for the stationarity checked in order to understand change in trend of the historical data. The series data fluctuates only around an average mean which is not depending on the time. The stationarity of obtained data can be checked using the mean discharge which shows that the generated series will be consisted of integrated technique regarding generation of ARIMA model (Fig. 1).

As the graph of mean discharge found to be as non-stationary, due to this non-stationarity in the raw series the integrated term 'I' has been introduced to ARIMA model. In order to the stationaries series the differencing of data has been performed and since data contained excessive negative values, hence an addition of maximum

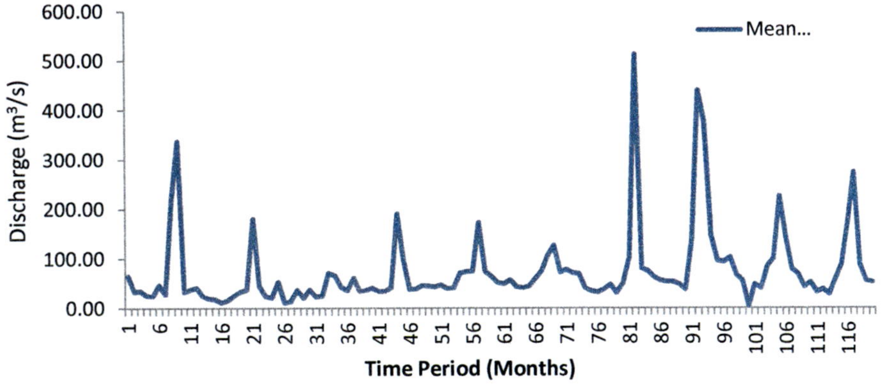

Fig. 1 Graph of mean discharge

value been done to differenced series. This would generate the positive values, and then log series was generated.

Models have been identified based on the ACFs and PACFs patterns; ACFs and PACFs are computed for the various lags of data series. There is autocorrelation between the Y_{t-1} and Y_t in the first lag, while ACFs and PACFs are computed between Y_{t-2} and Y_t for second time lag. The ACF and PACF methods for determining the lag length in an ARIMA model are commonly used, and there are other methods termed information criteria which can also be used. Mathematically:

$$\mathrm{ACF}(k) = \frac{\sum_{t=1+k} (Y_t - \bar{Y})(Y_{t-k} - \bar{Y})}{\sum_{t=1} (Y_t - \bar{Y})^2} = \frac{\mathrm{cov}(Y_t Y_{t-k})}{\mathrm{var}(Y_t)} \tag{1}$$

The additional correlation between Y_t and Y_{t-k} is presented through PACFs. The value of ACF and PACF is closely related and falls between -1 and $+1$. Mathematically:

$$\mathrm{PACF}(1) = \mathrm{ACF}(1) \tag{2}$$

$$\mathrm{PACF}(2) = \frac{\left[\mathrm{ACF}(2) - (\mathrm{ACF}(1))^2\right]}{1 - [\mathrm{ACF}(1)]^2} \tag{3}$$

4 Selection of ARIMA Model

According to the observed values of ACF and PACF, three models ARIMA (1, 1, 1), ARIMA (1, 1, 2) and ARIMA (2, 1, 1) were selected for the further diagnosis on the various parameters presented below:

Root-Mean-Square Error (RMSE)

$$\mathrm{RMSE}(\hat{\theta}) = \sqrt{\mathrm{MSE}(\hat{\theta})} = \sqrt{E((\hat{\theta} - \theta)^2)} \tag{4}$$

Mean Absolute Error (MAE)

$$\mathrm{MAE} = \frac{1}{n}\sum_{i=1}^{n}|f_i - y_i| = \frac{1}{n}\sum_{i=1}^{n}|e_i| \tag{5}$$

As the name implies, the mean absolute error is an average of the absolute errors $e_i = |f_i - y_i|$, where f_i is the prediction and y_i is the true value.

Mean Absolute Percentage Error (MAPE)

$$M = \frac{100\%}{n} \sum_{t=1}^{n} \left| \frac{A_t - F_t}{A_t} \right| \quad (6)$$

where A_t is the actual available value and F_t is the predicted value.

ME

$$\text{ME} = \overline{f - a} \quad (7)$$

Mean Percentage Error (MPE)

$$\text{MPE} = \frac{1}{n} \sum_{t=1}^{n} \frac{f_t - a_t}{a_t} \quad (8)$$

where a_t is the actual value of the quantity being forecast and f_t is the forecast.

Akaike Information Criterion (AIC)

$$\text{AIC} = 2k - 2\ln(L) \quad (9)$$

Bayesian Information Criterion (BIC)

$$\text{BIC} = n \cdot \ln\left(\hat{\sigma}_e^2\right) + k \cdot \ln(n) \quad (10)$$

where $\hat{\sigma}_e^2$ is the error variance.

The model is also tested on the basis of the autocorrelation factor (ACF) and partial autocorrelation factor (PACF) (Figs. 2 and 3).

The diagnostic checking has been performed on the selected model, which is presented as shown in Table 1.

Further after the assessment on the various criteria, performance of the model was tested AIC and BIC for the selection of most appropriate model. The model which has shown lowest value of AIC (8.63475) and SBIC (8.66234) has been emerged as the most suitable model. Therefore, this model has been used for synthetic generation of stream flow.

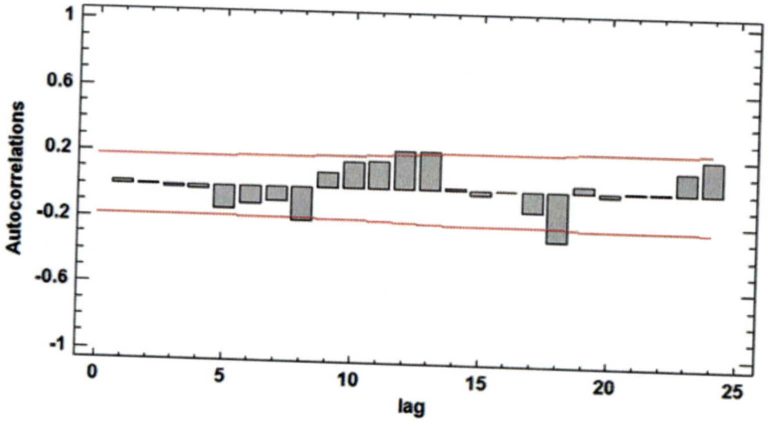

Fig. 2 ACF of the data

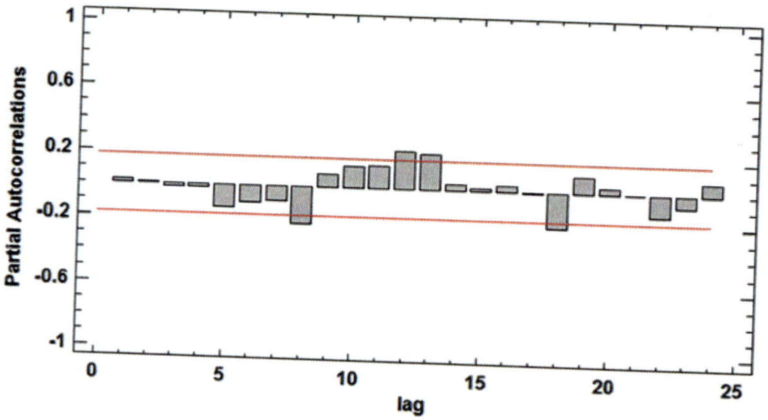

Fig. 3 PACF of the data

Table 1 Values of different tests

Model	RMSE	MAE	MAPE	ME	MPE	AIC	HQC	SBIC
(1, 1, 1)	73.0594	34.4823	54.0063	17.3684	−23.4770	8.6159	8.6348	8.6623
(1, 1, 2)	72.9638	32.2421	53.7621	17.8101	−22.3770	8.6299	8.6582	8.7000
(2, 1, 1)	73.0362	33.7302	54.6691	15.9794	−25.9700	8.6320	8.6602	8.7016

5 Result and Discussion

The result of the work is in the form of the mathematical model. The model which was generated is **ARIMA (1, 1, 1)** among the various generated models. This model was generated after the study of the discharge data collected from the gauge data.

Table 2 ARIMA model summary

Parameter	Estimate	Stnd. error	*t*	*P*-value
AR(1)	0.505815	0.0809405	6.24922	0.000000
MA(1)	0.979621	0.0156176	62.7254	0.000000

Model is selected based on the parameters like (i) constant mean = 3.97596, (ii) linear trend = 3.50374 + 0.00780529 t, (iii) simple moving average of 2 terms.

The model ARIMA (1, 1, 1) is chosen because from the calculated data of gauge site the AIC and the SBIC values are lowest among all the generated models. Thus, the model **ARIMA (1, 1, 1)** is selected for the forecasting of the synthetic stream flow of the river **Gomti at Lucknow, U.P., India** (Table 2).

$$Q_t = 0.505815\phi_{t-1} + Q_{t-1} + 0.979621\theta_{t-1} + \mu_v + \sigma_v\xi_t \quad (11)$$

Hence, developed model is used for the generation of twenty years' stream flow using the historical data of the ten years. The pattern depicting the trend is presented as shown in Fig. 4. It is very well seen from Fig. 4 that trend of actual and synthetically furcate discharge is matching very well all most in all the months. Even the extreme events are also captured to a significant extent.

The sample presentation of monthly mean and standard deviation of the synthetically generated series is presented in Figs. 5 and 6, which shows a convincing result. However, the mean of the furcated series is little bit higher than the actual one and the standard deviation curve is following very well to the actual stream flow series.

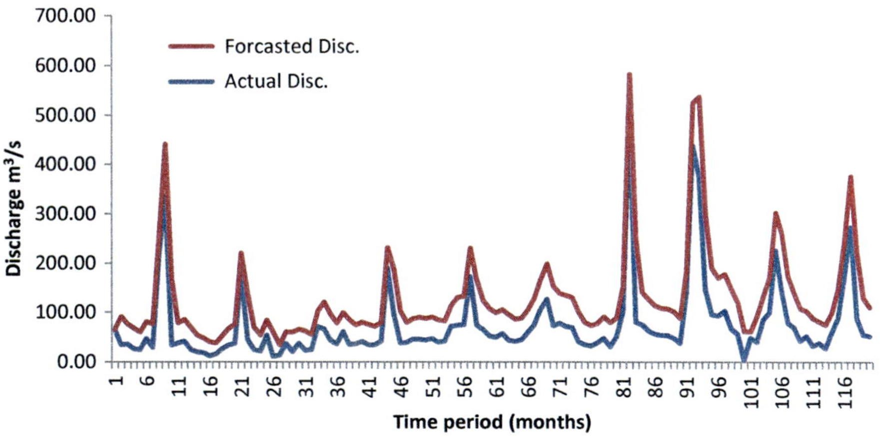

Fig. 4 Graph of forecasted data

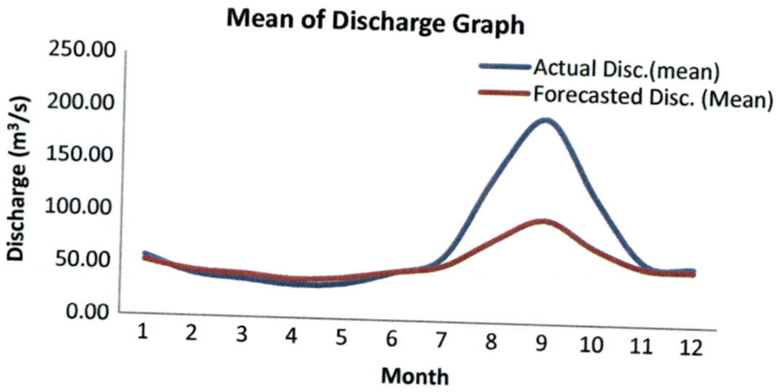

Fig. 5 Graph of mean discharge of main and forecasted data

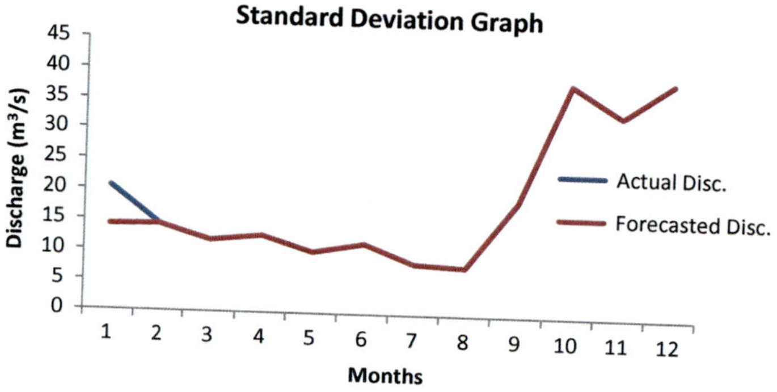

Fig. 6 Graph of standard deviation of main and forecasted data

6 Conclusion

The ARIMA model is developed with the help of the available data set of ten years from Gomti Barrage at Lucknow, U.P., India. AFC and PCF values for the different lags are computed, and using those ACF–PCF values the stream flow generation model is developed. After successful completion of various steps, **ARIMA (1, 1, 1)** is found the best model which can generate the synthetic stream flow. The model ARIMA (1, 1, 1) is used to generate the stream flow of about twenty years. The generated series is analysed with help of different verification criteria like RMSE, MAE, MAPE, ME and MPE, and model ARIMA (1, 1, 1) was found to be the best among other tentatively selected models. The results are presented graphically. The comparison of the actual series with the synthetically generated series depicts that the synthetically generated series is following the actual series quite a well in the most of the time period, and it found to be predicting less as compared to actual series in a few of monsoon periods.

References

1. Ahmed JA, Sarma AK (2007) Artificial neural network model for synthetic streamflow generation. Water Resour Manag 21(6):1015–1029
2. Box GEP, Jenkins GM (1976) Time series analysis forecasting and control. Holden-Day, San Francisco
3. Jianxin Li, Miyashita K, Kato T, Miyazaki S (2000) GPS time series modeling by autoregressive moving average method: application to the crustal deformation in central Japan. Earth, Planets and Space 52(3):155–162
4. Ong CS, Huang JJ, Tzeng GH (2004) Model identification of ARIMA family using genetic algorithms. Appl Math Comput 64(3):885–912
5. Mohammadi K, Eslami HR, Sh Mohammadi DK (2005) Comparison of regression, ARIMA and ANN models for reservoir inflow forecasting using snowmelt equivalent (a case study of Karaj). J Agric Sci Technol 7:17–30
6. Wurbs RA (1993) Reservoir-system simulation and optimization models. J Water Resour Plan Manage ASCE 119(4):455–472
7. Kim T-W, Valdes JBF (2005) Synthetic generation of hydrologic time series based on nonparametric random generation. J Hydrol Eng ASCE 10(5):395–404
8. Kumar D, Raju K, Sathish T (2004) River flow forecasting using recurrent neural networks. Water Resour Manag 18:143–161
9. Herman JD , Zeff HB, Lamontagne JR , Reed PM, Characklis G (2016) Synthetic drought scenario generation to support bottom-up water supply vulnerability assessments. J Water Resour Plan Manage 142(11). https://doi.org/10.1061/(ASCE)WR.1943-5452.0000701
10. Ghimire Bhola NS (2017) Application of ARIMA model for river discharges analysis. Nepal Physical Society JNPS 4(1):27–32
11. Rana MA, Yuan X, Kisi O, Curtef V (2017) Application of time series models for streamflow forecasting. Civ Environ Res 9(3):56–63
12. Wang DC, Salas JD (1991) Forecasting streamflow for Colorado River Sysyte. Completion report no. 164, Colorado Water Resources Research Institute, Colorado State University

Land Vegetation Change Detection Using Remote Sensing and GIS

Samshul Aarfin, Rishi Prakash, and Nitin Mishra

Abstract The study of vegetation cover change is one of the foremost arenas to understand the degree of interaction between man and environment. Changes in land vegetation cover affect biodiversity, water and other processes that come together to affect climate and biosphere. The study is done using remote sensing and GIS technique. The aim of this study is to detect vegetation cover change between 2003 and 2018 of Roorkee city region using Landsat image data. Vegetation changes have been detected by classification technique like unsupervised classification of both the images of year 2003 and 2018 using ERDAS Imagine 9.3. By monitoring of vegetation change, we can get the effects in biodiversity, water and other processes that come together to affect climate and biosphere the national development. The 15-year time period between 2003 and 2018 shows that a major change of vegetation cover change in the study area. The change in vegetation area observed in the duration of 15 years is 108.61 square.

Keywords Unsupervised classification · Vegetation change GIS · Remote sensing

1 Introduction

The study of vegetation cover change is one of the foremost arenas to understand the degree of interaction between man and environment. The changing nature of land vegetation depends on the existing ecological factors and the level of socioeconomic development [1, 2]. Information regarding land vegetation change cover is essential to outdo the problems arising from regular, uncontrolled progress and destruction of environmental quality, loss of basic water bodies and their associated lifeforms

S. Aarfin · N. Mishra (✉)
Department of Civil Engineering, Graphic Era Deemed to be University, Dehradun, Uttarakhand 248001, India
e-mail: nitinuag@gmail.com

R. Prakash
Department of Electronics and Communication Engineering, Graphic Era Deemed to be University, Dehradun, Uttarakhand 248001, India

L. M. Gupta et al. (eds.), *Advances in Civil Engineering and Infrastructural Development*, Lecture Notes in Civil Engineering 87,
https://doi.org/10.1007/978-981-15-6463-5_25

Fig. 1 Study area of Roorkee region

while planning changes in land vegetation cover affect biodiversity, water and other processes that come together to affect climate and biosphere the national development [3, 4]. The aim of this research paper is to estimate the land vegetation change detection of Roorkee region over time using satellite imagery for this area for the years 2003 and 2018.

2 Study Area

Roorkee is a city in north India and a municipal corporation in the Haridwar district of Uttarakhand state. Roorkee is located at 29.87° N 77.88° E. In Roorkee, the climate is warm and temperate. Roorkee is a city with a significant rainfall (Fig. 1).

3 Datasets Used

Landsat data is one of the most important and accurate datasets for understanding the global land cover status which is freely provided by the United States Geological Survey (USGS). Two Landsat images of Roorkee region were downloaded from USGS Web site on March 7, 2003 and March 8, 2018, of same path and row (146/39) for the change detection of vegetation cover of the study area. The image of 2003 is

Landsat 7 ETM and the image of Landsat 8 OLI. Landsat 7 is the seventh satellite of the Landsat program launched on April 15, 1999.

Landsat 8 is an American Earth observation satellite launched on February 11, 2013. It is the eighth satellite in the Landsat program; the seventh to reach orbit successfully. Providing moderate-resolution imagery, from 15 to 100 m, of Earth's land surface and polar regions, Landsat 8 operates in the visible, near-infrared, short-wave infrared and thermal infrared spectrums.

4 Methodology

The imagery which was used in this project was obtained from the USGS Web site. Landsat 7 image of date March 7, 2003 and Landsat 8 image of date March 8, 2018, were downloaded using earth explorer option of USGS. For the images which are downloaded from USGS Web site, there is no need for doing geo-referencing of the image. After extracting the image, it was imported into ERDAS Imagine. For proper outcome of the result from the downloaded images, it is necessary that both images have same row and path (146/39) (Fig. 2).

After importing the image into ERDAS Imagine, layer stacking process has been done in layer stacking; we combine all images and make them into one image. Landsat 8 has 11 bands, and we combine all those bands in layer stacking. The stacked images of year 2003 and 2018 can be seen in Fig. 3a and b.

Since the study area is a small part of the entire stacked image, sub-setting is performed to cut the study area (Fig. 4).

The image classification of both the images has to be done appropriately for getting the change detection. Analysis of digital image values, including one or more spatial, temporal, and spectral band relationships, to obtain categories of information about specific features [5]. Unsupervised classification is the method which is selected for this project unsupervised classification which is chosen because we do not know the exact area location, and training pixels cannot be accurately associated with the right class.

The number of desired classes was set to 50, and the number of iterations was set to 10. Several different outputs were obtained by setting different desired number of classes while performing the classification, and different iteration numbers were used to try and improve the accuracy of the classification process. After analyzing the different outputs that were obtained, the classes that fall under "Vegetation" category were identified, color-coded differently and renamed (Fig. 5).

Recoding is the process of reclassification of the classes. Recoding is done using thematic option on setup recode the water classes were recoded as 1, urban classes were recoded as 2 and vegetation classes were recoded as 3. The recoded images of year 2003 and 2018 (Fig. 6).

Change detection is defined as a process used to identify the change that occurred in a specific area over a span of time. By observing the same area at different time intervals using satellites or aerial photography, the user can identify the change of

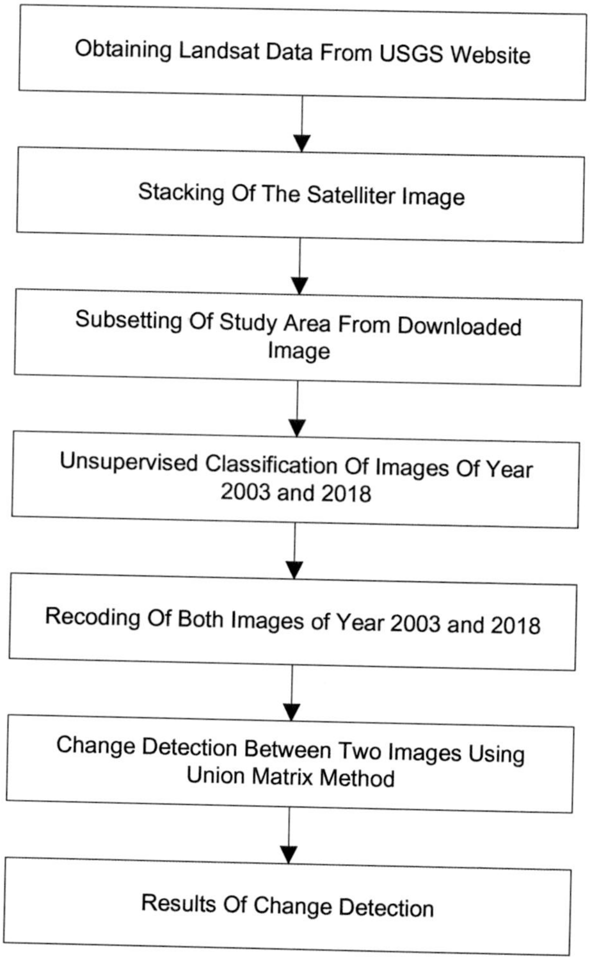

Fig. 2 Flowchart of the methodology

land use and land cover in that area. Since the images used for this study were already classified, the post-classification comparison method was used. This is a simple and more promising method which produces accurate results if the classification is done accurately. We have to open this file in Arc GIS software to generate a change detection matrix in Excel sheet from there which will display the change in pixel count between 2013 and 2018 and also the change in area. The Matrix Union function will create a matrix from two input thematic rasters. To open this dialog, go to the Raster tab, Raster GIS group, click the dropdown below Thematic and select Matrix Union. A Change Detection Matrix helps us to detect the change in pixel count during a given time duration from which we can compute the change in area in km^2. Attribute table of Matrix Union is open in ArcGIS (Table 1).

We can use this information to generate a change detection matrix from where we can also calculate the urban land change detection in km^2. We will create the Change

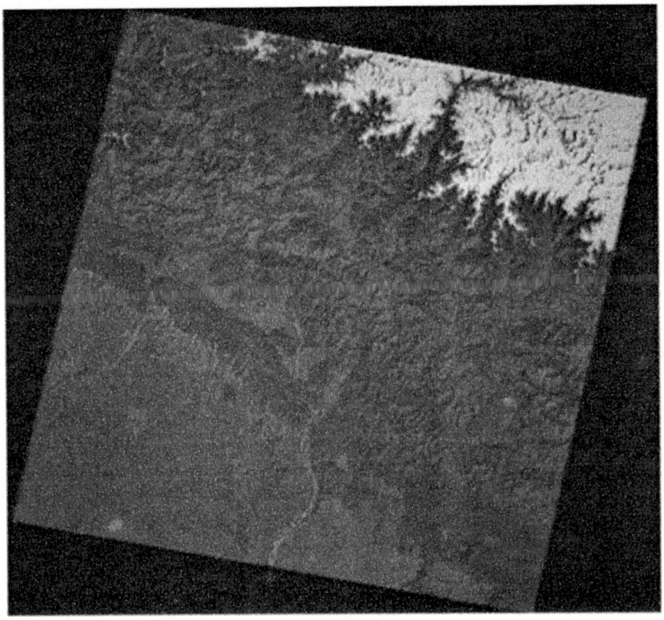

(a)

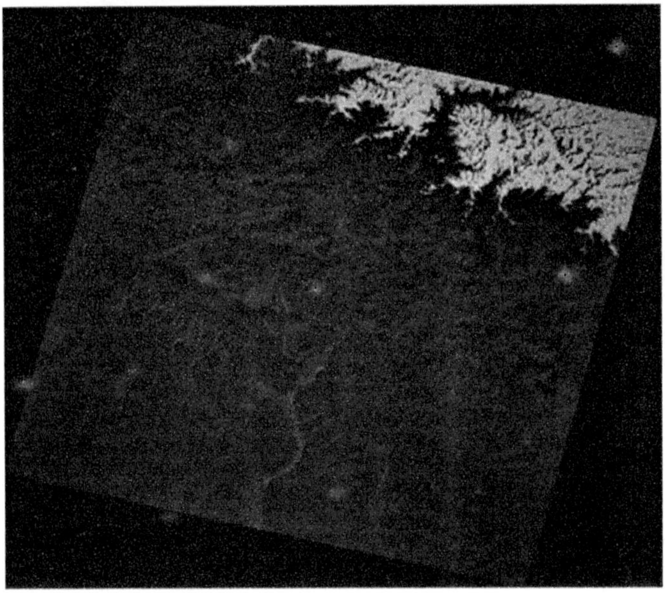

(b)

Fig. 3 **a, b** are Landsat image of date March 7, 2003 and March 8, 2018, respectively

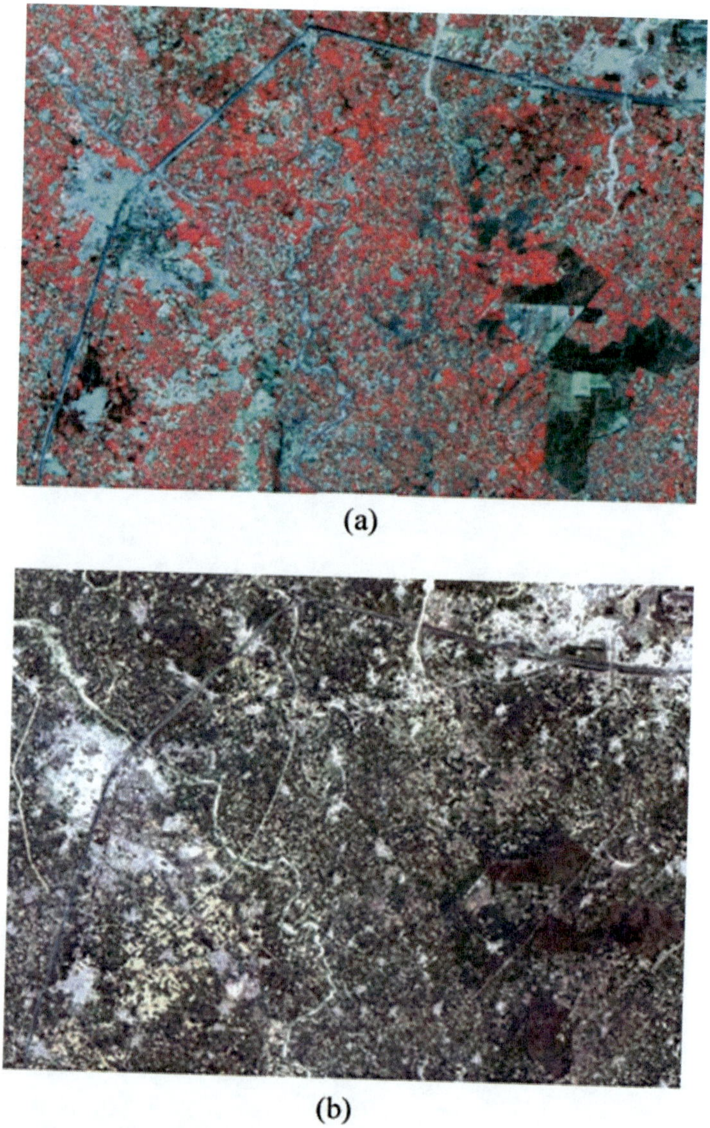

Fig. 4 **a, b** are subset image of date March 7, 2003 and March 8, 2018, respectively

Detection Matrix in Microsoft Excel 2016 for better reliability and performance. A Change Detection Matrix represents the change in pixel count between 2013 and 2018 of urban area, water bodies as well as in vegetation area. From the above table, it is shown that 6760 pixels are from urban land and 20,588 pixels are from vegetation land. It also shows that 21,867 pixels of water bodies are converted into urban land and 388,227 pixels of vegetation area are converted into urban area also. In order to

(e)

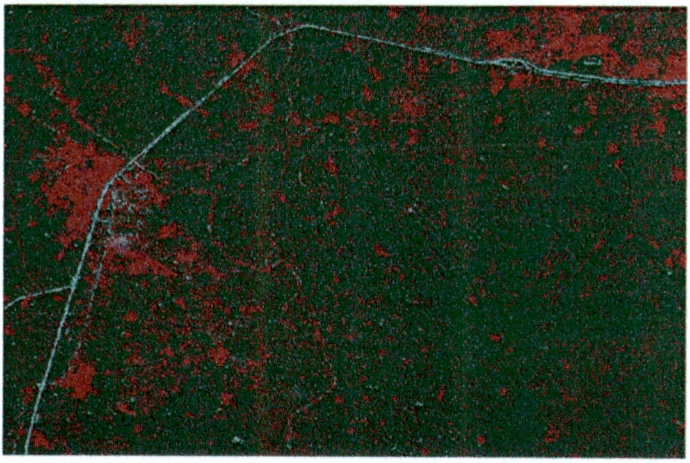

(f)

Fig. 5 **a, b** are classified image of date March 7, 2003 and March 8, 2018 respectively

convert pixel count into km^2, it is necessary to know about the cell size of the classes and then multiplying them by their corresponding pixel count to compute the area (Table 2).

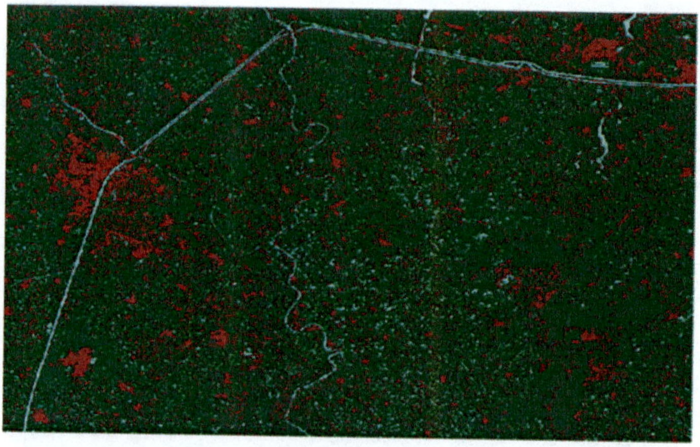

(a)

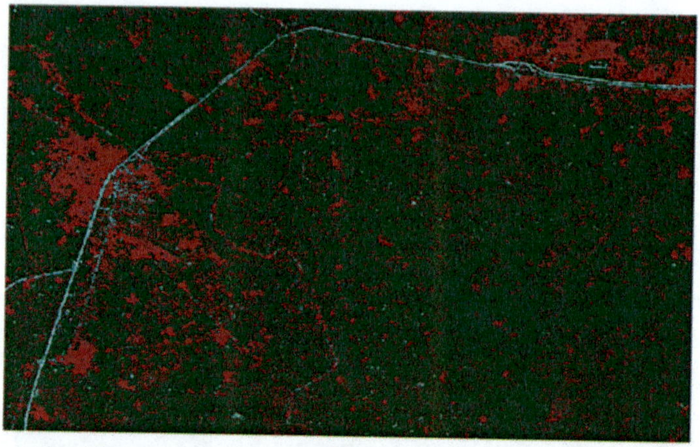

(b)

Fig. 6 **a, b** are recoded image of date March 7, 2003 and March 8, 2018 respectively

5 Result

The table given above shows the Change Detection Matrix depicting the change in area in km^2. It shows that the vegetation area was 1687.26 km^2 in 2003 and has been decreased to 1578.65 km^2 in 2018 (Table 3).

Table 1 Years 2003 and 2018 Matrix Union attribute table

OID	Value	Count	Recode 2003	Recode 2018
0	0	0	0	0
1	1	15,498	1	1
2	2	21,867	1	2
3	3	71,313	1	3
4	4	6760	2	1
5	5	151,018	2	2
6	6	216,823	2	3
7	7	20,588	3	1
8	8	388,227	3	2
9	9	1,465,921	3	3

Table 2 Change Detection Matrix for pixel count

	Water	Urban	Vegetation	Total 2018
Water	15,498	6760	20,588	42,846
Urban	21,867	151,018	388,227	561,112
vegetation	71,313	216,823	1,465,921	1,754,057
Total 2003	108,678	374,601	1,874,736	2,358,015

Table 3 Change Detection Matrix for area count (in km^2)

	Water	Urban	Vegetation	Total 2018
Water	13.95	6.08	18.53	38.56
Urban	19.68	135.92	349.40	505.00
Vegetation	64.18	195.14	1319.33	1578.65
Total 2003	97.81	337.14	1687.26	2122.21

6 Conclusion

Pre-processing operations of layer stacking and sub-setting are performed on both the images Landsat 7 and Landsat 8 of years 2003 and 2018. Unsupervised classification technique is used for classification, and accuracy for both the image is greater than 75%. A Change Detection Matrix is generated which depicts the change in pixel count from which change in km^2 can be computed during a span of 15 years. The change in vegetation area observed in the duration of 15 years is 108.61 square from 2003 to 2018.

Acknowledgements Authors are very thankful to Mr. Lakhwinder Singh, Ph.D. Research Scholar, IIT Roorkee, for his constant help regarding various GIS topics related to image processing and classification.

References

1. Kafi KM, Shafri HZM, Shariff ABM (2014) An analysis of LULC change detection using remotely sensed data: a case study of Bauchi city. In: International remote sensing and GIS conference and exhibition, vol 20, pp 1–9
2. Sharma K, Jalan S (2013) Change assessment of urban green spaces of Dehradun city using image derived parameters. Trans Inst Indian Geogr 35(1):63–74
3. Ahmadi H, Nusrath A (2012) Vegetation change detection of Neka river in Iran by using remote sensing and GIS. J Geogr Geol 2(1):58–67
4. Adhikari BS (2003) Ecological attributes of vegetation in and around Nanda Devi National Park, pp 30–35
5. Samaniego L, Schulz K (2009) Supervised classification of agricultural land cover using a modified k-NN technique (MNN) and Landsat remote sensing. J Remote Sens 1(9):875–895

Applications of Low Impact Development for Managing the Storm Water Surface Runoff in Urban Areas

Ruchika Dabas, Satish Kumar, and Munendra Kumar

Abstract Urbanization is one of the main important factors leading to urban flooding, which has caused major damage to the environment and society. Controlling urban flooding has become a new challenge for urban planners. There are many reasons of urban flooding in Delhi. It may occur due to increased urbanization, low rate of infiltration and poor infrastructure for storm water drainage network, improper drainage system. For controlling urban flooding, there is one of the control source solutions that help in reducing the urban flooding that is low impact development techniques. Low impact development (LID) is a new innovative technique for managing storm water at the source and helps to improve the water quality as well. It is most popular method which helps to reduce the hydrologic as well as water quality which effects by urbanization. The storm water management model (PCSWMM) is a widely used rainfall-runoff simulation model which has ability to model LID techniques. The LID used in this study is infiltration trench, green roof and rain water harvesting. The result obtained from the model (with LID) found to be satisfactory with the ground reality. In this present study, PCSWMM has been used for the first time for modelling storm water drains of the urban area (Delhi city, India), and the results obtained from simulation are in good consent with the observed data. Thus, the present study will help the flood managers to manage excess surface runoff during storm more efficiently in urban cities.

Keywords PCSWMM · Urban flooding · Storm water drainage systems · Low impact development (LID)

R. Dabas (✉) · M. Kumar
Delhi Technological University, Delhi, New Delhi 110042, India
e-mail: ruchikadabas23@gmail.com

S. Kumar
Indian Institute of Technology Delhi, Hauz Khas, New Delhi 110016, India

L. M. Gupta et al. (eds.), *Advances in Civil Engineering and Infrastructural Development*, Lecture Notes in Civil Engineering 87,
https://doi.org/10.1007/978-981-15-6463-5_26

1 Introduction

Urban drainage systems are planned to run out the surface runoff from urban areas (e.g. paved streets, parking lots, sidewalks and roofs) during storm events. However, excess storm water exceeding the drainage capacity can cause urban flooding and result in traffic interruption, economic loss, pollution and health issues. An increase in impervious land cover led to more surface runoff, faster runoff concentration and increase the peak flow rate. Thus, there is an increasing need to improve drainage capacity to reduce flooding in fast urbanizing areas. The conventional method to urban flooding becomes impractical nowadays because it requires more area. However, the conventional method can be designed in the catchment to collect storm water by adopting source control method. There are so many new innovative techniques like low impact development techniques (LID) is one of the new techniques which help to solve the urban flooding-related problems. Many new storm water management techniques have been developed to solve runoff problems such as green roofs and rainwater harvesting. These different techniques named as low impact development techniques.

For managing runoff generated in the urban catchments, several models are available like MOUSE (DHI), HEC-1(US army), Hydroworks (HRWL), SWMM (Storm water management model) and PCSWMM (CHI). Out of this, PCSWMM was identified as the GIS-based powerful model for modelling urban drainage networks because it precisely matches the simulated results with the observed data. PCSWMM used SWMM model for modelling storm water drainage systems. SWMM is a dynamic rainfall-runoff model for the simulation of quality and quantity complications related to catchments runoff in urban areas [12, 18]. EPA had developed storm water management model (SWMM) in the year 1791 which is extensively used to simulate all features of urban hydrologic and water cycles. The features include surface runoff, rainfall and drainage network's flow routing, snowmelt and concentrations of pollutants [12]. Further, simulation through single or continuous events can be executed for the catchments where combined sewer drains or natural drains or storm water drain exists. The model provides information like flooding locations as well as flood volume but unable to manage it.

There are several storm management methods have been developed to manage flood volume such as green roofs, infiltration trenches and rainwater harvesting. These technologies are known as sustainable drainage systems. LID has been recommended as an innovative solution for storm water management [1–6, 16, 17]. To understand more clearly why it is useful for communities, the watersheds are incorporated with LID techniques, and it is helpful to first understand the effects of traditional storm water management, which have marked land development and altered waterways and water quality for decades [7–11]. The main aim of this technology is to maintain rain water runoff and also help in minimizing flooding [12–15].

Thus, for managing the urban storm water system, distributed rainfall-runoff models are used. But the main difficulties in using these models are that developing of these models is a time taking the process and also about the model accuracy.

PCSWMM is the most dynamic rainfall-runoff model in which runoff is generated from the subcatchments goes directly to manholes or to the downstream subcatchments. Thus, spatial discretization of the catchments is required by the rainfall-runoff model for developing runoff model computationally.

In the present study, PCSWMM has been used for analyzing the storm water drainage system of Delhi city, India. Further, the LID is applied into the developed PCSWMM model for managing the excess storm water surface runoff. Thus, the study will help the flood managers to manage excess surface runoff during storm more efficiently in urban cities.

2 Study Area

The study area (Fig. 1) is Delhi. Delhi, the capital city of India, situated in the northern part between the latitudes of 28°35′02″ and 28°30′09″ N and longitudes of 77°15′53″ and 77°20′44″ E. Delhi has a huge variation in temperature and rainfall during summer and winter. The average temperatures in the city are around 29 °C that varies between 5 °C (in winter) to 48 °C (in summer). The average rainfall in Delhi is around 797.3 mm.

3 Model Specification: Storm Water Management Model (SWMM)

SWMM model is also capable of simulating the surface runoff and its quality along with the computable features of storm water. SWMM tracks the quantity and quality of runoff generated within each subcatchment, and the flow rate, flow depth and quality of water in each pipe and channel during a simulation period comprised of multiple time steps. SWMM was first developed in 1971 and has undergone several major upgrades since then. The newly added LID module (SWMMVersion5.1.006) is expected to have the capability of simulating the storm water management performance of various types of LID practices including permeable pavements. The SWMM model primarily developed for urban settlements is proficient in simulating the quality and quantity of water generating from the watershed for both single events as well as long-term (continuous) event. In the present study, dynamic wave method and Green-Ampt model is employed for computing flow routing and losses due to infiltration repectively. Rossman [18] can be referred for additional details on SWMM. Though in this study, inflows have not been assigned to any nodes nevertheless the feature of assigning inflow as a boundary condition to the nodes is also available in SWMM.

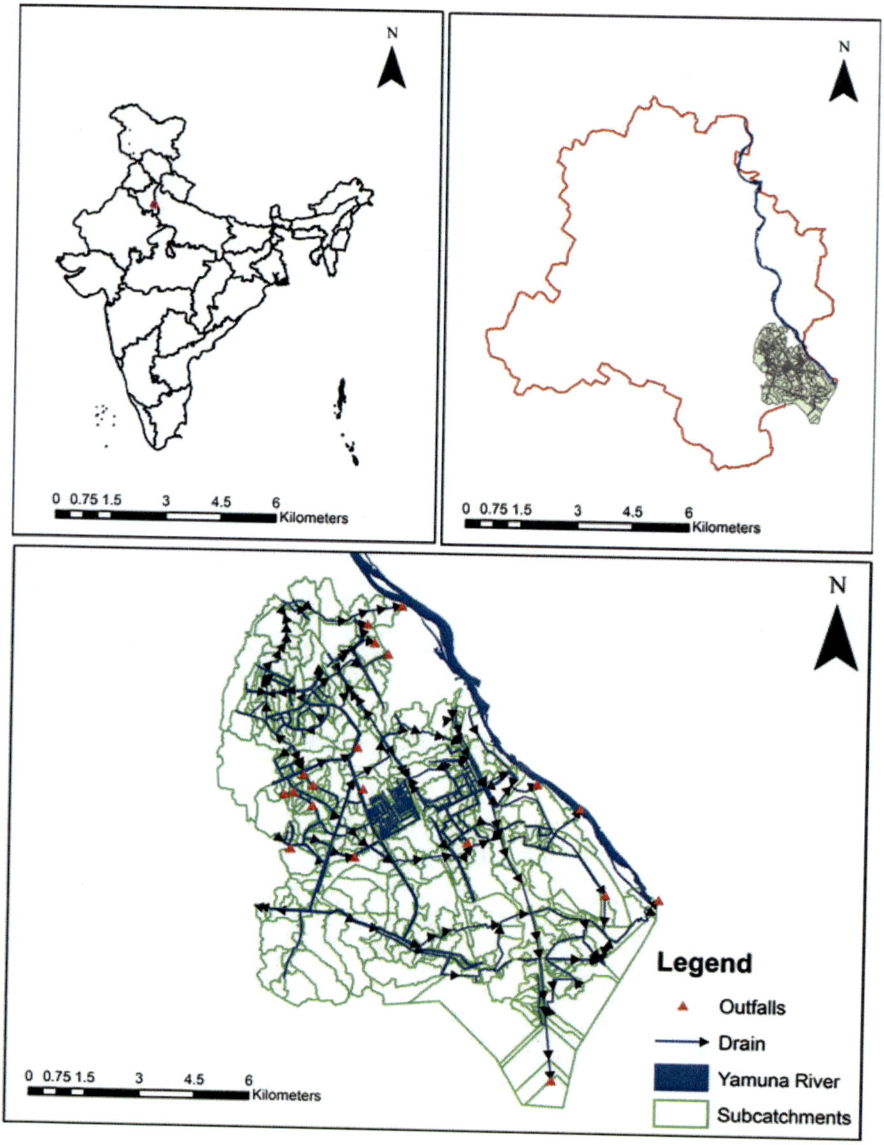

Fig. 1 Study area

4 Data Used

The data used to develop SWMM5 model have been collected from the various government agencies (Table 1).

Table 1 Description of the data used

S. No.	Spatial data	Resolution	Source
1	Digital elevation model (DEM)	30 m × 30 m	Shuttle radar topography mission (SRTM) of USGS
2	Landuse/land cover (LU/LC) of 2010	1:250,000	National Remote Sensing Centre, Indian Space Research Organization (ISRO)
3	Soil	1:250,000	National Bureau of Soil Survey & Landuse Planning (NBSS-LUP)
4	Drains cross section survey data	At 100 m RD (reduced distance)	Various agencies of Delhi Government (I&FC, PWD, MCD)

5 Model Setup

The surveyed data of the storm water drains for the study area was converted into conduits layers and manholes layers with the help of ArcGIS 10.1. The drains information like invert level, width, depth, latitude and longitude, conduits length was available in the attributes table. Then, the outfall point layers of the drains were generated in ArcGIS. All these layers were imported into the PCSWMM. In PCSWMM, watershed and subcatchments were generated from the DEM on the basis of manholes. The subcatchments was delineated in such a fashion that runoff generated from each subcatchments will contribute to single manholes. Further, the landuse and soil were given to the catchments. And finally, rainfall station information and rainfall data were given to the PCSWMM. After loading all the necessary input data required by the PCSWMM, the model was initially run for the day having peak rainfall. The methodology is explained by a flow chart as shown in Fig. 2.

6 Results and Discussion

The rainfall-runoff model was developed using PCSWMM is shown in Fig. 3. Precipitation data from the years 2015–2019 of 15 min interval received from Indian Meteorological Department, New Delhi, was used for the simulation. Between 2015 and 2019 years, the maximum rainfall recorded was 84.2 mm on 20 August 2016. Since, our objective is to store the excess storm water runoff overflowing from the existing drainage systems. So, the model was first simulated, and the results obtained were evaluated for the above days. Time series plot of the rate of flow of the drains, flooding rate and velocity of runoff is shown in Fig. 3a, b. For validating the model efficiency, the waterlogging locations identified by the model were checked with the

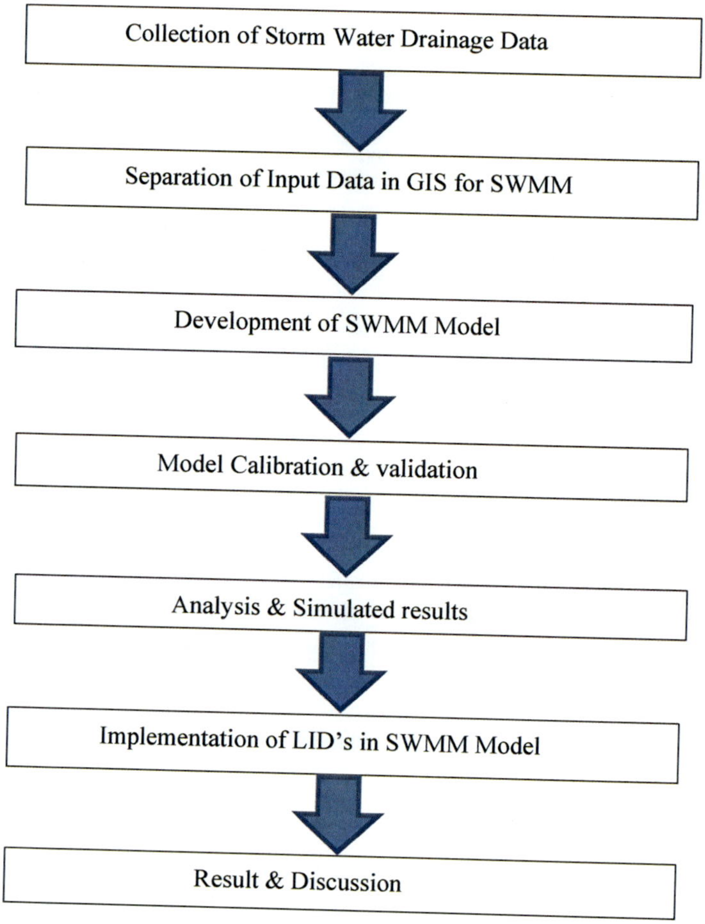

Fig. 2 Methodology

past records. The waterlogging locations information is available from Delhi government. It was found that the simulated results are in close agreement. The waterlogging may be caused due to various reasons like inadequate drainage infrastructure like drains are undersized due to which waterlogging takes place at the junctions or downstream. The other reasons are the rainfall intensity and duration of rainfall, improper gradient of drains, i.e. invert levels of the drains are not uniform due to which flow is not able to go downstream, and hence, backflow takes place causing waterlogging in the upstream. The most important reasons were the incapability of the existing drainage system to accommodate the excess runoff. Thus, further LID like infiltration trenches, green roof, rainwater harvesting was implemented in the study area has shown in given Fig. 4a, b. The results show huge reduction in the waterlogging locations and also in flood volume.

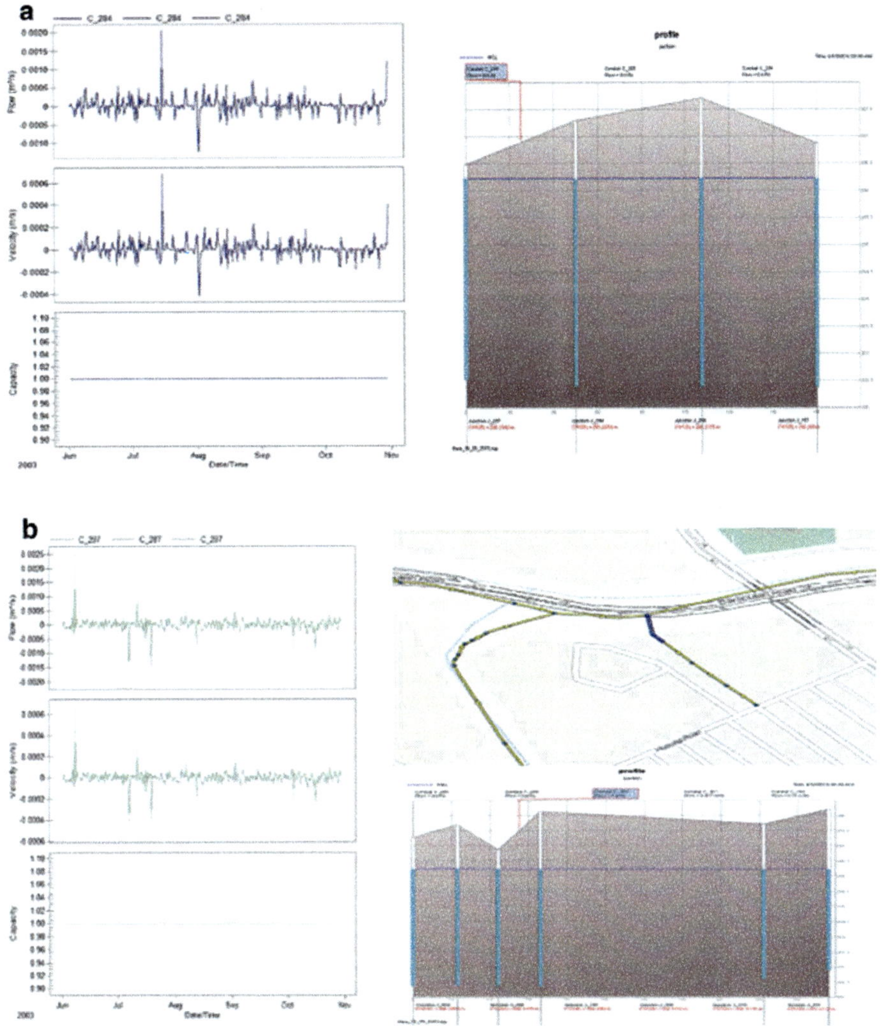

Fig. 3 Shows the time series plot of flow rate in the drains, flooding and velocity of runoff in a and b

7 Conclusions

Along with the rapid urbanization in Delhi, urban distress has occurred and has been great concerns of the society. Among these, urban problems give the negative impact of urban runoff on hydrology as well as the water quality of urban storm water. There are various conventional storm water models are available in the market, and all the models are ranging from simple processes for prediction of results. However, the conventional system can be designed in the catchment to collect storm water

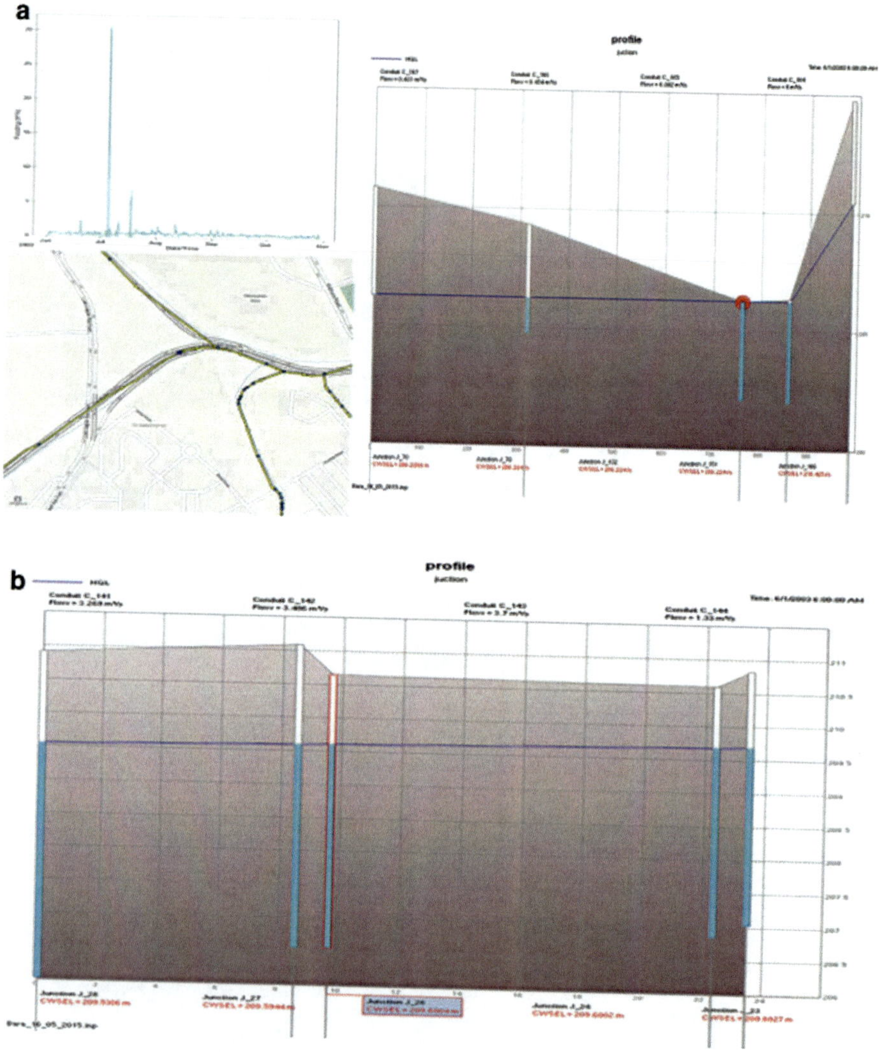

Fig. 4 Shows reduction in flooding by implementing LID like infiltration trenches, as given a and b

by adopting some source control. Many developed and developing countries adopt this LID's techniques to solve urban flooding-related problems. It is modern new innovative technique for management of storm water by controlling its source by achieving pre development conditions. The LID also helps in environmental issues like water quality and to minimize the quantity.

Acknowledgements The authors would like to thank Delhi Government, India, for providing the data to carry out the work. The authors also want to acknowledge the PCSWMM for providing free license for the research work.

References

1. Andoh R, Declerck C (1997) A cost effective approach to stormwater management, source control and distributed storage. Water Sci Technol 36(8):307–311
2. Berland A, Shiflett SA, Shuster WD, Garmestani AS, Goddard HC, Herrmann DL, Hopton ME (2017) The role of trees in urban stormwater management. Landscape Urban Plan 162:167–177
3. Chaffin BC, Shuster WD, Garmestani AS, Furio B, Albro SL, Gardiner M, Spring M, Green OO (2016) A tale of two rain gardens: barriers and bridges to adaptive management of urban stormwater in Cleveland, Ohio. J Environ Manag 183:431–441
4. Chang NB, Lu JW, Chui TFM, Hartshorn N (2018) Global policy analysis of low impact development for stormwater management in urban regions. Land Use Policy 70:368–383
5. Chen Y, Samuelson HW, Tong Z (2016) Integrated design workflow and a new tool for urban rainwater management. J Environ Manag 180:45–51
6. Dhakal KP, Chevalier LR (2016) Urban stormwater governance: the need for a paradigm shift. Environ Manag 57(5):1112–1124
7. Dhakal KP, Chevalier LR (2017) Managing urban stormwater for urban sustainability: barriers and policy solutions for green infrastructure application. J Environ Manag 203:171–181
8. Gogate NG, Kalbar PP, Raval PM (2017) Assessment of stormwater management options in urban contexts using multiple attribute decision-making. J Cleaner Production 142:2046–2059
9. Guizani M (2016) Storm water harvesting in Saudi Arabia: a multipurpose water management alternative. Water Resour Manag 30(5):1819–1833
10. Hoang L, Fenner RA (2016) System interactions of stormwater management using sustainable urban drainage systems and green infrastructure. Urban Water J 13(7):739–758
11. Hwang HM, Fiala MJ, Park D, Wade TL (2016) Review of pollutants in urban road dust and stormwater runoff: Part 1. Heavy metals released from vehicles. Int J Urban Sci 20(3):334–360
12. Huber WC, Dickinson RE (1992) The USEPA SWMM4 stormwater management model, version 4: user's manual, University of Guelph, Guelph, Ontario 174
13. Kirshen P, Aytur S, Hecht J, Walker A, Burdick D, Jones S, Mather L (2018) Integrated urban water management applied to adaptation to climate change. Urban Clim 24:247–263
14. Ma Y, Egodawatta P, McGree J, Liu A, Goonetilleke A (2016) Human health risk assessment of heavy metals in urban stormwater. Sci Total Environ 557:764–772
15. Martin C, Ruperd Y, Legret M (2007) Urban stormwater drainage management: the development of a multicriteria decision aid approach for best management practices. Eur J Oper Res 181(1):338–349
16. Montalto F, Behr C, Alfredo K, Wolf M, Arye M, Walsh M (2007) Rapid assessment of the cost-effectiveness of low impact development for CSO control. Landscape and urban planning 82(3):117–131
17. Palhegyi GE (2010) Designing storm-water controls to promote sustainable ecosystems: science and application. J Hydrol Eng 15(6):504–511
18. Rossman LA (2010) Storm water management model user's manual, version 5.0 (p. 276). Cincinnati: National Risk Management Research Laboratory, Office of Research and Development, US Environmental Protection Agency

Analysis of Parameters for Storm Water Management Model (SWMM) by Using GIS

Ruchika Dabas and Munendra Kumar

Abstract The most important and basic data required for assessment of water resource is Runoff. For this runoff water requires proper planning, sustainable management, water control Strategies. In this present investigation, the area of Yamuna river flowing through Delhi has been selected for the study. The study depicts development of watershed catchment delineation, flow direction, development of streams, Land use and cover map and Digital elevation model (DEM) by using Geographic Information System (GIS) applications. DEM is generated from satellite image. As a result of this, the Runoff of this catchment is discharged into the Yamuna River. By using the available data rainfall intensity analysis work has been carried out and the IDF curve has been generated.

Keywords Runoff estimation · GIS · Water-shed delineation · SWMM

1 Introduction

Runoff is very important data required for analysis of available water resources for optimum planning, management and new and innovative strategies developed such as waterways, storage facilities or erosion control structures. Knowledge of peak and total runoff due to rainfall is most important for planning and designing any hydrologic structure. Direct measurement of runoff is always good but in most of the cases, it is not possible at desired time and location and most of the developing countries infrastructure is not available for measurement of runoff and other hydrological parameters, as most of the watersheds are ungauged. Rainfall-runoff modelling in the urban catchment is quite different and complex as compared to the rural areas because of dynamic nature of hydrological process due to dynamic land use/cover, heterogeneous nature of urban catchment and more human interference etc. A rational of hydrograph method that can simulate the runoff corresponding to variable rainfall intensity (Smith and Lee 1984). Impervious area and rainfall are

R. Dabas (✉) · M. Kumar
Civil Engineering Department, Delhi Technological University, Delhi, New Delhi 110042, India
e-mail: ruchikadabas23@gmail.com

L. M. Gupta et al. (eds.), *Advances in Civil Engineering and Infrastructural Development*, Lecture Notes in Civil Engineering 87,
https://doi.org/10.1007/978-981-15-6463-5_27

the two main factors, which influence the runoff responses in urbanized watersheds. Hydrological cycle in urban areas can be assumed to start with rain falling over the ground, which is either impervious or pervious area. Some rainfall infiltrates into the sub-surface from the previous area and the remainder forms the surface runoff. Surface runoff will eventually flow into a watercourse and finally to a receiving water body. This is not the case for an impervious area, where nearly all the rainfall becomes runoff. A study being conducted by (Guo 2001) emphasized on improvement of the rational hydrograph method by introducing a new formula to compute the time of concentration.

Thus, rainfall-runoff phenomenon is dynamic in urbanized watersheds and the quantity of water available from corresponding water bodies' changes with time. Therefore, a simplified model/relationship is warranted to estimate the runoff generated from the urbanized watershed for planning level and design level applications. Such a model/relationship should account dynamic land-use/cover aspect of the urbanized watersheds and require minimum data. For the Indian conditions, such a model/relationship has not been reported in the literature. In urbanized watersheds, impervious areas contribute a major portion of runoff. So correct estimation of impervious areas or extraction of land use/cover information is very vital. Remotely sensed images and related techniques like digital image processing and GIS can be used to extract the land use/cover information. Further, we need to forecast the water demands and quantity of runoff available from local surface sources (lakes, ponds) in the near future, while planning the optimum use of available water resources. This, in turn, requires a relationship between urbanization (change in land use/cover information like pervious areas into imperious one) and runoff to be generated from urbanized watersheds for given average meteorological conditions. Therefore, the present study is aimed to investigate the various aspects of rainfall-runoff phenomenon for urbanized watersheds and This can be utilized for Rainfall-Runoff modelling for urbanized watersheds based on some of the easily available catchment and meteorological characteristics and suitable for planning level applications.

2 Literature Review

Many studies are available of rainfall-runoff modelling for rural areas, however, a very few attempts have been made to address the issue of urban areas. In the Indian context, no example of such deterministic rainfall-runoff model has been reported in the literature which is suitable for planning level applications. To find runoff depth of the basin area using SCS-CN (soil conservation service-curve number) method in integration with remote sensing and GIS, a well-developed technique with spatial and temporal distribution of hydrological parameters [1] but this method provides approximate results (Give the limitations of this method for urbanized catchments). Simple approach for calculating runoff and pollution loads in urban catchments, and discusses conceptual modelling methods for simulating daily runoff and pollution loads [2] but it has been found that this approach did not account spatial variability of urban hydrological process and catchment. So, further work can be certainly extended to develop a simple model into the deterministic model by incorporating catchment characteristics like slope, evapotranspiration based on land use/cover, evaporation, infiltration, etc.

The SCS model, which fully considers (e.g., topography, soil, and land use), to simulate the rainfall-runoff relationship of small watershed [3] but did not consider for the spatial and temporal variability of urban hydrological processes in which remote sensing and GIS technique will be more useful as it considers spatial and temporal of different hydrological processes can be addressed very efficiently and accurately. Some of the researchers have found that the true and detailed representation of the physical system considering spatial and temporal variability enhances the overall performance in the modelling process. Till now, few attempts have been made to model the various hydrological processes using RS and GIS techniques mostly for the urban estimation [4–6].

Rosmiller, 1982 [7] developed a rational hydrograph formula based on the assumptions of rational formula to compute hydrographs using a constant rainfall intensity deduced from the intensity-frequency-duration curves. Smith and Lee 1984 [8] developed a rational hydrograph method that can help to simulate runoff corresponding to variable rainfall intensity. A study being conducted by Guo, 2001 [9] gives importance on the improvement of the rational hydrograph method by introducing a new formula to compute the time of concentration.

Bennis and Crobeddu [10] presented runoff simulation model based on linear system theory for small urban catchment based on the improvement in the rational hydrograph (RH) method which mainly considers the pervious area & impervious area, time variability of rainfall, initial abstraction on impervious areas and infiltration on pervious area.

3 Materials and Methods

The different secondary data has been collected from various concerned departments in India and thereafter preparation of different thematic layers was done using remote sensing and GIS techniques. An understanding of urban dynamics and estimation of urban growth has been made to determine impervious area. Thereafter, model parameterization has been done by using GIS database in order to get model calibration and validation. Thereafter, model parameterization was done using GIS database in order to get model calibration and validation. Also, generation of rainfall-runoff responses with respect to space and time (runoff depth for different years at micro-watershed scale) for selected watershed has also been done. The formulation of the deterministic rainfall-runoff model suitable for planning level application has also been selected for the same. Thereafter the calibration of the proposed model has been done using watershed characteristics of the study area.

3.1 Data Used in Present Study

Required data has been collected from various departments like Central Water Commission (CWC), UP Irrigation department, Survey of India, Delhi Municipal corporation etc. The data being collected is observed flow data for model calibration and validation, Survey of India Topo sheets of the proposed area, Base Map, Satellite data/DEM, Soil Characteristics (Soil Map), Land use/land cover information, Drainage network Data, Meteorological data (Rainfall Data).

3.2 Present Study Area

To study the various aspects of rainfall-runoff modelling, an urbanized watershed is selected as the study area for which required data are available. After analysis and verification of available data, we have decided that Yamuna river data will be suitable for research work. The map shown through Fig. 1 in the catchment area contributed in Yamuna river in between old railway bridge in Delhi and Okhla barrage in U.P., India.

4 Results and Discussions

The results indicated in the form of a DEM, Land use/cover map, Soil map, flow direction, development of streams and preparation of drainage map.

4.1 Digital Elevation Model (DEM)

DEM is generated from satellite image. Digital Elevation Models are files that contain the elevation of the terrain over a particular area, generally used for fixed grid interval. DEM used for extracting terrain parameters, contour development, catchment area delineation and stream/Drain development by using GIS.

4.2 Soil Map

Geotechnical survey of India map has been collected and create a GIS database by using ARC GIS software. Soil map is used to find out the soil characteristics like permeability, porosity, rate of infiltration, etc. This will be required for the parameterization of rainfall-runoff model.

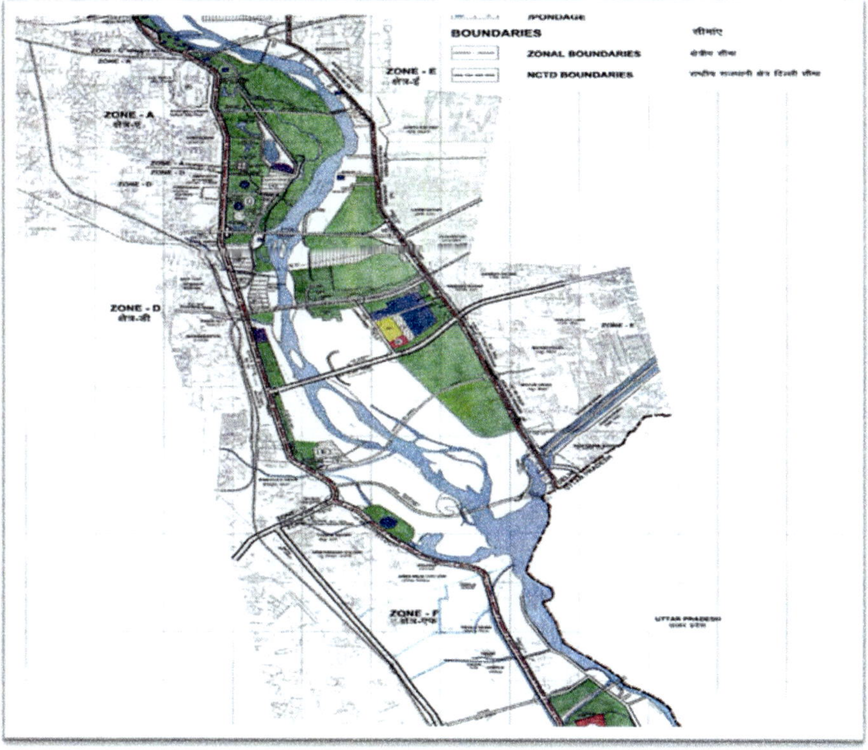

Fig. 1 Map showing the location of catchment area consider in the study

4.3 Land Use/Cover Information Map

Land use/cover map developed by ARC GIS software based on Delhi master plan as shown in Fig: 2 given below. Remote sensing and GIS related techniques are used to extract the land use cover information. Land use map is used for identification of catchment surface cover like a residential area, industrial area, water bodies, green area, open area, roads, public and semipublic facilities, etc. Land use map has been used for calculation of the percentage of previous area and drawing the drainage lines for base map of study area is shown in Fig: 3).

4.4 Drainage Network Map

Drainage map of selected study area is prepared with the help of topo sheets, base map collected from local authority, DEM using GIS technology, etc. as shown in

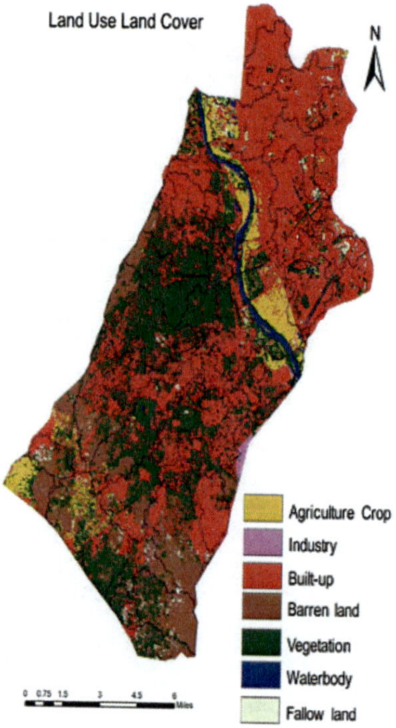

Fig. 2 Preparation of Land use/cover map of the study area

Fig. 4. General topography of the study area is flatter terrain, some low lying areas are situated along the Yamuna river and general slope of the study area is from west to east towards Yamuna river and north to south. All the major and minor drains have been marked on the map. Flow direction and demarcation of sub-catchment area are marked on the map. As shown in the map all the tributaries are connected to the main drains and main drains are finally discharge into Yamuna river. Catchment boundary of the selected watershed area has been marked on the base map and total catchment is about 14 km^2. Study area comprises of various areas like Okhla, Malviya Nagar, Mahrauli, Qutub Minar, Masud Pur, Chirag Delhi, Dhaula Kuon, India Gate, Connaught Palace, Sarojini Nagar, AIMS area, Delhi golf course, etc. (Fig. 5).

4.5 *Meteorological Data*

Rainfall data obtained from the meteorological department for the period 1975–2004. The annual average rainfall in Delhi is 555 mm. Most of the rainfall occurs between July to September. The maximum rainfall that occurs in a day during the last 30 years was 255.9 mm on 30 June 1998. The maximum rainfall that occurs one hour in a day during the last 30 years was 68 mm on 7 July 1993.

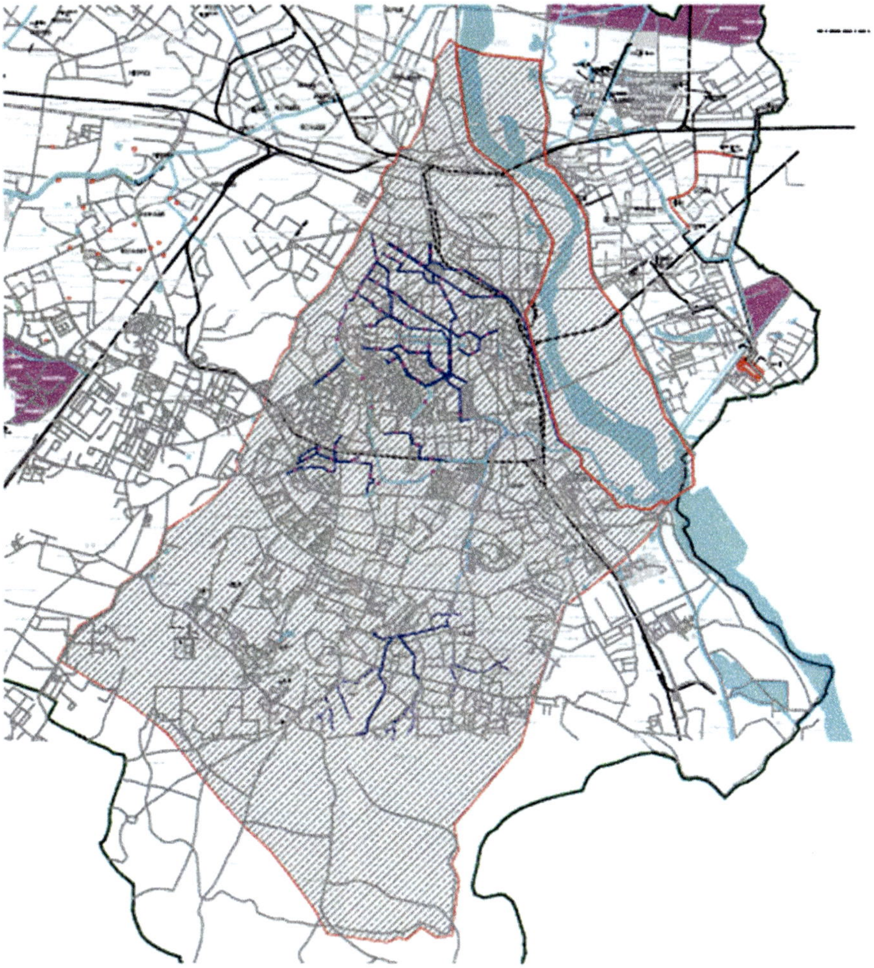

Fig. 3 Base Map of study area indicating drainage lines

4.6 Rain Fall Intensity Analysis

By using the above available data rainfall intensity analysis work has been carried out and IDF curve has been generated. As shown in figure Metcalf and Eddy equation does not represent the true rainfall characteristics of the region. The Talbot equation most closely represents the true rainfall characteristic of region (Fig. 6).

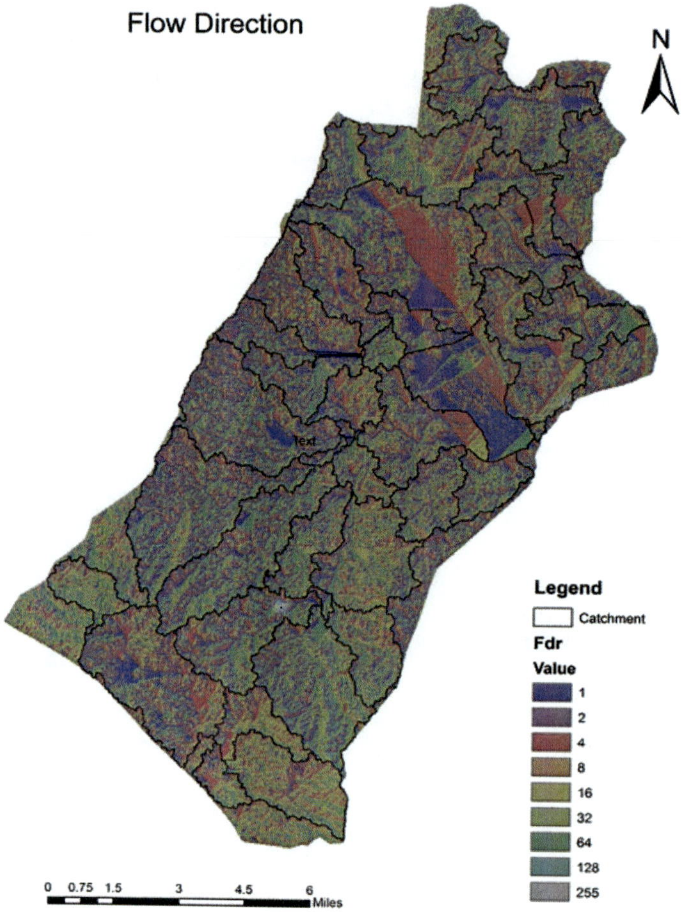

Fig. 4 Flow direction in watershed delineation of study

5 Conclusion

GIS database for SWMM which can be used for planning level applications and the above figures are indicated that development of DEM, land use/cover map, Drainage map, flow direction, development of streams, respectively. In this Arc GIS 10.2 is used, where DEM, Soil Type, Land use, Drainage Map, Flow Direction all are processed in GIS with the help of SWMM Model and then converted into usable Format of SWMM and Rainfall Data Is also used for this Study of this Area. In this study, it also discussed different parameters required for the SWMM model by using GIS. SWMM modelling approach as an efficient tool in catchment modelling of input parameters for catchment Hydrology. Therefore SWMM integrated with GIS have the capability to evaluate storm rainfall.

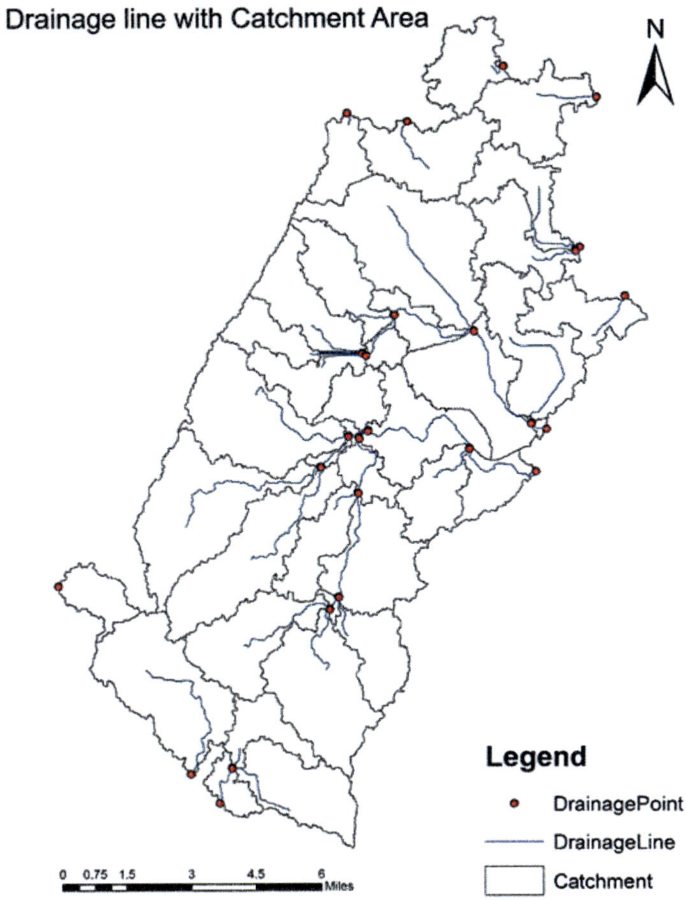

Fig. 5 Development of streams of the study area

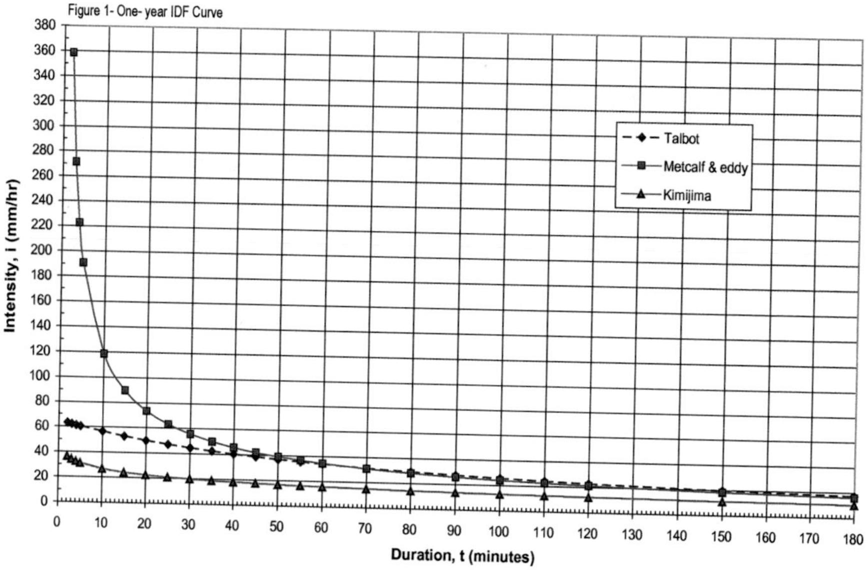

Fig. 6 Intensity Duration Frequency (IDF) curve

References

1. Coskun M, Musaoglu N (2014) Investigation of rainfall- runoff modelling of the Van Lake Catchment By using remoting sensing and GIS integration
2. Chiew FHS, McMahon TA (1999) Modelling runoff and diffuse pollution loads in urban areas. Wat Sci Tech 39(12):241–248
3. Liu X, Li J (2008) Application of SCS model in estimation of runoff from small watershed in Loess Plateau of China. Chin Geogra Sci 18(3):235–241 (College of Geography and Planning, Ludong University, Yantai 264025, China)
4. Ross LG, Mendoza QM, Beveridge MCM (1993) The application of geographic information systems to site selection for coastal aquaculture: an example based on salmonids cage culture. Aquaculture 112:165–178
5. Meyer SP, Salem TH, Labadie JW (1993) Geographical information system in urban storm-water management. J Water Resour Plan Manag 119(2)
6. Sample DJ, Heaney JP, Leonard TW, Koustas R (2001) Geographic information systems, decision support systems, and urban storm-water management. J Water Resour Plan Manage 127(3):155–161
7. Rossmiller RL (1982) Rational formula revisited. In: Proceedings of conference on stormwater detention facilities: planning, design, operation, and maintenance, engineering foundation urban water resources research council, ASCE, New England College, Henniker, N.H., Aug. 2–6, pp. 146–162
8. Smith AA, Lee K (1984) The rational method revisited. Can J Civ Eng (11):854–862
9. Guo JCY (2001) Rational hydrograph method for small urban watersheds. J Hydrol Eng 6(4):352–357
10. Bennis S, Crobeddu E (2007) New runoff simulation model for small urban catchments. J Hydrol Eng, ASCE, September-October-2007

Application of Six Sigma on RMC Plant

Viraj Parekh, Sagar Jotani, and Jay Patel

Abstract Six sigma is a quality improvement technique widely used in manufacturing industry. Application of six sigma is yet to be explored exhaustively in the field of construction. This study explores the possibility of application of six sigma in the process of manufacturing of concrete in the RMC plant situated at Navsari, Gujarat. This paper defines and analyzes process performance of RMC plant using six sigma technique. The data was collected in the form of batches and its mix proportion. Total data was collected for 469 batches of the said plant and analyzed using MS Excel. As per IS 4926:2003, tolerance limit for cement, aggregate, and water was set, and batch-wise upper limit and lower limit were calculated. These calculations were then compared with the target value set by the RMC plant. This comparison led to calculate the defects per million opportunity (DPMO) value for the concrete manufacturing process of the RMC plant. With the help of DPMO value, the sigma value is found out to be 1.77 which is not at par with the ultimate sigma level. This work provides valuable insights on the implementation of six sigma technique in the construction industry. Six sigma technique can evaluate the quality of the process and quantify the improvement goals if the process is under performance and less efficient. Improvement in quality of construction can be observed at large extent if six sigma technique is applied to more complicated, volatile, and multistep projects with a linkage to lean principle.

Keywords Six sigma · Quality management · Process performance · Ready-mix concrete

---

V. Parekh (✉) · S. Jotani · J. Patel
Institute of Technology, Nirma University, Ahmedabad, India
e-mail: viraj.parekh@nirmauni.ac.in

S. Jotani
e-mail: 17bcl171@nirmauni.ac.in

J. Patel
e-mail: 17bcl169@nirmauni.ac.in

L. M. Gupta et al. (eds.), *Advances in Civil Engineering and Infrastructural Development*, Lecture Notes in Civil Engineering 87,
https://doi.org/10.1007/978-981-15-6463-5_28

1 Introduction

The six sigma concept is derived from statistical distribution known as "standard normal distribution" illustrated by a symmetrical bell-shaped curve. Theoretically, this bell-shaped curve has been extensively studied and has been proven very useful as numerous natural continuous phenomena which seem to follow it or can be approximated by it [1]. Six sigma is used in manufacturing industry since last two decades to monitor and improve the quality aspect in the process. The six sigma concept emphasize on improving the process performance rather than focusing on final output only. Process performance can be evaluated by defects per million opportunities (DPMO). To achieve the sigma level at six, the value of DPMO is 3.4 [4].

The application of six sigma in manufacturing industry has been proven quite useful. So, it was applied in construction industry as well, but at small-scale projects such as a residential building to improve the quality of internal finishing work [3], this principle was applied to determine and improve the key input variables affecting the cracks in lightweight partition walls [2]. Application of this technique is yet to be explored for evaluating the process efficiency of RMC plant. Therefore, to assess the application of six sigma at this level, RMC plant of Navsari, Gujarat, was selected for the research purpose.

The selected plant for this study has a capacity to produce 30 cubic meter of concrete when all the four hopers are filled fully. The plant produces M30, M35, and M40 grade concrete mainly, and they transport in the surroundings for the various bridge construction sites.

2 Methodology

The data was collected for 469 batches of concrete. The sample of collected data is given in Tables 1 and 2 for three batches only. Table 1 shows the data for individual batches, and Table 2 shows the combined data of all the three batches. The collected data is provided with target value of each material which were used to manufacture the concrete. The checklist was prepared in MS Excel for all 469 batches of concrete. The checklist helped to verify whether the target value of all the materials which were used in manufacturing process of concrete were within the tolerance limit or not as per IS 4926:2003. Table 3 shows the sample checklist for the three batches. One which meets the standard requirement is marked with "√" sign and else is marked with "X" sign.

From the checklists of all 469 Batches, it was observed that total number of defects are 1108 out of total 2814 opportunities. Therefore, DPMO is found out as 393,745.557. DPMO is calculated as per following formula.

$$\text{DPMO} = (\text{No. of defects}/\text{No. of opportunities}) \times 1{,}000{,}000$$

Table 1 Sample of batch-wise collected data

Batch No.	Cycle time (sec)	White sand (kg)	20 mm aggregate (kg)	10 mm aggregate (kg)	Black sand (kg)	Cement (kg)	Water (kg)	Additive (kg)	Total (kg)
1	118	131	357	214	260	219.2	89.8	2.33	1273.33
2	59	130	365	226	238	215.2	85.1	1.76	1261.46
3	69	133	367	223	245	216.6	84.8	1.73	1271.13

Table 2 Sample of collected data for combined batches

	White sand	20 mm aggregate	10 mm aggregate	Black sand	Cement	Water	Additive
Moist. %	0	0	0	0	0	0	0
Actual total	394	1089	663	743	651.4	259.7	5.82
Target total	388.5	1107	670.5	753	649.5	255	5.25
Average value	131.33	363	221	247.66	217.13	86.56	1.94
Target value	129.5	369	223.5	251	216.5	85	1.75

Table 3 Sample of checklist prepared

	White sand	Black sand	20 mm aggregates	10 mm aggregates	Cement	Water
Target value	129.5	369	223.5	251	216.5	85
Upper control limit	133.385	380.07	230.205	258.53	220.83	87.55
Lower control limit	125.615	257.93	216.795	243.47	212.17	82.45
Batch 1	✓	✓	X	X	✓	X
Batch 2	✓	X	X	X	✓	✓
Batch 3	✓	X	X	X	✓	✓

Based on the sigma level table as given in Table 4, the sigma value was calculated with respect to DPMO value, and it came out to be 1.77 sigma level.

Table 4 Overview of sigma level and DPMO

Sigma level	DPMO
1	690,000
2	308,000
3	66,800
4	6210
5	320
6	3.4

3 Data Analysis

DMAIC, i.e., define, measure, analyze, improve, and control, is an integral part of six sigma process for continuous improvement. DMAIC is a five-step process where one has to define the key variables in process at first place, measure the process performance at second place, analyze the data collected, find out the root-cause at third place, improve the process after rectifying the root-cause at fourth place, and finally control the improved process.

To define, SIPOC analysis is used in the paper. SIPOC stands for supplier, input, process, output, and customer. SIPOC explains the entire process with all the activities and ingredients required in it as given in Table 5.

Next step of DMAIC is to measure the defect of the process, and for that Pareto chart is prepared and studied. Pareto chart is a bar and line graph, where data is represented by bar graph, and line graph represents cumulative total. It shows which factor is causing 80% defect by having 20% weightage in the process. Pareto chart for this study is presented as Fig. 1.

Fish-bone diagram is prepared to do root-cause analysis as a part of third step of DMAIC. Analysis of any data leads to a conclusion about what is to be improved and controlled as a part of fourth and fifth stage of DMAIC methodology. Figure 2 shows the fish-bone diagram for the manufacturing process of RMC plant.

Table 5 SIPOC analysis for RMC plant

Supplier	Inputs	Process	Outputs	Customers
RMC plant	1. Cement 2. Aggregate 3. Sand 4. Potable water 5. Admixture 6. Fly ash 7. Batching plant 8. Control cabin	1. Pumping of ingredients 2. Batching of ingredients based on grade of concrete 3. Transportation of ingredients through belt conveyor into mixture 4. Dry mixing of ingredients 5. Wet mixing of ingredients 6. Filling of concrete into trucks	Ready-mixed concrete (RMC)	Builder or owners

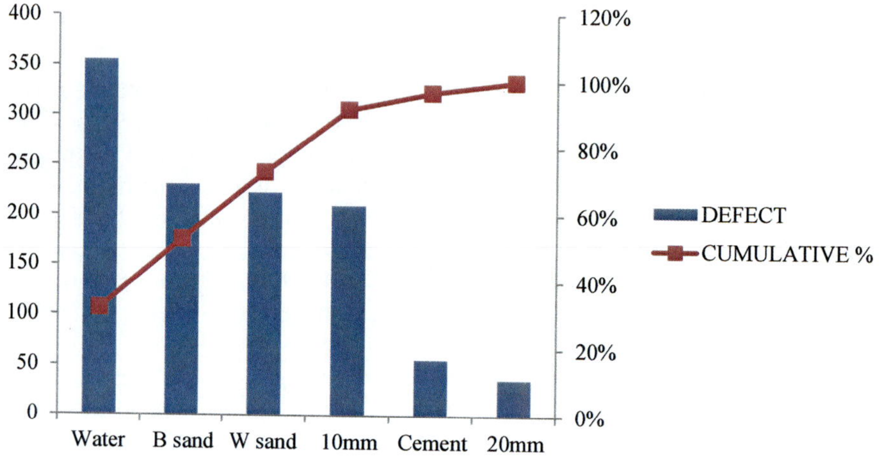

Fig. 1 Pareto chart for RMC plant

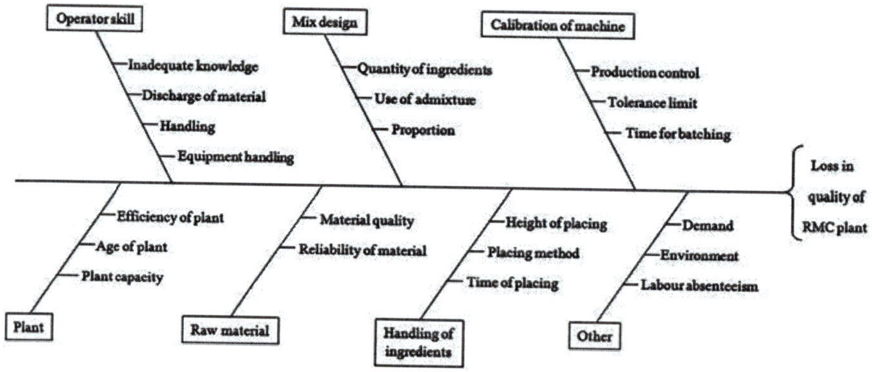

Fig. 2 Fish-bone diagram for RMC plant

4 Result

This work comprehend the application of six sigma technique in construction industry. The principle can be applied to RMC plant to evaluate the quality of the manufacturing process and to find out the scope of improvement.

The study carried out on the manufacturing process of RMC plant has given some valuable insights about the defect in the process, DPMO value, and sigma level. The sigma level was found out as 1.77 which shows that a lot of improvement is required in the process of manufacturing of concrete. The ultimate sigma level is 6 and compared to that, 1.77 is articulated as quite a low measure of quality. Pareto chart showed that it is the quantity of water which causes the major quality defect in the process. The same was found out from the fish-bone diagram too.

5 Conclusion

Quality control in construction projects focuses only on materials rather than on work flows. In present study, methodological and statistical approach is used to determine quality of manufacturing process of RMC plant. As of now, no research is available in the field of application of six sigma in RMC plant. So, this study helped us to apply well-known concept in comparatively new field of work. If quantity of water can be controlled for this study, it will ultimately increase the quality of produced concrete. Usage of that concrete will lead to the better quality of construction. However, quality of the concrete is ultimately considered based on the results of cube test but the objective here is to examine the quality of the process.

With the aid of the six sigma technique, the parameter which affects the quality can be controlled with scientific approach. This technique can be applied for each activity at micro-level for project improvement and process improvement. In future, it can be proven a very useful statistical tool to identify the defect and improve the output for any construction material manufacturing. This tool can be applied on construction projects as well to measure the quality of final outcome.

References

1. Aboelmaged MG (2010) Six sigma quality: a structured review and implications for future research. Int J Qual Reliab Manage 27(3):268–317
2. Kuo-Liang L, Yang S (2013) Applying six sigma to quality improvement in construction. J Manage Eng 29(4):464–470
3. Sawant S, Pataskar S (2014) Applying six sigma principles in construction industry for quality improvement. Int J Sci Eng Res 117(4):129–139
4. Seung H, Myung C et al (2008) Six sigma-based approach to improve performance in construction operations. J Manage Eng 24(1):21–31

Monitoring of a Construction Project in Its Execution Phase Using Project Delivery Success Factor (PDSF)

Shubhankar S. Pimplikar and Rohit R. Salgude

Abstract A construction project usually involves different types of activities and has number of factors affecting it. Such situations demand rigorous monitoring and fluent coordination on the project. The final output of any project must be that it is delivered successfully. In this study, the researchers have focused on application of a conceptual model based on KPIs and project constraints to monitor a real estate building project in Pune. To predict the project delivery success factor (PDSF) of a 13 storeyed commercial building, two additional parameters, project performance index (PPI) and project monitoring and coordination index (PMCI), have been defined and used. Data has been collected from on-site observations, collecting authentic documents related to the particular project, meeting various involved stakeholders, interviewing them and making a note of their opinion and expertise. It has been clearly observed through the results that the PDSF for the real estate project under study is less. Therefore, the particular project has been considered as a possible failure, and various measures have been recommended to improve the PDSF.

Keywords Project management · Project delivery success factor (PDSF) · Key performance indicators (KPIs) · Project performance index (PPI) · Project monitoring and coordination index (PMCI)

1 Introduction

A construction project usually involves hundreds of small tasks to be performed even if the project is of limited scope or is not having a huge budget. Also, when number of factors affecting is more, project management becomes a necessity on such construction projects. Every project is unique, and hence, it cannot be monitored and

S. S. Pimplikar (✉)
Dr. Vishwanath Karad MIT World Peace University, Kothrud, Pune, India
e-mail: shubhs1234@gmail.com

R. R. Salgude
Civil Engineering Department, Maharashtra Institute of Technology, Kothrud, Pune, India

L. M. Gupta et al. (eds.), *Advances in Civil Engineering and Infrastructural Development*, Lecture Notes in Civil Engineering 87,
https://doi.org/10.1007/978-981-15-6463-5_29

controlled in the same manner. In real estate sector, may it be either a residential or a commercial project, assuming that the project kicks off as decided along with the necessary and sufficient amount of revenue, what is left further is the real challenge—How to deliver the project successfully? Also to monitor whether the project is being delivered successfully is the next challenge to be faced.

1.1 Factors Affecting Project Success

When the target of successful delivery of the project is set, various factors affecting the project success and project performance need to be considered. The factors in favour of the project delivery success (PDS) need to be boosted, and adverse factors need to be monitored and controlled. For that, various problems faced on construction projects (only real estate projects as far as this study is concerned) need to be identified. A few of these are: (1) unrealistic planning, (2) construction delays, (3) cost overruns, (4) absence of proper communication among the employees and different stakeholders, (5) compromised quality of work, (6) improper inventory management, (7) blocked working capital and (8) high attrition rate.

1.2 Literature Review

From the existing literature reviewed, some following observations were noted:

1. To deliver a successful project outcome, an integrated approach towards project management is essential [1–5].
2. To improve the performance of a project delivery, a unified framework for construction project integration is helpful. This framework can be used by industry practitioners to determine where to allocate resources in order to have a more-integrated project [6].
3. Construction project performance relies on different dimensions of project management, out of which, project integration management has a strong impact on project management performance [2, 7–9].

2 Research Objectives and Methodology

Some of the above-mentioned factors affecting the performance of the project and further its successful delivery must be quantified by using project performance indicators to monitor and control the construction project. Therefore, main objective of the study was:

1. To determine the project delivery success factor (PDSF) based on calculating project performance index (PPI) and project monitoring and coordination index

(PMCI) of a real estate building project in its execution phase and analyse project status.
2. To recommend effective measures for project monitoring and control in order to increase the PDSF, thus making the project successful, in spite of various problems/issues faced during execution phase.

To achieve the targets as mentioned above the following methodology was adopted:

A real estate project was identified from which necessary data and information was collected. Different techniques such as personal interviews, on-field observations, expert opinion and collection of authentic documentation of the project under study through site visits were used. Once the data was obtained, it was organised and then analysed. It was quantified by giving ratings and weightages to different factors/parameters. The qualitative findings were also noted, and corresponding interpretation was done. Results were obtained from the data analysis. Recommendations were made to make the project successful.

3 Case Study

3.1 Basic Information and Problems Faced

Being an on-going project, the exact identity of the project has not been disclosed. Hence, the identified project was considered as 'P' by the researchers. It was located at Shivajinagar, Pune. The type of building occupancy was 'commercial'. The project scope was defined as a 13 storeyed building (G + 10 with two basements, B1 and B2). Various problems faced on the project are mentioned below:

(a) Location—Due to crowded place, problem was faced for equipment and machinery to enter site.
(b) Below the ground—Unstable soil mass was found, so retaining wall was required. Water table was encountered at less depth, so continuous dewatering was required.
(c) Above the ground—Accessibility issue. Cutting of some trees involved.

3.2 Determination of Project Performance Index (PPI)

Project performance is based on some of the key performance indicators (KPIs). The KPIs are measurable and hence could be used to determine the PPI. In this study, the researcher has identified six KPIs, namely speed, value, efficiency, innovation, impact and complexity. The relation of these KPIs with the project constraints, scope, time, cost and risk, have been used by the researchers to develop an equation for measuring PPI. Quality is a common parameter between project performance and project constraint. It could be considered as either of the two or both and hence finds its place as a separate identity (quality factor, QF) in equation.

$$\therefore \text{PPI} = \left[\frac{\text{Scope}^3 \times \text{QF}}{\text{Cost} \times \text{Time} \times \text{Ek}}\right] \quad (1)$$

Equation (1) represents the simplified equation for determination of PPI.

3.3 *Determination of Project Monitoring and Co-ordination Index (PMCI)*

Monitoring of project is done through various tools and techniques such as monthly review meetings, various reports such as DPR, MPR, quality reports and earned value analysis. The researchers have defined PMCI to check the level of monitoring and coordination on any project. The equation for calculating PMCI has been formulated as follows:

$$\text{PMCI} = \left[\frac{\text{Weighted Score of Issues, Disputes, NCs Resolved}}{\text{Weighted Score of Total No. of Issues, Disputes, NCs Occurred}}\right] \quad (2)$$

If PMCI lies within 0–0.2, then the project is very poorly monitored and coordinated.

If PMCI lies within 0.2–0.4, then the project is poorly monitored and coordinated.

If PMCI lies within 0.4–0.6, then the project is averagely monitored and coordinated.

If PMCI lies within 0.6–0.8, then the project is highly monitored and coordinated.

If PMCI lies within 0.8–1.0, then the project is very highly monitored and coordinated.

3.4 *Determination of Project Delivery Success Factor (PDSF)*

To determine the PDSF for any real estate project during its execution phase, the combined effect of PPI and PMCI must be considered. Project delivery success (PDS) is directly proportional to PPI and PMCI each. Also, both PPI and PMCI are interrelated in a way that if the monitoring mechanism on a project fails, it is definitely going to affect the performance of the project. Also, indirectly it means that if PPI is low, PMCI too might have been low, and hence, in return, the PPI has been adversely affected. This relation, interdependency, and direct proportionality between PPI and PMCI give the equation as follows:

$$\text{PDSF} = [\text{PPI} \times \text{PMCI}] \quad (3)$$

Equation (3) represents the exact final equation for determination of PDSF. The significance of PDSF as stated by the researchers is given below:

If PDSF ≥ 1, then the project is said to be delivered successfully.

If PDSF < 1, then the project's delivery is said to be a possible failure.

3.5 Data Collection

In order to meet the objectives of research, data was collected from the identified project 'P'. GFC project drawings, BOQ of the building (refer Fig. 2), detailed schedule with baseline and tracking (refer Fig. 1), minutes of meeting (MoM) for each month from May 2018 to December 2018 (refer Fig. 3), monthly project reports, daily project reports and daily labour reports (8th Sept, 2018--5th Nov, 2018), quality update reports and quality review reports (refer Fig. 4) were the data obtained from the above-mentioned project to determine its PPI, PMCI and further PDSF. From the available data, following observations were made:

1. Estimated duration for project 'P': 422 days (i.e., 14 months approx.).
2. Estimated cost of project (excluding GST): Rs. 20,61,57,924/–.
3. Project start: 29th March, 2018 (approx. April, 2018).
4. Project completed till status date (1st Nov, 2018): Ground parking floor slab (i.e., 1st slab in totality).
5. Amount spent till status date (1st Nov, 2018): Rs. 4,13,29,372/–.
6. Percentage of cash flow executed: 20.05%.

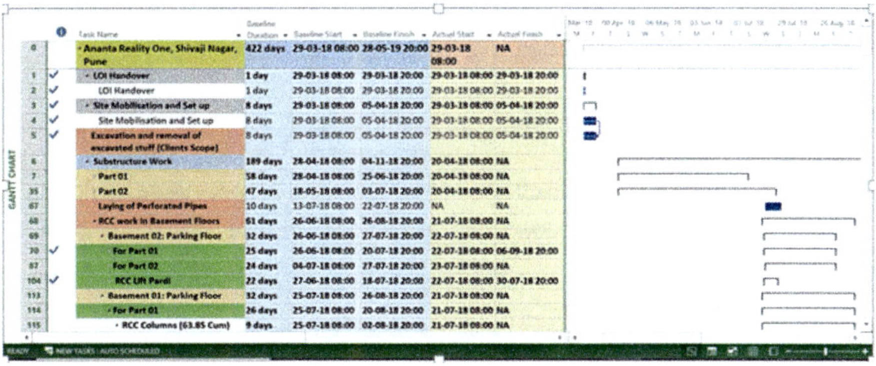

Fig. 1 Gantt chart (prepared in MSP) representing tracked schedule for project 'P'

COST SUMMARY SHEET		
SR. NO.	DESCRIPTION OF WORK	AMOUNT IN RUPEES
A	**Civil and finishes work**	
1	Earth work - Excavation,Filling , Anti-termite , Soling ,Surface dressing	4,507,915
2	RCC- PCC,Plum concrete,screeding , Shuttering , RCC ,Steel and PT Work	160,159,731
3	Internal Finishes - Internal Gypsum,celling,paints,AAC block work,Brick work and plaster	5,289,545
4	External finishes - External plaster,Texture paint	1,779,225
5	Waterproofing Works - Brick bat,chemical ,shabhad and crystalline waterproofing	760,470
6	Flooring Works- Flooring,Dado,skirting,sill,jambs,Concrete paver blocks,wash basin counters	5,906,722
7	Door and Window Works	96,750
8	Fabrication Works - Staircase railling	819,000
9	Misc Items - Column angle rubber guard,MS insert plate,Rock anchoring,Couplers,Manhole covers,Rebaring work,Dewatering work,Scaffolding work for Lift well.	1,601,597
B	**MEP Work**	
I	**Fire Fighting Works**	
1	SECTION - I - Fire hydrant/Wet riser system	1,281,277
2	SECTION - II - External fire hydrant	1,921,916
3	SECTION - III - Automatic sprinkler system	2,402,395
4	SECTION - IV - Portable Chemical Fire Extinguishers	960,958
5	SECTION - V - Pumps & Pump Room Accessores	2,882,875
6	SECTION - VI - Terrace booster pump & Accessories	1,201,200
II	**Internal Electrical Works**	4,804,791
III	**External Electrical Works**	3,203,195
IV	**Internal Plumbing Works**	171,972
V	**Duct Plumbing Works**	2,402,396
VI	**External plumbing Works**	4,003,994
	TOTAL FOR CIVIL ,FINISHES AND MEP WORKS	**206,157,924**

Fig. 2 BOQ summary (prepared in MS Excel) for project 'P'

Sr.#	Reference	Point of Discussion	Plan of Action	Person Responsible	Target Date
1	Oct MIS MOM	Poor finish product quality of concrete-Honeycomb,bulge,level of concrete	Action started after 15.11.2018 by increasing labor force said by Mr Bhor.	Shubham representative	24-Nov-2018
2	Oct MIS MOM	Concrete not cast in level.Less concrete quantity procurement issue.	Point is kept under observation.	Shubham representative	26-Nov-2018
3	OCT MIS MOM	Material quality, Workmanship,skilled labor, co ordination issues	Material quality not upto mark , Workmanship not improved and skilled labor is not yet deployed said by khade. Action started from 15.11.2018 said by Mr Bhor , as last meeting done on 20th Oct so till date no any assurance in work.	Shubham representative	

Fig. 3 Sample minutes of meeting for project 'P' reported by the PMC

Sr. #	Description	Implication	Frequency
Trade - Reinforcement Cement Concrete Work			
Subgroup - Resources Management			
1	Reinforcement steel stacking not done above ground level by at least 150mm (steel resting on ground) **Location/Remark:** (Commercial-1st Floor-Ground Floor-Directly rest on ground.	Major	1/1
Subgroup - Site Supervision and Quality Control			
2	Evidence of work found to be carried out without RFI clearance. **Location/Remark:** (Commercial-2 Wheeler Parking Floor-Column-Reinforcement RFI not cleared but shutter is done.	Very Severe	3/3
3	Side supports for column / wall shutters not as specified. **Location/Remark:** (Commercial-1st Floor-Column-Other side Column support taken.),(Commercial-1st Floor-Column-),(Commercial-1st Floor-Column-	Severe	3/3
4	Clear /nominal cover to rebars and links not as per dwg (0 to +10 mm tolerance for nominal cover). **Location/Remark:** (Commercial-1st Floor-Column.-Cover not found.),(Commercial-1st Floor-Column.-),(Commercial-1st Floor-Column.-	Severe	3/5

Fig. 4 Sample quality review report for six sigma analysis

3.6 Data Analysis and Interpretation

The project 'P' has been tracked in the month of November 2018. According to the researchers, whatever was shown in the GFC drawings as well as defined in the BOQ had to be achieved for successful completion of the project, and hence, it was identified as the 'scope' of work. Similarly, the 'time' parameter of the project was determined and measured using the detailed schedule of the project work. The 'cost' parameter was measured based on the BOQ and cash flow reports. The 'quality' of construction work was measured based on the various types of reports like quality review reports, quality update reports and also the minutes of meeting (MoM). Six sigma tool was used to determine the quality level of the project. 'Risk' factor was determined based on the complexities and challenges in project by assigning certain weightages to different parameters contributing to increment in risk. The data interpretation is done as shown in below (refer Table 1).

The project 'P' was monitored through continuous review meetings and MIS on construction site. No special tool or technique apart from this was used for monitoring and coordination of the project. Hence, the researchers studied the minutes of the meetings from MoM 1 (May 2018) to MoM 12 (Nov 2018), the details of which are given below in Table 2. The issues/disputes/non-conformities occurring on site due to lack of monitoring and improper coordination were pointed out from each of the meetings, and those were assigned with a weighted score based on not only the parameters involved in that particular issue/dispute/NC but also its seriousness and importance in accordance with the particular project. The weighted scores were also given to the issues resolved, and the results were tabulated as below (refer Table 2).

Table 1 Comparative study of expected versus actual work on site for project 'P'

Expected values	Actually values
Scope of the project-40%	Scope of the project-24%
Schedule of project-214 days	Schedule of project-129 days
Cost of project-32.2%	Cost of project-20.05%
Quality of project-sigma level 2 for general works for no rejection. Sigma level 3 for structural works	Quality of project-sigma level 1.6 for general works and sigma level 2.5 for structural works have been observed
Risk of project-30%	Risk of project-47%

Table 2 Calculation of weighted scores for issues/disputes/NCs on project

Meeting ID	Meeting type	Total no. of issues/disputes/NCs	Weighted score for total no. of issues/disputes/NCs	Weighted score for resolved issues/disputes/NCs
MoM 1	Review	12	12	6.25
MoM 2	MIS	16	16	9.5
MoM 3	Review	11	12	8
MoM 4	Review	13	13	7.5
MoM 7	MIS	13	27	9
MoM 9	MIS	17	25	11.5
MoM 10	Review	7	8	2.5
MoM 11	MIS	12	24	9.5
MoM 12	MIS	20	37	12
Total			174	75.5

4 Results and Discussion

The PPI is calculated by the researchers based on the analysed and interpreted data as given in Table 1.

$$\text{PPI} = \frac{\text{Scope}^3 \times \text{QF}}{\text{Cost} \times \text{Time} \times \text{Risk}} = \frac{\left(\frac{24}{40}\right)^3 * \left(\frac{1.6+2.5}{2}\right)/6}{\left(\frac{20.05}{32.2}\right) * \left(\frac{129}{214}\right) * (0.47)}$$

$$\therefore PPI = \frac{0.0738}{0.1764} = 0.4183 \cong 0.42$$

Therefore, the value of **PPI** for the project under study is **0.42** which is average.

The PMCI is calculated by the researchers based on the analysed and interpreted data as given in Table 2.

$$\text{PMCI} = \frac{\text{Weighted Score of Issues, Disputes,NCs Resolved}}{\text{Weighted Score of Total No. of Issues, Disputes, NCs Occurred}}$$
$$= \left(\frac{75.5}{174}\right) = 0.434(\text{lies in the range of } 0.4 \text{ to } 0.6)$$

Therefore, the value of **PMCI** for the project under study is **0.434** which implies that the project is averagely monitored and coordinated as in accordance with the scale set in Eq. (2)

The PDSF is calculated as

$$\begin{aligned} \text{PDSF} &= \text{PPI} \times \text{PMCI} \\ &= 0.42 \times 0.434 \\ &= 0.1822(\ll 1) \end{aligned}$$

If PDSF < 1, then the project is a failure as mentioned along with Eq. (3).

Therefore, in this case, as the value of **PDSF** for the project under study is **0.1822**, the project 'P' is identified as a possible *failure* by the researchers.

If at all the project 'P' would have been executed as planned, the PDS score for it would be calculated as:

$$\text{PPI} = \frac{\text{Scope}^3 \times \text{QF}}{\text{Cost} \times \text{Time} \times \text{Risk}} = \frac{\left(\frac{40}{40}\right)^3 * \left(\frac{2+3}{2}\right)/6}{\left(\frac{32.2}{32.2}\right) * \left(\frac{214}{214}\right) * 0.30} = \frac{0.4166}{0.3000} = 1.38$$

$$\text{PMCI} = 1(\text{Ideal Case})$$

$$\therefore \text{PDSF} = 1.38 \times 1 = 1.38$$

If PDSF > 1, then the project is a success.

Therefore, the researchers opine that the project 'P' could have been a proper success, if it would have been executed as per the approved plan and schedule.

4.1 Reasons for Project Failure

The problems/issues faced on project 'P' during its execution phase which resulted in low PDSF for that particular project are identified as below:

(1) Incorrect risk estimation.
(2) Unrealistic planning.
(3) Improper tracking and monitoring.
(4) Unsatisfactory quality due to improper supervision.
(5) Insufficient working capital due to overbilling by the contractor.
(6) Insufficient deployment of labour as well as site and office staff.

(7) Lack of coordination among the various stakeholders.
(8) Communication gap between client and the contractor.
(9) Less involvement of the project manager in controlling the project.

4.2 Measures to Make Project Successful

The measures that could be effective in improving the PDSF of the project and thus make it a success are as follows:

(1) Application of correct project management processes at right time through integrated management approach.
(2) Appointment of competent project manager.
(3) Correct cash flow management by avoiding overbilling.
(4) Proper planning of project before it actually starts by anticipating project delays and identifying precautionary measures to avoid them.
(5) Rigorous and constant follow-up for troubleshooting aspects of project activities is the essence of success of the project.
(6) Working out realistic timelines and keeping necessary and sufficient floats in scheduling of work.
(7) Ensuring speedy, clear communication and coordination among stakeholders.
(8) Achieving defect-free construction through effectively monitoring quality of work.
(9) Installing a separate mechanism for monitoring labour and productivity aspects.

5 Conclusion

1. Project performance index (PPI) of the project under study is 0.42, and hence, there is a need for improving the project performance by controlling time delays and achieving the targeted scope within the budgeted cost.
2. Project monitoring and coordination index (PMCI) value is 0.434. This value lies within the range of 0.4--0.6, so the project under study has not been properly monitored and controlled.
3. Based on the indices of PPI and PMCI, the PDSF has been determined. The value of PDSF (0.1822) has clearly indicated that the project 'P' may become a possible failure if not effectively monitored.
4. The reasons for project failure have been identified, and certain measures have been recommended in order to improve the PDSF.

References

1. Ahmed Taresh Ejel (2018) Impact of integration management on construction project management performance (Oil Refining Company). Int J Eng Technol 40–42
2. Sevilay Demirkesen and Beliz Ozorhon (2017) Impact of integration management on construction project management performance. Int J Project Manage ScienceDirect 35:1639–1654
3. Amir Naser Ghanbaripour, Craig Ashley Langston, Ariyan Yousefi (2017): implementation of 3D integration model for project delivery success: case study. J Constru Eng Manage ASCE 143(8):0733–9364
4. Y. Shiau, Mingming Wang, W. C. Wang (2002) Developing a construction integrated management system. Department of Construction Engineering Chung Hua University
5. A guide to PMBOK, Project Management Institute (2017), 6th edition.
6. Angelica Ospina-Alvarado; Daniel Castro-Lacouture, and James S. Roberts (2016) Unified framework for construction project integration. J Constru Eng Manage ASCE, 142(7)
7. Salwa Iqbal, Sheikh Kashif Raffat, Muhammad Sarim and Abdul Basit Shaikh (2014) Integration management—a report on the construction projects. Sci Int (Lahore) 26(3):1013–5316
8. Rita Mulcahy (2018) PMP exam Prep, 9th edn. Rmc Pubns Inc, Burnsville
9. Project Management Institute's Homepage. https://www.pmi.org. Last accessed 20 Sept 2018

Rainwater Harvesting at Vaageswari College of Engineering Karimnagar

Umank Mishra, Syeda Saba Huriya, Md. Areef, Ubaid Bin Hameed, and M. K. Upadhyay

Abstract According to the United Nations Population Fund's (UNFPA) State of the World population 2018 report, India's population grew at 1.2% a year between 2010 and 2019, marginally higher than the global average of 1.1.% a year and also double than China's population 0.5% a year. These will lead to high rate of consumption of most valuable natural resource "Water" resulting in augmentation of pressures on the permitted fresh and drinking water resources. Ancient method of damming river and transporting water to urban area has its own issues of eternal troubles causing socially and politically. In order to conserve and maintain the daily demand of water requirement, we need to think for alternative cost-effective and relatively simple technological methods of conserving water. Rainwater harvesting is one of the best methods fulfilling these requirements of water. The technical aspects of this paper are rainwater harvesting collected from rooftop which is considered to be catchment areas from institute's departmental building at Vaageswari College of Engineering (S_4 campus), Thimmapur, Karimnagar, Telangana (India). Data is collected, i.e., catchment areas & hydrological rainfall data. Water harvesting potential for the S_4 campus is calculated, and the tank capacity with suitable design is being considered. Volume of tank has been calculated with most appropriate method of estimation of water. Optimum location of tank on the basis of hydrological analysis and GIS analysis was done in the college campus. Finally, Gutter design, its analysis, first flush, and filtration mechanism are also carried out in detail.

U. Mishra (✉)
Shri Shankaracharya Technical Campus (Shri Shankaracharya Group of Institution), Bhilai, Durg, Chhattisgarh 490020, India
e-mail: umank17@gmail.com

S. S. Huriya · Md. Areef · U. B. Hameed
B.Tech Student, Vaageswari College of Engineering, Karimnagar, Telangana, India
e-mail: sabahuriya@rocketmail.com

M. K. Upadhyay
G.E.C. Bharatpir, Shyorana, Rajasthan, India
e-mail: newlifemanish@gmail.com

L. M. Gupta et al. (eds.), *Advances in Civil Engineering and Infrastructural Development*, Lecture Notes in Civil Engineering 87,
https://doi.org/10.1007/978-981-15-6463-5_30

Keywords Catchment · Rainwater harvesting · Recharging subsurface aquifers · Hydrological analysis · Filtration · GIS · Google earth

1 Introduction

Collecting rainwater harvesting means to make best use of rainwater at the location where it falls, i.e., protect it and not to allow to drain away and cause floods anywhere. The method of collecting rainwater is collecting and storing rainwater on the surface or in underground aquifers until it is lost in the form of surface runoff [1]. We can store it in tanks or use it to feed groundwater, depending on conditions and situation. This method helps to increase agricultural production by improving the availability of groundwater at a specific place and time and using rain, improving the quality of groundwater by diluting, without directly evaporating and polluting the environment of the area. increase of vegetation cover, etc. Potential areas comprise where groundwater levels are declining, where substantial amount of aquifer has been desaturated, where due to rapid urbanization, infiltration on rainwater in subsoil has reduced extremely, and recharging in groundwater has depreciated [2].

States that have adopted large-scale rainwater harvesting and groundwater recharge programs are Rajasthan, Maharashtra, Gujarat, Tamil Nadu, Talangana (TS), Karnataka, Andhra Pradesh (AP), Orissa, Madhya Pradesh (MP), and Chattisgarh (CG). Main portion of these areas is hidden by six water-scarce river basin schemes, namely Sabarmati rivers of Kachchh and Pennar, Saurashtra, Cauvery, rivers between Mahanadi and Godavari, east-flowing rivers among Kanyakumari and Pennar, which have less than 1000 m^3 of renewable water per annum [3].

The regulated Indian research on rainwater harvesting (**IRWH**) or artificial recharge up to now has absorbed on the engineering presentation of specific construct structures [4]. Although there are many unofficial data on economic and social benefits, there is little sympathetic based on experiential work.

- the impact of activities in water harvesting on the local hydrological system in terms of net water inflows
- the impact of the basin level on the total water balance of the basin; and also
- long-term economic imperatives.

Recently, researchers have raised queries about the thinkable unintended consequences of collecting water [2] and its economics [5]. One of the details for the lack of experiential research on the hydrological and economic facets of water collection schemes is the inability to obtain correct scientific data on hydrological, hydraulic, and meteorological limits that determine the performance and impact of water harvesting. The analysis of water collection systems also does not take into account the influence of the scale factor.

The detail that different periods of history are marked by the genesis, the emergence and fall of a new tradition of storing water [6], is also not valued.

Rainwater harvesting do many scientists study antique technique for various purposes throughout the year. Practical advice is available in books written by [7] titled "self-reliance in water" and [8], talks by Anupam Mishra. Other than this, two books, entitled first, Estimation and costing in civil engineering, by [9], and second, R.C.C. Designs, by [10], were referred for complete structural analysis of underground sump and for costing and estimation of underground sump.

In this paper, rainwater is being conserved or harvested for whole one campus area of Vaageswari College of Engineering, Karimnagar. And it used two methods of distribution of harvested rainwater (Rapid depletion method & Rationing method). Finally, the cost for construction of tank was calculated.

2 Study Areas

Thimmapur is located by 78.79 E longitude and 18.57 N latitude in Karimnagar district of Telangana at an elevation of about 275 m above mean sea level. Thimmapur has a tropical climate and receives high rainfall during June to September months. The rainfall here averages 907 mm. The precipitation varies 227 mm between the driest month and the wettest month. The monthly rainfall data of the Karimnagar district is given below in Table 1 which is expected to be same for the station of Vaageswari College of Engineering, Thimmapur campus. This data has been collected from the Web site www.en.climate-data.org.

Given below is a satellite picture that is showing the whole campus and the only campus considered for rainwater harvesting system at Vaageswari College of Engineering (Fig. 1).

Table 1 Monthly rainfall data of Karimnagar station (Web. P.2)

Month	Rainfall (mm)
Jan	9
Feb	5
Mar	9
Apr	15
May	23
Jun	140
Jul	232
Aug	199
Sep	181
Oct	73
Nov	16
Dec	5
Total	**907**

Fig. 1 Whole campus area of Vaageswari institutes taken from "Google Earth" (2019)

The rooftop area considered is the catchment area which obtains rainfall. Catchment area in our case is the whole S_4 campus measured. This measurement was done manually with the help of "reinforced fiber tape" which is the easiest method known as "tape survey." Before using the tape, it was checked that there was no error, and the length of the tape was also carefully checked to verify its accuracy. The places which are not accessible are measured by using the "Ruler" from tool box of "Google Earth" (Fig. 2).

Fig. 2 Rooftop area of the campus considered taken from "Google Earth" (2019)

3 Methodology

3.1 Components of Rainwater Harvesting System

The rainwater collection system contains several components that will be needed to transport rainwater through pipes or drains, filtration tanks, and storage tanks for collecting rainwater [11]. The system consists of following basic components—Catchment area, coarse mesh/leaf screen, gutters, down spout/conduit, first flushing device, filters, storage tanks, and recharge structures. The figure below shows the rainwater harvesting system (Fig. 3).

The various methods of recharging subsurface aquifers are through recharge pit, abandoned hand pump, abandoned dug well/open well, recharge trench, recharge through shaft, and recharge trench with bore (Fig. 4).

Design of rainwater harvesting system (calculations)

1. Catchment area

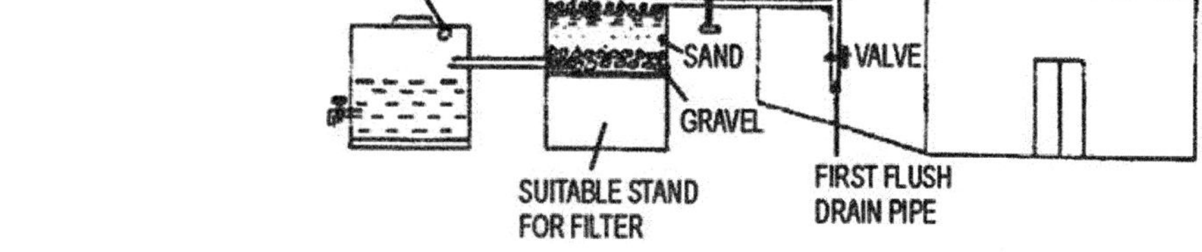

Fig. 3 Rainwater harvesting system

Fig. 4 Flowchart showing the usage of rainwater

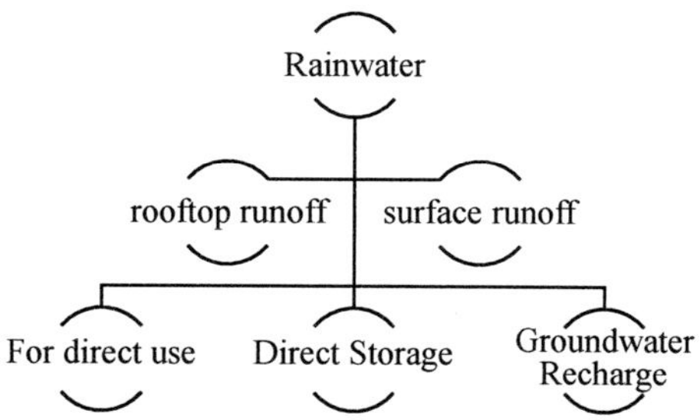

$$\text{Rooftop area} = 6600\,\text{m}^2$$

2. Average rainfall = 907 mm
3. Runoff coefficient

$$\text{For roof top area} = 1\ \text{(Totally impervious)}$$

Harvesting potential or volume of water received (m^3) =
Area of catchment (m^2) × amount of rainfall (mm) × runoff coefficient
Now, vȯlume of water received = 6600 × 907 × 1 = 5986 m^3 = 5900 m^3

By rationing method

In this method, the amount of water supplied to the student is limited which is equal to, say, 30 l/day per capita water demand.

- Therefore, let us suppose number of students at VGSE = 1500

  Then, total amount of water consumption per day = 1500 × 0.03 = 45 m^3/day
  Total no. of days we can utilize preserved water = stored water/water demand
  Hence, finally, no of days = 5900/45 = 131 days (or 4.4 months).

By rapid depletion method,

we assumed per capita water demand = 50 lt/day = 0.05 m^3/day

Total amount of water consumption per day = 1500 × 0.05 = 75 m^3/day

Total no. of days that preserved water can be utilized = 5900/75 = 79 days (or 2.63 months)

Finally, it is observed that if the amount of water stored is equal to 5900 m^3, then applying

1. RDM, consumer can only utilize the preserved stored water for about 79 days (2.63 months).
2. However, in RM, preserved stored water can be utilized for a period of 131 days (4.4 months).

So, finally for this volume of water received and usage of water in the campus, the recharge pit of dimensions 10 m × 5 m × 3 m can be constructed. One recharge pit can be constructed at the backside of the diploma campus.

By this, if we consider annual rainfall data, then we can obstruct 80–90% of rainwater.

The tank should be of the first class brick in 1:4 cement mortar foundation, and the floor should be cement 1:3:6. The inside of the septic tank should be finished with 12 mm cement plaster, and the floor should be treated with cement plaster with a thickness of 20 mm using solution 1:3 mixed with standard waterproofing compound. The upper and lower parts of the absorbing pit should be from the second class of

Table 2 Detail estimation of sump

S. No.	Particular	No.	Length (m)	Breadth (m)	Height/depth (m)	Quantity (m^3)
1	Earth work in excavation	1	10.80	5.80	3.3	206.712
2	Cement concrete 1:3:6 in foundation	1	10.80	5.80	0.3	18.79
3	I class brickwork in 1:4 cement mortar (i) Long wall (ii) Short wall	2 2	10.60 5.0	0.30 0.30	3 3	19.08 9
		Total				28.08
4	R.C.C work for slab cover	1	10.60	5.60	0.20	10.72
5	12 mm plastering inside with 1:2 cement mortar (i) Long wall (ii) Short wall	2 2	10 5	– –	3 4	60 30
		Total (Rs)				90

brick in cement mortar 1:6, and the middle part should be dry brick. Wall thickness is about 30 cm. Roof covering slabs shall be precast R.C.C.

4 Estimation and Costing

The total cost of tank was come out to be Rs. 491,726.4 (Tables 2 and 3).

5 Results and Discussion

Runoff potential = 5900 m^3, and for recharge, size of recharge pit or underground sump is taken as 10 m × 5 m × 3 m.

The estimated expenditure cost for the underground storage tank is Rs. 491,726.4 (Graphs 1 and 6).

Recharge pit/underground tank can be connected to borewell for borewell recharge.

For design purposes, the following data was considered constant. These values can be changed later depending on the other situation.

Therefore, it was assumed that the number of people who consumed water from any building was 1500 people.

The average annual rainfall was assumed to be 907 mm. The consumption rate here was set at 30 l/day (Table 4).

Table 3 Abstract of estimation cost

S. No.	Particular	Quantity	Rate	Cost (Rs.)
1	Earthwork in excavation	206.712 m^3	100 Rs./m^3	20,671.2
2	Cement concrete 1:3:6 in foundation with brick ballast	18.79 m^3	2700 Rs./m^3	50,733
3	I class brickwork 1:3 cement mortar	28.08 m^3	3000 Rs./m^3	84,240
4	R.C.C work for slab cover	10.72 m^3	2700 Rs./m^3	28,944
5	12 mm plastering with 1:2 cement mortar	90 m^3	2700 Rs./m^3	243,000
Total				427,588.2
6	Contingency + work charges establishment	(3% + 2% = 5%)	–	21,379.41
7	Engineering profit	10%	–	42,758.82
Grand total				491,726.4

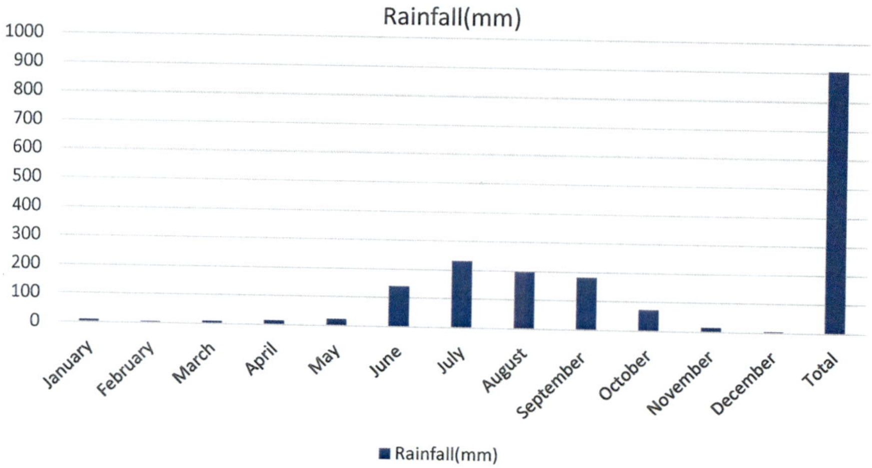

Graph 1 Graph showing amount of rainfall collected throughout the year

6 Conclusion

This paper covers all aspects of the fight against water scarcity in Vaageswari College of Engineering, Thimmapur, Karimnagar campus by applying ancient old method of rainwater harvesting. Two methods have been proposed for tank design, which have different approaches to consumption of harvested rainwater. For rooftop area having 6600 m^2 and volume of water received 5900 m^3, it was concluded that the water can be stored for a tank capacity of 6000 m^3 for 131 days by rationing method and 78.6 days by rapid depletion method. Therefore, it was finally concluded that

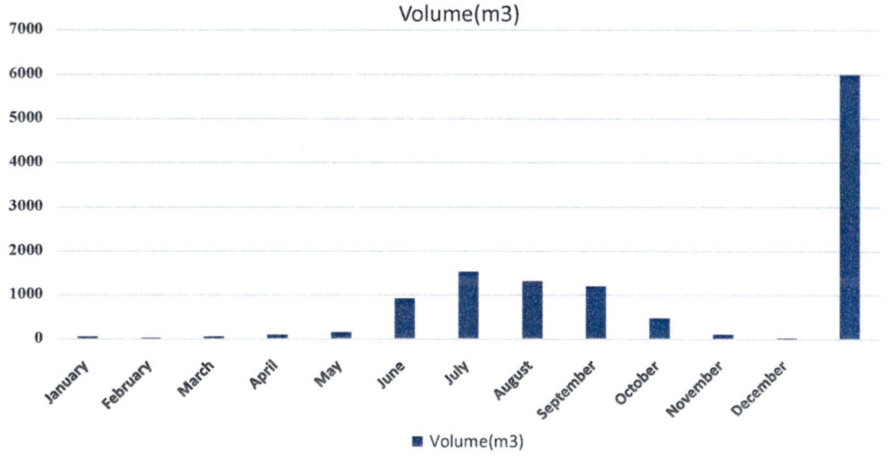

Grpah 2 Volume of water collected from rainfall throughout the year

Table 4 Analysis of distribution of stored harvested water by two methods

S. No.	Building name	Rooftop area (m^2)	Reservoir capacity	RM = R/30 (days)	RDM = R/45 (days)
1	S_4 Block	6600	6000 (R1)	131	78.6

the implementation of rainwater harvesting system in the S_4 campus of college will be the best approach to combat the current water scarcity scenario in all aspects, either from a financial point of view or from the optimal utilization of land surface. Therefore, water is a very valuable natural resource that is always in great demand on the Vaagesvari College of Engineering, Thimmapur, Karimnagar, in the state of Telangana (India).

References

1. Kumar MD, Ghosh S, Patel,A, Singh OP, Ravindranath R (2006) Rainwater harvesting in India: some critical issues for basin planning and research Land Use Water Resour Res 6:1–17
2. Badiger S, Sakthivadivel R, Aloysius N, Sally H (2002) Preliminary assessment of a traditional approach to rainwater harvesting and artificial recharging of groundwater in Alwar District, Rajasthan. Annual partners meeting, IWMI-Tata Water Policy Research Programme, Anand, February 19–20
3. Garg SK (2002) Hydrology and runoff computation, irrigation engineering & hydraulic structure, 16th edn. Khanna Publishers, Delhi
4. Muralidharan D, Athawale RN (1998) Artificial recharge in India: base paper prepared for Rajiv Gandhi National Drinking Water Mission, Ministry of Rural Areas and Development, National Geophysical Research Institute

5. Kumar MD (2004) Roof water harvesting for domestic water security: who gains and who loses? Water Int QW, 43–53
6. Pandey DN, Gupta AK, Anderson DM (2003) Rainwater harvesting as an adaptation to climate change: review article. Curr Sci 10–15
7. Ragade I (2012) Self reliance in water: a book by Indukanth Ragade. https://www.indiawaterportal.org/articles/self-reliance-water-book-indukanth-ragade
8. Pacey A, Cullis A (1998) Rainwater harvesting: the collection of rainfall and runoff in rural areas Administration Dev 8:119–121
9. Dutta BN (2012) Estimating and costing in civil engineering: theory and practice including specifications and valuation civil engineering book. UBS Publishers' Distributors (P) Ltd., New Delhi
10. Punmia BC, Jain AK (2015) R.C.C. Designs book. Laxmi Publications; Tenth edition (2015)
11. Sharma RK (2010) Rainwater harvesting at National Institute of Technology (N.I.T), Rourkela. BTech thesis

Web Portal Support (Internet)

12. http://as.ori.nic/balangir/rainfall/pubNormaldtl.asp
13. http://www.rainwaterharvesting.org
14. http://www.tn.gov.in/dtp/rainwater.htm
15. http://www.aboutrainwaterharvesting.com/
16. http://www.rainwaterharvesting.org/People/innovators-urban.htm#svis
17. Wikipedia.com
18. Google Earth
19. Bhuvan.com
20. www.en.climate-data.org

Elimination of Process Wastes in Construction by Using Last Planner® System

Mitesh K. Bhatt, S. S. Pimplikar, and Piyush Pandey

Abstract Recently, Government of India has decided vision of 100 smart cities in 2020, and for a better productivity (on time, within budget and promised quality of projects), construction industry should adopt new construction management technologies but due to the tendency to adopt new management technology and other more reasons almost 80% construction projects are either time overrun or cost overrun. This problem can be solved by the adoption of tools of lean construction in Indian construction sector. Last Planner® System is one of the most efficient tools of lean construction and with the use of this tool; process wastes and negative effect of traditional method can be mitigated. Thus, objective of this study is to overcome the limitations of traditional method of construction and process wastes by the implementation of Last Planner® System in residential project.

Keywords Traditional construction method · Cost overrun and time overrun · Lean construction · Last Planner® System

1 Introduction

Initially, lean concept was used in Venice for making of warship. They used continuous flow method for warship construction, and same concept was used by Henry Ford in 1914 for the production of ford car and then it failed because of not flexibility in design. German aircraft used Takt time concept, which means available production time by customer demand. Later Toyoto used the combination of Takt time concept and lean construction at same time and got best result in production. Koshleka was first who used lean concept in construction. Last Planner System is one of the best tools of lean construction, which contains the direct involvement of

M. K. Bhatt (✉) · S. S. Pimplikar
Construction, Engineering and Management, MIT-WPU, Pune 411058, India
e-mail: mkbhatt06@gmail.com

P. Pandey
Vedzen Institute, Pune 411040, India

L. M. Gupta et al. (eds.), *Advances in Civil Engineering and Infrastructural Development*, Lecture Notes in Civil Engineering 87,
https://doi.org/10.1007/978-981-15-6463-5_31

higher authority in project and better project delivery and planning system like master schedule, phase schedule, six-week look-ahead plan and weekly work plan. Last Planner® System creates a better communication platform between higher authority and lowest supervisor of the company about project constraints and progress status.

2 Literature Review

Through literature review, it was found that [1] explained that use of push schedule is the main reason for process waste in traditional method. Shingate and Hedaoo [2] explained the use of Last Planner System to overcome the limitations of traditional construction work. More et al. [3] explained that in residential buildings non-value adding activities can be reduced by 27% with the use of proper lean techniques. Abhiram and Asadi [4] identified process wastes in construction. Ballard [5] suggested use of Last Planner® System in Indian construction and benefits from it. Devaki and Jayanthi [6] identified the barriers to apply lean principles in Indian construction industry. Harsha et al [7] explained use of Last Planner® System has clear set of objectives rather than traditional construction method. Lindhard and Wandahl [8] suggested changes in each stage of Last Planner® System. Porwal et al. [9] explained that Last Planner® System is tool of lean construction which gives a better control on planning and monitoring of project. Seppanen et al. [10] explained the combination of Last Planner® System and location-based management system.

3 Objectives

a. To critically examine the owner's traditional organizational structure, for the functionality aspects considering coordination, communication, record keeping, stakeholder interest, teamwork aspects, contractor engagement hence determine the limitations and negative impacts on project performance.
b. To apply the concepts used in Last Planner® System (lean construction), to overcome the limitations of the traditional organizational project management practice on a residential project.
c. To analyse the existing Last Planner® System being currently used on a real estate project for its impact on project time and cost.

4 Methodology

Data has been collected from an ongoing building project, which is mentioned below. Subsequently, analysis of data has been done with the reference to the objectives set.

5 Limitations

The limitation of this study is that the project under construction was already partly completed using traditional construction management system; hence, the benefits of the Last Planner® System could not be used in early stage of project and another limitation is that the inferences obtained from the feedback analysis are based on limited responses from participants.

6 Data Collection

Data related to research work is collected data from Symphony Buildings B1 and B2. Developer of these properties is Pegasus Properties Pvt. Ltd., Hinjewadi, Pune.

(A) An online questionnaire survey has been done and on that, 41 persons have given their feedback. These include trainees and junior engineers, cost engineer, planning engineer, professor, valuation engineer and project manager.

(B-I) Constraint Management sheets and problem-solving sheets from Last Planner® System (duration between 27th December 2018 to 14th February 2019) have been collected of Symphony B1 building.

(II) Material procurement sheet (procurement schedule) of Symphony B1 building and steel procurement inventory sheet with the use of Last Planner® System from the duration of 15th November 2018 to 20th March 2019 are obtained.

(C) Site photographs of Symphony B1 from the period of April 2017 to April 2019 for better tracking of construction progress. Total expense sheet and planning sheet of construction (while the use of traditional method) are collected and also expense sheet, master schedule and six-week look-ahead plan (Last Planner® System) sheets have also been collected from Symphony B1 building.

7 Data Analysis

In the data analysis, the data of Last Planner® System is been compared with feedbacks which are received from the questionnaire survey as well as progress of symphony B1 building is also compared based of progress ratio with the use of traditional method and with the use of Last Planner® System.

(A) Based on questionnaire survey, the involvement of higher authority, maintenance of daily progress sheet and effect of resource planning in schedule can be scrutinized as:

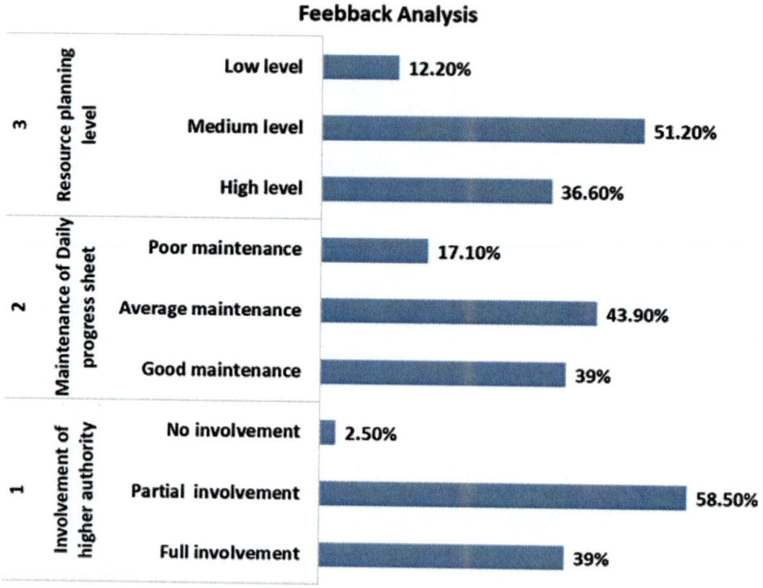

Fig. 1 Questionnaire survey feedback

As per Fig. 1, it can be identified that above-mentioned three factors are not getting efficiently used in traditional construction project. Comparison of traditional construction method and Last Planner® System is as shown in Table 1.

Based on these functionality aspects, limitations of traditional construction method can be identified as:

- Lack of top management involvement.
- Huge process wastes.
- Lack of material planning and delayed delivery.

Most negative effect of traditional construction method on project is time overrun and cost overrun.

(B) (i) Constraint management sheets and problem-solving sheets are used in Last Planner® System to better communicate about the issues. Constraint management sheets are reviewed to identify current issues on site and who will solve those issues, and at same time, futuristic constraints and their mitigation plans are also discussed among site engineers, project manager and higher authorities (Fig. 2).

With the use of constraint management sheets and problem-solving sheet at symphony site, it enabled the higher authority involvement. Therefore, constraints were mitigated quickly, hence preventing delay in project progress.

(ii) Material procurement system and material delivery are also the key problems of traditional construction method. In traditional method, material schedule is

Table 1 Comparison between traditional method and Last Planner® System

S. No	Functionality	Traditional method	Last planner® system	Researcher's critical observation
1	Coordination and communication	Have to wait for higher authority signature and approvals. (delay in process and problem cause solution)	Project manager (PM) is responsible for work and PM will directly communicate with higher authority If needed	In Last Planner® System, involvement of top management is high so any approval time will be fast (based on constraint management Sheet)
2	Record keeping	Not properly maintained	Well maintained	In Last Planner® System, each day-completed task has to be noted in daily progress sheet board. Thus, all data can be easily evaluated by higher authority
3	Schedule	Push (focused only on start and finish dates of activities)	Pull (focused on every possible aspect and make project Scheduling w.r.t it	In Last Planner® System, detailed project scheduling enables six-week look-ahead targeted work and hence possible to know all upcoming constraints
4	Stakeholders interest	Stakeholders' interests are not aligned	Stakeholders' interests are aligned	Audit is done on site (once in a week), and at that time, all internal stakeholders are supposed to be present, and based on performance, they get incentives
5	Team work	Functional	Cross-functional	Every morning GEMBA (at actual place—on site) meeting clears the daily targeted goal
6	Contractor's performance	Less effective	Highly effective	In Last Planner® System, contractors are aware of their roles and responsibilities. Which is highly effective for project progress

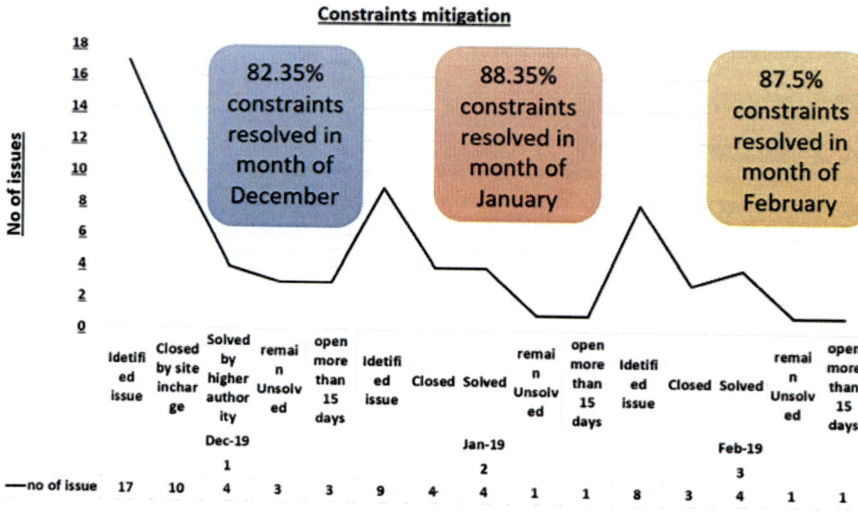

Fig. 2 Constraint resolution analysis at symphony B1 building

not done in proper manner because not used in detail scheduling. So most of the time at the end, material shortage is been recorded, and also, here material procurement takes much time for approvals from different departments, while in Last Planner® System 25% lead material stock is always available at site. In Last Planner® System, PM and store manager do not require higher authority permission for the procurement of material. So delay and over process kind of process wastes can be mitigated in material procurement (Table 2).

In Symphony B1 building, total quantity of received steel is 374,299 kg while actual issued is 374,501 kg. So 202 kg steel is used more than planned quantity but still it is just 0.05% of actual quantity. Thus by proper material procurement system, wastages like over-production and over-processing can be controlled.

(C) Symphony B1 project was started in April 2017 and decided end date was March 2019; but due to the failure of traditional construction method at symphony site, Last Planner® System has been implemented over there from November 2018 and estimated end date is March 2020. Analysis of cost and time overrun in both systems is given below:

As per Fig. 3 and Table 3, it is identified that in traditional construction method actual cost was much higher than expected cost while using Last Planner® System (1st floor to 8th floor), cost escalation was slightly higher than expected cost.

Table 2 Material procurement system tracking

Item	A	B	C	A + B + C
XYZ	Lead time	Avg. consumption	Safety stock 25%	Daily update

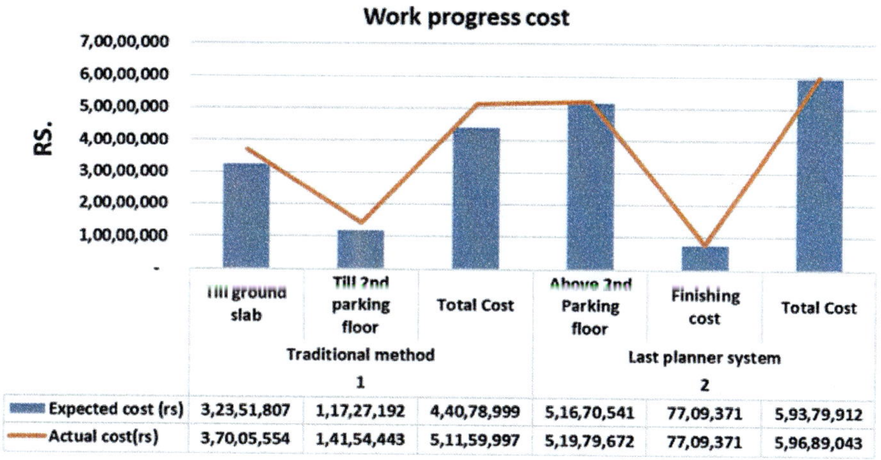

Fig. 3 Symphony B1 cost while using traditional const. method and Last Planner® System

Table 3 Expected versus actual cost percentage of Symphony B1

Percentage complete (lean construction method)			
1	Lean construction cost	5, 93, 79, 913.41	5, 96, 89, 044.74
2	Symphony B1 Estimated Cost	29, 52, 23, 059.00	29, 52, 23, 059.00
	Total Percentage	20%	20.21%

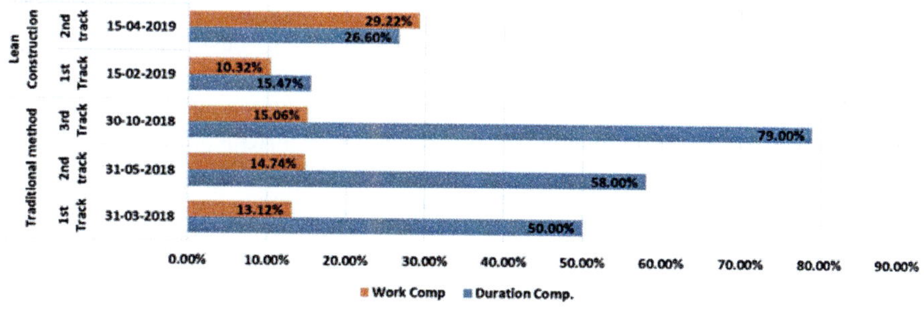

Fig. 4 Symphony B1 progress while use of traditional method and Last Planner® System

While tracking of project progress of Symphony B1, as per Fig. 4. it is identified that with the use of traditional construction method, project progress was highly delayed and while use of Last Planner® System, project progress was as per planned, and it seems that project might be completed on estimated time.

8 Conclusion

Distinct advantages of using the Last Planner® System (LPS) technique of lean construction over the traditional construction management system are summarized as:

(i) Significant involvement of top management is achieved. (ii) Daily monitoring of site progress is facilitated. (iii) Bottlenecks are visualized ahead of time and effective measures implemented due to six-week look-ahead schedule. (iv) Rework cost is reduced. (v) Effective control on time and cost overruns is attained. (vi) Actual cost of construction till second parking floor, using traditional method is Rs. 5.11 crores as against the expected cost of Rs. 4.407 crores, which is 16.08% more. (V) Actual cost of construction from above second parking floor [till the present stage of construction (8th floor) + finishing work (in progress)] using Last Planner® System is Rs. 5.969 crores as against the expected cost of Rs. 5.937 crores which are only 0.54% more.

Therefore, Last Planner® System enables better communication and coordination among the higher authorities than traditional method of construction and also enabling solving of constraints which cause delay; hence, it prevents time and cost overruns.

9 Future Scope

Last Planner® System enables consideration of all futuristic issues with the use of six-week look-ahead plan but still due to much uncertainty in construction sector, all issues are not getting solved, though steering committee and site in charge team know before six weeks but if steering committee as well site in charge team can prioritize them and solve them systematically than nearly all the futuristic constraints can be solved before they actually happen.

References

1. Bajjou M, Chafi A, En-nadi A (2017) A comparative study between lean construction and the traditional production system. Int J Eng Res Afr 29:118132
2. Shingate N, Hedaoo M (2017) Last planner system- theory to implementation. Int J Sci Res Dev 5:395–399
3. More V, Charhate S, Sinha M (2016) Lean construction techniques in indian construction industry: some analysis. Int J Civ Eng Res 7:59–65
4. Abhiram P, Asadi SS (2016) Implementation of lean methodology in indian construction. Int J Civ Eng Technol (IJCIET) 7(6):641–649
5. Ballard HG (2015) Bringing lean into India's construction industry Indian lean construction conference. In: 23rd Annual international conference of the international group for lean construction, pp. 01–10

6. Devaki and Jayanthi (2014) Barriers to implementation of lean principles in the indian construction industry. Int J Eng Res Technol 3:1189–1192
7. Harsha N, Suresh AV, Nagaraj N (2013) Implementation of lean concepts in the construction engineering project. Int J Eng Res Technol 2(5):1078–1085. ISSN: 2278-0181
8. Lindhard and Wandahl (2013) Looking for improvement in last planner system: defining selection criteria. ASCE 2013:27–35
9. Porwal V, Solis J, Lavy S et al (2010) Last planner system implementation challenges. IGLC, pp 548–556
10. Seppanen O, Ballard G, Pesonen S (2010) The combination of last planner system and location based management system. Lean Constr J 43–54

Runoff Volume Estimation by SCS-CN Method Through Arc-GIS Approach

Umank Mishra, Gowru Kiran Kumar, S. K. Gupta, P. Sarah Kiron, N. Vinay, and Lubna Ara

Abstract The salient resources of water for discharge of groundwater in a watershed or catchment area are rainfall and rainfall runoff. The primary source of groundwater recharge is rainfall. Runoff in the watershed or catchment area affected by geomorphologic feature principally, land use change affects the surface runoff volume over and rate of runoff considerably. This project aims to estimate the runoff volume over Gangadhara watershed (289.29 km^2), Telangana by one of the methods called Soil Conservation Services (SCS)–Curve Number (CN) method. In this method, alternate land use/land cover (LULC) and soil type shape files were first obtained and compiled using by remote sensing (RS) and geographic information system (GIS) techniques. Then, hydrologic soil group (HSG) maps will be developed by interpreting formative elements by soil taxonomy. Available rainfall data of the past 30 years (1988–2018) from rain gauge station of Gangadhara watershed is collected and screened to consistency and accuracy and is linked to GIS database. Based on soil and rainfall conditions of a watershed, the antecedent soil moisture condition (AMC) is found to be AMC-I, i.e. dry condition. By using relevant equations, the runoff volume over the watershed estimated is 0.81 km^3.

Keywords Runoff · SCS-CN method · LULC · HSG · RS and GIS · AMC

U. Mishra (✉)
Associate Professor, Shri Shankaracharya Technical Campus (Shri Shankaracharya Group of Institution), Bhilai, District Durg 490020, Chhattisgarh, India
e-mail: umank17@gmail.com

G. K. Kumar
Geophysicist, Groundwarer Department, Karimnager, India

S. K. Gupta
Civil Engineering Department, IIT (BHU), Varanasi, India

P. Sarah Kiron · N. Vinay · L. Ara
B.Tech Student, Vaageswari College of Engineering, Karimnager, India

L. M. Gupta et al. (eds.), *Advances in Civil Engineering and Infrastructural Development*, Lecture Notes in Civil Engineering 87,
https://doi.org/10.1007/978-981-15-6463-5_32

1 Introduction

SCS-CN method was developed by the USA in 1963, [1] There are numerous methods offered for surface rainfall runoff modelling, but **SCS-CN** method [2–4] is the simplest method to calculate the volume of runoff of the watershed. The method requires the numeric watershed characteristics which are on the base of a watershed runoff determination. The objective of this technique is to determine the accurate curve number of the watershed that defines the runoff potential. By using **RS and GIS,** the processing will be easier. [5] studied the hydrological reply of watershed to land use changes based on the geologic information system (GIS) and remote sensing (**RS**) approach. [6, 7] carried out hydrological modelling using **RS and GIS**. Accurate modelling is required for the estimation of spatial and worldly distribution of the water resources parameters. **Rainfall runoff estimation** and assessment are valuable and pertinent issues of hydrological in addition to geographical investigation. It is an important factor affecting the growth and progress of floods, soil erosion and other hydrological risks. There are many techniques offered for rainfall runoff modelling.

1.1 Objective of Study

The watershed hydrologic generations leading to the generation of surface runoff are governed by the interaction of precipitation with topological, land use and soil properties. Hence, the use of geographic information system is preferred over the traditional methods in proper quantification of runoff.

The following objectives are explored in this project:

(a) To develop **hydrological soil group** (HSG) maps for the watershed of the study region based on soil order, infiltration rate, soil depth and soil characteristics of that watershed.
(b) To detect the extent of **land use/land cover** (LULC) occurred in study region and examine its impact on CN.
(c) Cumulative rainfall data, soil parameters and hydrological groups of the watershed are modelled to determine the **curve number**.

2 Study Region

The Gangadhara watershed is a part of Mothevagu sub-basin and is located in upper part of Telangana. The Mothevagu flows in Gangadhara district and joins in Manair River. The area of watershed is 289.29 km^2. The highest elevation is Kachireddypalli (367 m) and the lowest elevation is Chegurthi (250 m). The drainage pattern is dendritic pattern with some percolation tanks (Fig. 1).

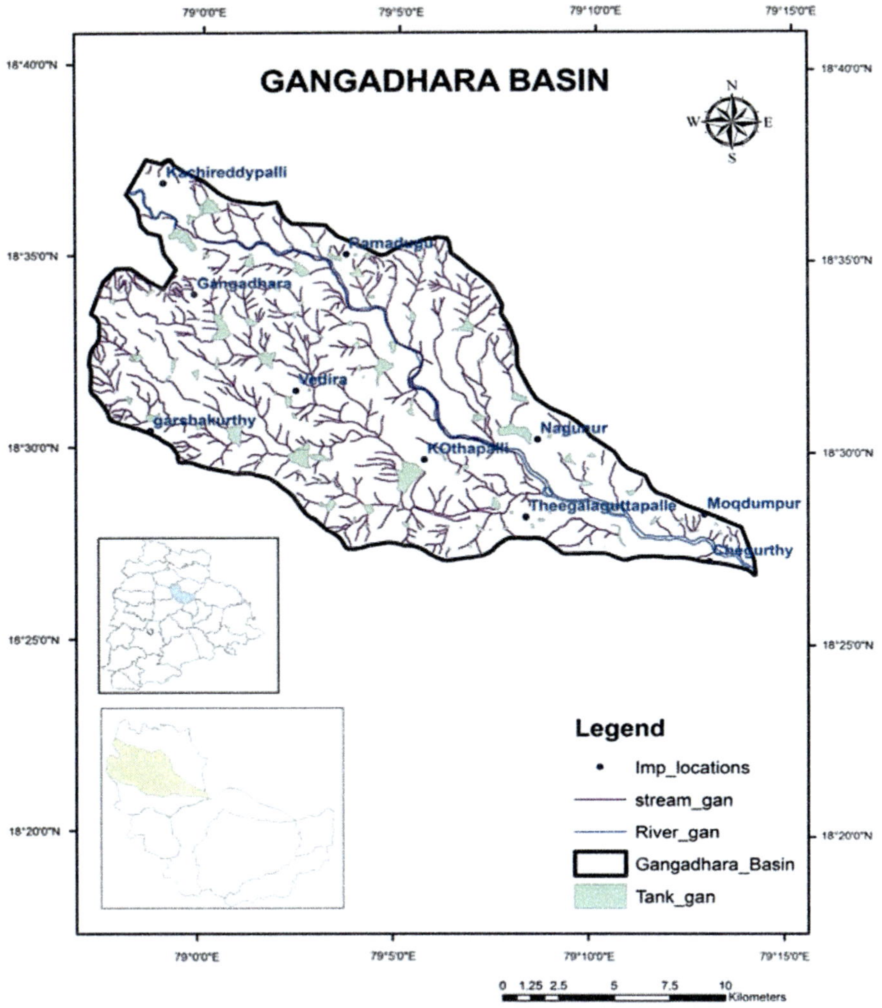

Fig. 1 Gangadhara watershed map

3 Methodology and Data Collection

Figure 2 shows the methodology that is adopted for this present study. The land use/land cover maps are extracted from the Landsat image downloaded from USGS. Soil information is collected from Telangana State Remote Sensing Application Centre (TRAC), and HSG of that watershed is determined according to infiltration capacity of that soil. Thirty years of rainfall data, i.e. 1988–2018, is collected from CPO, Karimnagar. By using all this data, Curve Numbers for the different HSG are found from LULC map and the weighted Curve Number is determined.

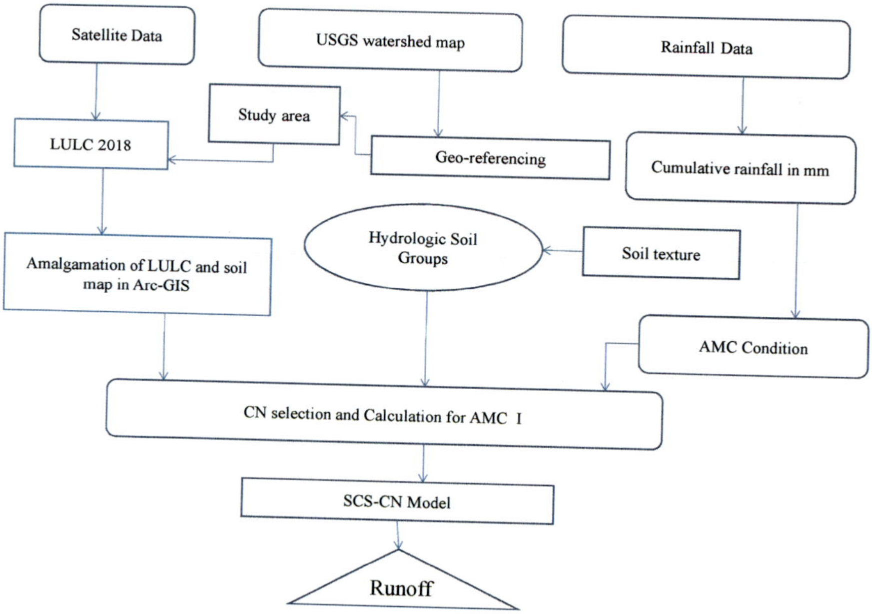

Fig. 2 Flowchart of methodology

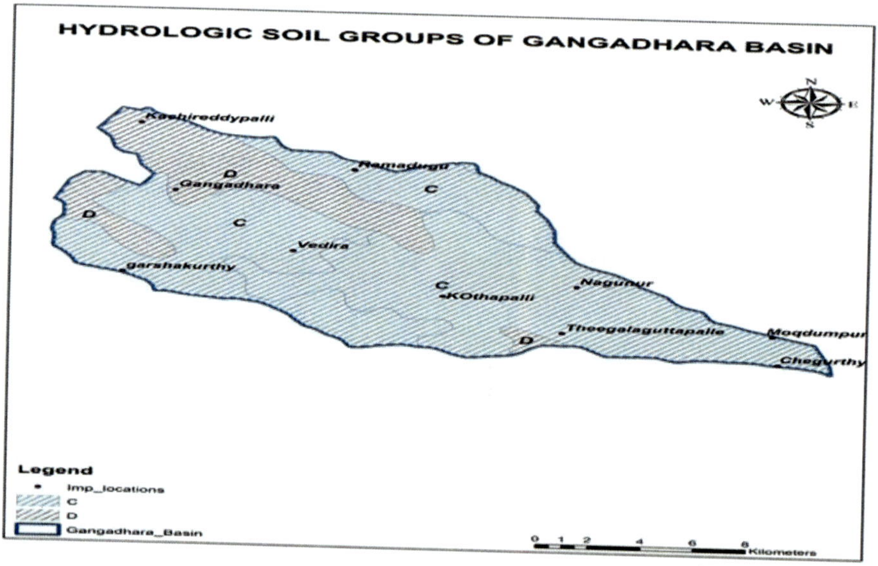

Fig. 3 Hydrologic Soil groups of the Gangadhara watershed

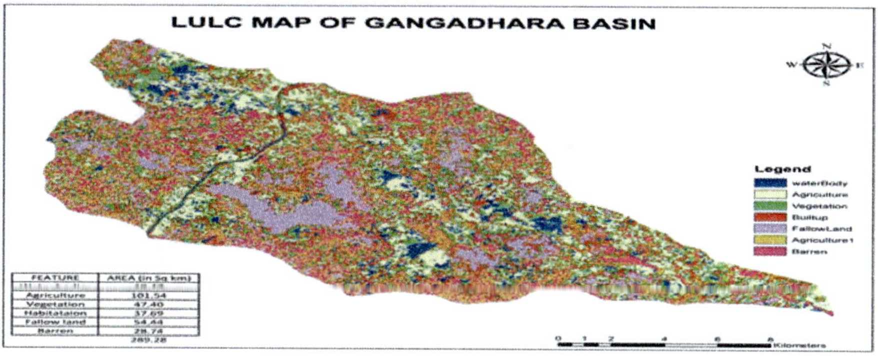

Fig. 4 LULC of Gangadhara watershed

Graph 1 Land use/land cover % of the Gangadhara watershed

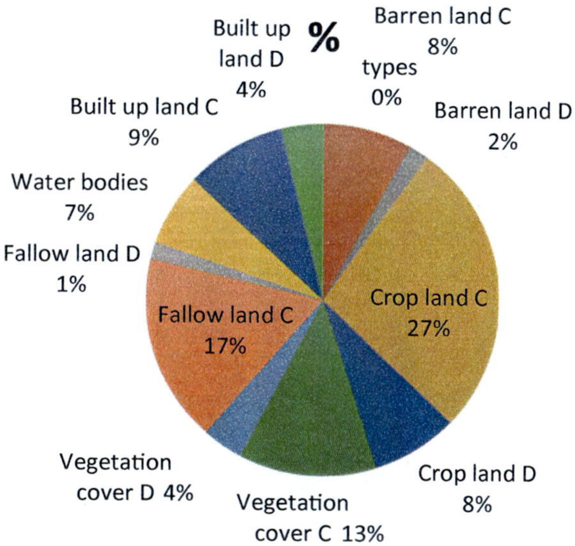

The above are the frequently used empirical methods to estimate the runoff value from the watershed. The following is the formula used to determine runoff.

$$Q = \frac{(P - 0.3S)^2}{P + 0.7S} \tag{1}$$

$$S = \left(\frac{25400}{\mathrm{CN}}\right) - 254 \tag{2}$$

where Q is discharge or runoff, P is the rainfall depth, S is a maximum possible retention after runoff begins, CN is Curve Number. Equation 1 is for Indian conditions.

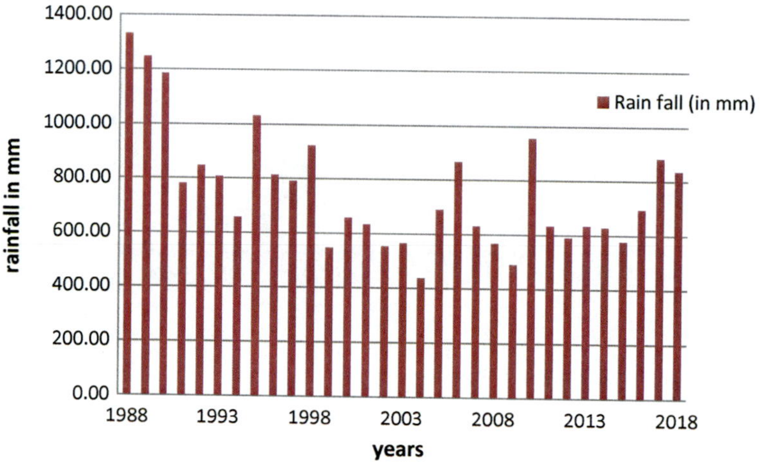

Graph 2 Bar graph between year and rainfall

Table 1 Hydrologic soil groups

Groups	Runoff potential	Soil texture
A	Low	Deep sand, deep loess and aggregated silt
B	Moderately low	Sandy loam, shallow loess, red loamy, red sand and red sandy loam
C	Moderately high	Clayey loam, shallow sandy loam, mixed red and black soils
D	High	Heavy plastic soils, certain saline soils and deep black soils

3.1 Hydrologic Soil Groups

According to SCS-CN technique, the soil types of a watershed are classified into four bands, i.e. A, B, C, D. But a watershed taken for this study has only C and D HSGs. Table 1 shows the hydrological soil groups (Fig. 3).

3.1.1 Antecedent Moisture Conditions

In SCS-CN method, AMC is one of the important factors to be considered for the calculation of runoff. AMC is again defined into three conditions, i.e. AMC-I, AMC-II and AMC III. But for this watershed, the soil condition is dry; therefore, it has AMC-I condition. Table 2 shows the soil conditions of AMC types.

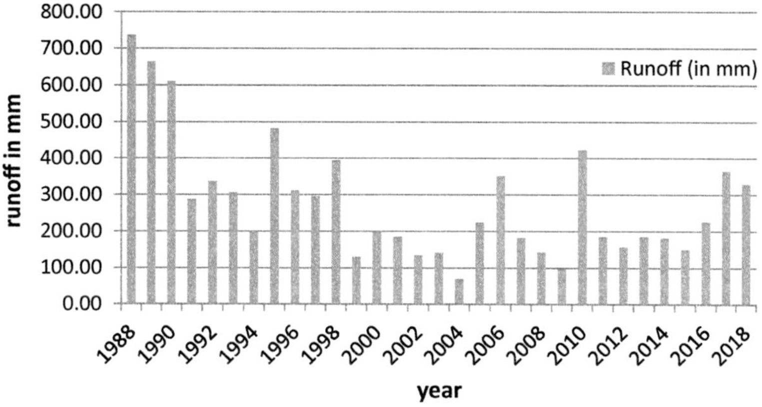

Graph 3 Bar graph between year and runoff

Table 2 Antecedent moisture conditions

AMC type	Soil conditions
I	Soil are dry
II	Average conditions
III	Saturated soil conditions

3.1.2 Land Use/Land Cover

LULC is the most significant factor in calculating the surface runoff of the watershed. The Gangadhara watershed is classified into many classifications. The classifications are barren lands, fallow lands, water bodies, built-up area, cropland and vegetation cover of the study region.

Figure 4 shows the map of LULC of Gangadhara watershed.

Calculation of weighted Curve Number CN_W:

$$CN_W = \frac{\%\text{area} \times \text{CN}}{\text{Totalarea}} \tag{3}$$

Table 3 shows the land use/land cover of the Gangadhara watershed.

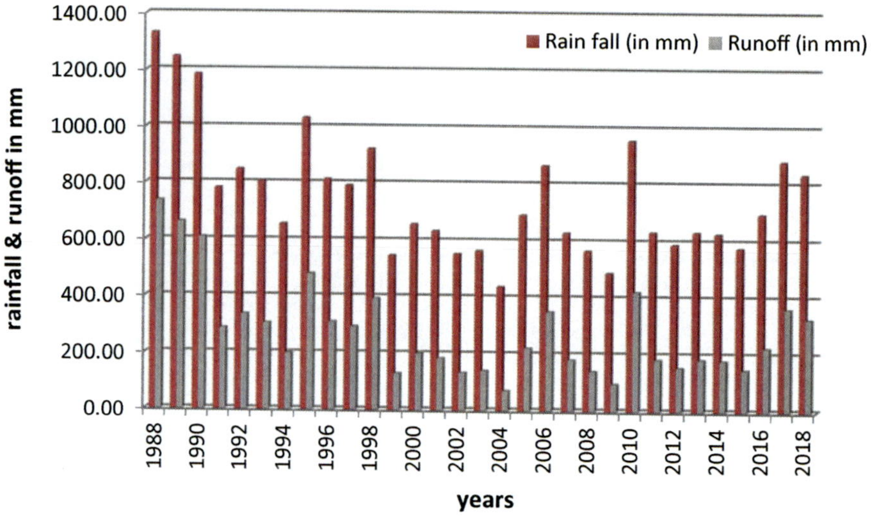

Graph 4 Bar graph between rainfall and runoff

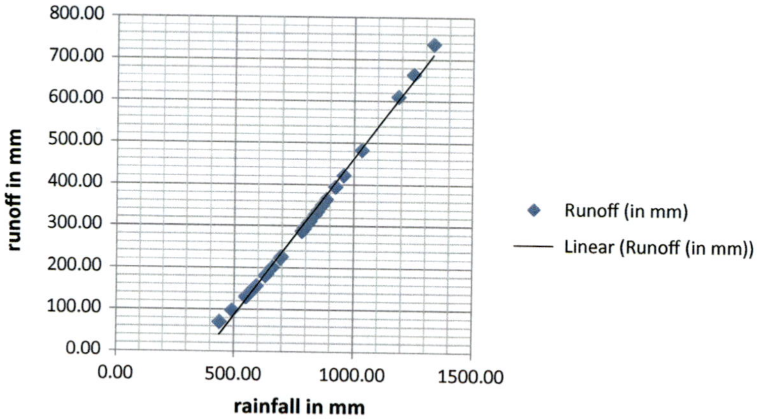

Graph 5 Scatter plot between rainfall and runoff

4 Result and Discussion

The average value of the runoff for the Gangadhara watershed is 280 mm. The average value of total losses for the **Gangadhara watershed** is 478 mm. The average value of volume of runoff for the Gangadhara watershed is 81037265 m^3, i.e. 0.810 km^3 for the last 30 years (1988–2018). Hence, there is a requirement for the construction of recharge structure (Table 4).

Table 3 LULC of Gangadhara watershed

Land use cover	HSG types	Area in km^2	CN	% Area	% Area × CN	Total area in km^2	CN_W
Barren land	C	23.44	77	8.10	623.92	289.29	2.16
	D	5.26	83	1.82	150.92	289.29	0.52
Cropland	C	78.58	88	27.16	2390.43	289.29	8.26
	D	22.9	91	7.92	720.37	289.29	2.49
Vegetation cover	C	36.46	77	12.60	970.49	289.29	3.35
	D	10.92	83	3.77	313.32	289.29	1.08
Fallow land	C	49.91	74	17.25	1276.74	289.29	4.41
	D	4.52	78	1.56	121.88	289.29	0.42
Water bodies		19.49	100	6.73	673.00	289.29	2.33
Built-up land	C	26.5	94	9.16	861.10	289.29	2.98
	D	11.31	55	3.91	215.03	289.29	0.74

5 Conclusion

The present work shows that remote sensing satellite images are very useful to determine runoff volume over a watershed. In this work, **SCS-CN model** was used to determine the runoff of **Gangadhara watershed**. From the Landsat images, Curve Number has been determined based on ***LULC and HSG soil types***. GIS arises as a well-organized tool for the preparation of most of the input data obligatory by the ***SCS-CN*** model and requires less time. The average runoff volume of the watershed computed was 0.810 km^3 as the soil conditions in that watershed are in dry condition and exist only Antecedent Moisture Condition-I. Since the conservation of water is possible in that watershed, no water is going to be wasted by joining the rivers. ***Soil Conservation Service–Curve Number*** (SCS-CN) method is efficiently established as a better method, which consumes ***less time and facility*** to handle extensive data set as well as larger environmental area to identify site selection of artificial recharge structure.

Table 4 Total losses acquired from 1998–2018

Years	Rainfall in mm	Runoff in mm	Runoff volume in m^3	Total losses in mm
1988	1331.5	736.78	213,143,301.9	594.72
1989	1247.8	664.16	192,136,178.8	583.64
1990	1184.9	610.35	176,566,779.5	574.55
1991	781.4	287.33	83,121,616.92	494.07
1992	847.4	336.72	97,408,404.42	510.68
1993	806.6	305.98	88,516,539.25	500.62
1994	657.2	199.81	57,803,493.62	457.39
1995	1032.0	482.75	139,655,788.1	549.25
1996	814.0	311.50	90,115,223.14	502.50
1997	792.0	295.14	85,381,467.28	496.86
1998	922.2	394.63	114,162,875.3	527.57
1999	548.0	130.48	37,745,824.32	417.52
2000	658.6	200.75	58,075,520.07	457.85
2001	635.0	185.06	53,536,197.55	449.94
2002	554.4	134.30	38,851,533.56	420.10
2003	566.8	141.80	41,020,808.96	425.10
2004	438.1	70.70	20,453,600.89	367.40
2005	691.0	222.83	64,462,910.87	468.17
2006	866.7	351.47	101,677,913.10	515.23
2007	630.9	182.37	52,757,861.69	448.53
2008	567.9	142.47	41,214,928.41	425.43
2009	491.0	98.00	28,350,099.6	393.00
2010	957.0	422.19	122,135,938	534.81
2011	634.1	184.47	53,365,076.83	449.63
2012	591.0	156.77	45,352,571.2	434.23
2013	634.3	184.93	53,498,157.81	449.87
2014	629.2	181.26	52,436,053.66	447.94
2015	578.8	149.17	43,152,951.15	429.63
2016	696.2	226.43	65,503,896.12	469.77
2017	883.9	364.74	105,515,128.8	519.16
2018	836.6	328.52	95,036,575.09	508.08

References

1. Subramanya K (2013) Engineering hydrology, textbook of IV edition. McGraw Hill, New York
2. Mishra SK, Jain MK, Singh VP (2004) Evaluation of the SCS–CN-based model incorporating, antecedent moisture. Water Resour Manag 18:567–589

3. Michel C, Andreassian V, Perrin C (2005) Soil conservation service curve number method: how to mend a wrong soil moisture accounting procedure? J Water Resour Res 41:W02011
4. Sahu RK, Mishra SK, Eldho TI, Jain MK (2007) An advanced soil moisture accounting procedure for SCS curve number method. J Hydrol Process. 21:2872–2881
5. Kiran ST, Singh PVS, Trevedi TP (2001) Hydrologic response of a watershed to land use changes a remote sensing and GIS approach. Int J Remote Sens 22
6. Satheeshkumar S, Venkateswaran S, Kannan R (2017) Rainfall–runoff estimation using SCS–CN and GIS approach
in the Pappiredipatti watershed of the Vaniyar sub basin, South India. Model. Earth Syst Environ 3:24 https://doi.org/10.1007/s40808-017-0301-4
7. Gangodagamage C, Agarwal SP (2001) Hydrological modelling using remote sensing and GIS. In: Asian conference on remote sensing, 5–9, Nov

Analysis and Design of Urban Water Distribution Network Using Hardy Cross Method and EPANET

R. K. Rai and P. S. Lingayat

Abstract Water distribution networks (WDNs) serve many purposes in addition to the provision of water for human consumption, which often accounts for less than 2% of the total volume supplied. The purpose of water distribution system is to supply required water quantity at adequate pressure. Water distribution networks play an important role in modern societies because its proper operation is directly related to the population's well-being. The analysis and design of hydraulic network of Waghapur Naka medical college (WNMC) in Yavatmal city were carried out which contains dead ends and modelling it in computer software EPANET. The hydraulic network was analysed for residual pressure of 7 m. Design for 12 m residual pressure for WNMC in Yavatmal city is given in this paper. The procedure developed for analysis and design in this paper has been explained and tested using a case study of WNMC in Yavatmal city as hydraulic network. It also compares between analytical solutions by Hardy Cross method using Hazen–Williams equation and analysis results using computer software EPANET. The aim of this paper is to develop a simple procedure for water distribution network analysis and design using Hardy Cross method and EPANET.

Keywords EPANET · Hardy cross method · Hydraulic network · Pressure · Water distribution network

1 Introduction

The demand of water required for agriculture, industry, sanitation and domestic consumption increases continuously along with the population and economic growth. Managing the water distribution system in a sustainable and integrated manner is

R. K. Rai
Department of Civil Engineering, Government College of Engineering, Amravati 444604, India

P. S. Lingayat (✉)
Environmental Engineering, Government College of Engineering, Amravati 444604, India
e-mail: plingayat5@gmail.com

L. M. Gupta et al. (eds.), *Advances in Civil Engineering and Infrastructural Development*, Lecture Notes in Civil Engineering 87,
https://doi.org/10.1007/978-981-15-6463-5_33

necessary to meet the growing demand of water for drinking, industrial and other necessities. In this regard, there is an urgent need to develop well-designed and optimized solution methods to achieve better control of water distribution networks using different solution approaches. The purpose of a system of pipes is to supply water at adequate pressure and flow.

The present-day water distribution networks are complex and required huge investments in their construction and maintenance. Engineers may not have enough time to monitor all the hydraulic parameters under different operating conditions. Hence, a number of modification attempts to the standard solution methods for the development of a powerful algorithm may help to assess both steady-state solution and particularly time-dependent simulations of water distribution systems when the nodal demands change on a daily basis.

EPANET has the capacity to analyse unlimited number of pipes and tanks. EPANET is a computer program that performs extended period simulation of hydraulic and water quality behaviour within pressurized pipe networks. A network consists of pipes, nodes (pipe junctions), pumps, valves and storage tanks or reservoirs. EPANET tracks the flow of water in each pipe, the pressure at each node, the height of water in each tank and the concentration of a chemical species throughout the network. This paper aims to develop a simple procedure for analysis of water distribution network using Hardy Cross method with the help of electronic spreadsheets and EPANET.

2 Literature Review

Anisha et al. [1] provided the information about various demands, losses and uses of the public. The design of a new network of supply will make the municipality be aware of the new demands and rate of increase in the demands.

Kakadiya and Mavani [2] presented the hydraulic analysis of pipeline network of Punagam area near Surat city using EPANET 2.0. It is difficult to provide safe water to the rural people in sufficient quantity, quality and at satisfactory pressure head with achieving economy constraint.

Rai and Sanap [3] reported that with rising population and increase in the living standards of people, demand for water in cities has increased. Water distribution network requires huge investments to be made, and hence, the network should serve its purpose. The analysis of complex hydraulic network is a time consuming and equally tedious task.

Swamee and Sharma [4] observed that designing a large water-distribution network as a single entity is difficult. The present practice of designing such a system is by decomposing or splitting it into a number of subsystems.

Lungariya and Katharotiya [5] presented a case study of the analysis of continuous water distribution system in Surat city using EPANET software and carried out the work with a specific objective of effective planning, development of water distribution network and analysis using EPANET software.

3 Objectives

Keeping in view the literature, the following objectives are planned.

1. To analyse the existing water distribution system using EPANET and to suggest some measures if the present network does not fulfil the present and future demand;
2. To analyse the data by using EPANET software; and
3. To check the discharge and pressure head in looped network.

4 Equations Used for Design and Analysis of Network

4.1 Hydraulic Equations

Hydraulic equations commonly used for design and analyses of water transmission network are as follows:

1. Darcy–Weisbach equation; and
2. Hazen–Williams equation.

In this paper, the design and analysis were worked out by using the most popular Hazen–Williams equation. This equation is conventionally acceptable equation for design of water conveyance system as it is simple to use. Hazen–Williams equation with hydraulic mean depth, slope and velocity is given by Eq. (1)

$$V = 0.852 C_H R^{0.63} S^{0.54} \tag{1}$$

where

C_H Hazen–Williams coefficient of pipe;
S Slope of hydraulic gradient line (m/m) and
R Hydraulic mean depth, m.

Substituting $V = 4Q/(\pi D^2)$, $C_H = 100$, $R = D/4$ and $S = h_f/L$ in Eq. (1) and after some algebraic manipulations, one can obtain equations.

$$h_f = \frac{10.68 L Q^{1.852}}{C_H^{1.852} D^{4.87}} \tag{2}$$

$$h_f = K Q^{1.852} \tag{3}$$

where K = Resistance coefficient of a pipe and given by

$$K = \frac{10.68 L}{C_H^{1.852} D^{4.87}} \tag{4}$$

The Hazen–Williams formula expressed in the forms of above equations can be used to compute the loss of head in a pipe flowing under pressure.

4.2 Method for Design of Branch Network

Most village water distribution networks are designed as dead-end network or branch network due to ease of design and economical aspects of the project. The design procedure for branch network is as follows:

1. As the demand at each node is known, then apply continuity equation at each node and find out the flows in each pipe;
2. The elevation at each node and HGL at reservoir is known;
3. Calculate the require heads at each node by adding design pressure and elevation at that node;
4. Consider the pipe connecting the reservoir and the first node;
5. Assume any suitable diameter for the given flow and find out pipe resistance constant K;
6. Find corresponding head loss and check whether the head loss is just less than the permissible head loss;
7. If this condition is not satisfied, decrease the diameter and calculate K and h_f and check again. Repeat this procedure until the proper diameter is obtained;
8. Now that the diameter and corresponding head loss for the first pipe is known, find out the new HGL at node 2; and
9. Now considering new HGL at node 2, find permissible head loss in pipe 2 connecting node 2 and 3. Repeat the procedure from Steps 5 to 7.

The above-mentioned procedure is utilized for obtaining suitable pipe diameters in the branch hydraulic network.

4.3 Steps for Analysis Using EPANET

1. Draw a network representation of your distribution system;
2. Edit the properties of the objects that make up the system;
3. Describe how the system is operated;
4. Select a set of analysis options;
5. Run a hydraulic/water quality analysis; and
6. View the results of the analysis.

5 Analysis and Design of Distribution Network for WNMC Outlet 1 Part 1

The water distribution network of WNMC outlet 1 part 1 is shown in Fig. 1, node, elevation and required head shown in Table 1, and its required HGL equals to 463.38 m. Elevation of source of supply is equal to 448.38 m, and hence, it is possible to provide such supply level. Also, pipe details are given in Table 1 like length, diameter and flow.

5.1 Calculations, Results and Discussion

WNMC of Yavatmal city water distribution network was modelled and analysed using HCM and EPANET. The water level of ESR is 15 m above from the ground level and calculate head loss in each pipe and determine the hydraulic gradient line. It was observed that the calculated pressure at each node is greater than 12 m, which satisfies the required condition. After satisfying designed requirement, then find out the required diameter and calculate the cost of each pipe for hydraulic network of WNMC. 15 m water level satisfies the pressure demand and required diameters for WNMC outlet 1 part 1. These nodes however meet pressure of 7 m in hydraulic network of WNMC outlet 1 part 1, and consumers at these nodes are expected to obtain flow by means of their own arrangements for pressure of 12 m. Residual pressure of different nodes in the hydraulic network is given in the tabular form. The cost analysis of WNMC outlet 1 part 1 was carried out; the total design cost

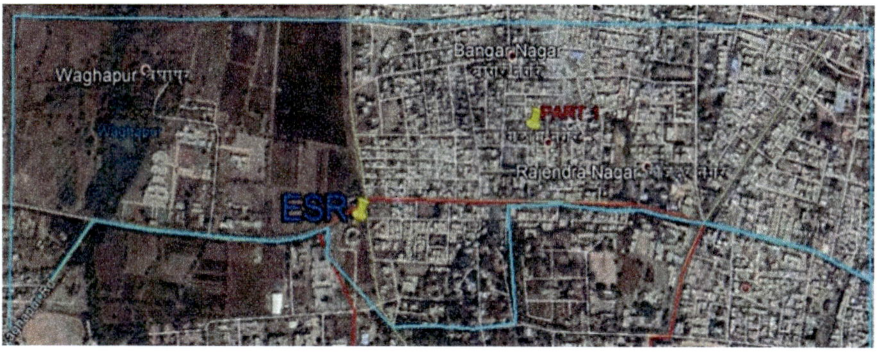

Fig. 1 WNMC outlet 1 part 1

Table 1 Pipe and node details for outlet 1 part 1

Pipe	L, m	*Q*, lps	*D*, mm	Node	Elevation, m	Req. Head, m
1–2	20	0.000	150	1	448.38	–
2–3	10	0.052	200	2	449.52	461.52
3–4	82	0.026	200	3	449.48	461.48
4–5	62	0.213	200	4	450.00	462.00
5–6	113	0.161	200	5	447.98	459.98
6–7	88	0.294	200	6	448.30	460.30
3–13	142	35.282	100	7	448.36	460.36
13–14	96	0.369	100	13	448.16	460.16
14–15	50	0.249	100	14	446.05	458.05
15–16	64	0.130	100	15	445.70	457.70
15–17	26	0.166	100	16	444.84	456.84
17–18	86	0.067	100	17	445.92	457.32
17–19	40	0.223	110	18	445.71	457.71
4–20	50	0.104	100	19	444.80	456.80
20–21	92	0.13	100	20	449.30	461.30
20–22	56	0.239	100	21	448.50	460.50
5–23	50	0.145	160	22	448.72	460.72
23–24	62	0.130	110	23	448.12	460.12
24–25	37	0.161	100	24	447.97	459.97
25–26	175	0.096	100	25	447.81	459.81

of WNMC outlet 1 part 1 distribution system was calculated and its find out that reduction in cost than conventional cost of WNMC outlet 1 part 1.

When the reservoir level is 15 m

Keep the water level 15 m above the ground level and calculate head loss in each pipe and determine the hydraulic gradient line. From the calculation, it can be observed that the calculated pressure at each node is greater than 12 m and the required diameter of each pipe is given in Table 2 (the minimum pressure is at node 4 (13.38 m), and maximum pressure is at node 80 (21.63 m), which satisfies required condition. Graph of node number versus nodal pressure is shown in Fig. 2.

Table 2 Node and pipe details for outlet 1 part 1

Node	Pressure, m	Pipe	*D*, mm
1	15	–	–
2	13.86	01-Feb	200
3	13.9	02-Mar	75
4	13.38	03-Apr	75
5	15.4	04-May	75
6	15.08	05-Jun	75
7	15.02	06-Jul	100
13	15.22	Mar-13	200
14	17.33	13–14	75
15	17.68	14–15	75
16	18.54	15–16	75
17	17.46	15–17	100
18	17.67	17–18	100
19	18.58	17–19	75
20	14.08	Apr–20	100
21	14.88	20–21	75
22	14.66	20–22	100
23	15.26	May-23	75
24	15.41	23–24	100
25	15.57	24–25	75
26	17.33	25–26	100

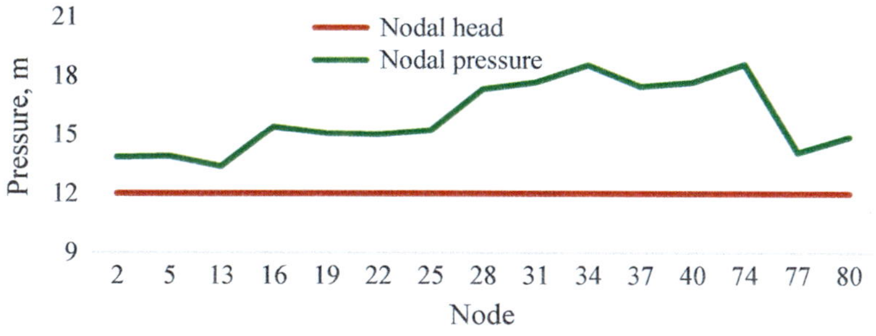

Fig. 2 Graph of node number versus nodal pressure

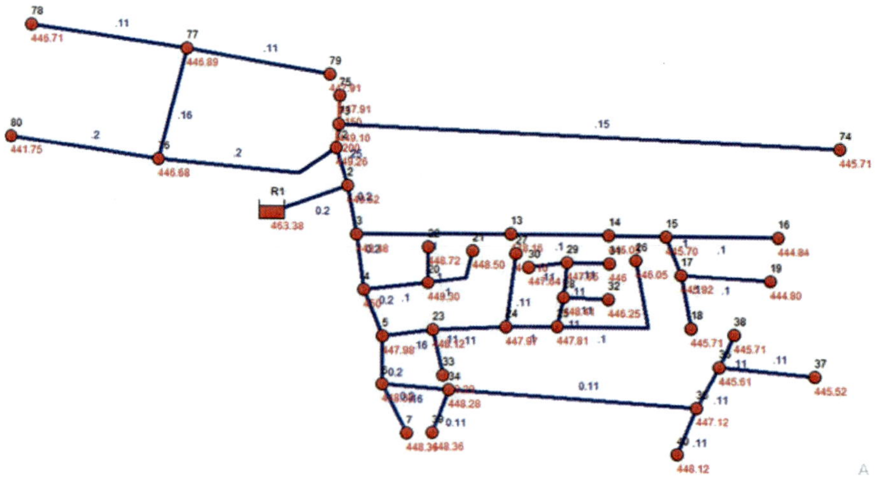

Fig. 3 Schematic view WNMC outlet 1 part 1 using EPANET

5.2 *Computational Analysis of Branched Network*

The step-by-step procedure for analysis of hydraulic network using EPANET was used.

Input
Elevations and diameters at all nodes and pipes in WNMC of Yavatmal city for water distribution network, respectively, are shown in Fig. 3.

5.3 *Results and Discussion*

WNMC of Yavatmal city water distribution network was successfully modelled and analysed in EPANET. Some nodes in hydraulic network did not meet requirements of residual pressure due to their high elevations. These are nodes 4, 6, 7, 34, 35, 39 and 40 in outlet 1 part 1. These nodes however meet pressure of 7 m, and consumers at these nodes are expected to obtain water by means of their own arrangements for pressure of 12 m from sump that can be provided at respective nodes. Results obtained from the computer tool, such as residual pressure of different nodes in the hydraulic network, are given in the table format obtained from EPANET.

Node report
It includes 43 junctions. The result obtained using EPANET is presented. The results obtained using EPANET software for WNMC outlet 1 part 1 were calculated. The error between actual pressure and the pressure computed using software is also compared.

The pressure was computed using Hazen–Williams approach. For WNMC outlet 1 part 1, pressures are negative at junctions 4, 6, 7, 34, 35, 37, 39 and 40, and hence, there is fluctuation in the pressure head.

Pipe report

Pipe report of WNMC outlet 1 part 1 includes 43 pipes. The result is obtained using EPANET software for WNMC outlet 1 part 1. The error between actual flow and flow computed using software was also compared. The error between actual head loss and head loss was also computed using EPANET software. The flow computed using EPANET is nearly equal to the actual flow, and the head loss computed using EPANET is nearly equal to the actual head loss.

5.4 *Cost Analysis*

Here, the cost analysis of WNMC outlet 1 part 1 was carried out. The item rates were referred from the current schedule of rates for the year 2018–2019. Using Equation ($C_m = K_m\, L\, D^m$), the cost of pipes was obtained and is given in Table 3.

The total amount for WNMC outlet 1 part 1 distribution system works out to be 5,178,543.99 Rs., and conventional cost of WNMC outlet 1 part 1 is 6,045,907.02 Rs., and percentage reduction of cost is 14.35%.

6 Conclusions

From the study, it can be concluded that:

1. At the end of the analysis, it was found that the resultant pressures at all the junctions and the flows with their velocities at all pipes are adequate enough to provide water within the study area.
2. The present study on the analysis of water distribution network using Hardy Cross method gives better emphasis on effective design and distribution of network pipes using EPANET.
3. Hardy Cross method is easy, but it requires more number of iterations to get acceptable results; and Hardy Cross method can be used for large networks easily.
4. Hardy Cross method can be easily applied for the analysis of dead-end systems with single source, if dummy pipes are included.
5. The analysis by using EPANET is less time consuming as compared to the excel programming;
6. Graphical details of demand, pressure at nodes and links can be obtained by EPANET without any tedious work; and

Table 3 WNMC outlet 1 part 1 design cost

Pipe	*D*, m	*L*, m	K_m	*m*	Total cost, Rs.
1	0.200	20	36876	1.8825	35,642.05
2	0.075	10	36,876	1.8825	2812.20
3	0.075	82	36,876	1.8825	23,060.06
4	0.075	62	36,876	1.8825	17,435.65
5	0.100	113	36,876	1.8825	54,616.29
6	0.100	88	36,876	1.8825	42,533.04
12	0.200	142	36,876	1.8825	253,058.60
13	0.075	96	36,876	1.8825	26,997.14
14	0.075	50	36,876	1.8825	14,061.01
15	0.075	64	36,876	1.8825	17,998.09
16	0.100	26	36,876	1.8825	12,566.58
17	0.100	86	36,876	1.8825	41,566.38
18	0.075	40	36,876	1.8825	11,248.81
19	0.100	50	36,876	1.8825	24,166.503
20	0.075	92	36,876	1.8825	25,872.26
21	0.100	56	36,876	1.8825	27,066.48
22	0.075	50	36,876	1.8825	14,061.01
23	0.100	62	36,876	1.8825	29,966.46
24	0.075	37	36,876	1.8825	10,405.15
25	0.100	175	36,876	1.8825	84,582.76
26	0.100	125	36,876	1.8825	60,416.25
27	0.075	44	36,876	1.8825	12,373.69
28	0.150	34	36,876	1.8825	35,254.49
29	0.100	37	36,876	1.8825	17,883.21
30	0.075	60	36,876	1.8825	16,873.21
31	0.075	60	36,876	1.8825	16,873.21
32	0.100	98	36,876	1.8825	47,366.345
33	0.075	55	36,876	1.8825	15,467.11
34	0.150	227	36,876	1.8825	235,375.61
35	0.075	82	36,876	1.8825	23,060.06
36	0.075	80	36,876	1.8825	22,497.62
37	0.100	130	36,876	1.8825	62,832.90
38	0.075	78	36,876	1.8825	21,935.18
39	0.100	75	36,876	1.8825	36,249.75
40	0.100	30	36,876	1.8825	14,499.90
41	0.075	20	36,876	1.8825	5624.40

(continued)

Table 3 (continued)

Pipe	D, m	L, m	K_m	m	Total cost, Rs.
42	0.200	1201	36,876	1.8825	2,140,305.56
43	0.100	255	36,876	1.8825	123,249.16
44	0.150	738	36,876	1.8825	765,229.95
45	0.10	290	36,876	1.8825	140,165.71
46	0.075	192	36,876	1.8825	53,994.29
47	0.075	155	36,876	1.8825	43,589.14
48	0.150	480	36,876	1.8825	497,710.54
					$\Sigma = 5{,}178{,}543.99$

7. Cost data from field for the conventional water distribution network was collected, and it was compared with design cost; the percentage reduction in the cost was found that indicates that the development of procedure for the analysis of water distribution network is successful.

References

1. Anisha G, Kumar A, Suvarna P (2016) Analysis and design of water distribution network using EPANET for Chirala municipality in Prakasam district of Andhra Pradesh. Int J Eng Appl Sci 3(4), 111–112. ISSN: 2394-3661
2. Kakadiya S, Mavani K (2016) Simulation of existing water distribution network by using EPANET: a case study of Surat city. Glob Res Dev J Eng Recent Adv Civ Eng Glob Sustain 2455–5703. e-ISSN
3. Rai RK, Sanap NG (2017) Analysis of hydraulic network using hardy-cross method and EPANET. Int J Innov Res Sci Eng 3(03), IJIRSE. ISSN: (O)2454-9665, ISSN: (P)2454-0663
4. Swamee PK, Sharma AK (1990) Decomposition of large water-distribution systems. J Environ Eng 116(2), ASCE, ISSN 0733-9372/90/0002-0269
5. Lungariya P, Katharotiya N (2016) Analysis of continuous water distribution in Surat city using EPANET: a case study. GRD J Eng Recent Adv Civ Eng Glob Sustain 2455–5703. e-ISSN

Estimation of Surface Runoff Potential Using SCS-CN Method and GIS for Parts of Doddahalla3 Watershed in Krishna River Basin

Bhavyata Jethva, Ashim Ghosh, and Lingaraju Yale

Abstract A study was conducted to estimate the runoff by using SCS-CN method and geospatial technology for part of Doddahalla3 watershed in Krishna River Basin. The study area falls under Gadag, Sirahati, and Mundargi Taluks of Gadag District, Karnataka, with an area of 752 km^2. Spatial data like raingauge station locations, land use and land cover, and soil texture for the study area was determined using Water Resources Information System (WRIS), Bhuvan, and Google Earth. Non-spatial data like daily rainfall was obtained from Karnataka State Natural Disaster Monitoring Centre (KSNDMC). The SCS-CN runoff curve number is one of the most widely accepted and frequently used techniques for the estimation of surface runoff. Once the yield is available, then further planning can be done for the river rejuvenation activities in the same area.

Keywords SCS-CN · Rainfall · Surface runoff · Gadag

1 Introduction

Gadag District is the dry region of Karnataka and is facing water crisis. To solve the problem of water unavailability in the area, the art of living along with the Government of Karnataka took up the river rejuvenation project for the entire Gadag District. Surface runoff availability is the first step in planning the river rejuvenation activities. Runoff is the part of rainfall, snowmelt, and/or irrigation water that runs over the soil surface toward the stream rather than infiltrating into the soil. For surface runoff to occur, the infiltration capacity of the soil has to be lower than the rainfall available. Runoff calculations become imperative for river rejuvenation projects (watershed management) as it will show the availability of rainfall which is available as a runoff to recharge the groundwater and harvest the rainwater. It has been explored

B. Jethva (✉) · A. Ghosh · L. Yale
River Rejuvenation Projects, The Art of Living, Bangalore 560082, India
e-mail: coordinator.rrp@projects.artofliving.org

L. M. Gupta et al. (eds.), *Advances in Civil Engineering and Infrastructural Development*, Lecture Notes in Civil Engineering 87,
https://doi.org/10.1007/978-981-15-6463-5_34

by various researchers that accurate measurements of runoff are time-consuming and costly. However, National Resources Conservation Service (NRSC), United States Department of Agriculture (JSDA) in 1969, developed SCS-CN method. It produces reasonably high accuracy, simple to carry out, and less time-consuming. It is widely accepted because of its reliability and stability in the scientific fraternity [1–3].

2 Study Area

The present paper deals with a project area of part of Doddahalla3 watershed that falls in Gadag District administrative boundary which drains into Tungabhadra River in the Krishna River basin. The watershed is spread across 752 km^2 covering parts of three taluks, namely Gadag, Sirahati, and Mundargi, of Gadag District. The study area extends between 15°05′53″ N-15°27′23″ N latitude to 75°36′20″ E-75°57′00″ E longitude. The highest land use and land cover are agricultural built up and water bodies. There are three raingauge stations which are influencing the study area, namely Gadag, Sirahati, and Mundargi raingauge stations.

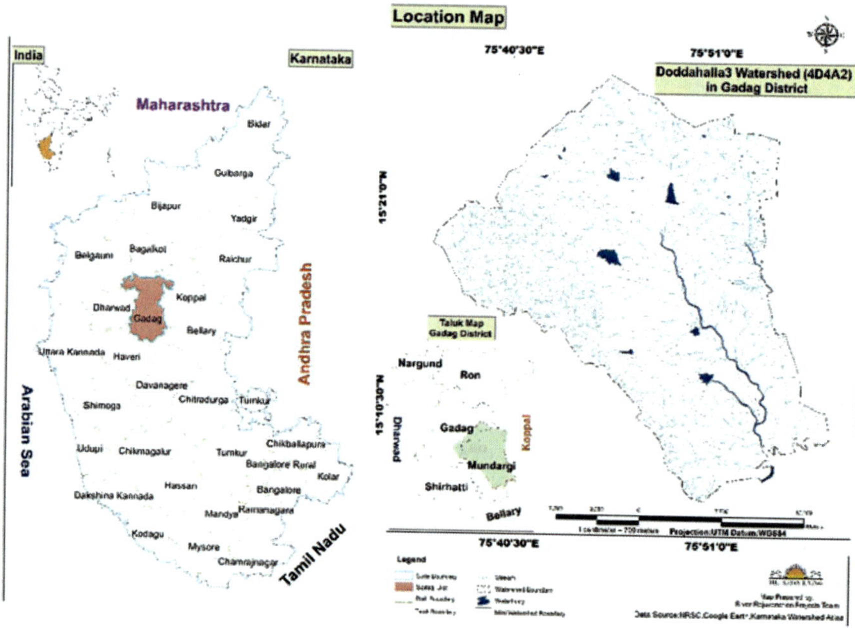

3 Materials and Methodology

Data is divided into two parts: spatial and non-spatial. It was collected from various sources. After data acquisition, the data was processed using Google Earth, MS-EXCEL, and ARC GIS software.

3.1 Data Collection and Software Used

1. Spatial Data:
   a. To prepare land use and land cover (LULC), watershed along with miniwatershed boundary and soil texture thematic layers, open sources like WRIS and Bhuvan WMS servers were used in ARC GIS.
   b. To locate the raingauge stations, Google Earth software was used.
2. Non-spatial Data:
   a. 30 years of monthly rainfall data were collected from KSNDMC.

3.2 SCS-CN Method and GIS

Soil layer
The soil texture map was prepared using WRIS WMS layer. After delineation of the map, it was further classified into hydrologic soil groups as Groups A, B, C, and D as per the definitions given by National Engineering Handbook (NEH) developed by USDA, based on the infiltration rate and other characteristics. The USDA-SCS soil classification is given in Table 1 [4].

Land use and Land cover Layer (LULC)
Various land use and land cover types were interpreted and prepared using WRIS WMS layer as per the USDA guidelines to determine the curve number [5].

Antecedent Moisture Condition (AMC)
The moisture content of the soil at the beginning of the rainfall event is showed by the AMC. Curve number depends on AMC type. There are three categories of AMC. According to the Engineering Hydrology book, AMC-I shows soil which is dry but not to wilting point. Soils where satisfactory cultivation have been taken place. AMC-II shows average conditions. AMC-III shows sufficient rainfall has occurred within the immediate past 5 days. That means, soil condition is saturated. AMC takes into consideration the season, whether it was in dormant or growing.

The following table shows the AMC class and its relation with the curve number (CN) [5] (Table 2).

Table 1 USDA-SCS soil classification

Hydrological soil group	Type of soil	Runoff potential	Final infiltration rate (mm)	Notes
Group A	Sands and gravel	Low	>7.5	Water transmission rate is high
Group B	Moderately deep with moderately fine to coarse textures	Moderate	3.8–7.5	Water transmission rate is moderate
Group C	Clay loams, shallow sandy loams, soils with moderately fine to fine textures	Moderately high	1.3–3.8	Water transmission rate is moderate
Group D	Heavy plastic and soils with a permanent high water table, clay soils that swell significantly	High	<1.3	Water transmission rate is low

Table 2 AMC for the determination of CN value [5]

AMC type	Total rain in the previous 5 days in mm	
	Dormant season	Growing season
I	Less than 13	Less than 36
II	13–28	36–53
III	More than 28	More than 53

Curve numbers to calculate the runoff were taken from the [5], which is given for the AMC-II type. For dry conditions of AMC-I or wet condition of AMC-III, equivalent curve numbers can be calculated using the following equations.

$$\mathrm{CN_I} = \frac{\mathrm{CN_{II}}}{2.281 - 0.0128\ \mathrm{CN_{II}}} \tag{A}$$

$$\mathrm{CN_{III}} = \frac{\mathrm{CN_{II}}}{0.427 + 0.00573\ \mathrm{CN_{II}}} \tag{B}$$

SCS–CN Method The runoff estimation is the most significant aspect in hydrological modeling. Out of available empirical methods for its estimation, SCS-CN is very popular due to its simplicity, flexibility, and requirement of a single parameter called curve number (CN) for computing the runoff.

It is based on the water balance equation of the rainfall in a known interval of time, which can be expressed as

$$P = Ia + F + Q \tag{1}$$

where

P total precipitation
Ia initial (early) abstraction
F cumulative infiltration excluding Ia and
Q direct surface runoff.

Two fundamental hypotheses are used in the above equation [5].

1. "The first suggestion states that the ratio of the real quantity of direct runoff to the maximum possible runoff is equal to the ratio of the amount of real infiltration to the quantity of the potential maximum retention".

   Hence,

$$\frac{Q}{P - Ia} = \frac{F}{S} \tag{2}$$

where S = potential maximum retention after runoff begins and is given by

$$S = \frac{25{,}400}{\text{CN}} - 254 \tag{3}$$

where CN is the curve number that is determined from land use and land cover and soil conditions.

2. The second hypothesis talks that the amount of early abstraction is some fraction of the probable maximum retention. Values of the constant range from $0.1 \leq$ constant ≤ 0.4 as documented in number of studies from various geographical locations, which include the USA and many other countries. For Indian conditions, the constant ranges from 0.1 to 0.3 subject to certain constraints of soil type [5].

$$Ia = (\text{Constant})\, S \tag{4}$$

Substituting values of Eqs. 2 and 4 in 1, we get

$$Q = \frac{(P - Ia)^2}{p - Ia + S} \tag{5}$$

Table 3 Raingauge stations details

Raingauge station name	Latitude	Longitude
Mundargi	15° 12′ 29.891″ N	75° 53′ 26.893″ E
Sirahatti	15° 13′ 50.493″ N	75° 34′ 56.660″ E
Gadag	15° 25′ 03.978″ N	75° 37′ 56.481″ E

For Indian conditions, some modifications were done and $Ia = 0.3S$ is to be considered. Hence, the above equation would be,

$$Q = \frac{(P-0.3S)^2}{(P-0.7S)} \tag{6}$$

Once, the value of CN is known, the value of S can be calculated using Eq. 3 and substituting value of S, we can find out value of Q using Eq. 6. The SCS-CN curve number is a function of type of the land use, the kind of antecedent moisture conditions prevailing and the ability of soils to allow infiltration of water. The depth of the volume of runoff from watershed with different types of hydrological soil groups and different land use and land cover can be obtained by using the above-mentioned equations.

3.3 Rainfall Data

The rainfall data was collected for 30 years for three raingauge stations mentioned below, which is influencing the study area (Table 3).

4 Results and Discussions

4.1 Soil and LULC

The most predominant categories as shown from the table and the map are agricultural cultivated straight-type land representing 84.12% of total area. The other categories like forest, residential, and water bodies constitute 12.87%, 0.92%, and 1.18%, respectively (Table 4).

Table 4 Land use and land cover classification as per SCS-CN type

S. No.	LULC type	Area (Sq km)	Area (percentage of total)
1	Cultivated straight	633.20	84.12
2	Forest scrub	72.99	9.70
3	Open forest	16.28	2.16
4	Dense forest	7.62	1.01
5	Residential, average 65% impervious	6.91	0.92
6	Fallow land	4.66	0.62
7	Wasteland	1.38	0.18
8	Hard surface areas	0.79	0.10
9	Water bodies	8.91	1.18
Total		752.75	100.00

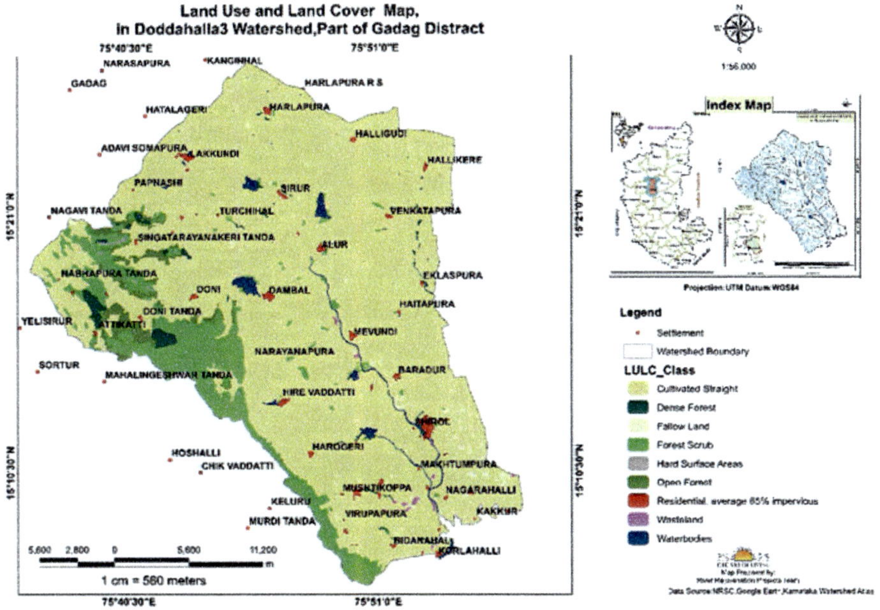

Soil layer was regrouped as hydrological soil groups (HSG), and the following table and map show the distribution, total area covered as well as percentage of HSG in the study area. Hydrological Soil Group C with moderately high runoff potential covers the maximum area followed by type B which shows moderate runoff potential (Table 5).

Table 5 Soil classification per hydrological soil group

S. No.	Soil texture	HSG type	Area (Sq km)	Area (percentage of total)
1	Clay	C	376.50	50.02
2	Sandy clay loam	B	152.85	20.31
3	Clay loam	C	121.82	16.18
4	Sandy loam	B	74.79	9.93
5	Loamy sand	B	10.44	1.39
6	Habitation mask	C	6.91	0.92
7	Rock outcrops	C	0.54	0.07
8	Water body mask	Waterbody	8.90	1.18

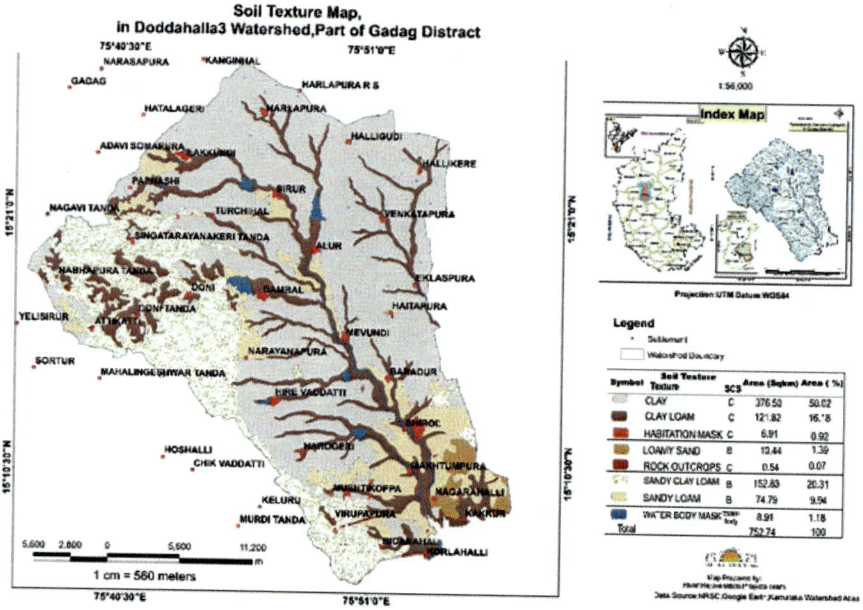

4.2 *Rainfall and Runoff Calculations*

The first step after collecting the rainfall data was to find out the probability of minimum available rainfall for 30 years (1984–2015) using semi-logarithmic graph of recurrence period versus annual rainfall. It shows us the probability of exceeding or equaling the rainfall for the given period [6]. Choosing the value of the year to be 2, i.e., once in every two year, the minimum rainfall that would definitely occur in the study area was computed for all three raingauge stations using this graph.

According to the weightage percentage of the area of influence (according to the Thiessen polygon), the rainfall was calculated for each miniwatershed.

AMC was computed using daily rainfall data of the last 30 years, considering those years for which the assumption of minimum rainfall requirement was met. It was found that the entire study area falls under AMC type I. By overlaying LULC- and HSG-type layers, we found the curve numbers for the average AMC, i.e., type II. Using Eq. A, the equivalent CN_I was determined.

The rainfall distribution for two years 2013 and 2014, pre-monsoon, southwest Monsoon, and northeast monsoon (post-monsoon) have been shown in the maps below. As it can be seen from the table, 2014 seems a good rainfall year where the total pre-monsoon and the NE monsoon rainfall are more than the previous year considering all three raingauge stations (Table 6).

Table 6 Monsoon distribution for 2013 and 2014

Raingauge station	2013				2014			
	Pre-monsoon (mm)	Southwest monsoon (mm)	North East monsoon (mm)	Annual (mm)	Pre-monsoon (mm)	Southwest monsoon (mm)	North East monsoon (mm)	Annual (mm)
Mundargi	90.00	416.00	73.00	579.00	147.00	350.50	167.00	664.50
Sirahati	77.50	354.50	0.00	432.00	285.00	442.00	136.50	863.50
Gadag	172.00	313.50	86.00	571.50	137.00	392.00	152.10	681.10

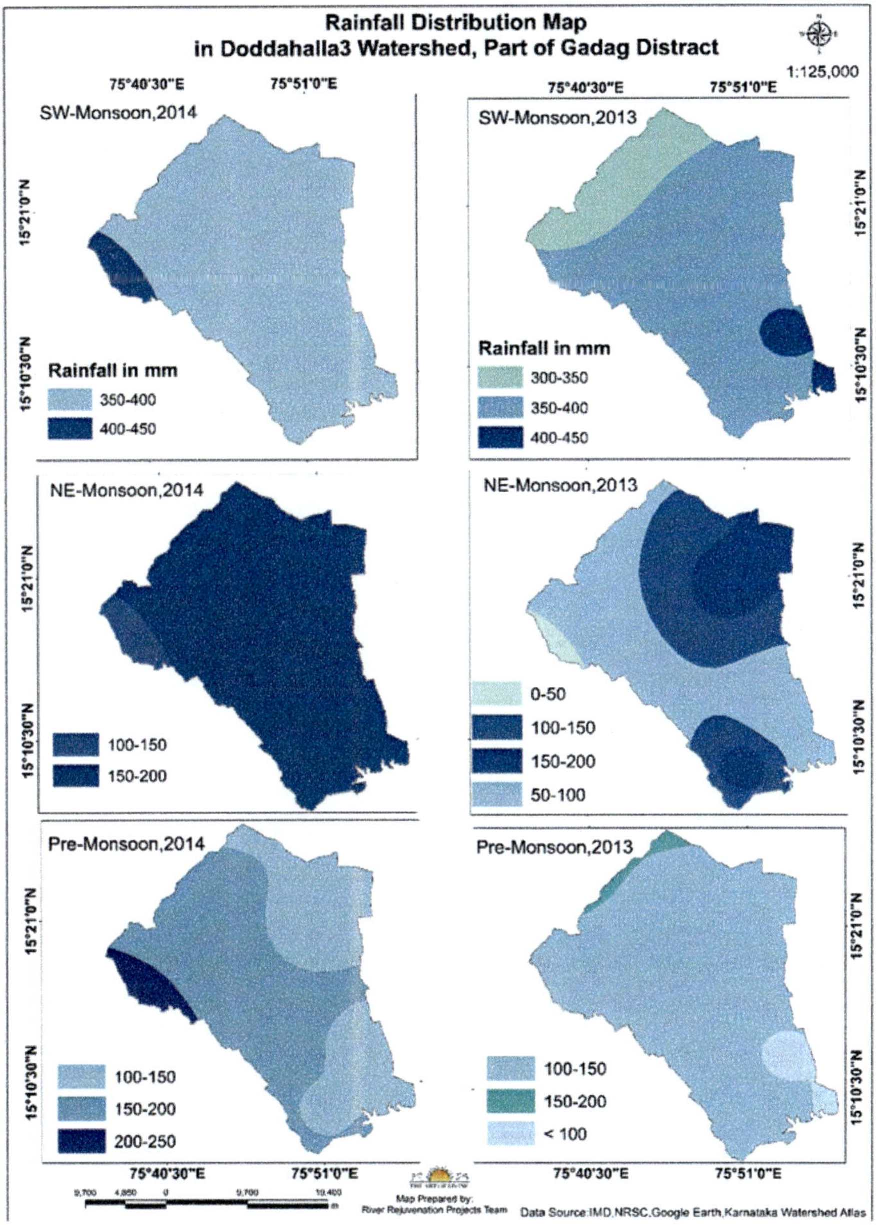

Overlaying thematic layers of LULC, soil texture, miniwatershed boundary in GIS environment using ARC GIS, the curve number for each miniwatershed was determined. Putting values in the above-mentioned equations for SCS-CN, and the final runoff value were calculated. The following table shows miniwatershed wise runoff calculations for the area (Table 7; Fig. 1).

Table 7 Miniwatershed-wise runoff calculations for the study area

S. No.	Name of miniwatershed	Rainfall (mm)	CN_I	Runoff (mm)	Area (Sq km)	Runoff volume (MCM)
1	Alur	566.86	78.00	489.15	33.11	16.20
2	Attikatti	562.73	55.46	331.96	18.20	6.04
3	Ayyanakerihalla	586.69	55.79	292.49	26.77	7.83
4	Baradur	584.99	71.90	501.08	22.19	11.12
5	Basapura (Bennihalli)	588.84	67.85	499.92	19.02	9.51
6	Bennihalli	592.90	75.31	499.35	18.24	9.11
7	Bidanahal	595.18	68.66	412.43	38.43	15.96
8	Dambal	595.24	61.06	414.33	24.14	10.00
9	Dindur	578.92	57.59	341.16	32.66	11.14
10	Doni	589.09	59.84	368.70	24.64	9.09
11	Eklaspura	582.36	76.76	502.84	25.18	12.66
12	Govankop	590.51	75.24	502.02	21.79	10.94
13	Gudigere	582.18	86.60	505.62	0.33	0.17
14	Gudigere North	582.60	86.60	505.67	1.69	0.85
15	Halligundi	582.73	73.37	503.65	19.50	9.82
16	Harlapura	552.88	70.01	475.13	38.03	18.07
17	Harogeri	589.95	63.47	452.80	19.48	8.82
18	Hatti	588.27	75.21	500.97	1.29	0.65
19	Hire Vaddatti	594.51	57.72	350.70	35.32	12.39
20	Kalkere	592.00	70.75	467.69	24.76	11.58
21	Kesalapura	588.23	77.90	501.15	22.98	11.52
22	Lakkundi	560.11	79.15	478.81	26.86	12.86
23	Mevundi	588.70	71.19	507.92	15.68	7.96
24	Mevundi (North West)	590.77	64.85	493.28	40.03	19.75
25	Mundargi	592.00	72.73	491.60	24.85	12.22
26	Murlapura	583.71	73.33	503.69	12.92	6.51
27	Nelogi	594.88	75.53	493.97	1.90	0.94
28	Ramenahalli	589.73	71.49	502.57	23.93	12.03
29	Shirol	589.60	71.39	509.60	20.69	10.54
30	Singatarayanakeri	592.12	67.59	492.98	16.36	8.07
31	Sirur Kere	563.28	57.56	466.03	26.90	12.54
32	Talavarthihalla	583.16	83.21	506.27	22.74	11.51
33	Turchihal	593.87	59.51	431.58	23.39	10.09
34	Venkatapura	582.64	75.72	504.08	28.45	14.34

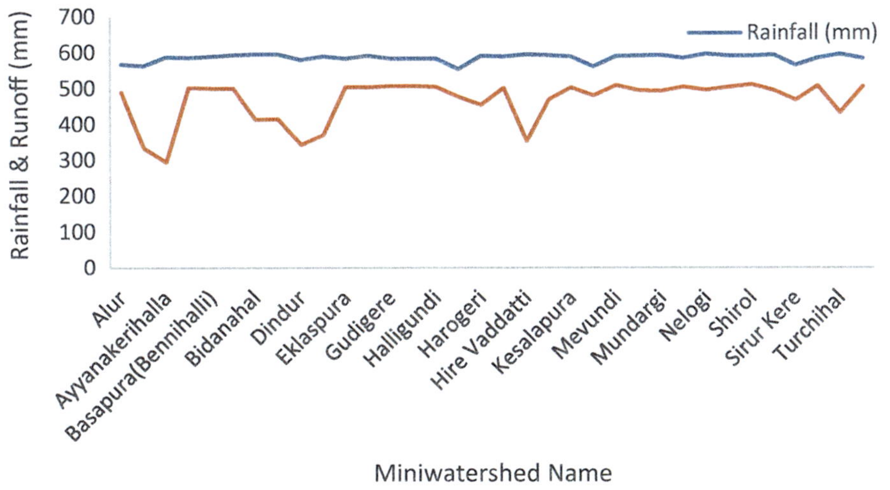

Fig. 1 Rainfall versus runoff graph for 34 miniwatersheds in Doddahalla3 watershed

5 Conclusion

With the integration of geospatial technology and SCS-CN model, the calculated estimated runoff shows the availability of water for harvesting and recharge. Hence, the watershed management becomes more efficient. Since it is showing runoff volume (yield), it can be concluded that the river rejuvenation projects team can go ahead and plan the engineering measures for the watershed management to recharge groundwater and harvest rainwater. It can further be concluded that the trend showed by different miniwatersheds is varying because of the LULC and soil themes being different in particular terrains. This method depicts that it is the powerful tool to estimate the runoff and hence, along with watershed management, land use planning [7] can also be done efficiently.

References

1. Tejaswini NB, Shetty A, Hegde VS (2011) Land use scenario analysis and prediction of runoff using SCS-CN method: a case study from the Gudguji Tank, Haveri district, Karnataka, India. International Journal of Earth Science and Engineering 4(5):845–853
2. Soulis KX, Valiantzas JD (2012) SCS-CN parameter determination using rainfall-runoff data in heterogeneous watersheds-the two-CN system approach. Hydrol Earch Sci 16:1001–1015
3. Dhawale AW (2013) Runoff estimation for Darewadi watershed using RS and GIS. Int J Recent Technol Eng 1(6):46–50
4. National Engineering Handbook, Part 630 Hydrology, Chapter 7: United States Department of Agriculture (2007)
5. Subramanya K (2013) Engineering hydrology, 4th edn. McGrawHill Education, New Delhi, pp 185–190

6. Nagarajan N (2012) Poongothai: spatial mapping of runoff from a watershed using SCS-CN method with remote sensing and GIS. J Hydrol Eng ASCE 17(11):1268–1277
7. Mishra SK, Singh VP Soil conservation service curve number (SCS-CN) methodology. Springer, India, Accessed as EBOOK

Use of BIM for Study of Life Cycle Cost Analysis of Residential Complex

Aditi A. Sathe and S. S. Pimplikar

Abstract The initial costs of a building are quite small in comparison to the life cycle costs, as it is estimated that they represent less than 30% of the total life cycle cost of a building. Accordingly, if we consider integration of the life cycle cost of a building into the estimation of investment values, then we would have a better and clearer idea of how much our total operation costs are during and after construction process. Did an overview of Life Cycle Assessment and Modelling tools and BIM integrated LCC evaluations. Used BIM objectives by Krieder for to generate Life Cycle Cost Analysis of the Residential Building. Findings of this study demonstrate the opportunities to manage and estimate a reliable and accurate information on a building's life cycle in real time by considering each element and its components into the calculation, which may change the metrics for a real estate economic evaluation in order to achieve sustainable and efficient property investment. The adoption of the building information modelling to generate relative LCCA databases directly benefits the real investment cost calculation for investors.

Keywords Life cycle cost analysis (LCCA) · Building information modelling (BIM) · Real estate · Revit · BIM

1 Introduction

Our construction industry has huge impact on our environment, and it contributes to significant amount of world's energy and resource consumption. The building materials manufacturing uses about 10% of the global energy, and operational phase produces at least 30% of all greenhouse gas emissions, and demolishing buildings are responsible for 40% of all solid waste. Building information modelling is dimensional tool which supports effective designs and construction technique. Due to increasing awareness and environmental concerns and pressure from several government bodies, clients, and other factors, many studies have been conducted to reduce building

A. A. Sathe (✉) · S. S. Pimplikar
Civil (Construction, Engineering and Management), MIT-WPU, Pune 411038, India
e-mail: aditia.sathe@gmail.com

L. M. Gupta et al. (eds.), *Advances in Civil Engineering and Infrastructural Development*, Lecture Notes in Civil Engineering 87,
https://doi.org/10.1007/978-981-15-6463-5_35

energy consumption and its environmental impacts. The cost-effective analysis of the process is termed life cycle costing (LCC). Current research in this area explores the use of building information modelling (BIM) tools for LCA of buildings, taking advantage of its integration of process, technology, and people. The ease of access to information and flexibility to early design changes that BIM tools offer probably make them more appealing for use for LCA of buildings. Life cycle costs give additional comprehensive assessment of the future cost effectiveness of a building than various alterative building analyses that importantly concentrate on the initial capital costs or on operating-related costs in the short term. Reduction in costs is significant to company profitability, and it is possible through action research that the life cycling costs of buildings can be reduced by design, often at no overall increase in capital cost. Notwithstanding the significance of life cycling costing, practical difficulties have prevented effective use of this technique in making decisions about buildings due to lack of rigorous techniques for life cycle costing and appropriate practical working tools in building design techniques.

2 Literature Review

Through the literature review, it was found that

1. The LCI analysis is the quantification of the inputs and outputs for each of the phases of the building's life cycle [2].
2. The parametric approach possesses the advantage that input parameters can be adjusted easily and quickly in BIM design process [4].
3. To support decision-making in the critical early design stages, we propose a workflow of using conceptual BIM models and visual scripting for life cycle cost assessment [6].

Abidin and Powmya [1], Rashid and Yusoff [7], Chau et al. [3], Bayer et al. [2] have given idea about the basic relation between the LCCA and BIM which helped in further analysis for the conclusions.

3 Objectives

a. To interpret how we can apply building information modelling (BIM) in the projects.
b. To show the data interoperability through building model.
c. To show how the adoption of the building information modelling to generate relative LCCA databases directly benefits the costing of project.

4 Methodology

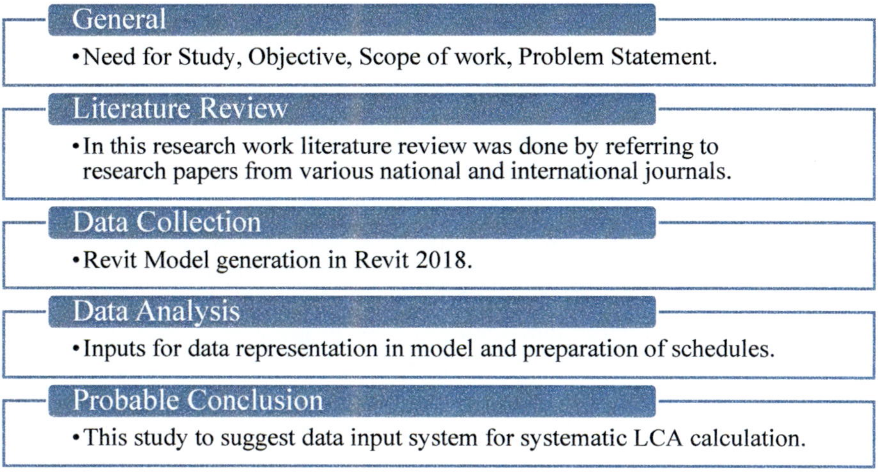

5 Data Collection

5.1 Prelude

The framework of LCA includes: the goal and scope definition, life cycle inventory (LCI) analysis, life cycle impact assessment, and interpretation of results (ISO 14040 2006). The goal and scope definition include the reasons to carry out the LCA and intended application, the intended audience and the use of the results, in addition to the definition of system boundary and functional units. The LCI analysis is the quantification of the inputs (such as materials and energy) and outputs (such as carbon emissions and wastes) for each of the phases of the building's life cycle [2]. Life cycle impact assessment is the classification and characterization (weighting) of the results from the LCI based on human and environmental impacts or effects [2, 7]. The last process in standard LCA framework is the interpretation of results. This phase uses an iterative procedure both within the phase and with the other phases of an LCA (ISO 14040 2006). At the interpretation phase, LCA results are reported in the most informative way possible, and the need and opportunities to reduce the impacts of the building on the environment are systematically evaluated [2]. Validation of results (comparing with results from other published articles) and sensitivity analysis (checking on the reliability of variables for LCA) may also be conducted during this process [7].

5.2 3D Model

3D BIM model was generated for representation of LCA data.

5.3 BIM Integrated LCC Evaluations

The estimation of LCC in the first stages of design is based on the comparison with the cost of similar buildings in the past, but as the design evolves in more detail, the life cycle cost can also be estimated more accurately. However, if the individual building components are identified, then the LCC of each of them can be estimated by assigning a certain price, life expectancy, and the frequency of maintenance and operating activities. Therefore, data aggregation by means of integrating all building components costs into the model database to estimate the efficiency and performance of the building can enable the investor to know the running cost of a design proposal. In this sense, the owner has the possibility to know how much the construction process, environmental performance, and future maintenance of a building would cost. To achieve an integrated evaluation, using parameterised modelling tools that associate geometric models of a building into databases can be considered as a practical solution. The BIM technology enables the possibilities to aggregate data from all sources into the model and simulate the building data for many different uses from the earliest conception to the demolition stage, in addition to simulations related to existing buildings. Based on the computer integrated construction research program at the University of Pennsylvania, which identified 25 BIM uses during the life cycle of a facility, this paper considers three of them to integrate into economic evaluation of LCC. The three selected uses are (1) cost estimation with quantity take-off methods, (2) energy consumption analysis, and (3) building (preventative) maintenance scheduling, where the integration and combination of these tools provide an instrumental integrated approach to run more accurate LCCA. These abovementioned uses can be summarised as follows.

5.3.1 Cost Estimation (Quantity Take-offs)

a. Precisely quantify modelled materials.
b. Generation of more cost estimates at faster rate.
c. Better visual representation of project and construction elements that needs to be estimated.
d. Easier exploration of different design options and concepts within the owner's budget.

5.3.2 Energy Consumption Analysis

a. Save time and costs by obtaining building and system information automatically from the BIM model instead of inputting data manually.
b. Improve building energy prediction accuracy by auto-determining building information, such as geometries and volumes precisely from the BIM model.
c. Optimize building design for better building performance efficiency and reduce building life cycle cost.

5.3.3 Building (Preventative) Maintenance Scheduling

a. Plan maintenance activities proactively and appropriately allocate maintenance staff.
b. Track maintenance history.
c. Reduce corrective maintenance and emergency maintenance repairs.
d. Evaluate different maintenance approaches based on cost.

6 Data Analysis

6.1 Prelude

BIM technology enables us to model and simulate the real construction process and its cost in a virtual environment. This process includes the information and performance of each building element, which provide an estimation possibility of accurate initial cost of a building, as well as the renovation cost. Building energy modelling (BEM) can be performed either in external applications or directly inside a BIM authoring program (i.e. Revit), in which it analyses the parametric model in terms of environmental behaviour and energy consumption. Accordingly, estimations of total cost per year based on systems and sources used in the building could be demonstrated. This would provide the total utility cost needed for the LCCA. The FM-BIM would demonstrate an accurate planning system connected with time and cost and with scheduling and tracking of service and maintenance events associated with their cost.

6.2 Case Study

The core value of BIM is the information attached to the model. This technology enables the modeller to collect and aggregate data from different sources and run a different series of simulations on them. This way, it is possible to produce an

Fig. 1 3D model of building

automatic and systematic report on final outcomes. This case study explores the use of autodesk Revit, a 30-year-old BIM authoring application, for life cycle cost analysis for an imaginary residential building project (Fig. 1).

The case study project is residential building of 2100 m^2 per floor with fourteen floor. The building is divided into two stories and is located on a hypothetical plot of vacant land near Kothrud without any nearby structure. To simplify the process and keep the focus of the study on workflow of the case study, this paper limits itself to this small building with simple functionality. The building has been virtually built in Revit through a collection of intelligent building elements, which constitute an important database. Throughout this case study, a simulation has been run to determine the building performance. The basic specifications of the construction elements and cost are defined in following Fig. 2.

6.3 *Data Analysis*

The methodological approach of this case study is based on the five major categories of purpose and objective of BIM that are defined by Kreider et al. [5]. To develop an integrated workflow, this study tries to use the five objectives as the path to generate a practical solution to integrate BIM into LCCA.

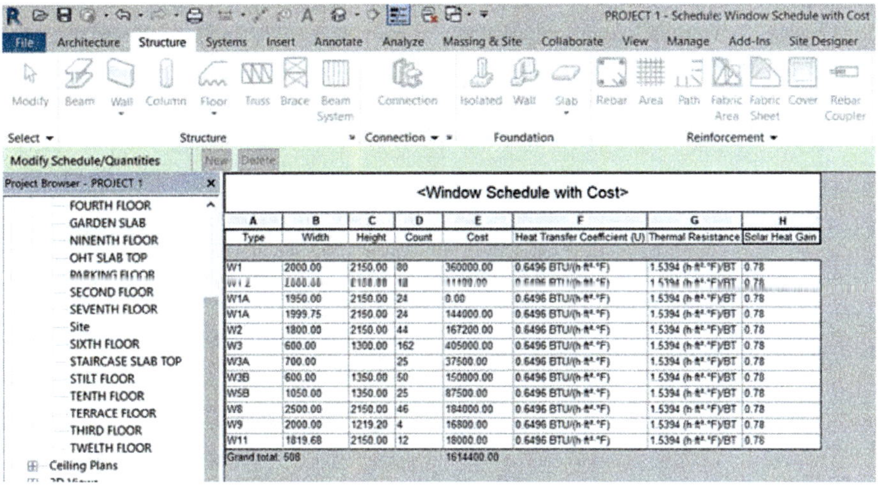

Fig. 2 Screenshot of Revit building materials properties that highlight the physical properties and cost in order to show how the information for simulation is entered to Revit

6.3.1 Gather: Collect or Organize Facility Information

There are two resources of data to collect and organize information about the facility. The first one refers to the different costs addressed to each building element, and the second one refers to the information needed to simulate the energetic performance of the facility to obtain its utility cost.

Cost Elements Resource Database

To be able to gather information for each building element, the cost breakdown structure that is used this way each building element can be connected to an external cost database. The BEDEC developed by the Institute of Construction Technology of Catalonia (ITEC) is a complex database where different levels of detail can be estimated accordingly, such as material, labour, machinery, and ancillary costs; this gives a direct connection of import and export to Revit as the BIM authoring application.

Energy Resource Database

To be able to run the energy analysis and estimate the utility cost of the project, four necessary sources have been identified (Table 1).

Table 1 Source of energy resource information

Sources	Information required
Construction elements of physical properties	Defined by the characteristics of their own materials, such as thermal conductivity, heat capacity, embodied energy, or embodied carbon
Environmental data	Location (latitude and longitude), soil type (thermal conductivity, density and heat capacity), wind protection, shading elements and climate data (air temperature, relative temperature, solar radiation, and wind speed)
Operational profile	Schedules of usage shall be set in order to have a more accurate energy analysis in different periods of time
Building systems	This indicates what kind of energy input and output systems are implemented. Sources of energy and cost of the purchased energy are inputted

6.3.2 Generate: Create Information Databases of a Facility

Once the resource databases are identified, it is time of aggregate the different sources of information together through the virtual construction of the facility inside Revit. The building materials have associated the physical properties, such as thermal conductivity, heat capacity, and embodied energy or embodied carbon of the same material used in the real construction. When the basic building structure is modelled, the rest of the building elements can be introduced into the model, such as doors, windows, and skylights. Therefore, as the model has more details, the analysis will be more accurate. Correspondingly, in the element cost breakdown defined for this case study, each building element can be linked with the external database (Fig. 3).

6.3.3 Analyse and Communicate: Examine Elements of the Facility to Gain a Better Understanding of It and Present Information of a Facility in an Exchangeable Format

To be able to simulate the future building performance and obtain the LLC costs, we need to internally analyse the energy efficiency of a facility in addition to communicating the information generated with external applications for more advanced databases or simulations. The final results are based on the combination of the initial cost information of each element and the quantity take-off list from the virtual construction made in Revit. Comparing this method with the traditional workflow, which should quantify the elements in a manual way, the quantity take-off process requires significant effort and time that may be associated with possible errors.

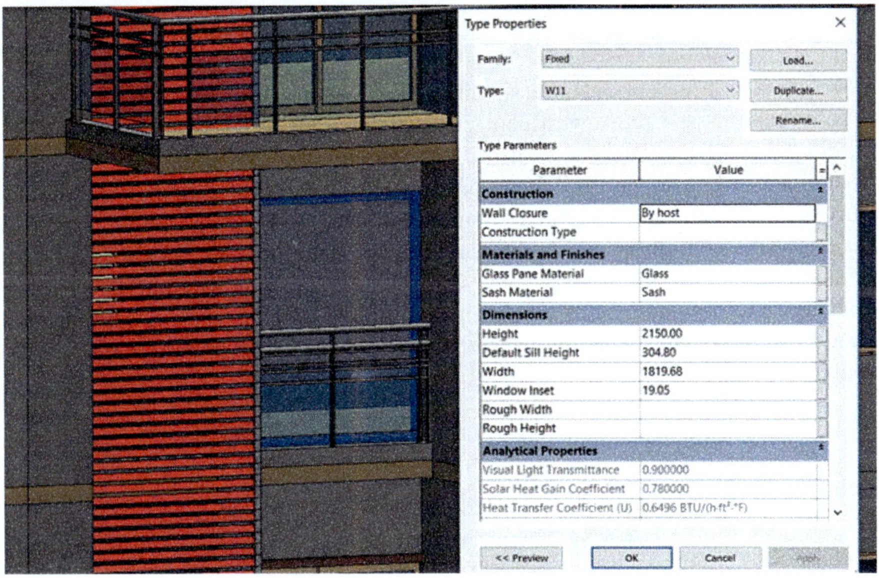

Fig. 3 Screenshot of Revit building element's physical properties

6.3.4 Realize: Make or Control a Physical Element Using Facility Information

In case of LCCA, the integrated database built throughout the virtual modelling process allows investors to realize and consequently control the total cost of an investment, including both initial and future running cost. In order to achieve sustainable and efficient property investment, the BIM model becomes a live database of an existing or future facility. This final stage, where all data has been realized through accurate reports and simulations, demonstrates the possibilities of the different choices that directly impact the total cost and also provides a more accurate vision of the future costs.

7 Conclusion

The findings of this study demonstrate the opportunities to manage and estimate reliable and accurate information on a building's life cycle in real time by considering each element and its components into the calculation via BIM tools and a series of simulations, which may change the metrics for a real estate economic evaluation in order to achieve sustainable and efficient property investment. In this study, the implementation of an integrated approach of BIM, BEM, and BIM-FM tools into LCCA is explored as an intelligent solution, which not only analyses the required aspects of a building but also simulates future behaviour and provisions. In this regard, the development of comprehensive databases is essential to serve as the basis for required future periodic assessments and to generate relative LCCA databases directly benefits the real investment cost calculation for investors. However, the validity of the BIM databases will be as accurate as the model; thus, the wrong modelling process can produce the wrong databases. Therefore, when the modelling process is correct, this integrated solution can assist private and public investors to achieve the goal of promoting the quality, sustainability, and competitiveness of the building stock. This study suggests an accurate data aggregation and series of simulations to provide a more accurate estimation of efficiency and performance of a building, where consequently investors are enabled to know the running cost of a design proposal.

References

1. Abidin NZ, Powmya A (2014) Perceptions on motivating factors and future prospects of green construction in Oman. J Sustain Dev 7(5):231–239
2. Bayer C, Gamble M, Gentry R, Joshi S (2010) AIA guide to building life cycle assessment in practice. The American Institute of Architects AIA, Washington, DC
3. Chau CK, Leung TM, Ng WY (2015) A review on life cycle assessment, life cycle energy assessment and life cycle carbon emissions assessment on buildings. Appl Energy 143:395–413
4. Hollberg A, Ruth J (2016) LCA in architectural design—a parametric approach. Int J Life Cycle Assess 21(7):943–960
5. Kreider R, Messner J, Dubler C (2010) Determining the frequency and impact of applying BIM for different purposes on building projects. In: 6th international conference on innovation in architecture, engineering and construction (AEC, University Park, Penn State University, PA)
6. Röck M, Hollberg A et al (2018) LCA and BIM: Integrated assessment and visualization of building elements' embodied impacts for design guidance in early stages. Procedia CIRP 69(2018):218–223
7. Rashid AFA, Yusoff S (2015) A review of life cycle assessment method for building industry. Renew Sustain Energy Rev 45:244–248

Water Resources Planning, Policy and Management—Experiences from Kumudvathi River Rejuvenation Project

Ravindra Desai, Anjali Kumari, Girish G. Shetty, and Bhavyata Jethva

Abstract Kumudvathi River Rejuvenation Project is an initiative taken to revive the natural process of water resource renewability, which existed in nature. The terrain analysis using geospatial technology revealed the geomorphological and geohydrological processes that had been operating since ages and the conventional human interventions which made the water resources sustainable [1]. It was possible to reveal the causative factors which distorted the pre-existing water resources renewability factors, which in turn helped in evolving a methodology to copy nature's method of cycling the water source between the surface and sub-surface with different layers of soil, vegetation, and subsurface aquifers, collecting the spatial and non-spatial information. It was possible to map, locate suitable sites for harvesting rainwater, recharge groundwater, revive the surface, and base flow with environment-friendly conditions. The structures constructed accordingly have been able to perform the function efficiently and yield good results of increased base flow, raising the water table, and improved groundwater potential. The pilot exercise proved efficient when concurrent impact assessment was done. This helped in project extension as stakeholder's confidence and involvement increased, resulting in an extension of the project to the entire basin. Many hindrances and constraints faced are discussed in this paper. It needs the attention of the policymakers and also the government to adopt such proven cost-effective, eco-friendly, and innovative methods elsewhere in the similar geological terrains (Sharachchandra in World Dev 19(6):607–621, 1991 [2]). This also brings awareness and provisions for regulations to protect and maintain such structures to sustain the augmented process of water resource renewability.

Keywords Spatial and non-spatial · River rejuvenation · Water resource management · Planning

---

R. Desai · A. Kumari (✉) · G. G. Shetty · B. Jethva
River Rejuvenation Project, Art of Living International Center, Bangalore 560082, India
e-mail: anjali.kumari@projects.artofliving.org

L. M. Gupta et al. (eds.), *Advances in Civil Engineering and Infrastructural Development*, Lecture Notes in Civil Engineering 87,
https://doi.org/10.1007/978-981-15-6463-5_36

1 Introduction

River basins of Karnataka have been undergoing increasing water stress for the last four decades resulting in drying up of rivers. Although the average annual available rainfall figure has not changed much, water is still not available throughout the year in many parts of Karnataka. This increases drinking water shortages, less yield of agriculture crops, and leads to migration of farmers toward cities and debt traps leading to farmer suicides [3]. The degradation of the rivers can be attributed to overexploitation of groundwater which results in sapping the base flow in rivers and also catchment degradation through mining, deforestation, and land-use changes [4]. Hence, there is an urgent need to work for reviving rivers.

One such pilot exercise to rejuvenate the river has been carried out by the volunteers of the River Rejuvenation Projects in the Kumudvathi watershed. The river catchment area is influenced by mainly three rain gauge stations, Thyamagondlu, Nelamangala, and Solur indicated the 30-year average rainfall to be 840.7 mm, 827.1 mm, and 785 mm, respectively, in the region. As the area receives good rainfall as observed, it creates good scope for harvesting the rainwater and recharges the aquifer [5]. For two decades, the process of natural groundwater recharge has decreased due to the removal of natural vegetation, siltation of the water bodies, increased evaporation, and overexploitation of groundwater resulting in water cycle imbalance [6]. The dependency on the groundwater made farmers indiscriminately drill deeper borewells, resulting in failure of borewells and/or more cases of borewells having fluoride/nitrate concentration [7]. Farmers are using water tankers and cans for drinking and other domestic needs [8].

The study of rejuvenating Kumudvathi watershed was started by comparing the historical and present conditions of groundwater availability and rainfall analysis for the calculation of available rainwater in the project area for recharge to the aquifer [9, 10]. Considering the basic concept of the hydrological cycle, drainage treatments were planned for rejuvenating the natural river system. The major objective was to restore the local water resources and sustainably recharge groundwater. Four important components of river rejuvenation to restore the water resources are:

- To bring back the hydrological balance of the area by integrating rainwater, surface water, and groundwater in harmony to make the river flow.
- To balance local water sources and user's requirements to ensure that water supplies are available when and where they are needed.
- To encourage the farmers for agroforestry/horticulture based plantations that help birds and bees to thrive, thereby restoring the ecosystem and increase greenness.
- Water Literacy: Awareness of Water Resources and its Management.

2 Study Area

Kumudvathi River is a tributary of Arkavathi River and originates from Shivagange hills in Nelamangala Taluk of Bangalore Rural District, Karnataka, India. The river catchment covers 460 km^2 area across 278 villages (18 mini-watersheds) of Nelamangala Taluk, Bangalore Rural District and Magadi Taluk, Ramanagara district, with a population of 1.92 Lakhs. Both Arkavathi and Kumudvathi flow into Thippagondanahalli Reservoir which was serving 30–40% water requirements of Bangalore city in the past. geologically the terrain as hard rocks. There was sufficient inflow to the reservoir but has drastically reduced for three decades due to deforestation, quarrying, soil erosion, siltation of the existing 223 irrigation tanks, and overexploitation of groundwater. The basin is declared as one of the overexploited and highly water-stressed areas by the Karnataka Groundwater Authority [11] (Fig. 1).

3 Water Resource Management Approaches and Methodologies

3.1 Planning

An initial and important step is to understand the natural processes of rivers and their flow based on the topography, geology, geomorphology, geohydrology, pedology, and biological framework inherent to the terrain. Survey and Site Suitability (Identification of area interest) is carried out by evaluation of the hydrogeological suitability of the site which largely depends on the ease of injecting, the aquifer storage capacity, and the aquifer's resistance to clogging should be carried out. The terrain, geological conditions, drainage, and soil scape are important aspects that decide the site suitability. With the help of GIS and remote sensing techniques, the preparation of thematic maps using remote sensing technology like soils, geomorphology, lithology and lineaments, and slopes is referred to locate the congenial sites for harvesting the rain waters and recharging the aquifer [12].

3.2 Mobilization, Capacity Building and Execution

3.2.1 Mobilization and Organization Hierarchy

It is imperative for any project of this size which needs a thoroughly thoughtout plan that addresses people mobilization, capacity building by consistent and periodic training sessions subjectively for every different role played and including different stakeholders. It was important that Art of Living as Nodal Agency created a platform for all stakeholders to perform their roles and ensured participation and ownership

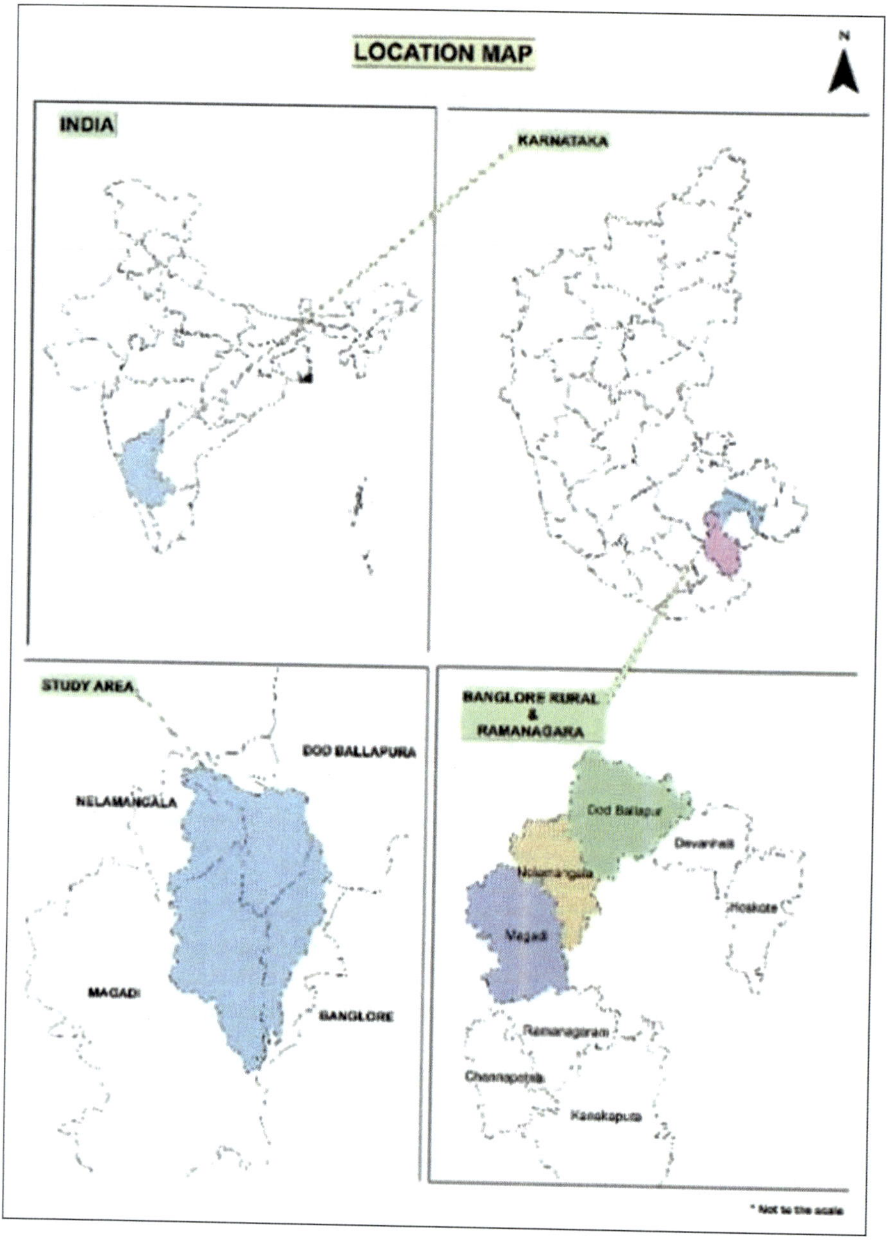

Fig. 1 Location of study area

along with facilitating labor mobilization, training and technically guiding the action plan implementation.

To ensure all processes are carried out effectively, a well-defined project management team was created. This involved many visits to each village/gram panchayats identifying qualified field engineers and needed contractors for implementation. A three-layer team was created, the first layer related field engineers interfacing with contractors on a day-to-day basis to assign and review action plan implementation by the contractors. This team will be sent to the target area to train the villagers and material suppliers in each village.

Roles for field engineers including contractors have been defined with clear job responsibilities, reporting hierarchies and information sharing. Each person in the team shall be trained on PRA techniques, river rejuvenation structure construction techniques and implementation.

The second layer is the technical team comprising of senior field experienced engineers who will select and award work to contractors. The field engineers will be trained, and many relevant educatory programs to all stakeholders are organized. This second layer continuously monitors work from the in every micro-watershed.

The third layer is the overall project management team, which will focus on ensuring the overall strategy, execution, monitoring, and documentation. Apart from these inputs from various experts, resource persons will be solicited. The whole technical and project management team is being trained through workshops and exposure visits [13–15].

3.2.2 Execution

The team outlined above will undertake the responsibility of executing the project.

Steps	Activity
Step 1	Micro-watershed-wise list of structures relayed to the second layer team
Step 2	Second layer team identifies and marks these points on ground and hands over it to field engineers to execute with contractors
Step 3	Second layer team will also work with gram panchayat elected members for intimation and needed approvals
Step 4	Field engineers will ensure work progress and reporting to the third layer and first layer

3.2.3 GPS Devices/Mobile Apps

Mobile monitoring devices with GPS and cameras with required mobile apps are used for continuous and effective monitoring of the work by field engineers. These data collected is analyzed at the central office and discussed with the second layer for improvements and course corrections. The third layer integrates various key performing areas and reports it to various stakeholders. Mobile apps are also used for baseline data collection (private bore wells data), farmers survey with the landholding and crop details, market place survey, PRAs, the water level at our observation borewells, etc.

3.2.4 Monitoring and Review

Roles and Responsibilities of Field Engineers: coordinating with Local Government Body (Panchayat Development Officer and Gram Panchayat President and Members) and ensuring contractors execute work on time and quality.

Roles and Responsibilities of the Second Layer: identification of the locations of the structures on the ground using the GPS data provided, regular visits to micro-watershed and ensuring and resolving field engineers' issues on the ground. Quality assurance and monitoring with appropriate authorization.

Roles and Responsibilities of the Third Layer: Planning, project survey, satellite mapping and structure identification; identifying physical locations of the structures planned using satellite imagery; regular review of the overall project and planning course corrections as need be. Reporting to various stakeholders including overall finance management and related reporting [16].

3.2.5 Convergence

Simultaneously, afforestation in forest area, other government lands, and in farmland, where our river rejuvenation activities are already in progress, is being planned to be taken up intensively to supplement the water harvesting and recharge structures. Through convergence with Integrated Watershed Management Program, the entire identified watersheds in these districts are being taken up for saturation; such works related to water conservation and watershed development are being identified through an intensive and participatory planning process, at the end of which Gram Sabha approvals are being obtained.

The creation of durable and productive assets through convergence with departments of agriculture, horticulture, sericulture, forest, animal husbandry, fisheries, minor irrigation, and Panchayat Raj Engineering Departments is being planned, executed, and monitored, to optimize the financial resources, physical infrastructures, and human resources of various government agencies to maximize the gain to the vulnerable sections of the society and contribute to the development of the state and its people.

3.2.6 Outcomes

The increase in the water yield in the borewells and thereby improvement in the agricultural crop production which thereby improves their economic status are the primary outcomes of this project. The third-party impact assessment carried out in the Kumudvathi River basin; based on the study of the recharge structures and the approach plan adopted by the Art of Living (AOL) Foundation and the data gathered, the following conclusions are drawn:

1. Rainwater recharge is one of the best strategies that AOL has adopted for quantitative and qualitative water management and adaptation to climate change in the Kumudvathi basin. The overall methodology adopted is found to be suitable to rejuvenate the river, particularly in Kumudvathi basin and in the hard rock terrain elsewhere also.
2. The background studies leading to site selection and the initial implementation on a pilot scale are yielding encouraging results. Availability of source for recharge, i.e., runoff, has been established by hydrological studies. Actions by AOL in adopting an integrated approach to conserve water within the hydrological boundary conceiving the concept of river rejuvenation need appreciation.
3. An increase in the water table has been observed in all the injection wells. The magnitude of the increase is from 20 to 137 ft. There are 23 injections wells that have recorded an increase in water level more than 50 ft. The increased water level in these non-pumping wells is a clear indication of augmented recharge. Overall improvement in the groundwater regime is observed (increased water levels within one year).
4. Augmented recharge is estimated approximately at 2% that needs to be increased by way of increasing the injection wells, i.e., intensification of recharge activities.
5. The time taken at the suggested rate of augmented recharge through intensification varies from 7 to 26 years for reaching an equilibrium status with groundwater draft and dynamic resources (natural and augmented).
6. Water pools have already exhibited useful results. These have been able to store water generated from the rainfall. Instances of livestock feeding on these pools reflect the actions by AoL in organizing water to cattle as well. If water pools can store 50% of the built-in capacity, a minimum of 900 cm of water gets recharged/percolated.

3.2.7 Community Participation

Water resource management activities can be classified under two major heads, namely supply-side activities and demand-side activities. AoL has specific and detailed programs designed to encourage and empower the commune to participate in the supply-side and demand-side activities based on the stakeholder's identification, understanding of their needs from and values for ecosystems.

Supply-side activities are:

- Increasing the groundwater recharge by constructing recharge structures along the length of streams,
- Maintaining the water bodies,
- Tree plantations, creating eco-restoration cells,

- Working on the area treatments like CCTs, contour bunds, farmer ponds, etc.

While the recharge structures and tree plantations are done based on the supply-side activities by the team specified in prior sections, it is imperative the Art of living (AoL) team which has to work on the demand side of the water to ensure that usage is understood, monitored, and contained based on the needs of the communes by communes itself. To bring the communes to this level, awareness is a key focus of AoL's demand-side activities.

Demand-side activities are:

- To undertake Water Literacy Programs.
- Plan and adopt water-use efficiency measures such as reducing water-intensive crops, drip/sprinkler irrigation methods, then flood-based irrigation, slowly and gradually adopting natural farming.
- Agroforestry and creating nurseries.
- Market creation to their current produce ensuring sustainable income.
- Creation of MSME for value additions on their produce.
- Ensuring linkages to all available and relevant incentives/support from GOVT. line departments.

On the above demand-side activities, approaches have been made on skill development (empowering farmers and their families with various types of skill development programs), leadership and entrepreneurship (leadership and entrepreneurship training, creation of local employment opportunities), value addition on animal husbandry, honey beekeeping, and other farm activities. Franchise retail outlets helping entrepreneurs to establish retail outlets in franchise business model for growth in every village. Also providing all necessary infrastructure support, technical skills, IT support, and marketing support to run the outlets. Providing transparency and fair opportunity to sell the farm products with end-to-end handling and marketing support. Technology usage for connecting various buyers and sellers (farmers) digitally has been planned and being implemented.

3.2.8 Rejuvenation Water Bodies

The rejuvenation of the water resources through cleaning the existing water bodies. For the same, volunteers have been called through the "Volunteers for Better India" program which resulted in assembling volunteers to clean the existing water bodies (stepwell, tanks). This attracted corporate, local communities to participate and take responsibility for the watershed.

3.2.9 Tree Plantation

All the trees planted are of native species that attract bees and birds for flowers and fruits. Creating eco-restoration cells around the existing water bodies accelerates the natural process of infiltration of rainwater to enrich groundwater, sustain surface water, and decrease temperature.

4 Need for Holistic Policy for River and Ecosystem Rejuvenation

4.1 Preamble for Policy

Water security is the key to India's continued economic growth and poverty reduction. India has 18% of the world's population, but only 4% of global renewable water resources within its territory. Parts of India are already considered water-scarce. Increased competition over fragile and finite resources poses risks to economic development, food and energy security, and livelihoods. Extremes of floods and droughts routinely strike the country, with disastrous consequences. Pressures on the water resource base and the frequency and intensity of extreme events are expected to be exacerbated by climate change. India has extensive groundwater resources, estimated at 30–40% of its annual utilizable renewable water resources. The physical characteristics of the groundwater resources vary considerably within India. Shallow, low-storage hard-rock aquifers in the basaltic and granitic systems of peninsular India have comparatively limited groundwater availability Approximately 2532 billion m^3 of groundwater is abstracted in India each year. This represents 25% of global groundwater withdrawals and makes India the world's largest user of groundwater. India's groundwater resources are under threat from overexploitation. Intensive and unregulated groundwater pumping in many areas has caused rapid and widespread groundwater decline, where an estimated 15% of India's food is being produced by groundwater mining.

Hence, it is quintessential that a large-scale intervention seeks to protect and revitalize our riverine ecosystem. Given the current state of affairs, what is imperative is a holistic policy framework, focusing on river rejuvenation that ensures the conservation of rivers by augmenting the recharge by artificial structures and effective tree plantations all along the catchment area and settlements. The focus has to be on the demand-side measures that emphasize water literacy, reduce consumptive groundwater use, and incentivize the participant to ensure de-risking. This needs to happen both at the national and state government levels.

Based on the current affairs as said above and AOLs experience implementing 12,000 structures and working various stakeholders and farmers at large as key and important benefactors across 5000+ km^2 area in four states reviving 41 rivers across nine main river basins, the following detailed recommendations can be made to address the current challenges

4.2 *Detailed Policy Recommendations*

4.2.1 River Rejuvenation

Policy framework that identifies key eco-sensitive zones in the very watershed and ensures a long-term and short-term revival is ensured in a planned approach. It should also regulate/ensure zero river pollution (industrial, domestic, and agricultural), stop sand mining, and regulate any detrimental land-use changes in context to the riverine and its ecosystems

1. Demarcation of key catchments area in every watershed and to be declared as key eco-sensitive zones. An afforestation drive in these areas by selecting native tree species those mostly are flower and fruit-bearing encouraging the birds/bees to help create these sensitive zones to key eco-restoration cells. This would be the key long-term measure to help revive the ecosystems related to the river and its entire basin. To ensure quality plantations are available key focus on cresting nurseries in each gram panchayat and/or block should be planned.
2. Declaration of these eco-sensitive zones to be chemical-free zones. STPs and micro-STPs must be commissioned with the involvement of local communities ensuring that the riverine system is safe from chemicals and gray water/sewage.
3. Afforestation through the key measure would take at least a decade and a half to two decades to reach the needed objective, but the crises being massive an immediate intervention should be made to ensure an increase in the rainwater harvesting. Governments should align various line departments such as Rural Development and Panchayat Raj, agriculture, minor irrigation, environment and forest, NABARD, Watershed Department, dairy, horticulture, etc., to increase fund source for creating recharge structures from catchments areas till valleys. Recharge structures must be planned in farmer lands as well [17].

4.2.2 Community and Stakeholder Participation

1. Engaging many civic societies, NGOs, and corporates for effective implementation of water literacy programs to all relevant stakeholders involving village, farmer, women, youth communes and industrial associations.
2. Identification of a key watershed manage(s) every village and or block/gram panchayat who is well trained on available water resources for her/his area and his management and should be responsible as keeper of important and key water bodies related to supply and demand side.

4.2.3 Farmers

1. Incentivize for effective usage of water
    - Open to smart water usage measurement and control tools.

- Drip irrigation (note: we have enough schemes to be made effective with awareness).
- Natural farming with NO fertilizers and NO pesticides (training and hand-holding/incentives to adapt).
- Support for improved livestock preferably Bos indicus if cows.
- Incentivize agroforestry and horticulture.

2. Market linkage and sustainability.

   - Identification of marketplace and introduction to the digital platform.
   - Effectively connecting through digitally connected logistics and storage godowns.
   - Associating farmer groups with NGOs/companies, for training and hand-holding so that farmer can be an entrepreneur that can brand and institutionalize his produce/product.

4.2.4 Water-Related Data and Its Dissemination

1. Evidence-based planning for groundwater management will have to be underpinned by investment in data collection, sharing, and use. Aquifer mapping is ongoing but has limited coverage, and water quality monitoring is nascent, and abstraction rates are inadequately monitored. Collaboration between central and state agencies on data sharing, resource assessments, and research needs to be strengthened.
2. The data must be made free of cost to all institutions working on these restoration and rejuvenation areas and is not used by these institutions for any commercial purposes.

5 Conclusion and Suggestion

This paper discusses Art of Living's water resource restoration (groundwater) by harvesting rainwater, its methods and outcomes. This has increased the rate of recharge and considerably changing groundwater table. The methodology being very holistic and effectively copying nature has yielded good results and can be made repeatable in similar terrain. Other terrains can also replicate process of understanding different geological and geomorphological situations and changing action plan suited to those areas. As it seems, it is quite important for effective handling on demand side similar to supply side that brings in effective usage or builds water usage efficiency. Based on experience on the technical and social side, many policy recommendations are made which will holistically bring about a drastic change in water supply and demand side [10].

Ethics Declaration
Statement of Human Right

This study has been approved by the Art of Living ethics committee and has been performed in accordance with the ethical standards as laid down in the 1964 declaration of Helsinki and its later amendments or comparable ethical standards. Also, the study deals with the data from the past event which was collected during the course of field survey; hence, for this study, formal consent of farmers and other human beings are not required. Also, all the data used in study has been authenticated by Art of Living.

References

1. Geo.hunter.cuny.edu Home page: http://www.geo.hunter.cuny.edu/~fbuon/GEOL_231/Lectures/Intro%20Basic%20Concepts.pdf. Accessed 1 May 2019
2. Sharachchandra ML (1991) Sustainable development: a critical review. World Dev 19(6):607–621
3. Revi A (2008) Climate change risk: an adaptation and mitigation agenda for Indian cities. Environ Urban 20(1):207–229
4. Mitropapas A, Koumantakis I, Vasileiou E, Perdikaki M, Kallioras A (2016) Overexploitation of groundwater resources in the coastal aquifer system of argolis. Bull Geol Soc Greece 50(2):865–873
5. Rahman MAU, Shafiqur Rahram ASM (2013) Rainwater harvesting in Bangladesh (published on 7 Jun 2013)
6. Ourgenerationusa.com Home page: https://www.ourgenerationusa.com/environment.html. Accessed 1 May 2019
7. Dcmetrosftp.org Home page: http://dcmetrosftp.org/appropriate%20tech/AT%20articles/Proceedings2ndInternationalConferenceOnAppropriateTechnologyZimbabwe2006.pdf. Accessed 1 May 2019
8. Elearning.reb.rw Home page: https://elearning.reb.rw/course/view.php?id=201&section=8. Accessed 1 May 2019
9. De Vries JJ, Selaolo ET, Beekman HE (2000) Groundwater recharge in the Kalahari, with reference to paleo-hydrologic conditions. J Hydrol 238(1–2):110–123
10. Atal bhujalyojana (Abhy) (2018) Document of the world bank. Revitalization of rivers in India, draft policy recommendation. Isha Foundation
11. Kumudvathi River Rejuvenation Project Report—phase I, Prepared by Dr. Lingaraju Y, M.Sc., Ph.D. Trustee SSIAST, Formerly Professor & Research Lead GAT, Director KSRSAC & Geomatics Centre, WRDO, Govt. of Karnataka, Assisted by Sri Janardhan, M.Sc., MI GIS World, Sri Abhilash, BE (M.Tech.) MI GIS World
12. Sandoval-Solis S, McKinney DC, Loucks DP (2011) Sustainability index for water resources planning and management. J Water Resour Plan Manag 137(5):381–390
13. Gadgil M, Prasad SN, Ali R (1983) Forest management and forest policy in India: a critical review. Soc Action 33(2):127–155
14. McHugh AN (2011) An assessment of sustainable water management at university campuses. Duke University

15. Grafton RQ, Hussey KS (2011) Water resources planning and management. Cambridge University Press, Cambridge
16. Verma S, Malpe DB (2017) Management and monitoring of watershed through remote sensing and GIS—a case study from eastern part of Yavatmal District, Maharashtra. Int J Innov Res Sci Eng Technol
17. Vijay KS, Girija BK (2014) Perspective on a water resource policy for India. Energy Resour Inst

Optimal Design of Sewerage Networks Using Swamee Algorithm

R. K. Rai and D. K. Madavi

Abstract Sewerage system is the significant infrastructural component of society which collects wastewater from various areas/localities and conveys the same toward the sewage treatment plant. Sewer line is the basic unit occurring repeatedly in design process of sewerage network. Huge amount of investment is required for the construction and maintenance of the sewerage system which includes cost of sewer pipe, cost of excavation, cost of manhole, etc. The reduction in the cost of sewer lines results in substantial savings in the total capital cost. In this paper, Swamee algorithm was used to design the cost-effective sewer line by determining the cost function and optimal solutions. Manning equation was used to obtain the set of feasible diameter, head loss, etc.

Keywords Cost function · Manning equation · Optimal design · Sewer line

1 Introduction

A set of sewers collecting discharges at their nodal points and emptying the same into another set of sewer lines is called as sewerage network. Sewage collection, treatment and disposal systems are embraced to ensure that the sewage released from colony/society is appropriately collected, transported, and treated without causing any environmental problems. Sewers must be designed of appropriate size and planned skilfully so that it is prevented from overflow and succeeding damages to properties and health.

In any society, sewer networks are important part of infrastructures, and investment required for construction and maintenance of such large-scale sewer line is huge. Therefore, it is mandatory to provide low-cost sewers as any saving in the cost of

R. K. Rai
Department of Civil Engineering, Government College of Engineering, Amravati 444604, India

D. K. Madavi (✉)
Environmental Engineering, Government College of Engineering, Amravati 444604, India
e-mail: dkmadavim@gmail.com

L. M. Gupta et al. (eds.), *Advances in Civil Engineering and Infrastructural Development*, Lecture Notes in Civil Engineering 87,
https://doi.org/10.1007/978-981-15-6463-5_37

these sewerage systems may result in considerable reduction in the total construction cost. Thus, the aim of optimal design is to find a cost-effective solution that reduces capital investment as well as ensures good system performance under given criteria and constraints.

In this paper, with the help of Manning's equation as hydraulic model the optimal design of sewerage network has been carried out using Swamee algorithm. The design processes have been explained and compared considering a small 3-link sewerage network, and the results obtained have been presented.

2 Objectives

Keeping in view the literature, following objectives was planned.

1. To study various design constraints, design criteria, and hydraulic design equations for design of partially full sewerage networks;
2. To develop and optimize the cost function for major component of sewerage network;
3. To develop a simple procedure for the optimal design of sewerage networks using Swamee algorithm;
4. To obtain optimal design of sewerage network using Swamee algorithm.

3 Equation Used for Design of Sewerage Network

3.1 Hydraulic Equations

The Manning equation is the most popular and reliable hydraulic model in practice around the world due to its simplicity [1].

Manning's equation is:

$$V = \frac{1}{N} R^{\frac{2}{3}} S^{\frac{1}{2}} \tag{1}$$

where

V = Flow velocity, m/s;
N = Manning's coefficient of roughness;
R = Hydraulic radius, m; and
S = Sewer pipe slope.

The value of Manning's coefficient of roughness for concrete sewers is 0.013–0.015. The method comprises of designing the sewers as circular concrete sections with sewer running partially full [2]. The formulae used for the design of sewer flowing partially full are as follows:

The central angle in terms of the relative depth of flow k_d is given by:

$$\theta = 2\cos^{-1}(1 - 2k_d) \tag{2}$$

The area ratio k_a:

$$k_a = \left(\frac{\theta - \sin\theta}{2\pi}\right) \tag{3}$$

The hydraulic mean depth ratio k_r:

$$k_r = \left(\frac{\theta - \sin\theta}{\theta}\right) \tag{4}$$

The velocity ratio k_v:

$$k_v = \left(\frac{\theta - \sin\theta}{\theta}\right)^{\frac{2}{3}} \tag{5}$$

The discharge ratio k_q:

$$k_q = \left(\frac{\theta - \sin\theta}{2\pi}\right)\left(\frac{\theta - \sin\theta}{\theta}\right)^{\frac{2}{3}} \tag{6}$$

For a given sewage flow, the design of a sewer can be obtained if the self-cleaning velocity and design relative depth are known, and these can be obtained only when the diameter is known using Eqs. (1)–(6).

3.2 Cost Function

For the optimal design of sewer line, it is necessary to describe the cost function of the various components of a sewer line. The main components are sewer pipes, manholes, and excavation [3].

3.2.1 Cost of Sewer Pipes

The capitalized cost of the sewer C_{mi} can be demonstrated as:

$$C_{\mathrm{mi}} = k_m L_i D_i^m \tag{7}$$

where

k_m and m = Cost parameters of pipe;
L_i = Length of pipe, m; and
D_i = Diameter of sewer, m.

The cost of RCC pipe for different diameters was taken and cost parameters found were $k_m = 3697$ Rs./m, and $m = 1.2148$. The cost of CI pipe for different diameters was taken and the cost parameters found were $k_m = 28{,}577$ Rs./m, and $m = 0.9961$ [4].

3.2.2 Cost of Excavation

The total cost of excavation is the summation of sheeting cost and shoring cost of sewer trenches and earthwork. Assuming sides of excavation trench to be vertical, the capitalized cost of earthwork for sewer C_{er} can be written as:

$$C_{\mathrm{er}}=\frac{1}{2}L_i w_i (d_{i-1} - d_i)c_e + \frac{1}{6}L_i w_i \left(d_{i-1}^2 + d_{i-1}d_i + d_i^2\right)c_r \tag{8}$$

where

C_{er} = cost of capital earthwork;
c_e = ground level unit earthwork cost, Rs./m^3;
c_r = per unit depth increase in unit earthwork cost, Rs./m^4;
d_{i-1} and d_i = invert depth at upstream and downstream nodes, respectively, m; and
w_i = width of sewer trench, m.

The surface area of side walls of excavation trenches influences the cost of sheeting and shoring of sewer trenches. The capital cost C_{es} can be demonstrated as:

$$C_{\mathrm{es}} = \frac{1}{2}L_i (d_{i-1} + d_i)c_s + \frac{1}{3}L_i \left(d_{i-1}^2 + d_{i-1}d_i + d_i^2\right)c_{\mathrm{rs}} \tag{9}$$

where

c_s = unit capital cost of sheeting and shoring at ground level, Rs./m^2; and
c_{rs} = increase in unit sheeting and shoring cost per unit depth, Rs./m^3.

Combining Eqs. (8) and (9), and rearranging the terms, the total capital cost of excavation C_e is given by:

$$C_{ei} = k_{ei} L_i (d_{i-1} + d_i) \tag{10}$$

where

k_e = earthwork cost coefficient given by

$$k_e = \frac{1}{2}w_i c_e + \frac{1}{6}w_i\left(\frac{d_{i-1}^2 + d_{i-1}d_i + d_i^2}{d_{i-1} + d_i}\right)c_r + \frac{1}{3}c_e\left(\frac{d_{i-1}^2 + d_{i-1}d_i + d_i^2}{d_{i-1} + d_i}\right)c_{\mathrm{rs}} \tag{11}$$

Costs of earthwork, and sheeting and shoring were taken. The cost parameters for excavation found were $c_e = 159.57$ Rs./m^3; $c_r = 15.33$ Rs./m^4; $c_s = 162.18$ Rs./m^2; and $c_{rs} = 21.2$ Rs./m^3 [4].

3.2.3 Cost of Manhole

The cost of a manhole is based on the depth of manhole. The capitalized cost of the manhole C_h can be demonstrated as:

$$C_{\mathrm{hi}} = K_h d_{\mathrm{hi}} \tag{12}$$

where

k_h = Manhole cost coefficients;
d_{hi} = Depth of manhole, m.

The cost parameters found were $k_h =$ 10,306 Rs./m [4].

3.2.4 Total Cost

The total cost C_T of a sewer line is the addition of the cost of prime components of the line together and is given by [5]:

$$C_T = C_{\mathrm{mi}} + C_{\mathrm{ei}} + C_{\mathrm{hi}} \tag{13}$$

3.2.5 Objective Function

To arrive at the objective function to be minimized, it is necessary to describe the cost function of the various components of a sewerage network. The objective function of sewerage network consists of (1) the cost of sewer pipe; (2) the cost of excavation of sewer trenches; and (3) the cost of manhole [6].

The cost function F for complete gravity sewerage network having n links is obtained by combining Eqs. (7), (10) and (12) and can be expressed as:

$$F = \sum_{i=1}^{n} k_m L_i D_i^m + \sum_{i=0}^{n} k_h d_i + \sum_{i=1}^{n} k_{ei} L_i (d_{i-1} + d_i) \tag{14}$$

After rearranging the terms, Eq. (14) can be written as:

$$F = k_h d_0 + k_{e1} L_1 d_0 + \sum_{i=1}^{n} k_m L_i D_i^m + \sum_{i=1}^{n} k_h d_i + \sum_{i=1}^{n} k_{ei} L_i d_i + \sum_{i=1}^{n-1} k_{ei+1} L_{i+1} d_i \tag{15}$$

Assuming k_{ei} to be constant with respect to D_i, and differentiating Eq. (15) with respect to D_i, equating it to zero and simplifying, the diameter equations were obtained as:

$$D_i = \left\{ \left(\frac{4^{\frac{16}{3}} Q^2 N^2}{3m\pi^2 k_m} \right) \left[(n - i + 1)k_h + k_{ei} L_i + 2 \sum_{j=i+1}^{n} k_{ej} L_j \right] \right\}^{\frac{1}{m+\frac{16}{3}}}, \quad i < n \tag{16}$$

$$D_n = \left\{ \left(\frac{4^{\frac{16}{3}} Q^2 N^2}{3m\pi^2 k_m} \right) [k_h + k_{en} L_n] \right\}^{\frac{1}{m+\frac{16}{3}}}, \quad i = n \tag{17}$$

3.3 *Procedure for Design of Sewerage Network*

The procedure for the design of sewerage network is as follows:

1. Determination of set of feasible diameters for each link;
2. Determination of head loss and slope for each diameter;
3. Determination of cost of the sewer link considering the cost parameters such as cost of pipe, cost of excavation, and cost of manhole;
4. Design of all the lines coming at junction manhole and selecting the line with minimum cost;
5. Extending the design with minimum cost and minimum downstream depth up to next junction point;
6. Select one main line and join the branch lines with minimum cost option and checking their downstream depth at junction; and
7. Repeat all the steps till outlet is reached.

4 Design of 7-Link 2-Junction Sewerage Network

The Swamee algorithm and dynamic programming have been adopted for the optimal design of a sewerage network. The process of optimal design of a sewerage network is illustrated using a design example of 7-link 2-junction sewerage network (Fig. 1; Table 1).

4.1 Results and Discussion

The design of 7-link 2-junction sewerage network design example was carried by the design procedure. The set of feasible diameters and the corresponding head loss for each link of the sewerage network were determined using design procedure. Feasible set of diameters with their corresponding head loss calculated for each link is given in Table 2.

Consider the sewer pipe 1-1 as a starting point. Slope and head loss for each diameter of line 1-1 were determined; invert depths and total costs for each option were calculated. Similarly, the invert depths and total cost for each option of the line

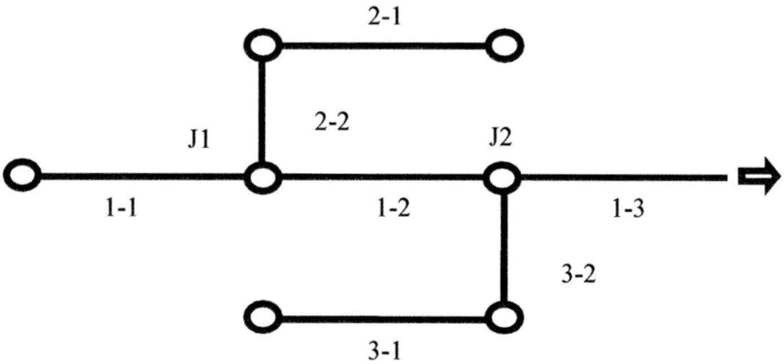

Fig. 1 7-link 2-junction sewerage network

Table 1 Design data for 7-link 2-junction sewerage network

Pipe	q, m^3/s	z_1, m	z_2, m	L, m
1-1	0.240	158.00	157.00	80
1-2	0.330	157.00	156.50	85
1-3	0.550	156.50	155.00	65
2-1	0.030	158.00	157.50	75
2-2	0.060	157.50	157.00	50
3-1	0.100	158.50	157.00	75
3-2	0.150	157.00	156.50	60

Table 2 Set of feasible diameters and head loss

Pipe	Set of feasible diameters (head loss), m
1-1	0.600 (0.175), 0.550 (0.280)
1-2	0.600 (0.355), 0.550 (0.560)
1-3	0.600 (0.745), 0.550 (1.185)
2-1	0.300 (0.210), 0.250 (0.765)
2-2	0.350 (0.190), 0.300 (0.430)
3-1	0.450 (0.205), 0.400 (0.385)
3-2	0.450 (0.370), 0.400 (0.690)

2-1 + 2-2 were determined. The line option having least cost was selected from both the line 1-1 and 2-1 + 2-2 and depending on the downstream invert depth of both the lines at junction J_1, line 1-1-2 having downstream invert depth equals to 1.710 m which is greater than the downstream invert depth of line 2-1-2 + 2-2-2. Therefore, line 1-1 was selected as main line and line 2-1 + 2-2 was selected as branch line. The total cost and downstream invert depth up to junction J_1 for each options of line 1-1 and 2-1 + 2-2 are given in Table 3.

The variation in the cost of different groups of line with the options up to junction J_1 is shown in Fig. 2. The variation in depth of two lines at junction J_1 is shown in Fig. 3.

The main line was extended up to next junction point and designed using the design procedure. Each option of the main line 1-1 was grouped with the options of link 1-2 thereby producing four options, and the invert depths and total cost for each option were determined. Similarly, the invert depths and total cost for each option of the line 3-1 + 3-2 were determined. The line option having least cost was selected from both the line 1-1 + 1-2 and 3-1 + 3-2 and depending on the downstream invert depth of both the lines at junction J_2, line 1-1-2 + 1-2-1 having downstream invert depth equals to 1.610 m which is greater than the downstream invert depth of line 3-1-2 + 3-2-2. Therefore, line 1-1 + 1-2 was selected as main line and line 3-1 + 3-2 was selected as branch line. The total cost and downstream invert depth for all the options of line 1-1 + 1-2 and 3-1 + 3-2 up to junction J_2 are given in Table 4.

Table 3 Cost options of line 1 and 2 at J_1

Line	Options	Total cost, Rs.	d at J_1, m
1-1	1-1-1	134,249	1.500
	1-1-2	114,125	1.710
2-1 + 2-2	2-1-1 + 2-2-1	246,035	1.350
	2-1-1 + 2-2-2	229,407	1.300
	2-1-2 + 2-2-1	227,118	1.350
	2-1-2 + 2-2-2	218,024	1.495

Fig. 2 Cost variation of line 1 and line 2 up to junction J_1

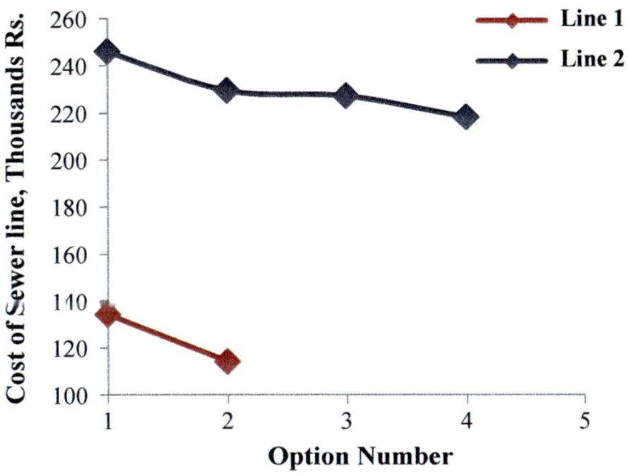

Fig. 3 Variation in depth of lines at junction J_1

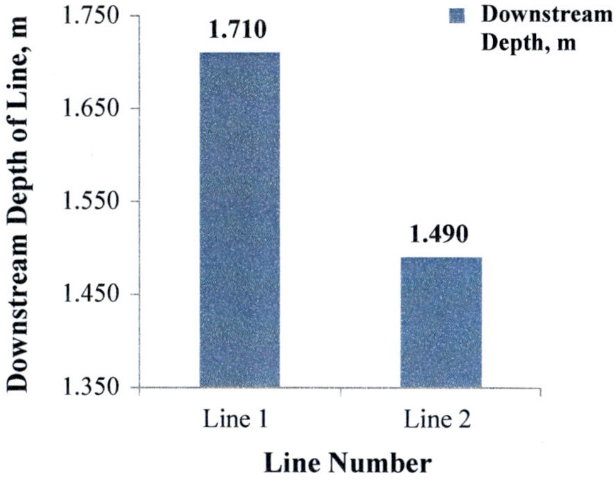

Table 4 Cost options of line 1 and 3 at J_2

Line	Options	Total cost, Rs.	d at J_2, m
1-1 + 1-2	1-1-1 + 1-2-1	516,445	1.710
	1-1-1 + 1-2-2	525,305	2.640
	1-1-2 + 1-2-1	488,026	1.610
	1-1-2 + 1-2-2	496,646	2.540
3-1 + 3-2	3-1-1 + 3-2-1	367,689	1.450
	3-1-1 + 3-2-2	358,411	1.690
	3-1-2 + 3-2-1	344,917	1.450
	3-1-2 + 3-2-2	331,170	1.590

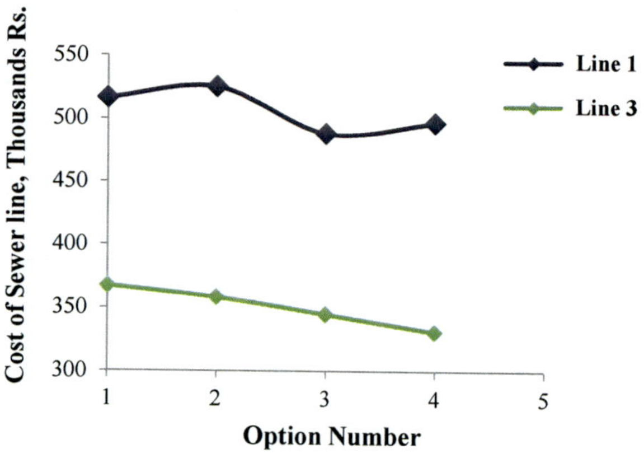

Fig. 4 Cost variation of line 1 and line 3 up to junction J_2

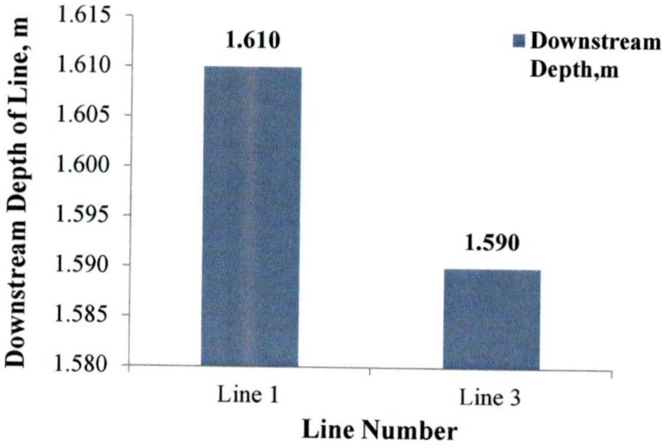

Fig. 5 Variation in depth of lines at junction J

The variation in the cost of different groups of line with the options up to junction J_2 is shown in Fig. 4. The variation in depth of two lines at junction J_2 is shown in Fig. 5.

The least cost solution of the extended line 1 was again checked for the downstream invert depth of branch line at junction J_1 and was found out to be ok. Thus, the design was continued considering line 1 as main line. The main line was extended up to outlet and designed using the design procedure. Each option of the line 1-1 + 1-2 + 1-3 was grouped thereby producing eight options and the invert depths and total cost for each option were determined. The line option 1-1-2 + 1-2-1 + 1-3-2 having least cost and greater downstream invert depth equals to 2.295 m at the outlet was selected. The total cost and downstream invert depth for all the options of line 1-1 + 1-2 + 1-3 up to outlet are given in Table 5.

The variation in the cost of main line with the options up to outlet is shown in Fig. 6. The variation in depth of two lines at outfall is shown in Fig. 7.

The least cost solution of the extended line 1 was again checked for the downstream invert depth of branch line at junction J_1 and J_2 which was found out to be ok. Thus,

Table 5 Cost options of line 1 at outlet

Line	Options	Total cost, Rs.	*d* at outlet, m
1-1 + 1-2 + 1-3	1-1-1 + 1-2-1 + 1-3-1	739,452	1.845
	1-1-1 + 1-2-1 + 1-3-2	737,446	2.335
	1-1-1 + 1-2-2 + 1-3-1	729,656	2.005
	1-1-1 + 1-2-2 + 1-3-2	722,848	2.395
	1-1-2 + 1-2-1 + 1-3-1	716,976	1.845
	1-1-2 + 1-2-1 + 1-3-2	714,971	2.335
	1-1-2 + 1-2-2 + 1-3-1	696,373	1.905
	1-1-2 + 1-2-2 + 1-3-2	689,501	2.295

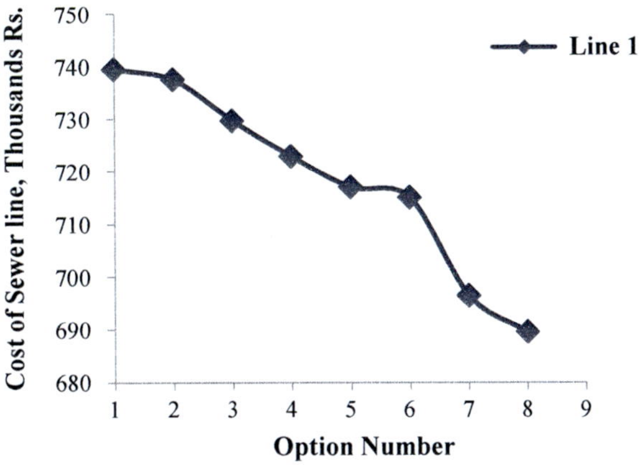

Fig. 6 Cost variation of line 1 up to outlet

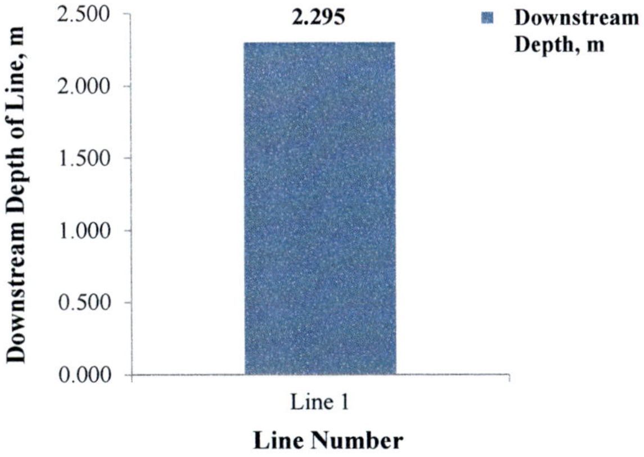

Fig. 7 Depth of line at outfall

Table 6 Design details for the optimal solution of 7-link sewerage network

Line	Pipe	Diameter, m	d_{i-1}, m	d_i, m	Total cost of line, Rs.
Main line	1-1-2	0.550	2.270	1.550	689,501
	1-2-2	0.550	1.550	1.610	
	1-3-2	0.550	1.610	2.295	
Branch line 1	2-1-2	0.250	1.250	1.515	218,024
	2-2-2	0.300	1.565	1.495	
Branch line 2	3-1-2	0.400	2.515	1.400	331,170
	3-2-2	0.400	1.400	1.590	
Total cost in Rs.					1,238,695

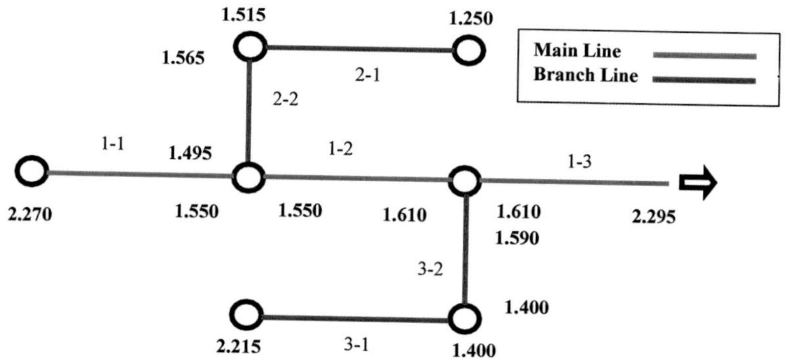

Fig. 8 Main and branch lines of 7-link 2-junction sewerage network

the design was ok considering line (1-1-2 + 1-2-1 + 1-3-2) as main line and line (2-1-2 + 2-2-2) and (3-1-2 + 3-2-2) as branch line. The ultimate optimal solution is as given in Table 6 having a total cost of Rs. 1,238,695.

At junction J_1, invert of incoming sewers is 1.550 and 1.490 m, and that of outgoing sewer is 1.550 m. At junction J_2, inverts of incoming sewers are 1.610 and 1.590 m, and that of outgoing sewer is 1.610 m. Since the terminal depth of branch line 1 at junction J_1 and of the branch line 2 at junction J_2 is less than the depth of main line at junction J_1 and J_2, the network design is feasible. Also, the invert depth at outlet is 2.295 m. The main line and branch lines of the optimal design of the 7-link sewerage network are represented in Fig. 8.

5 Conclusions

From the study, it can be concluded that:

1. Using Swamee algorithm, the diameters of each link were determined easily, since it gives only single value of diameter due to which set of feasible diameters is obtained in the increment and decrement of 50 mm;
2. Various constraints, design criteria, and hydraulic design equations for design of partially full sewerage network were studied;
3. Cost data from field for the major component was collected and its cost function was developed for sewerage network;
4. By solving the sewer lines coming at junction in parts using Swamee algorithm and developed procedure optimal solution of entire sewerage network was obtained easily;
5. The Swamee algorithm restricts the number of options at each stage which reduces numerous cost-effective grouping and thus, reduces error due to large data handling;
6. The upstream invert depth of the sewer pipe should be more than or equal to the downstream invert depth of the previous sewer pipe;
7. The design procedure was developed using illustrative design examples and successfully applied to the case study example for obtaining optimal solution; and
8. The optimal design of a sewerage network can be obtained easily using Swamee algorithm.

References

1. Manual (1993) Sewerage and sewage treatment, 2nd edn. CPHEEO, Ministry of Urban Development
2. Parate HA, Shetkar RV (2017) Design of sewer network using dynamic programming. Int J Innov Res Sci Eng Technol 6(5):8633–8641
3. Swamee PK (2001) Design of sewer line. J Environ Eng 127(9):776–781
4. Rai RK, Daine VV (2017) Optimization approach for sewerage network design using sewer line design algorithm. Int J Adv Res Sci Eng 6(3):191–199
5. MJP (2018) Schedule of rates for Maharashtra Jeevan Pradhikaran works for the year 2018–2019, Amravati Region, Amravati
6. Swamee PK, Bhargava R, Sharma AK (1987) Noncircular sewer design. J Environ Eng 113(4):824–833

Comparative Study of Designing of Common Effluent Treatment Plant by STAADX PRO Software and by Manual Calculations

Rishab Pardeshi, Abhishek Satputale, and Shraddha Admane

Abstract The increasing small and medium scale industries in India, environmental pollution and wastewater percentage increases. So waste minimization and cleaning technology are encouraged under Swacch Bharat Abhiyan. The government has many facilities to promote the CETP in small-scale industry. According to that Abhiyan, every industry should have their Effluent Treatment Plant, so that water will be treated before leaving it outside the industry. In this paper, the designing of a Common Effluent Treatment Plant (CETP) for 20 MLD is studied. The design parameters and the design verification of the plant are also done using certified design software of STAADX PRO. Therefore, the establishment of CETP plant is considered in saving rivers from adverse effects of industrial pollution. The treated water can also be used for boiler or gardening purposes. The calculations for all the equipment like sedimentation tank, precipitator, etc. have done. The simulation of CETP plant gives an overall idea about the minimization of pollution then according to treatment we can modify the design of a common effluent treatment plant.

Keywords Common effluent treatment plant (CETP) · Effluent-generating industries · Design and operational aspects · STAADX PRO

1 Introduction

India has so many water resources but that water is not used for drinking purposes. So India is now started to think that how can be water used for drinking or any purpose. Due to the increasing industrialization, government has made a compulsion to every industry that they should have their ETP plant so that they will not take out their wastewater directly into the river. The main aims of ETP plants are to decrease COD, BOD, DO, etc. all the parameters from the wastewater which are harmful to the environment. The wastewater plant should be more efficient and cost-effective

R. Pardeshi · A. Satputale · S. Admane (✉)
School of Chemical Engineering, MIT Academy of Engineering, Alandi (D), Pune, Maharashtra, India
e-mail: svadmane@chem.maepune.ac.in; shraddhaadmane@gmail.com

L. M. Gupta et al. (eds.), *Advances in Civil Engineering and Infrastructural Development*, Lecture Notes in Civil Engineering 87,
https://doi.org/10.1007/978-981-15-6463-5_38

[1]. The design will give the output of every process in a significant way. While designing the process first assume the overall ton over of company then by using software start designing the process [2]. So from the process, people will come to know about the plant layout of any industries. The common effluent treatment plant consists of various equipment like grit chamber, Primary settling tank, Secondary Settling tank, Trickling filter, etc. the actual implementation of common effluent plants in the industry will require the prior calculation of every equipment and process so the coast of pant reduces. The use of STAADX PRO software will provide the detailed designing of the water treatment plant [1]. The wastewater treatments are explained by preliminary treatment, primary treatment, secondary treatment, and tertiary treatment. By considering all these treatments the CETP plant is designed. By using this treatment the wastewater can be made environmentally friendly.

2 About Software

The STAADX PRO software is used for simulation of 20 MLD plant. For designing we have considered screen chamber, grit chamber, skimming tank, primary settling tank, aeration tank, trickling filter, secondary settling tank, digester units, and sludge drying beds, respectively [3–6].

It is a product of KOMVEE version 2016 of Sustaining Environment. It is commercially used software for quick estimation and verification of parameters of a common effluent treatment plant and also of each and every unit involved. Through this software, we can design the entire process to develop CETP and even verify manual calculations if necessary. This software also helps in evaluating and in turn verifying the credibility of calculations performed.

In this project, we have used STAADX PRO 2016 version software as a verification tool since it would not be feasible to implement a pilot plant and devise our calculations. This software provides the output based on basic and universally accepted theories and formulas using a pre-defined algorithm. This algorithm helps in providing output in a sequence similar to that of performing manual calculations, which makes it desirable for its use as a verification tool [7].

This can also be used for different water treatment methods and technologies like Sewage Water Treatment (STP-PD pro), Waste Water Treatment (WTP-PD pro), etc. with its different modules still in progress.

3 Calculations

See Table 1.

4 Discussion and Results

The table shows the comparative result of manual calculations as well as software results (Table 2).

After giving all the inputs to software the final design of the process is as given as.

The overall process diagram shows all the equipment required for designing common effluent treatment plants like screen chamber, grit chamber, Aeration tank, Primary settling tank, Trickling filter, Secondary settling tank, and digester tank. Figure 1 shows the overall dimensions of every equipment as well as total area required for the process. The design calculations are also shown. The comparative study of manual as well as software calculations is as follows.

5 Conclusion

The designing of common effluent treatment plants is highly affected by population density, population growth and type of industry. The study indicates that all major pollutants can be reduced in the wastewater after treatment on various levels. Design values obtained from manual calculations were compared with those obtained from software. Both the numerical values obtained were found to be nearly equal. The waste obtained in the form of sludge is also converted into useful Methane which can be used as fuel. Another by-product obtained after drying in sludge drying beds can be used as manure.

Table 1 All the conditions and general parameters for software

Screen chamber	Grit chamber
Plant capacity is 20 MLD Average flow = 20,000 m^3/day Maximum flow = 0.347 m^3/day Size of bars 9 mm × 50 mm, 9 mm facing the flow Inclination of bars with horizontal as 30° Assume velocity of flow normal to screen as 0.3 m/sec at average flow Assume velocity of flow normal to screen as 0.75 m/sec at maximum flow Net submerged area for average flow = 0.77 m^2 Therefore, gross submerged area of the screen = **0.38 m²** Therefore, velocity of flow in screen chamber = **0.6 m/sec** The gross width of the screen chamber will be = **1.2 m** Therefore, liquid depth at average flow = **0.316 m** Hence, total depth of the screen = **0.616 m** The size of the channel = 1.2 m (width) × 0.616 m (depth) Calculation for bed slope: Bed slope is nearly 1 in **4272 m**	$V_s = \frac{g}{18}\frac{[S-1]}{V} D^2 = \mathbf{3.15\ cm/s}$ Reynolds Number $R = V_s \cdot D/\nu = \mathbf{5.53} > 0.3$ Using $V_s = 3.15$ cm/sec, calculate R and Cd and then again V_s till it converges Subsequent trial $V_s = 2.4$ cm/sec, $R = 4.21$, $Cd = 7.5$ $V_s = 2.4$ cm/sec, Hence, O.K. Now for $\beta = 0.06, f = 0.03$, and $D = 0.02$ cm, $V_c =$ **22.76 cm/sec** Now $Q = 20$ MLD = 0.232 m^3/sec Therefore, C/S area $A = Q/V =$ **9.66 m²** Depth is taken as 3 m, So width becomes $W = 3.22$ m Since it's double chamber grit $W = 6.44$ m $L = 3 \times \frac{22.76}{2.4} = \mathbf{28.45\ m}$ **Aeration TANK** Flow to the aeration tank = 20,000 m^3/day Percentage recirculation = 0.25 Total flow to aerator = **25,000 m³/day** So volume of sludge returned to the aerator = 5000 m^3/day
Primary settling tank	

(continued)

Table 1 (continued)

Screen chamber	Grit chamber
Surface area of primary settling tank Design flow to the primary settling tank = 20,000 m^3/day Assume SOR = 20 m^3/m^2/day Area = $\frac{Q}{SOR}$ = 1000 m^2 Take number of tanks as 2 Area of each tank = 500 m^2 Flow through each tank = 10,000 m^3/day Take depth of tank as 3 m, Volume of tank = 3 × 500 = **1500 m^3** Check the overflow rate at average design flow The overflow rate = $\frac{Q}{Area}$ = 20 m/day This should not exceed 50. Hence satisfactory Average volume of primary settling tank = 1500 m^3 Detention time under average flow conditions = 0.15 day = **3.6 h** Calculation of tank diameter area of tank = 500 m^2 Diameter of tank = **25.22 m** Depth of primary settling tank Slope is 1:12 (V:H) Depth of settling tank at center = $\frac{1}{12}$ × Radius of tank = 1.05 m Provide a free board and safety of 0.75 m, Hence Total depth = **4.8 m** **Trickling filter** Influent BOD (S_o) = 150 mg/L Effluent BOD (S_e) = 20 mg/L Depth of filter (d) = 2 m No. of units = 6 Recycle ratio (R) = 1 Overall efficiency = $(1 - \frac{S_e}{S_o}) \times 100 = 86.66\%$ Applied BOD = $(S_o + RS_e)/(1 + R)$ = **85 mg/lit** Calculation of hydraulic loading rate (q)	Therefore, Area of aerator = 3125 m^2 Diameter of aerator = **63.09 m** **Secondary settling tank** Surface area of secondary settling tank Design flow to the secondary Settling tank = 20,000 m^3/day Take number of tanks as 2 Flow through each Tank = 10,000 m^3/day Take recycle ratio = 1 Total flow through each tank = **20,000 m^3/day** Assume SOR = 20 m^3/m^2/day Area = $\frac{Q}{SOR}$ = **1000 m^2** Area of each tank = 1000 m^2 Take depth of tank as 3.5 m, Volume of tank = 3.5 × 1000 = **3500 m^3** Check the overflow rate at average design flow The overflow rate = $\frac{Q}{Area}$ = **20** This should not exceed 50.Hence satisfactory. Detention time Average volume of secondary settling tank = 1500 m^3 Detention time under average flow conditions = **8.4 h** Calculation of tank diameter Area of tank = 1000 m^2 Diameter of tank = **35.68 m** Depth of secondary settling tank Slope is 1:12 (V:H) Depth of settling tank at center = $\frac{1}{12}$ × Radius of tank = 1.48 m Provide a freeboard and safety of 0.75 m, hence total depth = **5.24 m**

(continued)

Table 1 (continued)

Screen chamber	Grit chamber
	Sludge drying bed
$\frac{S_e}{S_o} = \exp(-2.94\ d\ q^{-0.5})$, $q = 16.52$ m^3/m^2/day Total flow to the trickling filter with recycle (Q) = 40,000 m^3/day Area of T.F (A) = **2421.3 m^2** Since there are 6 units area of each T.F = 403.55 m^2 Diameter of each T.F = **22.66 m** **Digester Unit** Volume of sludge from P.S.T = 19.456 m^3/day Volume of sludge from S.S.T = 14.755 m^3/day Volume of fresh sludge = 34.2113 m^3/day Volume of digester = **1539.5 m^3** Height of digester = 6 m So, area of digester = 1539.5/6 = **256.6 m^2** Diameter is therefore calculated from following equation- $A = \pi d^2/4$ Diameter of tank = **18.05 m**	Volume of fresh sludge to digester = 34.2113 m^3/day Volume of sludge out from the digester = **11.403 m^3/day** Depth of sludge application = 0.3 m Length of drying bed = 10 m Width of drying bed = **3.801 m** Area of drying bed = 38.01 m^2 Frequency of sludge removal = 10 days Number of sludge drying beds = 10

Table 2 Comparative study of software and manual calculations

Sr. No.	Parameters		Software calculations	Manual calculations
1.	Screen chamber	Gross submerged area of the screen	0.3859 m^2	0.38 m^2
		Gross width of screen chamber	1.2 m	1.2 m
		Slope of channel	4266.96 m	4272 m
2.	Grit chamber	Screening velocity	0.0315 m^2/sec	0.0315 m^2/sec
		Length of grit chamber	30 m	28.45 m
3.	Primary settling tank	Volume of tank	1500 m^3	1500 m^3
		Detention time	3.59 h	3.6 h
		Total suspended solid (depth)	4.8010 m	4.8 m
4.	Aeration tank	Capacity of tank	6250 m^3	6250 m^3
		Diameter of tank	63.09 m	63.09 m
5.	Trickling filter	Area of trickling filter	2412.0740 m^2	2412.3 m^2
6.	Secondary clarifier	Diameter of clarifier	35.6753 m	35.68 m
		Total depth at center of clarifier	5.4583 m	5.24 m
		Detention time	8.39 h	8.39 h
7.	Digester unit	Volume of digester	1520.5022 m^3	1539.5 m^3
		Diameter of digester tank	17.9591 m	18.05 m
		Quantity of total gas produced during digestion	278.85 m^2	256.6 m^2
8.	Sludge drying bed	Volume of sludge digester	11.4037 m^3	11.4037 m^3
		Number of drying bed	10	10

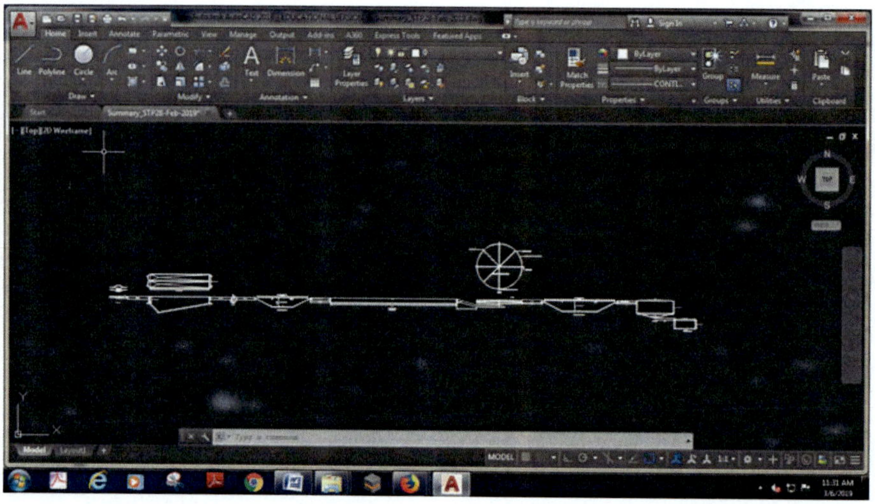

Fig. 1 Overall process flow diagram with designing calculation

References

1. Canter LW (1997) Environmental impact assessment, 2nd ed. McGraw Hill, University of Oklahoma
2. Central Pollution Control Board (October 2005) Performance status of common effluent treatment plants in India
3. Biswas AK, Kaul SN (2002) Sustainable environmental sanitation and water services. In: 28th WEDC conference prospects of common effluent treatment plant (CETP) in industrial sector. Kolkata, India
4. Prasad BGS, Gurunatha Rao BVS (1981) Studies on the treatment and disposal of tannery wastewater, Part 1. Leather Sci 28:221
5. Standard Methods for the Examination of Water and Wastewater (1995) American Public Health Association, Washington, DC
6. Charu K, Shukla NP (1990) Aerobic treatment of waste liquid effluents. Indian J Env Prot 10:276–280
7. Sangeeth Aiyappa S Common effluent treatments plants, technology and treatment process: the alternative strategies. Working Paper Series, No. 2

Experimental Modeling and Evacuation of Cr(VI) from Wastewater by Using Nanostructured Ceria

Harshada Deore, Mamta Sardare, and Parag Nemade

Abstract Ceria nanoparticle synthesizes by nanotechnology by using cerium chloride and sulfuric acid by the precipitations method which is evaluated as an adsorbent for the evacuation of Cr(VI) detachment from wastewater. Experimentation was done using the simple batch process where all the experiments were conducted to verify the efficiency of the nanoparticle for the removal of Cr(VI). Adsorption achieved balance inside at some point and was autonomous of starting chromium (VI) focus. The hybrid material was found to have maximum efficiency for the removal of hexavalent chromium with 94–95.6%, pH at 7, adsorbent dose of 0.1 mg and temperature of 28 ± 2 °C and initial concentration of 50 ppm. The mathematical modeling has done for the experimental data. The experimental data clearly fitted the Langmuir isotherm and Freundlich adsorption isotherm; the adsorption information was examined; and the adsorption instrument was explored utilizing X-beam diffraction by SEM.

Keywords Modeling · Evacuation · Ceria · Cr(VI) · Adsorption isotherm

1 Introduction

Water contamination is turning into the most significant issue that imperils and impacts on all living being and getting to be intense consistently. The impact of water contamination has consequences for the delicate equalization of nature. Overwhelming metals like mercury, chromium, nickel, lead, cadmium, copper are present in the debased water, and they present at low fixation in well equalization water. Anyway, these are available at high focus in terms of tainted water. Among these overwhelming metals, chromium hexavalent is viewed as a need. Cr(VI) turn out in the effluent of tanning, dye, fertilizer, electroplating, material assembling, jewels

H. Deore · M. Sardare (✉)
School of Chemical Engineering, MIT Academy of Engineering, Alandi, Pune 412105, India
e-mail: mdsardare@mitaoe.ac.in

P. Nemade
Department of Chemical Engineering, Institute of Chemical Technology, Nathalal Parekh Marg, Matunga, Mumbai 400019, India

L. M. Gupta et al. (eds.), *Advances in Civil Engineering and Infrastructural Development*, Lecture Notes in Civil Engineering 87,
https://doi.org/10.1007/978-981-15-6463-5_39

water, steel generation processing plants, and so on. The chromium hexavalent levels released might run from ten to several mg/l which are the most significant anthropogenic wellspring of chromium (VI) [1, 2]. Chromium is a heavy metal that is generally found in the environment. Natural chromium has three stable isotopes, namely ^{52}Cr, ^{53}Cr and ^{54}Cr. It happens for the most part in the chromium (III) from as chromite mineral, which contains oxygen, chromium (III) and different metals. Chromium occurs in environments as trivalent chromium structure. Hexavalent chromium is exceptionally lethal to vegetation. The system of danger is pH subordinate. It is just because of human exercises that significant measures of hexavalent chromium become present. Trivalent chromium mixes are not exceptionally harmful and are a fundamental supplement that encourages the body to utilize sugar, protein, fat and so on, and deficiencies may cause heart issues, interruptions of digestion systems and diabetes [3]. However, the take-up of an excessive amount of trivalent chromium can cause well-being impacts also, for example skin rashes. Hexavalent chromium mixes are partitioned up in water risk class 3 and are viewed as harmful. The unnecessary presentation to tidies or fogs of Cr(VI) mixes produces dermatitis, ulceration and puncturing of the nasal septum, just as liver and kidney harm. With long haul introduction to Cr(VI) mixes, human lung malignant growth increases.

Nanotechnology is getting created in a few dimensions: materials, gadgets and frameworks. The nanomaterials level is the most progressive at present, both in logical information and in commercial applications. Nanoparticles are not exactly a couple of 100 nm. This decrease in size achieves huge changes in their physical properties as for those saw in mass materials. They can be metallic, mineral, polymer-based or a blend of materials. A large portion of these progressions are identified with the presence of quantum impacts as the size reductions and are the birthplace of marvels, for example the super paramagnetism, Coulomb barricade, surface plasmon reverberation and so on. The expansion in the surface region to volume proportion is likewise a result of the decrease in size. It prompts the presence of surface impacts identified with the high number of surface iota's, just as to a high explicit territory, which is significant from the down to earth perspective.

2 Materials and Methods

Ceria synthesize using cerium chloride and sulfuric acid by the precipitations method [4]. Chemical and compounds utilized in this examination was of expository review and secured from E-Merck India Ltd and Aldrich are listed cerium chloride ($CeCl_2$), sulphuric acid (H_2SO_4), Potassium di-chromate ($K_2Cr_2O_7$), Nitric acid (HNO_3), Ammonium hydroxide (NH_4OH). 1:1 ratio was prepared for the synthesis of Ceria with a solution of various cerium chloride and sulfuric acid. 2 M cerium chloride solution is stirred with 300 rpm with 2 gm of ammonium bicarbonate in which dropwise 3 M sulfuric acid is added with pH adjusted by the base. The precipitate was kept for 24 h and washed with distilled water and dried at 50 °C for 3–4 h in the oven. The standard chromium (VI) solution of 10–100 ppm was treated with 0.02–0.1 gm ceria nanopowder. Adsorption experiments were done for different parameters contact time, the concentration of adsorbate and pH. The evolution of 220 UV-Visible spectrophotometers is used for the analysis of a sample at a wavelength of 540 nm. Evacuation of Cr(VI) from wastewater and uptake capacity was found as follows

$$(\%\text{Removal}) = \left(\frac{C_0 - C_t}{C_0}\right) \times 100 \tag{1}$$

$$\text{Uptake capacity} = (C_0 - C_t) \times \frac{V}{M} \tag{2}$$

where C_0 = initial concentration of Cr(VI), C_t = concentration of Cr(VI) at time t, V = volume of solution, M = mass of adsorbent.

2.1 *SEM Micrographs of Ceria Nanopowder*

The surface morphology of the material studied by SEM is shown in (Fig. 1). The SEM is made for understanding the surface morphology from figure that the material is having identifiable pores. It is done at 10.00 kV and at 50,000 magnification. This mainly shows the pores structures where adsorption of ceria nanopowder can take place.

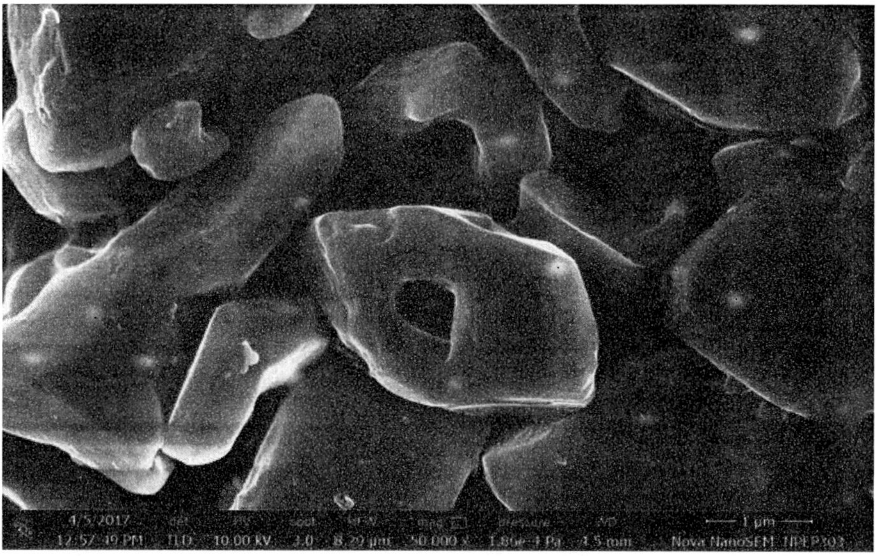

Fig. 1 SEM image of adsorbent before adsorption

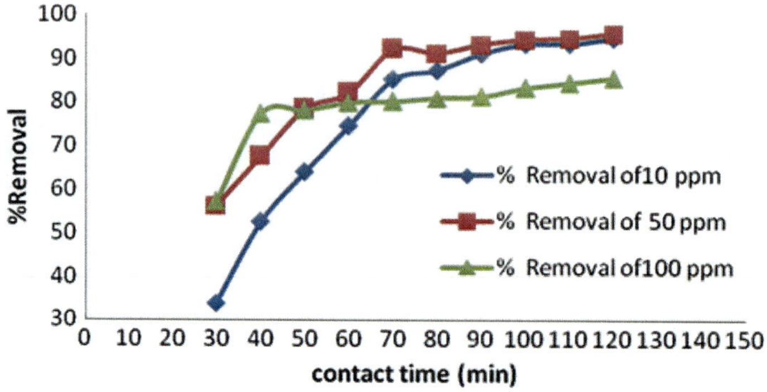

Fig. 2 Graph of percentage removal of Cr(VI) versus contact time

3 Result and Discussion

3.1 Effect of Contact Times

Batch investigation of Cr(VI) at various time intervals is considered for beginning chromium (VI) centralization of 10–100 mg/L at pH 7 and the adsorbent portion of 0.1 mg, keeping every single other parameter consistent. The outcome is spoken to in (Fig. 2). It is comprehended from (Fig. 3) that more than 95.5% evacuation happens in 120 min. The modification in the rate of expulsion may be because of the way that at first all adsorbent locales are idle, and in addition, the solute, center of attention inclination is soaring. Chromium (VI) uptake capacity rate by adsorbent is diminished provocatively, because of the reduction in the quantity of empty adsorption locales present in the nanomaterial which is decreases in the adsorption.

3.2 Effect of Adsorbent Weight

The impact of the adsorbent portion on the evacuation of chromium (VI) and uptake capacity is examined under the constant parameter (pH 7), at encompassing temperature (28 ± 2 °C) and contact time of 120 min for Cr (VI) centralization of 10–100 ppm. The outcomes are shown in Figs. 4 and 5. The expulsion of Cr(VI) is high at the 0.1 mg adsorbent dose of 95.5% at 50 ppm but the uptake capacity is decreases as adsorbent dose is increases. It is seen that after measurement of 0.07 mg/100 mL, there is stable change in the level of the evacuation of Cr(VI). It might be because of the covering of dynamic locales at higher measurements and the combination of the reciprocity. Along these lines, 0.1 mg/100 mL for 50 ppm is viewed as ideal portion and is utilized for more examination.

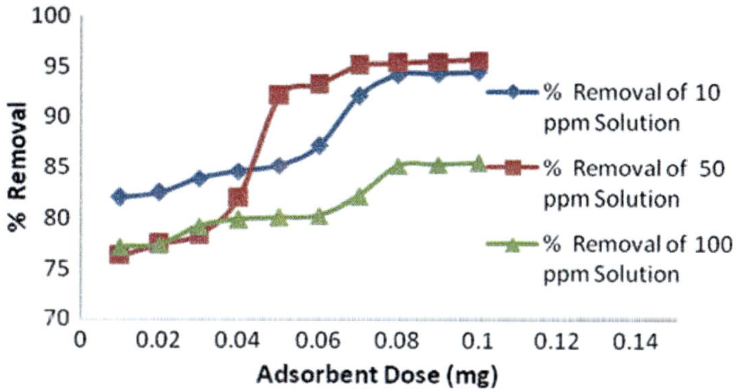

Fig. 3 Graph of percentage removal of Cr(VI) versus adsorbent dose of ceria nanopowder

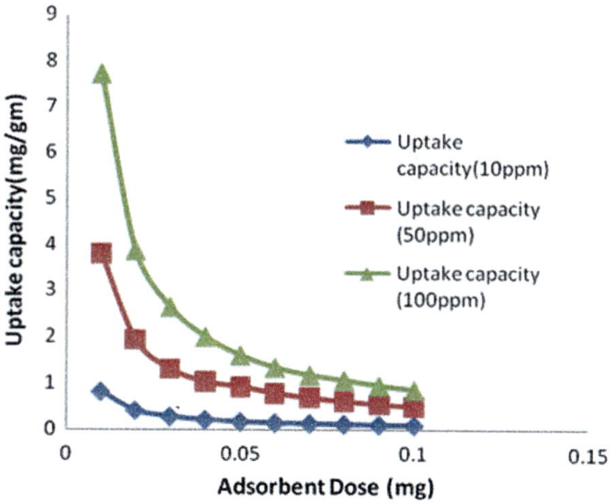

Fig. 4 Graph of uptake capacity versus adsorbent dose of ceria nanopowder

3.3 Adsorption Isotherms

3.3.1 Langmuir Adsorption Isotherm Model

Langmuir model points toward the character stability of adsorbate distribution on the adsorbent shell. During this research, we tried to investigate the obtained results using the Langmuir model. Langmuir model shows the good fit for the data found from the calculation. Langmuir isotherm model gives best fit than that of other. The following equation represents the Langmuir adsorption model.

$$\text{Non-linear equation}: q_e = (q_m \times k_a \times C_e)(1 + k_a \times C_e) \quad (3)$$

Table 1 Calculation for langmuir isotherm for a 0.1 mg adsorbent dose of ceria nanopowder

C_0	C_e	v/m	$q_e = v(C_0 - C_e)/m$	$1/C_e$	$1/q_e$
5	0.246	1000	4754	4.06	0.00021
10	0.406	1000	9594	2.46	0.000104
15	0.553	1000	14447	1.80	6.92E-05
20	0.609	1000	19391	1.64	5.16E-05
25	0.769	1000	24231	1.30	4.13E-05
30	1.335	1000	28665	0.74	3.49E-05
40	1.454	1000	38546	0.68	2.5E-05
50	1.562	1000	48438	0.64	2.06E-05

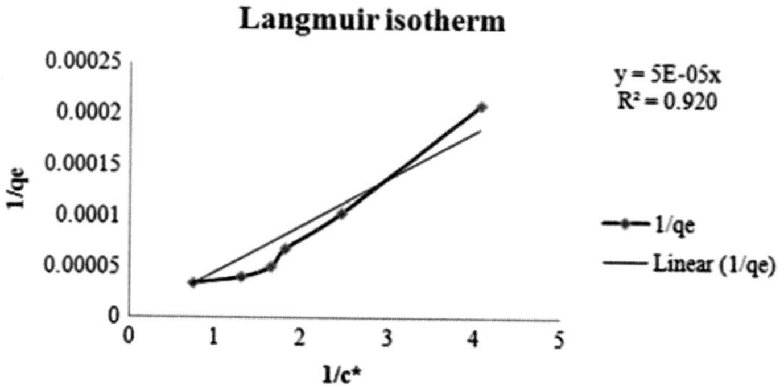

Fig. 5 Langmuir isotherm for 0.1 mg adsorbent dose

$$\text{Linear equation}: C_e q_e = q_m C_e + k_a q_e \tag{4}$$

where

- q_e Equilibrium adsorbed at the time (mg/g),
- C_e Equilibrium concentration (mg/L),
- q_m Maximum adsorption capacity in (mg/g),
- K_a Langmuir isotherm constant in (L/mg).

This is the Langmuir adsorption model for a 0.1 mg adsorbent dose of ceria nanopowder. At 5 to 50 ppm of chromium (VI) stock solution concentration at contact time of 120 min (Table 1).

3.3.2 Freundlich Adsorption Isotherm Model

Many faceted adsorptions are shown in the Freundlich model for heterogeneous adsorption surfaces. With this model, it is assumed that the adsorbate is incessantly

Table 2 Calculation for Freundlich isotherm for a 0.1 gm adsorbent dose of ceria nanopowder

C_0	C_e	v/m	$v(C_0 - C_e)/m$	$\log C_e$	$\log[v(C_0 - C_e)/m]$
5	0.246	1000	4754	−0.609	3.677
10	0.406	1000	9594	−0.391	3.982
15	0.553	1000	14447	−0.257	4.160
20	0.609	1000	19391	−0.215	4.288
25	0.769	1000	24231	−0.114	4.384
30	1.335	1000	28665	0.125	4.457
40	1.454	1000	38546	0.162	4.485
50	1.562	1000	48438	0.193	4.685

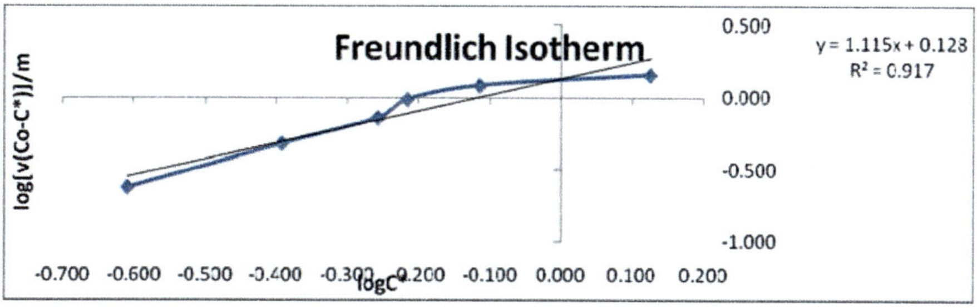

Fig. 6 Freundlich isotherm for 0.1 mg adsorbent dose

Table 3 Adsorption isotherm parameters for chromium adsorption on ceria nanopowder

Adsorbent	Langmuir parameters			Freundlich parameters		
0.1 mg ceria nanopowder	q_m (mg/gm)	K_a	R^2	K_f (mg/gm)	$1/n$	R^2
	1.832	1.631	0.920	1.1158	0.896	0.917

maintained obligatory to the adsorbent and obligatory energy required for adsorption varies exponentially distributed. This model can be represented by the below equations (Fig. 6, and Tables 2 and 3).

$$\text{Non-linear equation; } q_e = \left(\frac{X}{m}\right) = K_f \times C_e^{\left(\frac{1}{n}\right)} \tag{5}$$

$$\text{Linear equation; } \ln q_e = \ln K_f + \left(\frac{1}{n}\right) \ln C_e \tag{6}$$

where

q_e Equilibrium adsorbate at time (mg/g),
C_e Equilibrium concentration of adsorbate (mg/L),

k_f Capacity of the adsorbent (mg/g)
n Adsorption constant (L/mg).

4 Conclusion

Ceria is the impending adsorbent for the execution of Cr(VI) from aqueous systems for the purification of the wastewater. This nanopowder is allowed the traditional adsorption methods to eliminate the water contents. The ceria nanopowder is found to be partially amorphous in nature as marked in the XRD data. The adsorption model Langmuir isotherm gives the adsorbate molecular distribution on the adsorbent shell, and Freundlich isotherm model reflects the multilayer adsorption and applicable for heterogeneous adsorption surfaces. The adsorption data fitted best in the Langmuir isotherm than that of the Freundlich isotherm model. The execution efficiency of the Cr(VI) is optimized by varying the parameters. It is seen that ceria is appropriate material for evacuation of Cr(VI) from wastewater.

Acknowledgements Financial support from DST (SERB)-New Delhi under empowerment equality opportunities for excellence in science (Project Sanction file number SB/EMEQ-373/2014) is acknowledged.

References

1. Ramasubramaniam S, Govindarajan C, Gomathi T, Sudha PN Removal of chromium (VI) from aqueous solution using chitosan-starch blend. Scholar Res Library ISSN 0975–5071
2. Gallios GP, Vaclavikova M (2008) Removal of chromium (VI) from water streams: a thermodynamic study. Environ Chem Lett. https://doi.org/10.1007/s10311-007-0128-8
3. National Research Council (U.S.) (1974) Committee on biologic effects of atmospheric pollutants Chromium. Natl Acad Sci 155. ISBN 978-0-309-02217-0
4. Li R, Li Q, Gao S, Shang JK (2012) Exceptional arsenic adsorption performance of hydrous cerium oxide nanoparticles: part A. Adsorption capacity and mechanism. Chem Eng J 185–186:127–135

Impact of Micronutrients on Bioenergy Production with Addition of Animal Dung—A Pilot-Scale Study

Harshal M. Warade, Ramesh A. Daryapurkar, and Prashant B. Nagarnaik

Abstract Day by day energy issues are concentrating focus of the world's most of the nations due to limited production in crude oil [1, 2]. Bioenergy in the form of biogas is a suitable option for fulfilling this energy crises in rural areas because of the easy availability of raw substrates [3]. The focus of the current experimental study is to access in the generation of biogas and improve its production by adding some micronutrients along with its combination. Floating drum digester which has fixed in brick masonry chamber having a size of 2 m^3 in volume was used for the study. 20 kg cow dung along with 20 lit of water was mixed and filled into the reactor on daily basis, furthermore, addition of ferric chloride ($FeCl_3$), cobalt chloride ($CoCl_2$) and nickel chloride ($NiCl_2$) with doses of 10 mg/lit, 1 mg/lit, 1 mg/lit, respectively, added up to specific retention time. Daily gas generation in digester was measured with floating drum displacement. The result revealed that average gas production is 0.637 m^3/kg VS without micronutrient, 0.79 m^3/kg VS with adding ferric chloride, 0.781 m^3/kg VS with combination of cobalt chloride and nickel chloride and 0.874 m^3/kg (VS added) with a combination of ferric chloride, cobalt chloride, and nickel chloride.

Keywords Biogas production · Micronutrients · Anaerobic digestion · Methane · Anaerobic reactor

H. M. Warade (✉)
Assistant Professor, Department of Civil Engg., Yeshwantrao Chavan College of Engineering, Nagpur, India
e-mail: hrwarade@ycce.edu

R. A. Daryapurkar
CEO, Lars Enviro Pvt. Ltd, Nagpur, India

P. B. Nagarnaik
Professor, G. H. Raisoni College of Engineering, Nagpur, India

L. M. Gupta et al. (eds.), *Advances in Civil Engineering and Infrastructural Development*, Lecture Notes in Civil Engineering 87,
https://doi.org/10.1007/978-981-15-6463-5_40

1 Introduction

The global energy need is largely provided by fossil fuels (oil, natural gas, coal); and this storage is not at all infinite [4]. Due to overpopulation, the demand for energy also growing day by day ultimately heavy extraction of limited sources, thus the sustainable energy focus is diverted to renewable energy sources which are cost effective. Generation of biogas is achieved from the decomposition of highly organic substances in the absence of oxygen commonly known as anaerobic digestion. In anaerobic digestion process, active bacteria convert biological material into the stable matter without oxygen in a closed reactor [5, 6]. Normally biogas contains 55–65% pure methane, 25–35% carbon dioxide, and some traces of other gases like carbon monoxide, hydrogen sulfide, and oxygen [7]. The burning temperature of biogas is about 750 °C and it burns with blue color flame like LPG gas [8]. Cattle dung is the popular feeding material in most of the reactors constructed in the rural part of India due to its abundance availability. Cow dung is obtained about 12 kg per day from single cow whereas for buffalo produces about 22 kg per day [9]. The end product of anaerobic reactor is biogas which can be used as an alternative energy source also outlet manure which has greater nutrition value for cultivation of crop [10, 11]. Nowadays various advanced technologies are developed to utilize biogas for industrial purposes such as heating, electricity generation and as a vehicle fuel, etc. [12, 13]. The factors affecting biogas production are changes in atmospheric and reactor temperature, reactor design, organic loading rate and hydraulic retention time and bacterial activity inside the reactor [14, 15]. There are several environmental and socioeconomic benefits of bioenergy production which involves peoples and farmers who are directly or indirectly related to the production and utilization of feedstock for biogas generation [16, 17]. It also improves the economic capabilities, living standards, and related employment for poor families in rural areas. As a concern of urban areas biogas reactors can mitigate the various sanitation problems and may provide sustainable solutions on energy utilization as well as greenhouse gases [18, 19].

2 Experimental Setup

The model biogas plant on which the work was carried out is situated at Navegaon, Tq.-Umrer. Dist.-Nagpur (MS) India. This is a Floating drum digester type plant. The volume of digester is 2 m^3 and it is constructed in 2009. This biogas plant is allotted in farmer's scheme by Panchayat Samiti Umrer Dist-Nagpur. This Plant is running on 4 cattle manure approximately 15–20 kg/day and can fulfill the requirement of family of four members for cooking purposes. The height of RCC tank is 2 m and the height of floating drum is 1.2 m. During the project work LPG cylinder was arranged for the family so that biogas should not be utilized and for getting maximum production in 24 h. The outlet slurry of the digester is used as a fertilizer for the crop.

Fig. 1 Biogas plant setup

Figure 1 shows the floating dome biogas plant having inlet tank, liquid slurry digester, and Mild Steel floating dome. A pipe has attached to the gas collection chamber with outlet valve for escaping gas to the burner. Outlet slurry has managed in percolation pit where dewatering occurred through soil media.

3 Methodology

3.1 Material and Micronutrients Used

1. **Fresh Cow dung-**Fresh cow dung was collected and mixed with water thoroughly by manually and poured into the digester. Averagely 20 kg fresh cowdung was fed daily into the reactor. The ratio of cowdung to water was fixed at 1:1 so about 20 L of freshwater was used for blending with cowdung. A homogenous slurry was prepared by manual mixing and transfer daily into the reactor.
2. **Ferric Chloride-**Ferric or Iron chloride ($FeCl_3$) is a chemical compound has been used in various manufacturing process in the industrial sector. Color of ferric chloride depends on the angle of view, generally, it appears in green color in light reflection and purple or red [20]. It creates hydrolysis condition when blended with water and provides heat due to exothermic reaction. Ferric chloride combine a large portion of proteins when they are available in waste organic sludge resulting in stable organics [21, 22]. It also bind organic substances present in the reactor which breaks the attachment between proteins and iron substances and getting maximum efficiency in Volatile Solid reduction.

3. **Cobalt Chloride-**It is an inorganic substance ($CoCl_2$) obtained from cobalt and chlorine. The hexahydrate state of cobalt chloride generally used in the laboratory having dark purple color. Another state is anhydrous form having sky blue color [23]. Cobalt chloride is easily feasible in hydration and dehydration of substances also it is able to changes its color from blue to purple by variation in relative humidity in atmosphere [24].
4. **Nickel Chloride-**Nickel Chloride, i.e., $NiCl_2$ is available in yellow color with anhydrous and green color with hydrous nature, i.e., $Cl_2•6H_2O$ [25]. Nickel Chloride and its hydrate are generally used in synthesis of organic matter. They also accelerate the rate of reaction in the decomposition stage of organic substrates [26].

3.2 *Parameters to Be Checked*

During the experimental work, some important parameters were analyzed by the researcher, so that processes in the reactor can be monitor efficiently. The significant parameters like pH, Moisture Content, Total Solids, Volatile Solids, Methane Content, temperature, and hydraulic retention Time was determined on regular intervals.

3.3 *Doses of Micronutrients and Waste*

As per reference of various literature the micronutrients doses are decided on the basis of their quantity per lit. of water; therefore the amount of Ferric Chloride, Cobalt Chloride, and Nickel Chloride were decided to blend with slurry at the rate of 10 mg/L, 1 mg/L and 1 mg/L, respectively. The mentioned doses on micronutrients were added initially in 1 L of freshwater and mixed properly by manual stirring, subsequent to that this water uniformly mixed with cow dung slurry furthermore whole slurry was released into digester.

3.4 *Measurement of Gas*

As per addition of defined doses of micronutrients for a period of the experiment the average daily atmospheric temperature, Moisture Content, and Volatile Solids were calculated at an interval of 24 h up to 10 days. Biogas in the digester was measured in terms of height of floating drum so that gas is measured in m^3. After calculating Volatile Solids, gas quantity was expressed in m^3/kg VS.

3.5 Sample Collection

During the sampling process, produce biogas was collected in Urine Bladder having the capacity 1 L. To prevent air gaining access to the sample during sampling, about 10–15 times sweeping has been done by adding volumes of gas before taking the actual sample.

4 Results and Discussion

4.1 Gas Production

The total gas production was measured by calculating volume of floating dome and the average temperature on daily basis. The readings were taken in two seasons. A constant rate of water quantity added for slurry making i.e. 20 L. For all prescribed doses. The average temperature was recorded from daily publish newspapers.

The model reactor with a floating dome was used for the measurement of gas volume per day. It was operated on a 24 h interval basis so that the ultimate gas production should be obtained. Table 1 shows the daily gas production for all the prescribed doses with its combinations; also percentage increase with respect to without micronutrients.

As per graphical observation of Fig 2, it is noticed that the difference in gas production in the winter season by adding micronutrients and their combinations. In the winter three sets of reading were taken with micronutrients, in which the gas production is significantly increased by 19.37, 18.43, and 27.10% with the addition of $FeCl_3$, furthermore combination of $NiCl_3 + CoCl_3$ and $FeCl_3 + NiCl_3 + CoCl_3$, respectively.

As per graphical presentation of Fig. 3, it is justified that the difference in gas production was observed in the summer season by adding micronutrients and their combinations. In the summer season two sets of reading were taken with micronutrients, in which the gas production is considerably increased by 9.84% in combination of $FeCl_3 + NiCl_3 + CoCl_3$, also it is increased by 6.22% only feeding with cow dung. In this study period, due to blockages occurred in anaerobic reactor researcher unable to evaluate gas production in combination with $NiCl_3$ and $CoCl_3$.

The Fig. 4 clarifies the comparison of gas production in winter as well as in the summer season. In both the season three similar sets were taken with alone cow dung and micronutrients or its combinations. In this study period, it was observed that the gas production is slightly increased by 3.65 and 6.83% in the blending of $FeCl_3$ and cowdung and combination of $FeCl_3 + NiCl_3 + CoCl_3$, respectively.

Table 1 Increase of gas production for various doses

Season	Parameters	Without MN (control)	Addition of $FeCl_3$	Addition of $NiCl_3$ and $CoCl_3$	Addition of $FeCl_3$ $NiCl_3$ and $CoCl_3$
Winter	Gas production (m^3/day)	2.53	2.94	2.96	3.12
	Equivalent gas yield (m^3/kg VS) added at 25 °C	0.637	0.790	0.781	0.874
	% increase with respect to control		19.37	18.43	27.11
Summer	Gas production (m^3/day)	2.99	3.12	–	3.16
	Equivalent gas yield (m^3/kg VS) added at 25 °C	0.769	0.82	–	0.853
	% increase with respect to control	–	6.22	–	9.84

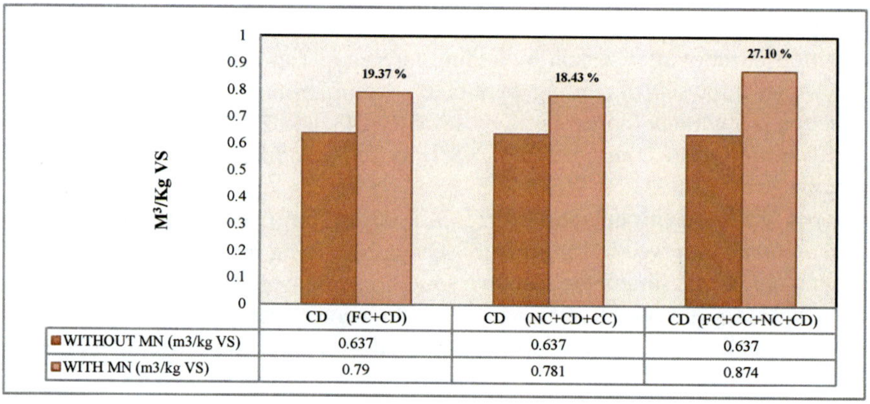

Fig. 2 Gas production in winter (with all doses combinations)

4.2 pH Variation

Accordingly, daily observations till continuous 75 days were taken for pH value of slurry in the reactor with the help of digital and portable pH meter.

During the 75 days of the observation, initially for a couple of days pH was 7–7.2. The observations were carried out continuously and after 27 days, pH was found to

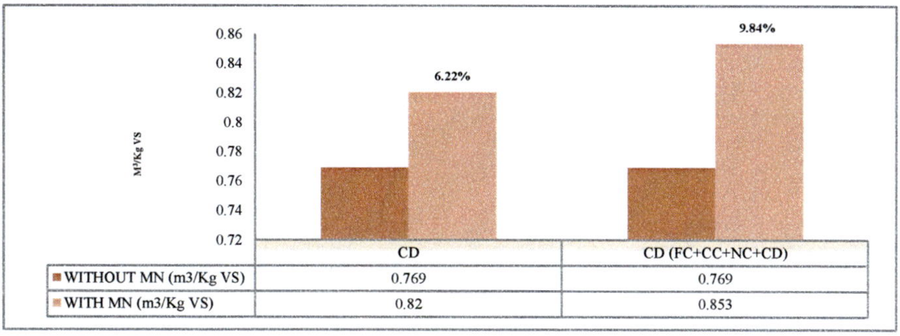

Fig. 3 Gas production in summer (with all doses combinations)

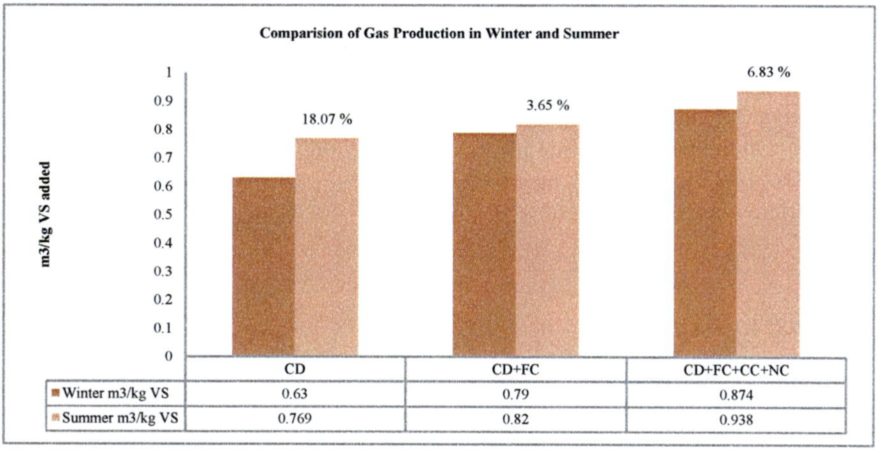

Fig. 4 Comparison of gas production in winter and summer

be 5.5. However, methane formation was taking place. In the later stage, pH started increasing and reached its normalcy 7.0; it further increased up to 7.9 in its alkaline condition. Thus, in this research it is observed that the pH value of the slurry in the reactor ranges from normal to acidic in 30 days approx.; thereafter from acidic to alkaline. In average, the pH value recorded was 7.1. From Fig. 5, it is clear that the growth of the anaerobic microorganism is normal and compatible.

4.3 *Organic Loading Rate*

At the initial stage when cow dung was used and the reactor was just initiated its processing, the gas production was low even if OLR was in 1.2–1.5 m^3/kg VS. Otherwise, gas production was justifiably in proportion with OLR.

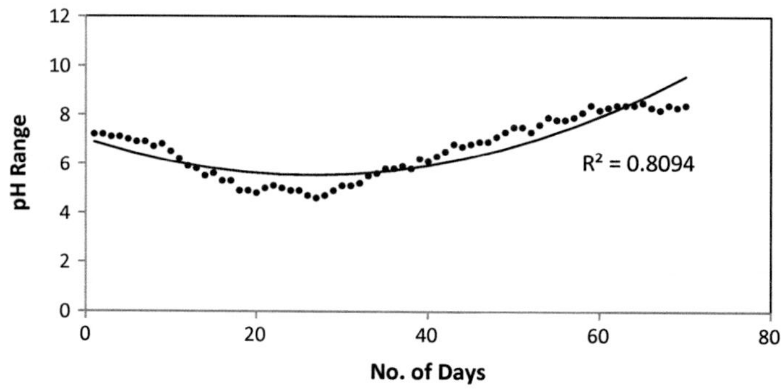

Fig. 5 pH variation

4.4 Methane (CH_4) and Carbon Di Oxide (CO_2)

In this study Methane (CH_4) and Carbon Di Oxide (CO_2) content was found by Orsat technique. On the basis of testing justified reading of CH_4 and CO_2 are considered. The content of CH_4 and CO_2 in produced biogas for all the determined doses of micronutrient was found to be 60.12% and 39.88%, respectively.

4.5 Cost to Benefit

In addition to that, the cost-benefit analysis has been done by considering the Calorific value of LPG and biogas. Researcher has calculated and compared the cost of biogas and LPG per m^3, resultantly proves that addition of $FeCl_3$ will give the highest value that is 68 times benefits while the combination of $NiCl_3 + CoCl_3$ will give 53 times benefits and combination of $FeCl_3 + NiCl_3 + CoCl_3$ will give 43 times benefits.

5 Conclusion

- Gas production is increased by 19.37% by the addition of Ferric Chloride dosed at the rate of 10 mg/L. It is increased 18.43% by a combination of Nickel and Cobalt Chloride dosed at 1 mg/L. When there was a combination of Ferric, Nickel, and Cobalt Chloride, the increase in biogas was 27.11%. Equivalent biogas production at 25 °C varied between 0.637 and 0.874 m^3/kg VS added for all the above cases.
- It was observed from gas production data that the cow dung feeding with ferric chloride showed better performance than the other feedings.
- During this project, another observation is made that gas production increased by 18.01% in Summer (38 °C) than that of winter (23 °C) season due to the effect of

temperature on biological process. It Is increased by 3.65% by addition of $FeCl_3$ with cow dung and 6.85% increased by the addition of $FeCl_3$, $NiCl_3$ and $CoCl_3$ with cow dung.

- From the cost-benefit analysis it has been observed that benefit to cost ratio of $FeCl_3$, combination of ($NiCl_3$ + $CoCl_3$) and Combination of ($FeCl_3$ + $NiCl_3$ + $CoCl_3$) are 68 times, 53 times and 43 times, respectively. The conclusion is therefore drawn that $FeCl_3$ is the best feeding micronutrient with cow dung in terms of cost-benefit ratio.

References

1. Hobson PN, et. al Methane production from agricultural and domestic waste, pp 181–217. Applied Science Publication Ltd., Microbiology Department, Rowett Research Institute UK
2. Warade H, Daryapurkar R, Nagarnaik PB (2019) Review of biogas production from various substrates and co-substrates through different anaerobic reactor. Int J Emerg Technol 10(2): 235–242. ISSN (Print): 0975-8364, (Online): 2249-3255
3. Ilaboya IR, et. al (2010) Studies on biogas generation from agricultural waste; analysis of the effects of alkaline on gas generation. World Appl Sci J 9(5):537–545, ISSN 1818-4952© IDOSI Publications
4. Warade H, Daryapurkar R, Nagarnaik PB (2019) Feasibility of biogas production from napier grass using animal manure as co-substrate on pilot scale reactor. Water Energy Int Central Board Irrig Power (CBIP) 62(3): 36–41. ISSN: 0974-4711
5. Momoh OL et al (2008) Effect of waste paper on biogas production from co-digestion of cow dung and water hyacinth in batch reactors. J App Sci Environ Manage 12:95–98
6. Ofoefule AU, et al (2010) Biogas production from paper waste and its blend with cow dung. J Adv Appl Sci Res 1(2):1–8
7. Rodionov DA et al (2006) Comparative and functional genomic analysis of prokaryotic nickel and cobalt uptake transporters: evidence for a novel group of ATP-binding cassette transporters. J Bacteriol 188:317–327
8. Chen JL et al (2016) Stimulation and inhibition of anaerobic digestion by nickel and cobalt: a rapid assessment using the resazurin reduction assay. Environ Sci Technol 50:11154–11163
9. Deshpande CV, et. al (August, 1997) Bioenergy production through biomethenation-a new approach. J Encol National Environ Eng Res Inst 12:23–29
10. Aquino SF et al (2007) Bioavailability and toxicity of metal nutrients during anaerobic digestion. J Environ Eng 133:28–35
11. Warade H, Daryapurkar R, Nagarnaik PB (2019) Assessment of biogas production from energy crop using animal manure as co substrate through portable reactor. In: International Conference on Smart Technologies for Energy, Environment and Sustainable Development, pp 165–174. Springer Nature. https://doi.org/10.1007/978-981-13-6148-7_18
12. Evranos B et al (2015) The impact of Ni, Co and Mo supplementation on methane yield from anaerobic mono-digestion of maize silage. Environ Technol 36:1556–1562
13. Zhang W et al (2005) Nanoscale iron particles for environmental remediation: An overview. J Nanopart Res 5:323–332
14. Shanableh A, et. al (June, 2005) Enhancement of biogas production from biomass using hydrothermal conditioning and partial oxidation. University Sharjah J Pure Appl Sci 2:59–73
15. Pecora V, et al (2006) Landfill biogas and its use for energy generation. In: International Scientific Conference of Mechanical Engineering, pp 7–9
16. Demirel B et al (2011) Trace element requirements of agricultural biogas digesters during biological conversion of renewable biomass to methane. Biomass Bioenerg 35:992–998

17. Murphy JD, et. al (2005) The optimal production of biogas for use as a transport fuel in Ireland. Renew Energy 2111–2127
18. Gustavsson J et al (2013) Bioavailability of cobalt and nickel during anaerobic digestion of sulfur-rich stillage for biogas formation. Appl Energy 112:473–477
19. Kida K et al (2001) Influence of Ni2+ and Co2+ on methanogenic activity and the amounts of co-enzymes involved in methanogenesis. J Biosci Bioeng 91:590–595
20. Fermoso FG et al (2008) Role of nickel in high rate methanol degradation in anaerobic granular sludge bioreactors. Biodegradation 19:725–737
21. Qiang H et al (2010) High-solid mesophilic methane fermentation of food waste with an emphasis on iron, cobalt, and nickel requirements. Biores Technol 103:21–27
22. Bianconi A et al (1980) Surface X-ray absorption spectroscopy: surface EXAFS and surface XANES. Appl Surface Sci 6:392–418
23. Zandvoort MH et al (2006) Induction of cobalt limitation in methanol feed UASB-reactors. J Chem Technol Biotechnol 81:1486–1495
24. Rodrigue A et al (2005) Identification of rcnA (yohM), a nickel and cobalt resistance gene in Escherichia coli. J Bacteriol 187:2912–2916
25. Van Hullebusch ED, et. al (2004) Sorption of cobalt and nickel on anaerobic granular sludges: isotherms and sequential extraction. Chemosphere 58:493–505. https://doi.org/10.1016/j.chemosphere
26. Barber WP et al (2000) Metal bioavailability and trivalent chromium removal in ABR. J Environ Eng 126:649–656

Study of Maximum Temperature Using Extreme Value Distributions

Shreenivas Londhe, Pradnya Dixit, and Shraddha Khandare

Abstract The study aims to analyse the maximum temperature in seven districts of Maharashtra using extreme value distributions. The total 100 years data (annual measured values) from 1901 to 2000 of maximum temperature is used for analysis in present study. The data is divided in two sets with first set from 1901 to 1950 and second set from 1951 to 2000. Maximum temperature for return periods 50 and 100 years is calculated using Gumbel and Log Person distributions. All seven districts have different meteorological aspects and have variety of temperature values. There is rise in temperature in all seven districts. Annual maximum temperature is highest rise in Gadchiroli, moderate in Nagpur, Chandrapur, beed and less rise in Pune, Satara, Nashik districts. There is considerable difference between annual maximum temperature values of each return period between both the sets for all the two distributions. This may also be contributed to effect of climate change over last 100 years.

Keywords Maximum temperature · Extreme value distribution · Gumbel · Log Pearson

1 Introduction

Extreme weather disturbance may cause substantial damage to our environment and ecosystem. This includes plants, animals, and all other living creatures on the

S. Londhe · P. Dixit
Department of Civil Engineering, Vishwakarma Institute of Information Technology, Pune 411048, India
e-mail: shreenivas.londhe@viit.ac.in

P. Dixit
e-mail: pradnya.dixit@viit.ac.in

S. Khandare (✉)
Water Resources and Environmental Engineering, Vishwakarma Institute of Information Technology, Pune 411048, India
e-mail: shraddha.218m0088@viit.ac.in

L. M. Gupta et al. (eds.), *Advances in Civil Engineering and Infrastructural Development*, Lecture Notes in Civil Engineering 87,
https://doi.org/10.1007/978-981-15-6463-5_41

Table 1 Shows details of studied districts and temperature

Sr. No.	Major region	District	Temperature min-max
1.	Marathwada	Beed	32.75–34.07
2.	Khandesh	Nashik	31.02–32.22
3.	Paschim Maharashtra	Pune	29.85–31.25
		Satara	30.11–31.57
4.	Vidarbha	Nagpur	33.01–34.47
		Chandrapur	32.33–34.61
		Gadchiroli	32.95–34.40

earth. Temperature is one of the very important parameters in weather forecasting. Greenhouse gases are trapping more heat in the earth's atmosphere, which is causing average temperature to rise all over the world. Gradually changing temperature affects climate change condition. Day-by-day increasing temperature is resulting to rise in sea water level, and this fully influences livelihood near it. Due to high temperature, evaporation rate increases and this creates scarcity of water in some places that is draught condition. When region experiences very high temperature for several days and nights, heat waves become common in those places. Ocean absorbs heat from temperature rise and ocean becomes warmer. Warmer oceans affect weather patterns, cause more powerful tropical storms, can impact many kinds of sea life as coral and fish. Hurricanes and other tropical storms get their energy from warm ocean water. These grow stronger with faster winds and heavier rain. This causes flooding, damage buildings, roads, other structures, harm crops and puts people's lives in danger. Study of variation in maximum temperature can give better ideas to plan the life pattern. It is necessary to be pre-analysed to reduce such hazardous effects.

2 Study Area and Materials

Analysis of annual maximum temperature data is performed in seven districts of Maharashtra state. Out of 36 districts, study is done in seven with variation in annual maximum temperature data. Study data of annual maximum temperature of 100 years is collected from Indian Water Portal. Table 1 shows details of study districts from major regions along with temperature variations. Figure 1 shows map of study areas, highlighted with yellow colour (http://www.indianwaterportal.org).

3 Methodology

As any hydrological parameter has developed its own trend across time span of many years, to know a particular for one parameter, minimum 30 years of data is

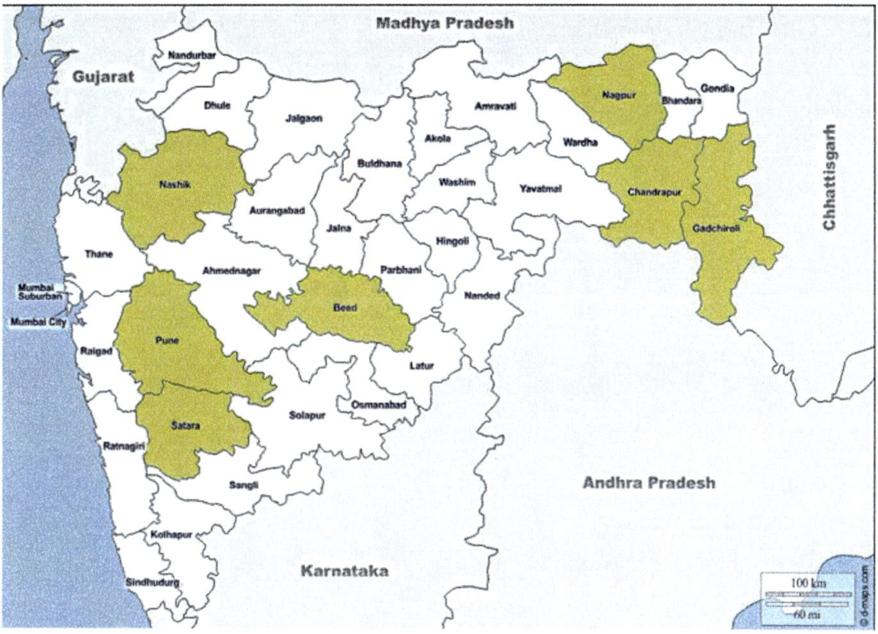

Fig. 1 Map of study area

to be observed [2]. Therefore, the available 100 years annual maximum temperature in degree Celsius data of each district is divided into two sets of 50 years each consequently, i.e. 1901–1950 and 1951–2000. In the present study, extreme value distributions are used to analyse the trend and pattern of annual maximum temperature in Maharashtra. Maximum temperature for return periods of 50 and 100 is calculated using extreme value distributions for each data set. Details about the extreme value distributions used are given below [2]. The fitness of a particular distribution to a particular data is judged using chi-squared test details of which are given in subsequent sections [1].

3.1 Extreme Value Distributions

Extreme value distributions arise as limiting distributions for maximums or minimums (extreme values) of a sample of independent, identically distributed random variables as the sample size increases. In hydrology, these distributions can be used to predict the extreme events such as maximum temperature, floods and droughts. In this analysis, Gumbel and Log Pearson distributions are used for analysis of maximum temperature.

3.2 Gumbel Distribution (1941) [2]

The following steps are used to analyse the annual maximum temperature probability distribution.

1. Means of annual maximum temperature.

   Standard deviation (Sn)

2. Reduced variate is calculated using the formula.

$$Yt = -\ln(\ln(T/(T-1))) \tag{1}$$

where

Yt Gumbel reduced variate, a function of "T"
T return period in years.

3. Frequency factor K for Gumbel distribution is calculated using the formula

$$K = \left(Yt - \overline{Yn}\right)/Sn \tag{2}$$

Both $\overline{Yn}$ (reduced mean) and Sn (reduced standard deviation) of the Gumbel variate and are functions of sample size N and their values are available in tabular form for several of N [2].

4. Magnitude of maximum temperature for the given return period is calculated by

$$Xt = \overline{X} + (K * Sn) \tag{3}$$

Xt magnitude of maximum temperature for given return period "T"
K Frequency factor
$\overline{X}$ Mean of annual
Sn standard deviation.

3.3 Log Pearson Distribution [2]

In this method, the sample (i.e. Z in this case) is first transformed into logarithmic form before analysing. For Log Pearson, Kz is a function of which is calculated by both the return period and the coefficient of skew (Cs). The values of Kz are given in the book [2] Following steps were followed for implementing Log Pearson distribution.

1. Log $X = \overline{Z}$ of all temperature data
2. Mean $(\overline{Z})$ of the log values

Standard deviation (Sn) of the log values coefficient of skew of the log values (Cs)

3. The value of frequency factor (Kz) was taken from the statistical table in Engineering Hydrology by K [2] corresponding to Cs to "T" (return period).
iv. Magnitude of temperature Xt for a given return period "T" is calculated by formula

$$Zt = Z + \left(\overline{Kz} * Sn\right) \tag{4}$$

where

Xt Antilog (Zt).

3.4 Test for Good Fit Chi-Square Test

The test intends to evaluate the difference between the sample data and the probability distribution. The formula for chi-square test is given by

$$\chi^2 = \sum_{i=1}^{N} \frac{(Oi - Ei)^2}{Ei} \tag{5}$$

where

χ^2 calculated chi-square
Ei frequency that is hoped regarding to the class division
Oi frequency on the same class
N number of class intervals.

The formula of Ei is as follow

$$Ei = \frac{n}{N} \tag{6}$$

where

n number of data
N number of class intervals.

The criterion for goodness is value of $\chi 2$ calculated should be less than value of χ^2 from the table [1]. The χ^2 value is read out from the table by knowing two parameters

1. Degree of freedom $= k - \text{h} - 1$

where

k number of class intervals
h number of estimated parameters.

2. Level of Significance.

4 Results and Discussions

Chi-square test

The test is carried at 5% significance level

1. Gumbel

   Degree of freedom $= k - \text{h} - 1 = 10 - 2 - 1 = 7$

k number of class intervals $= 10$
h number of derived parameters $= 2$ (Sn, X).

   Value of χ^2 from the table $= 14.067$

2. Log Pearson

   Degree of freedom $= \text{k} - \text{h} - 1 = 10 - 3 - 1 = 6$

k number of class intervals $= 10$
h number of derived.

parameters $= 3$ (Z, Sn, Cs) value of χ^2 from the table $= 12.59$

Table 2 shows consolidated results of all the locations while Table 3 gives results of χ^2 test which indicate whether the distribution is suitable for a particular data set or not (Table 4).

From 100 years maximum temperature data, rise in temperature for next 50 years and 100 years return period, using Gamble's method is ranging (0.1–0.4 °C) for Pune, Satara, Nashik (0.4–0.6 °C) for Nagpur, Chandrapur, Beed and greater than 0.6 °C for Gadchiroli. Using Log Pearson method, for next 50 and 100 years return period, rise in temperature ranging from (0.1 to 0.4 °C) for Pune, Satara, Nashik (0.4–0.6 °C) for Nagpur, Chandrapur, Beed and greater than 0.6 °C for Gadchiroli. Chi-square test is good fit for the Gumbel's method except at Gadchiroli, Chandrapur for period (1951–2000). Chi-square test is not good fit for the Log Pearson method.

Table 2 Shows consolidated results of temperature for 50 years and 100 years of return period using Gumbel and Log Pearson distributions method at all the seven districts

District	Gumbel's		Log Pearson	
Beed	50	100	50	100
1901–1950	33.9	34.08	33.66	33.76
1951–2000	34.36	34.55	34.08	34.16
Difference	0.46	0.47	0.42	0.40
	Gumbel's		*Log Pearson*	
Nagpur	50	100	50	100
1901–1950	34.47	34.66	34.24	34.34
1951–2000	34.92	35.13	34.58	34.67
Difference	0.45	0.47	0.35	0.33
	Gumbel's		*Log Pearson*	
Gadchiroli	50	100	50	100
1901–1950	34.28	34.45	33.99	34.09
1951–2000	34.93	35.15	34.51	34.57
Difference	0.66	0.7	0.52	0.49
	Gumbel's		*Log Pearson*	
Pune	50	100	50	100
1901–1950	31.37	31.55	31.13	31.23
1951–2000	31.58	31.74	31.36	31.44
Difference	0.21	0.18	0.23	0.21
	Gumbel's		*Log Pearson*	
Satara	50	100	50	100
1901–1950	31.49	31.66	31.22	31.31
1951–2000	31.77	31.93	31.55	31.63
Difference	0.287	0.28	0.33	0.32
	Gumbel's		*Log Pearson*	
Nashik	50	100	50	100
1901–1950	32.46	32.64	32.22	32.32
1951–2000	32.77	32.94	32.54	32.63
Difference	0.312	0.30	0.32	0.31
	Gumbel's		*Log Pearson*	
Chandrapur	50	100	50	100
1901–1950	34.21	34.39	33.95	34.04
1951–2000	34.77	34.98	34.39	34.47
Difference	0.559	0.58	0.45	0.43

Table 3 χ^2 calculated

District	Gumbel's		Log Pearson	
	1901–1950	1951–2000	1901–1950	1951–2000
Beed	8.8	10	64.4	95.6
Nagpur	8.8	10	74.8	70.8
Gadchiroli	6.4	27.4	64.4	59.2
Pune	11.6	10	56.8	59.2
Satara	10	4.8	87.2	72.4
Nashik	7.2	9	68.8	70.8
Chandrapur	8.2	24.3	64.4	52.2

Table 4 Chi-squared test results

Gumbel	Distribution suitable	Log Pearson	Distribution not suitable
χ^2 calculated <14.067		χ^2 calculated >12.59	

5 Conclusion

Observed annual maximum temperature values are inline with measured values in last 100 years. It is evident from calculated values using extreme value distribution (Gumbel's and log Pearson) trend of annual maximum temperature value that lie within observed pattern of earlier measured value for a 100 years. There is rise in maximum temperature for Gadchiroli district, moderate rise in temperature in Nagpur, Beed, Chandrapur and less rise in Pune, Satara and Nashik district. From above extreme value distribution method like Gumbel's and log Pearson, there is rise in temperature for 50–100 years return period within range between 0.2 and 0.6 °C. Due to rising temperature, increase in risk associated with extreme weather events is moderate within 0.2–0.6 °C, as temperature increases risk also increases. Ecological risk of widespread extinction is small at 0.6 °C rise but it will increase with increase in temperature. It is clear from chi-square test that Gumbel's method is good fit to do annual extreme value analysis of maximum temperature for seven districts. Rising temperature causes draught condition which leads to low crop productivity. Present study will definitely give input (provide path way) to know the upcoming effects of increasing temperature in these seven districts. It can give earth weather system parameter in climate change study. Furthermore, the applicability of the study will help to societal welfare to become aware for the climate change effect.

References

1. Reddy J (1997) Stochastic hydrology. Laxmi Publications Pvt. Ltd, New Delhi
2. Subramanya K (2008) Engineering hydrology. Tata McGraw Hill Education (India) Pvt. Ltd

Adsorptive Removal of Fluoride from Water Using Non-conventional Adsorbents

Disha Khandare, Ajay Tembhurkar, and Somnath Mukherjee

Abstract The rapid industrialization and ever-growing population have led to degradation of water quality and reduction in availability safe drinking water. Presence of priority and secondary pollutants in underground water is posing serious health concerns. Fluoride in drinking water is referred as a two-edge sword for its beneficial (up to 1 mg/L) and detrimental effects (>1.5 mg/L). In the present study, an attempt is made to assess the fluoride removal efficiencies of some waste material as well as the naturally occurring substances. In the present study, fly ash, modified neem bark powder and fish scale biochar were investigated for their feasible use as adsorbent for fluoride removal from water environment. Batch sorption experiments were employed to examine the effect of influencing parameters like adsorbent dose, contact time, pH, initial fluoride concentration and agitation speed, etc. The batch sorption data showed non-conventional adsorbents can be used effectively in fluoride removal and simultaneously also renders to reduce refuse disposal problem of human settlement.

Keywords Adsorption · Non-conventional adsorbent · Fluoride removal · Batch studies

1 Introduction

The presence of fluoride contamination of drinking water source is a serious concern worldwide. Many countries like Sri Lanka, parts of South Africa, China, Turkey

D. Khandare (✉)
Civil Engineering Department, Jadavpur University, Kolkata 700032, India
e-mail: dishakhandare@gmail.com

A. Tembhurkar
Civil Engineering Department, VNIT, Nagpur 440010, India
e-mail: artembhurkar@civ.vnit.ac.in

S. Mukherjee
Civil Engineering Department, Jadavpur University, Kolkata 700032, India
e-mail: mukherjeesomnath19@gmail.com

L. M. Gupta et al. (eds.), *Advances in Civil Engineering and Infrastructural Development*, Lecture Notes in Civil Engineering 87,
https://doi.org/10.1007/978-981-15-6463-5_42

and India are among the 23 nations around the globe where the occurrence of high levels of fluoride concentrations in ground water is responsible for skeletal and dental fluorosis [1, 2]. More than 260 million people around the world are being suffered owing to excess fluoride concentration in groundwater (>1.5 mg/L) [3]. As per Indian standards, an allowable fluoride concentration is 1.5 mg F^{-}/L for drinking water [4].

Various conventional processes are available for fluoride removal from water, i.e., ion exchange [5], electro-coagulation flotation [6], membrane filtration and reverse osmosis [7, 8], precipitation [9], nano-filtration [2] and adsorption [10]. However, many of these processes have limited application at small-scale treatment units because of their expensive operation and maintenance, requirement of skilled personnel, generation of toxic waste during regeneration and its disposal [8, 10]. Out of all the above treatments as mentioned, adsorption is considered as economically viable and most popular technological solution for defluoridation [11].

Recently, many researchers have explored different low-cost adsorbent materials such as bone-char powder [12, 13], fly ash [14], rice husk [14, 15], brick powder [16, 17], leaf powder [18], agricultural waste [19], egg shell waste [20], etc., for fluoride attenuation from water environment. The present study aims to investigate the adsorptive capacity of non-conventional materials like fly ash (FA), modified neem bark powder (MNBP) and fish scale biochar (FSB). Fly ash and fish scales are by-products of thermal and fishery industry, respectively, whereas neem bark is a natural product, and all the above materials are locally available to use as adsorbent. Fish scales constitute major part of fishery waste and usually be disposed as a waste with no commercial value. The feasibility of using fish scale derived biochar is investigated as a low-cost adsorbent for defluoridation since fish bone derived charcoal earlier was used as defluoriding adsorbent. In the present paper, an effort is made to assess the usefulness of these non-conventional, low-cost adsorbents to make a low cost an alternative of activated alumina for defluoridation of water.

2 Materials and Methods

2.1 Adsorbent Collection and Preparation

The thermal power plant rejects "fly ash" was used as one of the materials and was collected from Koradi Thermal Power Station (KTPS) located near Nagpur, Maharashtra. The fly ash was washed several times with tap water followed by double distilled water to clean the impurities and subsequently dried at 105 °C for 24 h in a hot air oven. The dried fly ash was then sieved through 100–150 μm mesh size and used for experimental purpose.

The bark of neem (*Azadirachta indica*) was collected from local woody area and pretreated to remove its organic content and color. The sun dried neem bark was crushed and grinded manually to produce powder form which was subsequently treated with acid and alkali. The chemical treatment incurred heating the neem bark powder with 1 N aqueous of solution HNO_3 and aqueous of solution 1 N NaOH for 20 min, respectively, followed by several times wash with distilled water after each treatment till the organic content and color of neem bark powder were physically removed. The chemically treated neem bark powder was sun dried for 2 days and also be oven dried and used as adsorbent material.

Fish scale waste of *Labeo rohita* (Rohu) was collected from local fish market of Nagpur. The scales were washed thoroughly with double distilled water to remove dirt, dust and soluble impurities. The cleaned fish scales were then sun dried for 2 days and carbonized at 600 °C in muffle furnace for 3 h for making necessary fish scale biochar as adsorbent.

2.2 Preparation of Adsorbate Solution

Stock solution of fluoride (strength) 1000 mg/L was prepared by adding 2.21 g of analytical grade (AR) sodium fluoride (NaF, Merck, India) in 1 l of double distilled water. Different solutions of varying initial concentrations were prepared by diluting above stock solution with double distilled water and stored in polythene containers for batch experimental studies.

2.3 Equipment

Measurement of fluoride ions and pH is done with Thermo Scientific Orion Fluoride Ion Selective Electrode Meter. Standard methods (4500-F-C) were used for measurement of fluoride. For agitating, adsorbent with 50 ml of synthetic fluoride solution in 100 ml polythene bottles in a horizontal shaker at 150 rpm speed was used.

2.4 Experiment and Sorption Studies

The influence of process controlling parameters, i.e., pH, adsorbent dose, contact time, initial concentration and agitation speed on efficiency of fluoride removal has been studied through different sets of batch adsorption with varied initial data sets. The adsorption study was done by conducting batch experiments in triplicate. All the experiments were carried out at ambient temperature. The removal of fluoride ion was calculated using the equation (1).

$$\%\text{Removal} = \frac{\text{Co} - \text{Ce}}{\text{Ce}} \tag{1}$$

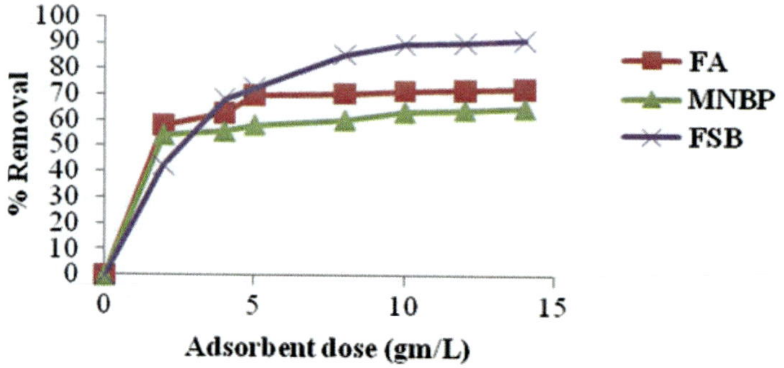

Fig. 3.1 Effect of adsorbent dose

where Co and Ce are the initial and equilibrium fluoride ion concentrations (mg/L), respectively.

3 Results and Discussion

3.1 *The Effect of Adsorbent Dose*

Figure 3.1 shows the effect of adsorbent dose on removal efficiencies of FA, MNBP and FSB. The effect of the amount of adsorbent was observed with various adsorbent doses between 2 and 14 g/L. The initial F-ion concentration was maintained as 5 mg/L, and solution was agitated at constant speed of 150 rpm. The highest fluoride removal (%) was observed as 72.5%, 64.8% and 91.2% for FA, MNBP and FSB, respectively. From the result, it is evident that efficiency of removal is dependent on adsorbent dose.

The percentage of removal of F^- was improved with an increase in the adsorbent amount. The ascending trend of removal efficiency at higher in adsorbent dose is due to the increase in surface area which yields more active site for adsorption of F^- [21]. However, no significant increase in fluoride removal was observed on further increasing the adsorbent dose. This might be due to attainment of equilibrium conditions of adsorption [22, 23]. Therefore, the optimum dose of 8 g/L, 12 g/L and 10 g/L for FA, MNBP and FSB, respectively, was used for further experimental studies.

3.2 *The Effect of Contact Time*

The effect of contact time was observed with initial F^- ions concentration = 5 mg/L and agitation speed of 150 rpm. The optimum dose obtained as 8, 12 and 10 g/L

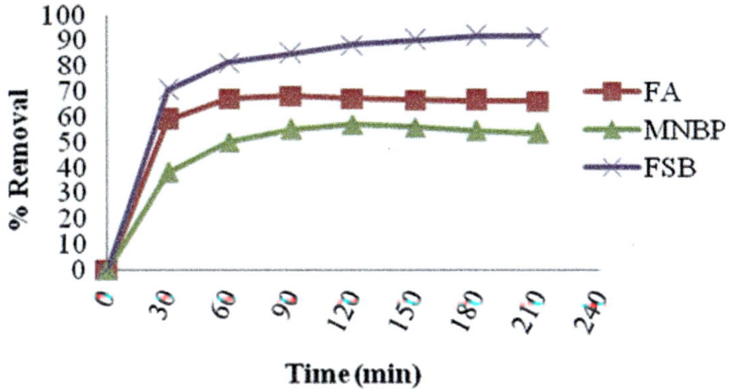

Fig. 3.2 Effect of contact time

for FA, MNBP and FSB, respectively. From Fig. 3.2, it is evident that contact time varies for different adsorbents depending upon properties of the adsorbent. It was noticed that the removal of F^- increases with increase in contact time, though, it gradually attained constant removal efficiency even in excess in contact time which was considered as equilibrium time [24]. The equilibration time for FA, MNBP and FSB was observed 60 min, 150 min and 180 min, respectively.

3.3 *The Effect of Initial Concentration*

The influence of initial fluoride concentration on its removal was examined by varying initial F^- ion concentration between 2 and 15 mg/L and with respective optimum dose and equilibrium time. As indicated in Fig. 3.3, it is apparent that the amount of fluoride ions adsorbed per unit mass of adsorbents decreases with rise in the concentration of fluoride. This is due to the reason that, at low initial fluoride concentration, the ratio of the fluoride ions to the number of available adsorption site was high, and with increase

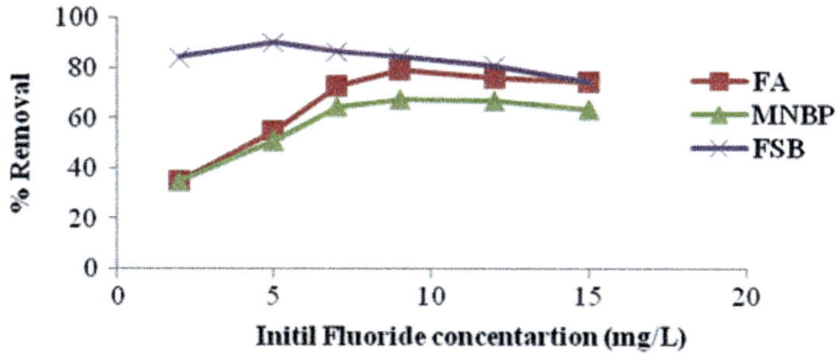

Fig. 3.3 Effect of initial F^- ion concentration

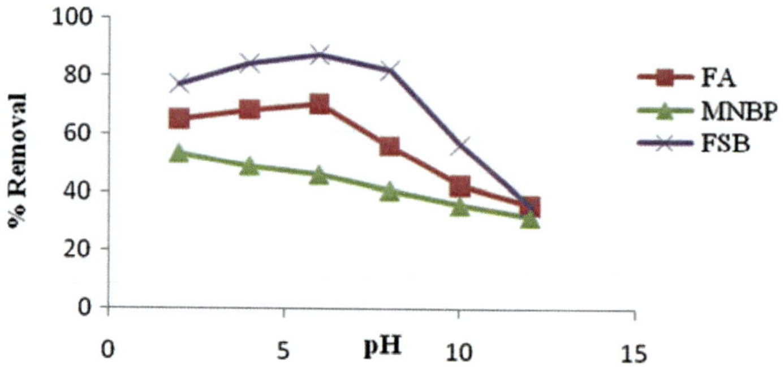

Fig. 3.4 Effect of pH

in fluoride concentration, the specific available sites were decreased [25, 26]. Such in decrease in amount of fluoride adsorbed per unit mass of adsorbent with increase in initial Fluoride concentration denoted the lesser capacity of adsorption at higher fluoride concentration. The maximum fluoride removal of 79 and 68% has been obtained at 9 mg/L for FA and MNBP, whereas for FSB, the % removal decreased after 5 mg/L concentration. Reduction in removal of F^- ions was due to availability of limited number of adsorption sites with a finite capacity on the adsorbent surface. Similar trend of observation has been reported by Zazouli et al. [27].

3.4 The Effect of PH

The effect of pH on removal of fluoride ions was also studied in the pH range of 2.0–12.0, and results are shown in Fig. 3.4. The result shows that the highest F^- ion removal of 53.21% was observed at pH 2.0 neem bark powder. Whereas for fly ash and fish scale biochar could remove fluoride up to 65.09% and 86% at pH 6, respectively. As the pH of solution increased, the % removal decreased rapidly. This is due to, at low pH level, positively charged H^+ ions on the surface enhance the binding of the negatively charged fluoride ions [25]. At higher pH, the relative adsorption decreases due to increased concentration of hydroxyl ions, and this retards adsorption of anionic fluoride ions [23].

3.5 The Effect of Agitation Speed

The effect of agitation speed on fluoride ion removal was examined over the range of 50–180 rpm. Figure 3.5 indicates the effect of agitation speed on adsorbent capacity of fluoride removal. The maximum removal of F^- was about 78.4%, 61.5% and 90% at 180 rpm was achieved for FA, MNBP and FSB, respectively. The results indicate that,

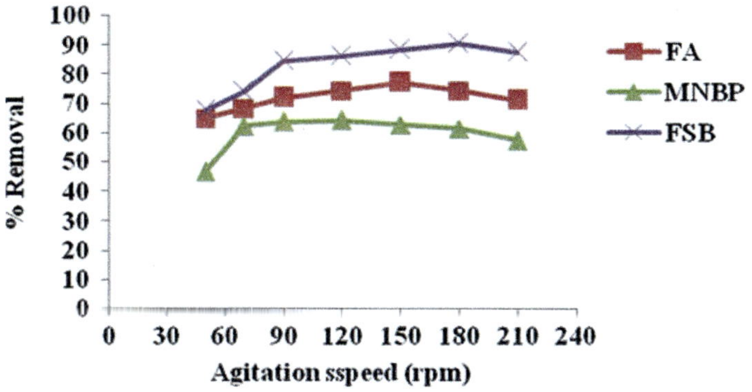

Fig. 3.5 Effect of agitation speed

the fluoride removal was increased with increase in agitation speed. The higher speed facilitates better contact between the adsorbent and adsorbate resulting in higher removal capacity [20]. At higher agitation speed, better mass transfer of F^- onto the surface of adsorbent was achieved. However, further increase in agitation speed resulted in reduction of fluoride removal efficiency due to increased possibilities of F^- ions to escape from surface of the adsorbent [22, 23].

3.6 *Advantage of Non-conventional Adsorbents Over Conventional Adsorbents*

- Non-conventional adsorbents used are available either as waste or naturally available material. Hence, they are relatively cheaper compared to conventional treatments for fluoride removal.
- The present study reveals that, the efficiencies of removal of fluoride ions of various non-conventional fly ash, modified neem bark powder and fish scale biochar as adsorbent mostly vary between 50 and 90%, so they can be effectively used as potential adsorbent for fluoride removal.
- The preparation of adsorbent is easy and requires very less maintenance and supervision.
- The reuse of fly ash and fish scale biochar as a filler material in various construction activities is possible.
- Though regeneration of the exhausted adsorbents is not needed as they are available abundantly and locally at very low cost, the regeneration of these adsorbent is possible with simple chemical treatment.

4 Conclusion

From the present study, it is concluded that fly ash, modified neem bark powder and fish scale biochar have potential to remove fluoride from water environment. The efficiency of these adsorbents in the removal of fluoride was assessed with varying dose of adsorbate, contact time, initial F^- ion concentration, pH and speed of agitation. From the batch sorption studies, it is evident that fluoride removal is greatly dependent on the adsorbent dose and contact time at a given initial solute concentration. The fish scale biochar shows better adsorptive potential as compared to fly ash and modified neem bark powder. The use of industrial by-product like fly ash and fish scale waste as adsorbent will help to reduce their possible environmental hazards. However, a detailed research may be required to improve the adsorption capacities and its safe disposal after use of as adsorbents.

References

1. Maheshwari RC (2006) Fluoride in drinking water and its removal. J Hazard Mater 137(1):456–463
2. Bhatnagar A, Kumar E, Sillanpää M (2011) Fluoride removal from water by adsorption—a review. Chem Eng J 171(3):811–840
3. Ram SM, Thakkar VP, Machale P (2017) Determination of fluoride level in drinking water from water samples in Navi Mumbai, Maharashtra". J Indian Assoc Publ Health Dent 15(4):395
4. Meenakshi S, Viswanathan N (2007) Identification of selective ion-exchange resin for fluoride sorption. J Colloid Interface Sci 308(2):438–450
5. Emamjomeh MM, Sivakumar M, Varyani AS (2011) Analysis and the understanding of fluoride removal mechanisms by an electrocoagulation/flotation (ECF) process. Desalination 275(1–3):102–106
6. Ndiaye PI, Moulin P, Dominguez L, Millet JC, Charbit F (2005) Removal of fluoride from electronic industrial effluent by RO membrane separation. Desalination 173(1):25–32
7. Hichour M, Persin F, Molénat J, Sandeaux J, Gavach C (1999) Fluoride removal from diluted solutions by Donnan dialysis with anion-exchange membranes. Desalination 122(1):53–62
8. Nawlakhe WG, Paramasivam R (1993) Defluoridation of potable water by Nalgonda technique. Curr Sci 65(10):743–748
9. Hu K, Dickson JM (2006) Nanofiltration membrane performance on fluoride removal from water. J Membr Sci 279(1–2):529–538
10. Sinha Sarita (2003) Removal of fluoride from aqueous solutions by Eichhornia crassipes biomass and its carbonized form. Ind Eng Chem Res 42(26):6911–6918
11. Viswanathan N, Sairam Sundaram C, Meenakshi S (2009) Removal of fluoride from aqueous solution using protonated chitosan beads. J Hazardous Mater 161(1):423–430
12. Brunson LR, Sabatini DA (2009) An evaluation of fish bone chars as an appropriate arsenic and fluoride removal technology for emerging regions. Environ Eng Sci 26(12):1777–1784
13. Chaturvedi AK, Yadava KP, Pathak KC, Singh VN (1990) Defluoridation of water by adsorption on fly ash. Water Air Soil Pollut 49(1–2):51–61
14. Deshmukh WS, Attar SJ, Waghmare MD (2009) Investigation on sorption of fluoride in water using rice husk as an adsorbent. Nat Environ Pollut Technol 8(2), 217–223
15. Deshmukh WS, Attar SJ, Waghmare MD (2009) Investigation on sorption of fluoride in water using rice husk as an adsorbent. Nat Environ Pollut Technol 8(2):217–223

16. Yadav AK, Kaushik CP, Haritash AK, Kansal A, Rani N (2006) Defluoridation of groundwater using brick powder as an adsorbent. J Hazard Mater 128(2–3):289–293
17. Chidambaram S, Ramanathan AL, Vasudevan S (2003) Fluoride removal studies in water using natural material. Water SA 29(3):339–344
18. Tomar V, Prasad S, Kumar D (2014) Adsorptive removal of fluoride from aqueous media using *Citrus limonum* (lemon) leaf. Microchem J 112:97–103
19. Yadav AK, Abbassi R, Gupta A, Dadashzadeh M (2013) Removal of fluoride from aqueous solution and groundwater by wheat straw, sawdust and activated bagasse carbon of sugarcane. Ecol Eng 52:211–218
20. Bashir M, Salmiaton A, Nourouzi M, Azni I, Harun R (2015) Fluoride removal by chemical modification of palm kernel shell-based adsorbent: a novel agricultural waste utilization approach. Asian J Microbial Biotech Env Sci 17(3):533–542
21. Vardhan VCM, Karthikeyan J (2011) Removal of fluoride from water using low-cost materials. In: Fifteenth International Water Technology Conference, IWTC-15, pp 1–14
22. Bhaumik R et al (2012) Eggshell powder as an adsorbent for removal of fluoride from aqueous solution: equilibrium, kinetic and thermodynamic studies. J Chem 9(3):1457–1480
23. George AM, Tembhurkar AR (2019) Analysis of equilibrium, kinetic, and thermodynamic parameters for biosorption of fluoride from water onto coconut (Cocos nucifera Linn.) root developed adsorbent. Chin J Chem Eng 27(1):92–99
24. Nemade PD, Vasudeva Rao A, Alappat BJ (2002) Removal of fluorides from water using low cost adsorbents. Water Sci Technol Water Supply 2(1):311–317
25. Khandare DA, Mukherjee S (2019) Fish scale waste: potential low-cost adsorbent for fluoride removal. J Indian Chem Soc 96:429–434
26. Gandhi N, Sirisha D, Shekar KC, Asthana S (2012) Removal of fluoride from water and waste water by using low cost adsorbents. Int J ChemTech Research 4(4):1646–1653
27. Zazouli MA, Mahvi AH, Mahdavi Y, Balarak D (2015) Isothermic and kinetic modeling of fluoride removal from water by means of the natural biosorbents sorghum and canola. Fluoride 48(1):37–41

Role of Meteorology in Seasonal Variation of Air Pollution

Avneesh Tiwari and A. K. Shukla

Abstract The meteorological parameters such as atmospheric temperature, wind speed and direction, and relative humidity are the governing factors for the concentration of air pollutants in the ambient atmosphere. The purpose of this study is to evaluate the impact of these meteorological parameters on ambient particulate matter ($PM_{2.5}$), nitrogen dioxide (NO_2), nitrogen oxide (NO), ozone (O_3), carbon monoxide (CO), sulfur dioxide (SO_2), and ammonia (NH_3) concentration during northeast monsoon, winter, summer, and southwest monsoon for Lucknow. Multiple linear regression analysis and Pearson's correlation matrix is prepared using data from October 2014 to October 2017. The monitored values by pollution control board and predicted values using regression equations are compared from October 2017 to October 2018. The result shows that in northeast monsoon, $PM_{2.5}$ is highly correlated with temperature, and O_3 and NH_3 are moderately correlated with relative humidity and wind speed, respectively. In summer, NO_2, O_3, and NH_3 are moderately correlated with relative humidity, and NH_3 is also moderately correlated with wind speed. In southwest monsoon, $PM_{2.5}$ is highly correlated with relative humidity and atmospheric temperature, and NO and O_3 are moderately correlated with wind speed and relative humidity, respectively. In winter, O_3 is highly correlated with relative humidity and $PM_{2.5}$ with wind speed and relative humidity, O_3 with atmospheric temperature and wind speed in moderately correlated, while rest of above pollutants in different seasons are low correlated with meteorological parameter.

Keywords Meteorology · Pollutants · Regression

A. Tiwari (✉)
Environmental Engineering, Institute of Engineering and Technology, Sitapur Road, Lucknow, India
e-mail: avneeshtiwari100@gmail.com

A. K. Shukla
Civil Engineering Department, Institute of Engineering and Technology, Sitapur Road, Lucknow, India

L. M. Gupta et al. (eds.), *Advances in Civil Engineering and Infrastructural Development*, Lecture Notes in Civil Engineering 87,
https://doi.org/10.1007/978-981-15-6463-5_43

1 Introduction

With the growth of population, urbanization, and industrial activities, air quality has been degrading and has become one of the most concerning issues. According to World Health Statistics, 2018, 91% of the world's population resides in regions exceeding WHO standards [1]. Air pollutants from various sources are released into the atmosphere, and the pollutants' concentration in the ambient air is not only governed by the quantity of emission but also depends upon atmospheres' ability in pollutants' dispersion or absorption [2]. Atmosphere acts as the medium for pollutants to move away from source, and atmosphere is governed by meteorological factors such as wind speed and direction, temperature, and relative humidity [3]. Hence, knowledge of meteorological parameters is important in understanding pollutants' concentration in ambient air. The concentration of pollutants in the urban environment is a function of mixing depth, wind speed, and size of the city. The average wind speed varies from time to time, i.e., month to month, morning to afternoon, and place to place [4]. Wind direction governs the movement of air pollutants in a particular direction across an area. The rate of dispersion increases with the increase in wind speed which dilutes pollutants in the atmosphere and reduces the ground-level concentration [5].

This paper investigates the influence of meteorological parameters on the variation in concentration of particulate matter, nitrogen dioxide, nitrogen oxide, ozone, carbon monoxide, sulfur dioxide, and ammonia in Lucknow among different seasons and generation of equation for predicting the concentration of pollutants using multiple regression model with the help of meteorological data.

2 Methodology

2.1 Study Area

Lucknow, the capital of Uttar Pradesh, situated in the heart of the state is selected for the study having latitude 26.8467° N and longitude 80.9462° E. According to the census 2011, Lucknow is at second position with urban population (66.21%) after Ghaziabad (67.55%) and at first position in terms of urban district with an area of 470.71 km^2 in Uttar Pradesh [6]. According to Indian Meteorological Department, the climate of Lucknow is tropical monsoonal type with four different seasons; northeast monsoon (October–November), winter (December–February), summer (March–May), and southwest monsoon (June–September). The temperature of the monitoring stations goes up to 40 °C in the summer. During winter, it goes below 10 °C [7].

Table 1 Location of air quality monitoring stations

S. No.	Monitoring station	Location	
		Latitude	Longitude
1	Central School, Lucknow—CPCB	26.882°	80.930°
2	Nishant Ganj, Lucknow—UPPCB	26.871°	80.957°
3	Lalbagh, Lucknow—CPCB	26.845°	80.936°
4	Talkatora District Industries Center, Lucknow—CPCB	26.834°	80.892°

2.2 Data Collection

The Uttar Pradesh Pollution Control Board (UPPCB) and Central Pollution Control Board (CPCB) are continuously monitoring ambient air quality at four continuous working stations in Lucknow. These continuous monitoring stations are at Gomtinagar, Nishant Ganj, Lalbagh, and Talkatora. Nishant Ganj monitoring station is operated by UPPCB, while other stations by CBCP have been considered for the study. Table 1 shows the location of the continuous monitoring stations. Hourly average pollutants' concentration data of NO, NO_2, SO_2, O_3, CO, $PM_{2.5}$, NH_3 from 2014 to 2018 were retrieved from CPCB (https://app.cpcbccr.com/ccr/#/caaqm-dashboard-all/caaqm-landing). The average of hourly values for each pollutant obtained from different monitoring stations was used in the study. The meteorological data were retrieved from National Oceanic and Atmospheric Administration (NOAA). Same data can be retrieved using this Web address.

https://www7.ncdc.noaa.gov/CDO/cdoselect.cmd?datasetabbv=GSOD&countryabbv=IN&georegionabbv=&resolution=40.

2.3 Data Analysis

In this study, pollutant data and meteorological data were collected for four years from October 2014 to September 2018. First, three-year data were used in the analysis followed by verifying the result with the help of last one-year data. The collected data were divided according to different seasons for doing analysis seasonally. The statistical analysis was carried out using R version 3.5.0; it includes multiple linear regression (MLR) model for predicting pollutant concentration with the help of metrological data and Pearson's correlation coefficient for investigating the relation between pollutants and meteorology in different seasons.

3 Result and Discussion

Meteorological parameters are highly variable due to different seasons and change in meteorological condition which leads to change in pollution level, and hence, to analyze the impact of meteorology and pollutants, Pearson's correlation analysis was done between pollutants and meteorological factors. Pollutants in the different season show significant positive/negative correlation with meteorology, R^2 values in between 0.5 and 1, 0.3 and 0.5, and below 0.3 shows high, moderate, and low correlations, respectively.

In northeast monsoon shown in Table 2, $PM_{2.5}$ is negatively correlated with temperature ($R^2 = -0.75$), NO_2 is positively correlated with NO ($R^2 = 0.69$), CO ($R^2 = 0.57$), and O_3 ($R^2 = 0.35$), NO is positively correlated with CO ($R^2 = 0.66$), and NH_3 ($R^2 = 0.32$), O_3 is positively correlated with CO ($R^2 = 0.49$), NH_3 ($R^2 = 0.40$) and negatively correlated with relative humidity ($R^2 = -0.32$); CO is positively correlated with NH_3 ($R^2 = 0.36$) and NH_3 is negatively correlated with wind speed ($R^2 = -0.35$).

In winter season shown in Table 3, $PM_{2.5}$ is negatively correlated with NO_2 ($R^2 = -0.32$) and wind speed ($R^2 = -0.41$), and positively correlated with relative humidity ($R^2 = 0.37$), NO_2 is positively correlated with NO ($R^2 = 0.65$), CO ($R^2 = 0.39$), O_3 ($R^2 = 0.37$), SO_2 ($R^2 = 0.34$), and NH_3 ($R^2 = 0.49$), NO is positively correlated with CO ($R^2 = 0.53$), and NH_3 ($R^2 = 0.31$), O_3 is positively correlated with CO ($R^2 = 0.37$), NH_3 ($R^2 = 0.57$), temperature ($R^2 = 0.33$), wind speed ($R^2 = 0.35$) and negatively correlated with relative humidity ($R^2 = -0.61$).

In summer shown in Table 4, NO_2 is positively correlated with NO ($R^2 = 0.50$), CO ($R^2 = 0.39$) and negatively correlated with CO ($R^2 = -0.42$) and relative humidity ($R^2 = -0.48$), NO is positively correlated with NH_3 ($R^2 = 0.39$), O_3 is positively correlated with NH_3 ($R^2 = 0.43$) and negatively correlated with relative humidity (R^2

Table 2 Pearson's correlation analysis for northeast monsoon between pollutants and meteorology

	$PM_{2.5}$	NO_2	NO	O_3	CO	SO_2	NH_3	Temp[a]	WS[b]	WD[c]	RH[d]
$PM_{2.5}$	1	−0.25	−0.12	−0.13	−0.03	−0.30	−0.01	−0.75	−0.09	0.16	−0.20
NO_2		1	0.69	0.35	0.57	0.22	0.29	−0.22	−0.24	0.09	−0.08
NO			1	0.28	0.66	0.18	0.32	−0.14	−0.26	0.01	−0.07
O_3				1	0.49	−0.01	0.40	0.03	0.05	−0.02	−0.32
CO					1	0.10	0.36	0.06	−0.11	−0.02	−0.16
SO_2						1	0.16	0.02	−0.14	0.06	0.04
NH_3							1	−0.17	−0.35	0.17	−0.06
Temp								1	0.07	−0.31	0.15
WS									1	−0.09	−0.05
WD										1	−0.45
RH											1

[a]Atmospheric temperature, [b]wind speed, [c]wind direction, [d]relative humidity

Table 3 Pearson's correlation analysis for winter season between pollutants and meteorology

	$PM_{2.5}$	NO_2	NO	O_3	CO	SO_2	NH_3	Temp	WS	WD	RH
$PM_{2.5}$	1	−0.32	0.06	−0.07	0.02	−0.02	−0.29	−0.24	−0.41	−0.11	0.37
NO_2		1	0.65	0.37	0.39	0.34	0.49	0.13	0.05	0.12	−0.06
NO			1	0.19	0.53	0.31	0.06	−0.03	−0.12	0.01	0.07
O_3				1	0.37	0.08	0.57	0.33	0.35	0.05	−0.61
CO					1	0.03	0.21	0.04	0.02	0.04	−0.11
SO_2						1	0.09	−0.001	−0.06	0.09	0.04
NH_3							1	0.23	−0.08	0.15	0.02
Temp								1	0.15	−0.22	−0.46
WS									1	0.05	−0.37
WD										1	0.02
RH											1

Table 4 Pearson's correlation analysis for summer season between pollutants and meteorology

	$PM_{2.5}$	NO_2	NO	O_3	CO	SO_2	NH_3	Temp	WS	WD	RH
$PM_{2.5}$	1	0.14	0.09	0.13	−0.28	−0.09	0.26	0.24	−0.24	0.12	−0.15
NO_2		1	0.50	0.39	−0.42	−0.20	0.07	0.06	0.06	0.26	−0.48
NO			1	0.13	−0.15	−0.08	0.39	−0.18	−0.19	0.11	−0.08
O_3				1	−0.24	−0.15	0.43	0.04	−0.04	0.12	−0.35
CO					1	0.52	−0.41	0.11	−0.21	−0.01	0.05
SO_2						1	−0.28	0.06	−0.14	0.01	−0.05
NH_3							1	−0.70	−0.37	0.19	0.33
Temp								1	0.07	−0.13	−0.54
WS									1	0.04	−0.19
WD										1	−0.44
RH											1

= −0.35), CO is positively correlated with SO_2 ($R^2 = 0.52$) and negatively correlated with NH_3 ($R^2 = -0.41$), and NH_3 is positively correlated with relative humidity ($R^2 = 0.33$) and negatively correlated with temperature ($R^2 = -0.70$) and wind speed ($R^2 = -0.37$).

In southwest monsoon shown in Table 5, $PM_{2.5}$ is positively correlated with temperature ($R^2 = 0.59$) and negatively correlated with relative humidity temperature ($R^2 = -0.63$), NO_2 is positively correlated with NO ($R^2 = 0.44$) and negatively correlated with CO ($R^2 = -0.4$), NO is negatively correlated with wind speed ($R^2 = -0.35$), O_3 is positively correlated with temperature ($R^2 = 0.31$) and negatively correlated with relative humidity ($R^2 = -0.38$).

The mathematical relation between pollutants and meteorological parameters was derived using multiple linear regression (MLR) model for predicting pollutant

Table 5 Pearson's correlation analysis for southwest monsoon between pollutants and meteorology

	$PM_{2.5}$	NO_2	NO	O_3	CO	SO_2	NH_3	Temp	WS	WD	RH
$PM_{2.5}$	1	0.01	0.01	0.22	0.01	−0.19	−0.10	0.59	0.10	0.09	−0.63
NO_2		1	0.44	−0.26	−0.4	−0.05	0.12	−0.05	−0.25	−0.03	0.21
NO			1	−0.24	−0.21	0.23	0.03	−0.19	−0.35	0.04	0.29
O_3				1	0.19	0.06	−0.06	0.31	0.19	0.13	−0.38
CO					1	0.13	−0.16	0.11	0.23	0.03	−0.13
SO_2						1	−0.04	0.001	0.22	−0.16	0.02
NH_3							1	−0.08	−0.11	0.02	0.11
Temp								1	0.26	0.13	−0.88
WS									1	−0.09	−0.31
WD										1	−0.24
RH											1

concentration with the help of metrological data in different seasons. Equations 1–7 represent pollutant concentration of $PM_{2.5}$, NO_2, NO, O_3, CO, SO_2, and NH_3 in northeast monsoon with R^2 values 0.65, 0.09, 0.08, 0.37, 0.06, 0.02, and 0.17, respectively, used to predict the pollutants concentration and compared with observed values shown in Fig. 1.

$$PM_{2.5} = 638.0778 - 1.0357 * RH - 5.4949 * WS - 0.0175 * WD - 16.1254 * TEMP \quad (1)$$

$$NO_2 = 77.7884 - 0.1318 * RH - 7.5146 * WS - 0.0029 * WD - 1.0362 * TEMP \quad (2)$$

$$NO = 48.2907 - 0.1675 * RH - 6.1737 * WS - 0.0127 * WD - 0.3798 * TEMP \quad (3)$$

$$O_3 = 61.6605 - 0.5771 * RH + 0.2647 * WS - 0.0260 * WD + 0.1275 * TEMP \quad (4)$$

$$CO = 1.8238 - 0.0155 * RH - 0.1397 * WS - 0.0006 * WD + 0.0186 * TEMP \quad (5)$$

$$SO_2 = 5.5139 + 0.0378 * RH - 0.9431 * WS - 0.0045 * WD - 0.0485 * TEMP \quad (6)$$

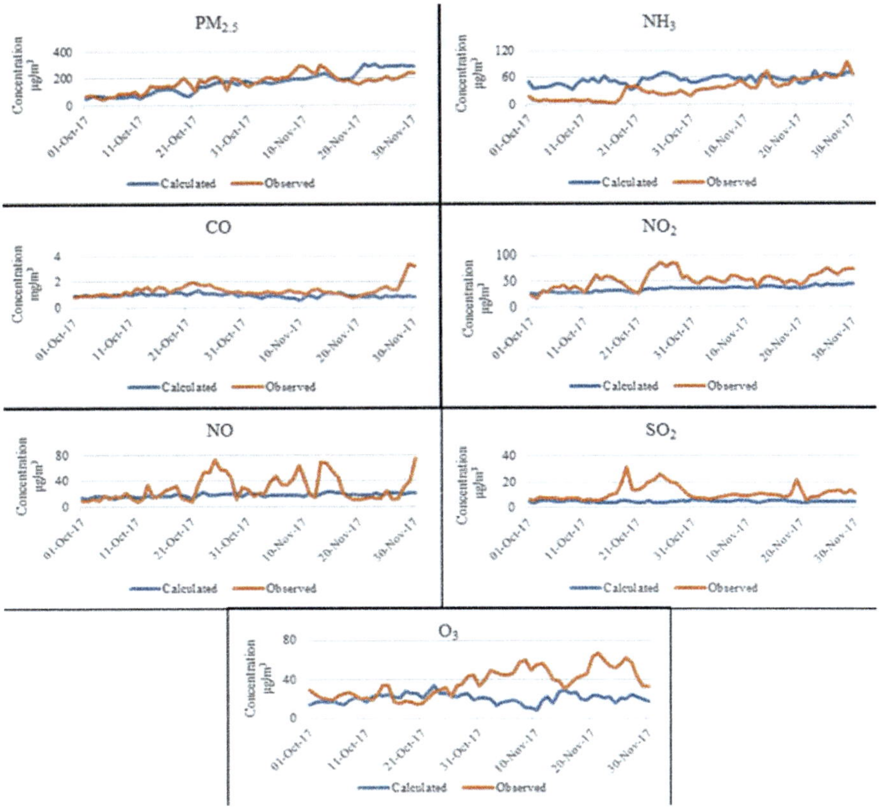

Fig. 1 Representation of observed and calculated pollutants concentration in northeast monsoon

$$NH_3 = 143.2269 - 0.7175 * RH - 33.1115 * WS + 0.0514 * WD - 0.7556 * TEMP \quad (7)$$

Equations 8–14 represent pollutant concentration of $PM_{2.5}$, NO_2, NO, O_3, CO, SO_2, and NH_3 in winter season with R^2 values 0.26, 0.05, 0.02, 0.41, 0.02, 0.02, and 0.11, respectively, used to predict the pollutants concentration and compared with observed values shown in Fig. 2.

$$PM_{2.5} = 250.1068 + 0.7289 * RH - 24.7457 * WS - 0.1214 * WD - 2.7780 * TEMP \quad (8)$$

$$NO_2 = 9.2954 - 0.0551 * RH - 0.0091 * WS + 0.0341 * WD + 0.8715 * TEMP \quad (9)$$

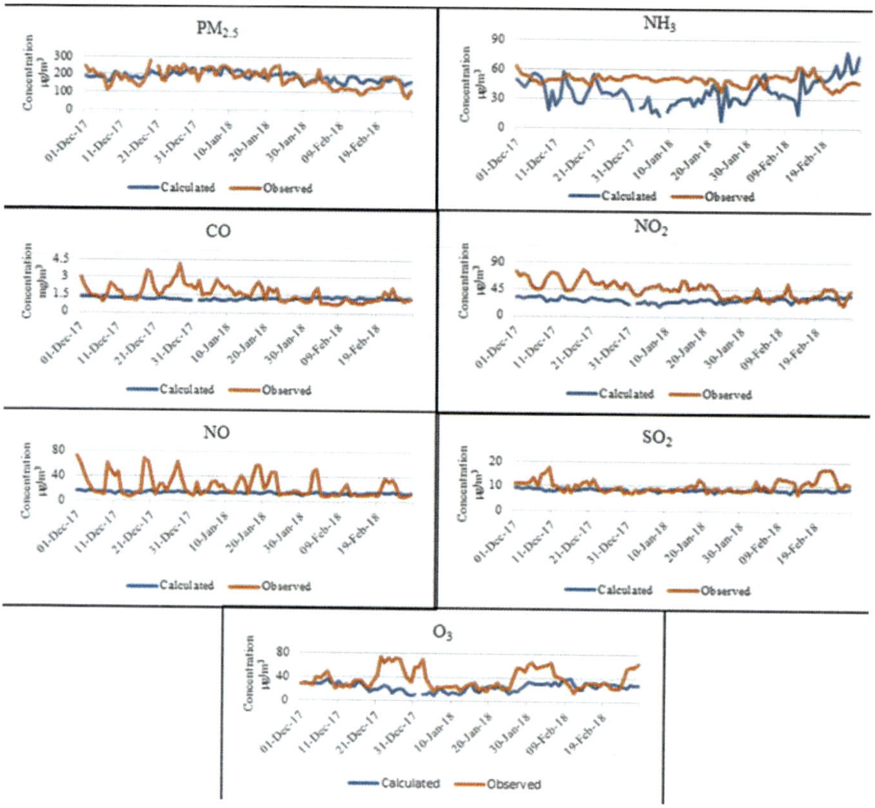

Fig. 2 Representation of observed and calculated pollutants concentration in winter season

$$NO = 16.2924 - 0.0077 * RH - 2.7711 * WS + 0.0039 * WD + 0.0681 * TEMP \tag{10}$$

$$O_3 = 48.7949 - 0.4849 * RH + 2.4773 * WS + 0.0102 * WD + 0.2934 * E \tag{11}$$

$$CO = 1.6673 - 0.0071 * RH - 0.0309 * WS + 0.0003 * WD - 0.0002 * TEMP \tag{12}$$

$$SO_2 = 6.7785 + 0.0034 * RH - 0.3923 * WS + 0.0048 * WD + 0.0570 * TEMP \tag{13}$$

$$NH_3 = -74.3352 + 0.2957 * RH - 4.9344 * WS + 0.1233 * WD + 4.1606 * TEMP \quad (14)$$

Equations 15–21 represent pollutant concentration of $PM_{2.5}$, NO_2, NO, O_3, CO, SO_2, and NH_3 in summer season with R^2 values 0.15, 0.32, 0.15, 0.20, 0.08, 0.02, and 0.58, respectively, used to predict the pollutants' concentration and compared with observed values shown in Fig. 3.

$$PM_{2.5} = 73.1173 - 0.1231 * RH - 12.1414 * WS + 0.0516 * WD + 1.9839 * TEMP \quad (15)$$

$$NO_2 = 116.7752 - 0.8668 * RH - 2.0311 * WS - 0.0218 * WD - 1.1988 * TEMP \quad (16)$$

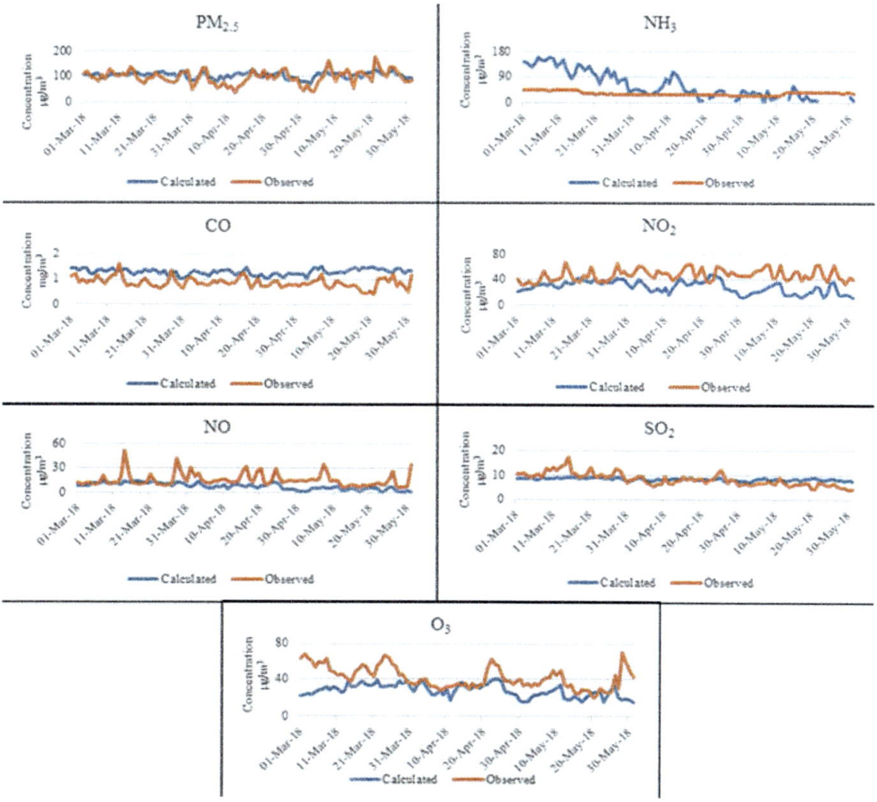

Fig. 3 Representation of observed and calculated pollutants concentration in summer season

$$NO = 44.1260 - 0.2239 * RH - 2.3795 * WS - 0.0076 * WD - 0.5933 * TEMP \quad (17)$$

$$O_3 = 98.5572 - 0.6562 * RH - 2.3054 * WS - 0.0285 * WD - 0.8939 * TEMP \quad (18)$$

$$CO = 0.2383 + 0.0086 * RH - 0.1237 * WS + 0.0007 * WD + 0.0274 * TEMP \quad (19)$$

$$SO_2 = 9.5741 - 0.0112 * RH - 0.5446 * WS + 0.0006 * WD + 0.0215 * TEMP \quad (20)$$

$$NH_3 = 285.1075 + 1.0237 * RH - 19.9270 * WS + 0.1686 * WD - 8.7139 * TEMP \quad (21)$$

Equations 15–21 represent pollutant concentration of $PM_{2.5}$, NO_2, NO, O_3, CO, SO_2, and NH_3 in southwest monsoon with R^2 values 0.41, 0.16, 0.19, 0.15, 0.06, 0.07, and 0.03, respectively, used to predict the pollutants concentration and compared with observed values shown in Fig. 4.

$$PM_{2.5} = 99.94 - 0.9865 * RH - 1.745 * WS - 0.00005 * WD + 1.108 * TEMP \quad (22)$$

$$NO_2 = -84.1154 + 0.4341 * RH - 1.9056 * WS + 0.0043 * WD + 2.3736 * TEMP \quad (23)$$

$$NO = -30.0636 + 0.1982 * RH - 1.7393 * WS + 0.0054 * WD + 0.8090 * TEMP \quad (24)$$

$$O_3 = 42.513 - 0.236 * RH + 0.937 * WS + 0.004 * WD - 0.370 * TEMP \quad (25)$$

$$CO = 0.7608 - 0.0026 * RH + 0.2326 * WS + 0.0003 * WD + 0.0029 * TEMP \quad (26)$$

$$SO_2 = 0.8292 + 0.0320 * RH + 1.0544 * WS - 0.0041 * WD + 0.1084 * TEMP \quad (27)$$

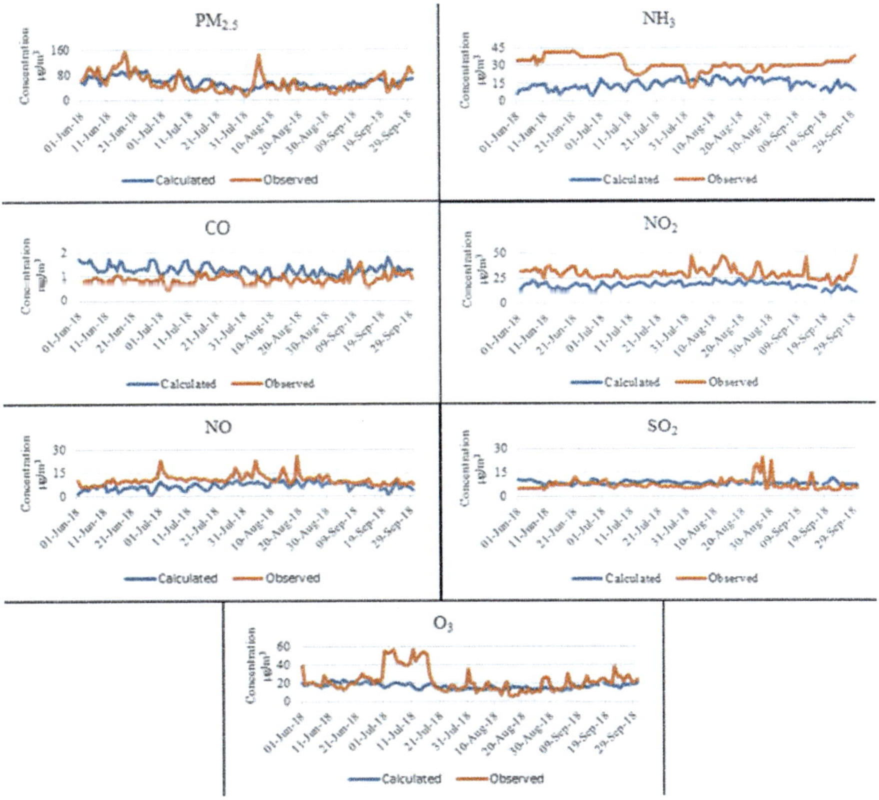

Fig. 4 Representation of observed and calculated pollutants concentration in southwest monsoon

$$NH_3 = -45.4073 + 0.3523 * RH - 2.6570 * WS + 0.0022 * WD + 1.2552 * TEMP \quad (28)$$

Figures 1, 2, 3, and 4 show the variation in concentration of pollutants calculated using equations generated by multiple regression analysis and concentration obtained from CPCB in northeast monsoon, winter season, summer season, and southwest monsoon, respectively.

4 Conclusion

On the basis of investigated relation between pollutants and meteorology, the pattern observed in Figs. 1, 2, 3, and 4 and by comparing the observed and predicted concentrations of pollutants, it can be concluded that the relation of meteorology with the pollutants is nonlinear. This nonlinear relation of pollutants and meteorology is also verified on the basis of R^2 value obtained during multiple regression analysis; as in most of the cases, the value of R^2 is very low except for $PM_{2.5}$ in northeast monsoon, winter, and southwest monsoon, for NH_3 and NO_2 in summer, for O_3 in northeast monsoon, and winter. Hence, nonlinear analysis and inclusion of other meteorological parameters like rainfall should be used in further studies to evaluate the role of meteorology in the seasonal variation of air pollutants.

References

1. World Health Organization: Status of the Health-Related SDGs. World Health Organisation (2017)
2. Yorkor B, Leton T, Ugbebor J (2017) The role of meteorology for seasonal variation in air pollution level in Eleme, Rivers State, Nigeria. J. Sci. Res. Rep. 17(3):1–17
3. Manju A, Kalaiselvi K, Dhananjayan V, Palanivel M, Banupriya GS, Vidhya MH, Panjakumar K, Ravichandran B (2018) Spatio-seasonal variation in ambient air pollutants and influence of meteorological factors in Coimbatore, Southern India. Air Qual. Atmos. Heal. 11(10):1179–1189
4. Holzworth, G.C.: Mixing depths, wind speeds and air pollution potential for selected locations in the United States. J. Appl. Meteorol. **6**(6):1039–1044
5. Jayamurugan, R., Kumaravel, B., Palanivelraja, S., Chockalingam, M.P.: Influence of temperature, relative humidity and seasonal variability on ambient air quality in a coastal urban area. Int. J. Atmos. Sci. **2013** (2013)
6. Directorate of Census Operations Uttar Pradesh: District Census Handbook, Lucknow, Series-10, Part XII-B (2011). Available: http://censusindia.gov.in/2011census/dchb/0926_PART_B_DCHB_LUCKNOW.pdf
7. Singh, O.P., Gupta, J.P., Warsi, A.H.: Climate of Lucknow. Indian Meteorological Department

Vehicular Emission Inventory of Lucknow

Tauqeer Alam and A. K. Shukla

Abstract Lucknow is amongst the most urbanized and rapidly growing cities in India. Based on the bottom-up approach, the COPERT 5 model was used to generate the emission inventory of the city. The pollutants CO, NO_X, HC, $PM_{2.5}$, SO_2, and CO_2 were considered, and emission for the year 2010–2018 was calculated. The vehicles' categories chosen for the study include cars, HCV, LCV, bus, motorcycle, and scooter. After the validation of the results with previous estimates of the city, the emission for the years 2020 and 2022 is predicted for different pollutants under each category of vehicles. CO and NOx are the major concerned air pollutants. It is predicted that besides the scheduled implementation of BS-VI emission standards from April 2020, the total emission in the city would increase due to the continuous growth of the city and thus increasing the number of two-wheelers and four-wheelers constituting almost 90% of the total vehicular population.

Keywords Lucknow · Emission · COPERT 5 · Bharat stage

1 Introduction

Air pollution is the release into the atmosphere of various gases, finely divided solids, or finely dispersed liquid particles at rates that exceed the natural capacity of the environment to dissipate and dilute or absorb them (Automotive Research Association of India 2011). With the consequences of significant impacts on the environment and human health, air pollution has received increasing research concerns [1]. The vehicular pollution is majorly caused due to the fuel combustion, which forms the

T. Alam (✉)
Environmental Engineering, Institute of Engineering and Technology, Sitapur Road, Lucknow, India
e-mail: alamtauqeer98@gmail.com

A. K. Shukla
Civil Engineering Department, Institute of Engineering and Technology, Sitapur Road, Lucknow, India
e-mail: arvind.shukla@ietlucknow.ac.in

L. M. Gupta et al. (eds.), *Advances in Civil Engineering and Infrastructural Development*, Lecture Notes in Civil Engineering 87,
https://doi.org/10.1007/978-981-15-6463-5_44

exhaust emission, as well as due to the evaporation of the fuel itself. The share of road transport in the total transport activity has been increasing in recent years. The biggest rise in road traffic has been noted with passenger cars. Consequently, air pollution emission from transport with their harmful effects on the man and the environment has also increased [2]. The pollutants like carbon monoxide (CO), oxides of nitrogen (NO_X), particulate matter (PM), methane (CH_4), carbon dioxide (CO_2), and hydrocarbons (HC) or volatile organic carbons (VOCs) are significantly present in emission from road transport [3]. Despite the implementation of Bharat Stage (BS) standards due to the increase in vehicle population of the country from 55 million in 2001 to 210 million in 2015, vehicular pollution is increasing continuously.

The objective of the study is to find the trend of major pollutants emitted from different vehicle categories in Lucknow from the year 2010 to 2018 and to evaluate the expected load of vehicular pollutants in the years 2020 and 2022 considering the implementation of BS-VI emission standards from April 2020. The pollutants included in the study are CO, NO_X, HC, $PM_{2.5}$, SO_2, and CO_2. The categories of vehicles, viz. passenger cars (petrol cars and diesel cars), heavy commercial vehicles (HCV), light commercial vehicles (LCV), diesel bus, motorcycle, and scooter have been considered for the study.

2 Study Area

The study area selected is Lucknow, the capital of Uttar Pradesh. Lucknow is one of the developing cities and is quickly rising marketing hub. Lucknow encounters a warm and moist subtropical meteorology alongside cool and dry autumn. In the late summer season, extreme temperatures go up to 45 °C. In winter, the normal day-by-day least temperatures go down to 7 °C in the month of January.

3 Methodology

COPERT 5 is used as the tool for the analysis of the vehicular emission of the city. It is one of the standard emission frameworks for vehicular emission developed by the European Environment Agency [4]. COPERT needs statistical parameters as its major input [5]. The fundamental parameters for the COPERT incorporate vehicle type, emission standards, annual vehicle kilometre travelled (VKT), fuel quality and annual fuel consumption, Reid vapour pressure, vehicle population and meteorological conditions, for example, temperature and humidity [6]. The basic equation used in developing COPERT model is stated in Eq. (1).

Quantity of pollutant m for vehicle n,

$$Q_{m,n} = \sum_i \sum_j \left(P_{i,j} \times \mathrm{VKT}_{m,i} \times \mathrm{EF}_{i,j,n}\right) \tag{1}$$

where i is the vehicle type; j is the local emission standard; $_{i,j}$ is the number of vehicles in category i with emission standard j; VKT_i is the annual average vehicle kilometres travelled (km) for vehicles category i; and $EF_{i,j,n}$ is the emission factor in g/km for pollutant n emitted from vehicle category i with emission standard j.

The vehicle registration data have been obtained from Regional Transport Office (RTO), Lucknow, and fuel quality and annual fuel consumption have been taken from the annual reports of the Council of Scientific and Industrial Research—Indian Institute of Toxicology Research (CSIR-IITR), Lucknow [7] and VKT, service life, obsolescence period, meteorological data from CPCB [8]. Annual VKT is studied in order to generate the total vehicular activity in the city over a particular year. It is assumed that VKT chosen for analysis remains constant over the year. In India, BS emission factors (EFs) have been used for analysis which is equivalent to Euro EFs. The detailed division of the vehicle with respect to engine classes and fuel type has been done, and it has been found that the two-wheeler covers 79% and passenger cars are 16% of the total vehicle population in the year 2018 almost same composition over the years as shown in Fig. 1. The bifurcation of passenger cars on the fuel basis has been done according to the report of CPCB [8].

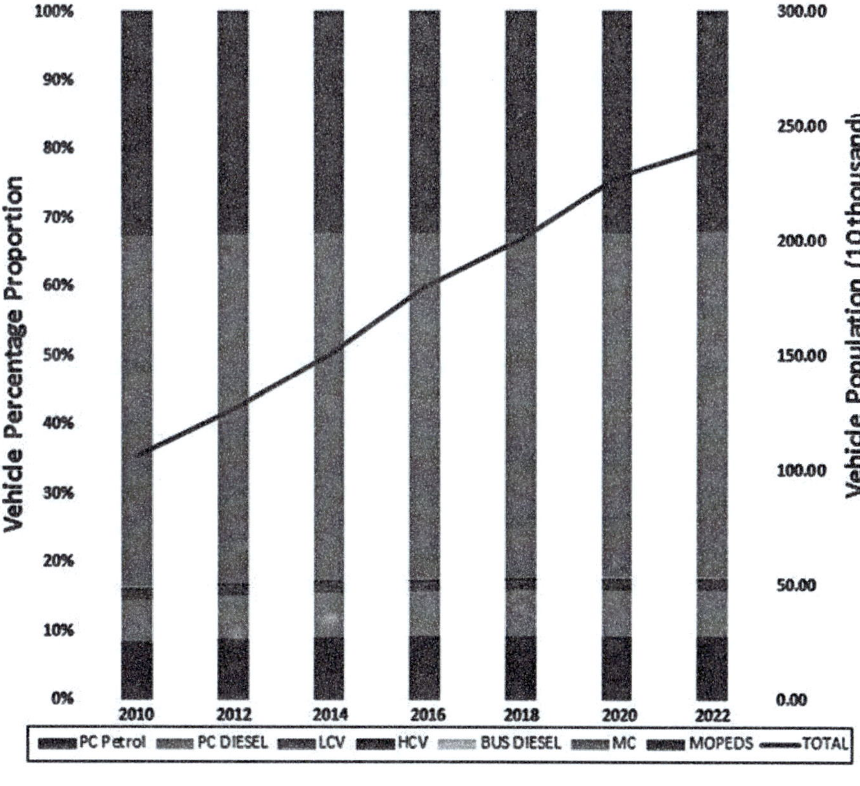

Fig. 1 Vehicle population and percentage proportion from year 2010 to year 2022

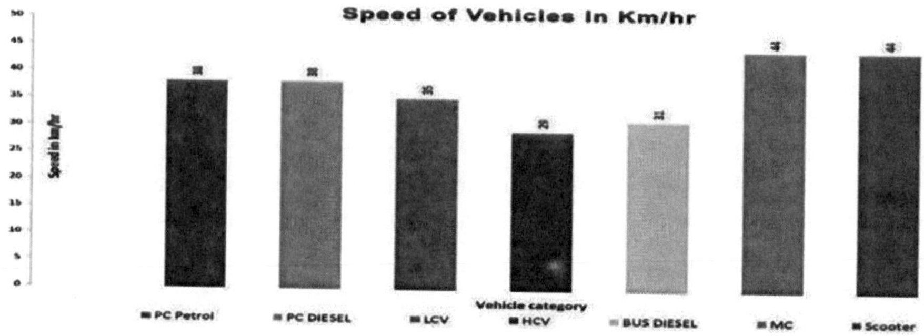

Fig. 2 Average speed of vehicles on roads in Lucknow

Table 1 VKT and service life of vehicles [8]

Vehicle	Motorcycle	Scooter	Car	Bus	LCV	HCV
VKT/day	49	39	54	113	90	49
Life (years)	10	10	15	12	11	11

To assess the properties of the fleet, the traffic survey has been done and the speed has been measured by using a radar gun at major roads of the city, and the average speed for each category of vehicles has been shown in Fig. 2.

The VKT is the distance travelled by a vehicle per day. The emission of a pollutant can be obtained by using the following equation:

$$\text{Pollutant}\,Q\ (\text{in g}) = \text{VKT} * \text{Emission Factor (in g/km)} \tag{2}$$

The parameters such as VKT, service life of the vehicle, minimum and maximum temperature, humidity, fuel properties, such as density, energy content, HC ratio, sulphur content, and total fuel sales have been used as input to the COPERT 5 model which has been shown in Table 1. Although over the years the VKT has been changed, it is assumed to be constant in the study and also the average of past 10 years data has been done for temperature and humidity.

The monthly data of minimum temperature, maximum temperature, and humidity are also used as input to the program which is shown in Fig. 3. It is presumed that the physical conditions of the roads will remain the same over the years. It is also assumed that the variation of exhaust with respect to temperature is neglected.

Some other inputs to the COPERT 5 program are fuel consumption, percentage vehicle with AC, AC usage, number of axles for commercial vehicles, Reid vapour pressure, etc. [7, 8].

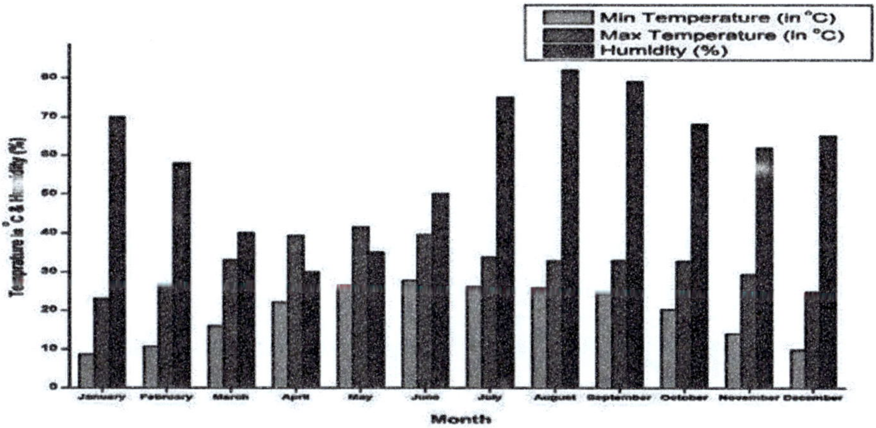

Fig. 3 Variation of minimum temperature, maximum temperature, and humidity

4 Results and Discussion

The emission results from the COPERT program can be obtained either in graphic user interface as shown in Fig. 4 or in form of Excel sheet by exporting it. The screenshot of one of the results obtained by COPERT 5 is shown in Fig. 4 order to authenticate the study.

The emission profile of each pollutant for the year 2010–2022 is shown in Fig. 5. It is observed that two-wheelers are the major emitters of CO followed by LCV. In 2010, CO was 12,000 ton/year which will raise up to 24,000 ton/year in 2022 due

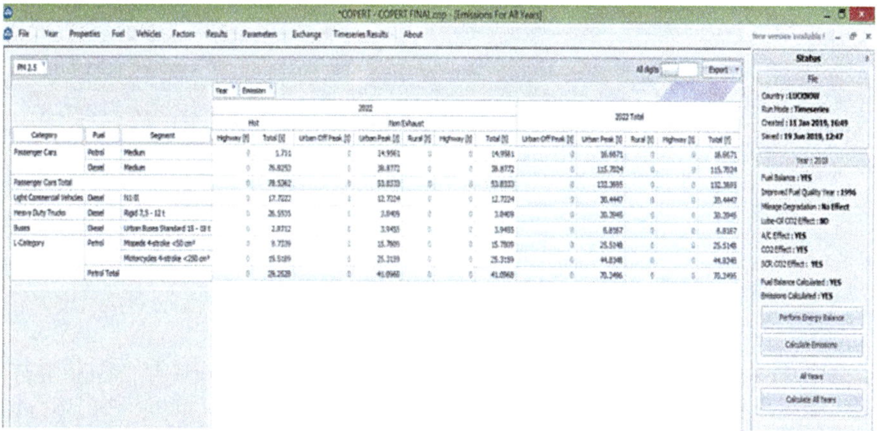

Fig. 4 Screenshot of one result obtained on COPERT graphic user interface

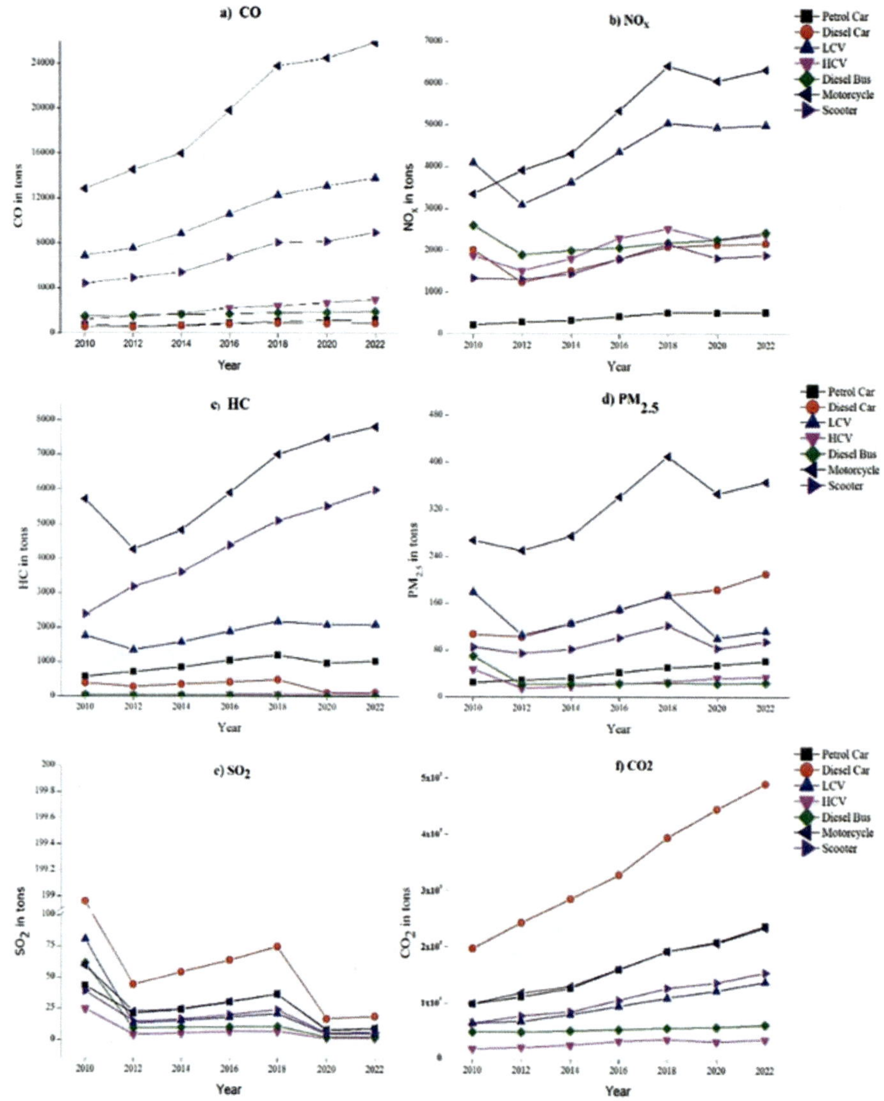

Fig. 5 Showing the variation of each pollutants **a** CO, **b** NO_X, **c** HC, **d** $PM_{2.5}$, **e** SO_2, and **f** CO_2 for different categories of vehicles for the years 2010–2022

to increase in two-wheelers population by 200%. It is also observed that the rate of increase in emission of CO would decrease due to the implementation of BS-VI norms from April 2020.

The predicted annual emission of each pollutant and their percentage contribution for the year 2022 under each category is shown in Table 2. Two-wheelers would be

Table 2 Predicted annual emission estimates of pollutants and percentage emission by each category for the year 2022

Pollutant	PC petrol (%)	PC diesel (%)	LCV (%)	HCV (%)	Bus (%)	Motorcycle (%)	Scooter (%)	Total emission 2022 (in tons)
CO	2	2	24	5	4	47	16	55,847
NO_X	2	10	24	12	11	31	10	20,698
HC	7	3	14	–	–	44	32	17,160
$PM_{2.5}$	5	18	18	3	2	42	12	928
SO_2	17	36	10	3	5	17	12	54
CO_2	17	36	10	3	5	17	12	22,609,953

the most prominent pollution source for the emission of CO. NO_X, HC, and $PM_{2.5}$ as it composes almost 71% of the total vehicle population in the city and higher VKT.

It is observed that 41% of the total NO_X emission is governed by two-wheelers followed by 35% from both LCV and HCV Fig. 6b. It is also expected that the total NO_X emission is going to remain almost constant during the year 2018–2022 due to the enhanced BS-VI technology and emission standards despite the increase in vehicle population by 20%.

Emission of HC is due to improper fuel burning due to continuous acceleration and deacceleration in the city traffic. Almost 78% of the total HC emission is constituted from two-wheelers alone. Passenger cars constitute 10–12% of the total HC emission. $PM_{2.5}$ is abundantly emitted from diesel cars and motorcycles due to unburnt fuel as exhaust emission (EE) and non-exhaust emission (NE) from the tyre wear, break wear, road surface wear, and resuspension of road dust. Due to the implementation of BS-VI, the EE is further going to be reduced in the year 2020, but the NE would

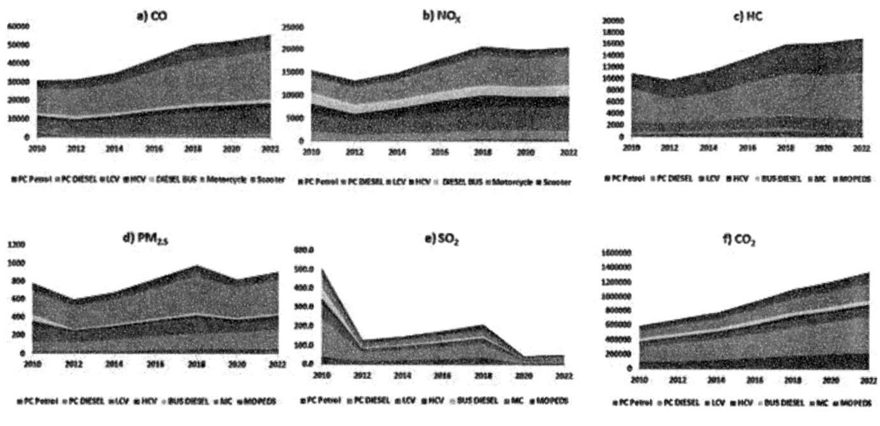

Fig. 6 Emission profile of **a** CO, **b** NO_X, **c** HC, **d** $PM_{2.5}$, **e** SO_2, and **f** CO_2 over the years from 2010 to 2022

be unaffected by BS-VI because it was assumed that the conditions of roads were unchanged in the study duration. The load of $PM_{2.5}$ would be increasing for diesel car rather than decreasing for the year 2020–2022 because of higher ratio (NE: EE) of non-exhaust emission in the total $PM_{2.5}$ emission with respect to that of the total emission in 2018–2020. 36% of the total SO_2 is generated from diesel cars alone followed by 18–22% by buses and commercial vehicles. The emission factor of SO_2 is 50 ppm in BS-IV, which is going to be cut by five times to 10 ppm in BS-VI in April 2020. Hence, the amount of SO_2 will be reduced by 60% in 2020. CO_2 is a potent greenhouse gas, 53% of it is being emitted from petrol and diesel cars alone followed by 29% from two-wheelers. It has increased rapidly by 2.5 times from 18,000 tons/year in 2020 to 45,000 tons/year in 2018 and is predicted to be 50,000 tons/year in 2022. The huge amount of CO_2 is generated due to the two-way catalytic converter used in vehicles in BS-IV standards.

4.1 Comparison of Emission Estimates with Previous Year Estimates in Lucknow

Comparison has been made with the estimation obtained by CPCB using IVE model for Lucknow as shown in Table 3. The CPCB emissions have been calculated by considering BS-III EFs, whereas in the study currently applicable BS-IV and BS-VI EFs are used. It can be seen in Table 3 that all the pollutants have increased except CO_2 due to the continuous growth in the number of vehicles from 2013 to 2018. The reason in the reduction of the concentration of CO_2 might be due to limited variables taken in the study, fewer resources, and lesser experimental data as compared to the detailed project and expertise methodology adopted by CPCB for IVE model.

Table 3 Comparison of our estimates to CPCB's previous estimates of Lucknow

Pollutants	CPCB estimates (2012-13) (tons/day)	Our emission estimates 2014 (tons/day)	Our emission estimates 2022 (tons/day)
CO	86.7	138.44	153
NO_X	37	57.33	56.7
HC	34.4	44.11	47
$PM_{2.5}$	3.6	2.69	2.54
SO_2	0.2	0.57	0.15
CO_2	4399.1	3033.41	6194.5

5 Conclusion

The emission inventory has been generated using the COPERT 5 program which is a bottom-up approach. It would be helpful in pollution modelling of developing cites like Lucknow. The inventory has been generated from the year 2010 to the year 2018, and the values for the year 2020 and the year 2022 have been predicted. The analysis shows that the prime pollutants deteriorating the environment of the city are CO and NO_X because 90% mobility of the vehicle is limited only to two-wheelers and passenger cars in the city. The most abundant pollutant CO, 63% of it, is emitted from two-wheelers; 53% of CO_2 is emitted from cars. Two-wheelers are substituting 76% of the total HC emission. From a fuel perspective, gasoline vehicles contributed to the largest share. It is observed that although the implementation of BS-IV standards contributed to the reduction in the rate of emission per vehicle, it could not meet an overall reduction in the total emission due to increasing motorized activity. Hence, the implementation of BS-VI would be helpful in reducing the rate of pollutants generating with the continuous growth of vehicles. If the emission of CO and HC is to be reduced, the two-wheelers would have to be minimized. Similarly, addressing NO_X and PM requires technological improvement in LCV, HCV, and bus.

References

1. Sharma AR, Kharol SK, Badarinath KVS (2010) Influence of vehicular traffic on urban air quality—a case study of Hyderabad, India. Transp. Res. Part Transp. Environ. 15(3):154–159
2. Kumari R, Attri AK, Gurjar BR (2011) Impact of CNG on emissions of PAHs and PCDDs/Fs from the road transport in Delhi. Atmospheric Pollut. Res. 2(3):394–399
3. Hu X, Xu D, Wan Q (2018) Short-term trend forecast of different traffic pollutants in Minnesota based on spot velocity conversion. Int. J. Environ. Res. Public. Health 15(9):1925
4. Ntziachristos, L., Gkatzoflias, D., Kouridis, C., Samaras, Z.: COPERT: a European road transport emission inventory model. In: Athanasiadis, I.N., Rizzoli, A.E., Mitkas, P.A., Gómez, J.M. (eds) Information Technologies in Environmental Engineering, pp. 491–504. Springer Berlin Heidelberg, Berlin (2009)
5. Ahlvik, S.E.P.: COPERT II Computer Programme to calculate Emissions from Road Transport: Methodology and Emission Factors. European Environment Agency, Technical Report
6. Gržini, D., Mihaljevi, G., Omeragi, A., Vinkovi, N.: COPERT III—the Effect of Estimated Parameters on the Total Emission of Road Transport in the Republic of Croatia, p. 11
7. Ahmed, M.T., Shukla, M.P., Ahmed, M.K., Pandey, M.B.M.: IITR Monsoon report, p. 39 (2018)
8. CPCB (2015) Status of pollution generated from road transport sector in six mega cities. Central Pollution Control Board, Government of India, New Delhi, India. https://www.google.com/urlsa=t&source=web&rct=j&url=http://www.indiaenvironmentportal.org.in/files/file/Report_Status_RoadTransport_SixCities.pdf&ved=2ahUKEwi6vv7h3cDqAhWPcn0KHe5OCNAQFjACegQIChAB&usg=AOvVaw1SITbynMdDiIV3jTelkEZQ

Inactivation of *E. coli* Present in Lake Water Using UV Reactor

Isha P. Khedikar and Abhijeet Ashok Paidalwar

Abstract Microbial contaminated water is the main cause for many water-borne diseases in developing countries. To make contaminated water microbial safe for drinking, disinfection is needed. Hence, this research work was carried out to disinfect lake water containing *E. coli* by using UV light. Four experimental setups were designed, i.e., UV reactor. Initially, the uncovered UV reactor (pair of 6 W UV light) was fabricated to disinfect the lake water for inactivation of *E. coli*, Later on, the UV reactors were modified three times by reflective mirrors (Modified 1), increased length and decreasing height and increased UV intensity (Modified 3). All experimentation were carried out at varying contact time, viz. 1:30 min, 2:30 min and 3:15 min. Raw lake water was characterized for and physicochemical and bacteriological parameters, viz. turbidity (1–6 NTU), temperature (23.9–25.4 °C), pH (6–8.5) and *E. coli*. Average *E. coli* in raw lake water was found to be 2400 MPN/100 ml, which exceeds the permissible limit as per IS:10500, (2012), Results of the investigation indicated that increased UV intensity (Modified 3) showed 99.99% inactivation, i.e., 3-log reduction.

Keywords Disinfection · *E. coli* · Inactivation · UV reactor · Ultraviolet radiation

1 Introduction

Provision of clean, safe drinking water is of paramount importance for human health. Lack of safe water, sanitation and health care affects billions of human beings [1]. It is essential to do disinfection to make water free from pathogenic microorganism and to make water safe for drinking. Various disinfection methods were attempted by the

I. P. Khedikar (✉)
Department of Civil Engineering, G. H. Raisoni College of Engineering, 440016 Nagpur, MS, India
e-mail: isha.khedikar@raisoni.net

A. A. Paidalwar
Solrun Energy Pvt. Ltd., Nagpur, MS, India
e-mail: abhijeetpaidalwar@gmail.com

L. M. Gupta et al. (eds.), *Advances in Civil Engineering and Infrastructural Development*, Lecture Notes in Civil Engineering 87,
https://doi.org/10.1007/978-981-15-6463-5_45

researcher. Worldwide chemical disinfectant chlorine is used [2], but it gives rise to disinfection by-products [3]. In the recent decades, UV disinfection is an established water treatment technology. Ultraviolet (UV) disinfection is effective in inactivating bacteria, protozoan parasites depending upon UV dosage [4]. It offers advantages like environmentally friendly, contact times are small, no unpleasant taste or odor. UV disinfection does not alter the water quality, inactivation of microorganism is independent of pH and temperature, no chemicals are needed, relatively inexpensive and easy to operate, no transportation, storage or handling of chemicals, small space requirement for installation [5].

In typical UV disinfection system, UV lamp is installed in the reactor through which contaminated water is passed which causes damage to genetic molecules [6]. Because of this damage to microorganisms, it prevents replicating in a human or animal body. UV light wavelength (100–400 nm) has lethal effect on microorganism; therefore, microbial inactivation is achieved. In this research work, lake water contaminated with *E. coli* (2400 MPN/100 ml) was considered for study. It is necessary to disinfect contaminated water as per IS code 10500:2012 to make it free from pathogenic microroganisms. This study is aimed to develop household UV reactor. Hence, an experimental setups were designed, i.e., uncovered UV reactor, modified UV reactor with reflective mirror (Modified 1), modified UV reactor with reflective mirror increased length and decreased height. (Modified 2), modified UV reactor with reflective mirror with increased intensity (Modified 3). For each set of experimentation percentage, inactivation was determined which subsequently helps to improve development UV reactor.

2 UV Light

According to photochemistry, in electromagnetic spectrum, UV light lies between X-rays and visible light. The spectrum can be divided into four ranges: UV-A (315–400 nm), UV-B (280–315 nm), UV-C (200–280 nm) and vacuum UV (100–200 nm) [5, 7], UV-A and UV-B reaches the earth's surface [8]. UV-C component of solar radiation does not reach the earth as it was absorbed by ozone within the stratosphere [9, 10]. Only UV-C radiation, at approximately 260 nm, has the greatest effectiveness because it corresponds to maximum absorption by DNA [2].

The UV-C range is so-called germicidal range since it is absorbed by proteins, ribonucleic acid (RNA) and deoxyribonucleic acid (DNA) and can lead to cell mutations or cell death; therefore, it is effective in inactivating pathogens [5]. The inactivation rate of microorganism increases with decrease in magnitude of wavelength as UV A < UV B < UV C [10].

3 Historical Background of UV Light

- Johann Ritter (Pharmacist) in the year 1801 discovered UV light by proving that silver chloride is decomposed by the invisible rays beyond the violet.
- Niels Finsen (Physician) in the year 1900 discovered a UV treatment for lupus vulgaris, a form of skin tuberculosis. He received the Nobel Prize in physiology or medicine for this discovery.
- Bernard and Morgan in the year 1903 found that for inactivation of bacteria around 250 nm is most sensitive wavelength.
- Henri et al. [11] described in Marseilles, France, the first commercial UV facility which utilized UV light to disinfect water.
- Gates [12] was the first to carry out detailed investigations of the action spectrum of *E. coli*; he showed that the optimal wavelength for inactivation was about 260 nm.
- Department of health, education and welfare (DHEW) in 1966 issued guidelines for disinfection by UV.
- In Norway, 1975, UV disinfection was introduced; because of this, concerns were expressed for disinfection by-products formed due to use of chlorine as disinfectant.
- N.W. Bergen wastewater treatment plant (1978) was installed successfully using UV system. In Europe (1985), over 1500 UV installations were done for the treatment of groundwater.
- Bolton et al. [13] found that UV was very effective against cryptosporidium.

4 Methodology

Water from an established Ambazari Lake, Hingna Road, Nagpur, Maharashtra, India, (21.1287° N, 79.0405° E) was collected for the research work. The lake water was supplied to the 13.5 MLD water treatment plant in MIDC, Hingna, to the industrial area. The characterization of raw lake water was done for different seasons of ten day's interval by standard equipments and methods as per APHA standards (1998). The physicochemical and bacteriological characterization of raw water was done for parameters turbidity, pH, temperature and *E. coli* before and after disinfection. Further, the obtained results were compared with the drinking water standard IS:10500 2012. The physicochemical parameters, viz. turbidity, pH and temperature were in permissible limit except bacteriological parameter, viz. *E. coli*. Initially, *E. coli* concentration of raw water samples was tested by a standard most probable number (MPN) as per APHA standards (1998). Lake water was passed to all the designed UV reactor with the help of peristaltic pump; water was collected at the outlet of the continuous flow reactor and tested for the final concentration of *E. coli* by MPN test for duration of November 1, 2015 to February 29, 2016. The complete setup of UV reactor is shown in Fig. 1.

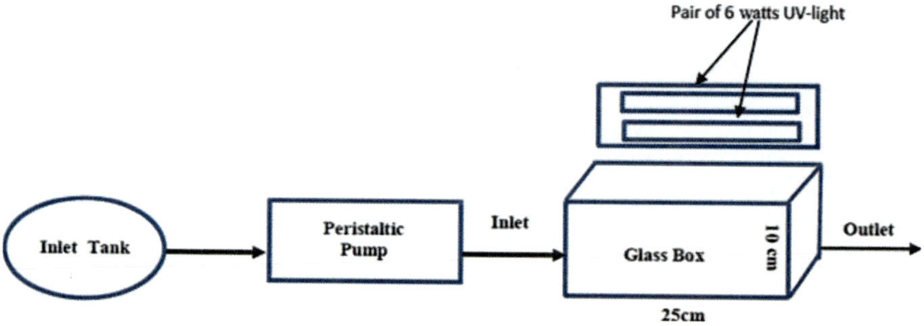

Fig. 1 Schematic diagram of experimental setup

5 Experimental Setups

A low-pressure (LP) Philips UV lamp of 6 and 11 watts is housed in a UV reactor connected in parallel on direct power supply. The lamp is of 15 cm long and 1.5 cm diameter. In this project, peristaltic pump is used to pass the water through the UV reactor at same flow rate. The UV reactor has been modified three times to achieve the maximum inactivation of *E. coli*. The experimental setup contained an uncovered glass box model of size 25 cm × 10 cm × 10 cm having two 6 W low-pressure UV germicidal lamp connected in parallel (Fig. 2), which emit monochromatic wavelength. The lamp is attached to the upper cover of the box with distance of lamp from the water level was 7 cm inside the UV reactor.

As the result is not satisfactory, the same experimental setup glass box of size 25 cm × 10 cm × 10 cm has been modified by providing reflecting mirror from inside the box to the whole area, i.e., covered UV reactor, provided with the same pair of UV lights of 6 W (Fig. 3). The distance of light from the water level is 7 cm. Lights emitted from UV-lamp was not properly concentrated on the water surface, and it gets pass through the glass box; hence, 60–70% lights were wasted, and due to this, the disinfection of water was not done effectively by UV lamps and *E. coli* not inactivated completely from water.

Fig. 2 Uncovered UV reactor

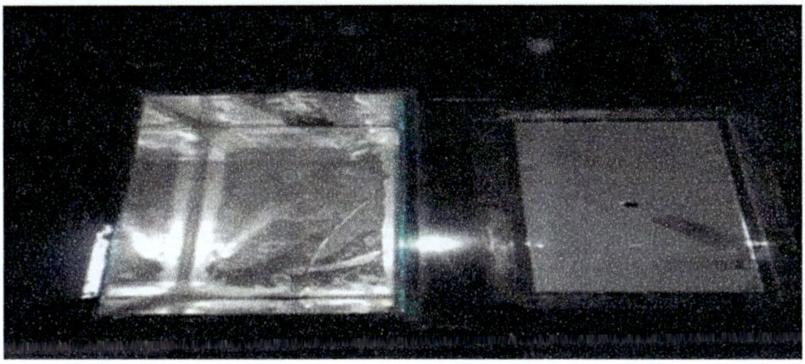

Fig. 3 Modified UV reactor with reflective mirrors (Modified 1)

Fig. 4 Modified UV reactor with increasing length and decreasing height (Modified 2)

As only few removal of *E. coli* has been observed, a new experimental setup has been fabricated with increasing length and decreasing height of setup. The new modified experimental setup is of size 32 cm × 10 cm × 5 cm with reflective glasses from inside to the whole area provided with a pair of 6 W UV light (Fig. 4). The distance of UV light inside the continuous flow reactor from water level was 2 cm.

As to achieve the 100% removal of *E. coli* from water, the same experimental setup of size 32 cm × 10 cm × 5 cm has been again modified by providing a two pair of UV lights of 6 and 11 W in series, to increase the intensity by UV light inside the UV reactor. The distance of UV light from water level is 2 cm (Fig. 5).

6 Results and Discussion

The secondary objective is to evaluate the effect on various physicochemical parameter by UV disinfection and effect caused by the physicochemical parameter on UV disinfection whose raw water characteristics are between 6 and 8, and as the water flows through the UV reactor, bacteria are exposed to 265 nm ultraviolet light radiation which causes damage to genetic molecules.

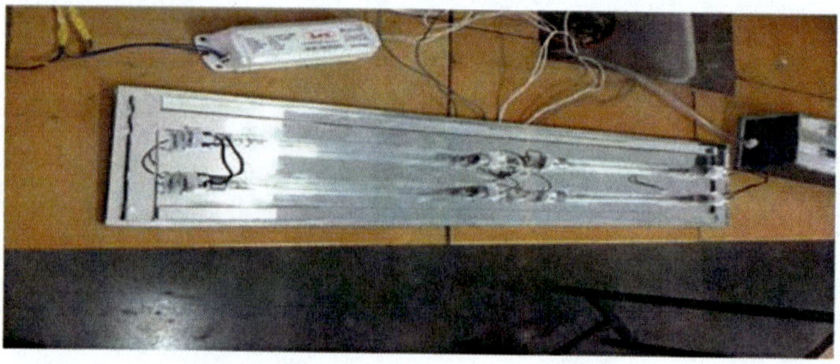

Fig. 5 Modified UV reactor with increasing intensity (Modified 3)

The characterization results obtained from water sample, i.e., lake water, turbidity (1–6 NTU), pH (6–8.5) and temperature (24.12 ± 1.2) were obtained within permissible limits as per IS: 10500, (2012), and hence, pre-treatment was not given to the raw water. The result from the uncovered UV reactor is not satisfactory as there is no inactivation of *E. coil* (Fig. 6) Most of the radiations from UV light get refracted outside the UV reactor. Therefore, the experimental setup is modified to increase the efficacy of inactivation of *E. coli.* The result from the modified UV reactor with reflective mirrors (Modified 1) showed some inactivation of *E. coli* at the contact time of 3:15 min (Fig. 7). The inactivation of *E. coli* comes down to 1600 MPN from 2400 MPN of raw water sample. The result does not show any inactivation for the contact time of 1:30 min and 2:30 min because of less detention time. The radiation in covered UV reactor is reflected back because of reflective mirrors, which increase the radiation and temperature inside the covered UV reactor. The modified UV reactor with increasing length and decreasing the height (Modified 2) shows good result with

Fig. 6 Inactivation of *E. coli* from uncovered UV reactor for the month of November, 2015

Fig. 7 *E. coli* in modified UV reactor with reflective mirrors (Modified 1) for the month of December, 2015

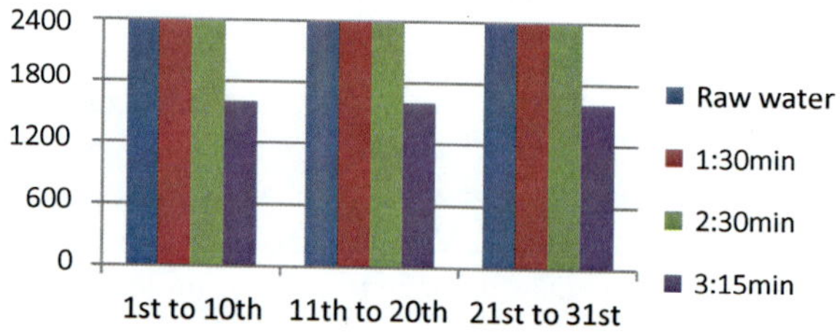

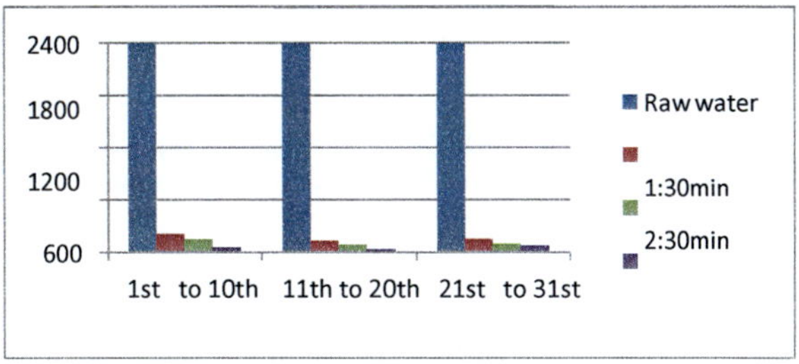

Fig. 8 *E. coli* from modified increased length and decreased height UV reactor (Modified 2) for the month of January, 2016

an average 90% inactivation of *E. coli* (Fig. 8). Due to increasing the length of UV reactor, the detention time increases and the water is subjected to longer time for inactivation, the height of the UV reactor is decreased by 5 cm, so that the radiation penetrate more in the water and increase the efficiency of inactivation. Modified UV reactor with increasing intensity (Modified 3) shows 99.99% inactivaiton of *E. coli* at the end of contact time 3:15 min.

7 Conclusion

This work shows that the UV disinfection by the UV light can be used for disinfection of water. The UV reactor performed well and disinfected 99.99% of *E. coli* under visible light irradiation at a contact time of 1:30 min, 2:30 min and 3:15 min of modified 3 UV reactors, i.e., increased intensity. Turbidity influences the disinfection efficiency of UV light and slightly decreases after disinfection. The variation of pH does not affect the disinfection by UV light. The pH slightly increases after disinfection. The increase in temperature increases the inactivation of *E. coli.*

References

1. Roman RMS, Soares AA, de Matos AT, Sediyama GC, DeSouza O, Mounteer AH (2007) Domestic wastewater disinfection using solar radiation for agricultural reuse. Am Soc Agric Biol Eng 50(1):65–71
2. Burch JD, Thomas KE (1998) Water disinfection for developing countries and potential for solar thermal pasteurization. Sol Energy 64(1–3):87–97
3. Bitton G (2005) Wastewater microbiology, 3rd edn. Wiley, New York
4. Rahmani AR, Samarghandi MR, Samadi MT, Nazemi F (2009) Photocatalytic disinfection of coliform bacteria using UV/TiO_2. J Res Health Sci 9:1–6
5. Clarke S, Bettin W (2006) Ultraviolet light disinfection in the use of individual water purification devices. U.S. Army Centre for Health Promotion and Preventive Medicine (USACHPPM)
6. Bhimate SJ, Kulkarni AD, Tolpadi A (2013) Disinfection of water containing *E. coli* by photocatalytic route using TiO_2/UV/visible light source. Int J Sci Res Manag Stud 1(3):80–87

7. McGuigan KG, Conroy RM, Mosler HJ, Preez MD, Jaswa EU, Ibanez PF (2012) Review solar water disinfection (SODIS): a review from bench-top to roof-top. J Hazard Mater 235–236:29–46
8. Chau A, Giang K, Leung M, Tam N (2008) Role of Rec A in the protection of DNA damage by UV-A in *Escherichia coli*. J Exp Microbiol Immunol 12:39–44
9. Navntoft C, Dawidowski L, Blesa MA, Fernandez-Ibanez P, Wolfram EA, Paladini A (2009) UV-A (315–400 nm) irradiance from measurements at 380 nm for solar water treatment and disinfection: comparison between model and measurements in Buenos Aires, Argentina and Almeria, Spain. Sol Energy 8:280–286
10. Wegelin M, Canonica S, Mechsner K, Fleischmann T, Pesaro F, Metzler A (1994) Solar water disinfection: scope of the process and analysis of radiation experiments. J Water SRT Aqua 43(3):154–169
11. Henri V, Helbronner A, Recklinghausen M de (1910) Nouvelles recherches sur la sterilization de grandes quantites d'Eau par les rayons ultraviolets. J Compt rend Acad Sci 151:677–683
12. Gates FL (1929) A study of the bactericidal action of ultraviolet light II: the effect of various environmental factors and conditions. J Gen Physiol 13:249–260
13. Bolton JR, Dussert B, Bukhari Z, Hargy TM, Clancy JL (1998) Inactivation of *cryptosporidium parvum* by medium-pressure ultraviolet light in finished drinking water. In: Proceedings of the AWWA annual conference. Denver, Colo.: AWWA

Evaluation of Fractal Growth Characteristic of Flocs for Aluminium Sulphate and Ferric Chloride Using Microscopy Method

P. S. Randive, D. P. Singh, A. G. Bhole, V. P. Varghese, and A. M. Badar

Abstract Coagulation and Flocculation processes are some of the important aspects of water treatment. The properties of flocs, such as size, structure, and strength have a significant effect on the solid and liquid separation process also it affects the basic operations of the industrial unit process. Floc structure is an important factor that influencing the coagulation effect and to identify the post-treatment load. The physical characteristics of the floc are therefore fundamental in determining their removal efficiency. Study of floc generated and its growth characteristics in the flocculation process plays an important role as it decides the removal policies, effect on subsequent processes, and ultimately the quality of water. The present investigation is to study the floc growth for two different aluminium and iron-based coagulants, Aluminium Sulphate and Ferric Chloride, respectively, using the microscopy method. This paper deals with the examination of two different kinds of flocs for strength, breakage, recovery, and floc growth rate.

Keywords Flocculation · Floc · Microscopy method

1 Introduction

In the water treatment plant, coagulation and flocculation is an important process for the removal of turbidity from water. The process of adding certain chemical called coagulants to water which reacts with impurity in order to form insoluble and gelatinous matter called flocs which can be precipitated quickly is known as coagulation. Addition of coagulants culminates in chemically charging collides so as to configurate substantially large molecule by accumulating all particles. In coagulation process, chemical coagulants are added into the water which reacts with alkalinity

P. S. Randive (✉) · D. P. Singh · V. P. Varghese · A. M. Badar
Department of Civil Engineering, KDKCE, Nagpur, India
e-mail: pallavi_randive@rediffmail.com

A. G. Bhole
Department of Civil Engineering, VNIT, Nagpur 440010, India

L. M. Gupta et al. (eds.), *Advances in Civil Engineering and Infrastructural Development*, Lecture Notes in Civil Engineering 87,
https://doi.org/10.1007/978-981-15-6463-5_46

to form gelatinous precipitated which is known as floc. Flocs are characterized as immensely porous, asymmetrical, saggy aggregate made up of smaller colloidal particles. Chowdhury et al. [1] studied the submicron colloid numbers, fluxes, and removals in two conventional water treatment plants. Also obtained the comprehensive study of supramicron (>1 p.m) particles and its behaviour. Concluded that removals of submicron colloids and supramicron particles ranged from 57 to 99% and 91 to 99%, respectively, considering two water sources from two conventional water treatment plants. The properties of flocs, such as size, structure, and strength have a significant effect on the solid and liquid separation process. The size, shape, and formation are the fundamental things that to be considered for the operation of industrial unit process. Dense flocs settle down easily. Therefore, knowing the physical characteristics of flocs subsequently increases its removal efficiency and results in low turbid treated water during settlement. To contemplate the necessity of floc structure and its behaviour in the flocculation process it is essential to analyze the floc structure, formation, and breakage in detail. There are various methods to measure floc size this is microscopy, light scattering, photography and image analysis, transmitted light, individual particle sensor, etc. Hurst [2] built an experimental apparatus to develop and understand floc blanket physics. In his study, he used visual insights floc blanket mechanics. Considered 1.3 cm thick section of floc blanket through which light intensity is transmitted. Experimental results show remarkable enhancement in particle removal efficiency and may drop operation and maintenance costs. Microscopy method is used to determine the floc parameter. For the determination of floc sizes microscopy is a widely used method. There are some limitations of microscopy method like it takes more time for sample preparation also required large sample size. Besides that, it is simple to perform, understand, and gives adequate information about the floc structure. This experimentation work aims to evaluate the characterize growth and regrowth ability of flocs with additional coagulant dosage strategy to explore the floc formation mechanism and floc characteristics using microscopy method. Ghanem et al. [3] studied ballasted flocculation processes to evaluate microscopic observations, Bench scale observations, density tests, and centrifugal settling tests. It gives information about the controlling mechanisms, interconnection between ballasting agent and coagulants for floc built up. Chakraborti et al. [4] used (DLA) mechanism and discussed the characteristics of flocs in flocculation process based on diffusion-limited aggregations. Microscopy is one of the commonly used floc sizing methods and carried out by precisely locating a selected sample particle from the suspension onto a microscope.

The floc strength factor is a measure of floc strength by exposing the floc to single level of increased shear rate within the carrying vessel and comparing the floc size before and after breakage. The floc recovery factor is an indication of the extent to which the floc size recovers during regrowth phase after breakage.

In the present study, two easily available and easily soluble coagulants are considered for experimentation and its sizes are measured for further calculations using microscopy method. General categorization of metal coagulants is aluminium-based and iron-based. For experimentation one from each category Aluminium Sulphate and Ferric Chloride coagulant is considered and floc strength, breakage, recovery factor as well as floc growth rate is determined and compared.

2 Experimental

2.1 Sample Preparation

Experimentation was conducted on artificial turbid water. Digital Nephelo Turbidity meter was used to measure the turbidity for each sample. After some trials, the required turbidity of 100 NTU was achieved by addition of 0.11 g/lit of kaolin clay in the water sample. Turbidity of 100 NTU was maintained throughout the experimentation for both the coagulants. Jar Test apparatus was used for determining the optimum coagulant doses for both Aluminium Sulphate and Ferric Chloride coagulant. Using pipette flocs formed during the jar test apparatus, procedure was collected steadily and observed under microscope. Diameter $d1, d2, d3$ of the above-collected sample is determined by placing them on glass slides and observing under a microscope.

2.2 Microscopy Method

Amongst various methods, microscopy refers to the most extensively used technique since long as a method for floc counting and sizing. [5, 6]. Pipette is widely used for collecting flocs from the treated sample. In this experimentation, the flocculated suspension was prepared in which a hollow glass tube was immersed and sealed at the top and then removed. The floc carried through the tube was slightly placed below the microscope for analysis. This method provides a good representative sample of the flocs without many ruptures. Microscopy method is the simplest method, at high magnification able to view the individual particles, examine, investigate and analyze it. This method is used to analyze floc shape and irregularity. Most popularly used method in many applications. Notable method in finding porosity and shape factor of flocs. Besides, microscopy is a comparatively inexpensive method to determine floc shape characteristics.

2.3 Coagulant Used

Aluminium Sulphate. Aluminium Sulphate is the most common coagulants used in waterworks. It chemical composition is $Al_2(SO_4)_3 . 18(H_2O)$. It requires the presence of alkalinity in water to form floc. Aluminium Sulphate (alum) added in water reacts

Table 1 Floc growth size of Aluminium Sulphate

Size 1 ($d1$) μm	Size 1 (d2) μm	Size 3 ($d3$) μm
200	120	173
266	133	186.2
293	146.3	213
Average = 253	Average = 133.1	Average = 191

with alkalinity present in it and leads to the formation of gelatinous precipitated of aluminium hydroxide which attracts fine suspended impurities over its surface does grow in size and easily get settled. It works on the pH range 6–8.5. Alum is widely used coagulant as it is more effective when pH of water ranges between 6.5 and 8.5. Its doses depend upon turbidity, colour, taste, pH value and temperature of water.

Ferric chloride. Ferric chloride is an inexpensive inorganic coagulant, as it is generated as waste material from steelmaking operations. However, it's by far the most corrosive and hazardous inorganic coagulant, and its use is limited to facilities equipped to handle it safely. pH sensitivity is somewhat less than alum. It works over a wide pH range 5–7. It is suitable for usage in the lime-softening process (pH 9).

Jar test apparatus is used for formation of flocs and also dosages were calculated considering 100 NTU turbidity for both the coagulants separately. After the experimental procedure of floc formation for both the coagulants diameter $d1$, $d2$, $d3$ is determined by using the microscopy method which is the simplest and accurate method of determining floc sizes. Details of diameter for Aluminium Sulphate are shown in Table 1 and images of flocs before, after breakage and recovery is shown in Fig. 1a–c.

Details of diameters for the coagulant Ferric chloride is shown in Table 2 and images of flocs before, after breakage and recovery is shown in Fig. 2a–c.

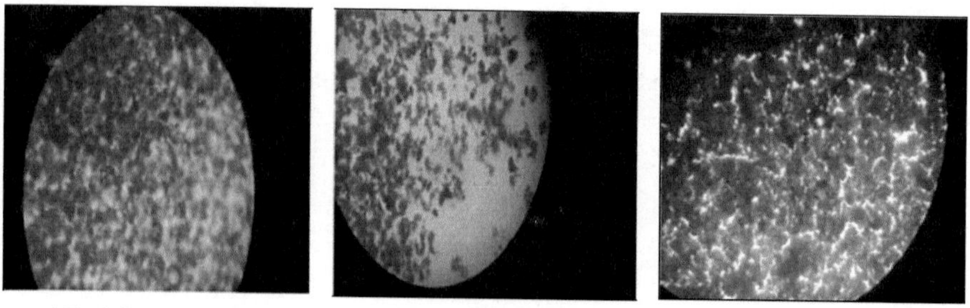

a.) Floc before breakage(d1) b) Floc after breakage (d2) c) Floc recovery (d3)

Fig. 1 Images of flocs of Aluminium Sulphate

Table 2 Floc growth size of ferric chloride

Size 1 ($d1$) μm	Size 2 ($d2$) μm	Size 3 ($d3$) μm
479	213	293
505.4	239.4	266
532	266	319.2
Average = 505.47	Average = 239.5	Average = 292.73

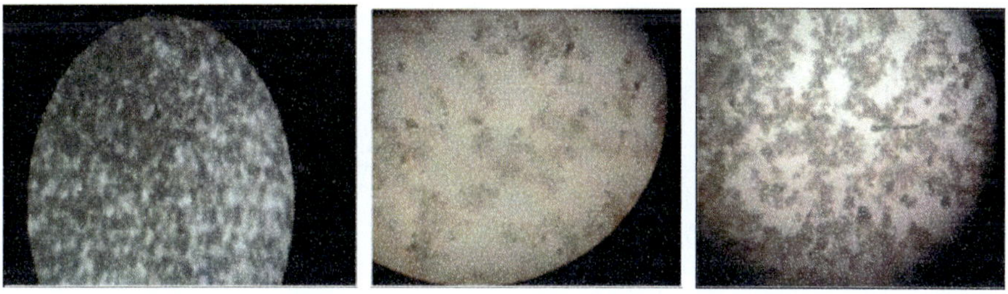

a) Floc before breakage(d1) b) Floc after breakage (d2) c) Floc recovery (d3)

Fig. 2 Images of flocs of ferric chloride

3 Results

3.1 Floc Strength Factor

The floc strength factor is a measure of floc strength by exposing the floc to single level of increased shear rate within the containing vessel and comparing the floc size before and after breakage. It is compared in Fig. 3.

From Fig. 3, it is clear that Aluminium Sulphate floc is more resistant to hydrodynamic shear force than that of ferric chloride flocs. As floc strength for Aluminium Sulphate floc is more than ferric chloride floc.

Fig. 3 Floc strength factor

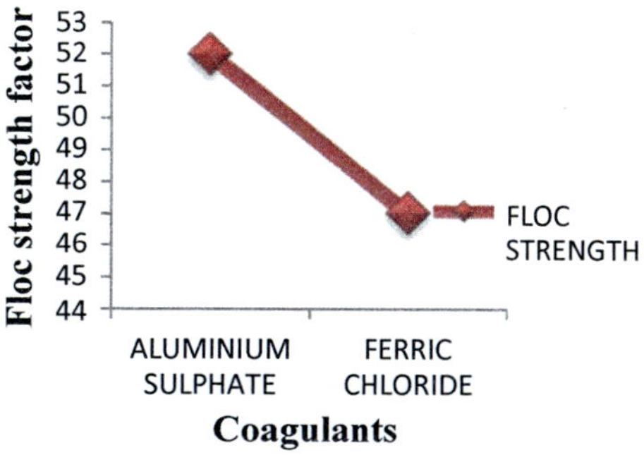

Fig. 4 Floc breakage factor

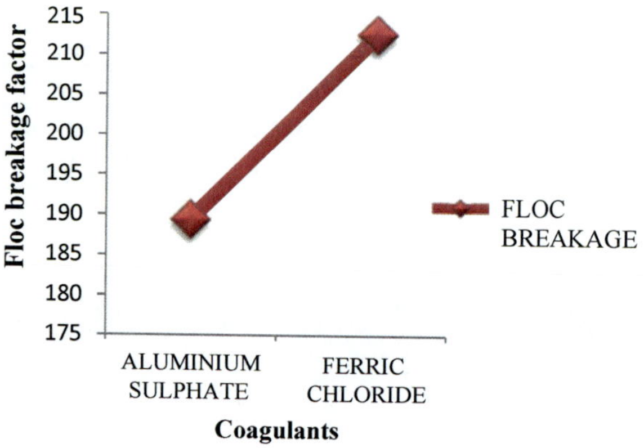

3.2 Floc Breakage Factor

Strength property of floc depends on the stability of floc in suspension and the number of bonds holding by the floc particles intogether. The floc breakage factor is compared in Fig. 4.

In the flocculation process, floc breakage plays a crucial role, the equilibrium of formation and breakage of floc results in the substantially stable floc size and structure. From Fig. 4, it is clear that ferric chloride flocs break more easily than Aluminium Sulphate flocs. As floc breakage factor is more for ferric chloride floc.

3.3 Floc Recovery Factor

The floc recovery factor is an indication of the extent to which the floc size recovers during the regrowth phase after breakage.

From above Fig. 5, it is clear that rate of recovery after breakage is more for Aluminium Sulphate flocs.

3.4 Floc Growth Rate

Floc growth rate is compared for both the coagulants as shown in Fig. 6.

From Fig. 6, it is clear that ferric chloride gives bigger and heavier floc than Aluminium Sulphate, hence the rate of settling will be more for ferric chloride flocs. It also shows that the rate of growth for floc is more for Aluminium Sulphate than ferric chloride floc. It also shows that floc growth time for ferric chloride is more than that of Aluminium Sulphate.

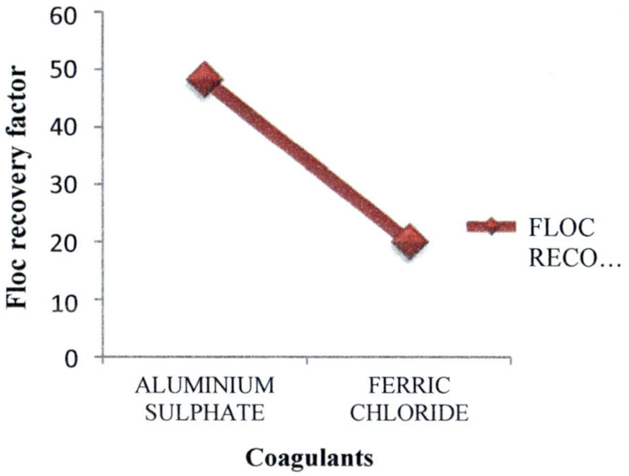

Fig. 5 Floc recovery factor

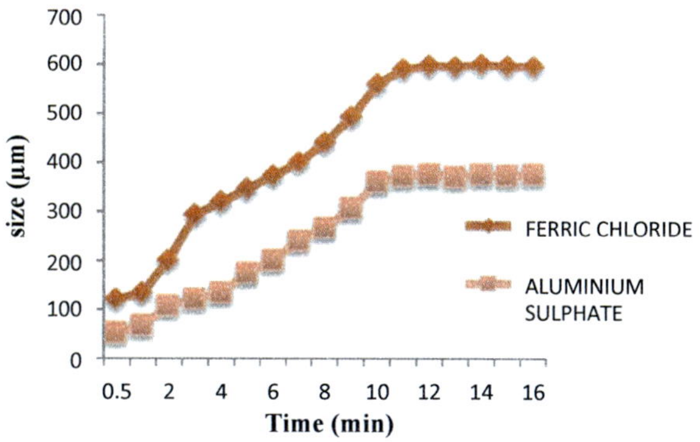

Fig. 6 Floc growth rate

4 Conclusion

Floc formation, breakage, and reformation were studied in detail which can give important indications for floc removal mechanisms. Experimental results were obtained from the microscopy method and they are compared considering both the coagulants Aluminium Sulphate and Ferric Chloride. The following conclusions were made.

1. The floc size for Ferric Chloride is more than Aluminium Sulphate floc. Therefore, the settling time for ferric chloride will be less than the Aluminium Sulphate floc. It will increase the efficiency of water treatment by reduction in settling time.
2. The floc strength and recovery is more for Aluminium Sulphate than Ferric Chloride coagulant. Hence Aluminium Sulphate flocs will be more resistant to hydraulic shear forces. Stability is more for Aluminium Sulphate floc.

3. The growth rate and growth time for Ferric Chloride are less than the Aluminium Sulphate. Hence Aluminium Sulphate floc requires less time for flocculation than ferric chloride, therefore, increase the efficiency for water treatment.
4. Cost of Aluminium Sulphate is less than that of Ferric Chloride. So Aluminium Sulphate is cost-efficient.
5. From the above results we can conclude that for floc removal, Aluminium Sulphate coagulant is more efficient than that of Ferric Chloride.

References

1. Chowdhury ZK, Amy GL, Bales RC (1993) Incorporation of submicron colloids into larger floc in Water Treatment. J Environ Eng 119(1):192–199
2. Hurst M (2014) Image analysis of floc blanket dynamics: investigation of floc blanket thickening, growth, and steady state. J Environ Eng 140(4):04014005
3. Ghanem AV, Young JC, Edwards FG (2007) Mechanisms of ballasted floc formation. J Environ Eng 133(3):271–277
4. Chakraborti RK, Atkinson JF, Van Benschoten JE (2011) Fractal growth characteristic of flocs in flocculation process in water treatment. In: ACS Publication environmental science and technology, vol 3
5. Aguilar MI, Saez J, Llorens M, Soler A, Ortuno JF (2003) Microscopic observation of particle reduction in slaughterhouse wastewater by coagulation-flocculation using ferric sulphate as coagulant and different coagulant aids. Water Res 37:2233–2241
6. Allen T (1997) Particle size measurement. In: Powder sampling and particle size measurement. Chapman and Hall, London

Estimation of Carbon Dioxide Emission from Vehicles in Lucknow

Ankit Kumar and A. K. Shukla

Abstract Emissions of greenhouse gases (GHGs) into the atmosphere have led to the rise in temperature of the planet. Increase in the concentration of GHGs is due to the emission from natural and anthropogenic activities. Various pollutants are emitted from the exhaust of vehicles including carbon dioxide (CO_2) which is a greenhouse gas, this has led to an increase in the concentration of anthropogenic CO_2 in the atmosphere. This study is conducted to determine the emission of CO_2 from the vehicles at four major roads of Lucknow. Traffic count survey is conducted to determine the traffic volume and its composition. Central Pollution Control Board (CPCB) emission factor is used for the estimation of emission. The result of traffic survey count shows that bikes are the major share of vehicles with 39% at Kanpur road, 47% at Faizabad road, 32% at Prayagraj road, and 37% at Sitapur road. The total concentration of CO_2 emitted is estimated to be 491,937 g/km/hr, 291,640 g/km/hr, 404,662 g/km/hr and 252,443 g/km/hr at Kanpur road, Sitapur road, Prayagraj road, and Faizabad road, respectively. Petrol cars and diesel cars are the major sources of CO_2 emission at all sites.

Keywords Greenhouse gas · Carbon dioxide · Vehicles · Emission · Lucknow

1 Introduction

The temperature around the world is rising. According to research conducted by scientists of the National Aeronautics and Space Administration (NASA) and Goddard Institute for Space Studies (GISS), the average temperature on Earth has

A. Kumar (✉)
Department of Environmental Engineering, Institute of Engineering and Technology, Sitapur Road, Lucknow, India
e-mail: ankitkumar1702@gmail.com

A. K. Shukla
Department of Civil Engineering, Institute of Engineering and Technology, Sitapur Road, Lucknow, India
e-mail: arvind.shukla@ietlucknow.ac.in

L. M. Gupta et al. (eds.), *Advances in Civil Engineering and Infrastructural Development*, Lecture Notes in Civil Engineering 87,
https://doi.org/10.1007/978-981-15-6463-5_47

increased by about 0.8 °C (1.4 °F) since 1880. Two-thirds of the warming has occurred since 1975, at a rate of roughly 0.15–0.20 °C per decade. Increase in concentration of greenhouse gases (GHGs) in the atmosphere is one of the major reasons for the rise in temperature as these gases absorb the heat [1].

A study conducted by the World Health Organisation (WHO) and the Joint Research Centre (JRC) in 2015 to determine the leading cause of air pollution. After analyzing data from over 50 cities around the world, vehicles were identified as a top contributor to air pollution. Pollutants emitted from vehicles include Carbon monoxide (CO), Hydrocarbons (HC), Nitrogen Oxides (NO_X), Particulate matter (PM), Carbon dioxide (CO_2) and Sulfur dioxide (SO_2) depending upon the type of fuel used in the vehicles [2]. CO_2 is a greenhouse gas that traps the earth's heat and contributes to the rising temperature of the planet resulting in climate change [3]. Vehicle exhaust is a major source of anthropogenic CO_2 in metropolitan cities [4]. With the introduction of the catalytic converter in the vehicles, the emission of CO_2 has increased significantly as it converts harmful emissions like CO and HC to less harmful emissions like CO_2 [5]. Traffic congestion is one of the reasons for the higher emission of pollutants in a city. A study conducted by researchers in Southern California proposes three methods to overcome traffic congestion which could reduce CO_2 emission by 7–12% [6]. This study was conducted on four major roads of Lucknow to determine the concentration of CO_2 emitted from the vehicles into the atmosphere.

2 Study Area

This study was carried out in the capital of Uttar Pradesh, Lucknow. It is the eleventh most populous city of India having a population of 2,902,900 (Census of India 2011). Four roads leading to Kanpur, Sitapur, Prayagraj, and Faizabad were selected for the study as shown in Fig. 1, they are marked by pins of the color black, red, yellow, and blue, respectively. Heavy traffic is observed on these roads throughout the day. Table 1 shows the coordinates of the location at which a traffic survey was conducted.

3 Methodology

A traffic survey was conducted at the interval of 15 min for 4 h in the morning, 4 h in the afternoon and for 4 h in the evening on each selected sites to determine the traffic volume and its composition. This survey was conducted on Tuesday, Wednesday, and Thursday making sure that there is no special event on that day [7]. Vehicles have been categorized as CNG buses, tractor, LCV, scooter, bikes, heavy commercial vehicles (HCV), three-wheelers, and cars. Further cars are classified according to their fuel type, i.e., petrol (56%), diesel (40%), and CNG (4%) [8]. The average number of vehicles per hour is calculated based on the observed vehicular count.

Fig. 1 Survey sites map

Table 1 Coordinates of traffic survey location

S. No	Site	Latitude	Longitude
1.	Kanpur road	26.800018	80.894775
2.	Sitapur road	26.793909	80.953492
3.	Prayagraj road	26.765684	80.946482
4.	Faizabad road	26.873539	81.008803

According to Auto Fuel Policy (2010), Bharat Stage (BS) IV norms were implemented in 13 cities which included Lucknow, therefore CPCB emission factor for BS-IV vehicles has been adopted for the estimation of emission as given in Table 2.

The following equation is used for the estimation of emission:

Table 2 Emission factor for BS IV vehicles as per CPCB standards

S. No	Vehicle type	CO_2 emission factor (g/km)
1.	CNG buses	806.5
2.	Tractor	799.95
3.	LCV	401.25
4.	Scooter	38.54
5.	Bikes	24.82
6.	HCV	762.39
7.	3 W	57.71
8.	Petrol car	126.5
9.	Diesel car	148.76
10.	CNG car	143.54

$$Ep = \sum (Vt \times Ft) \tag{1}$$

where

Ep Emission of CO_2 in g/km/hr
Vt Total number of vehicle of type (t) per hour
Ft Emission factor of CO_2 for vehicle of type (t) in g/km.

4 Result

Figure 2 shows the result of a traffic count survey of all the four sites. Y-axis shows the average, i.e., vehicles per hour at each site. It is observed that bikes are the major share of vehicles with 39% at Kanpur road, 47% at Faizabad road, 32% at Prayagraj road and 37% at Sitapur road followed by scooters at all the sites except Faizabad road where petrol cars (20%) has more share than scooter (12%).

Figure 3 shows the emission of CO_2 in g/hr at all the sites from a different category of vehicles. Kanpur road witnesses the maximum concentration of CO_2. Emission of CO_2 at Kanpur road, Sitapur road, Prayagraj road, and Faizabad road are 491,937 g/km/hr, 291,640 g/km/hr, 404,662 g/km/hr and 252,443 g/km/hr, respectively. Petrol and diesel cars contribute the most to CO_2 emission at all the sites. Even though the emission factor for bikes is very less but due to their presence in more quantity they have contributed significantly to the CO_2 emission. Emission through bikes at Kanpur road is 11%, Sitapur road is 12%, Prayagraj road is 10% and Faizabad road is 14% of their respective emission. Overall emission from tractors is very

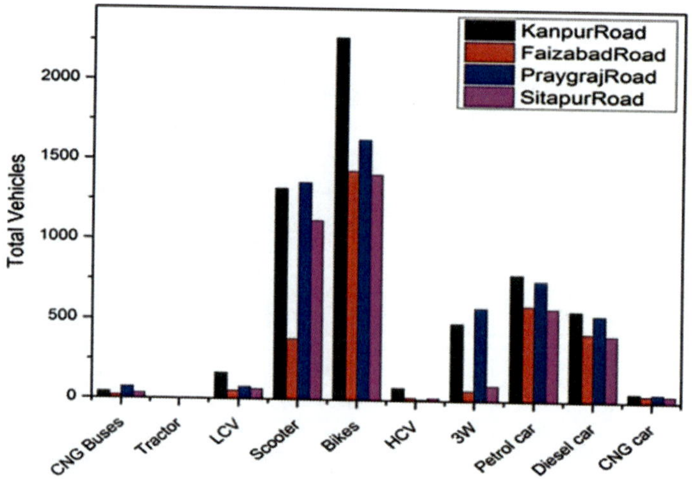

Fig. 2 Vehicular data of all sites

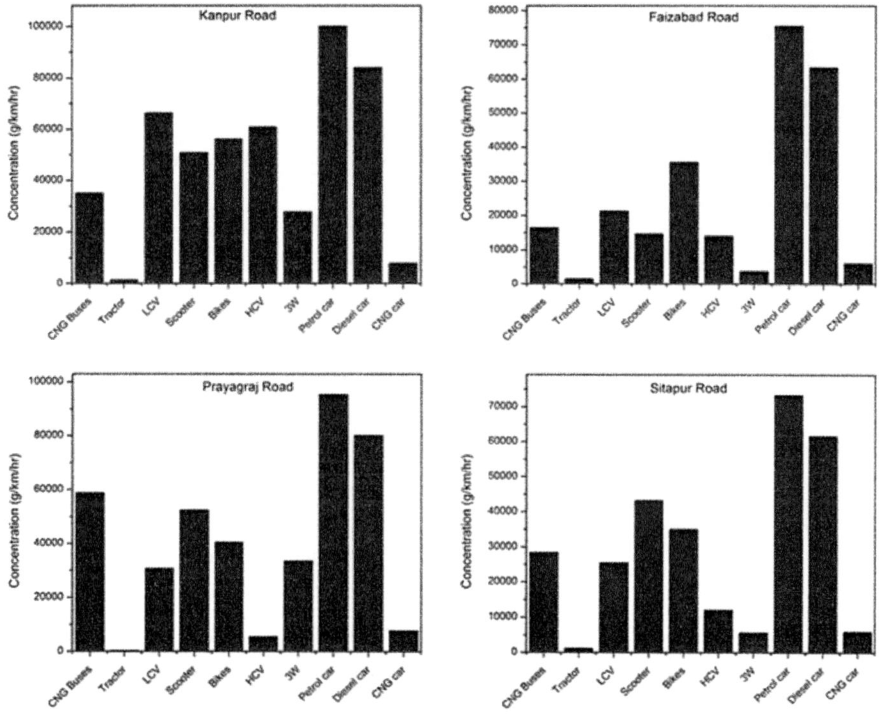

Fig. 3 CO_2 emission at all the sites

less as they were observed in less number during the survey. CNG cars and 3 W has contributed least in the overall CO_2 emission with 2% and 5%, respectively.

5 Conclusion

It was observed from the traffic count survey that bikes contribute to the major share of vehicles in Lucknow for about 38% followed by scooters 24%. These vehicles result in traffic congestion in many places due to poor traffic management and driving skills of people. Further, it is observed that the petrol car contributes 24% and diesel car contributes 20% to the total CO_2 emission from the vehicles at the sites. CO_2 is a greenhouse gas, therefore, it is required to lower down its concentration into the atmosphere to prevent climate change. Encouraging the use of public vehicles and improving traffic management can help deal with traffic congestion which can reduce CO_2 emission [6]. Transport sector contributes around 14% towards the global emissions of greenhouse gases [9]. The push for electric vehicles by the government is a noble step towards reducing pollutants emitted from vehicles.

References

1. Kumar A, Tiwari A, Alam T, Shukla AK (2019) Carbon footprint of municipal solid waste: a case study of Lucknow. Int J Tech Innov Mod Eng Sci 5
2. Bhandarkar S (2013) Vehicular pollution, their effect on human health and mitigation measures. VE 1(2):3340
3. Fogarty J, McCally M (2010) Health and safety risks of carbon capture and storage. JAMA 303(1):67–68
4. Kakouei A, Vatani A, Idris AKB (2012) An estimation of traffic related CO emissions from motor vehicles in the capital city of Iran. Iran J Environ Health Sci Eng 9(1):13
5. Amatayakul W, Ramnäs O (2001) Life cycle assessment of a catalytic converter for passenger cars. J Clean Prod 9(5):395–403
6. Barth M, Boriboonsomsin K (2008) Real-world carbon dioxide impacts of traffic congestion. Transp Res Rec 2058(1):163–171
7. Currin TR (2001) Turning movement counts. In: Stenquist B (ed) Introduction to traffic engineering: a manual for data collection and analysis, pp. 13–23. Wadsworth Group, Stamford, Conn
8. Central Pollution Control Board (CPCB) (2015) Status of pollution generated from road transport in six mega cities, p. 68
9. Central Pollution Control Board (CPCB) (2010) Status of the vehicular pollution control programme in India

Biosorption: Principles, and Applications

Poonam, Anju Rani, and Pradeep Kumar Sharma

Abstract In recent years, human and anthropogenic activities have generated elevated accumulations of toxic pollutants in the environment. These toxic pollutants become persistent in the environment and constitute major environmental problems by badly disturbing the ecosystems and human health. Pollutants including heavy metals, hydrocarbons, dyes, pesticides, and radionuclides are hard to remove from the environment given that most of them cannot be degraded by chemical and biological means and are eventually indestructible. Several approaches such as soil incineration, precipitation, or/and ion-exchange methods, have been extensively employed, but are expensive and detrimental to the environment. Apart from physicochemical methods, biological methods have also been acknowledged as an alternative remediation process. Fundamentals of these biological approaches are microbial activities through which degradation of pollutants depends. Moreover, the cosmopolitan distribution and significant properties of microorganisms in alteration and detoxification of pollutants make them an ultimate candidate. Among biological methods, biosorption is one of the promising technology for pollutant elimination/recovery from wastewater because of its effectiveness, simplicity, and easy biomass availability. This article critically reviews the mechanism, advantages, limitations, and the significance of biosorption for the removal of heavy metals, radionuclides, dyes, hydrocarbons, and pesticides from the environment.

Keywords Biosorption · Pollutants · Industralization

Poonam · A. Rani (✉)
Department of Life Sciences, Graphic Era Deemed to be University, Dehradun, Uttarakhand 248002, India
e-mail: anju.teotia@gmail.com

P. K. Sharma
Department of Environmental Science, Graphic Era Deemed to be University, Dehradun, Uttarakhand 248002, India

L. M. Gupta et al. (eds.), *Advances in Civil Engineering and Infrastructural Development*, Lecture Notes in Civil Engineering 87,
https://doi.org/10.1007/978-981-15-6463-5_48

1 Introduction

Rapid industrialization during the last decade has resulted in a considerable increase in wastewater generation, which contains many kinds of organic compounds, such as PCBs, pesticides, dyes, and hydrocarbons. Many of the present persistent organic compounds are toxic even at very low concentrations with serious effect on aquatic life and human health and represent global problems [1].

A number of physico-chemical methods have been explored for the removal of pollutants from wastewater. These methods have limited use in the treatment of contaminated water/soil due to high operational costs [2]. Moreover, some methods become ineffective if the concentration of contaminant reaches beyond the treatment level limit.

Biological methods which include biosorption and bioaccumulation may be a suitable strategy for the removal of contaminants when compared with existing technologies. In biosorption, biological material is used to eliminate (adsorb/absorb) contaminants from wastewater [3]. The use of microbes as biosorbents has various advantages such as high surface to volume ratio, widespread occurrence rapid kinetics of adsorption and desorption, and low cost. So far, the biosorption of heavy metals by microbes such as bacteria, fungi, and algae were extensively reported [4], however, its use for removing pesticides, dyes, hydrocarbons, and radionuclides has received considerably less attention.

2 Biosorption

The term sorption includes both absorption and adsorption. Absorption is the inclusion of a substance in one form into another different form while adsorption is the physical attachment of molecules/ions on the solid material's surface by means of bonding. The material which adsorbed at the interface is the adsorbate and the solid surface on which adsorption occurs is called as adsorbent [5]. The major advantages of biosorption above conventional treatment methods include:

- Economically favorable
- Efficient process (can take place at a wide range of pH and temperature)
- Reduction of biological sludge
- Optimum utilization of industrial waste
- Reuse of biosorbents
- No additional requirements (for example, nutrients)
- Probability of contaminants recovery for further use (in case of metals).

A broad range of naturally occurring biomaterials has been explored as biosorbent for the pollutant elimination. Of the different biosorbents, different organisms, i.e., bacteria, cyanobacteria, algae (including microalgae, macroalgae, and seaweeds), yeasts, fungi, and lichens have been demonstrated to possess good potential to remove contaminants. Generally, dead biomass is used as biosorbents owing to various properties (long storage time, no limitation by environmental conditions and adsorbate concentration, and no requirement of nutrients) when compared with the living cells as biosorbents [6].

3 Mechanism of Biosorbent-Sorbate Interactions

The binding of sorbate on the surface of biosorbent occurs via many physical or chemical interactions [7]. The cell wall composition plays an important role in biosorption interactions. The cell wall of different microbial groups is consists of complex polysaccharides, proteins, and lipids, which in turn have different functional groups (–OH, –COOH, –NH2, –SH, –CO, –COOR,). These functions groups have key involvement in biosorption of cations [8]. During the biosorption of metals, the naturally occurring cations and protons which remain attached with functional group of microbial surface are exchanged with metal ions [9].

The cell wall presents numerous active sites for the binding of metal ions. The amount and type of biosorbate (especially in case of metals and radionuclides) varies in different microbial groups due to difference in cell wall composition (Fig. 1).

Algal cells walls are made up of cellulose which consists of carboxylates, amines, imidazoles, phosphates, sulfhydryls, sulfates, and hydroxyls functional groups. Bacterial cells are mainly composed of peptidoglycans (polymer of N-acetylglucosamine –β 1, 4-N-acetylmuramic acid with peptide chains) with carboxylate as key functional group implicated in metal binding [10]. Fungal cells are composed of chitin and chitosan, which have the property to restore metal ions.

4 Factors Affecting Biosorption

Many factors control the process of biosorption. These factors take account of biomass properties and physio-chemical properties (Fig. 2).

Type and nature of Biomass: Broad array of microbial biomass has been explored in different biosorption studies as single and mixed organism biomass [11]. Microbial biomass when used in free suspended forms has several disadvantages that consist of small size, low mechanical strength, and complex biomass/effluent separation. Microbial cells can also be immobilized on inert materials support in the form of biofilm. Living cell biofilms also provide additional further removal of additional pollutants such as carbons, pesticides, and nitrates.

Growth, nutrition, and age of Biomass: Biomass growth, growth conditions (young or old culture), and nutrition is known to influence biosorption due to morphological change, cell wall composition of cells, and product formation (intracellular or extracellular), etc. [12]. Furthermore, surface area and surface area/volume ratio is another important factor in the case of individual cells and immobilized biofilms,

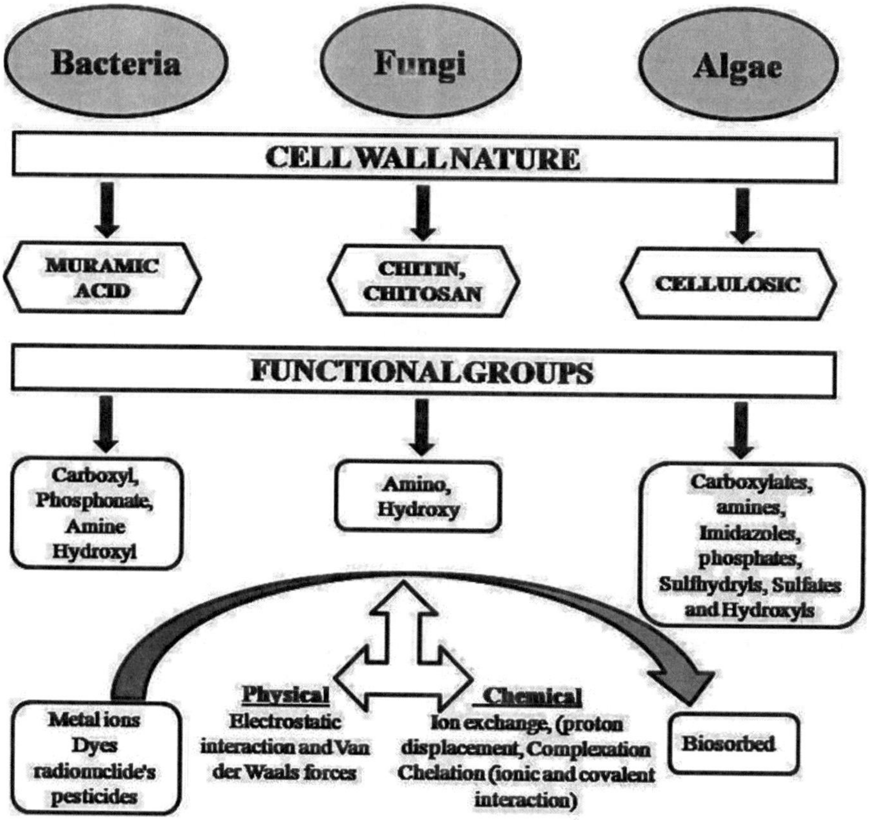

Fig. 1 Microbial cell wall nature and their interactions in biosorption

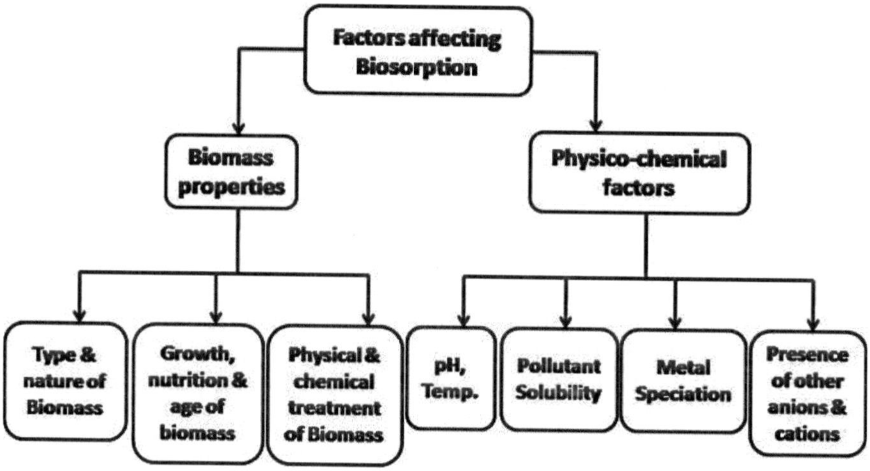

Fig. 2 Factors affecting biosorption

Table 1 Effect of different treatments on biosorption efficiency

Method	Effect	References
Heat, chemical sterilization, crushing	Offer a larger available surface area and expose the intracellular components and more surface binding sites due to the destruction of the cell membranes	[41–45]
Acidic pre-treatment	Availability of more binding sites and elimination of impurities and ions that block heavy metals binding sites	[46–48]
Chemical agents (Mineral acids, NaOH, Na_2CO_3, $CaCl_2$ NaCl, chloroacetic acid ethylenediamine, carbodiimide)	Opens new binding sites and enhance or modify binding sites	[49]
Detergent	Disruption of cell components increased cell permeability and exposed a wider variety of potential binding sites	[50, 51]

respectively. The amount of biomass may also influence biosorption effectiveness with a decrease in sorption/unit weight as biomass concentration increases.

Physical and chemical treatment of biomass: Biosorptive capacity of biomass can be altered by treating biomass with physical, chemical, and other methods which in turn modify surface properties [13, 14].

Many physical pre-treatment methods such as heat treatment, autoclaving, freeze-drying, and boiling have been used [15–18]. Chemical methods include the treatment with diverse organic as well as inorganic reagents. The magnitude of increase in biosorption capacity by chemical pre-treatment is dependent on the cellular components of the biomass, which occurs due to the modification of binding sites on biomass surface. The following table (Table 1) explains different treatment methods and their effect on biosorbent.

5 Biosorption of Different Pollutants

5.1 *Heavy Metals*

Various heavy metals released via industrial activities such as copper (Cu), cadmium (Cd), lead (Pb), mercury(Hg), cobalt (Co), chromium (Cr), and nickel (Ni) have deleterious effects on the environment. Bacteria, fungi, yeast and algae have been proved potential sorbents for heavy metal removal [19, 20]. We are not discussing heavy metals biosorption in detail here as a number of the biosorption studies have been done for heavy metal removal and literature is already available.

5.2 Dyes

Most of the industries involved in the manufacturing of paper, printing, textile and many other use dyes for coloring of their products [21]. Annual production of dyes across the world is about 7×10^5 metric tons [22], of which 70–80% is produced by textile industries. The dyes or their metabolites pose genotoxic effects on aquatic and human life. Therefore, the elimination of these colored pigments/dyes from water is crucial.

Various studies have proven that the effectiveness of dye removal is affected by the type of microorganisms and the process parameters. Apart from concentration and chemical structure of dye, other factors that influence biosorption are pH, temperature, oxygen level, added carbon and nitrogen sources [23–26].

Among the various microorganisms involved in biosorption of dyes, one of the important organisms is the fungal biomass, owing to its easy and cheap availability as a waste product from various fermentation industries [27]. The fungi such as *Aspergillus niger*, *Rhizopus arrhizus,* and *Rhizopus oryzae* have been reported to biosorb various kinds of dyes [28–30].

The well-known genera of bacteria discussed in the literature for the biosorption of dyes are *Pseudomonas, Bacillus, Sphingomonas, Aeromonas, Citrobacter, Escherichia, Desulphovibrio, Proteus, Schewanella and Alcaligenes* [24, 31].

5.3 Complex Hydrocarbons

Hydrocarbon pollution is by aliphatic compounds as well as aromatic compounds. Among different treatment methods, microbial degradation is the chief removal mechanism of hydrocarbons in soils. Numerous studies [32, 33] have reported good sorption of polycyclic hydrocarbons (naphthalene, phenanthrene, fluoranthrene, Benzo (a) pyrene and pyrene) on biomass-derived from algae, fungi, and bacteria. Among different microbes, *Phanerochaete chyrsosporium, Pseudomonas* sp., and *Bacillus* sp. have shown potential removal capacities.

5.4 Radioactive Elements

The anthropogenic activities and nuclear science developments have been contributing to an outstanding increase in the dispersion of radioactive elements in the environment [5]. In the natural environment, the most important concern is the oxidation state of radionuclide. Some radioactive elements pose more toxicity in higher oxidation state rather than low oxidation state, for example, uranium (U), technetium (Tc), plutonium (Pu), and neptunium (Np). The presence of these radioactive elements in high oxidation state create hazards to human, animals and aquatic flora and fauna due to high solubility and thus more mobilization in the aquatic environment. [34].

In this context, numerous studies have been done for the elimination of Uranium, Caesium, Thorium, and Strontium using bacteria, yeast, and fungi as biosorbents. Several factors affected biosorption of radionuclides. Among these, pH was observed as an important factor in different studies (maximum biosorption was in the range of 4–6). Biosorbent dosage also influenced the extent of biosorption. Sar and D'Souza reported maximum biomass loading of 541 mg Ug^{-1} dry mass or 430 mg Th g^{-1} dry mass by bacterial biosorbents. Several studies [35], Sar and D'Souza also reported interference in uranium sorption by *Pseudomonas sp.* in the presence of cations (Fe^{3+}, Fe^{2+}, Al^{3+}, Cu $^{2+,}$ and Ag^{2+}). Highest inhibition to U sorption was shown by Fe^{3+}. In another study [36] absorption rate for uranium decreased in the presence of Al^{3+}, Fe^{2+}, Fe^{3+}, Cu^{2+}, ${CO_3}^{2-}$ ions (at 0.1 M concentration) with maximum reduction noted in the presence of CO_3^{2-}.

5.5 *Pesticides*

The extensive use of pesticides in agriculture resulted in the accumulation of residues in the environment. These pesticides undergo a variety of transformations that provide a complex pattern of metabolites and create a risk to human health. In accordance with the EEC directive permissible limit of pesticide in drinking water should be below 0.1 g/L. Therefore, pesticide removal is of increasing concern.

Some studies have reported microbial biosorption of DDT, lindane, Malathion by bacteria, fungi algae and diverse culture of bacteria and fungi. Juhasz et al. [37] confirmed the sorption of organochlorine pesticide, DDT by the fungal biomass of *Cladosporium* sp. to a considerable level. Salam and Das [38] studied lindane biosorption by pretreated biomass of yeast *Cintractia sorghi.* In a similar study, the fungi *Rhizopus arrizhus* was also found effective against the biosorption of lindane [39]. Malathion is a synthetic, broad-spectrum organophosphorous pesticide that is toxic and carcinogenic in nature. The dry cells of *Bacillus* sp. were reported to be efficient in removing Malathion from the solution. Maximum biosorption of Malathion (81.4%) was observed under pH 6.5, temperature 25 °C [40].

6 Conclusion and Future Prospects

Although a number of studies have been carried out on heavy metal biosorption nevertheless there is a scope of improvement in the process by minimizing cost, better performance, and exploration of the process for other pollutants. To use industrial waste biomass as biosorbent serves two benefits, i.e., remedy for waste disposal and availability of biosorbent free of cost. Further investigation should be emphasized on modelling, reuse of biosorbent, and recovery of sorbate for further application.

Furthermore, consideration of other factors (pH, temperature, nature, and structure of contaminants and pre-treatment of biomass) and detailed study of a contaminated site in terms of prevailing environmental conditions is also a necessity to enhance biosorption. Exploitation of microbial consortia and immobilized microbial culture for biosorption will make the process more robust and economical. Furthermore, to make the process more pertinent by looking for the cheapest biosorbent in terms of minimum nutrient needs and amenable to the remarkably harsh environmental conditions.

References

1. Badawy MI, El-Wahaab RA, Moawad A, Ali MEM (2010) Assessment of the performance of aereated oxidation ponds in the removal of persistent organic pollutants (POPs): a case study. Desalin 251(1):29–33
2. Gadd GM (2010) Minerals and microbes: geomicrobiology and bioremediation. Microbiology 156:609–643
3. Ahmed M, Kibret M (2013) Recent trends in microbial biosorption of heavy metals: a review. Biochem Mol Biol 1(1):19–26
4. Mustapha MU, Halimoon N (2015) Microorganisms and biosorption of heavy metals in the environment: a review paper. J Microb Biochem Technol 7:253–256
5. Dos Marques AL, Santos W, Geraldo LP (2004) Direct measurements of radon activity in water from various natural sources using nuclear track detectors. Appl Radiat Isot 60(6):801–804
6. Baysal V, Cinar E, Bulut Y, Alkan H, Dogru M (2009) Equilibrium and thermodynamic studies on biosorption of Pb (II) INTO *Candida albicans* biomass. J Hazard Mat 161(1):62–67
7. Davis TA, Volesky B, Mucci A (2003) A review of the biochemistry of heavy metal biosorption by brown algae. Water Res 37:4311–4330
8. Talaro KP, Talaro A (2002) Foundations in microbiology, 4th edn. McGraw-Hill College, Blacklick, Ohio
9. Sar P, Kazy SK, Souza SFD (2004) Radionuclide remediation using a bacterial biosorbent. Int Biodeterior Biodegrad 54(2–3):193–202
10. Hassan SH, Kim SJ, Jung AY, Joo JH, Oh SE, Yang JE (2009) Biosorptive capacity of Cd (II) and Cu (II) by lyophilized cells of *Pseudomonas stutzeri*. J Gen Appl Microbiol 55(1):27–34
11. Muñoz R, Alvarez MT, Muñoz A, Terrazas E, Guieysse B, Mattiasson B (2006) Sequential removal of heavy metals ions and organic pollutants using an algal-bacterial consortium. Chemosphere 63(6):903–911
12. Abbas SH, Ismail IM, Mostafa TM, Sulaymon AH (2014) Biosorption of heavy metals: a review. J Chem Sci Technol 3(4):74–102
13. Garnham GW, Codd GA, Gadd GM (1993) Uptake of cobalt and caesium by microalgal- and cyanobacterial-clay mixtures. Microb Ecol 25:71–82
14. Treen-Sears ME, Volesky B, Neufeld RJ (1984) Ion ex-change/complexation of the uranyl ion by Rhizopus biosorbent. Biotechnol Bioeng 26:1323–1329
15. Gabr RM, Hassan SHA, Shoreit AAM (2008) Biosorption of lead and nickel by living and nonliving cells of *Pseudomonas aeruginosa* ASU 6a. Int Biodeterior Biodegrad 62:195–203
16. Göksungur Y, Üren S, Güvenc U (2005) Biosorption of cadmium and lead ions by ethanol treated waste baker's yeast biomass. Bioresour Technol 96:103–109
17. Mapolelo M, Torto N (2004) Trace enrichment of metal ions in aquatic environments by *Saccharomyces cerevisiae*. Talanta 64(1):39–47
18. Seki H, Suzuki A, Maruyama H (2005) Biosorption of Cr (VI) and As (V) onto methylated yeast biomass. J Colloid Interf Sci 281:261–266

19. Ahluwalia SS, Goyal D (2007) Microbial and plant derived biomass for removal of heavy metals from wastewater. Bioresour Technol 98:2243–2257
20. Oves M, Khan MS, Zaidi A (2013) Biosorption of heavy metals by *Bacillus thuringiensis* strain OSM29 originating from industrial effluent contaminated north Indian soil. Saudi J Biol Sci 20(2):121–129
21. Gupta VK, Ali I, Mohan D (2003) Equilibrium uptake and sorption dynamics for the removal of a basic dye (basic red) using low-cost adsorbents. J Colloid Interf Sci 265(2):257–264
22. Azmi W, Sani RK, Banerjee UC (1998) Biodegradation of triphenylmethane dyes. Enzyme Microb Technol 22:185–191
23. Koyani RD, Sanghvi GV, Sharma RK, Rajput KS (2013) Contribution of lignin degrading enzymes in decolourisation and degradation of reactive textile dyes. Int Biodeter Biodegr 77:1–9
24. Saratale RG, Saratale GD, Chang JS, Govindwar SP (2011) Bacterial decolourization and degradation of azo dyes: a review. J Taiwan Inst Chem Eng 42:138–157
25. Younes S, Bouallagui Z, Sayadi S (2012) Catalytic behavior and detoxifying ability of an atypical homotrimericlaccase from the thermophilic strain scytalidium thermophilum on selected azo and triarylmethane dyes. J Mol Catal B Enzym 79:41–48
26. Zabłocka-Godlewska E, Przystaś W, Grabińska-Sota E (2014) Decolourisation of different dyes by two pseudomonas strains under various growth conditions. Water Air Soil Pollut 225:1846
27. Kapoor A, Viraraghavan T (1995) Fungal biosorption—an alternative treatment option for heavy metal bearing wastewater: a review. Bioresour Technol 53:195–206
28. Fu Y, Viraraghavan T (2000) Removal of a dye from an aqueous solution by fungus *Aspergillus niger*. Water Qual Res J Canada 35:95–111
29. Fu Y, Viraraghavan T (1999) Removal of Acid Blue 29 from an aqueous solution by fungus Aspergillus niger. In: Nikolaidis N, Erkey C, Smets B (eds) Hazardous and industrial wastes—proceedings of the Mid-Atlantic industrial waste conference, pp. 510–519
30. Gallaghar KA, Healy MG, Allen SJ (1997) Biosorption of synthetic dye and metal ions from aqueous effluents using fungal biomass. In: Wise DL (ed) Global environmental biotechnology. UK, Elsevier, BV, pp 27–50
31. Wang J, Gao F, Liu Z, Qiao M, Niu X, Zhang KQ, Huang X (2012) Pathway and molecular mechanisms for malachite green biodegradation in exiguobacterium sp. MG2. PLoS ONE 7(12):e51808
32. Ding J, Chen BL, Zhu LZ (2013) Biosorption and biodegradation of polycyclic aromatic hydrocarbons by *Phanerochaete chrysosporium* in aqueous solution. Chin Sci Bull 58:613–621
33. Ke V, Luo L, Wang P, Luan TNF (2010) Effects of metals on biosorption and biodegradation of mixed polycyclic aromatic hydrocarbons by a freshwater green alga *Selenastrum capricornutum*. Bioresour Technol 101(18):6961–6972
34. Prakash D, Gabani P, Chandel AK, Ronen Z, Singh OV (2013) Bioremediation: a genuine technology to remediate radionuclides from the environment. Microb Biotechnol 6(4):349–360
35. Byerley JJ, Scharer JM, Charles AM (1987) U(VI) Bioadsorption from process solutions. Chem Eng J 36:B49–B59
36. Xu X, He S, Wang Z, Zhou Y, Lan J (2013) Biosorption of uranium by *Bacillus* sp.FB12 isolated from the vicinity of a power plant. Adv Environ Res 2(3):245–260
37. Juhasz A, Smith E, Smith J, Naidu R (2002) Biosorption of organochlorine pesticides using fungal biomass. J Ind Microbiol Biotech 29:163
38. Salam JA, Das N (2014) Lindane degradation by Candida VITJzN04, a newly isolated yeast strain from contaminated soil: kinetic study, enzyme analysis and biodegradation pathway. World J Microbiol Biotechnol 30:1301–1313
39. Cliff B, Weibel DE, Lockyer NP, Jungnickel H, Stephens G, Vickman JC (2003) Detection of chlorinated pesticides on the surface of fungus using ToF-SIMS. Appl Sur Sci 204:710–713
40. Adhikari S, Chattopadhyay P, Ray L (2010) biosorption of Malathion by immobilized cells of Bacillus sp. S14. Chem Spec Bioavailab 22(4):271–276
41. Bayramoglu G, Arica MY (2008) Removal of heavy mercury (II), cadmium(II) and zinc(II) metal ions by live and heat inactivated Lentinusedodes pellets. Chem Eng J143:133–140

42. Bayramoglu G, Celik G, Yalcin E, Yilmaz M, Arica M (2005) Modification of surface properties of Lentinussajor-cajumycelia by physical and chemical methods: Evaluation of their Cr^{+6} removal efficiencies from aqueous medium. J Hazard Mater 119:219–229
43. Brady JM, Tobin JM (1995) Binding of hard and soft metal ions to Rhizopus arrhizus biomass. Enzym Microb Technol 17:791–796
44. Robles LC, Feo JC, Aller AJ (2000) Selective pre-concentration of phenyl-mercury by living Escherichia coli and its determination by cold vapour atomic absorption spectrometry. Anal Chem Acta 423:255–263
45. Vilar VJP, Botelho CMB, Boaventura RAR (2006) Equilibrium and kinetic modelling of Cd(II) biosorption by algae Gelidiumand agar extraction algal waste. Wat Res 40:291–302
46. Abbas M, Nadeem R, Zafar MN, Arshad M (2008) Biosorption of chromium (III) and chromium (VI) by untreated and pretreated Cassia fistula biomass from aqueous solutions. Water Air Soil Pollut 191:139–148
47. Araujo GCL, Lemos SG, Ferreira AG, Freitas H, Nogueira ARA (2007) Effect of pre-treatment and supporting media on Ni(II), Cu(II), Al(III) and Fe(III) sorption by plant root material. Chemosphere 68:537–545
48. Vijayaraghavan K, Yun Y-S (2008) Bacterial biosorbents and biosorption-a review. Biotechnol Adv 26(3):266–291
49. Vijayaraghavan K, Yeoung-Sang Y (2007) Chemical modification and immobilization of corynebacterium glutamicum for biosorption of reactive black 5 from aqueous solution. Ind Eng Chem Res 46:608–617
50. Gadd GM (1993) Interactions of fungi with toxic metals. New Phytol 124:25–60
51. Jeon C, Holl WH (2003) Chemical modification of chitosan and equilibrium study for mercury ion removal. Water Res 37:4770–4780

Transportation and Geotechnical Engineering

Analysis and Management of Road Accidents Caused Due to Pavement Surface Conditions on Expressways

Rohit R. Salgude, S. S. Pimplikar, and Kunj R. Patel

Abstract Expressways are the highest class of roads in the Indian road network, and approximately, 1581.4 km of expressways are operational in India. Road accidents have rose over decades and are a significant contributor towards human deaths. It is increasing yearly with the increase of road users and construction of road pavement globally. In India, more than 150,000 people are killed each year in traffic accidents and that is about 400 fatalities a day. Also, India losses 3% of its GDP to road accidents. The objective of this study is to segregate critical accident blackspots on Mumbai–Pune Expressway, to map the accident blackspots using QGIS and to generate Karl Pearson correlation coefficient and regression model between number of accidents and pavement roughness index.

Keywords Expressway · Pavement roughness index · Rigid pavement · Road accidents · GIS

1 Introduction

An accident, also known as an unintentional injury, is an unfortunate incident that happens unexpectedly, typically resulting in damage or injury. Road accidents are the most frequent and endless. A total number of 1714 accidents both fatal and non-fatal occurred on Mumbai–Pune Expressway during the year 2018. Factors contributing towards road accidents on expressways may be distracted driving, drunk driving, reckless driving, night driving, design defects, drowsy driving, tire blowouts, deadly curves, pavement surface conditions, animal crossings and freedom of movement given to drivers, leading to over speeding, particularly on expressways [1].

R. R. Salgude (✉)
Department of Civil Engineering, Maharashtra Institute of Technology, Pune 411038, India
e-mail: rohit.salgude@mitpune.edu.in

S. S. Pimplikar · K. R. Patel
School of Civil Engineering, MIT World Peace University, Pune 411038, India

L. M. Gupta et al. (eds.), *Advances in Civil Engineering and Infrastructural Development*, Lecture Notes in Civil Engineering 87,
https://doi.org/10.1007/978-981-15-6463-5_49

Extensive research has been conducted to identify the various causes of accidents, and hence, remedial measures have been recommended. Shafabakhsh et al. [2], Selvasofia and Arulraj [3], Bhagyaiah and Shrinagesh [4], Jayan and Ganeshkumar [5] used GIS to map the blackspots; Alhasan et al. [6], Elghriany et al. [7], Serigos et al. [8], Lee et al. [9], Yao and Liu [10], Baek et al. [11] measured pavement distresses using automated and manual methods; Kassu and Anderson [12], Gebretensay and Juremalani [13], Mir et al. [14], Malik et al. [15], Shariff et al. [16], Ghosh and Paul [17] performed analysis to identify road accidents; and Li et al. [18], Gupta and Rokade [19], Sailaja and Raju [20], Agyemang et al. [21] generated accident analysis models to predict accidents.

Yet, for 30% of various accidents; their primary causes of accidents are unknown Ramaswamy [22]. Pimplikar et al. [23] observed severe distresses on the new constructed Mumbai–Pune Expressway, in asphalt as well as the concrete pavement. Geopathic stress, as one of the prime reasons, has been identified as being responsible for deterioration of materials at these locations. In fact, at few spots, same types of distresses were observed even after one year, resulting in frequent maintenance of the expressway, hence the motivation for this research.

One of the significant factors for road accidents may be the pavement surface condition as deterioration and defects on pavements lead to skidding, driving off track, improper manoeuvring to avoid the road defects and the like. Therefore, it is important to investigate whether the distresses in pavements are responsible for causing road accidents.

Most researchers have majorly focused on asphalt pavements, whereas only few researchers have studied rigid pavements. Apart from this, no study is done for evaluation of the pavement's riding quality as a possible cause of road accidents.

However, this study is restricted only to the Mumbai–Pune Expressway; also, the study is conducted at 29 accident spots only.

2 Need for Research

The IRC codes of design, IRC:37-2018: Guidelines for design of flexible pavements and IRC:58-2015: Guidelines for the design of plain jointed rigid pavements for highways considers that the road safety is fundamental requirement of any transportation system, is the basis of design, is the assumption that the entire pavement thickness of the particular material used and is the very good functional condition.

But the empirical findings from the literature review indicate that the reality is actually different, because of conditions mentioned above.

Thus, a variable dealing with the above aspects needs to be incorporated as a design parameter. The pavement condition rating (PCR) is based on the pavement condition index (PCI). Determining PCI is based on structural distresses as well as functional distress. It is empirically observed that on the roads with lesser values of

PCI, there are more accidents, Chafekar et al. [24], Pimplikar and Kharat [25]. Thus, the correlation between the technical aspects of the road and the number of accidents needs to be ascertained.

3 Objectives

a. To segregate the critical accident blackspots existing on Mumbai–Pune Expressway based on pavement roughness index values obtained.
b. To map critical accident blackspots thus obtained on Mumbai–Pune Expressway using QGIS (quantum GIS).
c. To determine Karl Pearson correlation coefficient between number of accidents per year (A) and pavement roughness index (PRI) at critical accident blackspots for Mumbai–Pune Expressway.
d. To formulate a regression model between number of accidents per year (A) and pavement roughness index (PRI) at critical accident blackspot for Mumbai–Pune Expressway.

4 Data Collection

4.1 Prelude

In this research work, the accident data and road roughness data was collected from MSRDC which documents monthly progress reports for short-term improvements, operation and maintenance of Mumbai–Pune Expressway. Road roughness survey was carried out by ROADBOUNCE. ROADBOUNCE is the surveyor which carries out an automated survey to analyse the road conditions of Mumbai–Pune Expressway.

4.2 Accident Data

KM-wise accident data of one year was collected from MSRDC for year 2018.

4.3 Road Roughness Data

As mentioned previously, road roughness was measured by ROADBOUNCE, an advance automated system which is speedy, accurate and reliable.

ROADBOUNCE Enterprise provides potholes detection, prioritize repair locations based on traffic and road roughness index estimation as per international standards. It identifies potholes and rough road patches just by driving a car on the road and measures an IRI/roughness index data and identifies those road sections which need immediate attention using smartphone vibrations while driving car. The data which is collected is stored on a secured cloud thus eliminating manipulation of data.

Data Collected:

The following data was collected for both Mumbai and Pune Corridor for each lane for every 100 m (Year 2018).

a. Pre-monsoon road roughness test data in mm/km
b. Post-monsoon road roughness test data in mm/km.

Recommended roughness values as per IRC-SP-16 2004 for cement concrete pavement is good if roughness index is <2200 mm/km, average if between 2200 and 3000 mm/km, poor if >3000 mm/km.

5 Data Analysis

5.1 *Prelude*

In the data analysis, the accident data was compared with the road roughness to determine the blackspots on Mumbai–Pune Expressway; later, accident blackspots were mapped using GIS, and finally, correlation and regression model was developed for comparing accidents with pavement roughness index.

5.2 *Accident Blackspots on Expressway Based on Pavement Roughness Index*

Police Department has considered a spot as accident blackspot if more than five accidents occur at that spot within 3 months.

On the expressway, there are a total number of 188 accident spots identified by the authorities considering the Pune Corridor and Mumbai Corridor. Typical causes of accidents such as over speeding, tyre bursting, drunken drivers, highway geometrics are identified as the primary causes. However, as mentioned previously, there are specific accident spots which have recurring accidents in spite of various remedial efforts implemented by expressway authorities based on national- and state-level safety audits. An attempt is made in this research to find the possible cause of accidents at such blackspots. Hence, in all, only 29 blackspots are considered for this analysis.

Table 1 lists a total of 29 critical accident blackspot Km wise on Mumbai–Pune

Table 1 Blackspots on Mumbai–Pune Expressway

Pune Corridor (towards Pune)			Mumbai Corridor (towards Mumbai)		
S. No.	Number of accidents	Blackspots (KM)	S. No.	Number of accidents	Blackspots (KM)
1	12	6–7	a.	13	20–21
2	13	15–16	b.	15	29–30
3	18	18–19	c.	10	30–31
4	12	35–36	d.	9	31–32
5	20	36–37	e.	12	44–45
6	28	37–38	f.	8	49–50
7	35	38–39	g.	7	50–51
8	27	39–40	h.	7	58–59
9	19	40–41	i.	8	67–68
10	33	41–42	j.	7	70–71
11	36	45–46	k.	16	76–77
12	11	46–47	l.	7	77–78
–	–	–	m.	16	78–79
–	–	–	n.	21	80–81
–	–	–	o.	9	81–82
–	–	–	p.	14	82–83
–	–	–	q.	9	92–93

Expressway based on pavement roughness index (Fig. 1).

5.3 Mapping of Accident Blackspots on Mumbai–Pune Expressway Using QGIS

QGIS (quantum GIS) is a free and open-source platform which functions as geographic information system (GIS) that supports viewing, editing and analysis of geospatial data to compose and export graphical maps.

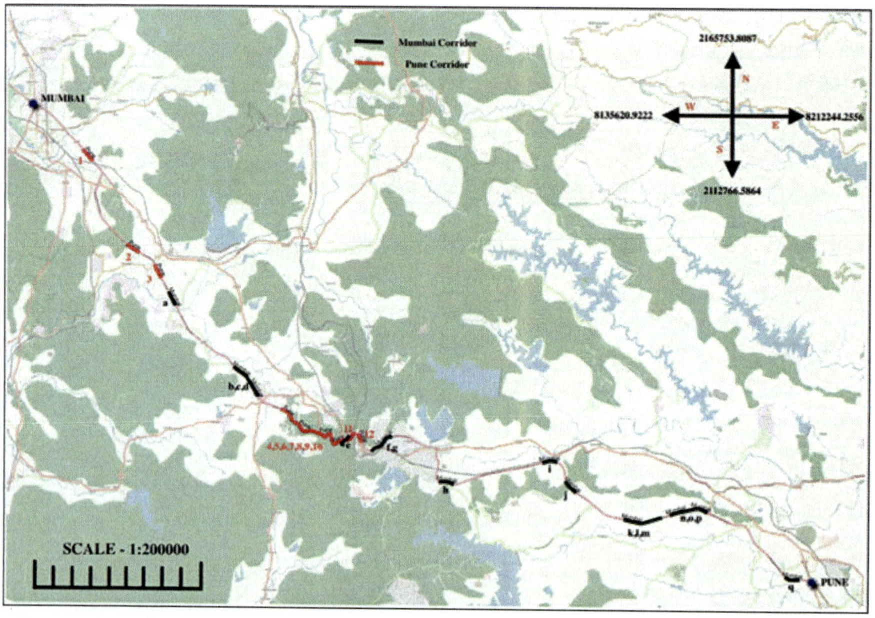

Fig. 1 Mapping of critical accident blackspots

5.4 Data Analysis for Karl Pearson Correlation Coefficient and Regression Model Between Number of Accidents Per Year(a) with Pavement Roughness Index (PRI)

5.4.1 For Pune Corridor

The Karl Pearson correlation coefficient between *A* and PRI is $r = +0.70$. From Fig. 2, it is interpreted that there is a linear, positive and strong correlation between number of accidents on Pune Corridor and average roughness index of three lanes. It

Fig. 2 Number of accidents versus pavement roughness index for Pune Corridor

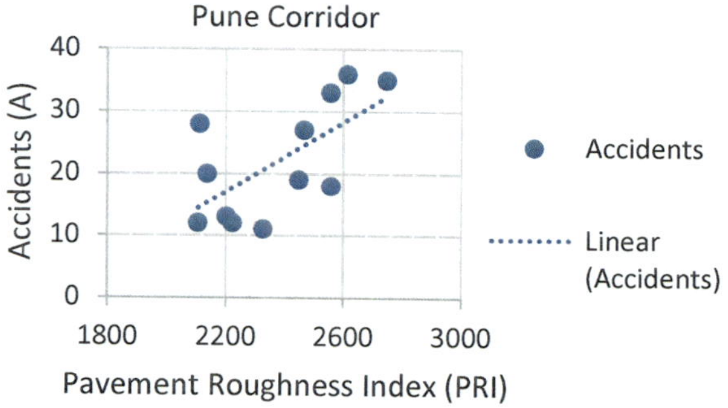

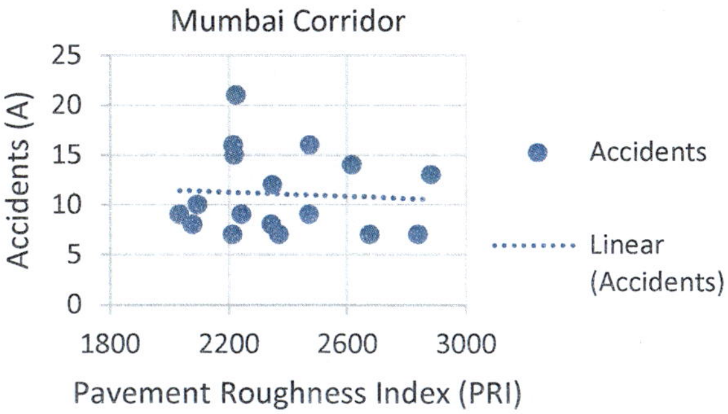

Fig. 3 Number of accidents versus pavement roughness index for Mumbai Corridor

is interpreted that as pavement roughness index increases, the number of accidents increases. It may therefore be statistically inferred that road accident is a function also of pavement roughness index. For accident spots of Pune Corridor, roughness index value varies from 2105 to 2748 mm/km, which as per IRC-SP-16:2004 shows that pavement condition has deteriorated to average roughness presently; further, deterioration may lead to poor roughness if it exceeds 3000 mm/km.

Based on this, linear regression equation gets formulated as

$$A = -44.58 + 0.0284(\text{PRI}) \tag{1}$$

Equation (1) enables the prediction of accidents at critical blackspots, provided that the pavement roughness index is known. For example, if the PRI is 2000 mm/km, then A will be 12, and if the PRI is 2800 mm/km, then A will be 34.

5.4.2 For Mumbai Corridor

The Karl Pearson correlation coefficient between A and PRI is $r = -0.1$. From Fig. 3, it is interpreted that there is a linear, negative and virtually no correlation. Thus, the number of accidents is independent of the pavement roughness for the Mumbai Corridor.

6 Conclusion

A total of 29 critical accident blackspots were segregated from 188 total accident spots based on the pavement surface conditions on Mumbai and Pune Corridor for Mumbai–Pune Expressway. A strong, positive correlation exists between the total number of accidents per year (A) and the pavement roughness index (PRI) along Pune Corridor. However, there is virtually no correlation between the total number of accidents per year (A) and the pavement roughness index (PRI) along the Mumbai

Corridor. The regression equation formulated for the Pune Corridor will enable the prediction of future road accidents; when the pavement roughness value is determined. However, further significant research regarding the existence of geopathic stress at these spots and its significance in accident studies needs to be carried out.

References

1. https://seriousaccidents.com/legal-advice/top-causes-of-car-accidents/. Last Accessed on 06 Sep 2018
2. Shafabakhsh GA, Famili A, Bahadori MS (2017) GIS-based spatial analysis of urban traffic accidents: case study in Mashhad, Iran. J Traffic Transp Eng (Eng Ed) 4(3):290–299
3. Selvasofia AS, Arulraj PG (2016) Accident and traffic analysis using GIS. Biomed Res (Special Issue)S103–S106
4. Bhagyaiah M, Shrinagesh B (2014) Traffic analysis and road accidents: a case study of Hyderabad using GIS. IOP Conf Ser: Earth Environ Sci 20:012026
5. Jayan KD, Ganeshkumar B (2010) Identification of accident hot spots: AGIS based implementation for Kannur District, Kerala. Int J Geomat Geosci 1(1):51–59
6. Alhasan A, Nlenanya I, Smadi O, MacKenzie CA (2018) Impact of pavement surface condition on roadway departure crash risk in Iowa. Infrastructures 3(14):1–16
7. Elghriany A, Yi P, Liu P, Yu Q (2016) Investigation of the effect of pavement roughness on crash rates for rigid pavement. J Transp Saf Secur 8(2):164–176
8. Serigos PA, Prozzi JA, de Fortier Smit A, Murphy MR (2016) Evaluation of 3D automated systems for the measurement of pavement surface cracking. J Transp Eng Am Soc Civil Eng 05016003-1–05016003-8
9. Lee J, Nam B, Abdel-Aty M (2015) Effects of pavement surface conditions on traffic crash severity. J Transp Eng Am Soc Civil Eng 04015020-1–04015020-11
10. Yao Q, Liu S (2014) Analysis of typical distress and cause on asphalt pavement for Liaoning highway. CICTP: Saf Smart Sustain Multimodal Transp Syst ASCE 919–924
11. Baek J, Yoo HM, Lee TH, Park YH, Kim MI (2014) Lessons from 20 years experiences of pavement management system on national highways in Korea: focus on distress survey. Des Anal Asphalt Mate Charact Road Air Pavements GSP 246 ASCE 170–177
12. Kassu Aschalew, Anderson Michael (2018) Determinants of severe injury and fatal traffic accidents on urban and rural highways. Int J Traf Transp Eng 8(3):294–308
13. Gebretensay FB, Juremalani J (2018) Road traffic accident analysis and prediction model: a case study of Vadodara City. Int Res J Eng Technol (IRJET) 5(1):191–196
14. Mir WA, Batra VS, Singla S (2018) Road traffic accident analysis and prediction model: a case study of Kashmir. Int J Tech Innov Mod Eng Sci (IJTIMES) 4(5):1343–1348
15. Malik FA, Jabbar SFSFFRMJ, Rashid IUHWSMSF (2017) Road accidents and prevention. IJEDR 5(2):40–46
16. Shariff SSR, Maad HA, Halim NNA, Derasit Z (2017) Spatial analysis of fatal road accidents along the north south expressway. J Eng Appl Sci 12(Special Issue 9):8764–8769
17. Ghosh Amit, Paul Suman (2013) Road accident scenario in Kolkata: a spatio-temporal Study. Euro J Appl Eng Sci Res 2(1):47–57
18. Li W, Zheng S, Lu Y (2017) The analysis of urban traffic accidents based on bayesian network. ASCE- CICTP, pp 4879–4888
19. Gupta H, Rokade S (2017) Development of crash prediction model using multiple regression analysis. Int J Curr Eng Sci Res (IJCESR) 4(6)82–86
20. Sailaja V, Raju SS (2015) Accident analysis on NH-18 by using regression model and its preventive measures. Int J Sci Res (IJSR) 4(4):2467–2470
21. Agyemang B, Abledu GK, Semevoh R (2013) Regression analysis of road traffic accidents and population growth in Ghana. Int J Bus Soc Res (IJBSR) 3(10):41–47

22. Ramaswamy (2003) Compendium on road accidents. FASCAO/CN, Bangalore
23. Pimplikar SS, Kharat AG, Raval PM, Dharmadhikari NP (2008) Effect of geopathic stress on the performance of roadways. Highw Res J Highw Res Bard Indian Road Cong 1(1):31–41
24. Chafekar BH, Jarad GP, Pimplikar SS, Dharmadhikari NP, Kharat AG, Sorate RR (2012) Effect of geopathic stress on pavement distresses. IOSR J Mech Civil Eng 1–8
25. Pimplikar SS, Kharat AG (2012) Effect of geopathic stress on the performance of roadways. IJECS 4:125–128
26. Baskara SN, Yaacob H, Hainin MR, Hassan SA (2016) Accident due to pavement condition—a review. Jurnal Teknologi (Sci Eng) 7(2):75–82. (Haryati Yaacob et al.)

Compact City and Related Impact on Sustainable Development in Urban Areas

Kaivalya Metre, Harshit Baghel, Gaurav Suman, Mohit Batra, and Sujesh D. Ghodmare

Abstract This article examines the city of Nagpur in India in the context of a compact city. The development concept is theoretical elements of the compact city. It tries to analyze the physiological aspects. The compactness of cities often refers to urban sustainability achieved by dense, mixed neighborhoods which make the city more livable. Today the developing cities have an adverse impact as they occupy large areas, convenient transport facilities are promoting the urban sprawl to spread over the large urban area. Compact city concept is one of the sustainable development approaches to restrict the impact of urban sprawl on the environment. There is a huge difference between the city characteristics around the world as compared to India. This paper reviews the compact city concept and tries to explore its impact on sustainable development in the Indian context.

Keywords Compact city · Sustainability · Urban sprawl · Modern planning

1 Introduction

Reduction in agriculture land, destruction of biodiversity, and ecosystem more fuel consumption for transport, loss of community life, ill-health, and less social cohesion are some of the ill effects of global urbanization. But as concluded by many researchers compact city has its own pros and cons of sustainable development. The pros are to reduce travel time and distance, efficient utility reduction of fuel consumption, and pollution. The cons are environmental quality; congest social acceptability the lack of urban green space, open space, privacy, and the effect of the urban form of ecology, wildlife nature resource, and economic well-being. Compactness can be evaluated using various parameters such as density, intensity, mixed land use. An

K. Metre (✉) · H. Baghel · G. Suman · M. Batra · S. D. Ghodmare
Civil Engineering Department, G. H. Raisoni College of Engineering, Nagpur, India
e-mail: kai.metre@gmail.com

S. D. Ghodmare
e-mail: sujesh.ghodmare@raisoni.net

L. M. Gupta et al. (eds.), *Advances in Civil Engineering and Infrastructural Development*, Lecture Notes in Civil Engineering 87,
https://doi.org/10.1007/978-981-15-6463-5_50

extended review of the literature and past-research findings are considered as the base for analyzing the compact city development. This paper focuses on selected criteria of compactness and that is highlighted in the methodology.

1.1 General Aspects of Compact City

1. Mixed land use: It refers to a combination in a building where it blends residential, commercial, cultural, institutional, and industrial functions. This promotes the reduction of overall working load on each individual worker.
2. Less complex land use: It is the process of regulating the use of land in an effort to promote more desirable social and environmental outcomes as well as more efficient use of resources.
3. Self-sufficiency in daily life: When the city provides the community with all the needs that the residents require then it is termed ads self-sustaining. Stores, employers, post offices, service providers, energy generation, waste disposal and processing, and small-scale agricultural production are included in this.
4. Efficiency in transportation: Reduction in area results in modicum transportation, thus directly affecting the fuel consumption of the vehicle which rapidly reduces.
5. Accessibility: It refers to the community's swiftness in providing amenities and supplies to the residents living there.
6. Open space ratio: It is the amount of open space required on a residential zoning lot in a non-contextual district, expressed as a percentage of the total floor area on the zoning area.

1.2 Compact City

The compact city also known as the city of short distances is a modern planning and new design concept. It emphasizes on promoting relatively high residential density which includes mix land uses. It hinges on an efficient public transport system. Compact city has an urban layout that incites walking, cycling, low energy consumption, and adding to it is benefits, the reduction of pollution. Opportunities for social interaction and a feeling of safety are instilled in residents as a result of high population.

2 Literature Review

Rajashri Kotharkar et al. [1] have worked and studied the compact city concept and its applicability in the Indian context and concluded that by increasing FSI, more people will be accommodated with lesser areas even if it may not have much impact on the overall density of the city. According to Krzysztof Rogatka et al. (2015) a brief intensification of land use which allows the use of existing facilities and encourages walking and public transport is given. Although the paper has worked on the same topic one is for the Indian scenario and other is for the foreign scenario [2]. Tomasz Bradecki has worked on 'Mapping Urban Open Space and Compact City-Research Methodology'. In this paper, research is introduced to describe the city compactness done with the help of a GIS-based system. It was concluded several policies that include maximum and minimum density ratio, urban open space ratio for development, and policies that promote the same [3]. Michael Neuman has worked on 'The Compact City Fallacy'. This paper concluded that the empirical data of whether compact cities are sustainable, considering the debates related to compact city and sprawl city. The paper also analyses whether compact city development is more sustainable than sprawl city development [4].

Jesper Ole Jensen et al. (2011) have worked on 'Sustainable Urban Development-Compact Cities or Consumer Practices?' This article evaluates compact cities while using lifestyle interpretation of urban forms to challenge the compact city approach. It analysis whether a compact city leads to a more sustainable city [5] Coorey et al. (2005) have worked on 'Urban Compactness and Its Progress towards Sustainability: The Hong Kong Scenario'. This paper evaluates the significant compact development adopted instead of sprawl development. It pointed out the positive as well as a negative impact, quantitative and qualitative attributes of a compact city. Chang et al. [6] have worked on 'Examine Sustainable Urban Space Based on Compact City Concept'. This article states that high compact cities show an improvement in public infrastructure and become most liveable. Hofstad [7] has worked on 'Compact City Development: High ideals and Emerging Practices'. This article's main aim is to search the environmental and social sustainable effect on the city area. The article assesses whether distinct institutional practices support the balancing of social, economic, and environmental goals [8].

3 Research Motivation

It is comparison between sprawl development and compact development. As in sprawl development, there is high utilization of fuel, time, increase in distance, pollution. But if we implement a compact city concept we can reduce the utilization of fuel, time, distance, etc. It also has some negative effects on the environment such as crowding, social acceptability, privacy, and urban greenery. These effects can be avoided by implementing the concept of a compact city.

4 Limitations

This concept cannot be implemented practically. In this work, the existing developments in particular areas are taken into consideration and cannot be made vacant. But here it is only assumed to be shifted in the NMC area nearer to the city Centre for the purpose of analysis. In this paper, all the parameters related to the compact city are not taken into consideration but has focused on only time, fuel consumption, and travel distance. The analysis is done only for selected locations around Nagpur City.

5 Study Area

The area selected for conducting the study is the city of Nagpur in Maharashtra. It covers the area of 227.36 km^2. The total population was 2,497,870 in the year 2011 with a density of 11,000/km^2. Nagpur has the metro rank of 13. Nagpur is the fifth fastest-growing city in the world from 2019 to 2035 with an average growth of 8.41%, this information is in accordance with the Oxford Economics reports (Fig. 1).

It ranks 1st among 100 cities in India and is also one of the Smart Cities in Maharashtra. Adjudged as the 20th cleanest city in India it is also a top mover in the western zone as per Swachh Sarvekshan 2016. Adding to its glory was the award that Nagpur received for the best city in terms of innovation and its practice in Swachh Sarvekshan 2018. Nagpur works progressively towards making it defecation free, its efforts paid off when it was declared as open defecation free under Swachh Bharat mission 2018.

6 Methodology

A methodology needs to be developed in order to describe cities compactness and urban open space. This paper has mainly focused on the issues relating to the public such as time distance, fuel consumption along with other resources. Such issues are considered as benefits of implementation of a compact city concept, which highly depends on the quality of life and sustainability.

For this purpose, we have considered the area which is outside the NMC area. The limit which has been considered was taken from zero miles square and diameter consider was of 10 km around the NMC area of the city. For this purpose, we have identified various locations with the help of a GPS system. Once the locations were identified, respected data were collected for the study. Next was to locate the vacant spaces in NMC region with the help of Google maps and collect the required data. With the help of all the data collected from both the places, the parameters were compared with the help of graphs. The graphs gave us the result of the comparison of the compact city and sprawl development.

7 Data and Analysis

Data is collected with the help of the GPS method. GPS method helps in giving the near about the same distance of the area from the zero miles to the considered 5 locations. The coordinates considered are Lata Mangeshkar Hospital, National Cancer Institute, Smt. Radhikatai Pandav College of Engineering, Bhavans B. P. Vidya Mandir, Mayur Lawns and Marriage Halls, Sez Mihan, Yashwantrao Chavan College of Engineering, Suryudaya College of Engineering, Raj Royal Lawns.

Data was collected by video recording during peak hours from 9 a.m. to 11 a.m. and in evening peak hours 5 p.m. to 7 p.m. which focus on the on-going traffic on that particular region. After recording, the traffic was calculated (only cars and bikes are considered). This gave us the count of people who travel a long distance. If these locations were able to build in the NMC area on the free spaces on those particular roads (or other vacant space in NMC area), the public would get relief. For approximate calculation, 60% of total bikes and 40% of total cars calculated by videography method of that area are considered to be going from zero miles to the respected co-ordinate. On that basis, the fuel consumption of the bikes and cars are calculated. The average mileage of cars is considered as 15 kmph and the bike is 40 kmph.

$$\text{Fuel Consumption} = (\text{total distance})/\text{mileage}$$

It is observed from Fig. 2 that the distance in sprawl development at location 3, which is the National cancer institute is 19.1 km whereas when it is converted into compact development it is observed to be 7.2 km. This reduces the total travel distance by 11.9 km. Also at location no. 9, Mayur lawn and marriage hall, the distance in sprawl development is observed to be 10.9 km whereas when converted into compact development it is observed to be 7.7 km. This reduces the total travel distance by 3.2 km.

It is observed in Fig. 3, the fuel consumption in sprawl development observed at location 3 which is National cancer institute is 1204.73 L whereas when it is converted into compact development it is observed to be 454.14 L. This reduces the total fuel consumption by 750.59 L. Also at location no. 9, Mayur lawn and marriage hall, the fuel consumption in sprawl development are observed to be 362.88 L whereas when converted into compact development it is observed to be 256.35 L. This reduces fuel consumption by 106.53 L.

It is observed in Fig. 4, the travel time in sprawl development observed at location 3 which is National cancer institute is 31 min whereas when it is converted into compact development it is observed to be 14 min. This reduces the total travel time by 17 min. Also at location no. 9, Mayur lawn and marriage hall, the travel time in

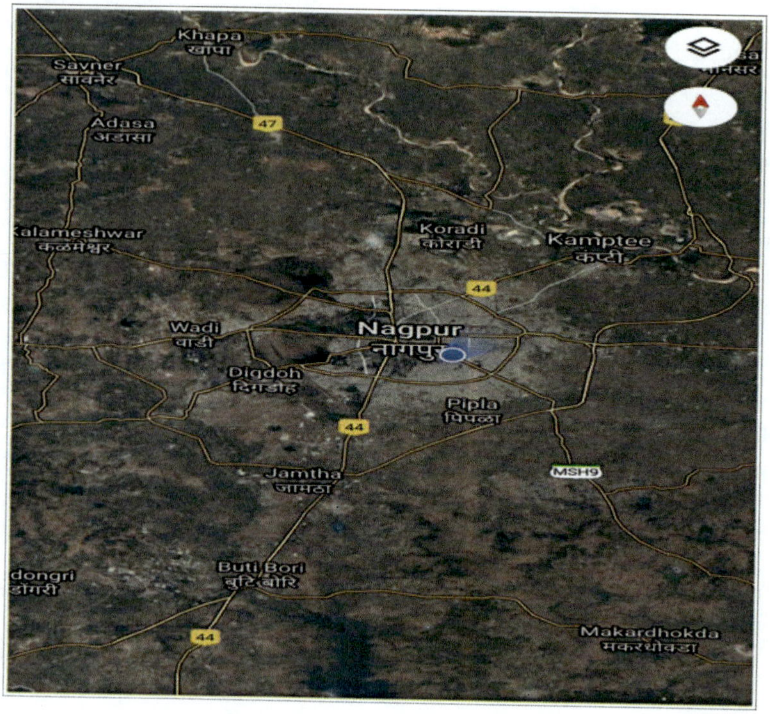

Fig. 1 Map of Nagpur (satellite view)

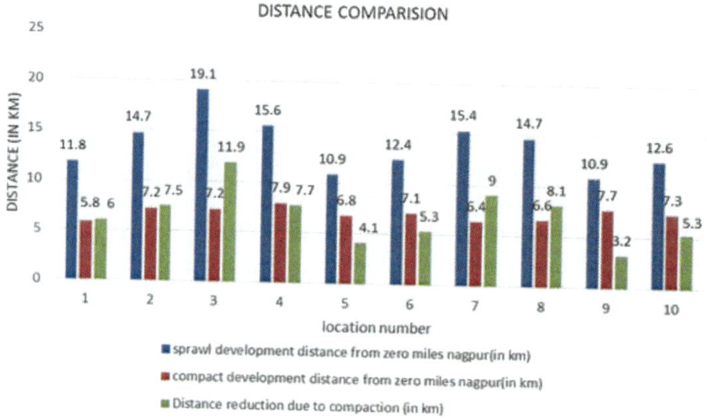

Fig. 2 Distance comparison chart

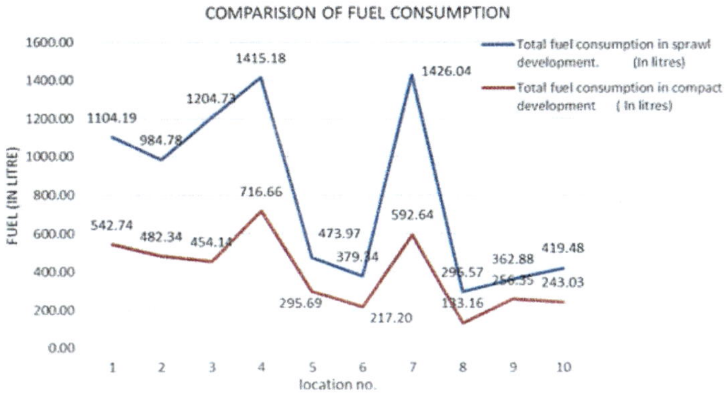

Fig. 3 Comparison of fuel consumption between sprawl and compact development

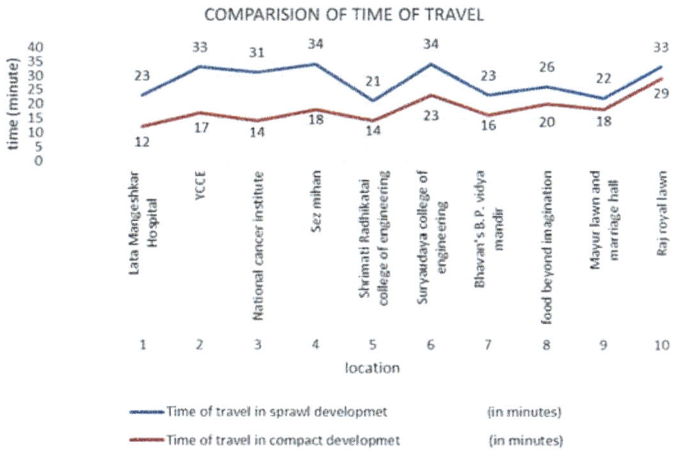

Fig. 4 Comparison of time of travel between the selected locations.

sprawl development is observed to be 22 min whereas when converted into compact development it is observed to be 18 min. This reduces the total travel time by 4 min.

8 Conclusion and Result

On the bases of the study conducted it can be observed that the time, distance, and fuel consumption values for an observed number of vehicles connected with the particular organization or the establishment were high. For example, the distance, time, and fuel consumption for National Cancer Hospital were 19.1 km, 31 min, and 1204.73 L, respectively with sprawl development. Whereas with the concept of the compact city after shifting, the distance, time, and fuel consumption are found to be 7.2 km, 14 min, and 454.14 L, respectively. Hence it is observed that the total saving of distance, time, and fuel are 11.9 km, 17 min, and 750.59 L respectively.

The study reveals a successful comparison between sprawl development and compact development which increases the efficiency of land use and resources. Reduce travel time encourages better social time and sustainable communities and also involves a reduction in distance and fuel consumption. Compact development also reduces pollution compare to sprawl development due to a reduction in fuel consumption. Compact development can be implemented in urban areas as well as in rural areas. It also reduces harmful impacts on nature and helps to maintain environmental balance. Compact city also has some negative effect on society like congest, open spaces, and privacy.

References

1. Rajashri Kotharkar PN (2012) PLEA 2012-28th conference, opportunities, limits & needs towards an environmentally responsible architecture Lima, Peru 7–9 November 2012; topic-compact city concept: its relevance and applicability for planning of Indian Cities
2. Krzysztof Rogatka RR (2015) Topic-a compact city and its social perception: a case study. Urbani izziv 26(1)
3. Bradecki T (2009) Mapping urban open space and the compact city research methodology
4. Neuman M (2005) The compact city fallacy, 25(1):11–26
5. Jensen JO, Christensen TH (2011) Sustainable urban development- compact cities or consumer practices? Dan J Geo-inf Land Manage, 46(1):50–64
6. Coorey SBA, Lau SSY (2005) Urban compactness and its progress towards sustainability: the Hong Kong Scenario. WIT Trans Ecol Environ 84. ISSN 1743-3641
7. Chang HS, Chen TL (2016) Examine sustainable urban space based on compact city space. Global J Inc (USA)
8. Hofstad H (2012) Compact city development: high ideas and emerging practices. Eur J Spat Dev

PM10 Dispersion Modelling at Unsignalised Intersection Using Caline4—A Case Study

Chintaman Santosh Bari and B. V. Khode

Abstract Air pollution caused due to human activities is one of the major problems in the whole world, and India is the third-largest contributor to the emission of greenhouse gases. Thus to control the air pollution, monitoring for the various pollutants as CO, PM10, PM2.5, SO_x and NO_x was carried out by CPCB in India. In Surat, daily monitoring was carried out at twelve sampling stations all around the city. The air quality index (AQI) was found to fairly harmful with a high concentration of particulate matter (PM). PM10 is a complex mixture of dry solid fragments, droplets of liquids or solid cores with coatings which is highly carcinogenic. The present study aims to study the dispersion of PM10 at an unsignalised intersection on urban roads. The temporal variation of the PM10 was studied compared with the traffic volume. It was observed that morning peak shows a high concentration of PM10 which goes on decreasing towards off-peak then again increases in the evening peak.

Keywords Caline4 · PM10 · Dispersion

1 Introduction

India a developing country is the second highest in the road network around the whole world. The vehicular population in India is about 250 million which also increases congestion and thus emissions. Millions of people get affected due to air pollution caused by human activities which causes great economic damage to ecosystems and society. India is the third largest in the emission of greenhouse gases after China and the USA. The severity of air pollution is so much that life expectancy among Indians on an average reduces by 3.4 years while among the residents of Delhi, it reduces by almost 6.3 years. It was observed that more than 1 lakh children were died due to the severe effect of air pollution in India (WHO). Particulate matter (PM) is a complex mixture tiny particle of metals, soot, soil and dust which vary greatly in shape, size and chemical composition. Particulates are highly carcinogenic and have

C. S. Bari (✉) · B. V. Khode
Transportation Engineering, G. H. Raisoni College of Engineering, Nagpur, India
e-mail: chintamanbari@gmail.com

L. M. Gupta et al. (eds.), *Advances in Civil Engineering and Infrastructural Development*, Lecture Notes in Civil Engineering 87,
https://doi.org/10.1007/978-981-15-6463-5_51

been classified in Group I carcinogen because of their ability to penetrate deep into the lungs and bloodstreams. They can also cause permanent DNA mutations, heart attacks and premature death. More than 80% of cities in India where air quality is monitored do not meet the standard of air quality prescribed by the Government of India. Various models have been applied for estimation, simulation and dispersion of gases such as Caline4 [3, 5, 9–11], MOBILE [1, 11], MOVES [6, 8]. Li et al. [5] have studied the simulation of emissions at a signalised intersection in China using Caline4 software and found that the Caline4 predicts good relation with the actual field data. Vincent et al. [9] have studied the spatial and temporal variation and dispersion of highway generated air pollution in residential urban area using Caline4 and found that the effect of pollution goes on decreasing away from the road Central Pollution Control Board [2] has studied the pollution level of six megacities with the use of IVE and action plans were taken for each city. Fallah-Shorshani et al. [4] have integrated Gaussian puff and street canyon model for the simulation of NO_x. Yazdi et al. [10] have studied the vehicular emissions near the highway in Tehran using tunnel experimental campaign and modelling procedure using Caline4, COPERT and found the emissions follow exponential decay pattern and are remaining up to 50–80% of the total in a distance between 100 and 150 m from highway. Tennyson and Kumar [7] have studied the carbon monoxide (CO) prediction and seasonal variation for Madurai City in India and found Caline4 to be predicting nearly equal values of actual data.

From the above literature, Caline4 was found to be most appropriate for the study. The aim of the present research was to study the dispersion pattern of the primary pollutant PM10 in an urban area. The study area was taken as an unsignalised intersection in the city of Surat. Surat is the eighth largest city and ninth largest urban agglomeration in Gujarat, India. The city is located 284 km, south of the state capital, Gandhinagar. The city centre is located on the Tapti River, close to the Arabian Sea. The Surat is located at 21°10′ N 72°50′ E. It has an average elevation of 13 m. The average maximum temperature in summer is 37 °C and in monsoon 32 °C and in winter it downs to 23 °C.

2 Methodology

The California Line Source Model (Caline4) is a line source Gaussian model which was developed by the California Department of Transportation (Caltrans). The Caline4 was used to estimate the dispersion of the pollutants such as carbon monoxide (CO), particulate matter (PM) and nitrogen oxides (NO_x) for around 500 m the links using receptors. The Caline4 is applicable for suburban, rural and Central Business District (CBD) areas and urban areas by changing the aerodynamic coefficient. The inputs for the Caline4 were ambient temperature, base pollutants, ambient velocity and stability class as meteorological parameters. The traffic volume, link size, link type, link height, emission factors and receptors position were given as inputs for traffic characteristics. The base data for the present study was used from the dataset obtained from the Gujarat Pollution Control Board (GPCB) for the last four years. The model for dispersion was formed for each temporal variation.

3 Data Collection

For the present study, the data was collected at an unsignalised intersection in Surat City on 27th September 2018. The videographic data was collected for one hour in the morning peak, off-peak and evening peak. The meteorological data were obtained by the meteorological department. The data were analysed using AVIDEMUX software on a large screen as shown in Fig. 1. Table 1 depicts the traffic volume count and weighted emission factors for morning peak, off-peak, and evening peak, respectively.

The Caline4 was used for dispersion modelling with the input data as shown in Table 2. The Caline4 was provided with only 20 receptors. So to increase the data points, the receptors are used in each quadrant, 20 at a time as shown in Fig. 2. Figure 2 shows the sample receptors in north-east quadrant, and same receptors were applied for other quadrants.

Figure 3 shows the yearly variation of three constituent pollutants for the last four years as obtained from GPCB. Figure 3 shows that there was a large variation in PM10. The least value was found to be 63 $\mu g/m^3$ and a maximum of about 107 $\mu g/m^3$. The concentration of PM2.5 was within range of 16–40 $\mu g/m^3$ and that of NO_x it was found to be 9.4–21 $\mu g/m^3$. The maximum values occurred in the year of 2017–2018 as shown in Fig. 3d. The annual average for PM10 was violating the

Fig. 1 Traffic volume count at unsignalised intersection, Surat

Table 1 Traffic volume and emission factors for each link

	Traffic volume (veh per hour)			Emission factor (g/km)		
	Morning peak	Off peak	Evening peak	Morning peak	Off peak	Evening peak
Northbound to southbound	3398	2934	4173	0.8073	0.7783	0.7808
Southbound to northbound	2179	1844	2234	0.8502	0.7740	0.8869
Eastbound to westbound	731	654	1163	0.8073	0.7783	0.7808
Westbound to eastbound	487	436	776	0.8502	0.7740	0.8869

Table 2 Base data for Caline4 for morning peak, off-peak and evening peak

	Time	Temperature (°C)	Wind direction (from north) (°)	Speed (m/s)	Base PM10 (µg/m^3)	Base PM2.5 (µg/m^3)	Base NO_x (µg/m^3)
Morning	9–10 a.m.	28	120	1.6666	81.05	26.05	18.09
Off	1–2 p.m.	37	210	1.6666	81.05	26.05	18.09
Evening	6–7 p.m.	34	290	1.9444	81.05	26.05	18.09

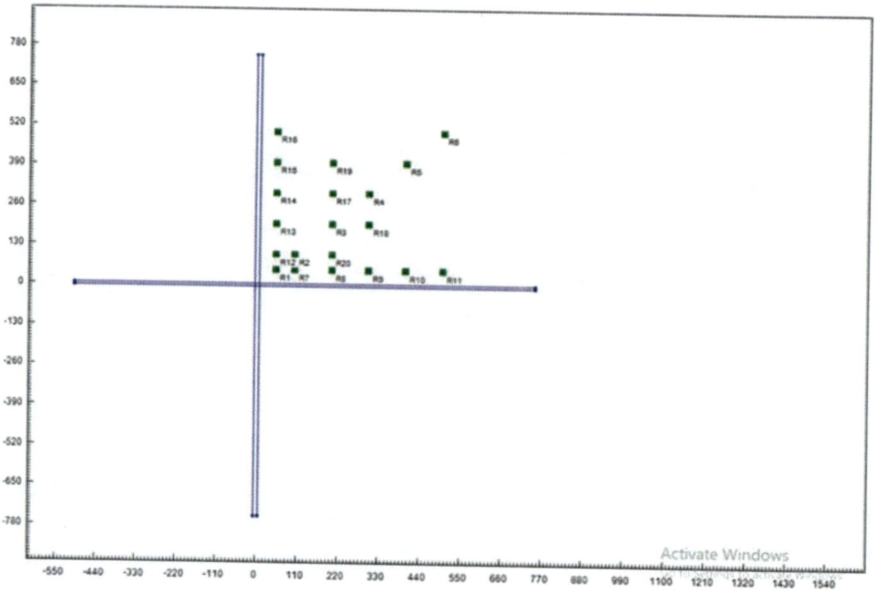

Fig. 2 Receptor position in Caline4 for north-east (NE)

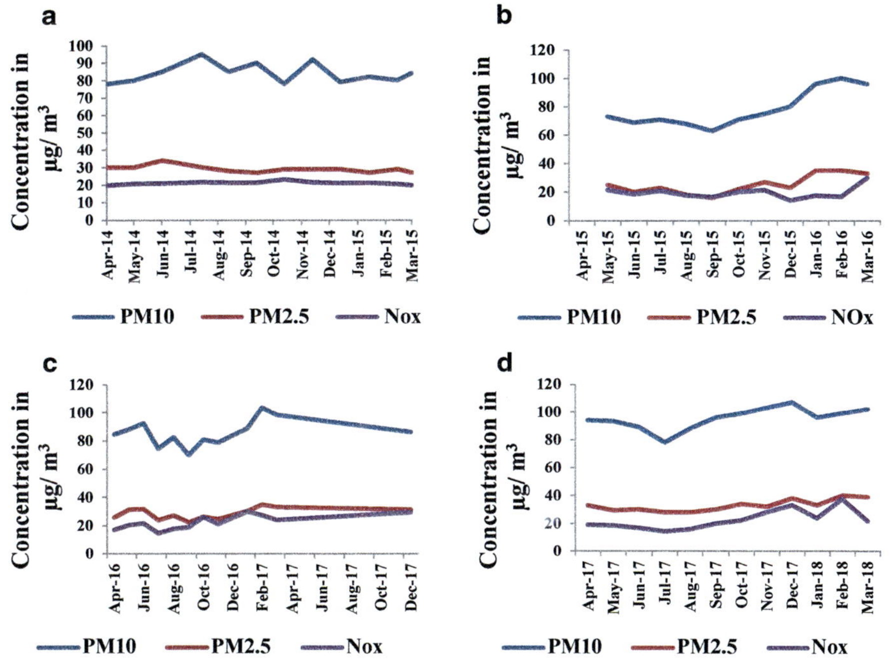

Fig. 3 Variation of PM10, PM2.5 and NO_x for last four year. **a** Variation of PM10, PM2.5 and NO_x for the year 2014–2015. **b** Variation of PM10, PM2.5 and NO_x for the year 2015–2016. **c** Variation of PM10, PM2.5 and NO_x for the year 2016–2017. **d** Variation of PM10, PM2.5 and NO_x for the year 2017–2018

prescribed standards by the Central Pollution Control Board (CPCB). The annual average values for PM2.5 and NO_x were within the limits given by CPCB.

The variation in concentration with respect to the season was also checked for base data. Figure 4 depicts the seasonal variation in pollutants. It was found that there is an overall increase in concentration per year for all three constituents. The PM10 concentration shows its maximum value in the season of summer as 101 $\mu g/m^3$. This value occurs may be due to the blowing of the wind and a lot of alteration in the environment as pressure zones. The base data also shows that the PM2.5 has a maximum value of 38 $\mu g/m^3$ in the season of summer. Contrary the NO_x shows the maximum value of 27 $\mu g/m^3$ in the month of winter. This may be due to a stagnant environment that helps the pollutant to stay within the environment which gets accumulates in the space.

Figure 5 depicts the variation in pollutants with respect to comparison in season within a year. It was observed that the year 2014 shows the decreasing trend for PM10 and PM2.5 from monsoon to summer. But from 2014 onwards, the maximum concentration was obtained in the season of summer that is increasing trend from monsoon to summer. But the NO_x shows the peak in winter as discussed earlier in Fig. 4.

Fig. 4 Seasonwise variation of pollutants

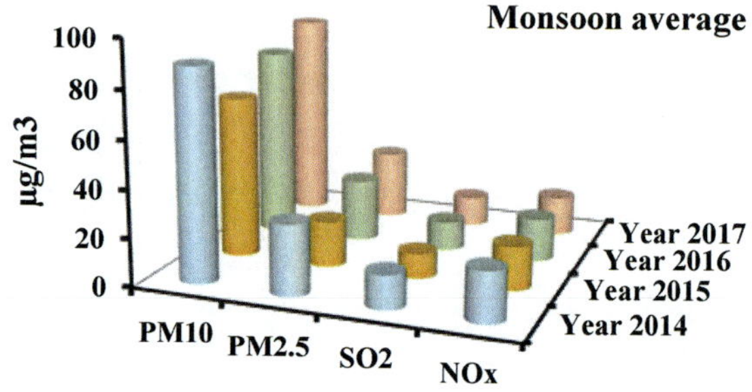

(a) Variation of pollutants in Monsoon season

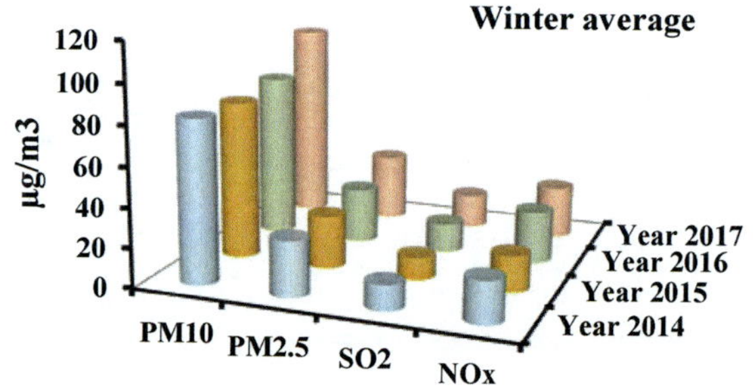

(b) Variation of pollutants in Winter season

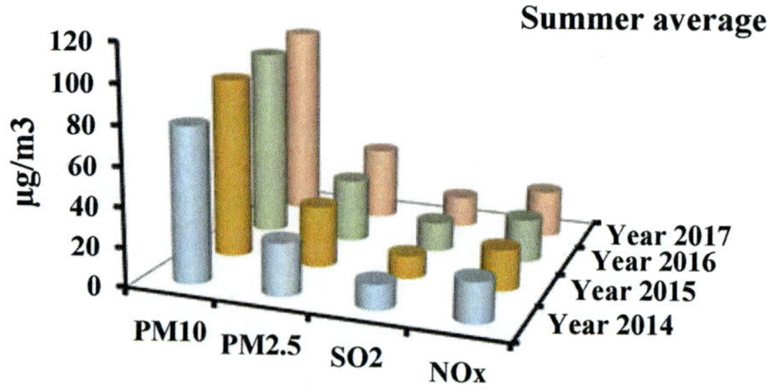

(c) Variation of pollutants in the Summer season

4 Results

The Caline4 model gives the predicted emissions for PM10 and PM2.5 as shown in Fig. 6. The results show that the PM10 concentration was maximum in morning peak of about 114.40 $\mu g/m^3$ and a minimum of about 85.80 $\mu g/m^3$. It may be due to low wind speed and less temperature which formed a calm environment for the accumulation of pollutants. Figure 6a shows that the concentration of PM10 pollutants was high near the road surface, and it goes on decreasing as distance from

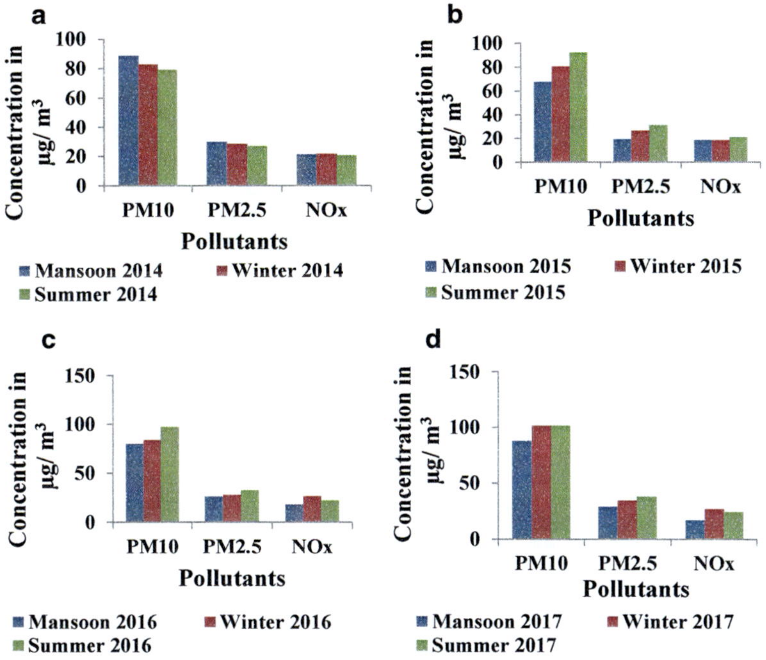

Fig. 5 Yearwise seasonal variation of pollutants. **a** Seasonal variation of pollutants year 2014. **b** Seasonal variation of pollutants year 2015. **c** Seasonal variation of pollutants year 2016. **d** Seasonal variation of pollutants year 2017

road increases, and it decreases up to 25% of the initial value. It was observed that at a particular region (−250, 500) in Fig. 6a, the concentration was high as the wind direction accumulates the pollutants at that point. Hence, it may be concluded that the green belt should be there for dispersion when wind direction is about south east (SE). Figure 6b shows the dispersion of PM10 for off-peak time, and it was found that the concentration was lower than both morning peak and evening peak. The concentration was high at the intersection, and some dispersion occurs towards the south west (SW) due to the prevailing wind. Figure 6c depicts the dispersion of PM10 for evening peak which shows maximum dispersion in north west (NW) that is in direction of the wind. Table 3 describes the descriptive statistics of PM10 emissions.

5 Conclusions

The air pollution is one of the major problems in the present condition and vehicular exhaust is a major source of emission. The health hazards due to air pollutions cause lung cancer, respiratory diseases and even death. Many times the pollutants in a calm environment with fog forms poisonous smog which is more harmful to health and

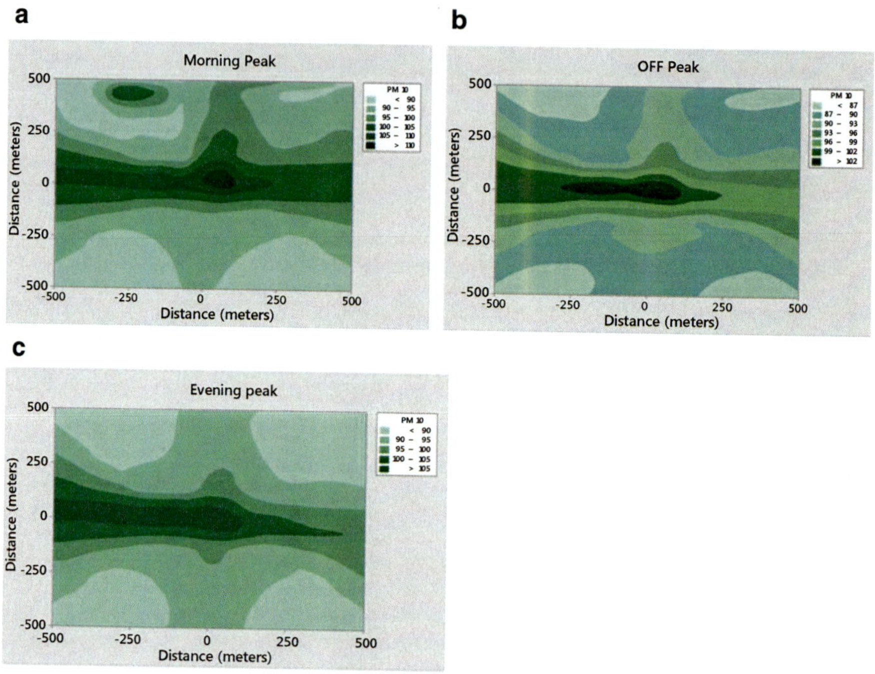

Fig. 6 Temporal variation in dispersion of PM10. **a** Morning peak variation in dispersion of PM10. **b** Off-peak variation in dispersion of PM10. **c** Evening peak variation in dispersion of PM10

Table 3 Descriptive statistics for PM10 output for morning peak, off-peak and evening peak

	Morning peak	Off peak	Evening peak
Mean (μg/m^3)	96.15	92.64	95.00
Minimum (μg/m^3)	87.00	86.00	85.80
Maximum (μg/m^3)	114.40	103.30	106.90
Standard deviation	6.58	5.06	6.16
Skewness	0.44	0.48	0.53

also for driving purpose. Thus, the present study aims to study the dispersion at unsignalised intersection for PM10 with the use of Caline4 software developed by Caltrans. The traffic volume was collected using video graphic data, and base data was obtained by GPCB. The model was developed with 20 receptors in each quadrant. The results show that the concentration of PM10 was high in the morning peak and found to be low in off-peak. It was also found that dispersion was a function of wind direction and its speed. Furthermore, the concentration of pollutants was found to go on decreasing as distance from the source that links go on increasing.

References

1. Boriboonsomsin K, Uddin W, Asce M (2006) Simplified methodology to estimate emissions from mobile sources for ambient air quality assessment. J Transp Eng 132:817–828
2. Central Pollution Control Board (2015) Status of pollution generated from road transport in six mega cities
3. Coats DM, Bushey RW, Mo ECS, Benson P, Nokes WA, Cramer RL (1980) Evaluation of the CALINE4 line source dispersion model for complex terrain application
4. Fallah-Shorshani M, Shekarrizfard M, Hatzopoulou M (2017) Integrating a street-canyon model with a regional Gaussian dispersion model for improved characterisation of near road air pollution. Atmos Environ 153:21–31
5. Li X, Yang X, Wang W (2002) Simulation on concentration of gas pollutants from motor vehicles near signalized intersection in Chinese city. Traffic Transp Stud
6. Liu H (2014) Validating MOVES PM2.5 emission factor empirically by considering accumulative emission effect
7. Tennyson D, Kumar R (2016) Seasonal trends and Caline4 predictions of carbon monoxide over Madurai city, India. IOSR J Environ Sci Toxicol Food Technol 10(9):77–85
8. Tu Y, Wang W, Xu C (2018) A real-time vehicle emission prediction model for freeway segments based on trajectory data. In: CICTP, pp 2376–2384
9. Vincent AS, Trull J, Zamore W, Brugge D, Durant J (2010) Spatial and temporal distribution of highway-generated air pollution in a residential urban neighbourhood: comparison of monitoring and dispersion modelling results, pp 594–601
10. Yazdi MN, Delavarra M, Arhami M (2015) Evaluating near highway air pollutant levels and estimating emission factors: case study of Tehran, Iran. Sci Total Environ 538:375–384
11. Zhang K, Batterman S (2010) Near-road air pollutant concentrations of CO and PM2.5: a comparison of MOBILE6.2/CALINE4 and generalized additive models. Atmos Environ 44(14):1740–1748

Rating and Condition Assessment of Flexible Pavements for Maintenance Decision

Shruti S. Wadalkar, R. K. Lad, and R. K. Jain

Abstract Pavement Management System (PMS) deals with the planning of maintenance, repair, and rehabilitation activities of road structure. Physical condition assessment is a critical step in pavement maintenance decisions. This decision made by identifying and measuring various distresses like cracking, potholes, raveling, depression, etc. In this work rating of flexible pavement in the Pimpri Chinchwad Municipal Corporation (PCMC) area of Pune region has done by measuring distresses. Based on the condition of the road maintenance decision is taken. For condition assessment and rating of road guidelines given by the Indian Road Congress (IRC 82-2015) has followed. The decision for support can be taken based on the rating of the road. The rating is provided from 1 to 3 for the condition of the poor to good. As per this process, the evaluation of the considered road segment in the PCMC area is in the range of 1.82–2.26, and the condition of the road is fair to good.

Keywords Pavement · Maintenance · Distresses · Condition assessment

1 Introduction

Pavement Management System (PMS) deals with the planning of maintenance, repair, and rehabilitation activities of road structure. A significant objective of PMS is to assist the engineer in making consistent and cost-effective decisions related to the construction, maintenance, and rehabilitation of pavements [1]. The First step towards the planning of maintenance operations is the evaluation of the existing pavement surface in terms of its physical condition as well as structural capacity [1].

S. S. Wadalkar (✉)
Department of Civil Engineering, DIT, SPPU, Pune 411018, India
e-mail: wadalkarshruti@yahoo.in

R. K. Lad
JSPM Technical Campus, Pune, SPPU, Pune, Maharashtra 411041, India

R. K. Jain
RSCOE, Pune, SPPU, Pune, Maharashtra 411033, India

L. M. Gupta et al. (eds.), *Advances in Civil Engineering and Infrastructural Development*, Lecture Notes in Civil Engineering 87,
https://doi.org/10.1007/978-981-15-6463-5_52

Physical condition of the pavement is assessed by identified and measuring various distresses on the pavement surface. Cracking, raveling, potholes, settlements, rut depth are some significant distress in flexible pavement. In this research work rating of flexible pavements by considering the above mentioned distresses has done. The methodology specified in IRC 82-2015 follows for the evaluation of flexible pavements IRC 82-2015 deals with mainly three types of maintenance viz. Routine, Preventive, and Periodic. Routine maintenance involves filling potholes, repairing cracks, and patchwork. Preventive maintenance is performed to improve or extend the useful life of the pavement surface while in good condition. And periodic maintenance includes regular maintenance operations like applying a renewal coat. In the case of urban roads, preventive maintenance is undertaken before pavement rating drops to 2 and periodic renewal may be conducted at a serviceability level of 2 [1].

2 Past Research Work

Boyapati and Kumar [2] prioritized the maintenance of pavements based on the Pavement Condition Index. Mannan and Tarefder [3] were analyzed the pavement performance of New Mexico's SPS-5 sections by analyzing five distress parameters; International Roughness Index (IRI), rutting, fatigue cracking, longitudinal cracking, and transverse cracking. Avinash et al. [4] had done performance evaluation of low volume flexible Pavements of an urban road network of Tumkur city, Karnataka, India. They have evaluated the functional and structural conditions in terms of CBR value of the subgrade soil, rebound deflection, International Roughness Index (IRI), and condition rating based on the extent of cracking, patching, rutting and potholes. Adu-Gyamfi et al. [5] had done a functional evaluation of pavement condition by using a complete vision system. Haider et al. [6] were estimated optimum timings for treatments on flexible pavements with surface rutting. Shaha et al. [7] developed a combined overall pavement condition index for the selected network of Noida urban roads. The four performance indices viz. Pavement Condition Distress Index, Pavement Condition Roughness Index, Pavement Condition Structural Capacity Index, and Pavement Condition Skid Resistance Index was developed individually. Then all these indices are combined to form an OPCI giving importance of each indicator. Ullas et al. [8] was evaluated the functional and structural condition of flexible pavements and developed performance prediction models using SPSS package. Five different roads types of urban pavement are considered for the study Suman et al. [9] presented as an approach to pavement treatment selection using a fuzzy logic inference system.

3 Methodology

In the present study, rating of flexible pavement is done. For the rating purpose methodology prescribed in IRC 82-2015 is adopted. Table 1 represents IRC 82-2015, criteria for the rating of pavements.

For the study purpose, 2.00 km segments of roads are considered from Pimpri-Chinchwad Pune. Cracking, raveling, potholes, settlements, rut depth are measured. Longitudinal and transverse cracks are measured in linear meter. Block and fatigue cracking are measured in square meter by selecting the appropriate area of developed cracks. Potholes are measured by measuring area and dept. Settlement is measured in area and depth. Rut depth is measured in linear meter. All the distress is expressed in percentage and shown in Table 2.

Based on Table 1 and the values of distresses obtained in Table 2, rating of all the road segments from 1 to 3 has been done which is shown in Table 3.

In IRC 82-2015, appropriate weights for the rating value of each parameter are given. The rating of each parameter multiplies by weight for calculating weighted rating value. The fixed weight for each parameter is given in Table 4.

Table 1 Pavement distress based rating for urban roads

Defects	Range of defects		
Cracking (%)	>15	5–15	<5
Raveling (%)	>10	5–10	<5
Potholes (%)	>0.5	<0.5	NA
Settlement (%)	>5	1–5	<1
Rut depth (mm)	>10	5–10	<5
Rating	1	1.1–2	2.1–3
Condition	Poor	Fair	Good

Source IRC 82-2015

Table 2 Values of distresses

Distresses	Road segment 1	Road segment 2	Road segment 3	Road segment 4	Road segment 5
Cracking (%)	21.24	3.74	6.41	0.58	14.66
Raveling (%)	0.082	0.59	0	0	0
Potholes (%)	0.092	0.048	0.065	0	0.15
Settlements (%)	0.07	0.0345	0.025	0	0
Rut depth in mm	0.89	0.58	0.36	0.4	0.32

Table 3 Rating of road segments

Distresses	Rating of road segments as per Table 1				
	Road segment 1	Road segment 2	Road segment 3	Road segment 4	Road segment 5
Cracking (%)	1	2.325	2.620	2.900	1.130
Raveling (%)	2.85	2.890	3.00	3.000	3.000
Potholes (%)	1.83	1.910	1.880	2.000	1.730
Settlements (%)	2.94	2.970	2.980	3.00	3.000
Rut depth in mm	2.84	2.740	2.940	2.930	2.940

Table 4 Fixed weightage

S. No.	Parameter	Weightage (fixed)
1	Cracking (%)	1.00
2	Raveling (%)	0.75
3	Potholes (%)	0.50
4	Shoving (%)	1.00
5	Patching (%)	0.75
6	Settlement (%)	0.75
7	Rut depth	1.00

Source IRC 82-2015

The average of the Weighted Rating Values of all parameters is the final rating value (IRC 82-2015). The final rating of road segment one is shown in Table 5.

Similarly, the final rating of all other road segments has done. Figure 1 represents the final rating values of all the road segments.

Table 5 Final rating value calculation for road segment 1

Distress type	Value of distress	Rating as per Table 1	Weightage	Weighted rating value
Cracking (%)	21.24	1	1.00	1.00
Raveling (%)	0.082	2.85	0.75	2.14
Potholes (%)	0.092	1.83	0.5	0.92
Settlements (%)	0.07	2.94	0.75	2.21
Rut depth in mm	0.89	2.84	1.00	2.84
Final rating value			1.82	
Condition			Fair	

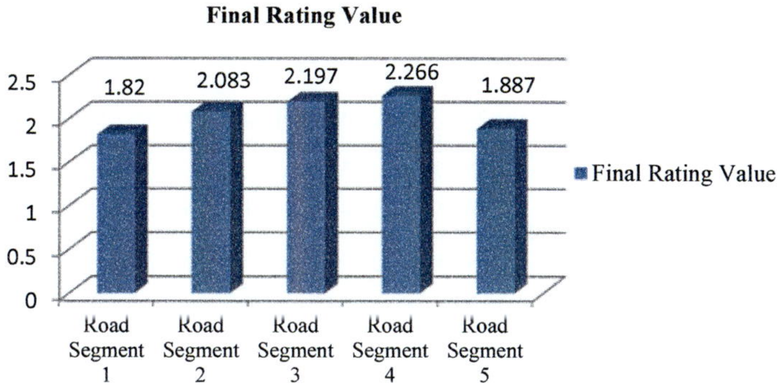

Fig. 1 Final rating value for road segments

Table 6 Maintenance decision for road segments

Road segment	Final rating range	Maintenance decision
1	<2	Periodic renewal
2	>2	Preventive maintenance
3	>2	Preventive maintenance
4	>2	Preventive maintenance
5	<2	Periodic renewal

4 Result and Discussion

From Fig. 1, it is observed that the final rating of the road for road segments one and five is between 1.1 and 2 so both the road segments are in fair condition. Road segment two scores are near about 2.1; hence, it is in fair to good condition. Road segment four is in good condition as the final rating is more significant than 2.1. Based on rating value, the maintenance decision for road segments is shown in Table 6.

5 Conclusions

In this study, rating of flexible pavements has been done by using the IRC 82-2015 method. Major distresses in the flexible pavements for urban roads are considered for the assessment. This method can be applicable for rating and condition assessment of flexible pavements efficiently. The rating of road segments is in the range of 1.82–2.26. Out of five road segments, two are in fair condition; one is in fair to good condition, and the remaining two are in good condition. Based on condition assessment, decisions for the maintenance of roads can be undertaken. Roads in this area required periodic or preventive measurements according to the condition of the road.

References

1. IRC 82-2015. Code of practices for maintenance of bituminous surfaces
2. Boyapati B, Kumar RP (2015) Prioritization of pavement maintenance based on pavement condition index. Indian J Sci Technol 8(14):64320
3. Mannan UA, Tarefder RA (2014) Evaluation of long-term pavement performance based on New Mexico LTPP SPS5 data. In: T & DI. ASCE, pp 269–279
4. Avinash NR, Vinay HN, Prasad D, Dinesh SV, Dattatreya JK (2014) Performance evaluation of low volume flexible pavements—a case study. In: T & DI. ASCE, pp 69–78
5. Adu-Gyamfi YO, Tienaah T, Attoh-Okine NO, Kambhamettu C (2014) Functional evaluation of pavement condition using a complete vision system. J Transp Eng ASCE 04014040-1–04014040-10
6. Haider SW, Baladi GY, Chatti K (2014) Impact of pavement surface monitoring frequency on pavement management decision making. In: T & DI Congress. ASCE, pp 656–666
7. Shah YU, Jain SS, Tiwari D, Jain MK (2013) Development of overall pavement condition index for urban road network. In: 2nd conference of transportation research group of India (2nd CTRG). Procedia Soc Behav Sci 332–341
8. Ullas S, Sreedevi BG, Sreelatha T (2013) Pavement performance modeling—a case study. Int J Innov Technol Explor Eng 3(3):166–170
9. Suman SK, Sinha S (2013) Pavement maintenance treatment selection using fuzzy logic inference system. Int J Eng Innov Technol 2(6):172–175

A Review on Improving Bearing Capacity of Soil by Effective Use of Geosynthetic Reinforcement

Anand Shrigondekar and Prabhuling Ullagaddi

Abstract With rapid enhancement in infrastructure development, the availability of competent land goes on reducing. By developing techniques of ground improvement, the weak soil can be converted into competent acceptable soil condition. A geosynthetic in the form of planar or three-dimensional or combination of both has been used in the field. In this manuscript, an effort has been made to study an effect of inclusion of geogrid, geocell or geotextile on bearing capacity of soil. It is found that there is considerable enhancement in load-carrying capacity of soil and also observed reduction in displacement of footing. The results of grid and anchorage-reinforced soil have found more sustainable solution than soil with geogrid because of large vertical confined zones and substantial improvement in reinforcement factor. The use of geosynthetic reinforcement has also been observed in embankment construction, railway track application, etc.

Keywords Geosynthetic · Soil reinforcement · Bearing capacity · Foundation · Settlement

1 Introduction

Nowadays due to increase in globalization, there has been deficiency of suitable or competent land. So, there is a need to develop the techniques which modify the properties of weak soil and convert soil into good acceptable condition. The sorts of soil improvement strategies, for example mechanical, hydraulic modification, physical and compound adjustment or improvement by reinforcement and confinement, have been developed to solve the problems. Among the various soil improvement methods, the inclusion of reinforcement in different forms is widely used in practice because of

A. Shrigondekar · P. Ullagaddi (✉)
Civil Engineering Department, Shri Guru Gobind Singhji Institute of Engineering and Technology, Vishnupuri, Nanded, Maharashtra 431606, India
e-mail: pbullagaddi@sggs.ac.in

A. Shrigondekar
e-mail: anandshrigondekar7@gmail.com

L. M. Gupta et al. (eds.), *Advances in Civil Engineering and Infrastructural Development*, Lecture Notes in Civil Engineering 87,
https://doi.org/10.1007/978-981-15-6463-5_53

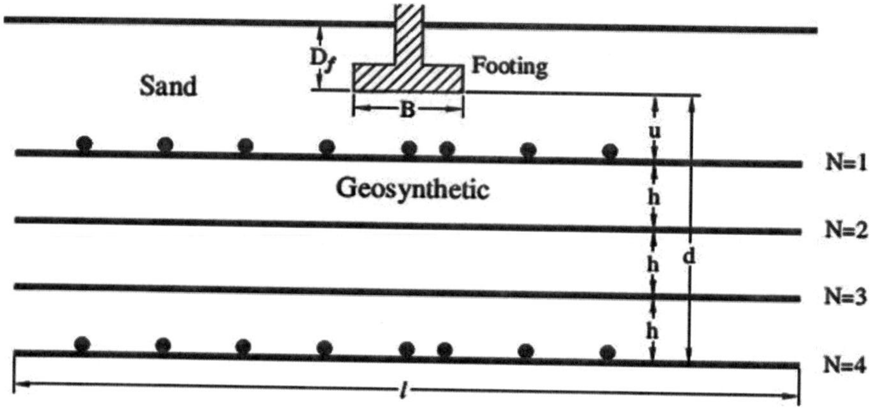

Fig. 1 Geometric parameters of reinforced soil [6]

cost-effective solution and simplicity in application [1]. Generally, the soil possesses characteristics of low tensile strength and it depends on environmental condition also. The tensile strength of soil is improved by providing reinforcement along the direction of tensile stress [2]. Geosynthetics as reinforcement have been effectively utilized in many areas of civil engineering such as airport runways, railways, roadways, retaining structures, embankments and landfills. Geosynthetics are polymeric products which include geogrids, geocells, geotextiles, geomembranes and geocomposites. Geosynthetic-reinforced soil (GRS) system has proved to be an effective way in improving bearing capacity of soil [3]. The load settlement behavior of footing can be modified considerably, and the general shear failure can be eliminated using geosynthetic reinforcement [4]. In this context, some of the researchers carried out experiments in different types of reinforced soil and found out suitable reinforcement configuration [5]. Figure 1 shows geometric parameters of strengthened soil framework.

Yetimoglu et al. [7] measured the bearing capacity of rectangular footing using geogrid and performed both numerical and experimental parametric analyses in single-layer as well as multilayered reinforced sand. For single-layer reinforced sand, it was seen that the load-carrying capacity is maximum for the embedded depth of reinforcement equal to 0.3 times footing width and in support of multilayered reinforced sand, the depth of the first reinforcement was observed as 0.25 times footing width. Das and Omar [8] analyzed maximum bearing pressure of surface strip foundation using laboratory test model on geogrid-reinforced sand as well as unreinforced sand. A fine-graded uniform sand and geogrid of BX1000 SSO were utilized for test and furthermore considered the effect of footing width and density index of sand on improvement factor known as bearing capacity ratio. Some researchers found that the improvement factor of geogrid-reinforced sand decreased with an increase in width of foundation and for a particular width of footing the improvement factor is slightly decreased with increase in relative densities.

1.1 Failure Mechanism in Reinforced Sand

Binquet and Lee [3] initially reported that the ultimate bearing capacity of compressible sand can be enhanced by using metallic reinforcement and the failure mechanism observed is as per the following:

1. Bearing capacity failure above upper reinforcement layer: When the ratio of depth of the first layer reinforcement (u) to width of footing (B) is more than $\frac{2}{3}$, the failure will occur above top layer reinforcement.
2. Pullout of reinforcement due to insufficient embedment length: The failure is likely to be occurred by pullout or tension, when the ratio of depth of the first layer reinforcement (u) to footing width (B) is less than $\frac{2}{3}$, $N < 2$ or 3 and presence of short ties.
3. Breaking of upper ties: The tensile failure or breaking of reinforcement occurs due to overstressing for u/B ratio $\left\langle \frac{2}{3}, N \right\rangle$ 4 and presence of long ties.

2 Previous Studies on Geosynthetic-Reinforced Soil

In this paper, an overview of geosynthetic reinforcement in a shallow foundation has been made based on some previous studies. The parametric study of geosynthetic-reinforced soil has been classified according to type of geosynthetic material used, like geogrid, geotextile, geocell or combination of two.

2.1 Studies on Geogrid-Reinforced Foundation System

Demir et al. [9] carried out a systematic research on bearing capacity of circular footing by using geogrid in compacted granular coarse fill layer overlying on clay deposit. It was seen that for granular fill layer of thickness $0.67D$ and for single layer of support, the depth of geogrid proved to be advantageous from strength perspective when it placed at $0.1D$ and $0.5D$ underneath the base of footing. The bearing capacity enhancement was around 90%, and the reduction in settlement amount was 53%. The effect of the second reinforcement found that bearing capacity expanded by up to 230% and vertical displacement reduced up to 60% when the second geogrid layer set in the range of $0.15D$ and $0.3D$ underneath the first geogrid layer. The outcomes from the finite element investigations were in very good agreement with test perceptions.

Durga Prasad et al. [10] made investigation on model of square footing to establish the load–displacement behavior of soil without reinforcement and strengthened granular soil. Load settlement response in three cases was obtained: (1) aggregate layer overlying sand layer with no reinforcement, (2) geogrid-reinforced sand layer and (3) an aggregate layer overlying sand by using geogrid. In the first case, the aggregate thickness overlying sand was differed from $0.1B$ to $0.5B$. It was observed that the bearing pressure increases with increment in thickness of aggregate. For 100-mm-deep aggregate layer overlying sand and settlement ratio of 10%, the bearing pressure proved to be increased by 81%. From the second case, it pointed out the effect of reinforcement width, i.e., 3*B*, 4*B*, 5*B*. At an optimum depth of reinforcement equal to 0.45*B*, the bearing weight expanded by about 3% and 10% when the reinforcement width increases from 3*B* to 4*B* and from 3*B* to 5*B* separately. The geogrid placed beyond 0.45*B* from bottom of footing was not efficient in improving load capacity of the footing. In the third case, the most suitable depth of geogrid for carrying maximum pressure was observed to be 0.3 times footing width and bearing pressure was increased by 27% compared to the first case. Load improvement factor in all three cases was calculated for the settlement ratio ranging from 5 to 15%, and it ranged from 1.3 to 1.9, 1.1 to 1.8 and 1.3 to 2.7 for the first, second and third cases, respectively.

Mirzaeifar and Ghazavi [11] taken in to account the geometry of foundation and the bearing capacity analysis of geogrid reinforced multi-edge shallow foundation were studied. The results of numerical analysis for square footing confirmed that the bearing capacity improves with addition of number of reinforcement layer up to 3. In case of plus-shaped cross footing for $\frac{B}{L} = 0.3\text{–}0.4$, more than two layers of geogrid were not shown any significant effect on bearing capacity and for $\frac{B}{L} = 0.5\text{–}0.6$, the improvement in bearing capacity was not significant for more than three layers of reinforcement. The result of bearing capacity ratio in H-shaped footing was found similar to that of plus-shaped cross footing. The results of parametric studies under multi-edged foundation showed that the spacing between reinforcement layers and first layer of reinforcement depth is 0.33 L.

Alawaji [12] observed the behavior of geogrid-reinforced sand over collapsible soil. Model tests were completed by utilizing circular footing plate of 100-mm diameter and Tensar SS2 geogrid. The effects on bearing capacity ratio and collapsible settlement were observed by varying width and depth of geogrid. Inclusion of geogrid causes decreased in collapsible settlement and improved the bearing capacity ratio. Elastic modulus increased with diameter of geogrid. From the results obtained, the most optimum depth of geogrid was 0.1 times diameter of loaded area.

Badakhshan and Noorzad [13] experimentally analyzed the ultimate bearing capacity of sand using geogrid. The effect of load eccentricities (7.5, 15, 22.5, 30 mm) was observed by utilizing a circular footing of 120-mm diameter. The maximum bearing pressure in both centrally and eccentrically loaded footings was observed at a depth of the first reinforcement equal to 0.42*D*, and the optimum distance between geogrids was also calculated as 0.42 times diameter of footing. A local shear failure observed in centrally loaded circular footing for reinforced and unreinforced sands whereas the failure in reinforced condition tends to general shear failure by increasing the load eccentricity. The results of experiments showed that bearing capacity ratio increases with increase in load eccentricity to the footing core boundary, beyond that BCR decreased.

Mosallanezhad et al. [14] observed the effect of anchorage system in geogrid-reinforced sand known as grid-anchorage (G-A) system. For this grid-anchorage system, the BCR was reported as 2.98. For the same soil using geogrid, it was almost 1.65. It is clearly understood that the bearing capacity ratio of the grid-anchorage system was roughly 1.8 times more than the traditional geogrid. The common geogrid improves the bearing load by restricting the downward movement of soil due to loading. In grid-anchorage framework, adding anchorage to geogrid can in fact increase the interface angle between soil and reinforcement. There was no significance difference noticed in improving initial and final stiffness factor due to provision of anchorage over traditional geogrid. The bearing capacity of grid-anchorage framework varies nonlinearly with width of footing, and from results obtained, it can be seen that the nonlinear relationship is more in G-A than ordinary geogrid.

Makkar et al. [15] utilized two sorts of three-dimensional geogrids with rectangular and triangular aperture. For rectangular and triangular pattern of three-dimensional geogrids, the improvement in BCR was observed as 3.05 and 2.7, respectively. Load-carrying capacity in this case was 1.85 times that of single layer of conventional reinforcement. For *3D* rectangular geogrid and spacing of $0.75B$ between two successive layers, surface heave of soil was completely eliminated. Certainly, *3D* geogrids performed superior to the usual geogrids in reducing surface deformation.

Sweta and Hussaini [16] completed a series of enormous direct shear test to examine the shear conduct of unreinforced and geogrid-strengthened sub-ballast interfaces. The normal stresses (σ_n) and rates of shearing (S_r) varied from 20 to 100 kN/m^2 and from 2.5 mm/min to 10 mm/min, respectively. The breakage of ballast occurred during shearing at the interface of ballast and sub-ballast was determined in terms of Marsal's breakage (B_g). For the unreinforced interface, it was observed that the breakage increased from 2.84 to 4.07% when σ_n was increased from 20 to 100 kN/m^2 at shearing rate of 2.5 mm/min. By using geogrid at ballast and sub-ballast interface, the value of B_g reduced from 2.84 to 2.08% at $\sigma_n = 20$ kN/m^2 and $S_r =$ 2.5 mm/min.

2.2 Studies on Geotextile and Geogrid-Reinforced Foundation System

Abu Farsarkh et al. [6] observed the settlement to be reduced by 20% at all balance weight levels when the laboratory test outcomes appeared with at least two layers of reinforcement. Performance of composite-reinforced sand (geogrid and geotextile) was superior to those strengthened by geotextile or geogrid alone. By appropriate arrangement of reinforcement, the footing load redistributed to progressively uniform way and stress fixation decreased, which results into reduced settlement. From model footing test results, it was discovered that an influence depth of $1.25B$ required for placing geosynthetic reinforcement irrespective of the geogrid or geotextile and embedment depth.

Ouria and Mahmoudi [17] examined bearing capacity geotextile-reinforced sand by considering impacts of cement treatment of interface between geotextile and sand. The test results calculated for single layer of geotextile the bearing capacity was increased in between 1.46–2.2 times unreinforced case depending upon the length of geotextile. By treating the interface zone of sand and geotextile, it was increased about 1.71–2.34 times without reinforcement. Also, it determined the bearing capacity ratio of geogrid or geocell-reinforced footing which was 1.1–1.15 times the bearing capacity ratio of geotextile-reinforced footings.

Sridhar and Prathapkumar [18] conducted the model test on geotextile-reinforced sand by varying no. of coir geotextile layers from one to four. A graph of peak stress versus % strain for various layers of geotextile clearly indicated that with increase in layers of geotextile from $N = 1$ to $N = 4$, the peak stress increased from 140 to 1060 kN/m^2. The addition of geotextile layer in sand showed that peak stress increases for all values of strain. Subsequently, BCR additionally improved with geotextile layers but rate of increment of BCR reduced with increasing number of geotextile layers up to three beyond that it was insignificant. Due to greater mobilization of shear strength, it was observed that reduction in settlement improves by increasing stress.

Deb and Konai [19] compared the ultimate bearing capacity of sand with and without geotextile by varying percentages of fines (5, 10, 20, 30% by mass). The results obtained from stress–displacement graph clearly indicate that the load-carrying capacity of soil in both cases increases with adding % fines up to 10% fines. The density of the sand also increased as the amount of fine added in sand up to 21% fines. From the graph of % improvement in bearing capacity due to application of geotextile layer versus % fines, it was noticed that percentage improvement of ultimate load-carrying capacity of reinforced soil is more in case of sand mixed with 5% of fines as compared to the pure sand having 65% density index. It was seen that at 5% fines mixed with sand, the interface friction and adhesion were higher. Hence, the effectiveness of the geotextile reinforcement was more in the presence of 5% fines and enhancement in bearing capacity was found more due to the application of reinforcement. The improvement in load due to the application of reinforcement was more in case of pure sand as compared to the 10 and 20% fine mixed soil. This is because of higher density observed in case of 10 and 20% fines mixed in sand compared to pure sand. The effectiveness of reinforcement decreases for dense soil compared to the loose soil.

Tavangar and Shooshpasha [20] performed the plate load tests utilizing 27 cm × 27 cm and 35 cm × 35 cm square plates. The maximum bearing capacity was discovered when first layer of geotextile at $0.2B$ and $0.15B$ for 27 cm × 27 cm and 35 cm × 35 cm plates separately. The ideal number of geotextile layers was resolved for geotextile-strengthened sand with estimation of $\frac{U}{B} = 0.15$, $\frac{W}{B} = 4$ and $\frac{h}{B} = 0.3$. The utilization of geotextile layers up to impact depth underneath the footings can considerably increase bearing capacity of soil. However using more than 4 layers of geotextile, have not major effect on improving bearing capacity.

2.3 Studies on Geocell and Geogrid-Reinforced Foundation System

Hegde and Sitharam [21] modeled a square footing on geocell-strengthened soft clay by using "FLAC-3D". The model was approved with the exploratory test results, and it was discovered the outcomes were in great concurrence with one another. For the soft clay without reinforcement, ultimate bearing capacity in the range of 30 kPa was noted. The results obtained were confirmed that the load-carrying capacity of geocell and geogrid reached to about 150 kPa, whereas the individual effect of geogrid and geocell foundation bed on bearing capacity was observed to be about 70 and 110 kPa, respectively. The most extreme bearing pressure was noticed when the foundation bed reinforced with both geogrid and geocell.

Biswas et al. [22] observed the effect of planar geogrid and three-dimensional geocell reinforcements on bearing capacity in a layered foundation system of clay and sand. In the first case, the sand bed overlying clay subgrade with geogrid at the interface, the most extreme bearing weight decreased from 315 kPa at $H = 0.63D$ to 203 kPa at $H = 2.19D$ when $\frac{s}{D} = 24\%$. For geocell-reinforced sand bed overlying clay subgrade, it was noticed that the pressure settlement responses appeared as higher bearing capacity against the footing settlement than that of in the geogrid-strengthened foundation systems. The maximum bearing pressures were noted 412, 510, 538 and 559 kPa at depth $H = 0.63D$, $1.15D$, $1.67D$ and $2.19D$, respectively, at $\frac{s}{D} = 24\%$. When foundation system contains geocell-reinforced sand bed and base geogrid overlying clay, it was observed that maximum bearing pressure was higher around 720 kPa for $H = 1.15D$ and $\frac{s}{D} = 24\%$. Geocell was observed to be the most beneficial soil fortification method, giving highest degree of improvement.

Dash et al. [23] performed the experiments on a strip footing resting on sand with geocell mattress. The geocell made up of biaxial grid and NP-2 grids was observed to have an equal improvement in bearing capacity up to settlement of about $0.2B$. At higher settlement values, the performance of geocell made up of biaxial grid was better due to high stiffness. From the load settlement behavior of sand with geocell mattress, it was noticed that the increase in bearing capacity equal to 8 times that of unreinforced soil. The geocell formation effect shown that chevron pattern was more beneficial than diamond pattern. The most suitable depth of placing geocell mattress found out was at a depth of $0.1B$ from bottom of footing. Up to a geocell height equal to 2 times the footing width, the performance improvement was remarkable. Beyond that, there was no significant improvement observed. The ideal perspective proportion of geocell pockets for supporting strip footings was observed to be about 1.67.

Table 1 An overview of results obtained in geosynthetic-reinforced soil (GRS) system

S. No.	Author	Footing type	Reinforcement type	Results
1	Yetimoglu	Rectangular	Geogrid	Max load-carrying capacity at a depth of reinforcement = 0.3*B*, depth of the first reinforcement u = 0.25*B*
2	A. Demir	Circular	Geogrid	90% increase in bearing capacity and 53% reduction in settlement
3	B. Durga Prasad	Square	Geogrid	Optimum depth of reinforcement = 0.45*B*, load improvement factor = 1.1–1.8
4	M. Ghazavi	Multi-edge foundation	Geogrid	h = 0.33L and u = 0.33L
5	Ehsan Badakhshan	Circular	Geogrid	h = 0.42*D* and u = 0.42*D*
6	M. Mosallanezhad	Square	Geogrid and anchorage	BCR of G-A system = 2.98 BCR of GRS system= 1.65
7	Femy M. Makkar	Square	Rectangular 3*D* geogrid	Bearing capacity ratio = 3.05
			Triangular 3D Geogrid	Bearing capacity ratio = 2.7
8	Murad Abu Farsarkh	Square	Geotextile and geogrid	Influence depth of reinforcement = 1.25*B*
9	Ahad Ouria	Rectangular	Geotextile	Bearing capacity ratio = 1.46–2.2
10	Yashar Tavangar	Square	Geotextile	u = 0.15*B*, w = 4*B*, h = 0.3*B*
11	Sujit Kumar Dash	Strip	Geocell	Optimum depth of placing geocell = 0.1*B* from bottom of footing

3 Conclusion

This paper mainly discussed the parametric study of different geosynthetic-reinforced soil systems. The failure mechanism of reinforced soil systems is explained in brief. Also, an effort has been made to study the effect of geosynthetic configuration on load–displacement curve and properties of soils. After reviewing the previous literature on geosynthetic-reinforced soil systems, the following inferences can be stated

- The bearing capacity increases up to 3–4 layers of reinforcement depending upon the influence depth of stress distribution. Beyond that, adding reinforcement is not significant.
- The bearing capacity ratio of the grid-anchorage framework is about 1.8 times higher than that of conventional geogrid because adding anchorage to geogrid can effectively increase the interface frictional angle between soil and reinforcement.
- The depth of the first reinforcement and width of reinforcement are 0.1–0.3 times and 2–4 times width of footing, respectively, to get more efficient results in geosynthetic-reinforced soil system.
- Geogrid with rectangular pattern performs superior to that of triangular pattern due to high amount of soil confinement within a cell of 3D rectangular geogrid compared to 3D triangular geogrid.
- The influence depth of placing geosynthetic reinforcement is in the range of 0.8–1.6 times width of footing.
- The pattern of geocell formation shows that chevron pattern is more beneficial than diamond pattern.
- For a constant width of footing, the improvement in bearing capacity is slightly decreased with increase in relative densities.

References

1. Biswas A, Krishna AM (2017) Geocell-reinforced foundation systems: a critical review. Int J Geosynth Gr Eng 3(2):1–18
2. Sarant S (2005) Engineering aspects of reinforced soil. Ind. Geotech. J. **35**(1)
3. Binquet J, Lee KL (1975) Bearing capacity tests on reinforced earth slabs. ASCE J Geotech Eng Div 101:1241–1255
4. Mandal JN, Manjunath VR (1995) Bearing capacity of strip footing resting on reinforced sand subgrades. 14–15
5. Al-Sinaidi AR, Ali AH (2006) Improvement in bearing capacity of soil by geogrid—An 6. experimental approach. In: *IAEG2006*, no 240, pp 1–5
6. Abu-Farsakh M, Chen Q, Sharma R (2013) An experimental evaluation of the behavior of footings on geosynthetic-reinforced sand. Soils Found 53(2):335–348
7. Yetimoglu T, Wu JTH, Saglamer A (1994) Bearing capacity of rectangular footings on geogrid-reinforced sand. J Geotech Eng 120(12):2083–2099
8. Das BM, Omar MT (1994) The effects of foundation width on model tests for the bearing 9. capacity of sand with geogrid reinforcement. Geotech Geol Eng 12(2):133–141
9. Demir A, Yildiz A, Laman M, Ornek M (2014) Experimental and numerical analyses of no. 4, circular footing on geogrid-reinforced granular fill underlain by soft clay. Acta Geotech 9:711–723
10. Durga Prasad B, Hariprasad C, Umashankar B (2016) Load-settlement response of square footing on geogrid reinforced layered granular beds. Int J Geosynth Gr Eng 2(4):1–10
11. Mirzaeifar H, Ghazavi M (2015) Bearing Capacity of Multi-edge Shallow foundations on Geogrid reinforced sand. In: The 4th international conference on Geotechnical Engineer and soil mechanics, December 2015

12. Alawaji HA (2001) Settlement and bearing capacity of geogrid-reinforced sand over collapsible. Soil. Geotext Geomembr 19(2):75–88
13. Badakhshan E, Noorzad A (2017) Effect of footing shape and load eccentricity on behavior of geosynthetic reinforced sand bed. Geotext Geomembr 45(2):58–67
14. Mosallanezhad M, Hataf N, Sadat Taghavi SH (2016) Experimental and large-scale field tests of grid-anchor system performance in increasing the ultimate bearing capacity of granular soils. Can Geotech J 53(7):1047–1058
15. Makkar FM, Chandrakaran S, Sankar N (2017) Behaviour of model square footing resting on sand reinforced with three-dimensional geogrid. Int J Geosynth Gr Eng 3(1):1–10
16. Sweta K, Hussaini SKK (2019) Behavior evaluation of geogrid-reinforced ballast-subballast interface under shear condition. Geotext Geomembr 47(1):23–31
17. Ouria A, Mahmoudi A (2018) Laboratory and numerical modeling of strip footing on geotextile-reinforced sand with cement-treated interface. Geotext Geomembr 46(1):29–39
18. Sridhar R, Prathapkumar MT (2017) Behaviour of model footing resting on sand reinforced with number of layers of coir geotextile. Innov Infrastruct Solut 2(1):1–8
19. Deb K, Konai S (2014) Bearing capacity of geotextile-reinforced sand with varying fine fraction. Geomech Eng 6(1):33–45
20. Tavangar Y, Shooshpasha I (2016) Experimental and numerical study of bearing capacity and effect of specimen size on uniform sand with medium density, reinforced with nonwoven geotextile. Arab J Sci Eng 41(10):4127–4137
21. Hegde AM, Sitharam TG (2015) Three-dimensional numerical analysis of geocell-reinforced soft clay beds by considering the actual geometry of geocell pockets. Can Geotech J 52(9):1396–1407
22. Biswas A, Krishna AM, Dash SK (2016) Behavior of geosynthetic reinforced soil foundation systems supported on stiff clay subgrade. Int J Geomech 16(5):04016007
23. Dash SK, Krishnaswamy NR, Rajagopal K (2001) Bearing capacity of strip footings supported on sand with geocell mattress. Geotext Geomembr 19(4):235–256

Effect of Geotextile-Reinforced Sand on Pore Water Pressure Using Shake Table

Aarti Patil and S. M. Nawghare

Abstract Pore water pressure is a key parameter that governs the strength of soil. The effect of geosynthetic reinforcement to investigate the dissipation of pore water pressure generated due to dynamic load was studied. In the present study, shake table was used for assessing the effect of geotextiles on pore water pressure in case of sand. The tests were conducted on sand with and without reinforcement at relative densities of 30 and 60%. The frequency of base shaking was 5 Hz and was subjected to sinusoidal excitation of 0.5 g. Tests were carried out on woven and nonwoven geotextiles. The effect of variation in the number of geotextile layers on pore water pressure was observed. It was observed that excess pore pressure reduced due to inclusion of more number of reinforcement layers. The time to build up maximum excess pore pressure increased with the number of woven and nonwoven geotextile layers. The excess pore water pressure generated was observed more in case of 30% relative density than 60% dense soil. The change in pore water pressure was more significant in case of nonwoven geotextile than woven geotextile. The percentage increase in reduction of excess pore water pressure in case of nonwoven geotextile for four layers was around 54% and 32% at a relative density of 60 and 30%.

Keywords Pore water pressure · Geotextiles · Shake table

1 Introduction

Earthquakes are one among the most destroying natural disasters causing huge loss of life and severe damage to various buildings and other structures too. This damage is caused due to excess pore water pressure built up in soil strata which results in reduction of shear strength of soil. When saturated loose sands are subjected

A. Patil (✉) · S. M. Nawghare
College of Engineering, Pune, India
e-mail: patila17.civil@coep.ac.in

S. M. Nawghare
e-mail: smn.civil@coep.ac.in

L. M. Gupta et al. (eds.), *Advances in Civil Engineering and Infrastructural Development*, Lecture Notes in Civil Engineering 87,
https://doi.org/10.1007/978-981-15-6463-5_55

to dynamic loading, propagation of shear waves takes place. Loose saturated sand and silts behave like a liquid when shaken by an earthquake as soil grains lose their contact and cause increase in pore water pressure. Hence, some improvement method to increase the strength and to reduce excess pore water pressure is required.

A number of studies under static loading have been conducted on behaviour of geosynthetic-reinforced soil, whereas less research studies have been done under dynamic loading. The effect of reinforcement on liquefaction resistance of Solani sand was studied. They conducted tests on Solani sand reinforced with geogrid sheet, geosynthetic fibre and natural coir fibre. Coir fibres were more effective in improving liquefaction resistance of sand [3]. Various researchers studied the influence on pore water pressure due to variation in frequency and amplitude. Experiments were conducted using uniaxial geogrid as a reinforcing material, and effects of variation in acceleration and relative density on reinforced soil were studied. There was 25% increase in liquefaction resistance [5]. Pore water pressure ratio increased with acceleration amplitude under building pressure. A small change in acceleration influences the changes in pore water pressure developed [8]. The effect of varying ranges of relative density and acceleration on saturated sand was also studied. It was concluded that for higher magnitude of shaking and denser soils, initiation of liquefaction was observed for pore pressure values greater than one [6]. The effect of geosynthetics sandwiched between gravel layers was also studied. This gravel layer proved to be effective in dissipation of pore water pressure and also restrained the shear deformation under embankment due to high permeability. This improved layer restrained the deformation and prevented embankment failure [4]. The mechanical response and pore pressure generation in granular filter were studied. This was subjected to uniaxial cyclic loading. It was commented that cyclic load induces excess pore water pressure which results in internal erosion of fines. The excess pore water pressure is direct function of loading frequency and reduction in permeability [2]. Proper excess pore pressure dissipation can help in consolidation. Consolidation is faster when excess pore pressure is distributed more towards drain and less at impermeable boundary [1].

In this paper, the results of experiments conducted on a shake table for sand reinforced with woven and nonwoven geotextile are reported. The effect of varying number of geotextile-reinforced sand and sand density on pore water pressure is investigated. A series of shake table tests were carried out on saturated sand with different reinforcement material, varying density and number of geotextile layers.

2 Material Used in Study

2.1 Sand and Its Properties

The sand collected from seashore in Goa was used for testing. Laboratory tests were performed to determine the index properties of sand and are presented in Table 1.

Table 1 Index properties of sand

S. No.	Properties	Value
1	Soil type	Poorly graded sand
2	Specific gravity	2.67
3	Uniformity coefficient (C_u)	4.69
4	Coefficient of curvature (C_c)	0.44
5	Grain size D_{50}	0.019 mm
6	D_{10}	0.083 mm
7	Maximum void ratio (e_{max})	0.744
8	Minimum void ratio (e_{min})	0.449
9	Void ratio for $D_r = 30\%$	0.656
10	Dry unit weight for $D_r = 30\%$	15.82 kN/m^3
11	Void ratio for $D_r = 60\%$	0.567
12	Dry unit weight for $D_r = 60\%$	16.72 kN/m^3

Reinforcement Woven and nonwoven geotextile

The ultimate tensile strength of nonwoven and woven geotextile is 3.1 kN/m and 17.8 kN/m, respectively. The thickness of nonwoven geotextile was 3.5 mm, and its mass per unit area was 170 GSM. The aperture opening size of woven and nonwoven geotextile was less than 0.075 mm and 0.07–0.2 mm, respectively.

3 Experimental Investigation

The test program involves a total of 14 tests conducted on shake table apparatus. Shake table is used to simulate earthquake conditions. From the testing program, 2 tests were performed on sand without reinforcement and 12 tests were performed using a different number (i.e., 2, 3 and 4) of woven and nonwoven geotextile reinforcements along with sand at relative density of 30 and 60% and acceleration of 0.5 g.

3.1 Test Model

Experimental set-up includes three main components: a vibrating platform, control panel and motor. The equipment used for the work is unidirectional horizontal shake table. Vibrating platform vibrates in horizontal direction with the soil model attached to it. The size of platform is 1000 mm × 1000 mm and is made of cast iron coated with silver paste. A control panel is an important component as it controls the frequency of shaking. A standard frequency and amplitude can be given to produce the desired acceleration. A motor of 3 HP capacity with a three-phase connection is used. Soil

Fig. 1 Soil model

model of size 400 mm × 400 mm × 400 mm as shown in Fig. 1 was used. The model was mounted on a unidirectional shake table. The model was rigid and 12 mm thick. The pore pressure transducers were connected to soil model at 0.1 and 0.2 m from the base of the model as shown in Fig. 1. The capacity of pore pressure transducers was 100 kPa with a least count of 0.1 kPa. Pore pressure transducers are connected to digital indicator having a capacity of 100 kPa and least count of 0.1 kPa. The pore water pressure ratio (r_u) is defined as

$$r_{\mathrm{u}} = \frac{U}{\sigma'_{\mathrm{vo}}}$$

where U is excess pore water pressure and σ'_{vo} is effective overburden pressure.

4 Test Procedure

One of the most important aspects of test procedure is the method of sample preparation. The basic purpose was to obtain a homogeneous sample with a desired relative density and degree of saturation. The raining technique was adopted to fill the soil in mould using Hooper as it assures uniform relative density. The soil mould was divided into 5 layers up to 350-mm depth from the base of mould. All layers were of 7 cm each.

The sample was prepared by following the procedure:

1. The void ratio (e) analogous to a relative density (D_r) of sand was calculated from

$$e = e_{\max} - D_{\mathrm{r}}(e_{\max} - e_{\min})$$

where e_{max} is maximum void ratio, e_{min} is minimum void ratio and e is void ratio of sand for desired relative density. The values of void ratio (e) at relative density of 30% and 60% were 0.656 and 0.567, respectively.

2. From the above value of void ratio (e), dry unit weight of sand (γ_d) was determined by the following equation:

$$\gamma_d = \frac{G}{1+e}\gamma_w$$

where γ_w is unit weight of water (taken as 9.81 kN/m^3). The values of dry unit weight γ_d for D_r of 30% and 60% were 15.82 kN/m^3 and 16.72 kN/m^3, respectively.

3. Assuming the constant height of mould to be filled up to as 350 mm, the volume occupied by the sample in the mould was determined using the plan dimensions of mould, i.e., 400 mm × 400 mm.
4. The dry weight of sand (W_d) is determined by the following equation:

$$W_d = \gamma_d V$$

The quantity of dry sand (W_d) calculated was filled into the mould using Hooper.

A desired value of amplitude is set with the help of crank shaft arrangement, and frequency was set using control panel before filling the mould. The pore pressure transducers were connected to the mould, and other end was connected to the digital indicator. The knobs of pore pressure transducers were opened at the instant of shaking, and pore water pressure values were recorded. The tests were performed at acceleration value of 0.5 g and at a frequency of 5 Hz. Degree of saturation used was 90%. The shake table was accelerated till the pore water pressure is decreased or shows a constant value after reaching a peak value. The depth at which geotextile sheets were placed is shown in Fig. 2.

5 Permeability Test

The constant head permeability tests were conducted as per IS 2720: Part 17, on sand and sand reinforced with a varying number of woven and nonwoven geotextile sheets.

Coefficient of permeability of sand alone at relative density of 30 and 60% was 1.05×10^{-5} m/s and 2.5×10^{-6} m/s, respectively. From Figs. 3 and 4, it can be observed that the coefficient of permeability was more for both woven and nonwoven geotextiles reinforced with sand at relative density of 30%. The permeability of nonwoven geotextile-reinforced sand was comparatively more because of large aperture opening size as compared to woven geotextile. It can also be observed that permeability of woven geotextile reduced for more number of layers at both the relative densities.

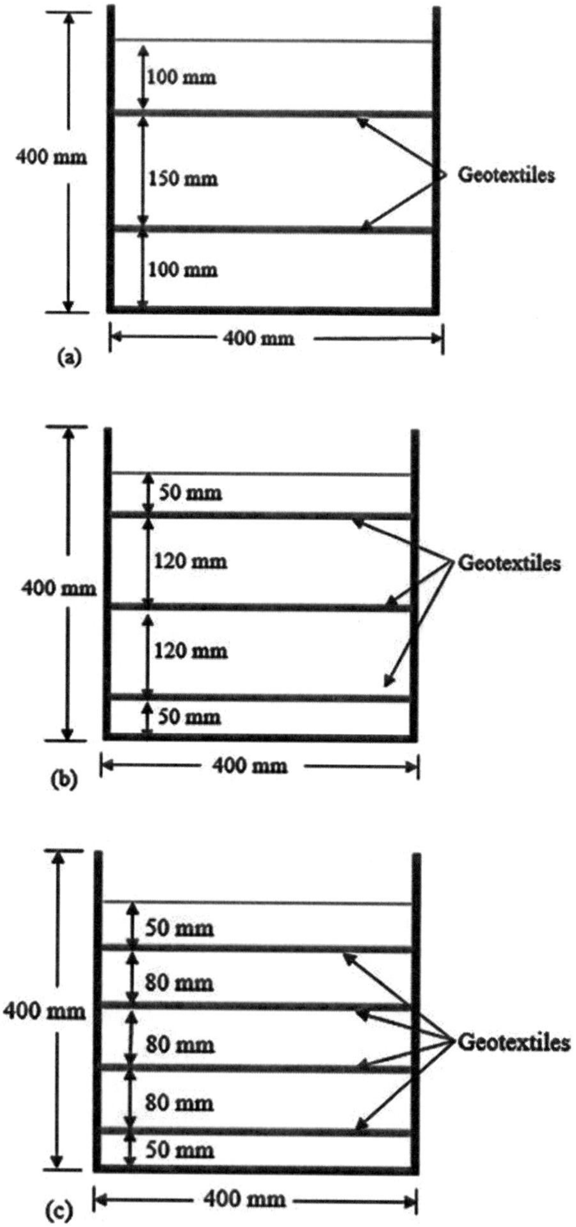

Fig. 2 Position of different geotextile layers. **a** Two layers; **b** three layers; and **c** four layers

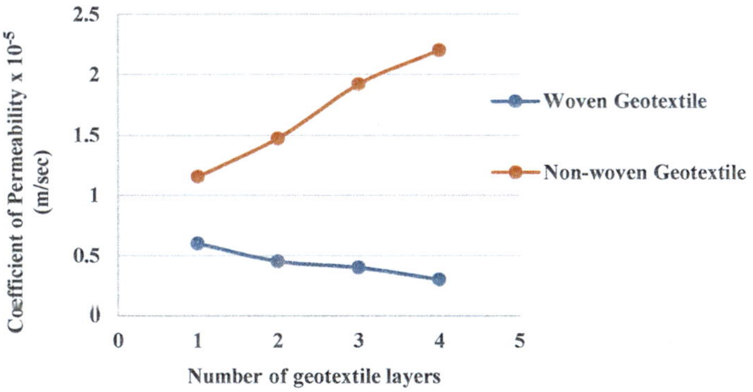

Fig. 3 Variation of coefficient of permeability with geotextile layers at 30% relative density

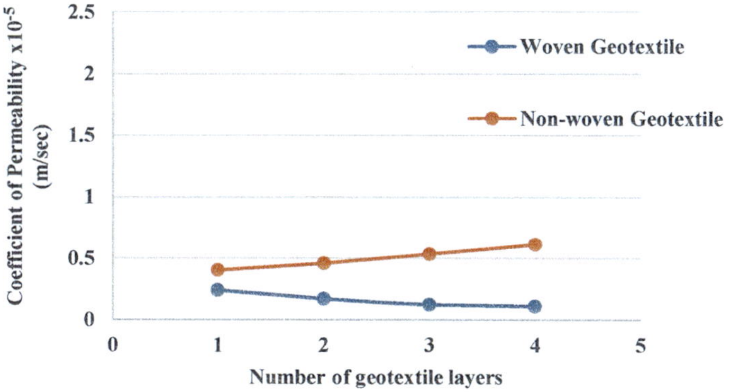

Fig. 4 Variation of coefficient of permeability with geotextile layers at 60% relative density

6 Observation and Results

In each test, the variation of excess pore water pressure with time has been recorded with pore pressure transducer and digital indicator. The pore water pressure was recorded at two depths in model: one pore pressure transducer at bottom (PPB) and another pore pressure transducer at middle (PPM).

Sand Only

Figure 5 shows the results for sand only. The variation of excess pore pressure with time at middle and bottom point is shown.

It can be observed from Fig. 5 that pore water pressure is more at bottom transducer as compared to the middle transducer. This may be due to more effective overburden pressure at bottom. As the depth increases, effective overburden pressure also increases.

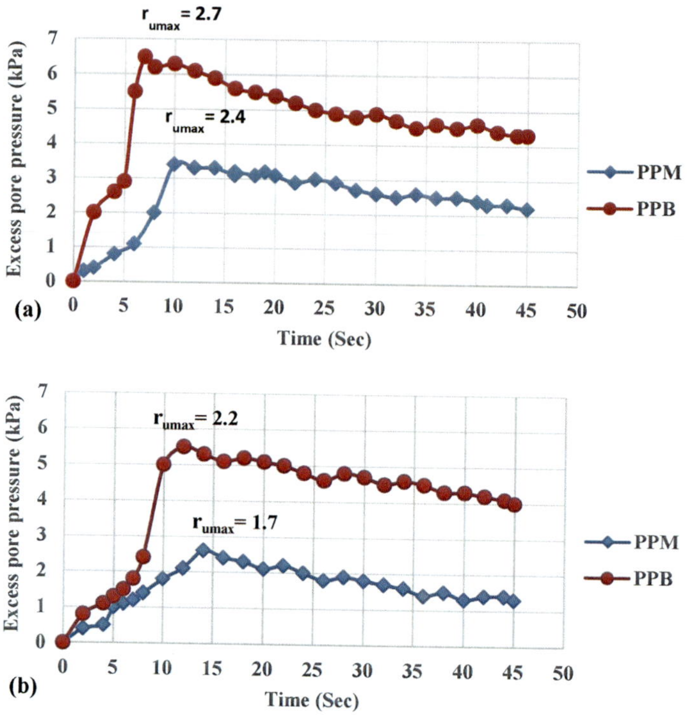

Fig. 5 Variation of excess pore water pressure with time at a relative density of **a** 30% and **b** 60%

The value of excess pore pressures is less at a relative density of 60% as compared to 30% relative density. This may be due to the state of denseness of the soil sample. As relative density increases, the values of pore water pressure decrease. The number of cycles to reach peak value pore pressure was less at 30% relative density than 60% relative density. The pore water pressure ratio was greater than one. Liquefaction was observed in the sample after shaking.

Sand with Woven Geotextile

Variation of excess pore pressure with time at relative density of 30% for two, three and four number of woven geotextile layers was observed.

From Fig. 6, it can be observed that as the number of woven geotextile layers increases excess pore water pressure reduces. But the change in excess pore water pressure is not much significant. The change in pore water pressure was less, and this may be due to permeability of woven geotextile with sand being less. The inclusion of woven geotextile increases the strength and stiffness of sand. Due to addition of four layers of geotextile sheet, the time required to attain peak pore pressure value increased than two geotextile sheets.

The variation of excess pore pressure with respect to time at relative density of 60% for two, three and four number of woven geotextile sheets is shown in Fig. 7.

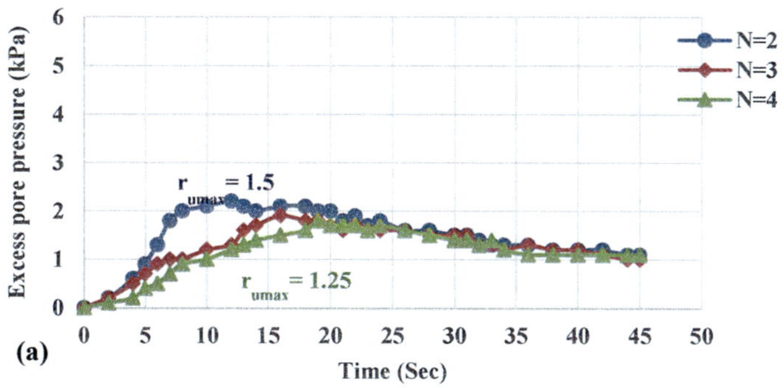

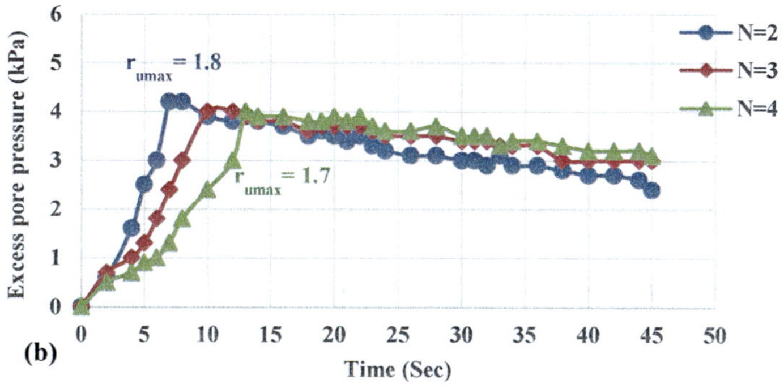

Fig. 6 Variation of excess pore pressure with time at relative density of 30% for **a** middle transducer and **b** bottom transducer

It can be observed from Fig. 7 that excess pore water pressure has reduced slightly at a relative density of 60% which can be due to the denseness of sand. As there was increase in the number of reinforcement layers, the time required to build up excess pore pressure also increased. Inclusion of geotextile layers improves the shear strength of soil. The excess pore water pressure was lower at relative density of 60% than that at 30% relative density. Pore water pressure was more at bottom point than middle point which can be due to the greater effective overburden pressure. The change in excess pore pressure is not much appreciable.

Sand with Nonwoven Geotextile

Variation of excess pore pressure with time at relative density of 30% for two, three and four number of woven geotextile layers was observed.

The change in excess pore water pressure was more significant in case of nonwoven geotextile sheets as shown in Fig. 8. The permeability of nonwoven geotextiles was more as compared to woven geotextiles. This may be due to the apparent opening size being larger in case of nonwoven geotextiles than woven geotextile.

From Fig. 9, it can be concluded that the excess pore pressure generated was less at relative density of 60%. The change in pore pressure was much significant than the results observed in case of woven geotextile.

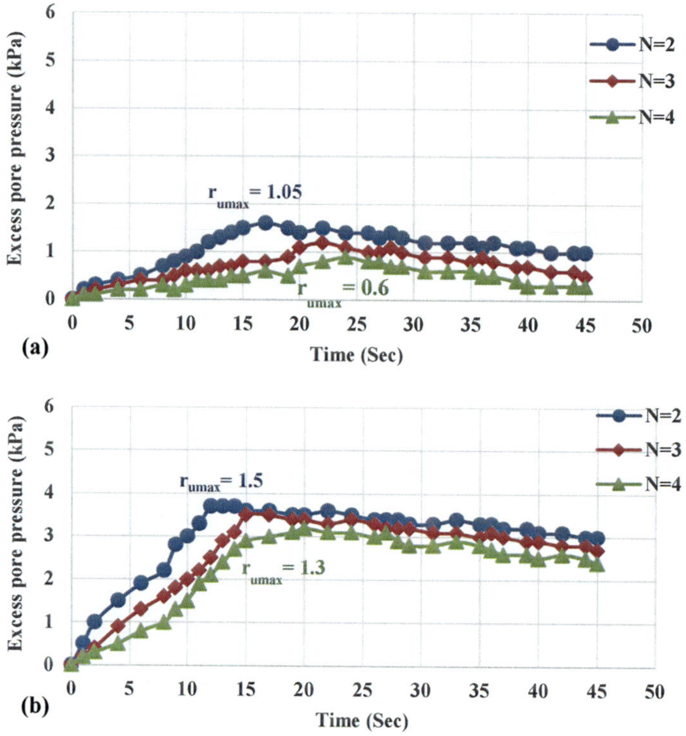

Fig. 7 Variation of excess pore pressure with time at relative density of 60% for **a** middle transducer and **b** bottom transducer

The use of nonwoven geotextile sheets is a better choice than woven geotextiles. The time taken to build up excess pore water pressure was more in case of nonwoven geotextile sheets. The pore water pressure ratio was below one when four number of nonwoven geotextile layers were used and even liquefaction was not observed in the sample, while in rest of the cases where pore water pressure ratio was greater than one liquefaction was seen.

7 Conclusions

The use of geotextile to increase liquefaction resistance was more effective at a relative density of 60%.

The percentage increase in reduction of excess pore water pressure in case of nonwoven geotextile for four layers was around 54 and 32% at relative density of 60 and 30%.

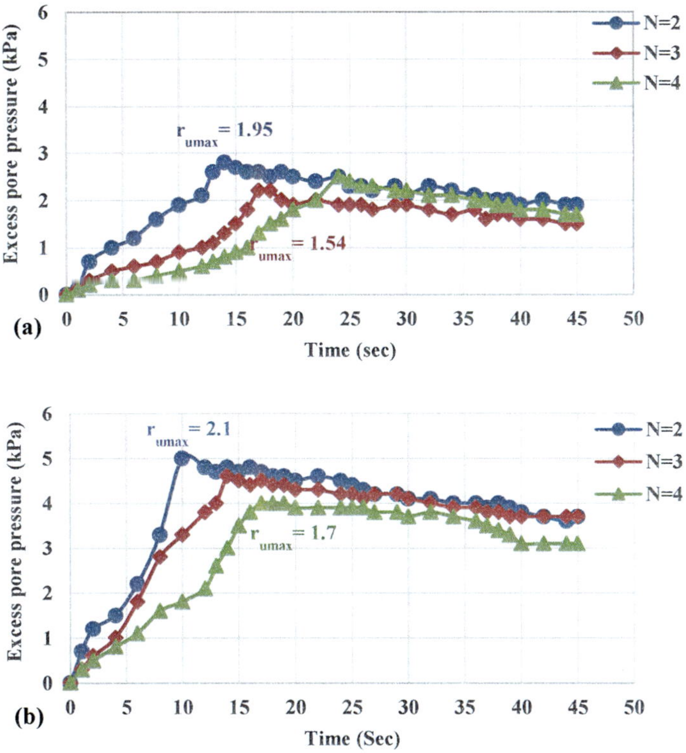

Fig. 8 Variation of excess pore pressure with time at relative density of 30% for **a** middle transducer and **b** bottom transducer

The time required to reach excess pore water pressure was more in case of nonwoven geotextiles and hence more reliable as compared to woven geotextiles. The change in excess pore pressure was more significant in nonwoven geotextiles.

On increasing the number of geotextile layers, the pore pressure value reduces in both woven and nonwoven geotextiles.

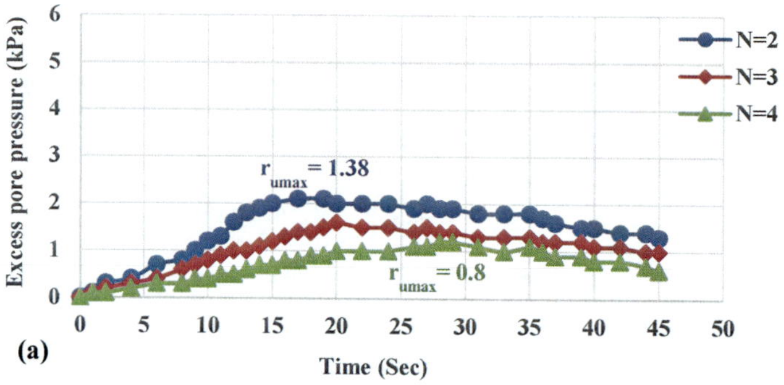

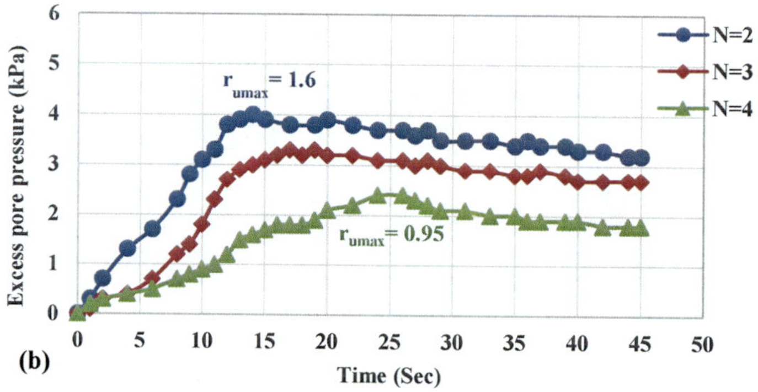

Fig. 9 Variation of excess pore pressure with time at relative density of 60% for **a** middle transducer and **b** bottom transducer

References

1. Ganesalingam D, Read W, Nagaratnam S (2013) Consolidation behavior of a cylindrical soil layer subjected to nonuniform pore-water pressure distribution. Int J Geomech 665–671
2. Israr J, Buddhima I (2018) Mechanical response and pore pressure generation in granular filters subjected to uniaxial cyclic loading. Can Geotech J 1–47
3. Maheshwari BK, Singh HP, Saran S (2012) Effects of reinforcement on liquefaction resistance of Solani sand. J Geotech Geoenviron Eng 138(7):831–840
4. Masaho Y, Hashimoto, R, Kurachi Y (2016) Shaking table tests on a deformation mitigation method for road embankment during liquefaction by using gravel and geosynthetics. Int Collab Lifeline Earthq Eng 384–389
5. Mittal S, Chauhan R (2013) Liquefaction behavior of reinforced saturated sand under dynamic conditions. Int. J. Geotechn. Geoenviron Eng 7(1):109–114
6. Pathak SR, Kshirsagar MP, Joshi MS (2013) Liquefaction triggering criteria using shake table test. Int J Eng Technol 5(5):4439–4449
7. Siddharthan R, Norris GM (1990) Residual pore water pressure and structural response. Soil Dyn Earthq Eng 9(5):265–271
8. Varghese RM, Madhavi Latha G (2014) Shaking table studies on the conditions of sand liquefaction. In: Geo-congress 2014 technical papers, GSP 234, pp 1244–1253
9. Waseim RA, Farouk AA (2016) Performance of reinforced sandy soil under uniform shaking. Electron J Geotech Eng 21(04):1575–1586

Effect of Polypropylene Fiber Length on Geotechnical Properties of Fly Ash

S. M. Nawghare and J. N. Mandal

Abstract In this study, an attempt has been made to improve the geotechnical properties and check the suitability of fly ash for civil engineering applications like slope stability, backfill material for retaining walls, and road construction. Fly ash was reinforced with different lengths (6, 12, 24, 36, and 40 mm) of polypropylene fibers. The fibers were added 0.6% by weight of the fly ash. The various effects of fiber lengths on dry density, moisture content, angle of internal friction, cohesion, unconfined compressive strength, and California bearing ratio value were studied. It was observed that different lengths have different effects on each property. Overall observation was that the 12 mm fiber length has shown good results for angle of internal friction, maximum dry density, and 24 mm fiber length was good for unconfined compressive strength value. The CBR value was increasing for higher length.

Keywords Fly ash · Fibers · Ground improvement

1 Introduction

The fly ash is a byproduct produced by the burning of coal in thermal power station. It causes environmental pollution. Many studies have been reported in the literature for various beneficial use of fly ash and to check feasibility of fly ash in various engineering applications. The various test was performed on fly ash by many researchers to assess the fly ash properties by mixing it with soil, sand, and reinforced with polypropylene fibers.

Polypropylene is flexible and tough in nature. This makes polypropylene applicable in engineering applications. It has good fatigue resistance. The density of polypropylene fibers normally ranges between 0.895 and 0.92 g/cm^3. Many

S. M. Nawghare (✉) · J. N. Mandal
Indian Institute of Technology Bombay, Mumbai, India
e-mail: sariputcr@gmail.com

J. N. Mandal
e-mail: cejnm@iitb.ac.in

L. M. Gupta et al. (eds.), *Advances in Civil Engineering and Infrastructural Development*, Lecture Notes in Civil Engineering 87,
https://doi.org/10.1007/978-981-15-6463-5_56

triaxial test were carried out on cohesionless soil samples reinforced with randomly distributed fibers and found that fibers inclusion increases the peak shear strength and residual strength of cohesionless soil [1]. Same tests were conducted on fiber-reinforced cemented soil. It was found that fiber reinforcement has increased peak strength. It was more considerable for soil without cement. On the other hand, increase in residual strength was more considerable for soil with cement. As long as fibers are not degraded, the fibers have a bridging effect on the crack [2]. The individual and combined effect of both, fiber addition and stabilization using cement on soil properties were studied. It was mentioned that fiber addition changes the brittle behavior of soil to ductile behavior [3]. Direct shear tests were carried on sand reinforced with fibers. It was mentioned by the author that the initial stiffness of the fiber-reinforced sand was not affected much by fiber reinforcement. Even considerable change in peak shear strength of fiber-reinforced sand was not observed [4]. The polypropylene fibers can reduce the spalling and it can improve the residual strength [5]. It is reported that fiber reinforcement can increase the angle of internal friction at the interface of clay barrier and a geomembrane which is smooth in nature [6]. The fiber reinforcement benefit depends on bond strength and angle of internal friction. It is the controlling mechanism for the fiber reinforcement [7].

The fiber reinforcement improves the unconfined compressive strength value of highly compressible clay. This is due to the crimping of fibers. The fiber length and percentage inclusion of fiber also affected the unconfined compressive strength value of clay which is highly compressible in nature [8]. The coir addition and fly ash in combination were used in soil. It increases the dry density and optimum moisture content of soil. It was also found that 10% of fly ash content is sufficient to improve the unconfined compressive strength value and California bearing ratio value [9]. It was found that bearing capacity of silty sand mix with fly ash can be improved by reinforcing it with randomly distributed fibers [10]. Researchers studied the effect of polypropylene fibers and fly ash mixed with high as well as low plastic clays on unconfined compressive strength and CBR. It was concluded that at 10% fly ash content the unconfined compressive strength of both clayey soils increased to a maximum value and at 0.75% polypropylene fibers content mixed with clayey soils the unconfined compressive strength and CBR also increased [11]. Effects on high plastic clay stabilized with fly ash and different size and type of randomly distributed fibers were studied. It was observed that compressive strength was maximum when the soil was stabilized with 25 mm length of fibers for black and white types of fibers. Apart from compressive strength, tensile strength of fly ash stabilized soil also increased up to 70% in the case of black fibers and 80% for white fibers. Authors also commented that fiber length has an appreciable effect on the ductility of soil [12].

Although many researchers have worked on fiber-reinforced soils, fibers reinforced sand and fly ash-soil reinforced with polypropylene fibers mixtures. It was observed from the literature that very less amount of work is found in respect to the effect of length of polypropylene fibers on fly ash properties without mixing with soil and sand. The present paper focuses on the effect of various lengths of polypropylene fibers on fly ash properties. The experiments further can be analyzed to recommend the application of fly ash in civil engineering, particularly geotechnical engineering.

2 Experimentation

2.1 *Standard Proctor Test*

The standard Proctor test was conducted by mixing fly ash with polypropylene fibers. For fiber reinforcement, 0.6% polypropylene fibers by weight of dry fly ash were taken for sample preparation. The fly ash was mixed with fibers in dry conditions before adding water. The fly ash and fibers were mixed by hands manually to prepare the sample. Proper care was taken to obtain homogeneous mixture. The water was added by 4% increment every time to compact fly ash and fiber mix into the mould. The homogeneous mixture so formed was tested for standard Proctor test to find optimum moisture content and maximum dry density. A standard Proctor compaction test was carried out on different lengths of fibers (i.e., 6, 12, 24, 36, and 40 mm) mixed with fly ash. The standard Proctor test was conducted as per the IS:2720 Part 7.

It can be observed from Fig. 1, that 12 mm length of fibers have the highest dry density as compared to other fiber lengths. The 24 mm length and 36 mm length fibers have almost same dry density and 40 mm length fibers have low density as compared to all the fiber lengths. From the observations, it can be mentioned that 12 mm length fibers may be used for higher density requirements. The increase in the dry density of fly ash reinforced with fibers may be due to very less thickness of the fibers and the honeycomb structure of fibers after stretching. Further when the length of fibers was increased the dry density of fiber-reinforced fly ash has decreased.

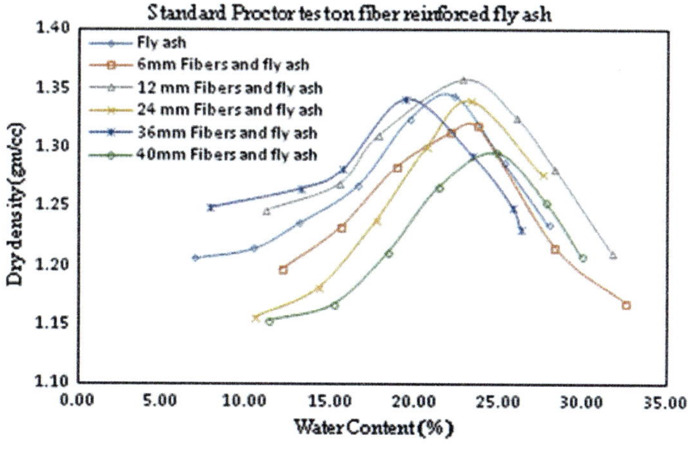

Fig. 1 Standard Proctor test on fiber-reinforced fly ash

Table 1 Standard Proctor test on fiber-reinforced fly ash

S. No.	Fiber length (mm)	Maximum dry density (gm/cc)	Optimum moisture content (%)
1	0	1.34	22.28
2	6	1.31	23.65
3	12	1.35	22.78
4	24	1.33	23.31
5	36	1.34	19.42
6	40	1.29	24.74

In the process of sample preparation, it was observed that as the fiber length was increased in the fly ash the clog formation took place. It leads to difficulty in sample preparation and compaction of the sample. Due to this maintaining uniformity in the fiber distribution was also not possible. Table 1 shows the maximum dry density and optimum moisture content values.

Optimum moisture content for all fiber lengths was almost the same. Fibers absorb less amount of water due to this there may not be much changes in the optimum moisture content in all lengths of fibers mixed in the fly ash.

2.2 *Direct Shear Test*

The direct shear test was conducted on 60 mm × 60 mm × 25 mm sample according to IS:2720 (Part 13)-1986. The samples were prepared by mixing 0.6% fibers with fly ash. The samples were compacted at their corresponding optimum moisture content and maximum dry density obtained from the compaction test. Figure 2 shows the

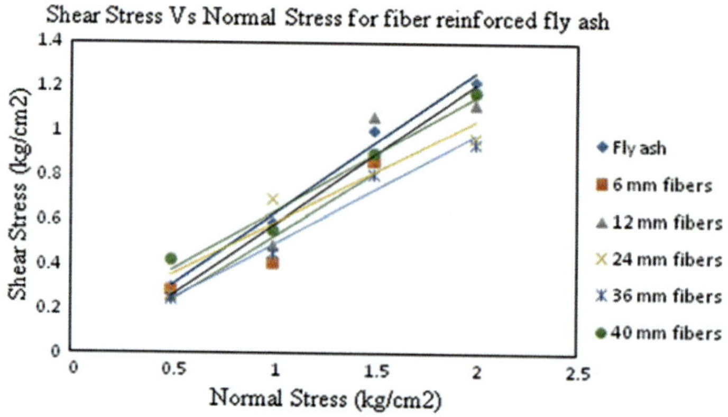

Fig. 2 Variation of shear stress with normal stress for different lengths of fibers

Fig. 3 Sample after failure

variation of shear stress with normal stress for different lengths of fiber-reinforced in the fly ash.

From Table 2 it can be observed that as fiber length increases initially up to 12 mm in size, angle of internal friction also increases and reaches a maximum value. The maximum value obtained was 36° for 12 mm fiber length. As the polypropylene fibers had honeycomb-like structure, when it is stretched, while preparing samples it was observed that the fibers got stretched during compaction. This makes the fibers to hold the fly ash in the structure, which can develop the friction between fibers and fly ash. This could be the reason for increase in the angle of internal friction for the fly ash reinforced with fibers. Further, when the fiber length was increased up to 24 mm there was decrease in angle of internal friction. As shown in Table 2, for 36 and 40 mm lengths of fiber, small increase in the angle of internal friction was observed while for further lengths of fiber there was slight increase in angle of internal friction. It is necessary to mention that for longer fiber length than 12 mm it was difficult to prepare the sample particularly, for 36 and 40 mm lengths of fibers in direct shear test.

Table 2 Angle of internal friction for different length of fibers

S. No.	Fiber length (mm)	φ (degree)
1	0	29
2	6	33
3	12	36
4	24	26
5	36	28
6	40	29

2.3 Unconfined Compressive Strength Test

The unconfined compressive strength test was conducted on the sample as per the procedure given in IS 2720-10: "Methods of tests for soil, Part 10: Determination of unconfined compressive strength". The sample was prepared with the maximum dry density and optimum moisture content. This was obtained through the standard Proctor test as mentioned earlier.

The fly ash and fibers were mixed in dry condition. The water then added equals to optimum moisture content. The unconfined compressive strength value goes on increasing gradually up to addition of 24 mm fibers in the fly ash and then goes on decreasing on further addition of 36 mm and 40 mm length of fibers in fly ash as shown in Fig. 4. Table 3 shows the unconfined compressive strength values of fly ash mixed with different lengths of fibers. The unconfined compressive strength was 186.54 kPa at 24 mm length of fiber. Particularly for 24 mm length of fibers bridging effect was observed in the sample as shown in Fig. 3. The bridging effect of fibers is nothing but the specimen holding capacity of fibers while cracks were developed in the specimen. The fiber reinforcement may resist the failure of the sample during the test as observed in Fig. 3. It was observed that sample preparation was difficult for large lengths of fiber as mentioned in Sect. 2.2.

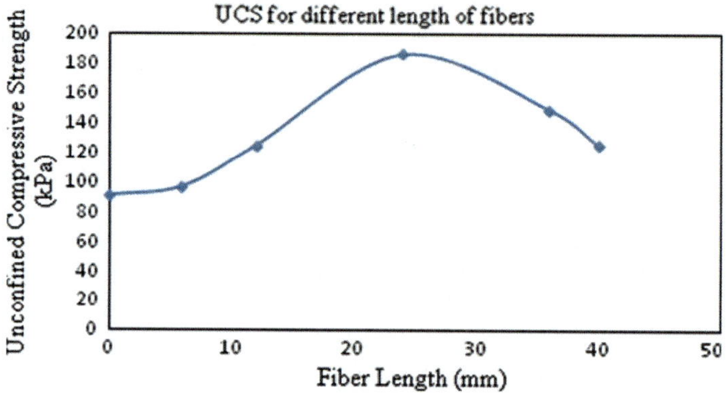

Fig. 4 Variation of unconfined compressive strength with fiber length

Table 3 Unconfined compressive strength (UCS) for different fiber lengths

S. No.	Fiber length (mm)	UCS (kPa)
1	0	90.54
2	6	96.40
3	12	124.36
4	24	186.54
5	36	149.12
6	40	124.9

2.4 California Bearing Ratio(CBR) Test

The California bearing ratio sample was prepared by considering maximum dry density and optimum moisture content. This was obtained as explained earlier through the standard Proctor test. The procedure for the determination of California bearing ratio value was followed as per the IS 2720-16: "Methods of test for soils, Part 16: Laboratory determination of CBR".

From Table 4 we can observe that as we reinforce the fly ash with different lengths of fibers California bearing ratio increases. The California bearing ratio was maximum in case of 40 mm length of fibers. This shows the strength of fly ash increases on the addition of fibers of different lengths. It was particularly observed that all the tests have shown good results for 12 and 24 mm fiber length, but for California bearing ratio value, it was observed that 40 mm fiber length has a good effect. In fact, the CBR value was increasing continuously with increasing the length of fibers as observed in Fig. 5. In the present study, the highest length used was 40 mm, hence the maximum value of CBR observed for 40 mm length fibers. Further, it can be mentioned that if length of fiber is more it will have more friction resistance against the penetration which may lead to more California bearing ratio value. The increase

Table 4 California bearing ratio (%) for different fiber sizes

S. No.	Fiber size	CBR (%)
1	Plain fly ash	30.52
2	6 mm	32.73
3	12 mm	33.73
4	24 mm	34.22
5	36 mm	35.21
6	40 mm	35.96

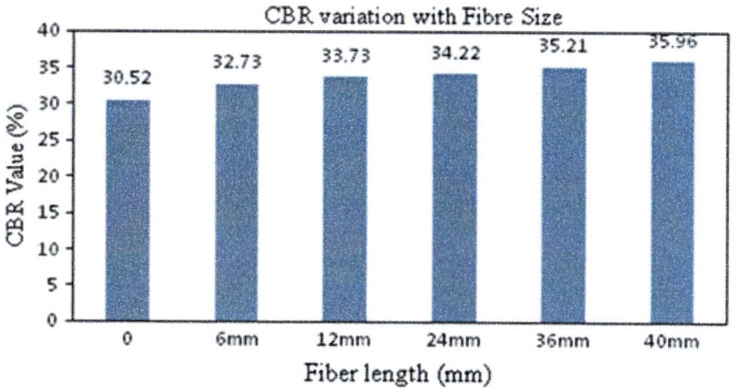

Fig. 5 California bearing ratio value variation with fiber length

in the value of California bearing ratio can also happen due to the interlocking of fibers with fibers and fly ash with fibers.

3 Conclusion

- It was observed from all the tests in the present study that different fiber lengths have different effects on the fly ash property. It was particularly observed that 12 mm fibers have good effect as compared to all other fiber lengths. As the fibers are lightweight, there was no much effect observed in the dry density of fly ash due to addition of fibers with different lengths.
- For the fiber-reinforced fly ash the angle of internal friction was increased by 24% as compared to unreinforced fly ash with 12 mm fiber length.
- The unconfined compressive strength value of fiber-reinforced fly was increased twice as compared to unreinforced fly ash with 24 mm fiber length.
- California bearing ratio value of fly ash was increased up to 18% for fiber reinforcement of 40 mm length.

References

1. Ranjan G, Vasan RM, Charan HD (1996) Probabilistic analysis of randomly distributed fiber-reinforced soil. J. Geotech. Eng. (ASCE) 122(6):419–426
2. Consoli NC, Prietto PDM, Ulbrich LA (1998) Influence of fiber and cement addition on behavior of sandy soil. J Geotech Geo Environ Eng (ASCE) 124(12):1211–1214
3. Kaniraj SR, Havanagi VG (2001) Behavior of cement stabilized fiber-reinforced fly ash-soil mixtures. J Geotech Geo Environ Eng (ASCE) 127(7):574–584
4. Yetimoglu T, Salbas O (2003) A study on shear strength of sands reinforced with randomly distributed discrete fibers. Geotext Geomembr 21(2):103–110
5. Xiao J, Falkner H (2006) On residual strength of high-performance concrete with and without polypropylene fibers at elevated temperatures. Fire Saf J 41(2):115–121
6. Zornberg JG (2005) Geosynthetic reinforcement in landfill design: US perspectives. In: Zornberg JJ, Gabr M, Bowders JJ (eds) In: Proceedings of geo-frontiers, geotechnical special publication 141 [CD-ROM]. ASCE, Reston, VA
7. Tang CS, Shi B, Gao W, Chen F, Cai Y (2007) Strength and mechanical behavior of short polypropylene fiber reinforced and cement stabilised clayey soil. Geotext Geomembr 25(3):194–202
8. Kumar A, Walia B, Mohan J (2006) Compressive strength of fiber reinforced highly compressible clay. Constr Build Mater 20(10):1063–1068
9. Rout SS (2017) Influences of fly ash and coir fiber on strength properties of soft soil. Res J Eng Sci (2017)
10. Nithin, S., Sayida MK (2012) Stabilization of silty sand using fly ash and coir fiber. In: Proceedings of recent advances in civil engineering, Kerala, India
11. Senol A, Etminan E, Yildirim H, Olgun CG (2012) Improvement of high and low plasticity clayey soils using polypropylene fibers and fly ash. In: ICSDEC, pp 500–509

12. Shafique SB, Gupta, SD, Huang J, Rezaeimalek S (2017) The effect of fiber type and size on the strength and ductility of fly ash and fiber stabilized fine-grained soil subbase. Geotech Front 19–29

Review of Experimental Techniques for Evaluating Unsaturated Shear Strength of Soil

P. B. Pande, S. R. Khandeshwar, and S. P. Bajad

Abstract Shear strength is the noteworthy engineering property of soil governing various geotechnical terminologies. Conventional shear strength apparatus and procedures are merely suitable for testing saturated or dry soil. Researchers and practitioners brought the usage of unsaturated shear strength in the respective arenas. Numerous experimental procedures are developed in last few decades to evaluate the shear strength parameters including suction of an unsaturated soil. Most of these methods are complicated, time consuming and expensive. In this concern, the present paper reviews various equipment and experimental procedures incorporating different combinations for measuring shear strength and suction. This study is emphasized on the combination of shear test apparatuses and suction measuring techniques. Moreover, soil used and drainage conditions for testing of unsaturated soil are reviewed and summarized. The combination of conventional triaxial shear test and filter paper method is not trialed in Indian context. This combination of measuring shear strength unsaturated soil would be easy and cost-effective.

Keywords Triaxial shear test · Direct shear test · Unsaturated Shear strength

1 Introduction

In geotechnical engineering, the shear strength is the most essential engineering property and fundamental state variables of soil required for analyses and prediction slope stability, earth pressures and foundation design [52]. In case of saturated soil, conventional direct and triaxial shear test apparatus are normally used to determine the shear strength in the laboratory. Various experimental procedures are developed in the previous few decades to evaluate the shear strength parameters of

P. B. Pande (✉) · S. R. Khandeshwar
Department of Civil Engineering, Yeshwantrao Chavan College of Engineering, Nagpur, India
e-mail: prashantbpande21@gmail.com

S. P. Bajad
Regional Office, Maharashtra State Board of Technical Education, Pune, India

L. M. Gupta et al. (eds.), *Advances in Civil Engineering and Infrastructural Development*, Lecture Notes in Civil Engineering 87,
https://doi.org/10.1007/978-981-15-6463-5_57

an unsaturated soil. In recent decades, several authors like Fredlund and Rahardjo [24], Wheeler and Sivakumar [57], as well as Cui and Delage [16] validated the fundamental theories essential for unsaturated soils with the help of suction-controlled oedometer, modified triaxial as well as direct shear testing apparatus. These apparatuses are not easily available, difficult to operate and costly.

Moreover, extensive engineering judgment is required to apply undrained strength analyses on saturated soils. This type of analysis can easily be applied because it does not involve determination of pore water pressure. However, the use of total strength parameters is not satisfactory because its pore pressure response is unknown.

In Indian context, it is necessary to develop the methodology to involve unsaturated conditions in shear strength analysis. And so, it is essential to measure suction along with shear strength of soil. In this concern, the present paper reviews various equipment and experimental procedures incorporating different combinations for measuring shear strength and suction. This study is emphasized on the combination of shear test apparatuses and suction measuring techniques. Moreover, various strain rates adopted, soil used and drainage conditions for testing of unsaturated soil are reviewed and summarized.

2 Shear Strength of Saturated Soil

Evaluation of shear strength of saturated soils is based on Mohr–Coulomb theory. Terzaghi [43] proposed the effective stress equation for evaluating the shear strength of saturated soils. The effective stress variable (σ') proposed by Terzaghi [43] is effectively used in Mohr–Coulomb's theory. Mohr–Coulomb suggested the two-dimensional $\sigma - \tau$ plot as shown in Fig. 1. The saturated shear strength equation proposed by Terzaghi [43] is given as follows:

$$\tau = c' + (\sigma_n - u_w) \tan \phi' \tag{1}$$

where

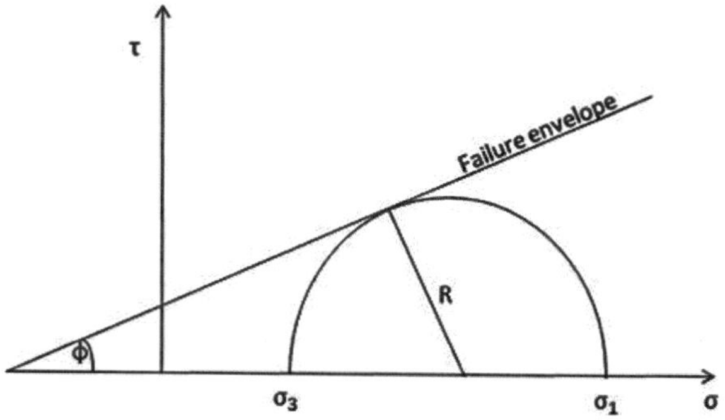

Fig. 1 Mohr–Coulomb failure envelope [43]

τ = shear strength
σ_n = total normal stress on plane of failure
$\sigma_n - u_w$ = effective normal stress
u_w = pore water pressure
c' and $\text{Ø}'$ = saturated effective shear strength parameters.

3 Shear Strength of Unsaturated Soil

Formerly, Biot [4] proposed a consolidation theory for an unsaturated soil. Hilf [29] was the first investigator who studied the effect of suction on compacted cohesive soils and proposed Eq. (2) established from Mohr–Coulomb strength envelope helps in defining the shear strength of unsaturated soils:

$$\tau = C + \sigma_n \tan \phi \tag{2}$$

where

τ = shear strength
c = cohesion
Ø = internal friction.

The most commonly used two stress state variables are the net normal stress $(\sigma - u_a)$ and the matric suction $(u_a - u_w)$. Fredlund et al. [23] described the behavior of unsaturated shear strength of soils in the form of Eq. (3) resulting from experimental results for d as follows:

$$\tau_f = c' + (\sigma_n - u_a) \tan \phi' + (u_a - u_w) \tan \phi^b \tag{3}$$

τ_f is the unsaturated shear strength of soil.
c' is the effective cohesion of saturated soil.
Ø′ is the effective angle of shearing resistance for a saturated soil.
Ø^b is the angle of shearing resistance with respect to the matric suction.
σ_n is the total normal stress on plane of failure.
u_a is the pore air pressure.
u_w is the pore water pressure.
$(u_a - u_w)$ is the matric suction of soil at failure.
$(\sigma_n - u_a)$ is the net normal stress on the plane of failure.

In the case of an unsaturated soil, the Mohr circle and strength envelope at failure are extended on a three-dimensional Mohr–Coulomb's plot as shown in Fig. 2.

Ø^b indicates the increase rate in shear strength with matric suction. From experimental and theoretical studies, it is stated that Ø^b is the nonlinearly decrease in the matric suction. The modified Mohr–Coulomb's failure envelope is observed between the shear stress (τ) and matric suction $(u_a - u_w)$. Here, φ^b can also be defined as an

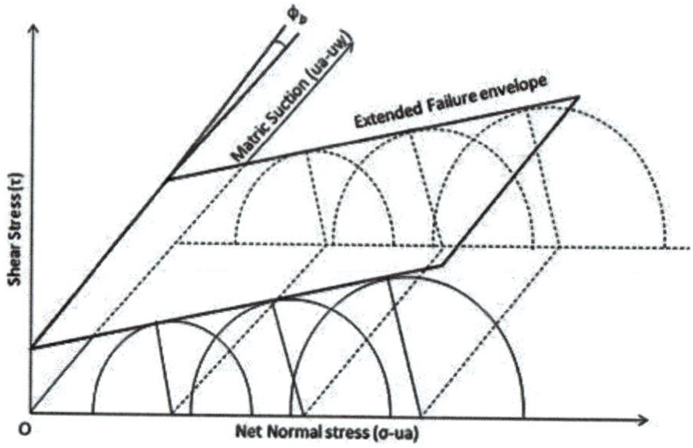

Fig. 2 Three-dimensional Mohr–Coulomb's plot Fredlund and Rahardjo [24]

angle of internal friction with respect to $(u_a - u_w)$ variation under a constant $(\sigma_n - u_a)$.

4 Experimental Techniques for Measuring Shear Strength of Unsaturated Soil

The unsaturated soil testing techniques are not analogous to the saturated soil owing to suction. Measuring suction is the most accentuating mechanism in unsaturated soil testing techniques. Therefore, the brief review on suction measuring techniques is essential.

The various methods measuring soil suction are broadly categorized into direct and indirect methods and presented in Fig. 3. Suction plate, tensiometer, pressure plate and suction probe are the direct methods of measuring suction. Indirect methods are again classified as primary, secondary and tertiary methods. Thermocouple psychrometers, transistor psychrometers and chilled-mirror psychrometers are primary methods. The filter paper technique is the secondary method; the tertiary methods comprise the thermal and electrical conductivity sensor techniques [36].

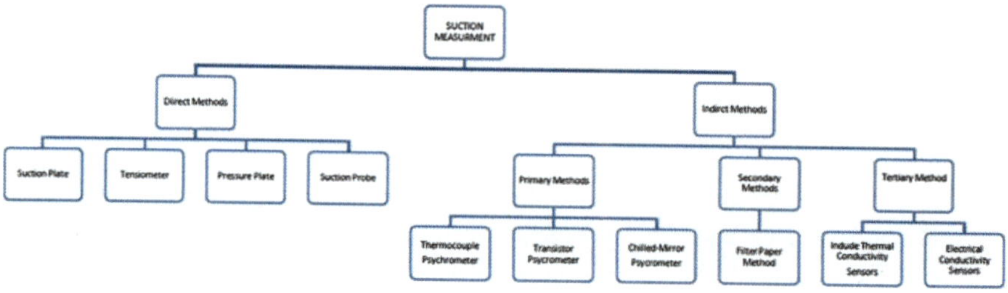

Fig. 3 Suction measurement techniques

Till date, numbers of models are proposed to validate the fundamental theories of unsaturated soils with oedometer controlling suction, triaxial and direct shear testing [1]. Fredlund and Rahardjo [24], Wheeler and Sivakumar [57] and Delage [17] anticipated theoretical parameters to study unsaturated soil. Suction-controlled equipment is difficult to operate and expensive; thus, it is necessary to develop and establish the specific methodology and protocol for evaluating shear strength parameters of unsaturated soil in combination with conventional equipment which are easily available and cost-effective.

In the 1950s, the modifications were suggested and commenced in shear strength testing apparatuses to accommodate unsaturated soils. The modification of conventional equipment is necessary to quantify pore air and pore water pressure of unsaturated soil separately, by means of porous disks with high air entry value (HAEV) [7]. Triaxial and direct shear test apparatuses are conventionally used for testing saturated soil and are modified and extensively used in the determination of the shear strength parameters of unsaturated soils well along. Nonetheless, the modified equipment for testing unsaturated soil is difficult to operate and costly. Unsaturated shear strength of soil can be evaluated without considering pore water and pore air pressure at failure [10].

In the 1980s, most of the researchers performed either direct or triaxial shear test for evaluating the unsaturated shear strength of soil [29]. Some investigators had chosen the direct shear apparatus to study suction effect on the shear strength parameters of unsaturated soils. Fredlund and Morgenstern [23], Escario and Saez [21], Gan and Fredlund [27], Vanapalli [51] and Vaunat et al. [54] used direct shear tests for controlling and measuring the suction on unsaturated soils using the axis-translation technique. Others selected triaxial shear tests to investigate the role of suction in controlling and measuring the shear strength of different soils.

Richards [40] worked on controlling the suction using axis-translation technique, and further investigators used this technique to modify the conventional testing equipment for evaluating the unsaturated shear strength of soil.

Initially Kjellman [32] experimented the triaxial test apparatus with the plane strain condition to study the unsaturated shear strength of soil. Subsequently, he performed the series of modified direct shear tests with controlled suction on fine sands and coarse silts using axis-translation technique. Donald [18] was the first investigator who modified the conventional direct shear testing apparatus and determined the SWCC and the shear strength.

Bishop et al. [8] presented test results obtained from the triaxial compression tests conducted on saturated sand under undrained condition and stated that apparent angle of shearing resistance is observed with respect to negative pore water pressure during shear. Several techniques were proposed by Bishop et al. [8] for evaluating the unsaturated shear strength of soil with triaxial equipment. In 1961, the first suction-controlled triaxial apparatus was developed by Bishop and Donald [5] and performed test on loose silt under consolidated drained condition. Coleman [15] suggested to measure the pore air pressure as an isotropic stress state variable [14].

Bishop and Blight [9] confirmed the suitability of the axis-translation technique used for testing unsaturated shear strength of soils [9]. Blight [10] performed unsaturated shear strength test on silty soil under consolidated drained condition, wherein shear strength escalation was observed owing to matric suction along with increasing net confining pressure. Lee [33] conducted triaxial test under drained and undrained conditions under plain strain. Consolidated drained tests were performed by modified triaxial cell using axis-translation technique to study the shear strength behavior of unsaturated Dhanauri clay and Madrid gray clay by Satija and Gulhati [42] and Escario [19], respectively. Ho and Fredlund [30] performed a series of unsaturated multistage triaxial tests on Hong Kong residual soils from by making the necessary modifications on a conventional triaxial cell. Chantawarangul [13] conducted the series of six undrained and unconfined compression tests on unsaturated compacted clayey soil with 52% sand, 18% silt and 30% clay.

Mahalinga-Iyer and Williams [34] applied liquid infiltration technique and measured the shear strength of unsaturated soil with triaxial shear test apparatus in combination with the filter paper method. Unconsolidated, undrained triaxial tests on unsaturated soil were conducted with pore pressure measurement and measured suction with the filter paper method explicated by McQueen and Miller [35]. Escario and Saez [21] observed nonlinearity in the shear strength envelope of unsaturated soils for the complete suction range from the results of the modified direct shear test.

Fredlund [22] reanalyzed the triaxial test results of Satija [41] and revealed some nonlinearity in the failure envelope of shear stress versus matric suction. Gan and Fredlund [27] performed a modified conventional direct shear apparatus by applying matric suction greater than 101 kPa through axis-translation technique. Gan and Fredlund [27] also observed nonlinearity in the failure envelope of shear stress versus matric suction, and they have made a suggestion for handling the nonlinearity behavior from a practicing engineer's point of view. Escario and Juca [20] designated elliptical failure envelope for suction value lower than the air entry value by performing the test on Guadalix red silty clay with the application of 120–600 kPa net normal stress. Fredlund and Rahardjo [24] arranged the number of state-of-the-art reports and keynote lectures together and published the textbook on "unsaturated soil mechanics." Further, Gan and Fredlund [27] modified direct shear test and triaxial test apparatus and performed experimentation on decomposed granite from season hill, Hong Kong. Again Fredlund and Xing [25] observed the nonlinear shear strength envelope of unsaturated soils with the data tested on the direct shear box over a wide suction range. Fredlund and Xing [25] introduced a tool called as the SWCC used to predict the shear strength of unsaturated soils from saturated soil parameters.

Vanapalli [50] started studying the relationship between SWCC and the shear strength unsaturated soil with respect to matric suction. He conducted the consolidated drained direct shear test on statically compacted glacial till specimen under saturated and unsaturated condition and proposed empirical procedures for evaluating the unsaturated shear strength of soils.

Vanapalli and Lane [49] used the conventional direct shear equipment for testing Indian Head till under a net normal stress ($\sigma - u_a$) of 25 kPa by compacting the soil at optimum moisture content for a matric suction ($u_a - u_w$), range of 0–500 kPa. Then, the achieved results were compared with data obtained by Vanapalli et al. [50] from the modified direct shear tests performed on the same soil under identical conditions and found reasonably good comparison between both methods. Then, Vanapalli and Lane [49] proposed combination of experimental procedure and semiempirical procedures for appraising the unsaturated shear strength of fine-grained soils.

Moreover, Vanapalli and Fredlund [48] worked on the concept of residual state and the observed increase nonlinearly in shear strength until the residual suction value reached and negative value of the strength of suction values beyond the residual state. Khalili and Khabbaz [31] focused on methods for evaluating the shear strength parameters of unsaturated soil using the effective stress concept and SWCC. The saturated test using the conventional triaxial equipment was conducted for determining the shear strength with pore pressure measurements [2].

Vanapalli and Fredlund [53] provided theoretical method for evaluating and interpreting the undrained shear strength of unsaturated soil in terms of normal stress $(\sigma - u_a)$ and matric suction $(u_a - u_w)$. Nishimura and Fredlund [38] studied the behavior of unsaturated silty soil and kaolin by developing the SWCC from triaxial test data. Vanapalli and Fredlund [53] compared the measured and predicted shear strength values from four different procedures for three different soils with limited and large suction range. Vanapalli and Fredlund [53] also noticed shear strength behavior of silty soil over the entire suction range (0–1,000,000 kPa). Rahardjo et al. [39] carried out the consolidated drained triaxial tests on granitic Bukit Timah formation. Caruso and Tarantino [12] accompanied constant water content test by using a shear box developed at the University of Trento. Boso [11] modified shear box by introducing osmotic technique. Zhan and Ng [37] performed suction-controlled direct shear test on unsaturated expansive clay. Wanatowski and Chu [56] studied biaxial shearing tests and comparisons with triaxial tests. Vilar [55] proposed hyperbolic function-based procedure for evaluating shear strength envelope. Thamer Ahmed [44] carried out a test on an unsaturated granitic residual soil and described shear strength–SWCC relationship. Tse [46] obtained shear strength parameters for suction-controlled direct shear tests.

Hamid and Miller [28] studied the shearing behavior of low-plasticity fine-grained unsaturated soils by conducting direct shear tests. Uchaipichat [47] tested compacted kaolin using modified conventional triaxial equipment and developed some constitutive models for unsaturated soils with effective stress approach. Bai and Liu [3] performed extensive experimental program on Nanyang expansive soil using a combination of conventional direct shear test (unconsolidated and undrained) and filter paper method. Khalili and Khabbaz [31] evaluated shear strength parameters and matric suctions with axis-translation technique in the undrained triaxial test.

The summary of various testing techniques reviewed, for measuring shear strength of unsaturated soil along with the suction measuring arrangement, is presented in Tables 1 and 2.

Table 1 Summary of direct shear test on unsaturated soil

Researchers	Soil	Test	Condition	Suction measured by
Donald [18]	Fine sands and coarse silts	Modified direct shear test	–	Maintained constant negative head to control negative pore water pressure
Escario and Saez [21]	Compacted red clay of Guadalix	Modified direct shear test	–	Axis-translation technique
Gan and Fredlund [27]	Regina clay	Modified direct shear test	CD	Axis-translation technique
Vanapalli et al. [50]	Glacial till (clayey till)	Multistage direct shear	CD	Pressure plate apparatus
Boso [11]	Reconstituted partially saturated clayey silt	Modified direct shear test	–	Axis-translation technique
Bai and Liu [3]	Nanyang expansive soil	Conventional direct shear test	UU	Filter paper method

5 Summary and Conclusion

In the mentioned literature review, only Mahalinga and Williams [34] and Bai and Lui [3] used the combination of conventional shear test and filter paper method. Unconsolidated, undrained conventional triaxial compression tests and filter paper method were performed on lateritic soils. Toll's theory was used to interpret the unsaturated shear strength parameters of soil [34]. Bai and Lui [3] conducted the unconsolidated, undrained conventional direct shear test along with filter paper method on Nanyang expansive soil. The obtained results were interpreted using soil–water characteristic curve.

The researchers are normally focused on framing the shear strength equation of unsaturated soil; nevertheless, few researchers premeditated the influence of initial matric suction on the shear strength of soil under undrained condition.

Combination of conventional triaxial shear test and filter paper method is not witnessed under unconsolidated, undrained condition in Indian context. Thus, the research program to evaluate undrained, unsaturated shear strength parameters can be undertaken on Indian soil.

Table 2 Summary of triaxial shear test on unsaturated soil

Authors	Soil	Test	Condition	Suction measured by
Bishop and Eldin [6]	Saturated soil specimen	Triaxial compression tests	CU	De-aired measuring system
Hilf [29]		Triaxial compression tests	–	Axis-translation technique
Bishop et al. [8]	Compacted shale	Triaxial compression tests	CD	Constant water (CW) content
Satija [41]	Dhanauri clay	Triaxial compression tests	CW and CD	Axis-translation technique
Ho and Fredlund [30]	Residual soils from Hong Kong	Modified conventional triaxial compression tests	–	Axis-translation technique
Mahalinga-Iyer and Williams [34]	Lateritic soils	Triaxial compression tests	UU	Filter paper method
Escario and Juca [20]	Madrid gray clay	Triaxial compression tests	CD	Axis-translation technique
Escario and Juca [20]	Guadalix red clay	Triaxial compression tests	CD	Axis-translation technique
Gan and Fredlund [27]	Decomposed granite from Shouson Hill, Hong Kong	Modified direct shear test and triaxial test apparatus	–	Axis-translation technique
Zerhouni [58]	–	Triaxial compression tests	–	Psychrometers
Nishimura and Fredlund [38]	Silty soil	Triaxial compression tests	–	Pressure plate technique
Chavez and Alonso [14]	Compacted specimens of Pancrudo shale	Triaxial compression tests	–	Axis-translation technique
Futai [26]	Sandy clay	Triaxial compression tests	CD	Axis-translation technique
Rahardjo et al. [39]	Granitic Bukit Timah formation	Triaxial compression tests	CW and CD	Axis-translation technique
Vilar [55]	–	Triaxial compression tests	–	Axis-translation technique
Rahardjo and Thu [45]	Silt (coarse kaolin)	Triaxial compression tests	CW and CD	NTU mini suction probe
Uchaipichat [47]	Compacted kaolin	Modified conventional triaxial equipment	CD	Axis-translation technique

(continued)

Table 2 (continued)

Authors	Soil	Test	Condition	Suction measured by
Khalili and Khabbaz [31]	Limestone powder	Triaxial tests	UU and CD	Axis-translation technique

References

1. Alonso EE, Gens A, Josa A (1990) A constitutive model for partly saturated soils. Geotechnique 40(3):405–430
2. Al-khazaali M, Vanapalli SK (2017) Experimental model to investigate the axial force-displacement behavior of a pipeline in unsaturated sandy soil. In: 70th Canadian Geotechnical Conference, GeoOttawa, Ottawa, Canada
3. Bai FQ, Liu SH (2013) Measurement of the shear strength of an expansive soil by combining a filter paper method and direct shear tests. Geotech Test J 35(3):451–459
4. Biot MA (1941) General theory for three-dimensional consolidation. J Appl Phys 12(2):155–164
5. Bishop AW, Donald IB (1961) The experimental study of partly saturated soil in the triaxial apparatus. In: Proceedings of the Fifth International Conference on Soil Mechanics and Foundation Engineering, Paris, vol 1, pp 13–21
6. Bishop AW, Eldin AKG (1950) Undrained triaxial tests on saturated sands and their significance in the general theory of shear strength. Geotechnique 2:13–32
7. Bishop AW, Henkel DJ (1962) The measurement of soil properties in the triaxial test. 2nd edn. Edward Arnold, London
8. Bishop AW, Alpan I, Blight GE, Donald IB (1960) Factors controlling the shear strength of partly saturated cohesive soils. Paper presented at the research conference on shear strength of cohesive soils, ASCE, University of Colorado, Boulder, CO, pp 503–532
9. Bishop AW, Blight GE (1963) Some aspects of effective stress in saturated and unsaturated soils. Geotechnique 13(3):177–197
10. Blight GE (1967) Effective stress evaluation for unsaturated soils. J Soil Mech Found Eng Div ASCE 93(SM2):125–148
11. Boso M (2005) Shear strength behavior of a reconstituted partially saturated clayey silt. Ph.D. thesis University of Trento
12. Caruso M, Tarantino A (2004) A shear box for testing unsaturated soils at medium to high degrees of saturation. Geotechnique 54(4):281–284
13. Chantawarangul K (1983) Comparative study of different procedures to evaluate effective stress strength parameters for partially saturated soils. M.Sc. thesis, Asia Institute of Technology, Bangkok, Thailand
14. Chávez C, Alonso EE (2003) A constitutive model for granular aggregates which includes suction effects. Soils Found 43(4):215–227
15. Coleman JD (1962) Stress/strain relations for partly saturated soils. Geotechnique 12(4):348–350
16. Cui Y, Delage P (1996) Yielding and plastic behaviour of an unsaturated compacted silt. Geotechnique 46(2):291–311
17. Delage P (2002) Experimental unsaturated soil mechanics. In: Proceedings of 3rd international conference on unsaturated soils, UNSAT, 1961(3), 973–996
18. Donald IB (1956) Shear strength measurements in unsaturated non-cohesive soils with negative pore pressures. In: Proceedings of the second Australia-New Zealand conference on soil mechanics and foundation engineering, Christchurch, New Zealand, pp 200–205
19. Escario V (1980) Suction–controlled penetration and shear tests. In: Proceedings of the fourth international conference on expansive soils, vol 2. American Society of Civil Engineers, Denver, pp 781–797

20. Escario V, Juca F (1989) Strength and deformation of partly saturated soils. In: Proceedings of 12th international conference soil mechanics and foundation engineering 1. Rio, Balkema, Rotterdam, pp 43–46
21. Escario V, Saez J (1986) The shear strength of partly saturated soils. Geotechnique 36(3):453–456
22. Fredlund DG (1987) Slope stability analysis incorporating the effect of soil suction, in MG Anderson and KS
23. Fredlund DG, Morgenstern NR (1977) Stress state variables for unsaturated soils. J Geotech Eng Div ASCE 103(GT5):447–466
24. Fredlund DG, Rahardjo H (1993) Soil mechanics for unsaturated soils. Stress Int J Biol Stress 517
25. Fredlund DG, Xing A (1994) Equations for the soil-water characteristic curve. Can Geotech J 31(3):521–532
26. Futai MM (2002) Theoretical–experimental study of the behavior of non-saturated tropical soils applied to a gully erosion. Ph.D. thesis, COPPE-Federal University of Rio de Janeiro, Rio de Janeiro, Brazil, 559 p
27. Gan JKM, Fredlund DG (1988) Multistage direct shear testing of unsaturated soils. Geotech Test J ASTM 11(2):132–138
28. Hamid TB, Miller GA (2009) Shear Strength of Unsaturated Soil Interfaces. Can Geotech J 46:595–606
29. Hilf JW (1956) An investigation of pore-water pressure in compacted cohesive soils. PhD thesis, technical memorandum. No. 654, U.S. Department of the Interior, Bureau of Reclamation, Design and Construction Division, Denver, CO
30. Ho DYF, Fredlund DG (1982) Increase in shear strength due to soil suction for two Hong Kong soils. In: Proceedings of specialty conference on engineering and construction in tropical and residual soils, ASCE, Honolulu, HI, pp 263–295
31. Khalili N, Khabbaz MH (1998) A unique relationship for χ for the determination of the shear strength of unsaturated soils. Geotechnique 48(5): 681–687
32. Kjellman W (1936) Report on apparatus for consummate investigation of mechanical properties of soils. In: Proceedings of 1st international conference soil mechanics, vol 2, pp 16–20
33. Lee KL (1970) Comparison of plane strain and triaxial test on sand. J Soil Mech Found Div 96(3):901–923
34. Mahalinga-Iyer U, Williams DJ (1985) Unsaturated strength behaviour of compacted lateritic soils, Tech Note Geotech 45(2):317–320
35. McQueen IS, Miller RF (1968a) Determination of a soil moisture potential, water in the unsaturated zone. In: Rijtema PE, Wassink H (eds) International association of science and hydrology Publication No 82, pp 147–155
36. McQueen IS, Miller RF (1968) Calibration of a wide-range gravimetric method for measuring moisture stress. Soil Sci 106(3):225–231
37. Ng WW, Tony LT (2006) Shear strength characteristics of an unsaturated expansive clay. Can Geotech J 43:751–763
38. Nishimura T, Fredlund DG (2000) Relationship between shear strength and matric suction in an unsaturated silty soil. In: Proceedings of the Asian conference on unsaturated soils, UNSAT-ASIA 2000, Singapore, pp 563–568
39. Rahardjo H, Aung KK, Leong EC, Rezaur RB (2004) Characteristics of residual soils in Singapore as formed by weathering. J Eng Geol 73(1–2):157–169
40. Richards LA, Fireman M (1943) Pressure-plate apparatus for measuring moisture sorption and transmission by soils. Soil Sci J 56:395–404
41. Satija BS (1978) Shear behaviour of partly saturated soils. PhD thesis, Indian Institute of Technology, New Delhi
42. Satija BS, Gulhati SK (1979) Strain rate for shearing testing of unsaturated soil. In: Proceedings of the sixth Asian regional conference on soil mechanics and foundation engineering, Singapore, pp 83–86

43. Terzaghi K (1936) The shear strength of saturated soils. In: Proceedings of the first international conference on soil mechanics and foundation engineering, vol 1. Cambridge, pp 54–56
44. Thamer Ahmed M, Ali FH, Ali SH, Huat BBK (2006) Relationship between shear strength and soil water characteristic curve of and unsaturated granitic residual soil. Am J Environ Sci 2(4):142–145
45. Thu TM, Rahardjo H, Leong EC (2006) Shear strength and pore-water pressure characteristics during constant water content triaxial tests. J Geotech Geoenviron Eng ASCE 132(3):411–419
46. Tse EYM, Ng CWW (2008) Advances in geo-engineering. Taylor and Francis Group, London, pp 481–486
47. Uchaipichat (2012) Variation of undrained shear strength of unsaturated clay with suction, In: International conference on ecological, environmental and bio-sciences, 13–15, Pattaya
48. Vanapalli SK, Fredlund DG (1999) Empirical procedures to predict the shear strength of unsaturated soils. In: Proceedings of the eleventh Asia regional conference on soil mechanics and geotechnical engineering, Seoul, vol 1, pp 93–96
49. Vanapalli SK, Lane JJ (1998) A simple technique for determining the shear strength of fine-grained unsaturated soils using the conventional direct shear apparatus. In: 2nd Canadian specialty conference on computer applications in geotechnique at: Winnipeg, Canada, P7B5E1
50. Vanapalli S, Fredlund D, Barbour S (1996) A rationale for an extended soil-water characteristic curve. In: 49th Canadian geotechnical conference, New Foundland
51. Vanapalli SK (1994) Simple test procedures and their interpretation in evaluating the shear strength of unsaturated soils. Ph.D. thesis, University of Saskatchewan, Saskatoon
52. Vanapalli SK (2010) Shear strength of unsaturated soils and its applications in geotechnical engineering practice. In: Unsaturated soils: theoretical and numerical advances in unsaturated soil mechanics—proceedings of the 4th Asia Pacific conference on unsaturated soils, pp 579–598
53. Vanapalli SK, Fredlund DG (2000) Comparison of different procedures to predict unsaturated soil shear strength. In: Proceedings of the GeoDenver conference. https://doi.org/10.1061/40510 (287)
54. Vaunat J, Romero E, Marchi C, Jommi C (2002) Modeling the shear strength of unsaturated soils, 3rd. In: International conference on unsaturated soils, vol 2. Balkema, pp 245–251
55. Vilar OM (2006) A simplified procedure to estimate the shear strength envelope of unsaturated soil. Can Geotech J 43:1088–1095
56. Wanatowski D, Chu J (2006) Stress strain behavior of a granular fill measured by new plain strain apparatus. Geotech Test J 29(2):149–157
57. Wheeler SJ, Sivakumar V (1995) An elasto-plastic critical state framework for unsaturated soil. Geotechnique 45(1):35–53
58. Zerhouni (1995) Triaxial testing using Psychrometers (in French). In: Proceedings of 1st international conference unsaturated soils, pp 673–678

The Role of Artificial Intelligence in Industry 4.0 and Smart City Development

Sujesh D. Ghodmare, B. V. Khode, and Sachin M. Ladekar

Abstract Transition is the reconstruction of the field from new fundamentals which changes the field's common elementary theoretical concepts along with methods and applications. During transition, there will be the partial overlap between existing and new paradigms due to decisive differences. Artificial intelligence and Internet of things are unlocking the huge opportunities due to its advancements, especially in relation to smart city and its development. Artificial intelligence signifies the era of evolution in science and technology. Nowadays, the tasks which were once performed manually by human beings have been overtaken by artificial intelligence. Sixty-two percentage of enterprises will be using AI by 2018 as conformed by Narrative Science. Many researchers have quoted the opinion that artificial intelligence may have the impact more than Industrial Revolution on overall world. In this paper, it is tried to focus on industrial and economic development trend due to advancements in the urban scenario. It is tried to consolidate the fact knowledge related to technological advancements across the world. The fundamental concepts of artificial intelligence and Internet of things as a game changer are tried to extract in order to understand their suitability for proper implementation of smart city concept.

Keywords Smart city · Artificial intelligence · Internet of things · Industrial Revolution · Game changer

1 Introduction

During the first Industrial Revolution (eighteenth to nineteenth century), the world's estimated population was 1.8 billion. The concepts notified were use of water and

S. D. Ghodmare (✉) · B. V. Khode
G. H. Raisoni College of Engineering , CRPF Gate, No.3, Hingna Rd, Digdoh Hills, Nagpur, Maharashtra 440016, India
e-mail: sujesh.ghodmare@raisoni.net

S. M. Ladekar
NIT Polytechnic, Nagpur, India

L. M. Gupta et al. (eds.), *Advances in Civil Engineering and Infrastructural Development*, Lecture Notes in Civil Engineering 87,
https://doi.org/10.1007/978-981-15-6463-5_58

steam energy. The main focus was on innovations related to the agriculture production and searching of raw materials. The notable shift was rural–urban migration in search of different facilities and needs. Development of new localities and related conflicts was also a major issue during that era of development. The entire revolution was more or less bounded by new product development and trading. The many changes were observed during 1870–1914 (second Industrial Revolution). The key concepts were electricity and fossil fuels along with the mass production. The notable inventions related to assembly lines and mass production were witnessed along with fast urbanization and expansion in infrastructural developments.

The issues and conflicts were common as the world population noted was around 4.4 billion. The third Industrial Revolution which is well known as the digital revolution was assumed to be started from 1960 with the concept of globalization. The approximate estimation of population was around 7.6 billion. As stated above, renewable energy, nuclear energy and geothermal energy concepts were prominent along with electricity and fossil fuels. The value addition concepts such as Internet and intellectual property services also came into picture during this period. Due to computing and digitalization, the notable gems during the said period till date are personalization and value products based on the information and innovation. The micro-level impact of the above revolutions on society can be summarized on the basis of focus area and the technological outcomes (Table 1).

According to Porter, the four forces connected with competitive rivalry which impacts the business environment are threat of entrance, substitute products, bargaining power of suppliers and bargaining power of buyers. This had emerged with queries like what about the e-waste?, is it sustainable?, etc. The industrial and economic revolution was having perspectives to deal with the issues like fast urbanization, industrialization, effect on environment, disasters, e-waste, waste management, slums, conflicts, congestion, child labor and health-related issues.

1.1 Transformation from Linear Economy to Circular Economy

Then, the trend of circular economy came into picture. Earlier, the linear economy was personal oriented with many restrictions and limitations. But the circular economy emerged as society-oriented system having benefits to all at large and of open nature. This transformation was really the cause for the paradigm shift in both business and industry modeling. The circular economy has number of unique benefits like raw material conservation, innovations, durable products, new jobs, efficiency, reduction in CO_2 emissions, etc. The advantages which can be fetched out of circular economy are reusing, remanufacturing, recycling and reducing the use. The output of this can be summarized in terms of societal issues as below (Table 2).

Then, the new era of people and society came into picture due to advancements and globalization. It had led to the expectations of sustainable coexistence and solutions

Table 1 Micro-level impact of Industrial Revolutions on society

Issues	Impact on society	
	Positive	Negative
Social and economic	Formation of custom unions and trade blocs	Unequal wealth distribution
	Cashless economy	Regionalization
	Entrepreneurship and innovations	Trade barriers
	Consumer power	Shortage of resources
	Globalization and diversity	Lifestyle diseases
	Connectivity	Poverty and limited access to basic needs
	Cultural fusion	Crime and unemployment
	Media and social networking	Urban crowding and increased slums
	Skill learning and education	Illegal migrations, communication barriers, racial and cultural conflicts
Technological	E-commerce	Social isolation
	Robotics	Energy consumption at exponential pace
	Internet and smartphones	Breakdowns in relation
	Digitalization	Internet addiction
	Fintech	Cybercrime
Environmental and ethical	Green movement	Pollution, global warming, e-waste
	Ethical consumption	Deforestation and illegal mining
	CSR	Child labor, human trafficking and exploitation
		Increased cybercrime rate

Table 2 Outputs of circular economy concept

Outputs	People attractiveness	Profit—business competitiveness	Planet—society
Parameters	Smart living, basic needs, health, safety and quality	Fair business practices, effective value chain and industry sectors	Recycling, environmental conservation and reusability
Views	Smart cities	Sustainable growth	E-waste management

for smart cities. The main intension was to create the healthy environment so as to allow the people to coexist with the quality of life along with superior living experiences. This had resulted in the need of transition of conventional cities into smart cities.

2 Methodology Adopted

In this paper, the various benefits and challenges of AI are tried to focus. For the purpose, the deep review regarding the applications of AI and related aspect is carried out. Different case studies are tried to elaborate. The fundamentals of AI are also tried to elaborate for understanding the basic concepts. On the basis of the different case studies, the existing and future scope of introducing and adopting AI is analyzed. At the same time, the different challenges related to the AI-related applications are focused. Finally, the recommendations regarding feasibility of adopting AI related to smart city development as a future requirement are tried to summarize.

3 Literature Review

As stated by Albert Einstein, we cannot solve our problems with the same thinking we used when we created them.

3.1 Highlights of Industry 4.0 (Fourth Industrial Revolution)

The notable point during this fourth Industrial Revolution is connectivity of technologies. This connectivity is in the form of extended digital capabilities, scale by emerging and embedding in life of individuals. This transformation at each level of industry and society is ultimately the result of the confluence of data based on huge massive storage and cognitive power. This has created the unimaginable opportunities in almost all sectors from education, health, agriculture, manufacturing to services. The entire Industry 4.0 Revolution is based on the core technological concepts such as AI, IOT, quantum and cloud computing, block chain and distributed ledgers, advanced materials, robotics, additive manufacturing and multidimensional (3D) printing, neural technologies, biotechnologies, virtual and augmented realities, geo-engineering, space technologies, energy capture, energy store and energy transmission. These all concepts are playing the vital role in shaping the fourth Industrial Revolution. In this paper, it is tried to focus mainly on the core technological concepts of artificial intelligence (AI) and Internet of things (IOT).

3.2 *Artificial Intelligence*

Basically, AI is a computer-based program or a software enabled with the mechanism to learn on the basis of knowledge to make decisions in a new situation similar to human beings [1]. For this software, inventors and researchers are trying to create the codes capable to read images, texts, pictures, videos or audios so as to learn something out of it. Artificial Intelligence is totally based on the concept that if all knowledge is stored in the machine itself then it can be used anywhere as per need and requirement. Intelligence is the ability to adopt the changes. Hence, AI is nothing but the data science working on the basis of stored data [3].

As shown in Fig. 1, USA is leading in terms of share in AI-related contributions followed by EU28 group and China. As far as Indian status is considered, the contribution is about 3% and 0.38 in reference to Nr. Players/GDP. AI has many applications necessary to deal with the various new emerging issues with availability and scarcity of resources.

The fast growing applications of AI are static image recognition, classification and tagging, algorithmic trading strategy performance improvements, efficient scalable processing of patient data, predictive maintenance, object detection and classification, content distribution on medias, cybersecurity protection and chatbots [4]. AI has the capability to adopt the existing creative processes like force field analysis, brainstorming, rapid prototyping, TRIZ, mind mapping [3]. It can also create its own creativity paradigm.

As shown in Fig. 2, it can be observed that AI contributes in almost all fields of new era of development, out of which static image reorganization, classification and tagging are the major parameter which contributes about 8097.9 million USD

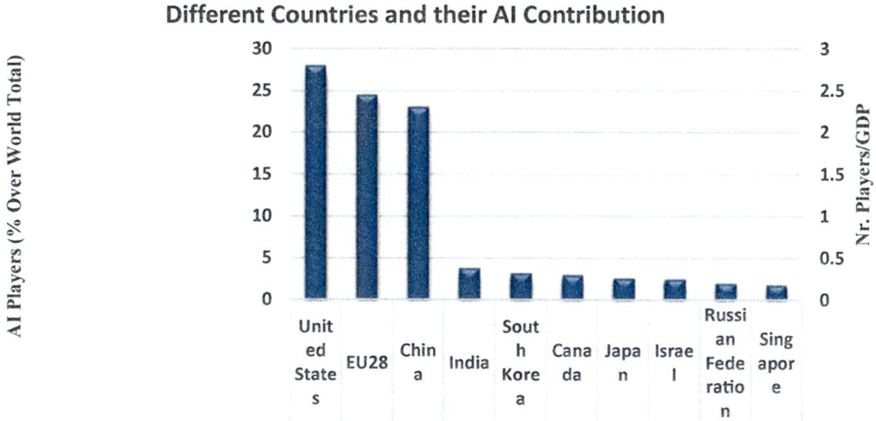

	United States	EU28	China	India	South Korea	Canada	Japan	Israel	Russian Federation	Singapore
AI Players (% Over World Total)	27	24	23	3	2.7	2.5	2.5	2.2	2	1.5
Nr. Players/GDP	2.8	2.45	2.3	0.38	0.32	0.3	0.26	0.25	0.2	0.18

Fig. 1 Major player in AI (% over world total)

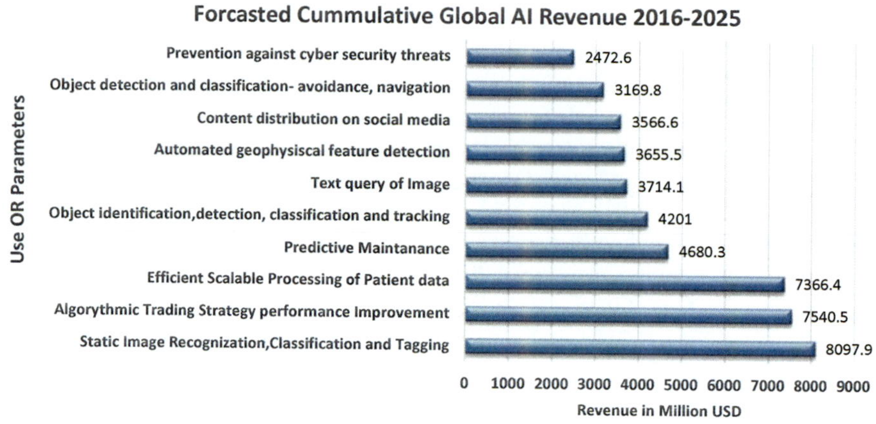

Fig. 2 Future of AI. *Source* Tractia, Statista Charts

revenue. Object identification, prediction, performance improvement and scalable processing are some other aspects which contributes the major share of the global revenue as forecasted for period of 2016–2025.

AI includes the ability of machine to use algorithm to learn from data and use what has been learned to make the decisions like human beings. It also includes machine learning which can be considered as the primary approaches to AI where machine has the ability to learn without being explicitly programmed. It consists of a subset machine learning that has networks which are capable of bearing unsupervised from data that is unstructured or unlabeled.

The graph in Fig. 3 clearly indicates the probable rise in the revenue from AI for enterprise application market from 2016 to 2025 across the world. It can be observed that the quantity which was nil during 2014–15 has reached to about 200 million USD in 2018. It is predicted that the growth will be on rise and may reach up to 24,000 million USD in 2024 and up to 32,000 million USD in 2025. This ultimately implies the importance of adoption of AI for various activities in scenario of globalization.

3.3 *Smart City*

Smart city uses the ICT in an integrated, collaborated and sustainable approach across all vertical segments to cater the needs of citizens, organizations and all other stakeholders while maximizing its triple bottom line. The basic challenges with the conventional city ICT were non-IOT, partial integration of vertical sectors, fragmented business chains without API integration and bureaucratic skepticism to change within some stakeholders. Due to this transition in view of cost-controlled scalable solution, fourth Industrial Revolution (Industry 4.0) connected with the

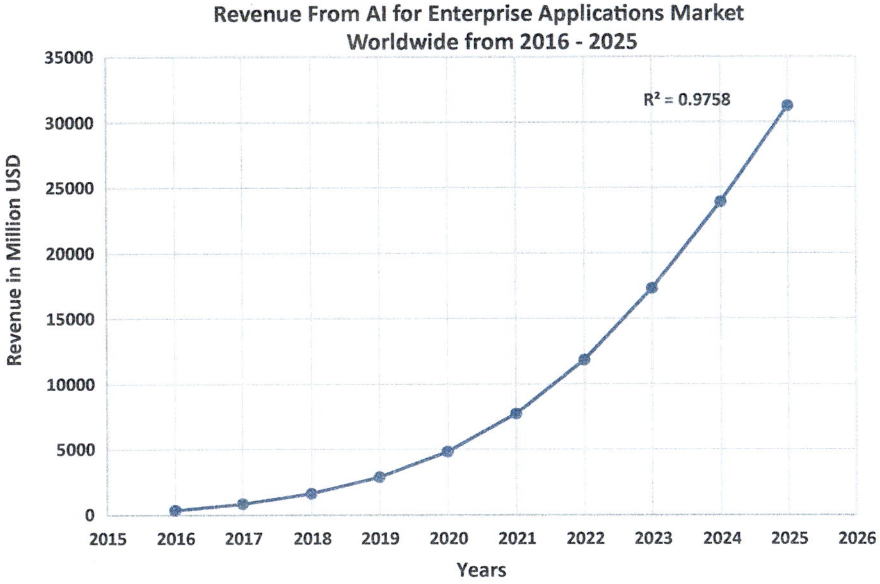

Fig. 3 Revenue from AI for enterprise application market worldwide from 2016 to 2025. *Source* Tractia, Statista Charts

various technologies at the heart proposed the different core technological concepts which can be termed as the cyber-physical system.

3.4 Internet of Things

The interconnection of uniquely identifiable embedded computing devices is termed as Internet of things (IOT). The devices which can be connected are sensors and monitors along with computers. This connectivity can be with the fixed line, mobile broadband, WiFi or any other technologies. The various changes in the word related to smart technologies such as different appliances, homes and cities are the output of IOT. It has the capability to deliver the better life in the future through digitalization. This movement will also lead to the concept, Internet of everything in the future. The devices supported by IOT include home electronics and appliances, smartphones and wearable, security and surveillance, automotive, aircraft and drones, monitoring implant heart and pacemakers, biochip implants, infrastructure, cloud system, cities and nation along with entire globe.

Table 3 Fundamentals of AI and IOT

Fundamentals of AI	Parameters and concepts	Fundamentals of IOT	Parameters and concepts
Deep and machine learning at the core	Artificial intelligence	Interconnectivity with the other network of the devices	Better performance than the individual device
	Machine learning		
	Deep learning		
Modes of machine learning	Supervised (task-driven regression and classification)	Ambient intelligence	Communicating devices act in background for serving
	Unsupervised (data-driven clustering)		
	Reinforcement (algorithm learns to react to an environment		
Cognitive and sensory abilities	To see (computer vision)	Big data concept	Data collection, sensor monitoring, measuring and reporting environmental elements
	To hear (speech recognition)		
	Comprehend (natural language processing)		
Interpretation of data	Structured data	Transmission and action implementation	Data available is transmitted and actions implemented in controlled patterns with self-correction cycles
	Unstructured data		

3.5 *Benefits of AI*

The aspects such as smart adaptive advertising, smart travel experience, smart power grid, renewable energy and optimization, disaster prevention, smart agriculture, smart health, smart security and surveillance can be achieved if proper implementation of AI and IOT is introduced in smart city development program. As indicated in Fig. 4, AI is currently playing the vital role in enhancing the features [4], functions and performances of the various products in the industry with about 51% of share. Similarly, optimization of operations (36%), automation (36%), decision making (35) and new product development (32%) along with the reduction in head count (22) [2] are the major benefits which are experienced by various executives.

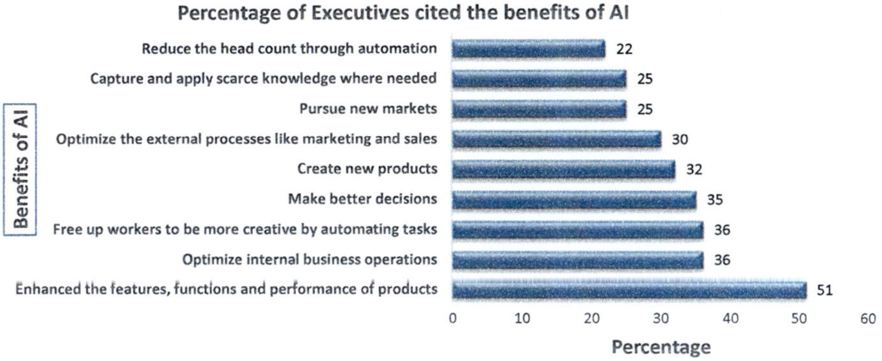

Fig. 4 Benefits of AI. *Source* Deloitte 2017, from—artificial intelligence for the real world—Thomas H-2018

3.6 Challenges in AI

Along with the number of positive points, there are wider challenges or the cross-industry questions as follows are also in picture with increased applications of AI and IOT (Fig. 5).

The major issue related to AI is its over-publicity in wrong and confusing manner [3]. In other words, it can be stated that on behalf of AI the trend to oversell the technologies in market places which contribute about 18% is the major challenge. It can also be noted that 47% of executives faced the difficulties while integrating the cognitive projects with the existing processes and systems. Similarly, expensive technologies (40%), not understanding the cognitive technologies (37%), scarcity of AI knowledge (35%) and immature technologies (31%), occupy the major share of

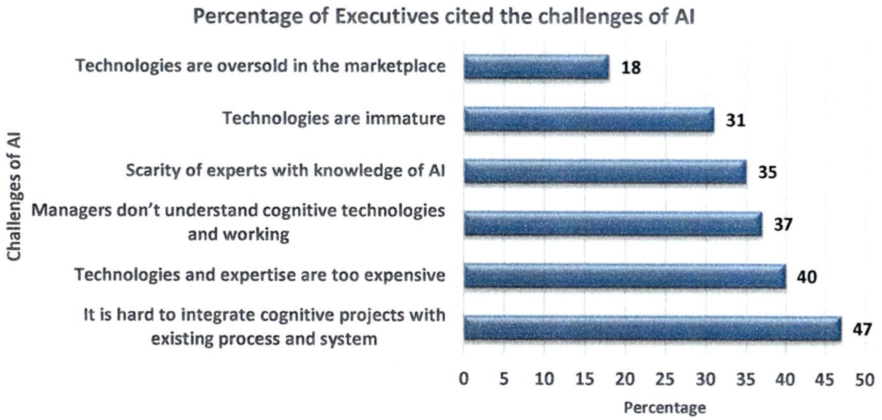

Fig. 5 Challenges in AI. *Source* Deloitte 2017, from—artificial intelligence for the real world—Thomas H-2018

Table 4 Challenges in AI

Issues	Description
Unemployment	As machines will take over more tasks, it will become difficult to prepare displaced humans to fill the roles created by new technologies
Inequality	Distribution of the wealth created by machines be very difficult
Humanity	The interactions with machines will affect behavior of human beings which will be unpredictable
Artificial stupidity	It will become out of control to guard the mistakes committed by the machine at particular stage
Racist robots	It will be very difficult if AI bias issue will come in existence
Security	It will be very difficult to keep AI safe from adversaries
Evil genies	Unintended consequences if occurred will not be possible to protect
Singularity	To maintain control over complex, intelligent systems will be the new issue to deal with
Robot rights	Treatment to AI components humanely will not be possible to implement

notable challenges due to adoption of AI-related advancements. Table 2 shows the additional aspects which describes the various AI-related issues and the probable challenges covering socioeconomic issues (Table 4).

4 Applications of AI and IOT to Industry and Case Study

In this section, it is tried to focus on the different case studies along with the industrial applications of AI [4]. For the purpose, industries like automotive and logistics, retail, health care and professional services are taken into consideration.

4.1 Industry Automotive and Logistics

Applications of AI and IOT Road traffic accidents predicted to be 5th leading cause of death by 2030 kill about 1.25 million people and injure about 20 million to 50 million, which cost about 518 billion USD. Assisted-driving vehicles along with crash avoidance technology are lowering accident rates nowadays. This may again plummet if AI-driven autonomous cars will come in action [4]. One of the implications related to the insurance aspects due to this transformation is not clear till date but liability shift from driver to manufacturers of sensors and devices, the software designer and the company along with operators and passengers.

Case Study Self-driving truck company—Otto (by Uber)—is testing self-driving trucks on roads within Northern California. The robot trucks are taking control on highways partially and if more automated, totally functional rollout by 2020 as per their goal. Trucking is a 700 billion USD industry in USA only, and it is predicted that self-driving trucks will be on the market before self-driving cars.

4.2 Industry Retail

Applications of AI and IOT Retailers are using supply-chain cost optimization at present. If AI is deployed for deeper understanding of consumer behavior, it may result in increased revenues [4]. In the future, AI-enabled digital assistants will be used to seamlessly and automatically find out order to deliver the most suitable option to customers. This will enhance the satisfaction level of consumer.

Case Study Humanoid robot (Pepper) in Japan—Nescafe—deployed about one thousand Pepper robots to retail appliance stores to help consumers pick out Nespresso coffee machines in Japan. Telco SoftBank also experimented by opening a pop-up mobile phone store run entirely by robot (Pepper) sales clerks capable of speaking 19 languages along with facial expressions and conversation recognition. In the future, testing of the robots to deploy them across Tokyo to serve foreign visitors during the 2020 Olympic Games is going on priority basis.

4.3 Industry Health Care

Applications of AI and IOT Digitization of clinical information through the implementation of electronic medical records (EMRs) resulted in the generation of significant amounts of real-time data by billions of connected devices. This has led in lower-cost access to genomic information which is the wealth of information captured on the Internet.

This information can be used by next-generation analytics technologies on concepts such as big data, cognitive computing and machine learning. This will be useful in improving the methods of cancer treatments, predict chronic diseases along with personalized medical interventions and drive behavioral change.

As per the studies of AI, it takes about 12–14 years and 2.6 million USD to bring cancer-combing drug in market which will be possible within the half of that with AI. Thus, AI is also helpful in reducing tome to introduce new drugs to society [4].

Case Study AI and machine learning help physicians as used by IBM. According to IBM, medical images make up at least 90% of all medical data. The computing giants Watson Health and Merge Healthcare arms recently partnered with the Radiological Society to demonstrate some AI-driven solutions to help clinicians extract insights from imaging data. Watson Health has developed cognitive tools for peer review, data summarization and physician support. The MedyMatch brain bleed application has been designed to help emergency room physicians diagnose stroke or brain bleed on the basis of relevant evidence in patient records. Merge's Marktation augments the work of physicians by raising image-reading speeds and accuracy.

Cloud application for eliminating common causes of errors in medical imaging, and a lesion segmentation and tracking module is also implemented in many treatment processes.

4.4 Industry Professional Services

Applications of AI and IOT AI is emerging as an unprecedented opportunity to augment humans' abilities to 'do,' 'think,' 'learn' and 'feel' making possible to automating routine tasks and allowing humans to focus to solve more important problems [4].

This is resulting in enhancing the digital innovations along with workplace productivity which is expected to rise by about 22% till 2020. This will also reduce the labor needed to complete a job by about 4% in coming years.

Case Study (the AI lawyer)—Law firm BakerHostetler recently employed its first AI lawyer, ROSS, for its bankruptcy practice which is built on IBM's cognitive computing platform Watson. ROSS is designed to read and understand language, postulate hypotheses when questioned, conduct research and generate responses (along with references and citations) to back up its conclusions. ROSS (a robot) also learns from experience, gaining speed and knowledge the more lawyers it interacts with. It also constantly monitors current litigation so that it can notify colleagues about recent court decisions that may affect their cases.

To summarize with the case studied, it can be stated that AI has huge applications in all aspects as far as the current scenario is considered. The companies like Facebook, Amazon, Alphabet, IBM and Microsoft launched a partnership to research and promote best practices around AI in 2016 for self-regulation with intension to provide a formal structure of communication. Figure 6 clearly indicates the estimated economic impact of AI in relation to the major Asian industries. It can be noted that at present the estimated amount ranges from 300 billion USD (financial services and transportation) to 550 billion USD (manufacturing). It can also be seen from the graph below that this will definitely increase to the range of 480 (health care), 590 (transportation), 600 (financial services), 620 (retail) and 770 (manufacturing). This shows that in the future in almost all sectors there will be the major role played by AI and related technologies.

5 Conclusion and Recommendation

From above discussion, it can be stated that AI and IOT have the capability to support the current Industry 4.0 developments which are ongoing. As discussed above, almost all fields are in need of efficient performance in view of optimized use of available resources. This efficiency can be attained if AI and IOT fundamentals are incorporated for smart city development. Smart city concept has the huge parameters to be

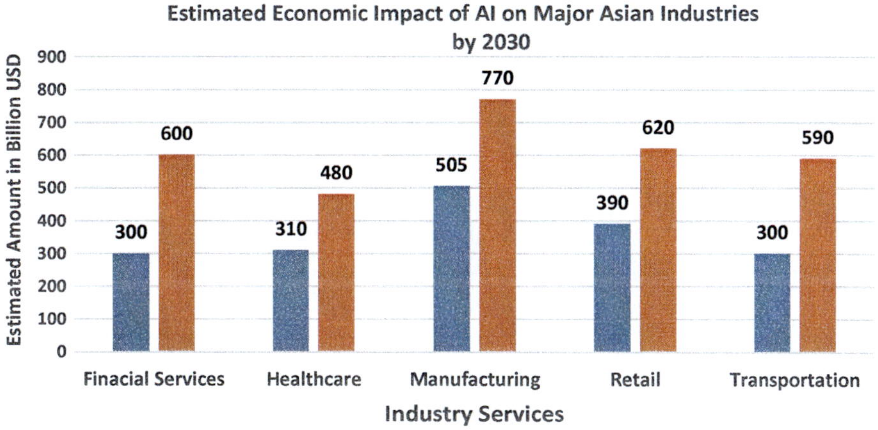

Fig. 6 Estimated economic impact of AI on major Asian industries by 2030. *Source* World Bank, UBS estimates as of January 2017

implemented to overcome the conventional and traditional methods of development. It has to be noted that World Economic Forum on the Digital Transformation Initiative has uncovered a variety of possible societal impacts stemming from the adoption of AI by different industries. It can also be stated that integration of predictive intelligent technologies during the smart city development can improve the public safety. Considering the expectations by industries in future, virtual reality, machine learning, augmented reality, natural language question answering and autonomous vehicles etc., representing AI are becoming the power for many innovations. It is also necessary to educate the maximum number of individuals regarding the benefits and need of AI and IOT. From above discussion and as proposed by many leading players, use of AI and its annual economic growth rates will double in 2035 with increase in productivity by about 40%. It can also be suggested to include the nonprofit organizations, ethicists and activists in the AI-related activities to accommodate diverse sources of expertise and to tackle with the related issues. In view of this, in Indian region also it is expected that the activities related to AI and IOT should be promoted and motivated for attaining the expected outcomes so as to match up with the requirements of Industry 4.0 Revolution. At the same time, the solutions for facing the unexpected difficulties which may arise without intimation should be prepared as way of defending. AI has the potential to drive significant productivity increases that improve the quality of lives around the world, but its development and use are not without risk and it must be implemented responsibly.

References

1. Rouhiainen L (2009) Artificial intelligence: 101 things you must know about our future (Kindle Edition)
2. Tavani D (2019) Artificial intelligence, the future of work, and inequality. Colorado State University
3. Lu P et al (2012) Artificial intelligence in civil engineering. Hindawi Publishing Corporation mathematical problems in engineering, vol 2012, Article ID 145974, 22 pages. https://doi.org/10.1155/2012/145974
4. Inclezan D et al (2017) Viewpoint: a critical view on smart cities and AI. J Artif Intell Res 60(2017):681–686
6. Allam Z et al (2019) On big data, artificial intelligence and smart cities, 80–89. 0264-2751/© 2019 Elsevier Ltd.

Analysis and Design of a Flyover Over a Railway Cross

Sonu Johny and Grace Mary Abraham

Abstract Heavy traffic congestion is observed in Guruvayur town. The main reason for this is the closure of railway gate for the passage of trains at Guruvayur–Choondal road. In order to solve this problem, planning and constructing a flyover bridge over railway crossing may be a viable option. A traffic flow survey is done in order to determine the traffic volume in the area, and a questionnaire survey is carried out to check the necessity of the flyover. Analysis of the flyover is performed in SAP 2000 software. In addition, a comparative analysis of the deflections produced in a pier when reinforced with steel and shape memory alloy (SMA) is done.

Keywords Flyover · Railway crossing · Guruvayur · SAP 2000 · Traffic survey · SMA

1 Introduction

Guruvayur is a municipal temple town in Thrissur district, Kerala. It is also where the Guruvayur Sri Krishna Temple resides, the fourth largest temple in terms of the number of devotees visiting per day [1]. Heavy traffic congestion is observed in the town of Guruvayur during the rush hours of the day. A railway crossing at the Guruvayur–Choondal is the major cause of the traffic congestion in the area. The roads are closed almost 32 times a day for about 5–10 min leading to a great amount of traffic congestion. Accidents have been also recorded in the area due to rushing of the driver to cross the railway line during the gate closing time. The timing of the train cannot be changed, so even emergency ambulances are required to wait for a long time.

A flyover is proposed in the area to reduce the traffic congestion based on the traffic and questionnaire survey that is conducted in the area to obtain the opinion of

S. Johny (✉) · G. M. Abraham
Amal Jyothi College of Engineering, Koovapally P.O, Kanjirappally, Kottayam, Kerala 686518, India
e-mail: sonujohny95@gmail.com

L. M. Gupta et al. (eds.), *Advances in Civil Engineering and Infrastructural Development*, Lecture Notes in Civil Engineering 87,
https://doi.org/10.1007/978-981-15-6463-5_59

the public [2]. The flyover is proposed over the existing road, and it is modelled and analysed using the software SAP 2000.

2 Data Collection and Assessment

Data was collected from the field in order to understand the actual problems faced by the people in the area. It helped in finding the feasibility of constructing a flyover across the railway cross to solve the problem [3].

Questionnaire survey was conducted for both the road users and the shopkeepers in the area. The questions were asked which helped in identifying the situation in the area. It also helped in finding the problems and to find feasible solutions for the problems faced by people.

From the study, it was found that about 65% of the users were daily users who went to schools or offices. There were also people who came to visit the temple which constituted about 25%, and the rest were occasional users. There were also a large number of shops in the area in which 52% included automobile shops, 26% restaurants and the rest grocery and medical shops.

The public were asked as to what they felt was the reason of the congestion, to which most of them responded that it was because of the railway cross. About 17% felt that it was because of the less width of roads, and 10% felt it was due to the poor management of the railway cross.

When asked for a solution, 80% responded that a flyover would solve the problem. About 14% were of the view that an underpass would solve the existing problem. This category mainly consisted of the shopkeepers as they felt that with construction of flyover their sale would greatly reduce and they would also lose their land.

A traffic survey was also conducted to find the amount of traffic in the area. The most common method of collecting traffic flow data is the manual method, which consists of assigning a person to record traffic as it passes. Traffic volume count was taken from Guruvayur to Choondal and Choondal to Guruvayur road. Traffic volume count was taken manually and by using video photography. As the traffic is composed of different classes of vehicles, it is converted to equivalent passenger car units. The flow is then expressed as PCU per hour or PCU per day. The distribution of different classes of vehicles along the Guruvayur–Choondal road stretch was found, and the peak hour of traffic volume was identified. It was seen that during 8.45 am to 9.45 am and 4.15 pm to 5.15 pm maximum number of vehicles occupied the roads. Hence, it was termed as peak hours of morning and evening, respectively.

During the morning peak, the two wheelers contribute the highest percentage of 42% followed by passenger cars 31%. There were also a few number of buses and vans during this peak time of about 2–4% carrying both students and office workers. A large number of auto-rickshaws about 20–23% were also noted. Few trucks and large vehicles run along Choondal to Guruvayur road and Guruvayur to Choondal road.

Survey was also conducted during the evening peak hours which showed similar trends to that of the morning peak hour. The two wheelers contributed the highest percentage of 49% followed by passenger cars 28–30%, and buses and vans 3–4% carrying both students and office workers. A large number of auto-rickshaws about 16–19% and trucks and large vehicles of 0–2% run along Choondal to Guruvayur road and Guruvayur to Choondal road.

Form both the questionnaire survey and the traffic survey, it was concluded that there is a problem of traffic congestion in the area and it can be reduced by constructing a flyover. A two-lane flyover for medium traffic is proposed.

Soil report was collected from boreholes conducted in the nearby area. From the borelogs, it is observed that the top stratum is soft to medium sandy/clayey soils. Considering the loads coming from the superstructure, deep foundation shall be a feasible system. Bored cast in situ RCC piles by DMC method installed as per the relevant clauses of IS 2911 part 1 and taken up to a depth of 13–18.5 m from the existing ground level and penetrating 30 cm into the rocky strata shall be provided as the foundation.

3 Modelling

The structure of the bridge is modelled using the finite element software SAP 2000. The grade of concrete that is used for the superstructure which includes the deck slab, longitudinal and cross girders is M_{25}. The grade of concrete that is used for the substructure consisting of pier, pier cap, pile, pile cap is M_{20}.

After the model is created, then the loads are applied on the structure as per the IRC specification [4, 5]. The loads include dead loads and live loads. The dead loads of the structure consist of the self-weight of the various components. The bridge loads are assigned in the form of moving loads and impact loads as per IRC: 6-2014 [6].

4 Analysis and Results

A static structural analysis is done which determines the displacements, stresses, strains and forces in structure caused by loads that do not induce significant inertia and damping effects. Figure 1 shows the bridge model, and Figs. 2 and 3 show the deflected shape and bending moment diagram.

Fig. 1 Two spans of the bridge modelled

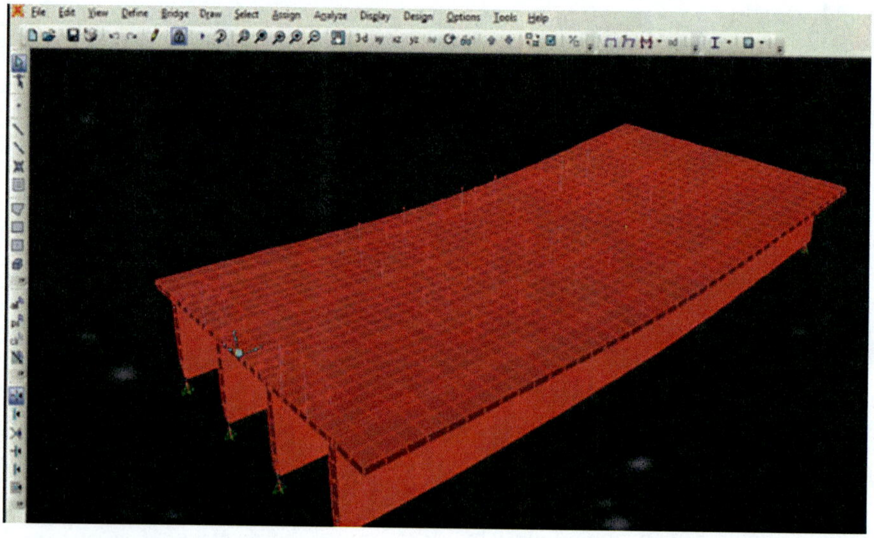

Fig. 2 Deflected shape of the structure

4.1 Shape Memory Alloy

A shape memory alloy abbreviated as SMA is an alloy that remembers its original, cold-forged shape and returns to the pre-deformed shape by heating. Studies show that this material is a lightweight, solid-state actuator alternative to conventional actuators such as hydraulic, pneumatic and motor-based systems [7]. There are many

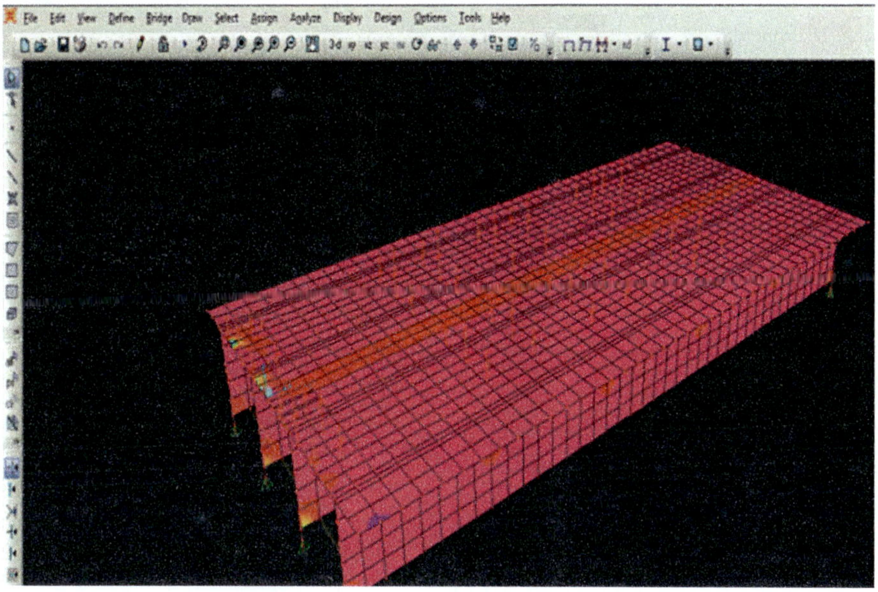

Fig. 3 Bending moment for deck slab

applications of SMA in different areas due to their high power density, solid-state actuation, high damping capacity, durability, corrosion resistance and fatigue resistance [8]. When used in civil structures, SMAs can be passive, semi-active or active components that reduce damage caused by environmental impacts or earthquakes. There are mainly three main types of shape memory alloys, which are copper–zinc–aluminium–nickel, copper–aluminium–nickel and nickel–titanium (NiTi) alloys, but SMAs can also be created by alloying zinc, copper, iron and gold.

Here, the application of shape memory alloy wire in seismic control of a bridge to reduce deflections is investigated. SAP 2000 engineering software was utilized to investigate and compare the performance of a steel-reinforced bridge column with SMA-reinforced bridge column under seismic loads and earthquake ground excitation.

The most common type of shape memory material is an alloy of nickel and titanium called Nitinol. This particular alloy has a very good electrical and mechanical properties, long fatigue life, durability and high corrosion resistance. As an actuator, it is capable of up to 5% strain recovery and 3515.348 kN/mm^2 restoration stresses with many cycles. A Nitinol wire 0.508 mm in diameter can lift as much as 7.3 kg. Nitinol can be used as an actuator, sensor and heater all in one material. Nitinol is also very popular as it is available in the form of wire, rod and bar stock, and thin film [9]. The properties of Nitinol, concrete and steel that is used are given in Table 1 [8].

Table 1 Properties of concrete, steel and SMA

Concrete	Compressive strength (N/mm^2)	40
	Density (kN/m^3)	24
	Modulus of elasticity (kN/m^2)	210×10^6
	Poisson's ratio	0.2
	Tensile strength (N/mm^2)	3
Shape memory alloy (SMA)	Density (kN/m^3)	64.5
	Modulus of elasticity (kN/m^2)	8.7×10^4
	Poisson's ratio	0.3
	Damping capacity	0.05 (5%)
	Diameter used (mm)	25
	Ultimate tensile strength (N/mm^2)	900
Steel reinforcement	Density (kN/m^3)	78.5
	Modulus of elasticity (kN/m^2)	210×10^6
	Poisson's ratio	0.3
	Diameter used (mm)	25
	Ultimate tensile strength (N/mm^2)	410

4.2 *Response Spectrum Analysis*

In the response spectrum method, the peak response of a structure during an earthquake is obtained directly from the earthquake response spectrum or design spectrum. This procedure is quite accurate for structural design applications. In this approach, multiple modes of response of a building to an earthquake are taken into account. For each mode, a response is read from the design spectrum, based on modal frequency and the modal mass. The responses of different modes are combined to provide an estimate of total response of the structure using modal combination methods such as complete quadratic combinations (CQCs), square root of sum of squares (SRSS) or absolute sum (ABS) method. Response spectrum analysis is carried out using the design spectrum specified in IS Code—1893:2002(Part 3) [10]. As per IS 1893, the design base shear (V_B) shall be compared with a base shear (V_{BX}) calculated using a fundamental time period T. If V_B is less then V_{BX} then all the response quantities (for example member forces, displacements) shall be multiplied by $\frac{V_{BX}}{V_B}$.

Seismic zone of the proposed structure is Zone 3, and zone factor is 0.16 as per IRC 6.

The correction factor is found as per the codes and applied in the load case of response spectrum. Value of 2.83 for EQ-X load case and 3.47 for EQ-Y load case is found as the scale factor and is provided as input in software.

4.3 Results

After application of load, analysis is carried out. The deflection at the nodes of the bridge column is found, and the values obtained for static and dynamic analysis are compared when reinforcement of steel and SMA is provided. This shows a positive result as in the case where SMA was introduced in T-beams [11]. The deflection of each point on the pier is noted from the deflection obtained after analysis. It was observed from the static analysis of the column frame that the maximum deflection reduced from 48 to 15 mm which gives a reduction of 68.75% in the deflection of the column as shown in Fig. 4. It was also observed from the dynamic analysis of the column frame that the maximum deflection of the column reduced from 115 to 75 mm which gives a reduction of 34.78% in the deflection as shown in Fig. 5.

The pier acts as a column, so the deflections are found to be maximum at the middle portion.

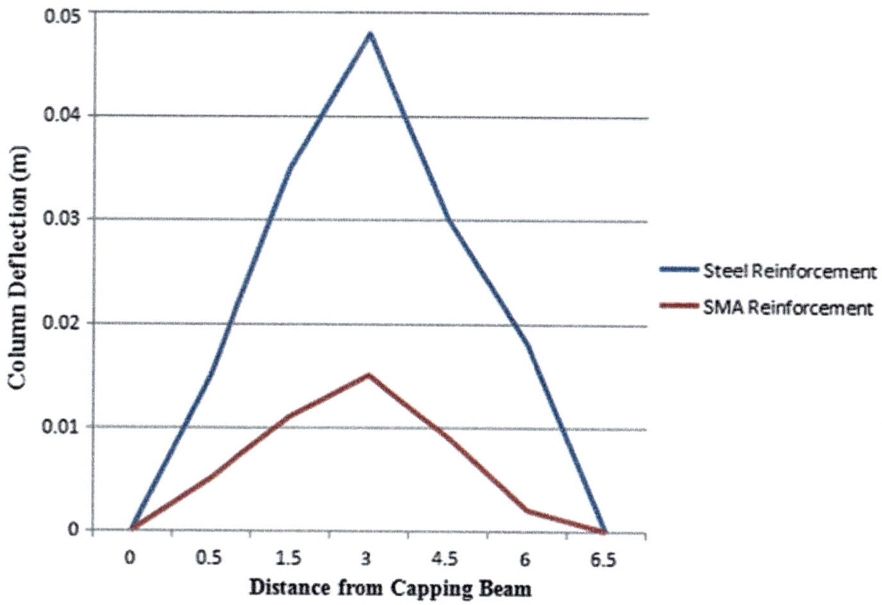

Fig. 4 Column deflection under static analysis

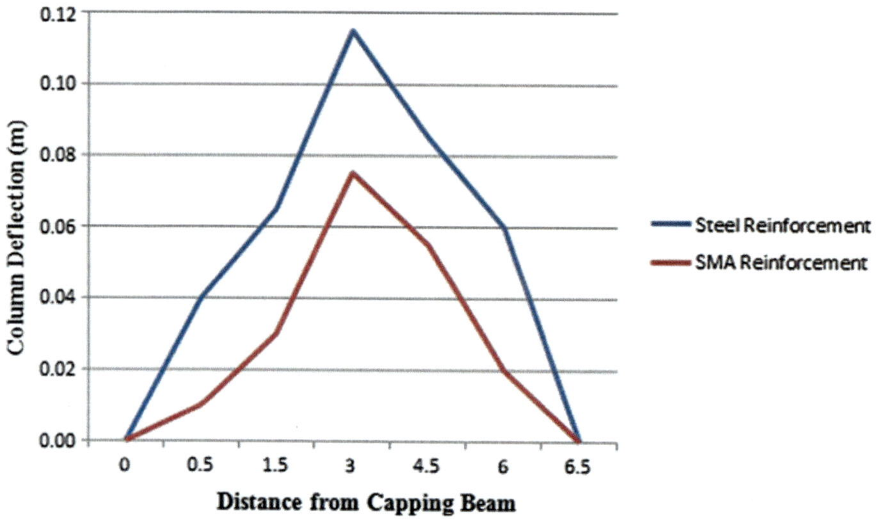

Fig. 5 Column deflection under dynamic analysis

5 Design of Flyover Elements

The design of flyover elements is done according to the IS codes [12] and IRC standards. The effective span of the T-beam is taken as 20 m, and the width of the road is 7.5 m. Width of footpath and kerb is 1.125 m on each side. The thickness of wearing coat is 0.08 m. The live load is taken as IRC class AA tracked vehicle [13]. The longitudinal girders provided are 4 in number and at a spacing of 2.5 m. The cross girders are provided at a spacing of 4 m.

5.1 Design of Interior Slab Panel

The slab that is taken into consideration is of size 4 × 2.5 m. Total dead load is found as 6.56 kN/mm^2. The dead load bending moment $M_B = 2.688$ kNm and $M_L = 1.174$ kNm. Dead load shear force is calculated as 7.216 kN.

The live load is taken as class AA tracked vehicle. One wheel is placed at the centre of the panel, and then live load bending moment is: $M_B = 31.01$ kNm and $M_L = 45$ kNm. And the overall depth is assumed as 200 mm.

Main reinforcement of 12-mm diameter bars at 100 mm c/c distance and 10-mm diameter bars at 100 mm c/c distance for secondary reinforcement is provided.

5.2 Design of Longitudinal Girder

The design of longitudinal girder is done using Courbon's theory. The depth of the girder is taken as 1.6 m and width of 0.3 m. IRC class AA tracked loading is considered for maximum loading. IRC class loads are arranged eccentrically for class AA tracked with minimum clearance = 1.625 m.

The corresponding reactions on the girder are found out, and the dead load that is shared by each girder is found to be 20.05 kN/m.

An impact factor of 10% is considered, and the live load is placed centrally in the span. Live load bending moment with impact and reaction for outer girder is 1338.83 kNm, and the live load bending moment with impact and reaction for inner girder is 1030.02 kNm.

The depth of the girder is assumed as 1600 mm. Total dead load per metre on girder is 30.13 kN/m. Maximum bending moment at the centre is calculated as 1808.5 kNm.

Main reinforcement of 12 bars of 36-mm diameter in three rows and 10-mm diameter four-legged stirrups at 250 mm spacing at support and 300 mm spacing towards the centre of span for shear reinforcement is provided.

5.3 Design of Cross Girder

The dimension of cross girder is taken similar to that of the longitudinal girder with depth 1.6 m and width 0.3 m. Total load on cross girder is 18.28 kN/m, and the reaction per girder is 34.27 kN/m. So, total bending moment is 438.01 kNm.

Main reinforcement of 4 bars of 25-mm diameter and shear reinforcement of 8-mm diameter two-legged stirrups at 350 mm c/c spacing throughout the length of the cross girder are provided.

5.4 Design of Pier Cap

The dead load and live load moments are calculated using IRC Class AA tracked vehicle loading and the reinforcement detail is found and is given as Top reinforcement 30 numbers of 25 mm diameter bars in 2 layers. Side reinforcement of 10 numbers of 16-mm diameter bars on each face is provided. Inclined reinforcement of 10 numbers of 16-mm diameter bars on each face and shear reinforcement of 12-mm diameter four-legged stirrups @ 150 mm spacing are provided.

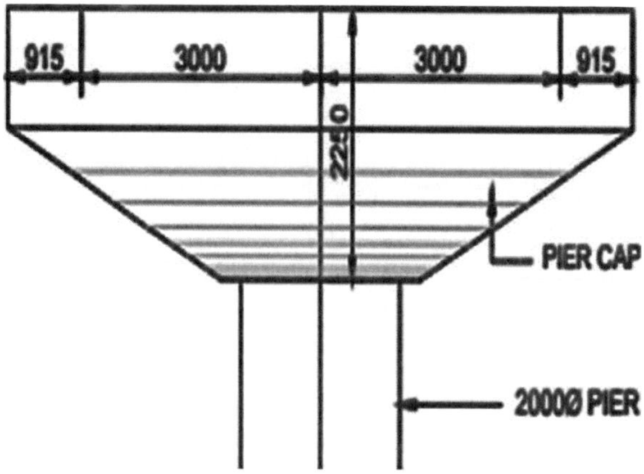

Fig. 6 Pier cap and pier

5.5 Design of Pier Column

The diameter of the pier is taken as 2 m and height 6.5 m. The dead load from all the above structures is calculated and added and designed as a column structure.

Main reinforcement of 32 numbers of 25-mm diameter bars and shear reinforcement of 10-mm diameter bars of lateral ties at 300 mm c/c distance are provided.

5.6 Design of Pile

The diameter of pile is 1 m. The design of pile is done using limit state method, and the reinforcement of 18 bars of 40-mm diameter bars is provided as main reinforcement. Lateral reinforcement of 12-mm diameter tie bars at a pitch of 190 mm at centre, 12-mm diameter tie bars at a pitch of 60 mm at pile head and 12-mm diameter tie bars at a pitch of 40 mm at pile end is provided.

5.7 Design of Pile Cap

A three-pile system is considered. The centre-to-centre spacing of piles is taken as 3 m. The design of pile cap is done as two beams. First beam connecting two pile and second beam connecting the third pile with the first beam.

Reinforcement of 26 bars of 25-mm diameter bars is provided for first beam and 26 bars of 25-mm diameter bars for second beam.

Figure 7 shows the position of the piles placed in the pile cap.

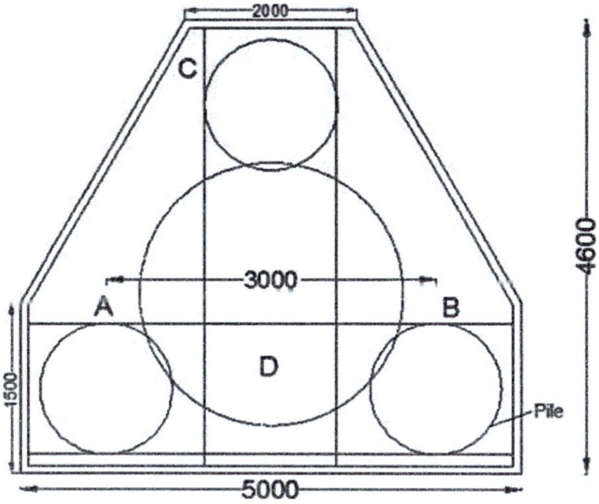

Fig. 7 Pile cap

6 Conclusion

The paper discusses the design and analysis of flyover at Guruvayur. A comparative study of the deflections in a pier column when reinforced with steel and SMA is also conducted. The analysis of the flyover has been done using the software SAP, and the design is also done manually.

From the comparative study of pier reinforced with steel and SMA, it has been observed that static analysis showed reduction of 68.75% in the deflection of the column and the response spectrum analysis gave a reduction of 34.78% in the deflection when SMA was used. Thus, the introduction of SMA reduced the deflection of the column in both analysis cases.

All procedures performed in studies involving human participants were in accordance with the ethical standards of the institutional and/or national research committee and with the 1964 Helsinki declaration and its later amendments or comparable ethical standards. For this type of study, formal consent is not required.

References

1. Guruvayur Temple, https://en.wikipedia.org/wiki/Guruvayur_Temple. Last accessed 2019/06/20
2. Sapariya AD, Patel KB (2015) Feasibility study for planning a fly- over bridge over railway crossing at Vijalpore Road, Navsari. Int J Adv Eng Res Dev 1:175–179
3. Batra U, Sarode MV (2013) Traffic surveying and analysis. Int J Appl Innov Eng Manag (IJAIEM) 14:123–128
4. IRC 21-2000, Standard specification and code for practice for road bridges, section III, cement concrete (plain and reinforced)
5. IRC 83- 1987, Part II standard specification and code of practice for road bridges

6. IRC 6-2000, standard specification and code of practice for road bridges
7. El-Hacha R (2018) Strengthening/reinforcing concrete structures using shape memory alloy. In: 5th world congress on smart and emerging materials, vol 5, pp 23–25
8. Tazarv M, Alam MS (2018) Shape memory alloy for bridge columns eleventh U.S. national conference on earthquake engineering, Los Angeles, California
9. Jaureguizahara SM, Chapettib MD, Yawnya AA (2016) Fatigue of NiTi shape memory wires. Procedia Struct Integrity 2:1427–1434
10. IS 1893 Part -3, earthquake resistant design of bridges and retaining walls
11. Paul NK, Saha S (2017) Improvement of load carrying capacity of an RCC T- beam bridge longitudinal girder by replacing steel bars with SMA bars. Int J Civil Environ Eng 7:1008–1012
12. IS 456:2000, Indian standard specification and code of practice for plain and reinforced concrete
13. Kavitha N, Jaya Kumari R, Jeeva K, Bavithra K, Kokila K (2015) Analysis and design of flyover. In: National conference on research advances in communication, computation, electrical science and structures (NCRACCESS-2015)

Smart Materials and Construction Management

Experimental Study on Pervious Concrete Using Coconut Shell Aggregate

Vilobh Vijayrao Ingale

Abstract Nowadays, pervious concrete is one of the best materials used in the construction industry to control the ground water table in the pavement system at their sources. In addition, to increase in the production of the waste materials increased the interest in utilizing of waste material for environmental and technical benefits. This paper presents the strength characteristics of pervious concrete with coconut shell-based concrete specimen of M20 grade with ordinary Portland cement of 53 grade. To study the experimental study of this work is to contribute sufficient data the characteristic strength of the pervious concrete by using the coconut shell aggregate with replacement of the conventional coarse aggregate in the proportion of 5, 10, and 15%. The result of this study states that the partial replacement of coconut shell gives good strength at up to extent but after that it will decreases.

Keywords Concrete · Cement · Coconut shell aggregate · Compressive strength · Flexural strength · Split tensile strength

1 Introduction

Due to the rapid growth in the population in the society, the production of the waste material is also increased rapidly. This waste material is damped in the soil; it will not be properly decomposed in the environment and it remains as it is for several years. The material which will not decomposed in the soil it create harmful effect on the environment so this material can be used as recycle material in other uses which lead to reduce the pollution. The better option to improve the waste disposal management is to reuse, recycle, and reduce.

Concrete may be a widely used construction material for numerous styles of structures thanks to its structural stability and strength. Due to rapid construction activities, most of the places are covered with impervious concrete. Due to this,

V. V. Ingale (✉)
Department of Civil Engineering, Vishwakarma Institute of Information Technology, Pune 411048, India
e-mail: Vilobhingale99.vi@gmail.com

L. M. Gupta et al. (eds.), *Advances in Civil Engineering and Infrastructural Development*, Lecture Notes in Civil Engineering 87,
https://doi.org/10.1007/978-981-15-6463-5_60

ground water table decreases. Concrete with pores (pervious) pavement is an effective solution to minimize this issue. Pervious concrete (PC) is open cell structure that allows the water to pass through it. Pervious concrete structure could also be a structure with interconnected voids through that rain and storm water is allowed to pass into the lowest. The gradient of pervious concrete is cement, aggregate, and water. Pervious concrete is associate environmental friendly artifact and is known as the best management follow for watershed management (M. Uma Maguesvaria and V. L. Narasimha).

The main property of concrete is compressive strength. This property of the concrete is mostly due to the coarse aggregate present in the concrete. Nowadays, there is rapid growth in the construction; due to this, the cost of the material is increased and for this construction natural material like stone is used most. And this creates a harmful effect on the environment. This harmful effect can be reduced by using some amount of agriculture waste in concrete like coconut shell. Coconut shell (CS) is the best material to substitute for the conventional aggregate in the concrete. Coconut shell is formed with polyose, hemi-cellulose, and polymer as major composition that shield it from decay. In India, the production of the coconut shell is about 11,930,000 tones (Vijay Kumar Shukla). If we partially replace coarse aggregate with the coconut shell aggregate in the concrete, it will save some amount of the conventional aggregate. The aim of this paper is to find the optimum proportion with the combination of natural aggregate to its strength properties.

2 Objectives

1. To test the effectiveness of pervious concrete made by partial replacement of coarse aggregate with coconut shell aggregate.
2. To find the optimum percentage of coconut shell in pervious concrete so that maximum strength can be achieved.

3 Methodology

The material was used to make the pervious concrete is cement, coconut shell aggregate, and coarse aggregate. The grade of cement is 53 grade of Super Birla. The size of coconut shell and the coarse aggregate was 20 mm passing and 12.5 retaining for the mix proportion. The mix design is made according to the IS code (IS 10262-2009).The maximum water cement ration is 0.45. The specimens are casted for the compressive strength, flexural strength, and split tensile strength.

Table 1 Properties of cement [1]

S. No.	Test	Standard value	Obtained value
1	Normal consistency	–	29%
2	Specific gravity	3.10–3.20	3.15
3	Initial setting time	>30	42 min
4	Fineness	<10%	0.8%

Table 2 Properties of course aggregate

S. No.	Tests	Obtained values
1.	Specific gravity	2.56
2.	Water absorption	1.46%
3.	Grading zone	II
4	Impact value	18.64%
5	Crushing value	24.5%

4 Material and Method

4.1 Cement

Cement could be a binding material in concrete for creating concrete normal hydraulic cement grade fifty three, factory-made by Birla Super confirming to IS 12269-2013 is employed. The properties of cement square measure given in Table 1.

4.2 Coarse Aggregate

The coarse mixture of size passes through 20 mm and retentive on twelve. 5 mm normal sieve is employed as coarse mixture. 20 mm size coarse aggregates confirming to IS 383-1970 the varied properties of coarse mixture square measure bestowed in Table 2.

4.3 Coconut Shell (CS)

The coconut shell aggregate of size passing through 20 mm passing and retain on the 10 mm sieve is used. The properties of coconut shell are determined same as conventional properties is given as in Table 3 (Fig. 1).

Table 3 Properties of coconut shell aggregate

S. No.	Tests	Obtained values
1	Specific gravity	1
2	Water absorption	24%
3	Impact value	8%
4	Crushing value	1.48%

Fig. 1 Coconut shell aggregate

4.4 Mix Design

Cement: aggregate: water

1: 3.21: 0.45.

Quantities of materials per m^3 concrete:

Cement: 405 kg/m^3
C.A.: 1490 kg/m^3
Water: 186 kg/m^3.

4.5 Experimental Procedure

The specimen of standard cube of 150 mm × 150 mm × 150 mm and standard cylinder of 300 mm × 150 mm and beam of size 150 mm × 150 mm × 700 mm were used to determine the compressive strength, flexural strength of concrete, and split tensile strength.

Table 4 Results of compression test

S. No.	% Coconut replacement (%)	Compressive strength (MPa) @28 days
1	0	16.46
2	5	6.42
3	10	5.95
4	15	3.11

5 Test Results and Discussion

5.1 Compression Test

The compressive test was done after the 28-day curing using the IS 516-1959. The cubes were tested on the compressive testing machine (CTM) of capacity of 2000 kN (Table 4; Fig. 2).

5.2 Flexural Test

The flexural test was done after the 28-day curing using the IS 516-1959. The beams were tested on the universal testing machine (UTM) of capacity of 1000 kN (Table 5; Fig. 3).

5.3 Split Tensile Test

The split tensile test was done after the 28-day curing using the IS 516-1959. The cylinders were tested on the compressive testing machine (CTM) of capacity of 2000 kN (Table 6; Fig. 4).

6 Conclusion

From the experimental study, we can conclude that the optimum percentage replacement of coconut shell with aggregate for the pervious concrete is 5–10% that leads to increase in strength whereas the 15% replacement leads to decrease in strength of concrete for compressive strength, split tensile strength, and also for flexural strength.

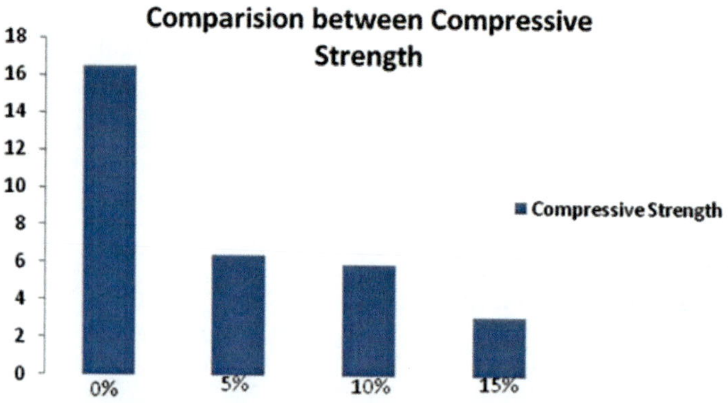

Fig. 2 Compressive testing using CTM

Table 5 Results of flexural test

S. No.	% Coconut replacement (%)	Flexural strength (MPa) @28 days
1	0	5
2	5	2.81
3	10	2
4	15	1.75

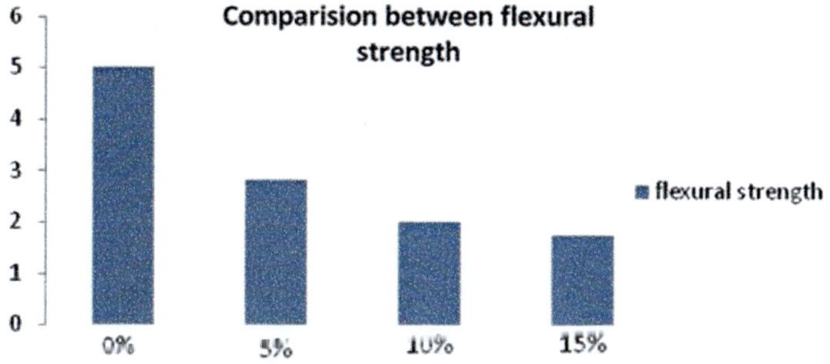

Fig. 3 Flexural testing using UTM

Table 6 Results of split tensile test

S. No.	% Coconut replacement (%)	Split tensile strength (MPa) @28 days
1	0	2.70
2	5	1.80
3	10	1.59
4	15	0.75

Coconut shell is more water absorber than coarse aggregate; so it decreases the water cement ratio and also increases the workability of the concrete and achieves more porosity as compared to conventional concrete so that the maximum water will be run-off through it.

By replacing the coarse aggregate with coconut waste, it reduces the weight of the concrete; also, the concrete is more economical than conventional concrete.

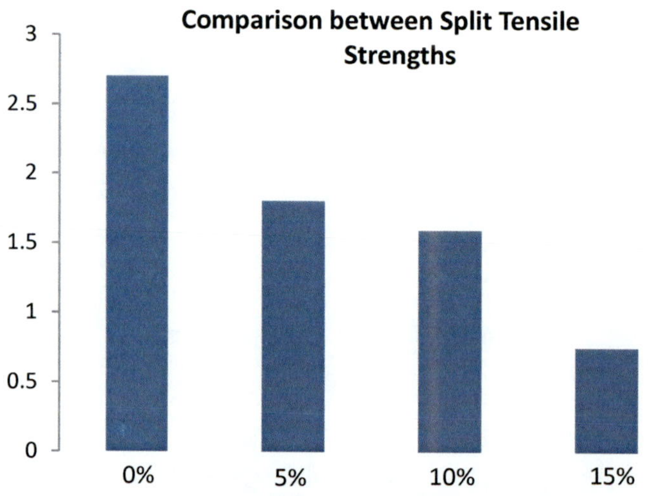

Fig. 4 Split tensile using CTM

Reference

1. Admute A, Nagarkar V (2017) Experimental study of green concrete. Int Res J Eng Technol (IRJET) 04(04)

Experiments on the Workability of Steel Fiber Reinforced Concrete

Swapnil Kore and Shardul G. Joshi

Abstract Despite various advantages and increased demand for steel fiber reinforced concrete in the construction industry, its drawback about low workability remains the same. This causes time-consuming construction and higher labour efforts. This study aims to perform experiments on the workability of SFRC by adding various admixtures. Fly Ash, Alkaline solution (NaOH solution), and viscosity-enhancing agent (poly-carboxylate ether) are the additives used to study their effect on the workability of SFRC. Different methods and contents were used for the study of these additives. The workability of fresh concrete was examined by a slump cone test. The effect of additives on the 7 days compressive strength was also studied.

Keywords Steel fiber reinforced concrete · Workability · Slump cone test

1 Introduction

Concrete is the composite construction material. Comparatively low tensile strength and ductility of the plain concrete is restrained by the incorporation of steel reinforcement having higher tensile strength and ductility. Conventionally this reinforcement is in the form of continuous and properly placed steel bars. Short and discontinuous fibers are randomly distributed throughout the concrete to produce a composite construction material known as fiber reinforced concrete (FRC). Fibers are introduced to increase the structural integrity of concrete. Fibrous material controls the cracking due to plastic and drying shrinkage [1].

Generally, the quantity of fibers in the concrete mix is expressed as the percentage of steel fibers by the total volume of concrete along with the fibers compositely. It is termed as volume fraction. Aspect ratio of the fibers is expressed as the ratio

S. Kore (✉) · S. G. Joshi
Department of Civil Engineering, Vishwakarma Institute of Information Technology, Pune, India
e-mail: swapnilkore3@gmail.com

S. G. Joshi
e-mail: shardul.joshi@viit.ac.in

L. M. Gupta et al. (eds.), *Advances in Civil Engineering and Infrastructural Development*, Lecture Notes in Civil Engineering 87,
https://doi.org/10.1007/978-981-15-6463-5_61

of fiber length and its diameter [2]. Steel fibers are the most used type of fiber in FRC which is known as Steel Fiber reinforced Concrete (SFRC). Along with the prevention of cracking due to plastic and drying shrinkage, steel fibers also increase some other properties of concrete such as flexural toughness, ductile behaviour prior to the ultimate failure, the energy absorption capacity, and durability [3]. Mostly, Steel fibers act as secondary reinforcement while hand-tied continuous rebars act as primary reinforcement. Only in the case of high volume fraction of steel fibers (where steel fibers volume fraction is more than 2%), they are used without other continuous reinforcement.

Despite various advantages of SFRC, steel fibers reduce the workability of fresh concrete. Workability is the property of fresh concrete or mortar which defines the ease and the homogeneity, with which it can be properly mixed, transported, placed, compacted and finished [4]. The inclusion of steel fibers results in the time-consuming construction and inconvenient handling of the labours. Concentration of steel fibers at surface level also causes bad finishing of the concrete. Steel fiber content and aspect ratio are the important factors influencing the workability of SFRC. The dimensional compatibility between the steel fibers and the coarse aggregates affect the relative mobility of the coarse aggregates. Therefore, higher fiber content and longer fibers cause difficulties in the relative movement among coarse aggregates and mobility of the mixture consequently [5, 6].

2 Objectives

The study aims to perform the experiments on the workability of steel fiber reinforced concrete. It seeks improvement in the workability of SFRC. This research studies and tries to find an alternative for the superplasticizer so that the fresh steel fiber reinforced concrete become flowable.

The objectives of this research include the experiments on workability of steel fiber reinforced concrete by using various admixtures other than the usual superplasticizers. The effect on the compressive strength of steel fiber reinforced concrete due to the addition of admixtures is also evaluated.

3 Methodology

The experiments on the workability of SFRC were done in this study. Fly ash, Alkaline solution (NaOH solution) and viscosity-enhancing agent (poly-carboxylate ether) were selected to do the experiments with the rheology of the steel fiber reinforced concrete. Previous studies suggest that alkaline solutions used in geopolymer concrete act as activation chemicals [7]. It may also affect the workability of concrete. Poly-carboxylate ether acts as viscosity-enhancing agent to retain the workability properties [8].

Different concrete mixes were designed as per the available literature. Four trials were made to test the workability of SFRC. First of all the normal steel fiber reinforced concrete without any extra ingredient were casted, then SFRC with fly ash, NaOH solution, poly-carboxylate ether were casted, respectively.

3.1 Materials

The materials and admixtures used in this research are discussed below:

Steel Fibers. The steel fibers used were hooked type and short in length. These are fibrex type of steel fibers. These are cold drawn steel wire fibers crafted from high quality, low carbon steel having tensile strength 1100 MPa. The specific gravity of this material is 7850 kg/m^3. The length of steel fibers is 30 mm and the aspect ratio is 25.

Fly Ash. This is Pulverized Fuel Ash confirming IS 3812:2003 Part 1. The fineness of fly ash affects the workability and strength of the concrete [7]. Fineness (retention on 45 micron sieve) of fly ash is 18%. Loss on ignition is 1.3% and the moisture content is 0.3%.

Alkaline Solution. The NaOH solution is used to improve the workability of fresh SFRC. The steel fibers were immersed in 1 N NaOH solution (as shown in Fig. 1) so that the relative mobility among the fibers gets enhanced. The equivalent weight of NaOH is 40. Therefore, 20 g of dry NaOH were dissolved in 0.5 L of water to make 1 N NaOH solution. 7.5 M NaOH solution is used in a sample concrete mix. 300 grams ($7.5 \times 40 = 300$) of dry NaOH were dissolved in 1 L of water to make 7.5 M NaOH solution (as shown in Fig. 2). As this reaction is exothermic, the Solution gets hot. Therefore, proper precautions should be taken while handling this solution.

Viscosity-Enhancing Agent. Poly-carboxylate ether is a viscosity-enhancing agent (Fig. 4). It has the specifications confirming IS 9103:1999. It works as super-

Fig. 1 Steel fibers immersed in NaOH solution

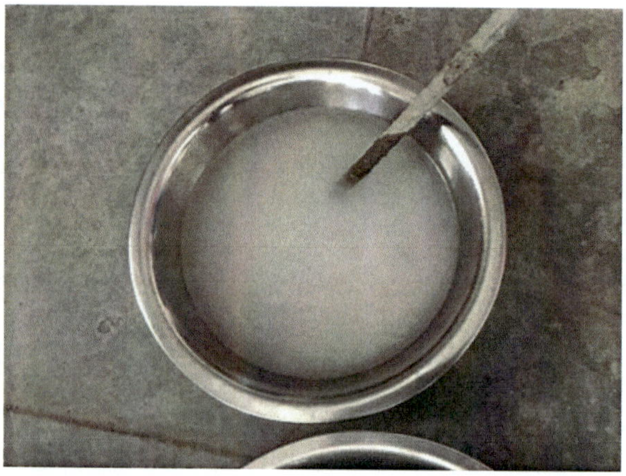

Fig. 2 7.5 M NaOH solution

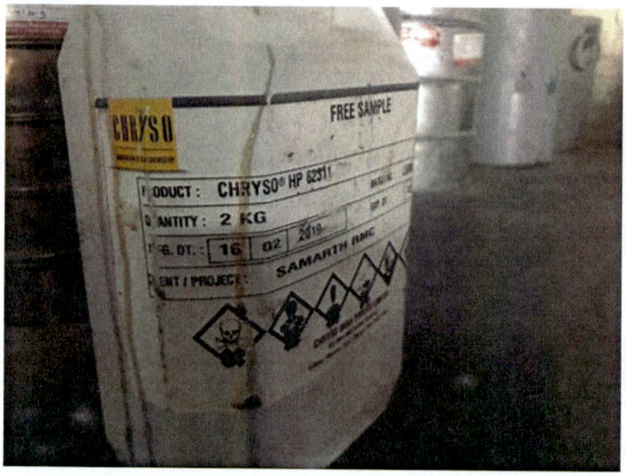

Fig. 3 Super-plasticizer

plastizer with retention for high-performance concrete and self-compacting concrete. It has specific gravity 1.11 and has no chloride content (Fig. 3).

3.2 *Experimental Execution*

The quantities of materials required for 3 cubes, i.e., 0.01125 m^3 of concrete were calculated as per mix design. The quantities for 1 m^3 of concrete are given in Table 1. All the materials were weighed as per the requirement. These materials were dry mixed properly in the mixer. After adding water and admixture the concrete mix was prepared.

Oil was applied thoroughly to the clean slump cone and its platform. Slump cone test [9, 10] was carried out on the concrete (Fig. 5). Workability was noted in mm.

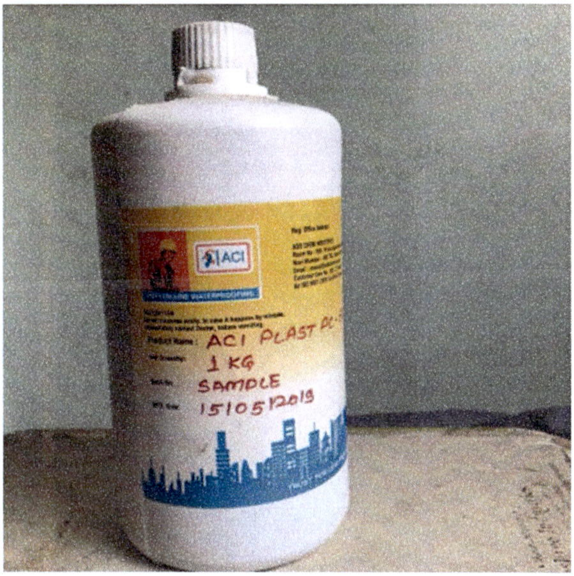

Fig. 4 Poly-carboxylate ether

Table 1 SFRC mix design material quantities for 1 m^3 of concrete

Materials	Trial 1	Trial 2	Trial 3	Trial 4
Cement	273.24 kg	218.59 kg	273.24 kg	273.24 kg
Fly ash		54.65 kg		
Water	190.75 kg	190.75 kg	103.25 kg	190.75 kg
7.5 M NaOH solution			87.5 L	
Coarse aggregates	1333.73 kg	1333.73 kg	1333.73 kg	1333.73 kg
Fine aggregates	828.05 kg	828.05 kg	828.05 kg	828.05 kg
Chemical admixture	2475 mL	2475 mL	2475 mL	2475 mL
Poly-carboxylate ether				4.44 kg
Steel fibers	78 kg	78 kg	78 kg	78 kg
Water-cement ratio	0.5	0.5	0.5	0.5

The concrete was poured in three oiled cube moulds of size 15 cm × 15 cm × 15 cm. While filling these cubes; were placed on the mechanical vibrator. Cubes were filled in three layers along with the external compaction. These cubes were allowed to set for 24 h and then they were demoulded. These cubes were immersed in water tank for 6 days. Compression Tests [9, 11] were carried on these cubes to determine the 7 day compressive strength (Fig. 6).

For trial 3, the Steel fibers were allowed to get soaked in 1 N NaOH solution for 24 h before casting of the concrete. As the reaction of dry NaOH with water is exothermic, solution gets hot. Hence proper precautions must be taken. At the time of casting, 7.5 M NaOH solution was used in the mix which also needs proper

Fig. 5 Slump cone test

Fig. 6 Compression test

handling. For trial 4, three different mixes were designed with different contents of poly-carboxylate ether. 35 mL, 45 mL, and 50 mL of viscosity-enhancing agent were used, respectively.

The successful trial was also checked for long steel fibers. The percentage of steel fibers was kept 0.5% by volume fraction.

4 Result and Discussion

Three different additives were studied under this research. Workability and compressive strengths were determined for each of them. Their impact and results are discussed as below:

- The workability of conventional SFRC is 10 mm by slump cone test and the flowability is zero by slump flow test. The average compressive strength after 7 days was found to be 33.97 MPa which is 113.23% of the design compressive strength.
- The workability of SFRC with fly ash is 15 mm by slump cone test and the flowability is zero by slump flow test. The average compressive strength after 7 days was found to be 25.93 MPa which is 86.42% of the design compressive strength.
- The workability of SFRC with NaOH solution is 12 mm by slump cone test and the flowability is zero by slump flow test. The average compressive strength after 7 days was found to be 14.32 MPa which is 47.7% of the design compressive strength.
- The work abilities of SFRC with poly-carboxylate ether are 30 mm, 43 mm and 55 mm by slump cone test and the flowability is zero by slump flow test for 35 mL, 45 mL and 55 mL of poly-carboxylate ether, respectively. Segregation of fresh concrete was observed for the 55 mL of poly-carboxylate ether. Reduction in water content was unable to provide required workability. Hence, we are considering the trial mix with 45 mL of poly-carboxylate ether. The average compressive strength after 7 days was found to be 25.48 MPa which is 84.93% of the design compressive strength (Figs. 7 and 8).

5 Conclusion

Here are some conclusions made from the analysis of results:

- The workability of steel fiber reinforced concrete can be improved by adding viscosity-enhancing agent. Comparatively it is the most effective additive in the study. The optimum quantity of poly-carboxylate ether in steel fiber reinforced

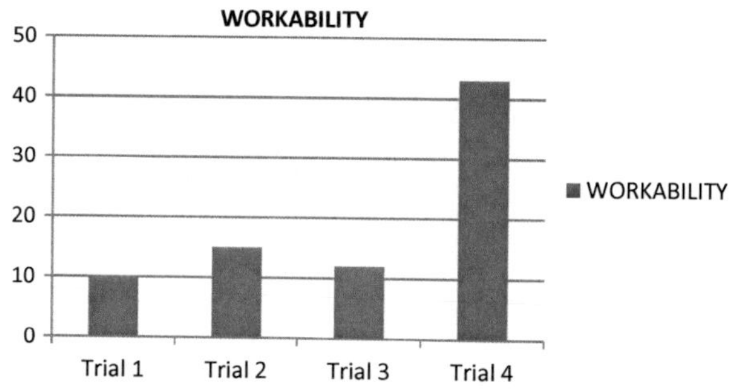

Fig. 7 Workability values for different trials

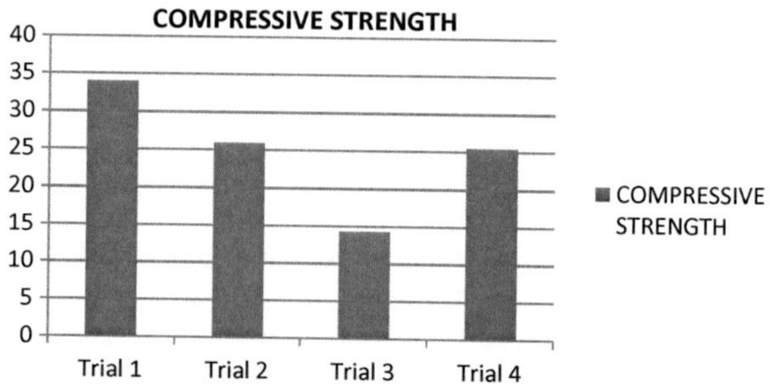

Fig. 8 Compressive strength values for different trials

concrete enhances the viscosity as well as affects the workability, flowability, and consistency of fresh concrete.

- The compressive strength of steel fiber reinforced concrete also remains unaffected by the addition of viscosity-enhancing agents (poly-carboxylate ether).

References

1. Behbahani HP, Nematollahi B, Farasatpour F (2011) Steel fiber reinforced concrete: a review. In: International Conference on Structural Engineering Construction and Management. Kandy, Sri Lanka
2. https://theconstructor.org/concrete/fibre-reinforced-concrete/150/
3. Altun F. et al (2006) Effects of steel fiber addition on mechanical properties of concrete and RC beams. Constr Build Mater 654–661
4. Indian Standard Code, IS6461 (Part VII) (1973)
5. Maidl B (1991) Stahlfaserbeton. Ernst & Sohn Verlag für Architektur und technische Wissenschaften, Berlin
6. Figueiredo AD, Ceccato MR (2015) Workability analysis of steel fiber reinforced concrete using slump and Ve-Be test. Mater Res 1284–1290
7. Patankar SV, Ghugal YM, Jamkar SS (2014) Mix design of fly ash based geopolymer concrete. Adv Struct Eng 1619–1636. Springer India
8. Akhlaghi O, Menceloglu YZ, Akbulut O (2017) Poly (carboxylate ether) based superplasticizer achieves workability retention in calcium aluminate cement. Sci Rep
9. Shetty MS (2005) Concrete technology: theory and practice, 6th edn. S. Chand Publications

10. ASTM C 143
11. IS 516: 1959 (2006) Methods of test for strength of concrete

Development of Concrete Canvas for Structural Applications

Kaustubh A. Kadam

Abstract Concrete canvas being an innovative technique is adopting a better stand in recent trends in civil engineering field. Being a thin membrane and flexible fundamental behaviour, application of concrete canvas is raising. The material is being transported in batch rolls which makes installation easier and faster for applications such as slope stabilization, canal lining, weed suppression and dust suppression, pipe protection, emergency shelters, waterproofing agent over warehouses and protection over sand bunkers in military works. This study aims the manufacturability, properties and the performance of laboratory processed concrete canvas and to compare the results with authentic concrete canvas record. The tests include flexural test, compression test and tensile test for the four variations in cast of test specimens. Equivalent sets of casts were produced by introducing steel wire mesh to examine improvement in performance of laboratory manufactured specimens which can benefit the increase the use of concrete canvas in additional applications which can parallelly used as structural elements.

Keywords Concrete canvas · Manufacturability · Reinforced concrete canvas · Mechanical properties

1 Introduction

Concrete is termed to be most commonly used construction material in current developing construction stream. Concrete covers most advantages and benefits, but flexibility boundaries of concrete are restricted [1]. The enhancement of the concrete for its improvement in versatility is carried out widely. In spite of this concrete is failing to prove itself in few implementations, there is scarcity in actual repairs of potholes in concrete pavements. The absolute and active shotcreting is unavailable. The actual efficient canal with minimum absorption is still unstirring. The world is still running for such snags.

K. A. Kadam (✉)
Department of Civil Engineering, Vishwakarma Institute of Information Technology, Pune, India
e-mail: kaustubhkadam21.kk@gmail.com

L. M. Gupta et al. (eds.), *Advances in Civil Engineering and Infrastructural Development*, Lecture Notes in Civil Engineering 87,
https://doi.org/10.1007/978-981-15-6463-5_62

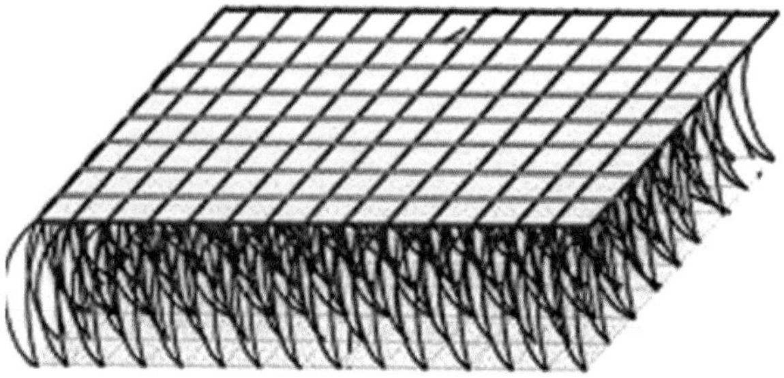

Fig. 1 Sectional view of concrete canvas [4]

The concrete field has come up with fresh technology. As its existence has not gained much duration in construction segment it has not been much familiar, but its achieving demand expeditiously [2]. Concrete canvas is a flexible material manufactured in dry state and hardens when hydrated. CC his holographic cast/grid which holds impermeable layer on bottommost layer, dry concrete mix in centrum and fibrous surface on top. The concrete canvas is packed in batched rolls, transported and laid as per requirement [3]. This can also be designated as precast medium (Fig. 1).

Concrete canvas is easier, faster and cost effective in installation as compared to conventional concrete. It also has benefited the environment by reducing the environment impact by 95%. As compared to conventional concrete, the installation speed of concrete canvas is 10% faster [5]. The dry density of concrete canvas is 1500 kg/m^3 and is increased to 2050 kg/m^3 post hydration [6]. Concrete canvas is available in three manners namely CC5, CC8 and CC13 which designates the thickness 5 mm, 8 mm and 13 mm, respectively. The hydration can be done by spraying or complete immersion in water. The installation can be carried out easily by nailing with specific spacing's of pegs or with adhesives for smooth bond with different surfaces. This technology was not commercialized for supplementary works in civil industry but was initiated for emergency shelters. Later, the applications raised for the concrete canvas [7].

2 Problem Statement

There is scarcity in actual repairs of potholes in concrete pavements. The absolute and active shotcreting is unavailable. The actual efficient canal with minimum absorption is still unstirring. The world is still running for such snags. The manufacturing of concrete canvas is not processed in India, and as a result, the initial cost of product is too high. There are no codal requirements for manufacturing of this material.

3 Objectives

The main objective is to produce workable un-reinforced and reinforced concrete canvas in laboratory. Secondly, to determine the mechanical properties of laboratory concrete canvas samples. Later, to study the site preparation, installation and hydration of concrete canvas and finally to identify the potential structural applications for concrete canvas.

4 Prototype Assembly

Three layers were assembled with polypropylene fibre sheet, cement and synthetic top surface. Several alterations were opted to eliminate the snags. The ultimate trial was comprised with tailored polypropylene fibre sheet, cementitious mix and cotton cloth sheet.

5 Experimental Analysis

To interpret over the strength characteristics of laboratory manufactured concrete canvas, there are significant vital experiments to be conducted over the test specimens. The examination of the results was done over three-point bending test, compressive test and pull-out test for flexural strength, compressive strength and tensile strength, respectively.

5.1 Test Specimen Preparation

The ultimate probation from foregoing casts is to be operated for the test specimens for experiments. The test specimens are followed by four variations in specified test.

1. Polypropylene sheet tailored with fibre sheet, Birla super OPC 53 grade and cotton cloth as pervious medium.
2. Polypropylene sheet tailored transversely by threads, Birla super OPC 53 grade and cotton cloth.
3. Polypropylene sheet tailored with fibre sheet, replacing 1/4th part of OPC 53 with Dr. Fixit 202 crack X powder and cotton cloth.
4. Polypropylene sheet tailored transversely by threads, replacing 1/4th part of OPC 53 with Dr. Fixit 202 crack X powder and cotton cloth.

(a) (b)

Fig. 2 Specimen under flexure test. **a** Test setup before application of load and **b** post-loading form of specimen

5.2 Flexural Test

Three-point bending test on the specified size of sample is conducted in accordance with IS 14862-2000 [8] (Fig. 2).

5.3 Compression Test

Compressive test is accomplished quoting to IS 2542-1981 Part II [9]. Results are further normalized for specimen as per loading disc size (Fig. 3).

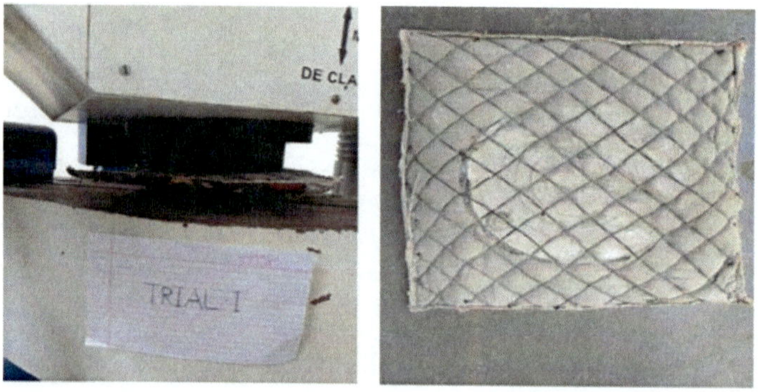

(a) (b)

Fig. 3 Specimen under compression test. **a** Test setup before application of load and **b** post-loading form of specimen

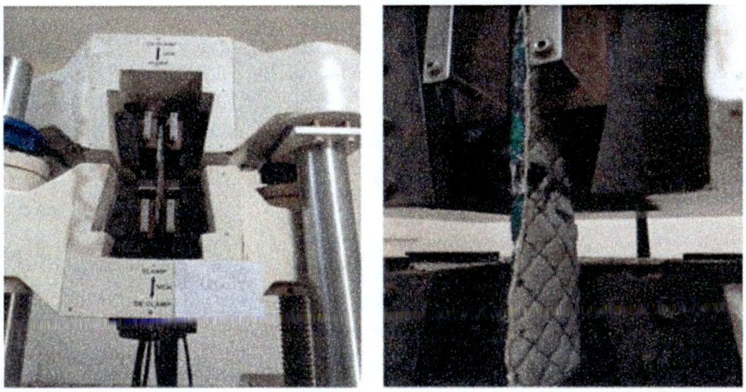

(a) (b)

Fig. 4 Specimen under tensile test. **a** Test setup before application of load and **b** post-loading form of specimen

5.4 Tensile Test

Pull-off test is carried out on the specimens as per ASTM D-5035 [10] (Fig. 4).

6 Results and Discussion

By experiencing the specified test for the specimens, the following results were obtained for the opted variations;

- The ultimate results of three-point bending test of un-reinforced specimens in terms of modulus of rupture (MOR) are termed as 4.347 MPa, 5.542 MPa, 6.258 MPa and 4.682 MPa, respectively.
- The ultimate results of three-point bending test of reinforced specimens in terms of modulus of rupture (MOR) are termed as 3.822 MPa, 6.306 MPa, 8.504 MPa and 5.255 MPa, respectively.
- The ultimate results of compression test of un-reinforced specimens in terms of compressive strength are termed as 50.311 MPa, 52.140 MPa, 52.768 MPa and 53.661 MPa, respectively.
- The ultimate results of compression test of reinforced specimens in terms of compressive strength are termed as 47.09 MPa, 54.398 MPa, 54.328 MPa and 52.049 MPa, respectively.
- The ultimate results of pull-off test of un-reinforced specimens in terms of tensile strength are termed as 0.606 MPa, 0.497 MPa, 1.109 MPa and 0.709 MPa, respectively.

- The ultimate results of pull-off test of reinforced specimens in terms of tensile strength are termed as 1.115 MPa, 0.958 MPa, 1.527 MPa and 0.939 MPa, respectively.
- The values opted from the tests that were executed are formed in the graphical representation of below in the terms of strength (MPa) vs specimen probation (Fig. 5).

The results obtained by a holographic cast made by locally available materials lead is sustainability and performance in comparison with actual concrete canvas. As we have analysed the results of un-reinforced specimen, by introducing the steel wire mesh as reinforcement, the further results with similar forms were obtained (Fig. 6).

As steel wire mesh has been introduced in the specimen, the increased variation in flexural strength and tensile strength was observed.

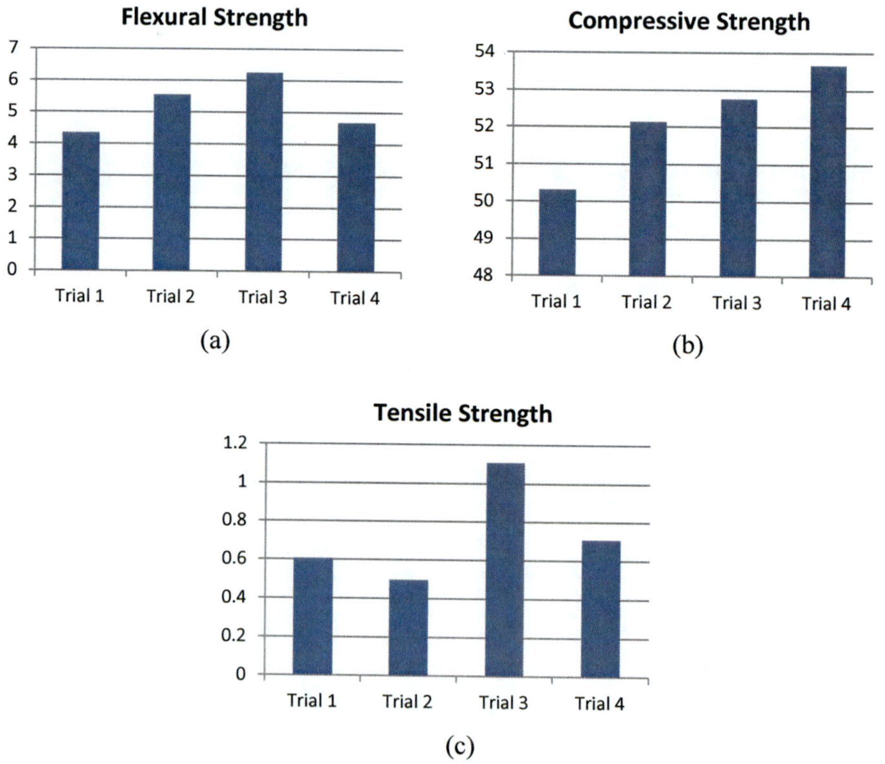

Fig. 5 Strengths of un-reinforced laboratory manufactured concrete canvas. **a** Flexural strength, **b** compressive strength and **c** tensile strength

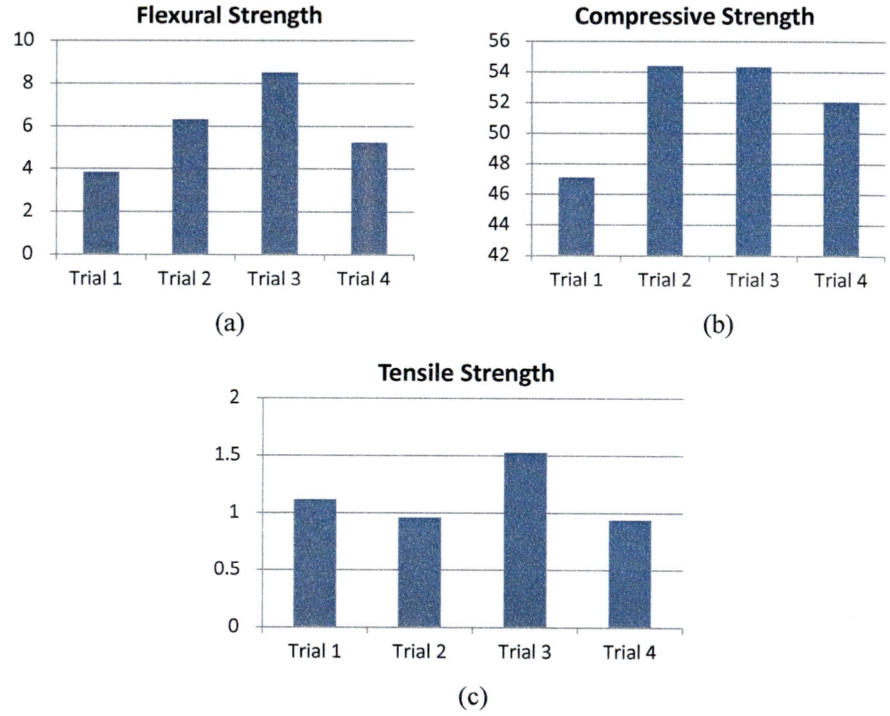

Fig. 6 Strengths of reinforced laboratory manufactured concrete canvas. **a** Flexural strength, **b** compressive strength and **c** tensile strength

7 Conclusion

The holographic cast formed equivalent to concrete canvas presented successive results to sustain in remarkable operations alike concrete canvas. The maximum number of manufacturability parameters is being managed to produce the desired cast in laboratory. The use of heavy equipments is not necessary. It can be manufactured in bulk and transported to workplace and installed faster than other conventional methods.

References

1. Ansari UA, Pasnur PK (2018) Experimental study of the mechanical behaviour of aluminium mosquito sheet on concrete canvas panels. Int J Adv Res Eng Technol 9(4):154–161
2. Akhtar V, Tyagi A (2015) Study of Canvas concrete in civil engineering works. Int Res J Eng Technol 2(9):591–594
3. Milliken Infrastructure (2017), "Concrete Cloth", geosynthetic cementitious composite mat
4. Influences of geometric patterns of 3D spacer fabric on tensile behavior of concrete canvas

5. https://www.slideshare.net/mobile/SandeepKumar1637/concrete-cloth-concrete-canvas
6. https://theconstructor.org/concrete/concrete-cloth-properties-applications-process/17070/
7. Anjaneyulu G (2017) study of concrete cloth (CC) in Civil engineering construction works. In: International conference on recent trends in engineering science and management
8. IS 14862-2000 (2003) Fibre cement flat sheets-specification
9. IS 2542-1981 (1991) Methods of test for gypsum plaster, concrete and products
10. ASTM D-5035 (2006) Standard test method for breaking force and elongation of textile fabrics

Experimental Study of Fresh and Harden Properties of Concrete Infused with Carbon Dioxide

Puja Patil

Abstract Concrete is a key material in the construction industry but the cement major ingredient of concrete is alone responsible for 5% of the world's annual carbon dioxide (CO_2) emissions. Carbon dioxide has been found as a beneficial admixture to freshly hydrating cement concrete. One recent technology infuses CO_2 during concrete production itself. The carbonation reaction between CO_2 and the hydrating cement forms nanosize calcium carbonate partials. The CO_2 can be infused into conventional concrete production through a simple gas cylinder and pressure control system. The feasibility of using CO_2 at the time of concrete production was investigated. A detailed experimentation evaluation of CO_2-infused concrete was conducted. The effect of the CO_2 on fresh properties of concrete was assessed in terms of workability, and effect on harden properties was assessed through compressive strength and split tensile strength. The comparison between the conventional concrete batch and batch infused with CO_2 (by weight of cement) was made to understand the behaviour of CO_2-infused concrete. The CO_2-infused concrete technology delivers valuable input to the concrete production while reducing the concrete's environmental impact.

Keywords Concrete · Carbon dioxide · Carbon dioxide-infused concrete · Slump cone test · Compressive strength

1 Introduction

Cement is the most widely used man-made material on earth. It produces concrete when mixed with water and aggregates and is used in the construction of everything from roads and buildings to dams and bridges and all kinds of other infrastructure. But while cement has widely framed the modern built environment, it is also major source of carbon dioxide emission to atmosphere. It is responsible for roughly 7% of the world's greenhouse gas emissions, according to International Energy Agency

P. Patil (✉)
Department of Civil Engineering, Vishwakarma Institute of Information Technology, Pune, India
e-mail: pujapatil8241@gmail.com

L. M. Gupta et al. (eds.), *Advances in Civil Engineering and Infrastructural Development*, Lecture Notes in Civil Engineering 87,
https://doi.org/10.1007/978-981-15-6463-5_63

[1]. Cement production alone contributes approximately 5% of the world's annual carbon dioxide emissions [2]. As global population grows, some estimates suggest that cement production may increase 23% by 2050. Concrete is one of the few materials in the world that has capacity to absorb carbon dioxide naturally. It is slow process, but over the time, it may be able to absorb a substantial amount of carbon dioxide emissions. There is need to find out techniques to speed up absorption process. Carbon capture and storage (CCS) can capture CO_2 emissions of cement industry before they are released and store them permanently [3]. If cement industries were retrofitted with carbon capture and storage (CCS) technology, then a substantial amount of emissions could be reduced from entering the atmosphere. One potential method is to upcycle CO_2 in concrete mixture prior to the end of their production [4]. If an industrial process could successfully use CO_2 in the production of concrete, there would be substantial carbon utilization that can close the loop for some of the CO_2 emitted during the production and additionally producing sustainable building material. Recent advances have been made globally, by infusing CO_2 in concrete production to produce sustainable and enhanced concrete.

A more recent technology infuses carbon dioxide in concrete during its production. Addition of carbon dioxide during its production results in formation of calcium carbonate particles which actually makes concrete stronger [1]. The strength improvement can then be leveraged in the optimization of the mix design such as early strength gain, 28 days strength or efficiency of binder. The process also results in less energy and cement during the production phase of concrete. It has been found that by adopting this recent technology, concrete products may rapidly gain required timely strength and achieve durability. Additionally, the resulting concrete products serve as storage units for undesirable carbon dioxide, which is drawn from industrial smoke-stacks resulting in sustainable building products [4]. The technology of CO_2-infused concrete is, however, not yet implemented by the concrete industries at national or local levels, due to lack of knowledge and application understanding related to this new technology in India. In experimental study, a feasibility study is made to use carbon dioxide in fresh concrete mixture during its production. The addition of CO_2 in fresh concrete accelerates the hydration and strength development without affecting the fresh properties of concrete [5].

2 Objectives

The objective of this paper is to prepare practically useful carbon dioxide-infused concrete in laboratory and to determine behaviour of this enhanced concrete with varying dosage of carbon dioxide. Characterize this special concrete by testing its fresh and hardened properties.

3 Methodology

1. Conceptual understanding (literature review) of carbon dioxide-infused concrete technology and carbonation behaviour of conventional and CO_2-infused concrete.
2. Arrangement of required concrete materials and industrial grade CO_2.
3. Based on the outcomes and findings of literature survey, a test matrix was developed aimed to ascertain the performance of fresh and hardened CO_2-infused concrete mix.
4. Casting of conventional and CO_2—infused concrete mixtures.
5. Laboratory testing of fresh properties (workability) and hardened properties (compressive strength and split tensile strength) of produced conventional and CO_2-infused concrete.
6. Finding out the behaviour of modified CO_2-infused concrete mix.

4 Experimental Analysis

The concept was to test behaviour of the modified CO_2-infused concrete and therefore there was need to provide controlled CO_2 gas to get absorbed in fresh concrete. The CO_2 used in the experiments was purchased as liquid CO_2 from deluxe industrial gases, Pune.

4.1 Gas Mixing and Handling

A tank of liquid CO_2 was placed near concrete mixer, and it was connected to gas controlled system and injector pipe. The liquid was weighed for injection into concrete mixer whereupon it converted into CO_2 gas and solid carbon dioxide snow. The mixing started with placing aggregates; this was followed by addition of cement and then water. The gas was injected once all the ingredients of concrete had been added and mixed [6]. The gas was injected with the help on injector pipe in the concrete mixing drum once all the ingredients of concrete had been added and mixed for around 3 min. The pressure maintained during injection process was 60 kg/cm^2. The pressure should be adjusted properly with the help of gas controlled system. The early age carbonation reactions involve reaction of main calcium silicate phases of hydrating cement with CO_2 in the presence of water to form calcium carbonate and calcium silicate hydrate gel as shown in following equations [6]:

$$3CaO \cdot SiO_2 + (3_{-x})CO_2 + yH_2O \rightarrow {}_xCaO \cdot SiO_3 \cdot yH_2O + (3_{-x})CaCO_3$$

$$2CaO \cdot SiO_2 + (2_{-x})CO_2 + yH_2O \rightarrow {}_xCaO \cdot SiO_3 \cdot yH_2O + (2_{-x})CaCO_3$$

Fig. 1 Carbon dioxide cylinder

Further any calcium hydroxide present in the cement paste will react with carbon dioxide, as shown in following equation:

$$Ca(OH)_2 + CO_2 + H_2O \rightarrow CaCO_3 + 2H_2O$$

These reactions are spontaneous such that carbonation was achieved within 120 s of CO_2 injection during concrete mixing. The reaction proceeds in the presence of water when Ca^{2+} ions from the cement paste meet CO_3^{2-} ions from the injected gas. The addition of CO_2 results in formation of nanoscale homogenously distributed carbonate reaction particles which contribute to stronger concrete (Fig. 1).

The carbon dioxide injection was synchronized with default mixing time so that the process cycle time remained unchanged. The CO_2 was injected into concrete mixer drum at specified flow rate for fixed injection time period, consequently it reacted with the fresh hydrating cement during initial mixing. According to literature review, the CO_2 in the liquid form was supplied at the rate of 1% by weight of cement and 1.5% by weight of cement. To observe variation in properties, two different batches of concrete were casted: conventional batch and modified CO_2 concrete batch. For the laboratory experimentation purpose, concrete cubes of M30 grade were casted and tested for its fresh and hardened properties. Compression testing was carried out at 3, 7, 14 and 28 days, respectively. For each doses, 3 cubes were casted at each test day. Split tensile strength of concrete cylinders were carried out at 14 and 28 days. Slump cone test was carried out on fresh concrete sample for workability of concrete. The samples for analysing behaviour of CO_2 concrete were created by taking fresh sample from mixing drum.

Table 1 Mix design

S. No.	Type	Unit	Quantity
1	Cement	kg	34.3
2	Coarse aggregates	kg	86.47
3	Fine aggregates	kg	54.4
4	Water	L	18.45

4.2 Material Specification

To satisfy the mix design as per the IS10262:2009 code [7], the materials were selected as coarse aggregates, fine aggregates confirming to IS383:1970, ordinary Portland cement 53 grade confirming to IS12269 (Table 1).

5 Results and Discussion

The process of CO_2 injection did not include any alterations in mix design or concrete production process. To understand the behaviour changes in fresh concrete, slump cone test was carried out [8]. The overview of fresh properties for the concrete produced for each mix type is presented in Table 2 (Fig. 2).

Table 2 Slump values for each mix type

S. No.	Mix type	Slump (mm)	Temp (°C)
1	Conventional concrete	75	32
2	Concrete with 1% CO_2	72	32
3	Concrete with 1.5% CO_2	68	32

Fig. 2 Slump cone test

Table 3 Average compressive strength of conventional concrete

S. No.		Day	Average comp. strength (MPa)
1	Conventional concrete	3	12.19
2		7	19.52
3		14	26.35
4		28	30.52

Table 4 Average compressive strength for 1% CO_2 addition

S. No.		Day	Average comp. strength (MPa)
1	Modified concrete with 1% carbon dioxide	3	14.42
2		7	24.15
3		14	34.43
4		28	43.04

Table 5 Average compressive strength for 1.5% CO_2 addition

S. No.		Day	Average comp. strength (MPa)
1	Modified concrete with 1.5% carbon dioxide	3	16.23
2		7	25.92
3		14	37.87
4		28	44.26

The average compressive strength measured for each dosage of CO_2 at 3, 7, 14 and 28 days, respectively. The concrete cubes were tested for compressive strength in accordance with IS516:1959 [9]. The compressive strength data is summarized in Tables 3, 4 and 5. The tasting measured the strength of three cubes per day for each CO_2 dosage (Figs. 3, 4, 5, 6 and Table 6).

The strength of carbon dioxide-infused concrete was found to be better than conventional concrete. Compressive strength results of the 1% CO_2-infused concrete showed a 19.29% improvement for the cubes tested at 3 days 23.71% at 7 days, 30.66% at 14 days and 41% at 28 days. For 2% CO_2 addition, average compressive strength enhancement was found to be 33.14% at 3 days, 32.78% at 7 days, 43.7% at 14 days and 48% increase at 28 days. Split tensile strength results of the 1% CO_2-infused concrete showed 10.1% improvement for the cylinders tested at 14 days, 14.98% at 28 days. For 1.5% addition of CO_2, average increase in tensile strength was 19.13% for 28 days and 3.92% for 14 days. Based on the slump test conducted on fresh mix from production line, the CO_2 infusion process had minimal effect on workability of concrete. For 1% CO_2 infusion, reduction in slump was found to be 3 mm. By increasing the dosage of CO_2 to 1.5%, reduction in slump was 7 mm than conventional concrete mix. To avoid reduction in workability, the addition of CO_2 was kept smaller in amount up to 1.5% by weight of cement.

Fig. 3 Compressive strength test

Fig. 4 Split cylinder tensile strength test

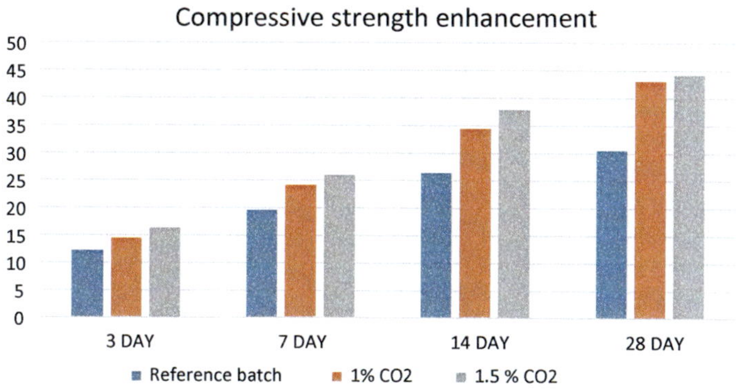

Fig. 5 Comparison of comp. strength of conventional concrete with CO_2-infused concrete

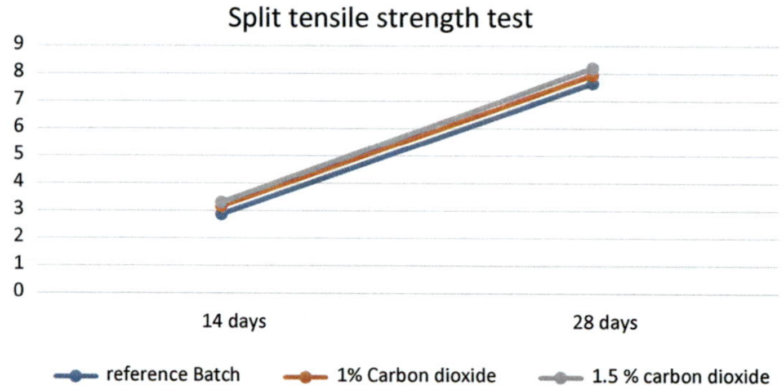

Fig. 6 Comparison of split tensile strength of conventional and CO_2-infused concrete

Table 6 Average split cylinder tensile strength

S. No.	Day	Conventional concrete mix	1% CO_2	1.5% CO_2
1	14	2.87	3.16	3.30
2	28	7.64	7.94	8.20

6 Conclusion

By infusing CO_2 into freshly mixing concrete, a series of concrete mixtures was produced and tested for fresh and hardened properties. The results of slump test showed that the CO_2 infusion process had a neutral effect on workability of concrete for the lower dosage. The compressive strength benefit was observed for the concrete that was infused with CO_2, by just adding small amount of CO_2 (1% by weight of cement) increase in compressive strength was 41% at 28 days and 48% for 1.5% CO_2 addition. The results of split tensile strength showed neutral to positive effect on tensile strength of concrete. The observed increase in strength with all dosage of CO_2 may result from reaction of CO_2 with calcium ions from hydrating cement and formation of nanoscale carbonation reaction products.

References

1. Carboncure homepage. https://www.carboncure.com/
2. Shao Y, Monkman S, Boyd AJ (2010) Recycling carbon dioxide into concrete: a feasibility study. In: Concrete sustainability conference
3. Monkman S, Shao Y (2006) Assessing the carbonation behaviour of cementations materials. J Mater Civ Eng
4. Monkman S, MacDonald M (2016) Carbon dioxide up cycling into industrially produced concrete blocks. J Constr Build Mater

5. Monkman S (2014) Carbon dioxide utilization in fresh industrially produced ready mixed concrete. In: International concrete sustainability conference. National Ready Mixed Concrete Association, Boston
6. Monkman S, MacDonald M, Hooton RD, Sandberg P (2016) Properties and durability of concrete produced using CO_2 as an accelerating admixture. Cem Concr Compos 74:218–224
7. IS 10262:1982 (2009) Recommended guidelines for concrete mix design
8. Shetty MS (2005) Concrete technology: theory and practice, 6th edn. S. Chand Publications
9. IS 516: 1959 (2006) Methods of test for strength of concrete

Experimental Study of Use of Flyash as Retarding Agent in Black Cotton Soil

Vishal Gajghate and Prakash Patil

Abstract Black cotton soil is found in various parts of our country. It considerably swells and shrinks due to variation in moisture content. Therefore, such soil is very treacherous for undertaking any construction work. Flyash is known to have pozzolanic action which can possibly have some beneficial effect in countering welling in expansive soils. It is proposed to use flyash as retarding agent and to overcome the problems of swelling in black cotton soil (Mishra in Int J Sci Res (IJSR), 2015 [1]). It has been concluded that flyash reduces the swelling pressure. The mode of application is more important aspect in controlling the swelling.

Keywords Black cotton soil · Flyash · Swelling

1 Introduction [2]

Black cotton soil is found in various parts of our country. In India, it is extensively found in regions of the Deccan Plateau. Black cotton soil is highly expansive, sticky plastic clays formed from residual weathering of deposits derived from volcanic rocks like basalts. Typical behaviour of three soil under different climatic conditions has made the construction and maintenance of structure not only expansive, but also difficult.

V. Gajghate
G. H. Raisoni University, Saikheda, India
e-mail: vishal.gajghate@ghru.edu.in

P. Patil (✉)
G. H. Raisoni College of Engineering, Nagpur, India
e-mail: prakash.patil@raisoni.net

L. M. Gupta et al. (eds.), *Advances in Civil Engineering and Infrastructural Development*, Lecture Notes in Civil Engineering 87,
https://doi.org/10.1007/978-981-15-6463-5_64

1.1 Characteristic of Black Cotton Soil [3]

Dark cotton soil contains montmorillonite clay minerals which show high partiality towards water. The colloidal earth content in dark cotton soil is found up to 40% and the part passing 0.075 mm sifter ranges from 70 to 100%. As far as possible, pliancy list esteems range from 40 to 100 and 20 to 60% individually. Dark cotton soil has low shrinkage limit (10–18%) and high OMC (20–30%). Every one of these rates renders the dirt very touchy to moisture changes, compressible and plastic in nature.

1.2 Problems of Black Cotton Soil [4]

From stabilization point of view, the following are the main problem in case of black cotton soil.

- It is very difficult to pulverize the soil, as the dry lumps are difficult to break due to high dry strength and wet soil is too sticky and unmanageable.
- There is large variation is volume and stability with variation in water content.
- They show considerable shrinkage on drying resulting in the formation of extensive cracks. Black cotton soil compacted at OMC will also shrink when dried as shrinkage limit is much lower than OMC.
- The black cotton soil exerts high swelling pressure on being soaked.
- Conventional construction materials like sand aggregate may not be available within easy reach. The black cotton soil is very poor and undependable subgrade materials, therefore such unsuitable as construction material.

1.3 Purpose of Present Work [5]

Keeping in mind the huge quantity of flyash which has been lying unutilized this project work is undertaken to study the effect of flyash as retarding agent is swelling of black cotton soil. Flyash is known to have pozzolanic action which can beneficially use for altering the properties of swelling soils. The present project work is undertaken to study the effect of flyash on the swelling characteristics of black cotton soil.

2 General Work Procedure for Constant Volume Method for Determination of Swelling Pressure [6]

The standard proctor mould was used for the constant volume method tor determination of swelling pressure. In this method, the soil to be tested for swelling pressure was taken in dried state and then added with the moisture constant corresponding to optimum moisture content of the soil. The result we mixed thoroughly and left for maturity period of one hour. This sample was compacted with standard specifications of standard proctor test. Filter paper was placed at the top and bottom of specimen and the porous stone was placed above the filter paper at top and below the filter paper, at bottom of soil sample. This assembly was placed in a water tank with the connections made to the proving ring (for swell pressure measurement which was fixed to load frame). The expansion if any, of the soil samples was measured by displacement dial gauge. Before adding water in the tank, the proving ring dial displacement reading, was initialized to zero. A seating pressure of 0.05 kg/cm^2 was applied and water was added to the tank. The constant volume was maintained by bringing displacement dial reading to its initial value by rotating handle of load frame and corresponding swell pressure being measured by proving ring. This procedure was repeated every 24 h till displacement dial did not show any marked changes. The final swell pressure was recorded from the proving ring.

3 General Work Procedure for Variable Method for Determination of Swelling Pressure Using Consolidometer

The using general work procedure for determination of soil pressure by variable volume method is as follows:

Two porous stones were taken, soaked in water, taken out of water, and any traces of water were wiped away. One stone was placed in the central seating of the water through. The guide rings were attached to both ends of the specimen ring and it was placed gently on the porous stone inside the water through with a filter paper against each face of the specimen. Then, the other porous stones were placed on top of the specimen, followed by the pressure pad and the steel base over it.

The consolidation cell was placed in position on the bed of loading machine. The counterbalance system of loading beam was checked and adjusted carefully. The counterbalance beam was adjusted in level position with approximate load transmitting member in contact with pressure pad through the ball seating, mould and finally the dial gauge was adjusted. The initial reading was noted before water was added. Seating pressure of 0.05 kg/cm^2 was added and the assembly was kept for 2 days to allow swelling to take place.

The loads were increased in the same manner and the corresponding dial gauge reading was noted. A semi-log plot of load versus settlement was plotted. From the graph, the pressure corresponding to the initial sample thickness gave the swell pressure.

Table 1 Differential free swell test values

S. No.	Soil type	Volume in water		Volume in kerosene		DFS (%)	Avg. DFS (%)
		Initial	Final	Initial	Final		
1	Black cotton soil	12	12.5	8	8	56.25	40.9
		13	14	9	10	40	
		12	12.5	9	9	38.89	
2	Black cotton soil + 5% flyash	16	17.5	12	12.5	40	39.08
		13	14.5	10	10.5	38.11	
		13	16	11	11.5	39.12	
3	Black cotton soil + 10% flyash	15	15	11	11	36.36	35.45
		10.5	13.5	10	10	35	
		10.5	13.5	10	10	35	
4	Black cotton soil + 15% flyash	13	13	10	10	30	33.33
		13	13.5	10	10	35	
		13	13.5	10	10	35	
5	Black cotton soil + 20% flyash	13	13	10	10	30	31.87
		13	13.5	10	10	35	
		13	13	10	10	30	

4 Preparation of Sample for Swell Test [2]

For this test, about 2.5 kg of oven-dried soil was taken. To this soil 23.14% (i.e., OMC of soil) of water was added and mixed thoroughly. The above soil was compacted in the mould of the 100 mm diameter and height 127.3 mm in three layers. Each layer was given 25 blows with a hammer weighing 2.6 kg with free drop of 310 mm. Then the mould was placed on soil extruder machine and the specimen cutter was put on the top of the mould. The handle of machine was taken rotated to press the outer into the compacted soil. The excess soil was removed and the specimen cutter was taken out for performing test. Some soil was taken from this sample and kept in oven for water content determination. The procedure for preparation of sample for black cotton soil with admixture (flyash) was same as that described above with only difference that flyash is added as an admixture (Tables 1, 2, 3, 4, 5, 6, 7, 8, 9 and Figs. 1, 2, 3, 4, 5, 6, 7, 8).

5 Differential Free Swell Test

The DFS test conducted were divided into two heads:

Table 2 Observation sheet for variable volume swell test (black cotton soil)

S. No.	Load on hanger (*w*) (g)	Lever ratio	Pressure on soil (kg/cm^2) $w \times 10/1000 \times 28.27$	Final dial reading ($\times 10^{-2}$ mm)
1	0.0	1:10	0.0	100
2	142		0.05	170
3	565		0.2	168
4	2261		0.8	143
5	4521		1.6	94

Table 3 Observation sheet for variable volume swell test (BC soil + 5% flyash)

S. No.	Load on hanger (*w*) (g)	Lever ratio	Pressure on soil (kg/cm^2) $w \times 10/1000 \times 28.27$	Final dial reading ($\times 10^{-2}$ mm)
1	0.0	1:10	0.0	0.0
2	142		0.05	25.0
3	565		0.2	23.5
4	1130		0.4	22.0
5	2261		0.8	18.0
6	4521		1.6	−11.0

Table 4 Observation sheet for variable volume swell test (BC soil + 10% flyash)

S. No.	Load on hanger (*w*) (g)	Lever ratio	Pressure on soil (kg/cm^2) $w \times 10/1000 \times 28.27$	Final dial reading ($\times 10^{-2}$ mm)
1	142	1:10	0.05	00.00
2	142		0.05	61.5
3	1130		0.4	9.0
4	2261		0.8	2.0
5	4521		1.6	−35.0

Table 5 Observation sheet for variable volume swell test (BC soil + 15% flyash)

S. No.	Load on hanger (*w*) (g)	Lever ratio	Pressure on soil (kg/cm^2) $w \times 10/1000 \times 28.27$	Final dial reading ($\times 10^{-2}$ mm)
1	142	1:10	0.0	0
2	142		0.05	80
3	1130		0.4	57.5
4	2261		0.8	7.0
5	4521		1.6	−18.0

Table 6 Observation sheet for variable volume swell test (BC soil + 20% flyash)

S. No.	Load on hanger (w) (g)	Lever ratio	Pressure on soil (kg/cm^2) $w \times 10/1000 \times 28.27$	Final dial reading ($\times 10^{-2}$ mm)
1	142	1:10	0.0	0
2	142		0.05	40
3	1130		0.4	12
4	2261		0.8	−44

Table 7 Observation sheet for variable volume swell test (BC soil + 10% flyash) sandwich layer

S. No.	Load on hanger (w) (g)	Lever ratio	Pressure on soil (kg/cm^2) $w \times 10/1000 \times 28.27$	Final dial reading ($\times 10^{-2}$ mm)
1	142	1:10	0.0	0
2	142		0.05	69
3	1130		0.4	5
4	2261		0.8	−16

Table 8 Observation sheet for variable volume swell test (BC soil + 20% flyash) sandwich layer

S. No.	Load on hanger (w) (g)	Lever ratio	Pressure on soil (kg/cm^2) $w \times 10/1000 \times 28.27$	Final dial reading ($\times 10^{-2}$ mm)
1	142	1:10	0.0	0
2	142		0.05	31.5
3	1130		0.4	−32

Table 9 Observation sheet for variable volume swell test (only flyash as padding layer)

S. No.	Load on hanger (w) (g)	Lever ratio	Pressure on soil (kg/cm^2) $w \times 10/1000 \times 28.27$	Final dial reading ($\times 10^{-2}$ mm)
1	142	1:10	0.5	12
2	1130		0.4	−64

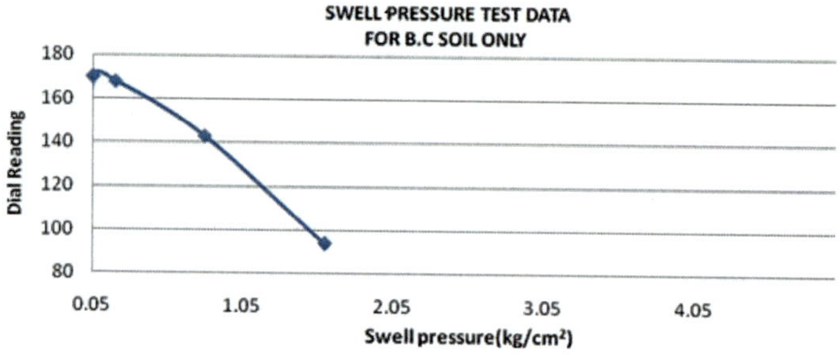

Fig. 1 Swell pressure from graph: 1.6 kg/cm^2

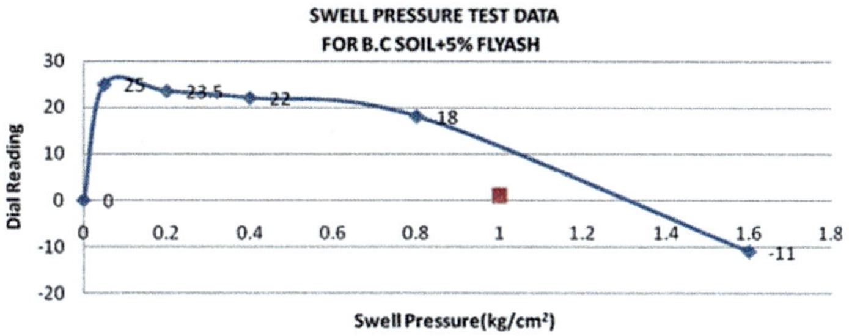

Fig. 2 Swelling pressure from graph: 0.85 kg/cm^2

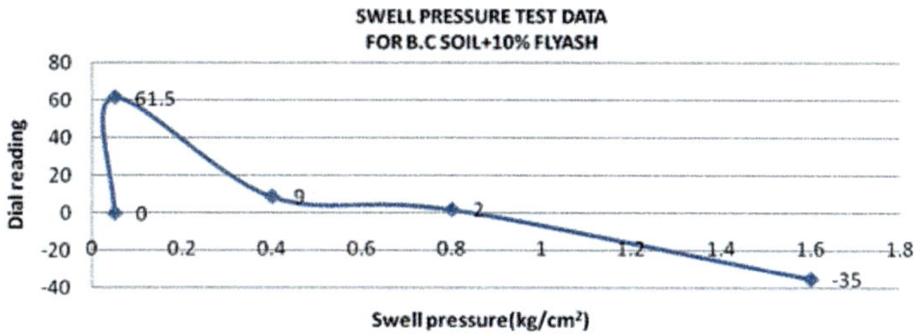

Fig. 3 Swell pressure from graph: 0.8 kg/cm^2

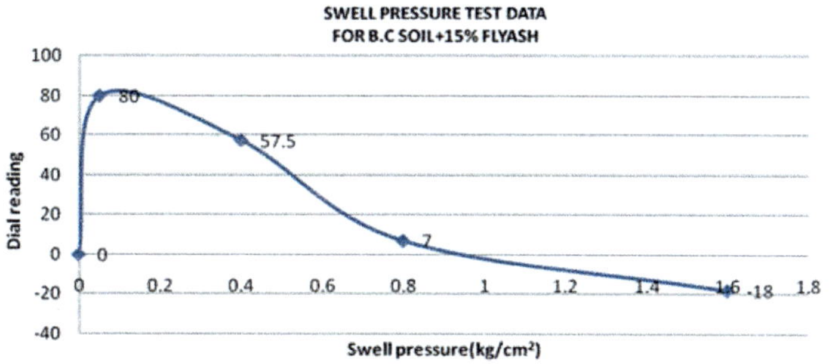

Fig. 4 Swell pressure from graph: 0.65 kg/cm^2

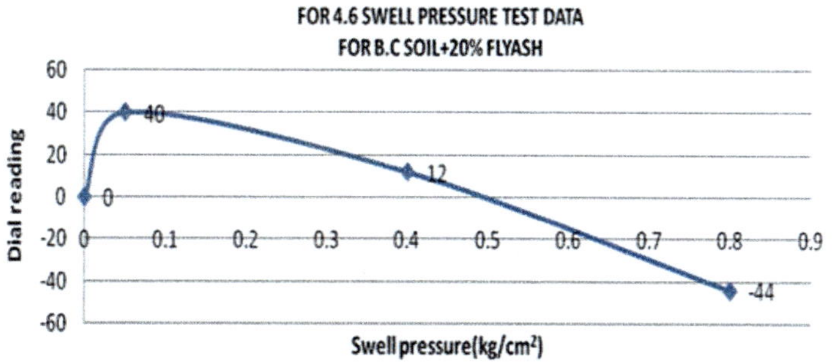

Fig. 5 Swell pressure from graph: 0.50 kg/cm^2

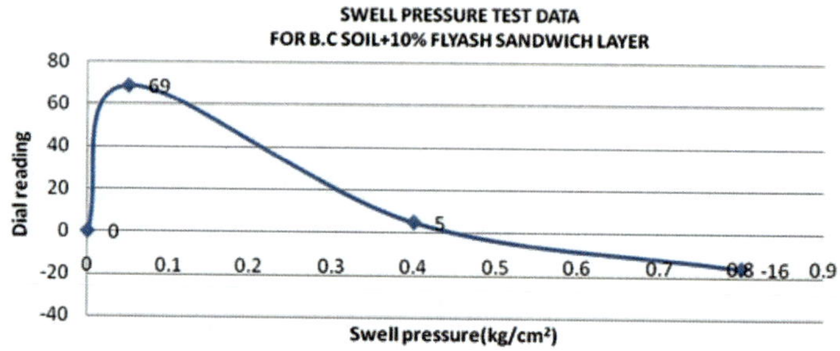

Fig. 6 Swell pressure from graph: 0.41 kg/cm^2

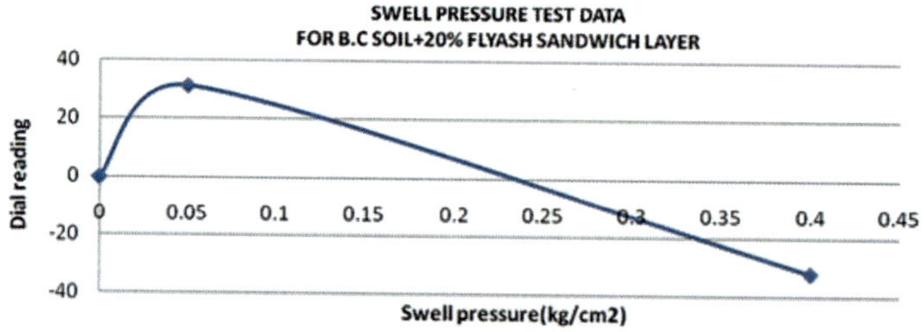

Fig. 7 Swell pressure from graph: 0.20 kg/cm^2

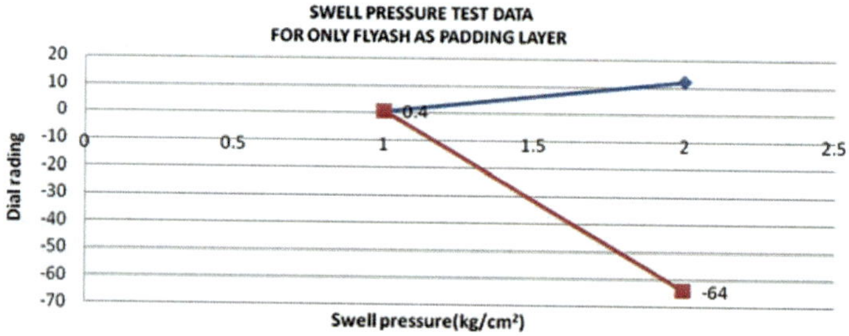

Fig. 8 Swell pressure from graph: 0.12 kg/cm^2

5.1 DFS Test for Black Cotton Soil Without Admixture

In this test, only black cotton soil was taken and its swelling was observed. It was found that the swelling high which is dangerous for civil engineering structures to remove this pressure of black cotton soil was very much disadvantage, and this soil was mixed with flyash.

5.2 DFS Test for Black Cotton Soil with Admixture

DFS test for black cotton soil with admixture (flyash): In this test, the admixture, i.e. flyash was added in the black cotton soil in varying ranges of 5, 10, 15 and 20%. It was observed that the swelling was considerably reduced which is desirable.

6 Results

The swelling and shrinking of expansive soils have always posed problems to civil engineering structures. This project work involves the use of flyash as a counteracting material against the highly expansive nature of the black cotton soil highly considering the future aspect.

The various results obtained for differential free swell values and the swell pressure values are summarized in Tables 10 and 11.

Table 10 DFS value

S. No.	Soil type	DFS (%)	Remarks
1	Black cotton soil	40.9	By increasing the percentage flyash, the degree of expansion reduces from high to moderate
2	Black cotton soil + 5% flyash	39.08	
3	Black cotton soil + 10% flyash	35.45	
4	Black cotton soil + 15% flyash	33.33	
5	Black cotton soil + 20% flyash	31.67	

Table 11 Swell pressure values

S. No.	Soil type	Swell pressure
1	Black cotton soil	1.6
2	Black cotton soil + 5% flyash	0.85
3	Black cotton soil + 10% flyash	0.8
4	Black cotton soil + 15% flyash	0.65
5	Black cotton soil + 20% flyash	0.5
6	Black cotton soil + sandwich layer (BC soil +10% flyash)	0.41
7	Black cotton soil + sandwich layer (BC soil + 20% flyash)	0.2
8	Black cotton soil + sandwich layer (pure flyash)	0.12

References

1. Mishra B (2015) A study on engineering behavior of black cotton soil and its stabilization by use of lime. Int J Sci Res (IJSR). ISSN (Online): 2319-7064. Index Copernicus Value (2013): 6.14 | Impact Factor (2014): 5.611
2. Gaikwad KS et al (2014) Analysis of engineering properties of black cotton soil & stabilization using by lime. Int J Eng Res Appl 4(5) (Version 3):25–32. www.ijera.com. ISSN: 2248-9622
3. Sudhakar P, Ramesh Babu V, Ramesh Babu B (2016) A study on subgrade characteristics of black cotton soil treated with lime and phosphogypsum. Int Res J Eng Technol (IRJET) 03(12)
4. Fulzele UG, Ghane VR, Parkhe DD (2016) Study of structures in black cotton soil. Int J Adv Sci Eng Technol 4(4). ISSN: 2321-9009
5. Jaya Prakash Babu V, Satyanarayana PVV, Manikantha S, Moin A (2016) Engineering properties of black cotton soil modified with fly ash and cement. Int J Eng Trends Technol (IJETT) 35(10)
6. Ramesh HN, Manjesh L, Vijaya Kumar HA (2014) Evaluation of engineering properties of black cotton soil treated with different stabilizers. Int J Eng Res Technol (IJERT) 3(12)

Performance of Concrete with Partial Replacement of Coarse Aggregate with Tyre Chipped Rubber

Sulagno Banerjee, Aritra Mandal, and Jessy Rooby

Abstract Transfer of tyre rubber suits a tremendous difficulty in India step by step. Analysts are attempting to utilize waste rubber in structural building venture from numerous days back. Crumb rubber replaced fine aggregate is a typical practice nowadays. Up to 20% chipped rubber with coarse aggregate had been replaced earlier and found that 5% replacement is optimum but that lacks some strength from conventional concrete. In this research programme to minimize this gap, extra 5% microsilica of the weight of cement had been added and also 40% of cement had been replaced by GGBS. Here, cubes, cylinders and prisms were cast to test compressive strength, tensile strength, flexural strength and durability against heat and were observed after 28 days and 56 days.

Keywords Tyre · Rubberized concrete · Chipped rubber concrete

1 Introduction

Presently, sustainability is the fundamental factor for research. For environment impact, researchers had tried to use waste products as much as they can and reusing of waste item is the primary key for research. In this specific research programme,

S. Banerjee (✉) · A. Mandal · J. Rooby
Department of Civil Engineering, Hindustan Institute of Technology & Science, Padur, Chennai, Tamil Nadu, India
e-mail: connect2sulagno@rediffmail.com

A. Mandal
e-mail: amiaritra@yahoo.co.in

J. Rooby
e-mail: jessyrooby@gmail.com

S. Banerjee
Department of Civil Engineering, Elitte College of Engineering, Sodepur, Kolkata, India

A. Mandal
Department of Civil Engineering, Techno International Batanagar, Maheshtala, West Bengal, India

L. M. Gupta et al. (eds.), *Advances in Civil Engineering and Infrastructural Development*, Lecture Notes in Civil Engineering 87,
https://doi.org/10.1007/978-981-15-6463-5_65

waste tyre chipped rubber is reused as coarse aggregate which goes about as 5% substitution of ordinary coarse aggregate. For being more eco-friendly, 40% cement is replaced by GGBS and additional 5% microsilica is being added to improve the strength. Here, cubes, cylinders and prisms were casted to test compressive strength, tensile strength, flexural strength and durability against heat and were observed after 28 days and 56 days.

2 Objective and Past Research

Concrete is the most utilized material in the development obligated for the consumption of regular assets and builds the shortage of the fixings, for example, cement, steel and aggregates, thus there is an interest for these materials in the business part. Further mining of waterway sand causes extreme ecological harm by bringing down groundwater table and breaking down of shake strata causes avalanche and seismic tremor. Architects are on edge to conquer this issue with different choices. Numerous scientists have endeavoured to recognize the backup utilization of the conventional materials. Emiroglu et al. [1] discovered slump that relies upon rubber substance and progressive diminishing in quality with the expansion of rubber. Gammel et al. [2] tried concrete with 10–25% crumb rubber supplanting alongside silica fume and rubcrete. Tayeh [3] found satisfactory performance against impact load and bending load with increase in percentage of sand replacement by the crumb rubber. Helme [4] prescribed 25% substitution that demonstrated compressive quality inside admissible range for most utilizations of concrete of the control blend plan. Naito et al. [5] discovered unit weight of CRC diminishes linearly. Richardson et al. [6] discovered concrete strength reduction means that air void/crumb separating which offers freeze/thaw resistance. Richardson et al. [7] concluded addition of 0.5% and 1% rubber crumb by mass of concrete to replicate levels of air entrainment that will provide freeze thaw resistance. Naik et al. [8] found that it is conceivable to make moderately high-quality rubber concrete utilizing magnesium oxychloride cement, which gives better holding attributes to rubber and altogether improves the presentation of rubcrete. Vadivel and Thennozhi [9] discovered evaluation of grade of concrete assumes the real job in the ductility performance of rubber replaced concrete.

Plate 1 Rubber aggregate

3 Experimental Investigation

3.1 Materials Used

3.1.1 Cement and Aggregates

In the present study, ordinary Portland cement of grade 53, confirming to IS: 12269-1987 [10], was used for preparing the concrete. The specific gravity of cement was 3.15. Fine aggregate—natural river sand passing through 4.75 mm IS sieve is used for making concrete. As per IS: 383-1970 [11], natural river sand was categorized under grading zone I. The specific gravity and fineness modulus of sand are found to be 2.65 and 3.05. Coarse aggregate—coarse aggregate was passed through 80 mm sieve and retained on 4.75 mm sieve confirming IS: 383-1970 [11] was used for concreting. The specific gravity and fineness modulus of coarse aggregate are found to be 2.695 and 7.7.

3.1.2 Water

Clean potable water free from suspended particles, chemical substances, biological elements, etc. is used both for mixing of concrete and curing.

3.1.3 Rubber Aggregate

This study has concentrated on the performance of a single gradation of rubber prepared by manual cutting (Plate 1). In this study, 5% of coarse aggregate is replaced by this chipped rubber. The maximum size of the rubber aggregate was 40 mm. The properties of the rubber used as aggregate are given in Table 1.

3.1.4 GGBS

GGBS which is near white in colour is a high-quality product, manufactured from a by-product of the steel or iron making industry. In this study, 40% of cement is replaced by GGBS. The physical properties of GGBS sample are shown in Table 2.

Table 1 Rubber properties

Parameters	Unit	Standard specs
Acetone extraction	%	5–10
Ash content	%	4 Max
Bulk density	gm/cc	0.30–0.45
Sieve analysis passing 40 mm sieve	%	99
Sieve analysis passing 2 mm sieve	%	1

Table 2 GGBS properties

S. No.	Characteristics	Value
1	Specific gravity	2.90
2	Bulk density (kg/m^3)	1220
3	Surface area (m^2/kg)	416
4	Insoluble residue (%)	0.14
5	Moisture content (%)	0.14
6	Loss on ignition (%)	0.19

3.1.5 Glenium 51

Glenium 51 superplasticizer (Plate2) is used for higher workability. In this study, 0.5% of cementitious material is used as glenium. It is an admixture of a new generation based on modified polycarboxylic ether.

Typical properties:

Aspect	Light brown liquid
Specific gravity	1082-1142 kg/lt at 20 °C
PH	6–7
Chloride Content	≤0.10% by mass
Alkali Content	≤3.0% by mass.

3.1.6 Microsilica

Microsilica is a very fine pozzolanic, amorphous material, a by-product of the production of elemental silicon or ferrosilicon alloys in electric arc furnaces. The specific gravity is 2.63. It is odourless white coloured powder with a pack density of 0.76 gm/cc. In this study, microsilica as 5% by weight of cement is being extra added.

Table 3 Mix proportions

Grade of concrete	Target mean strength (N/mm^2)	W/C ratio	Mix proportion
M 25	31. 60	0.45	1:2.20:2.72

4 Mix Design (as Per IS 10262-2009)

Based on the trial mixes, the final design mix was prepared for M25 grade of concrete as per IS 10262:2009 [12]. The concrete mix proportions were shown in Table 3.

5 Tests for Properties

The workability test, compression test, split tensile test, flexure strength test and heat tests were carried out to determine the strength and workability.

6 Results

6.1 Workability Test

Slump test was conducted using slump cone apparatus (Plate 2) to determine the workability and it is shown in Table 4.

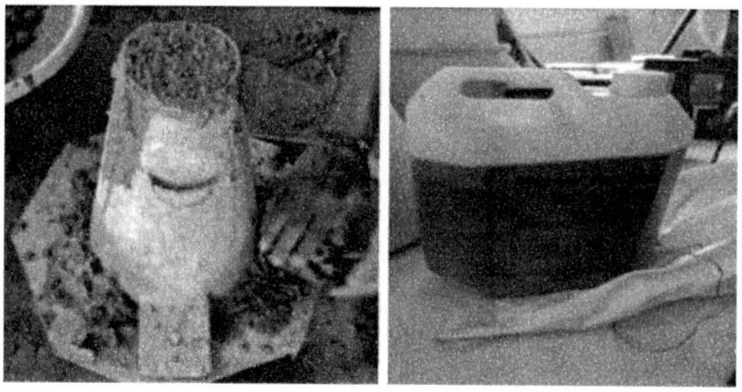

Plate 2 Slump test and glenium

Table 4 Result of workability test

Specification	Dosage of superplasticizer (% weight of cement)	Slump (mm)	W/C ratio
CC (control concrete)	0	100	0.45
CCR5 (Control concrete + 5% rubber)	0.5	90	
SGR0 (40% GGBS replaced cement concrete with no rubber and silica)	0.25	85	
SGR5 (40% GGBS replaced cement concrete + 5% rubber + 5% added silica)	0.5	95	

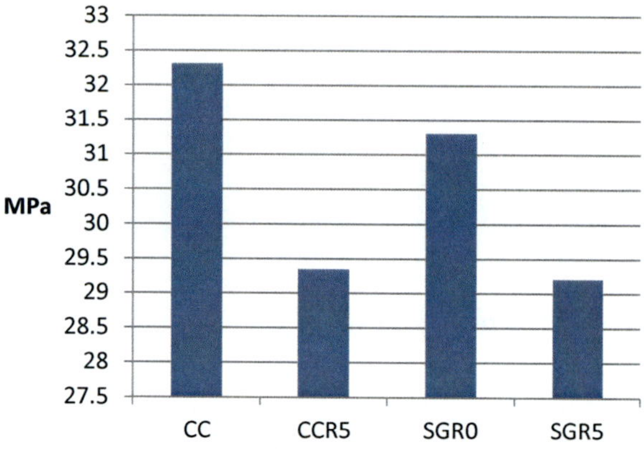

Fig. 1 Compressive strength (28 days)

6.2 Compressive Strength

The compressive strength of the specimens was determined in a universal testing machine of 200 tones capacity for 28 and 56 days, respectively, and they are tabulated in Figs. 1 and 2.

6.3 Split Tensile Strength

The 56 days split tensile strength of the specimens was determined and it is tabulated in Fig. 3.

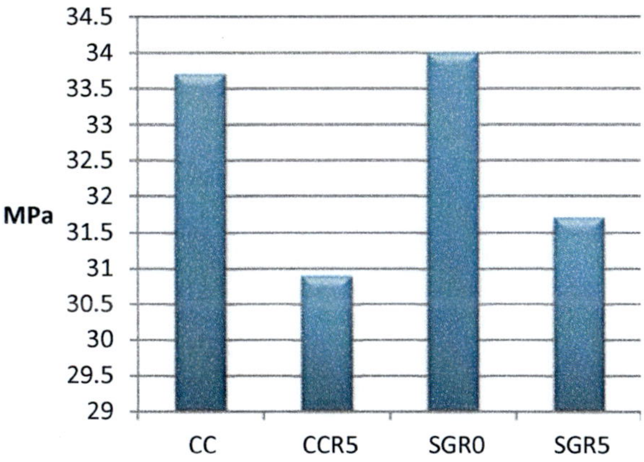

Fig. 2 Compressive strength (56 days)

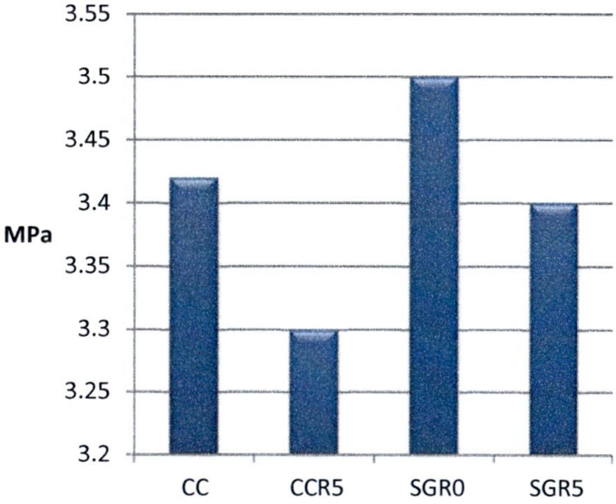

Fig. 3 Split tensile strength (56 days)

6.4 Flexural Strength

The 56 days flexural strength of the specimens was determined and it is tabulated in Fig. 4.

6.5 Heat Study

Cubes and cylinders are cast and after 56 days of curing in water they are dried and put into oven for 3 hours at a temperature of 150 °C. Then, compressive strength and split tensile strengths were obtained which are tabulated in Figs. 5 and 6.

Fig. 4 Flexural strength (56 days)

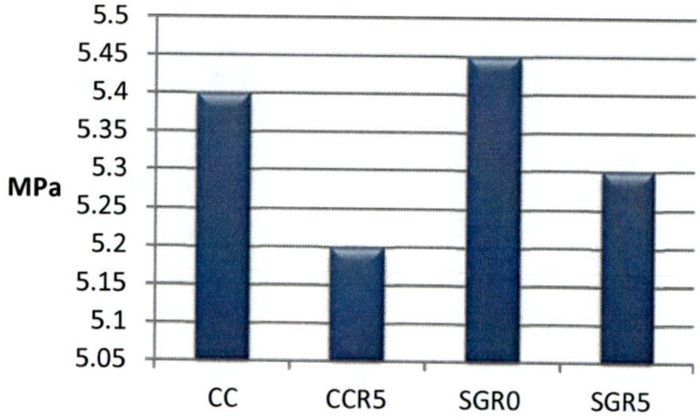

Fig. 5 Compressive strength (56 days) (At 150 °C for 3 h)

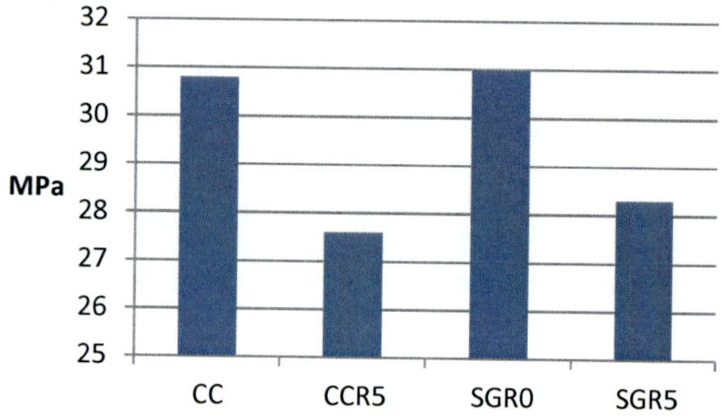

Fig. 6 Split tensile strength (56 days) (At 150 °C for 3 h)

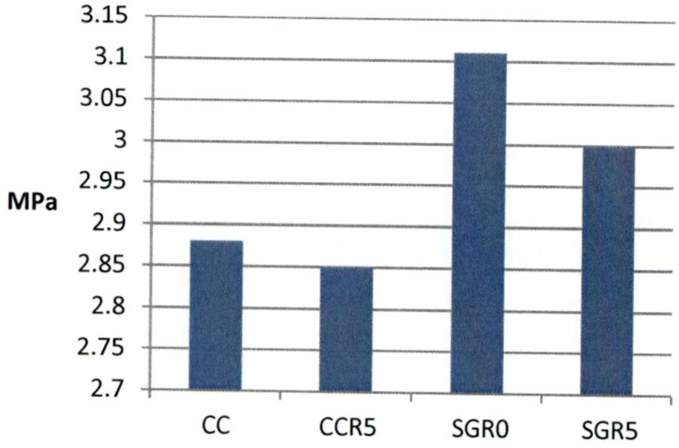

7 Discussions

Microsilica in cement improves its strength and durability as it gives progressively uniform dissemination and a more noteworthy volume of hydration items and diminishes the normal size of pores in the cement paste. With the addition of rubber, slump value decreases and so we have to add some superplasticizer in order to get the required slump value. The GGBS replaced concrete showed better performance after 56 days rather than 28 days. After 56 days, concrete with 5% replaced rubber aggregate shows only 5.9% reduction in compressive strength, 5.8% reduction in split tensile strength and 1.9% reduction in flexural strength. The heat study showed that rubber concrete shows almost equal heat resistance compared to normal concrete at 150 °C for 3 h.

8 Conclusions

- By adding microsilica and replacing cement with GGBS, we achieved the similarity in strength of conventional concrete and rubber concrete.
- When rubber is added the mixture becomes dry and so slump value decreases and hence we have to add superplasticizer in order to get the desired slump value.
- The GGBS replaced concrete with and without rubber shows better performance after 56 days rather than after 28 days. In fact, they reached the target strength after 56 days and are showing more strength than the conventional concrete after 56 days.
- There is very negligible decrease in compressive, split tensile and flexural strength in rubber concrete than conventional concrete.
- From the heat study, we can see negligible amount of decrease in compressive as well as split tensile strength. So, we can conclude at around 150 °C for 3 h the rubber concrete is getting least affected.

References

1. Emiroglu, M, Yildiz S, Kestemur O, Kelestemur MH (2012) Bond performance of rubber particles in the self-compacting concrete. Publisher Creations. ISBN: 978-88-907078-3-4
2. El Gammel A, Abdel-Gawad A-K, El Sherbini Y, Shalaby A (2010) Compressive strength of concrete utilising waste tire rubber. J Emerg Trends Eng Appl Sci (JETEAS) 1(1):96–99
3. Tayeh A (2013) Effect of replacement of sand by waste. Civ Environ Res 3(13). ISSN 2224-5790 (Paper) ISSN 2225-0514 (Online)
4. Helme JL (2010) Concrete, a use for use tires. CSSF Projects 2010
5. Naito C, Asce M, States J, Jackson C, Bewick B (2014) Assessment of crumb Rubber concrete for flexural structural member. AESC J Civ Eng 26(10)
6. Richardson A, Coventry K, Pienaan J (2011) The potential for achieving freeze/thaw protection in concrete through addition of rubber crum. Paper's no. BUE-FISC-2
7. Richardson A, Coventry K, Dave U, Piemar J (2011) Freeze/thaw performance of concrete using granulated rubber crumb. J Green Build 6
8. Naik T, Siddique R (2002) Properties of concrete containing scrap tire rubber—an overview. Report no. CBU-2002-06 REP-459. Centre for bi product utilization, Feb 2002
9. Vadivel S, Thennozhi R (2012) Experimental study on waste tyre rubber replaced concrete-ecofriendly construction materials. J Appl Sci Res 3966
10. IS 12269-1987 Indian Standard of OPC 53. BIS, New Delhi

11. IS383-1970, Indian Standard of aggregate, BIS, New Delhi
12. IS 10262-2009 Indian Standard of concrete mix design. BIS, New Delhi

Estimation of Design Shear Strength of Concrete Using Genetic Programming

Preeti Namjoshi and Shardul Joshi

Abstract Many codes provide empirical formulation for design shear strength of concrete which greatly vary from code to code. Moreover, many investigations into the shear problem that have been carried out have led to numerous empirical or semi-empirical formulae. These formulae usually agree quite well with the corresponding test result but not applicable for general use. The researchers have made use of experimental data set or analytical data set obtained from nonlinear finite element analysis. The equations are derived using nonlinear regression technique in which the form of the equation is required to be initially assumed. The present study investigates the application genetic programming (GP) in predicting the design shear strength of concrete. It is concluded that the values obtained by the equations derived from GP models estimate the design shear strength of concrete fairly close to the actual values.

Keywords Design shear strength · Genetic programming · Reinforced concrete

1 Introduction

Shear transfer mechanisms are very complex and are influenced by complex stress distribution which occur in the concrete beams after cracking. These mechanisms include shear resistance by uncracked compression zone of the beam, shear transfer due to aggregate interlock and dowel action of the longitudinal shear reinforcement. ACI-ASCE committee reports that shear strength of the beam is influenced by strength of the concrete, the percentage of longitudinal reinforcement and shear span-to-depth ratio of the beam. It is observed that shear strength of the beam increases with increase in compressive strength of concrete and it deceases with increase in shear span-to-depth (a/d) ratio as well as depth of the member. Beams are classified as deep beams (a/d between 0 and 1), short beams (a/d ratio between 1 and 2.5) and normal beams (a/d ratio greater than 2.5). Several methods which are proposed till today in

P. Namjoshi · S. Joshi (✉)
Department of Civil Engineering, Vishwakarma Institute of Information Technology, Pune 411048, India
e-mail: shardul.joshi@viit.ac.in

L. M. Gupta et al. (eds.), *Advances in Civil Engineering and Infrastructural Development*, Lecture Notes in Civil Engineering 87,
https://doi.org/10.1007/978-981-15-6463-5_66

the literature are limited to a particular type of beam only. Kani [1] reported 40% decrease in the shear strength of concrete when the depth of the beam increases from 150 to 1200 mm. This size effect of the shear strength of the beam is also studied by Bazant and Kim [2] and Bazant and Sun [3]. Appa Rao et al. [4] performed nonlinear regression analysis on the 612 data set points in which (i) compressive strength of concrete, percentage of longitudinal reinforcement, shear span-to-depth ratio and depth of the beam were considered as the influencing parameters. The authors have proposed a simple model to predict the design shear strength of the concrete beam. Kim and Park [5] have proposed a rational and mechanics-based equation to predict the design shear strength of the concrete beams even for high strength concrete up to 100 M Pa. In the era of artificial intelligence (AI), the researchers have applied data-driven techniques in the form of artificial neural networks (ANN) and adaptive neuro-fuzzy inference system (ANFIS) to predict the shear strength of the concrete. Goh [6] has developed ANN model to predict the shear strength of deep beams. Sanad and Saka [7] have also applied ANN tool for prediction of shear strength of concrete using 111 experimental data set. The other researchers Seleemah [8], Kim et al. [9] and Nehdi et al. [10] have developed ANN models to predict the shear strength of the concrete. Jung and Kim [11] developed two ANN models to estimate the shear strength of R. C. beam. Amani and Moeini [12] have used ANN and ANFIS models to predict the shear strength of concrete beams. Cladera and Mari [13] have taken review of EC-2 provisions for shear strength of concrete and concluded that EC-2 procedure is very easy to use by practising engineers but it presents a great scatter of results.

The codal provisions and the equations suggested by the other researchers are indicated in Table 1.

The foregoing paragraph discussed the importance and necessity of estimating design shear strength of concrete beam. The aim of the research work is to apply one more data-driven tool of genetic programming (GP) to estimate the design shear strength of concrete beam. Genetic programming technique yields the equation to estimate the design shear strength of concrete beam using the influencing parameters as defined by input parameters.

2 Genetic Programming

The concept of genetic programming (GP) is based on the evolution process which take place according to the principle of 'survival of the fittest'. The GP gives its output in the form of either a computer program or an equation [14].

Table 1 Emperical equations for design shear strength of concrete beam

Code/researcher	Empirical equation for shear capacity of concrete
ACI with web reinforcement	$V = \frac{A_{sw}}{s} Z f_{ywd}$
ACI with without web Reinforcement	$V_c = (0.167\sqrt{f_c}) b_w d$
EURO with web reinforcement	$V_R = 0.9 \rho_v f_{ywd} b_w d$
EURO without web reinforcement	$V = [0.0525 f_{ck} K(1.2 + 40\rho)] bd$
IS with web reinforcement	$V_{us} = \frac{0.87 f_y A_{sv} d}{S_v}$
IS without web reinforcement	$V_c = s\left(\frac{0.85}{6\beta}\right)(0.8 f_{ck})^{\left(\frac{1}{2}\right)}\left((1+5\beta)^{\frac{1}{2}} - 1\right)$
Appa Rao et al.	$V = \left(0.56 + \frac{4.0}{(a/d)^{1.5}}\right)\left[f_{ck}^{0.33} \rho^{0.5} d^{-0.44}\right]$
British code	$V_c = \frac{0.79}{\gamma_m}\left(\frac{100 A_s}{b_v d}\right)^{0.33}\left(\frac{400}{d}\right)^{0.44}\left(\frac{f_{cu}}{25}\right)^{0.33}$
Niwa	$Vc = 0.2\left(\rho_w f_{ck} \frac{d}{a}\right)^{0.33}\left(d^{-0.44}\right)\left(0.75 + 1.4\frac{d}{a}\right) b_w d$
SNI	$V_c = \frac{1}{7}\left(\sqrt{f_{ck}} + 120 \rho_w \frac{d}{a}\right) b_w d$
Zsutty	$V_c = 2.17\left(\rho_w f_{ck}\left(\frac{d}{a}\right)\right)^{0.33} b_w d$

2.1 The Primitives of Genetic Programming

Evolution of GP solution is assembled from two sets of primitives nodes, terminals and functions. The terminal nodes cater for the input to the GP system while the function nodes cater for the actual processing of data values. During the evolution process, GP randomly selects the nodes from either of the sets. While taking the trials, only a relatively simple node set is initially provided and then nodes are usually added only if required [15].

2.2 Tree-Based Genetic Programming

GP creates a structure by assembling the function and terminal nodes. The three main types of structure available are tree, linear and graph.

2.3 Algorithm of Genetic Programming

The genetic programming develops either computer programs or equations to solve problems by executing the following three steps:

1. Generating initial set of computer programs or equations considering random compositions of the functions and terminals of the problem

2. Performing iterations till the defined termination criteria is reached:
   (a) Each program or equation is evaluated and fitness value is assigned
   (b) New set of programs and equations are created following two primary operations.
      (i) Copy existing computer programs or equations to the new population.
      (ii) Create new computer programs or equations by applying genetic operations on two existing programs.
3. The equation or computer program which gives better fitness value is considered as the result of GP. This result may be a solution (or an approximate solution) to the problem [16].

Advantages of Genetic Programming A key advantage of GP over the other traditional modelling approaches is that it does not start with assuming an equation in any specified format. In a typical regression method, the model structure is specified in initially and the coefficients are determined. For neural networks, the time-consuming task of initially defining the network structure has to be undertaken and then the weights are found by the learning algorithm. On the other hand, in GP, input and output parameters and set of functions are defined initially and the learning method subsequently finds both the optimal structure of the model and its coefficients. It is observed that handling of constants in the equation is trouble-some in case of GP. The selection of appropriate input parameters is extremely important in case of GP as well because any inappropriate input parameter may hamper the accuracy of the model.

3 Methodology

The present work has made use of the experimental data set summarized by Cladera and Mari [13]. Original data set was composed of 202 beam specimens. The database was developed by 23 different researchers. The database was reduced to 122 beam specimen by removing the beam samples failed in flexure. All these beams were simply supported and tested under 3 or 4 point loading. The shear span-to-depth ratio (a/d) for all the beams was kept more than 2.49. The database used for the present study has been shown in Annexure 1. Out of the data set of 122 beam specimens, data set of 85 beam specimens was used to train the GP model, 18 beam specimens were used for validation and 19 beam specimens were used for testing the GP model. The input parameters used for developing the GP models are shown in Table 2.

Table 2 Input parameters

Input no.	Parameter	Notation
1	Compressive strength of concrete	fc
2	Percentage longitudinal reinforcement	Ptl
3	Percentage transverse reinforcement	Ptv
4	Yield strength of steel	fy
5	Shear span-to-depth ratio	a/d

The output parameter was design shear strength of the concrete (τ). Table 3 shows the data set used for validation and testing of the model.

Table 3 Data set for validation and testing

S. No.	fc (N/mm^2)	Ptl	Ptv	fy (N/mm^2)	a/d	τ (N/mm^2)
1	125.00	2.89	0.23	464.00	3.00	3.365553
2	73.00	2.80	0.16	569.00	2.50	3
3	25.00	1.76	0.10	350.00	3.95	1.56128
4	76.00	3.50	0.18	543.00	2.50	5.269433
5	36.00	2.49	0.07	525.00	3.10	1.350406
6	64.00	2.80	0.16	569.00	2.50	3.068493
7	64.00	2.80	0.16	569.00	2.50	3.123288
8	67.00	2.79	0.10	632.00	2.49	3.12628
9	49.00	1.95	0.28	430.00	2.88	2.85289
10	73.00	2.80	0.13	569.00	2.50	3.191781
11	24.00	1.80	0.10	330.00	3.92	1.628664
12	87.00	2.80	0.16	569.00	3.01	3.315068
13	89.00	4.46	0.16	569.00	2.50	3.319728
14	50.00	2.99	0.24	540.00	3.08	3.504274
15	74.00	3.50	0.13	543.00	2.50	5.889366
16	89.00	4.46	0.16	569.00	3.30	3.360544
17	75.00	2.80	0.13	569.00	2.50	3.438356
18	64.00	2.80	0.16	569.00	2.50	3.465753
19	73.00	2.80	0.16	569.00	2.50	3.465753
20	50.00	2.28	0.11	530.00	3.06	2.521246
21	67.00	2.79	0.10	632.00	2.49	2.397306
22	61.00	2.23	0.14	530.00	3.06	2.549575
23	42.00	2.99	0.12	530.00	3.30	3.10231

(continued)

Table 3 (continued)

S. No.	fc (N/mm^2)	Ptl	Ptv	fy (N/mm^2)	a/d	τ (N/mm^2)
24	38.00	2.90	0.21	500.00	3.25	3.110013
25	45.00	2.93	0.16	500.00	3.28	3.179834
26	45.00	2.86	0.16	500.00	3.21	3.205128
27	51.00	1.95	0.21	430.00	2.88	2.493178
28	41.00	2.93	0.21	500.00	3.28	0.332485
29	27.00	2.34	0.15	340.00	3.97	1.900998
30	67.00	2.80	0.16	569.00	2.53	2.780822
31	69.00	2.28	0.14	530.00	3.06	2.889518
32	89.00	4.46	0.13	569.00	3.30	2.789116
33	64.00	2.80	0.16	569.00	2.50	2.821918
34	80.00	2.81	0.18	543.00	2.50	4.935622
35	64.00	2.80	0.16	569.00	2.50	2.849315
36	26.00	1.17	0.10	350.00	3.97	1.482537
37	26.00	1.71	0.10	350.00	4.01	1.54249

4 Results and Discussions

As stated in Sect. 3, the genetic programming models were developed changing the genetic parameters and the three equations are shown in Table 4, obtained from the study. The accuracy of these models is assessed from the coefficient of determination (COD or r^2) and root mean squared error (RMS).

Equations 1 and 3 indicate the influence of compressive strength of concrete and percentage longitudinal reinforcement over the design shear strength of concrete, whereas Eq. 2 shows the influence of percentage transverse reinforcement and yield strength of concrete in addition to the compressive strength of concrete and percentage longitudinal reinforcement. The effect of a/d ratio is not seen in any of the equations yielded by GP technique. Equation 2 has maximum value of COD and least value of RMS, it may estimate the design shear strength of concrete close to the experimental values. Table 5 shows the comparison of the design shear strength values obtained by these three equations over the experimental values.

Table 4 Hypothesis of GP models

S. No.	COD	RMS	Hypothesis	GP equations
1	0.702	0.675	τ = (sqrt(ptl) * sqrt(sqrt((0.145604536 * fc))))	Equation 1
2	0.745	0.62611	τ = (0.1386998 * sqrt((sqrt((sqrt((pow(fc, 2) * ptv)) * fy)) * ptl)))	Equation 2
3	0.701	0.675011	τ = (sqrt(sqrt((ptl * (ptl * fc))))/1.61222219)	Equation 3

Table 5 Comparison of design shear strength values obtained from hypothesis

S. No.	fc (N/mm^2)	Ptl	Ptv	fy	a/d	τ (N/mm^2)			
						Expt	Equation 1	Equation 2	Equation 3
1	125	2.89	0.23	464	3	3.37	3.51	3.04	3.52
2	73	2.8	0.16	569	2.5	3.00	3.02	2.63	3.03
3	25	1.76	0.1	350	3.95	1.56	1.83	1.33	1.84
4	76	3.5	0.18	543	2.5	5.27	3.41	2.98	3.43
5	36	2.49	0.07	525	3.1	1.35	2.39	1.84	2.40
6	64	2.8	0.16	569	2.5	3.07	2.92	2.54	2.94
7	64	2.8	0.16	569	2.5	3.12	2.92	2.54	2.94
8	67	2.79	0.1	632	2.49	3.13	2.95	2.50	2.96
9	49	1.95	0.28	430	2.88	2.85	2.28	1.99	2.29
10	73	2.8	0.13	569	2.5	3.19	3.02	2.56	3.03
11	24	1.8	0.1	330	3.92	1.63	1.83	1.32	1.84
12	87	2.8	0.16	569	3.01	3.32	3.16	2.75	3.17
13	89	4.46	0.16	569	2.5	3.32	4.01	3.49	4.02
14	50	2.99	0.24	540	3.08	3.50	2.84	2.57	2.85
15	74	3.5	0.13	543	2.5	5.89	3.39	2.85	3.40
16	89	4.46	0.16	569	3.3	3.36	4.01	3.49	4.02
17	75	2.8	0.13	569	2.5	3.44	3.04	2.57	3.05
18	64	2.8	0.16	569	2.5	3.47	2.92	2.54	2.94
19	73	2.8	0.16	569	2.5	3.47	3.02	2.63	3.03
20	50	2.28	0.11	530	3.06	2.52	2.48	2.03	2.49
21	67	2.79	0.1	632	2.49	2.40	2.95	2.50	2.96
22	61	2.23	0.14	530	3.06	2.55	2.58	2.17	2.59
23	42	2.99	0.12	530	3.3	3.10	2.72	2.24	2.73
24	38	2.9	0.21	500	3.25	3.11	2.61	2.28	2.62
25	45	2.93	0.16	500	3.28	3.18	2.74	2.31	2.75
26	45	2.86	0.16	500	3.21	3.21	2.71	2.28	2.72
27	51	1.95	0.21	430	2.88	2.49	2.31	1.94	2.31
28	41	2.93	0.21	500	3.28	0.33	2.68	2.34	2.69
29	27	2.34	0.15	340	3.97	1.90	2.15	1.64	2.16
30	67	2.8	0.16	569	2.53	2.78	2.96	2.57	2.97
31	69	2.28	0.14	530	3.06	2.89	2.69	2.27	2.70
32	89	4.46	0.13	569	3.3	2.79	4.01	3.39	4.02
33	64	2.8	0.16	569	2.5	2.82	2.92	2.54	2.94
34	80	2.81	0.18	543	2.5	4.94	3.10	2.71	3.11

(continued)

Table 5 (continued)

S. No.	fc (N/mm^2)	Ptl	Ptv	fy	a/d	τ (N/mm^2)			
						Expt	Equation 1	Equation 2	Equation 3
35	64	2.8	0.16	569	2.5	2.85	2.92	2.54	2.94
36	26	1.17	0.1	350	3.97	1.48	1.51	1.10	1.52
37	26	1.71	0.1	350	4.01	1.54	1.82	1.33	1.83

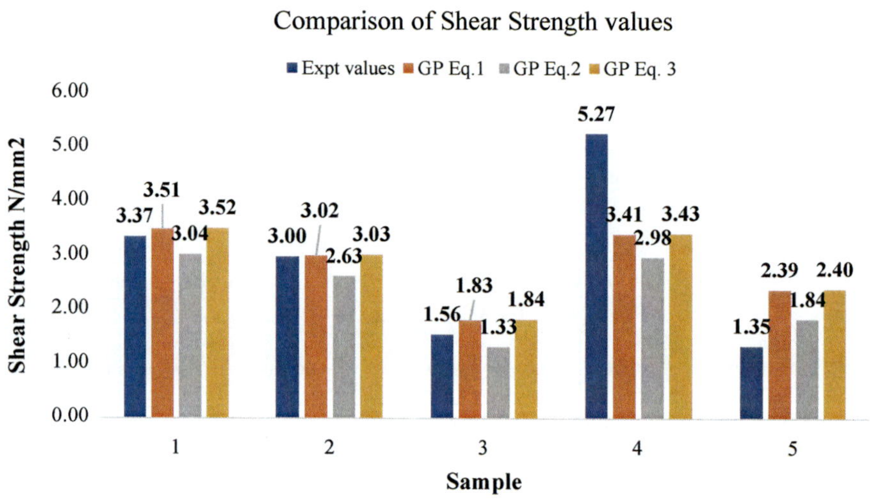

Fig. 1 Comparison of design shear strength values

Figure 1 shows graphical comparison for few of the beam specimen.

5 Conclusions

The conclusions obtained from the study are summarized below.

1. The genetic programming has correlated design shear strength of the concrete to the characteristic strength of concrete, yield strength of steel as well as the percentage reinforcement in longitudinal and vertical direction.
2. The equations obtained from genetic programming tool agrees with the theoretical knowledge. The GP2 equation fairly matches with the experimental values.
3. The genetic programming has given simple expressions for determining the shear strength of the concrete.

Annexure 1

b: mm	d: mm	fc: MPa	Ptl: %	Ptv: %	fy: MPa	a/d	Vfail: KN
307	466	24	1.8	0.1	330	3.92	233
305	460	25	1.69	0.1	350	3.98	168
229	457	24	2.28	0.15	340	4.01	173
305	457	26	1.71	0.1	350	4.01	215
305	462	25	1.76	0.1	350	3.95	220
231	460	27	2.34	0.15	340	3.97	202
305	460	26	1.17	0.1	350	3.97	208
240	1200	25	1.26	0.15	440	3	468
152	272	34	4.16	0.21	276	3.6	117
152	272	31	4.16	0.21	276	3.6	112
76	95	29	1.97	0.16	275	3	16
76	132	28	3.95	0.12	258	3	25
76	132	26	3.95	0.34	179	3	28
76	132	28	3.95	0.12	258	4	20
152	298	22	3.36	0.12	292	3.6	76
152	298	28	3.36	0.26	269	3.6	95
152	298	47	3.36	0.26	269	3.6	121
152	298	69	3.36	0.26	269	3.6	151
152	298	82	3.36	0.26	269	3.6	116
152	298	47	3.36	0.38	271	3.6	133
152	298	83	3.36	0.38	271	3.6	150
305	539	36	2.49	0.14	525	3.1	338
305	539	36	2.49	0.07	525	3.1	222
305	539	56	2.49	0.14	525	3.1	383
406	385	29	2.31	0.39	549	2.65	460
457	871	72	1.88	0.16	445	3	788
457	762	125	2.35	0.16	483	3	749
457	762	125	2.89	0.23	464	3	1172
180	233	40	2.23	0.09	844	4	115
180	233	75	2.23	0.09	844	4	123
180	233	76	2.81	0.09	844	4	138
180	233	70	3.5	0.09	844	4	147
180	233	80	2.81	0.18	543	2.5	207
180	233	74	3.5	0.13	543	2.5	247
180	233	76	3.5	0.18	543	2.5	221

(continued)

(continued)

b: mm	d: mm	fc: MPa	Ptl: %	Ptv: %	fy: MPa	a/d	Vfail: KN
127	203	41	3.2	0.49	322	3	87
127	198	98	4.54	0.51	324	3	102
127	198	90	4.54	0.65	323	3	108
127	198	103	4.54	0.78	324	3	123
203	419	57	3.03	0.34	426	3.27	267
203	419	56	3.03	0.34	426	3.27	267
375	655	36	2.8	0.08	430	3.28	457
375	655	36	2.8	0.08	430	3.28	363
375	655	36	2.8	0.11	430	3.28	483
375	655	67	2.8	0.11	430	3.28	552
375	655	67	2.8	0.16	430	3.28	689
375	655	87	2.8	0.14	430	3.28	598
375	655	87	2.8	0.23	430	3.28	721
250	292	64	2.8	0.157	569	2.5	228
250	292	64	2.8	0.157	569	2.5	208
250	292	64	2.8	0.157	569	2.5	206
250	292	64	2.8	0.157	569	2.5	278
250	292	64	2.8	0.157	569	2.5	253
250	292	64	2.8	0.157	569	2.5	224
250	292	73	2.8	0.105	569	2.5	260
250	292	73	2.8	0.126	569	2.5	233
250	292	73	2.8	0.157	569	2.5	253
250	292	73	2.8	0.157	569	2.5	219
250	292	73	1.65	0.209	569	2.5	282
250	297	67	2.79	0.101	632	2.49	178
250	293	67	2.79	0.101	632	2.49	229
250	292	67	2.8	0.101	632	2.49	175
250	198	67	2.78	0.157	569	2.5	258
250	292	67	2.8	0.157	569	2.53	203
250	292	87	2.8	0.157	569	3.01	242
250	292	87	2.8	0.157	569	2.74	260
250	294	89	4.46	0.157	569	2.5	244
250	294	89	4.46	0.126	569	3.3	205
250	294	89	4.46	0.157	569	3.3	247
250	294	89	4.46	0.196	569	3.3	274
250	294	75	4.46	0.224	569	3.3	304
250	292	75	2.8	0.262	569	3.3	311

(continued)

(continued)

b: mm	d: mm	fc: MPa	Ptl: %	Ptv: %	fy: MPa	a/d	Vfail: KN
250	292	75	2.8	0.105	569	2.5	272
250	292	75	2.8	0.126	569	2.5	251
250	292	75	2.8	0.157	569	2.5	266
250	292	75	2.8	0.196	569	2.5	289
250	292	75	2.8	0.227	569	2.5	284
200	260	26	1.47	0.12	267	2.77	89
200	260	26	1.47	0.16	269	2.77	89
200	260	26	1.47	0.25	256	2.77	93
200	260	26	1.96	0.13	262	3.46	85
295	920	75	1.36	0.16	500	2.5	583
169	459	74	1.03	0.13	500	2.72	139
169	459	74	1.16	0.13	500	2.72	152
295	920	71	1.03	1.16	500	2.5	516
300	925	21	1.01	1.791	508	2.92	282
300	925	38	1.01	1.791	508	2.92	277
300	925	47	0.76	1.791	508	2.92	342
290	278	49	1.95	0.11	430	2.88	158
290	278	49	1.95	0.18	536	2.88	169
290	278	49	1.95	0.28	430	2.88	230
290	278	51	1.95	0.214	430	2.88	201
110	443	55	2.58	0.48	499	2.82	155
110	398	74	5.8	0.48	538	3.14	265
110	463	43	1.23	0.333	555	2.7	105
150	310	75	2.59	0.23	255	3	156
150	310	73	3.08	0.2	255	3	143
150	310	82	4.43	0.13	425	5	97
150	310	75	4.43	0.18	425	5	119
150	310	82	4.43	0.27	425	5	125
200	353	50	2.28	0.109	530	3.06	178
200	351	50	2.99	0.239	540	3.08	246
200	353	61	2.228	0.141	530	3.06	180
200	353	69	2.28	0.141	530	3.06	204
200	351	69	2.99	0.239	530	3.08	255
200	351	50	2.99	0.239	540	3.08	267
127	216	45	2.07	0.378	421	3	63
127	198	88	4.54	0.65	421	3	107
127	198	87	4.54	0.78	421	3	121

(continued)

(continued)

b: mm	d: mm	fc: MPa	Ptl: %	Ptv: %	fy: MPa	a/d	Vfail: KN
127	198	83	4.54	0.51	421	4	95
200	303	42	2.99	0.166	530	3.3	177
200	303	42	2.99	0.118	530	3.3	188
199	307	38	2.9	0.21	500	3.25	190
199	306	39	2.92	0.16	500	3.27	151
195	306	39	2.99	0.12	500	3.27	128
200	312	45	2.86	0.16	500	3.21	200
200	302	44	2.95	0.12	500	3.3	150
200	306	42	2.91	0.16	500	3.27	177
201	306	39	2.9	0.12	500	3.27	164
199	3053	41	2.93	0.21	500	3.28	202
199	305	45	2.93	0.16	500	3.28	193
199	307	43	2.91	0.12	500	3.25	147

References

1. Kani GNJ (1967) How safe are our large reinforced concrete beams? ACI J 64(3):128–141
2. Bazant ZP, Kim JK (1984) Size effect in shear failure of longitudinally reinforced beams. Proc ACI Struct J 81(5):456–468
3. Bazant ZP, Sun HH (1987) Size effect in diagonal shear failure: influence of aggregate size and stirrups. ACI Mater J 84(4):259–272
4. Appa Rao G, Injaganeri SS (2011) Evaluation of size dependant design shear strength of reinforced concrete beams without web reinforcement. Sadhana 36(Part 3):393–410 (Indian Academy of Sciences)
5. Kim JK, Park YD (1996) Prediction of shear strength of reinforced concrete beams without web reinforcement. ACI Mater J Title no. 93-M24, 213–222
6. Goh ATC (1995) Prediction of ultimate shear strength of deep beams using neural networks. ACI Struct J 92:28–32
7. Sanad A, Saka MP (2001) Prediction of ultimate shear strength of reinforced-concrete deep beams using neural networks. J Struct Eng 127:818–828
8. Seleemah AA (2005) A neural network model for predicting maximum shear capacity of concrete beams without transverse reinforcement. Can J Civ Eng 32:644–657
9. Kim KS, Jung S, Han SE (2005) Prediction of shear strength using artificial neural networks for reinforced concrete members without shear reinforcement. J Comput Struct Eng Inst Korea 18:201–211
10. Nehdi M, El-Chabib H, Said A (2006) Evaluation of shear capacity of FRP reinforced concrete beams using artificial neural networks. Smart Struct Syst 2006(2):81–100
11. Jung S, Kim KS (2007) Knowledge based prediction of shear strength of concrete beams without shear reinforcement. Eng Struct 30:1515–1525
12. Amani J, Moeini R (2012) Prediction of shear strength of reinforced concrete beams using adaptive, neuro fuzzy inference system and artificial neural networks. Sci Iran A 19(2):242–248
13. Cladera A, Marı AR (2007) Shear strength in the new Eurocode 2. A step forward? Struct Concr 8(2):58–66

14. Gaur S, Deo MC (2008) Real-time wave forecasting using genetic programming. Ocean Eng 35(11–12):1166–1172
15. Shaw D, Miles J, Gray A (2004) Genetic programming within civil engineering. In: Organization of the Adaptive Computing in Design and Manufacture 2004 Conference, Engineers House, Clifton, Bristol, UK, April 20–22
16. Koza JR (1992) Genetic programming on the programming of computers by means of natural selection. A Bradford Book, MIT Press

Advance Technique of Precast Concrete Production

Uddesh U. Gaude and K. G. Guptha

Abstract The rate of infrastructure development growth and the necessity of producing better quality construction material demand shifting from conventional method of the production to certain advanced techniques. Water–cement ratio plays an important role in the quality of concrete product. High water–cement ratio produces lower strength concrete and vice versa. Normally, water–cement ratio required for hydration of cement is as low as 0.23 as compared to what is being practised during actual mixing considering the workability factor. Concept of extracting water after mixing and filling the mould can yield a higher strength. One such technique of precast concrete production is the “wet pressing”, in which wet concrete is pressed within the moulds using hydraulic pressure to expel out the excess water present in the concrete. Pressing time generally varies from 10 to 30 s based on the quality of materials used and quantity of water added. Pressed products are then immediately lifted and stacked in the shade for overnight curing. These products are then cured only for 3 days by sprinkling water. The study focuses on wet pressing technique of precast concrete production and comparison of products produced by this technique and old conventional methods. Wet pressed products provide better compressive strength, hardness, surface finish, durability and also reduce permeability and curing time. Method also proves to be faster and economical.

Keywords Wet press · Precast · Concrete

U. U. Gaude (✉) · K. G. Guptha
Department of Civil Engineering, Goa College of Engineering, Farmagudi, Goa, India
e-mail: udgaude77@gmail.com

K. G. Guptha
e-mail: kgg@gec.ac.in

L. M. Gupta et al. (eds.), *Advances in Civil Engineering and Infrastructural Development*, Lecture Notes in Civil Engineering 87,
https://doi.org/10.1007/978-981-15-6463-5_67

1 Introduction

The conventional method of precast concrete production is very slow, and the products imparted are also of lower quality. These methods lack in fulfilling the current production demand and also in terms of quality. Process of conventional production starts with mixing the material in required proportion, filling the moulds of required shape and then de-moulding and stacking for curing. This overall process is time-consuming and also requires larger space for stacking moulds.

Now from quality perspective as we know water–cement ratio play an important part in improving the quality of the concrete. Higher water–cement ratio will yield lower strength and vice versa. Basically, water required by concrete is for hydration of cement only. But if we add only that much quantity of water than the concrete won't be workable at all, and it will be difficult for filling moulds without honeycombing. In order to make the concrete workable, we have to add the excess water to the concrete. This excess water does not participate in hydration and evaporates forming voids in the concrete. Formation of such voids reduces the strength and durability of the concrete. So high water–cement ratio proves to be detrimental to the concrete. This issue can be addressed if we are able to extract the extra water immediately after filling up the moulds. Extraction of water after filling will handle the workability issue as well as impart the better strength. As the water evaporating from the concrete will be very less almost negligible, resulting less pores formation and highly dense concrete. One such method of extracting extra water after filling up the moulds is "wet press technique". This process of manufacturing concrete unit with better strength and durability has become popular in overcoming such deficiencies. Unlike conventional concrete water quantity is not critical in these types because extra water is drained off during pressing. While a required quantity of water for cement hydration and to ensure mix flows into the moulds easily without vibrations is added, immediately after pressing the product is strong enough to be handled and is removed from the mould and stacked.

This property of removing from mould immediately after pressing makes the mould available for next unit, which increases the production rate. The pressing time generally ranges from 10 to 30 s. This means that moulds can be reused within 50 s, whereas in conventional type it requires time to harden and de-moulded only on the next day of casting. To achieve highest productivity within confined manufacturing space, wet press manufacturers can employ turntable style station machines (carousel) for accomplishing the mould-filling, pressing/drainage and de-moulding phases. Lifted products can be stacked on edge close to each other (not touching). Property of stacking on the vertical edge also makes it possible to cure in minimum space. These concrete units are stacked in dry area for overnight curing and then transferred to stacking area for normal completion of curing [1]. The current study focuses on wet press technique and comparison of products produced by wet press technique and our conventional methods.

Table 1 Mix proportion

Cement (kg)	20 mm (kg)	10 mm (kg)	6 mm (kg)	Fines (kg)	*W*/*C*
90	100	130	100	300	0.53–0.56

2 Experimental Study

2.1 *Materials*

Concrete materials, i.e. ordinary Portland cement, 43 grade confirming to IS 8112:2013, coarse aggregate of 6, 10, and 20 mm size, fines and water, were blended together for the formation of concrete [2].

2.2 *Mix Proportion*

For comparative study of both the methods, same mix proportions, i.e. cement, fines, 6 mm aggregate, 10 mm aggregate, in the proportion 1:3.33:1.11:1.45:1.11, respectively, were used. Water–cement ratio for both the methods was also kept the same during mixing. The quantities of materials used in the trials are given in Table 1.

2.3 *Influence of Water–Cement Ratio*

Water–cement ratio is the ratio of mixing water to that of cement in the mixture. It plays an important role in governing the strength and durability of the concrete products. Lower the water–cement ratio will give higher strength and better durability. Water required in the concrete is for mixing the materials and hydration of cement. IS 456: 2000 specifies certain maximum water–cement ratio ranging from 0.4 to 0.6 for different grades of concrete and different exposure conditions [3]. Basically, cement requires a water of about 38% for complete hydration of cement over a life period. Since mixing is done for a short period of time, only 23% of water participates in the process of the hydration of cement.

If only 23% of water is added, it won't be workable at all and will be very difficult for placement. Considering workability and easy placement, we have to increase the water content. This excess water over 23% will retain in the concrete even after hardening and will get evaporate upon atmospheric exposure leaving the voids in the hardened concrete [4, 5]. Higher water–cement ratio affects the following concrete properties [6].

1. Compressive strength reduction.
2. Drying shrinkage or cracking due to the reduction in tensile strength.

3. Flexural strength.
4. Increase in permeability.
5. Lower abrasion resistance.
6. Reduced durability.

2.4 Wet Pressing Technique

As discussed above, higher water–cement ratio reduces the strength and durability of the concrete products. But we have to add the excess water considering workability aspect. This excess water has nothing to do with hydration of cement and has no role to play after placing the concrete. The issue of higher water–cement ratio can be solved if we are able to drain out the excess water present in the concrete immediately after placing in the moulds. Wet press exactly does the same thing. It uses the hydraulic pressure on to the mould to expel out the excess water present in the concrete immediately after placement. Extraction of excess water reduces the formation of void and increases the strength and durability of the concrete. Complete process of wet pressing the concrete within the moulds is explained in further section.

3 Machinery

Wet pressing technique requires special machineries for production as discussed below:

1. Batching system for accurately weighing the materials.
2. Conveyer belt and trolley arrangement for transferring weighed material into the mixer.
3. Cement and water dispenser for adding cement and water, respectively.
4. Electrically operated mixer for concrete mixing fitted with discharge nozzle for filling the moulds directly.
5. The special machinery used for pressing unit.

3.1 Hydraulic Press Machinery

Hydraulic wet press machinery comes with a complete assembly of volumetric material dispenser, single-stage or three-stage mould, pressing system, ejector station and handling system (see Fig. 1) [7].

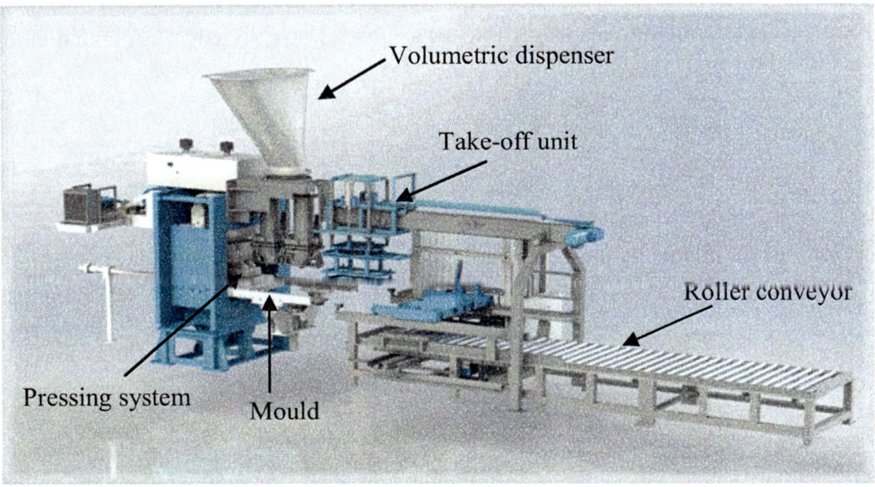

Fig. 1 Single-mould system

3.2 *Volumetric Measuring Dispenser*

These dispensers are attached to the bottom of the mixer for filling mould with wet concrete. Volume to be dispensed into the mould is pre-fixed based on type and size of the product.

3.3 *Moulds*

These machines are available in two varieties, viz. one with a single-mould system and the other with three-mould system. The single-mould press gives more cost-effective tooling or set up option to that of triple-mould system. It is still a 400-ton-power machine and is capable of producing any product that the triple mould can output. Single moulds are used for lower production demand. Triple mould pressing process has the rotary table system for filling and pressing operations. These are used for higher production requirements.

3.4 *Pressing System*

Filled moulds are pressed with hydraulic pressure of 400 ton to expel out the water from the concrete. Pressing ram automatically starts pressing on wet concrete for the time duration set by the operator. Pressing time varies based on size and type of the product and generally ranges from 10 to 30 s. Quick water drainage from the mould is achieved through vacuum units.

3.5 Ejector Arm

Ejector arm automatically pushes the concrete out of the mould. Product is then lifted creating vacuum on the top surface and moved to the waiting pallet for stacking.

3.6 Handling System

This system consist of a take-off unit which may be manually operated, semi-automatic or fully automatic unit. When the ejector arm presses the product out of the mould, at this point the take-off-unit takes over, lowering the vacuum head plate down onto the ejected concrete product creating a vacuum and picks up the product. The carrier then travels over to the waiting pallet where the products are neatly stacked along with previously made products and will continue until the pallet is full.

4 Methodology

The manufacturing process is as shown in the flow diagram (see Fig. 2).

4.1 Process

As discussed above, two types of systems are available single-stage mould and three-stage mould system. Process is basically same in both the cases except three-stage system uses rotary table system for filling the mould and pressing. Main steps are filling moulds, pressing the concrete to squeeze out extra water, ejecting, stacking and curing the finished product.

Usually, a concrete mix design for the product is required using locally available material as far as possible for economical production. Cement, fines, 6 mm aggregate, 10 mm aggregate, 20 mm aggregate are accurately weighed in the proportion 1:3.33:1.11:1.45:1.11, respectively. All these materials except cement are precisely weighed and transferred to a mixer using electrically operated trolley.

Cement, chemical admixture if any and water required for cement hydration and to form slurry that flows into the mould without any vibration is directly added into the mixer. Mixer is run till a good slurry is formed which is inspected with inspection

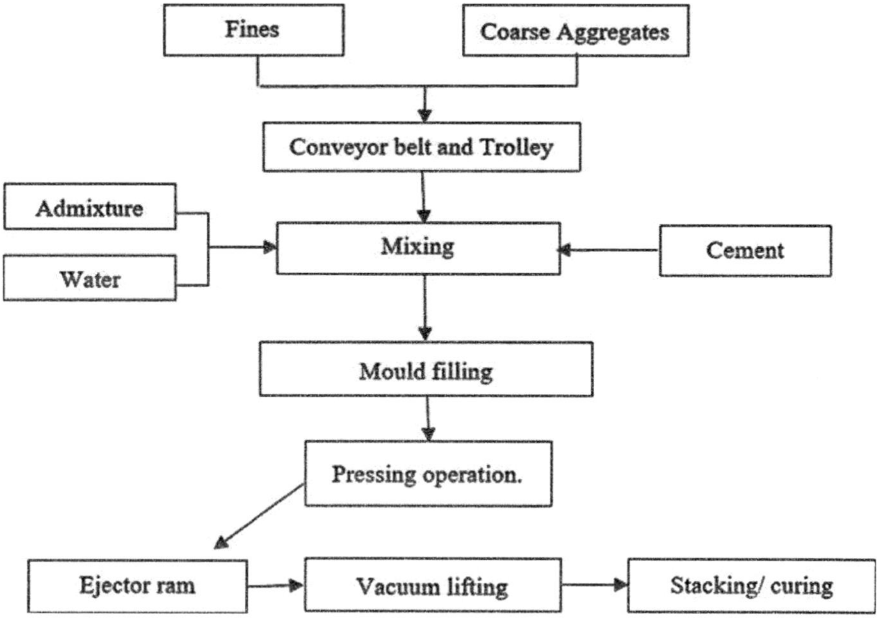

Fig. 2 Manufacturing process flow diagram

window provided on the mixer. The powder or fines is used to obtain close texture and smooth finish on the surface. To prevent the escape of fines as the free water is drawn off by the pressure, a sheet of porous paper is placed at the bottom of the mould.

A measured amount of wet mix is then dispensed through hopper into the mould. The mix easily runs and fills the mould without vibration, and another filter paper is placed on the top. The press is then actuated, and the hydraulic ram presses down into the mould squeezing out the excess water immediately after pressing. The product is strong enough to be handled at the end of this stage. The ejector ram automatically pushes the pressed product out of the mould. At this point the take-off machine takes over, lowering the vacuum head plate down onto the ejected concrete product creating a vacuum and picks up the product. The carrier then travels over to the waiting pallet where the products are neatly stacked along with previously made products and will continue until the pallet is full. The pallet when full travels along a roller conveyer to a point where it can be lifted off by a forklift truck and stacked within a dry atmosphere of the building for overnight curing. The products are then stacked out into the stacking area for normal completion of curing [1].

5 Tests

Different tests were conducted on different specimens from conventionally casted products and wet pressed products, and results were compared as discussed below.

5.1 Hardness Test

Test specimens from both methods were tested for finding surface hardness of the specimen. Samples were scratched with nails to observe the hardness of the product. Wet pressed shows a better resistance to scratching as compared to conventionally casted products.

5.2 Compressive Strength Test

Compressive strength test as per IS 516:1959 was conducted on both the specimens. Concrete cubes of 15 × 15 × 15 cm with mix proportion as discussed earlier were casted. These specimens were then cured and tested for 28 days compressive strength in a compression testing machine [8]. Test results for both the methods are tabulated.

Conventionally casted products Compressive strength results for conventionally casted concrete cubes are given in Table 2.

Wet pressed products Core cut specimen was taken from wet pressed products. These specimens were then tested as per IS: 516-1959 in a compression testing machine to find equivalent 28 days cube compressive strength. Test results are given in Table 3.

Table 2 Compressive strength

S. No.	Weight (kg)	Load (kN)	Compressive strength (N/mm^2)	Density (kg/m^3)
1	8.195	480	21.33	2428
2	8.025	512	22.75	2378
3	8.144	451	20.04	2413
4	8.102	424	18.84	2401
5	8.108	458	20.35	2402
6	8.051	475	21.11	2385

Table 3 Compressive strength

S. No.	*L* (mm)	*D* (mm)	Load (kN)	Compressive strength (N/mm^2)	Corrected strength (N/mm^2)	Equivalent cube strength (N/mm^2)	Density (kg/m^3)
1	152	123	319.9	26.93	24.28	30.73	2411
2	152	123	336.7	28.35	25.88	32.35	2405
3	152	123	325.8	27.40	25.02	31.28	2421
4	152	123	321.7	27.08	24.72	30.90	2386
5	152	123	324.9	27.35	24.97	31.21	2398
6	152	123	333.8	28.10	25.65	32.06	2428

6 Results and Discussion

When products from both the methods were tested, wet press technique proves to be better in several ways. These products impart better compressive strength and improve overall durability of the concrete as compared to our conventional method of casting. Better compressive strength also increases the abrasion resistance. Graphical representation of compressive strength test results for both the methods is shown (see Fig. 3).

As discussed in wet press technique, water is expelled out after mixing by applying hydraulic pressure. It compresses the product to the maximum possible level making highly dense and impermeable concrete products. Pressing the wet concrete within the moulds at initial stage avoids the formation of voids in the hardened concrete.

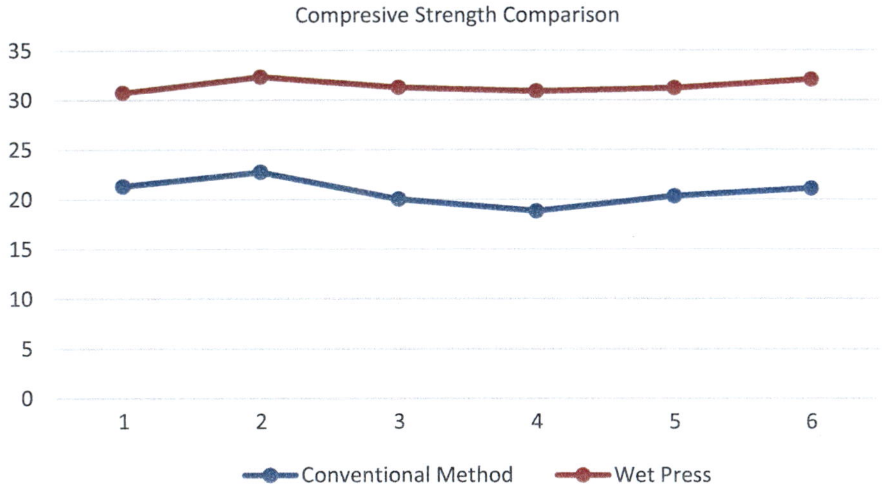

Fig. 3 Compressive strength comparison

This also improves the tensile strength and reduces the formation of surface cracks. Lower water content avoids the dusting on the surface due to the segregation of aggregates.

These products require lesser curing period as it starts gaining strength from the instant of pressing itself. These products are normally cured in shed for 1 day to avoid any cracks and then transferred to stacking area and cured by sprinkling water for the next 3 days, whereas in conventional method products are cured for at least 14 days of casting. Due to the extraction of water in the green state of concrete, it has many advantages over conventional method.

Advantages Wet press technique imparts the following advantages over conventional methods:

1. Better surface finish, well-defined edges.
2. Higher compressive strength.
3. Highly dense and impermeable products.
4. Rate of strength gain is higher.
5. Lesser curing period.
6. Lower cost of production.
7. Fully automated process.
8. Forms reusability within 50 s.
9. Lower cement consumption.

7 Conclusions

The following conclusions were drawn:

1. Wet press technique is a good option for improving the strength and durability of the concrete without affecting the workability and ease of concrete placement.
2. Draining off excess water in green state by applying hydraulic pressure increases the compressive strength by about 35% for the same mix proportion.
3. No or very less free water is retained in the hardened concrete and reduces the voids formation due to evaporation under atmospheric exposure.
4. Reduced voids formation within the concrete lowers the concrete permeability.
5. As abrasion resistance is directly proportional to the compressive strength, increase in compressive strength also increases the abrasion resistance.
6. Lower water content prevents segregation and dusting of top surface.
7. Lesser drying shrinkage improves tensile strength reducing the surface cracks.
8. This method proves to be better in quality and productivity over conventional method of casting.

References

1. Rieder KA et al (2015) Wet press concrete slab manufacturing
2. IS 8112-2013 ordinary Portland cement 43 grade—specification
3. IS 456-2000 specifications for plain and reinforced concrete
4. The constructor. https://theconstructor.org/concrete/vacuum-concrete-techniques-equipments-advantages/6867/. Last accessed 07 June 2019
5. Water cement ratio. http://www.buildingresearch.com.np/services/ct/ct8.php. Last accessed 18 June 2019
6. Effects of excess water in concrete mix. The Constructor. https://theconstructor.org/concrete/excess-water-effect-concrete-mix/29135/. Last accessed 18 June 2019
7. Forest press hydraulics FPH—single mould press. http://www.forestpresshyd.com/FPH_HOMEPAGE.html. Last accessed 07 June 2019
8. IS 516:1959—method of tests for strength of concrete

Performance of High Density Concrete with Mill Scale Waste

Vishnu Gavandi and K. G. Guptha

Abstract Concrete is being used on a very large scale throughout the world in an indiscriminate way thereby calls for an environmental friendly solution to produce sustainable concrete by judicious use of resources. Studies are carried out to find alternatives for the natural aggregates which are being depleted. Steel industry during its production is known to generate different types of solid waste, one of which is mill scale waste. This waste was used to develop a concrete mix by completely replacing natural aggregates. A high density mix was achieved with a density of 3.1 t/m^3 (intended to be used in the production of ballasts for washing machines). Early strength requirements of 35 MPa are met within 7 days of curing. This paper emphasizes on the resistance of the mix to acid attack. The performance of concrete was studied by immersing the cubes in sulphuric acid of 1 and 2% concentration. Investigations carried out have showed that the loss of compressive strength was minimum by the introduction of corrosion inhibiting admixture. Loss of mass was also considerably low.

Keywords Acid attack · Ballast · High density concrete · Mill scale waste

1 Introduction

It is a known fact that concrete is the second most consumed material only after water [4]. Aggregates form the major constituents of concrete. Hence, this calls for sustainable concrete by replacing the aggregates with wastes. Steel industry yields millions of tons of waste during its production [9]. Mill scale waste (MSW) is one of the wastes produced by it. During steel manufacturing, the specific production of mill scale can go up to 35–40 kg per ton [6]. Hence, it can be used in concrete, thereby

V. Gavandi (✉) · K. G. Guptha
Department of Civil Engineering, Goa College of Engineering, Farmagudi, Goa, India
e-mail: vishnugavandi01@gmail.com

K. G. Guptha
e-mail: kgg@gec.ac.in

L. M. Gupta et al. (eds.), *Advances in Civil Engineering and Infrastructural Development*, Lecture Notes in Civil Engineering 87,
https://doi.org/10.1007/978-981-15-6463-5_68

avoiding depletion of natural aggregates. The waste sample used has a well-graded particle distribution and hence serves as all in aggregates. This ensures that mill scale waste completely replaces both fine and coarse aggregates.

The primary aim of this research is to incorporate this design mix in forming high density concrete. Apart from the multiple uses of high density concrete, the intended use of this mix will be to produce washing machine ballasts. Since mill scale waste possesses a high specific gravity, it serves as an appropriate aggregate choice to attain heavy weight concrete [8]. The main purpose of ballasts in washing machine is to counteract the spinning forces acted on it during operation [7]. Ballasts manufactured with high density ensure that it occupies lesser volume in the assembly of the machine for a given weight.

The ballasts are designed to retain a density of 3.1 t/m^3 and are required to gain a minimum strength of 35 MPa at 7 days of curing. A number of trials were casted, and a final mix was decided. Apart from the physical requirements, it is necessary to ensure that the ballasts are durable too.

1.1 Tests on Resistance to Acid Attack

Ordinary Portland cement is an alkaline material with pH value of over 12 and hence has a very little resistance to acid attacks [5]. When in contact with acid, its pH value decreases, and when it decreases further, lower than 7, the components of cement start to break, and this is known as acid attack. Sulphuric acid (H_2SO_4) is one such acid with pH lower than 3 and is known to be quite harmful because of acid attack and also due to sulphate attack [3]. Hence, sulphuric acid serves as a standard to measure the durability of concrete. It reacts with calcium hydroxide present in concrete and produces gypsum causing increase in volume [1]. It also leads to corrosion and cracking, thereby reducing the mechanical strength of concrete. As mill scale waste is constituted by high iron content, it is of prime importance that corrosion does not affect it in damp conditions.

2 Materials and Methodology

2.1 Materials Used

Binders Cement serves as the primary binder, and ground granulated blast-furnace slag (GGBS) is used to enhance the durability of the concrete. The amount of GGBS is limited to avoid noticeable changes in the density of hardened concrete. The specific gravity of cement and GGBS was determined to be 3.14 and 2.85, respectively.

Water Clean potable water is used for all concreting works from mixing to curing.

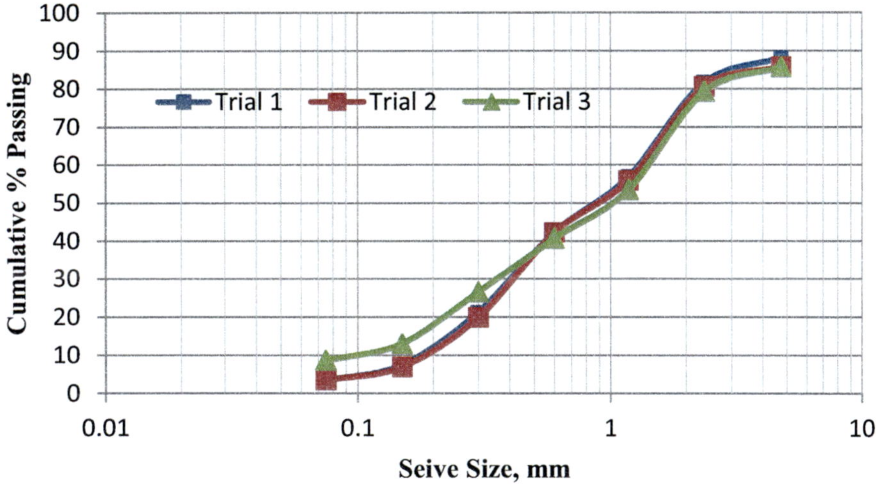

Fig. 1 Particle size distribution curve for mill scale waste

Mill Scale Waste A well-graded sample of MSW was used as a replacement for aggregates. The specific gravity of the sample was 6.20, and dry loose bulk density (DLBD) obtained was 2865 kg/m^3. Particle size distribution curve for the sample is as shown in Fig. 1.

Fibres Bajaj Fibretuff synthetic macro fibres were used in the design mix. These help in preventing shrinkage cracks and add flexural strength and also give considerable resistance to vibration effects. It has an embossed pattern finish which enables better bond with concrete and thereby increasing its performance. They also provide protection against acidic and alkaline environments and protection against corrosion [2]. It has a specific gravity of 0.90–0.92.

Chemical Admixtures Dynamon SP 508 is used as a super plasticizer for better workability and early gain of strength. Mapei Mapefluid IF 328 is a blister reducing admixture and improves the overall durability of the concrete. Most importantly, a corrosion inhibiting chemical EPKO KP 200 is used to prevent corrosion due to acid attack. Inclusion of this chemical reduced the loss in compressive strength and mass of the test cubes under the action of sulphuric acid.

2.2 *Methodology*

After mixing all the ingredients, density corrections are made to ensure that the hardened concrete has a fixed density of 3.1 t/m^3. For density corrections, DLBD vessel was used having a volume of 3 L. Thus, knowing the volume, mass of the

Table 1 Mix proportions of various constituents of concrete

Material	Cement	GGBS	MSW	Water	Fibretuff fibres	Dynamon SP 508	Mape fluid IF 328	EPKO KP 200
Mass (kg/m^3)	1047	83.33	2141	333	1.88	10.44	4.22	4.22

mix is checked for any variations, and accordingly, changes are done in proportions to either increase or to reduce the density to 3.1 t/m^3. The mix proportions of the materials are as shown in Table 1.

Cubes casted were demoulded after 24 h. Samples are immersed in sulphuric acid having concentrations of 1 and 2% for acidic curing. Cubes are also kept in a standard curing tank to compare the deterioration as control. Initially, cubes were casted without the use of a corrosion inhibiting admixture. After observing its performance in acidic environment, the admixture was added, and the results were compared to check the effectiveness of the inhibitor. Cubes were tested for 1, 3 and 7 days for its compressive strength and its mass in both standard and acidic environment and both with and without EPKO KP 200 serving as the corrosion inhibitor.

3 Results and Discussions

For each type of curing, tests were carried out for compressive strength and mass of the cubes. A density of 3.129 t/m^3 is obtained with the design mix. The results of compressive strength for different concentrations of exposure and for both, with and without use of corrosion inhibitor, are tabulated in Table 2.

Table 2 Compression strength data for different conditions for design mix

Type of casting	Days of curing	Compressive strength achieved (MPa)		
		Concentration of sulphuric acid		
		0%	1%	2%
Without EPKO KP 200	1	7.69	7.69	7.69
	3	33.96	33.42	33.03
	7	39.50	37.93	36.72
With EPKO KP 200	1	9.88	9.88	9.88
	3	38.42	38.11	37.78
	7	43.50	42.50	41.25

3.1 Performance in Acidic Environment Without EPKO KP 200

Although EPKO KP 200 does not contribute to the strength of concrete, it can be seen that 7 day strength for this mix is slightly lesser than the one with the said admixture. This is due to the fact that mill scale waste consists of high iron content up to 70%, and without this admixture, corrosion takes place and affects the mechanical properties. Even for normal curing, minor corrosion marks are visible on the surface. It can be seen that a loss of strength of 1.59% and 3.58% has taken place for 3 days and 7 days, respectively, for 1% concentration of H_2SO_4. Similarly, for 2% concentration, these values are higher at 2.74 and 7.04%. The comparison for different concentrations of acid exposure and the percentage loss of strength for each is as shown in Fig. 2. The linear regression equations for strength loss are shown in Eqs. (1) and (2) for 1% and 2% concentrations, respectively, and regression values obtained are 0.99 for both. Where x is the number of days of curing and y is the subsequent loss of strength in percentage.

$$y = 0.653x - 0.5393 \tag{1}$$

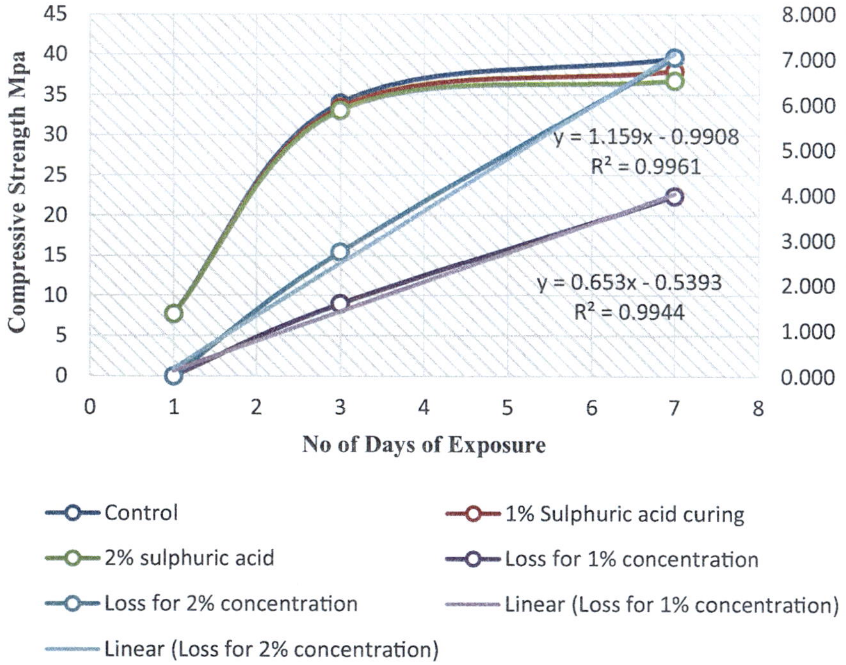

Fig. 2 Comparison of strength obtained without the use of EPKO KP 200

$$y = 1.159x - 0.9908 \quad (2)$$

Loss of mass is an important attribute to be considered for the ballasts. As it is expected that the variations in density be not too high, loss of mass due to acid attack needs to be kept as low as possible. Without the corrosion inhibitor, mass loss is noted to be 0.64% for 1% sulphuric acid curing and goes as high as 0.94% for 2% concentration.

3.2 *Performance in Acidic Environment with EPKO KP 200*

With the introduction of EPKO KP 200 corrosion inhibitor, the reduction in strength was lesser. The reduction in strength for 1% concentration was noted to be 0.81% and 2.29% for 3 days and 7 days, respectively. Furthermore, the loss of strength for 2% concentration was observed to be 1.67 and 5.17%. This indicates that the resistance to acid attack is improved by the introduction of corrosion inhibitor. Also visually, the test cubes were affected less severely in this case. The comparison for different concentrations of acid exposure and the percentage loss of strength for the trial with EPKO KP 200 is as shown in Fig. 3. The linear regression equations for strength

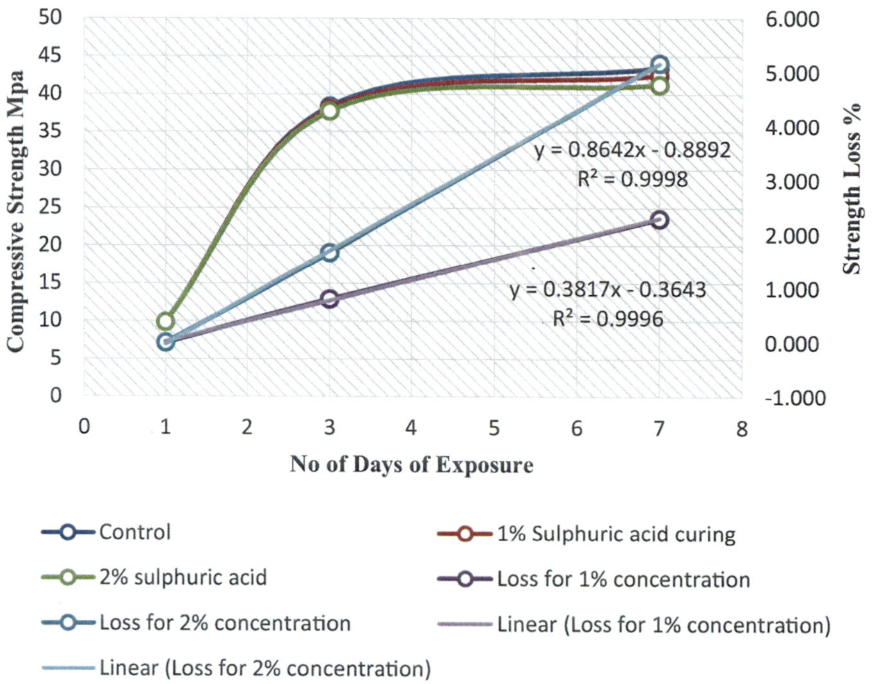

Fig. 3 Comparison of strength obtained with the use of EPKO KP 200

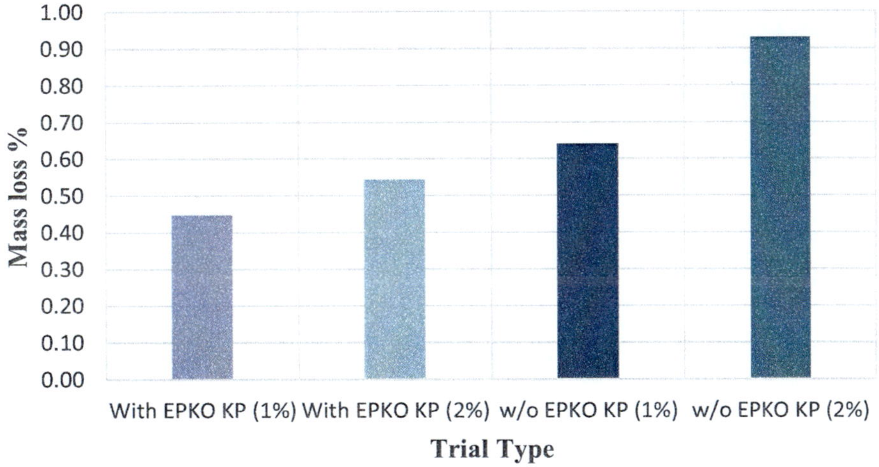

Fig. 4 Comparison of loss in mass of concrete

loss are shown in Eqs. (3) and (4) for 1% and 2% concentrations, respectively, and regression values obtained are almost equal to 1.

$$y = 0.3817x - 0.3643 \tag{3}$$

$$y = 0.8642x - 0.8892 \tag{4}$$

Loss of mass is observed to be 0.45 and 0.54% for 1 and 2% concentrations. These values are significantly less than those without the inhibitor. Although variations up to 1% are allowable, lower loss of mass due to acid attack ensures room for minor variations in density. The comparison of mass loss is as shown in Fig. 4.

4 Conclusions

From the experimentations, the following inferences are drawn.

1. High density concrete mix with mill scale waste as a complete replacement for aggregates achieves a density of 3.129 t/m^3 and gives 7 day strength of 43.5 MPa meeting the requirements of the ballasts.
2. It is observed that the loss of compressive strength is significantly reduced with the introduction of EPKO KP 200, a corrosion inhibiting admixture.
3. Loss of mass is also reduced due to the corrosion inhibitor.

4. Apart from surface degradation, there is no presence of corrosive marks in the cube despite high iron content of mill scale waste indicating no damage in the interior surface due to lower penetration of sulphuric acid.
5. The losses in strength and mass are minimal, and hence, it is observed that the mix possesses enough resistance to acid and sulphate attack.

References

1. Ahmed S, Munirudrappa N (1998) Effect of sulphuric acid on plasticized concrete. Indian J Eng Mater Sci 5:291–294
2. Bajaj Reinforcement. http://www.bajajreinforcementsllp.com/fibretuff.html. Last accessed 23 May 2019
3. Barbhuiya S, Kumala D (2017) Behaviour of a sustainable concrete in acidic environment. Sustainability 9(9):1556, 1–13
4. Gagg C (2014) Cement and concrete as an engineering material: an historic appraisal and case study analysis. Eng Fail Anal 40:114–140
5. Jamal H. Acid attack on concrete. https://www.aboutcivil.org/acid-attack-on-concrete.html. Last accessed 23 May 2019
6. Martín M, Lopez F, Torralba J (2012) Production of sponge iron powder by reduction of rolling mill scale. Ironmak Steelmak 39(3):155–162
7. Nice K. How washing machines work. https://home.howstuffworks.com/washer2.htm. Last accessed 19 May 2019
8. Ouda A (2015) Development of high-performance heavy density concrete using different aggregates for gamma-ray shielding. Prog Nucl Energy 79:48–55
9. Singhal A, Bhunia D, Pandel B (2015) Effect of sand replacement by mill scale on the properties of concrete. Int J Eng Technol Sci Res 2:60–67

Performance Evaluation of Rubberized Concrete with the Use of Steel Fibers

Dhiraj Agrawal, U. P. Waghe, and S. P. Raut

Abstract The worldwide use of concrete is second only to the water, as the demand of concrete increases; the requirement of its constituents also increases. Most of the developing countries like India are facing acute shortage of coarse and fine aggregates as they are natural resources to be used in concrete. To fulfill the demand of these materials, there is need to invent an alternative for coarse and fine aggregates but while inventing the alternative, it should be kept in mind that the alternatives must be from industrial or agricultural waste available at a very cheaper rate or available for free. Due to the evolution in automobile industries, there is remarkable increase in the production of waste tires which are made up of rubbers. The waste problem associated with used tires is one of the major problems facing the world today as a source of environmental pollution and health hazards. Currently, these waste tires are being utilized for recycling for rubber goods, pyrolysis, and in road construction. In the present study, sand is partially replaced from 0 to 20% by crumb rubber. Concrete specimen with M-25 grade is checked and compared with control mix for compressive strength, split tensile strength, flexural tensile strength and for acidic action. After using crumb rubber, it is found that the strengths mentioned above are achieved up to 10%. For further enhancement in the strengths, 0.2% steel fibers are used.

Keywords Crumb rubber · Rubberized concrete compressive strength · Split tensile strength · Flexural tensile strength · Acidic action test

1 Introduction

The natural resources like sand, gravel, and aggregate are depleting rapidly with increase in construction activity. To fulfill the need of construction activity, we have to find out new alternative that can fully or partially replace the construction materials.

D. Agrawal (✉) · U. P. Waghe · S. P. Raut
Department of Civil Engineering, Yeshwantrao Chavan College of Engineering, Nagpur, India
e-mail: erdhiraj007@gmail.com

L. M. Gupta et al. (eds.), *Advances in Civil Engineering and Infrastructural Development*, Lecture Notes in Civil Engineering 87,
https://doi.org/10.1007/978-981-15-6463-5_69

A large number of waste materials are produced and discarded as a waste. River sand is the universally accepted constituent of concrete as a fine aggregate, but almost the every part of world is facing the acute shortage of sand as it is the natural resource. Extraction of sand from river beds also affects the environment and the aquatic life of that specific river. To avoid this, various governing authorities in India are implementing different rules, regulations, and taxes on the collection of river sand from the river bed.

At the same time, global municipal solid waste production is increasing day by day with a rapid rate. India generates 0.1 million tonnes municipal solid waste per day, which amounts to be almost 36.5 million tonnes for a year. Along with this generation of discarded tires due to a significant increment in the vehicles causes environmental threat. These discarded tires can be utilized as viable construction material with partial replacement of aggregates in concrete.

Concrete is known as an excellent structural material and reflected as essential for the modern development and social humanity. The use of waste tires in concrete has become technically achievable, and the concrete is being measured as lightweight concrete. The probability of using crumb rubber derived from waste tires in concrete to improves the strength as well as to avoid damage to the environment [1]. Rubberized concrete reinforced with steel fibers is studied for load deflection and proved that the merger of crumb rubber receives the elastic strength to concrete; however, the supplementary gain in elasticity from the increased strength due to the addition of steel fiber as a secondary component [2]. Use of crumb rubber in concrete for structural columns is experimented by evaluating the use of fiber-reinforced polymer confinement as a means of overcoming the material deficiencies and indicated that the use of fiber-reinforced polymer to confine rubberized concrete effectively increased ductility that arises from rubberized concrete; hence, it gives the likely potential for structural column applications, particularly in seismic zones [3]. The use of waste tire in concrete had increased toughness, which is much greater than unmodified concrete. Absorption of energy for modified concrete was higher than the control sample. The fibers bridging over the cracks, the crack opening width can be controlled [4]. The dynamic and loads were applied to plain and hybrid rubberized-normal concrete beam, and the results demonstrated that the impact, inertial load, and bending load of hybrid rubberized-normal concrete beam increased with the increase in the percentage of sand replacement by recycled fine crumb rubber [5]. The fracture toughness and fracture energy can be increased up to certain rubber content [6].

2 Experimental Procedure

In this study, initially the materials were tested for their physical and mechanical properties in accordance with Indian Standards.

2.1 Properties of Crumb Rubber

Crumb rubber is produced by grinding in the size of 2–4.75 mm. Crumb rubber is checked for its specific gravity and it is found to be 0.626 with density 430 kg/m^3.

2.2 Properties of Fine Aggregates

Sand used in this study as fine aggregate has particle size 75 μm–4.75 mm, with fineness modulus as 3.76 and with a density of 1715.52 kg/m^3. The water absorption of sand is 1.2% and specific gravity as 2.65.

2.3 Properties of Coarse Aggregates

Coarse aggregates used have impact value as 34, hardness as 12, and specific gravity as 2.78. All these properties are within the permissible limits.

2.4 Tests on Specimen

In this section, mix design is carried out using IS 10262-2009 for M-25 grade concrete. Concrete specimen of size 15 * 15 * 15 cm was casted for determining the compressive strength test cured for 28 and 56 days. Split tensile strength was determined using cylinders of 150 mm diameter and 300 mm long. Flexural strength was determined by preparing concrete beams of size 100 * 100 * 500 mm.

For acidic action, test cube of size 15 * 15 * 15 cm was casted and kept for curing for 28 days in water; after drying it for 01 day, they were immersed in tank, which contains sulfuric acid in an amount of 3% to the volume of water in the tank. They were kept for 60 days. The procedure for acidic action test was referred from ASTM C267.

Table 1 shows the variation of percentage replacement of sand by crumb rubber and addition of steel fibers.

Table 1 Table for abbreviations used

% of crumb rubber added	Abbreviation used	% of crumb rubber and steel fiber added	Abbreviation used
0% rubber	R0	0% rubber + 0.2% steel fiber	R0 + 0.2S
5% rubber	R5	5% rubber + 0.2% steel fiber	R5 + 0.2S
10% rubber	R10	10% rubber + 0.2% steel fiber	R10 + 0.2S
15% rubber	R15	15% rubber + 0.2% steel fiber	R15 + 0.2S
20% rubber	R20	20% rubber + 0.2% steel fiber	R20 + 0.2S

R0 is the control concrete, i.e., without any replacement of sand by crumb rubber

3 Results and Discussion

3.1 *Compressive Strength for Rubberized Concrete with and Without Steel Fibers*

From Table 2, the highest compressive strength is obtained at 10% rubber content. The increase in strength was 12.36% as compared to control mix, wherein due to the further increase in rubber content, compressive strength is decreased. In the same manner, the compressive strength for 56 days is increased up to 10% rubber content but the percentage increase in strength is found to be 13.95%.

From Table 3, the highest compressive strength is obtained at 10% rubber content. The increase in strength was 13.18% as compared to control mix, which is higher than the compressive strength obtained in the mix without the addition of steel fibers. Similarly, the compressive strength for 56 days with the addition of 0.2% steel fibers is increased by 13.95%.

Figures 1 and 2 represent the comparison between compressive strength with and without steel fibers, and it is found that the strength is increased for each mix due to the addition of steel fibers along with crumb rubber.

Table 2 Compressive strength of cubes 28 and 56 days for rubberized concrete without steel fibers

Days	28 days	56 days
Type of mix	MPa	MPa
R0	28.3	30.1
R5	30.7	33.2
R10	31.8	34.3
R15	29.3	31.5
R20	27.3	29.9

Table 3 Compressive strength of cubes 28 and 56 days for rubberized concrete

Days	28 days	56 days
Type of mix	MPa	MPa
R0 + 0.2S	32.3	33.6
R5 + 0.2S	33.8	35.3
R10 + 0.2S	35.2	36.9
R15 + 0.2S	34.3	34.9
R20 + 0.2S	33.4	33.7

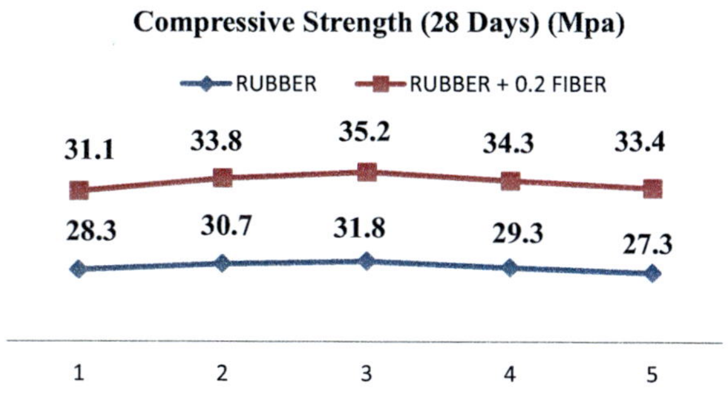

Fig. 1 Comparison of compressive strength of 28 days for rubber and rubber with 0.2% steel fiber

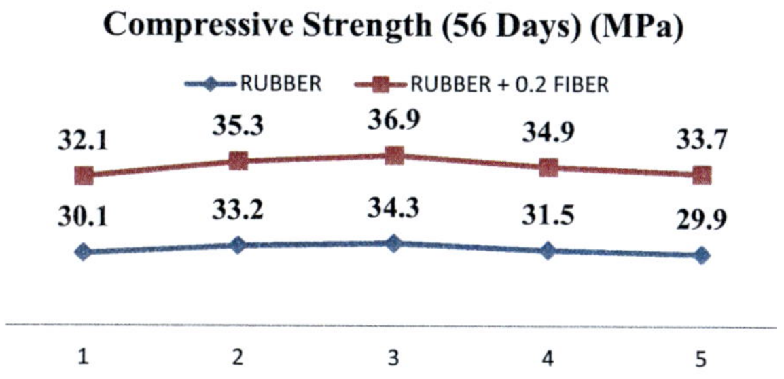

Fig. 2 Comparison of compressive strength of 56 days for rubber and rubber with 0.2% steel fiber

3.2 Flexural Tensile Strength for Rubberized Concrete with and Without Steel Fibers

From Table 4, the highest flexural tensile strength is obtained at 10% rubber content. The increase in strength was 48.48% as compared to control mix.

From Table 5, the highest flexural tensile strength is obtained at 10% rubber content. The increase in strength was 52.63% as compared to control mix.

Figures 3 and 4 represent the comparison between flexural tensile strength with and without steel fibers, and it is found that the strength is increased for each mix due to the addition of steel fibers along with crumb rubber. Flexural strength is increased up to 15% replacement of sand with crumb rubber.

Table 4 Flexural tensile strength of cubes 28 and 56 days for rubberized concrete without steel fibers

Days	28 days	56 days
Type of mix	MPa	MPa
R0	3.3	4.5
R5	4.3	5.1
R10	4.9	5.8
R15	3.8	4.9
R20	3.2	3.9

Table 5 Flexural tensile strength of cubes 28 and 56 days for rubberized concrete with steel fibers

Days	28 days	56 days
Type of mix	MPa	MPa
R0 + 0.2S	3.8	4.9
R5 + 0.2S	4.2	5.6
R10 + 0.2S	5.8	6.3
R15 + 0.2S	4.9	5.2
R20 + 0.2S	4.2	4.8

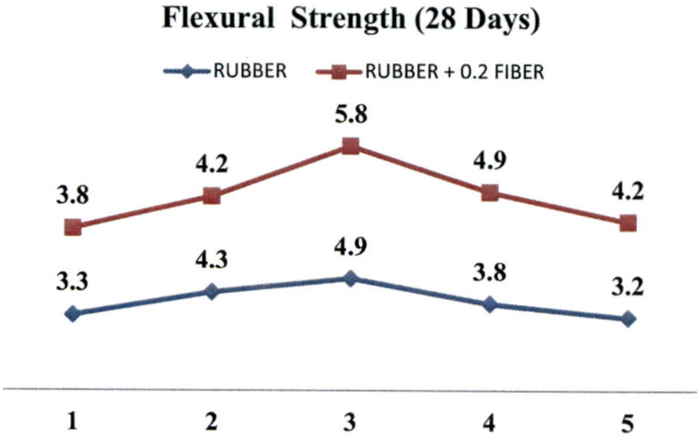

Fig. 3 Comparison of flexural tensile strength of 28 days for rubber and rubber with 0.2% steel fiber

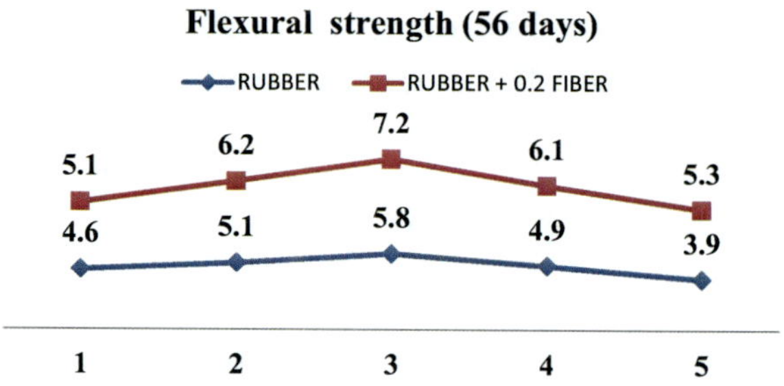

Fig. 4 Comparison of flexural tensile strength of 56 days for rubber and rubber with 0.2% steel fiber

3.3 Split Tensile Strength for Rubberized Concrete with and Without Steel Fibers

From Table 6, highest split tensile strength is obtained at 10% rubber content. The increase in strength was 33.33% as compared to control mix. There is no decrease in split tensile strength up to 20% replacement.

From Table 7, highest split tensile strength is obtained at 10% rubber content. The increase in strength was 40.90% as compared to control mix.

Figures 5 and 6 represent the comparison between split tensile strength with and without steel fibers, and it is found that the strength is increased for each mix due to the addition of steel fibers along with crumb rubber. Split tensile strength is increased up to 15% replacement of sand with crumb rubber.

Table 6 Split tensile strength of cubes 28 and 56 days for rubberized concrete without steel fibers

Days	28 days	56 days
Type of mix	MPa	MPa
R0	3.9	4.8
R5	4.9	5.4
R10	5.2	6.1
R15	4.7	5.7
R20	4.1	4.8

Table 7 Split tensile strength of cubes 28 and 56 days for rubberized concrete with steel fibers

Days	28 days	56 days
Type of mix	MPa	MPa
R0 + 0.2S	5.1	5.9
R5 + 0.2S	5.8	6.2
R10 + 0.2S	6.2	7.2
R15 + 0.2S	4.8	6.1
R20 + 0.2S	4.1	5.3

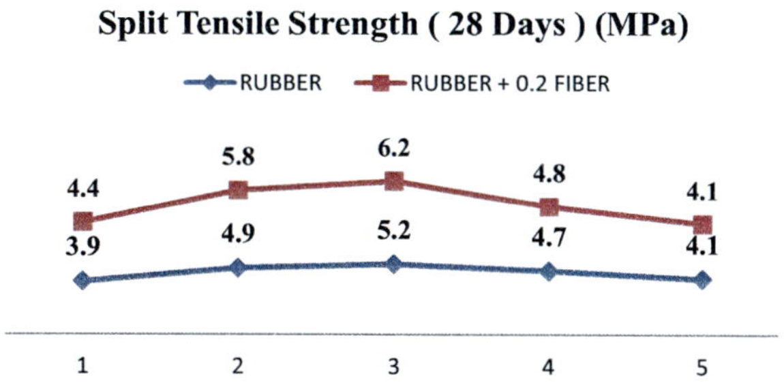

Fig. 5 Comparison of split tensile strength of 28 days for rubber and rubber with 0.2% steel fiber

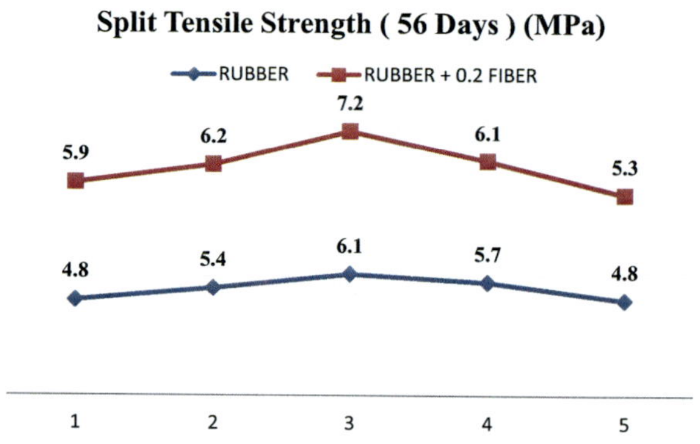

Fig. 6 Comparison of split tensile strength of 56 days for rubber and rubber with 0.2% steel fiber

Table 8 Split tensile strength of cubes 28 and 56 days for rubberized concrete without steel fibers

Days	60 days
Type of mix	MPa
R0	23.50
R5	25.10
R10	26.66
R15	22.40
R20	21.60

3.4 *Compressive Strength for Rubberized Concrete with and Without Steel Fibers After Acidic Action*

From Table 8, it is seen that compressive strength for acidic action is constantly decreasing even for the control mix. Rate of decrease is slow with percentage replacement of rubber.

From Table 9, it is seen that compressive strength for acidic action is remained constant with slight increase as compared to control mix up to 10% replacement.

Table 9 Split tensile strength of cubes 28 and 56 days for rubberized concrete with steel fibers

Days	60 days
Type of mix	MPa
R0 + 0.2S	27.12
R5 + 0.2S	28.90
R10 + 0.2S	29.33
R15 + 0.2S	28.10
R20 + 0.2S	27.13

4 Conclusion

- It is recommended to use crumb rubber as a partial replacement of sand for up to 10%.
- As rubber content in concrete is increased, workability is decreased.
- Use of rubberized concrete can lead to reduction in carbon emission.
- The decrease in compressive strength of concrete due to weak bonding between cement paste and crumb rubber.
- Rubberized concrete can be used for the construction of partition walls, kerbs on bridges and highway, parking lots, boundary walls and architectural components.
- The split tensile strength of rubberized concrete is more compared to flexural strength with the same ratio of replacement.
- After acid attack, more losses in weight and compressive strength were observed in the control mix than rubberized concrete.

References

1. Kumaran GS, Mushule N, Lakshmipathy M (2008) A review on construction technologies that enables environmental protection: rubberized concrete. Am J Eng Appl Sci 1941-7020
2. Hussein ZS (2011) Flexural behavior of steel fiber-reinforced rubberized concrete. Proquest LLC. Umi Number: 1507492
3. Youssf O, ElGawady MA, Mills JE, Ma X (2014) An experimental investigation of crumb rubber concrete confined by fibre reinforced polymer tubes. Constr Build Mater 53:522–532
4. Garrick GM (2001) Analysis and testing of waste tire fiber modified concrete. Master of Science thesis
5. Al-Tayeb MM, Abu Bakar BH, Ismail H, Akil HM (2013) Effect of partial replacement of sand by recycled fine crumb rubber on the performance of hybrid rubberized-normal concrete under impact load: experiment and simulation. J Clean Prod 59:284–289
6. Guo YC, Zhang JH, Chen G, Chen GM, Xie ZH (2014) Fracture behaviours of a new steel fiber reinforced recycled aggregate concrete with crumb rubber. Constr Build Mater 53:32–39

Feasibility of Fly Ash for Manufacturing of Self Compacting Concrete

Swapneel Satone, Dhananjay Parbat, Devendra Pratap Singh, Dipali P. Jasudkar, and Manmohandas Goel

Abstract Self-compacting concrete is able to flow and consolidate under its own weight and is deaerated almost completely while flowing in the formwork. It is cohesive enough to fill the formwork of any size and shape without segregation or bleeding. SCC is found useful for heavily-reinforced concrete members or in complicated work forms. The objective of this work is to study the different test adopted on self-compacting concrete-like slump flow test, U-tube, L box, J ring and compare the same with conventional concrete. And also to compare the Compressive Strength values of self-compacting and normal concrete specimens at different ages and to study the effect of superplasticizer, VMA and fly ash on concrete. All SCC mixtures exhibited greater values in compressive strength after being tested, compared to normal concrete. The compressive strength of SCC is found to be higher than that of normal concrete. Use of mineral and chemical admixtures, which usually improve the bonding between aggregate and cement paste, thus increasing the strength of concrete.

Keywords CC · VMA · Superplasticizer · SSC

S. Satone (✉) · D. P. Singh
Civil Engineering Department, K.D.K. College of Engineering, Nagpur 440009, India
e-mail: ssatone1980@yahoo.com

D. P. Singh
e-mail: dp_singh123@yahoo.com

D. P. Jasudkar
Chemistry Department, K.D.K. College of Engineering, Nagpur 440009, India
e-mail: dipjasudkar@gmail.com

D. Parbat
Government Polytechnic, Bramhapuri 441206, India
e-mail: parbatdk@gmail.com

M. Goel
Civil Engineering Department, Visvesvaraya National Institute of Technology, Nagpur 44010, India
e-mail: mdgoel@gmail.com

L. M. Gupta et al. (eds.), *Advances in Civil Engineering and Infrastructural Development*, Lecture Notes in Civil Engineering 87,
https://doi.org/10.1007/978-981-15-6463-5_70

1 Introduction

Concrete is that the second most generally consumed material within the globe owing to its good looks, strength and sturdiness. The thought of Self Compacting Concrete was projected in 1986 by academician Hajime Okamura in 1997, however, academician Ozawa in 1989 at the University of Japan developed the prototype [1]. Durability of concrete structures can be improved by using SCC. Chemical makeup of fly ash is silicon dioxide, alumina, and iron, based. Colour usually varies from dim gray to yellow-tan for fly ash consumed for concrete. Besides, it has pozzolanic characteristics. A pozzolana could be a siliceous or aluminosiliceous based material that is finely divided type and within the existence of wetness, with chemically add with the soluble calcium hydroxide. Thus, fly ash based concrete becomes toughened, stronger and additionally sturdy compared to straight long-standing when matched to conventional concrete mixtures made without using pozzolanic stuff.

The focus of this study is to point of how the concrete trade will increase its conformity with the stress of green initiative.

- Increased consumption of supplementary cementious stuff
- Improved Concrete Hardened properties.

1.1 Chemistry Behind Consumption of Fly Ash as a Supplementary Cementious Stuff

Reaction of Ordinary Portland cement in concrete after Fly ash addition is as below.

Ordinary Portland cement made up four major mineralogical phases characteristically diagrammatic by C_3S, C_2S, C_3A and C_4AF. The hydration reaction of major chemical compounds may be shown as [2]

$$2(3\,CaO.\,SiO_2) + 6H_2O \rightarrow 3CaO.\,2SiO_2.3H_2O + 3Ca(OH)_2$$
$$2(2\,CaO.\,SiO_2) + 4H_2O \rightarrow 3CaO.\,2SiO_2.\,3H_2O + Ca(OH)_2$$

Fly ash reacts with calcium hydroxide in concrete,

$$\text{Calcium Hydroxide} + \text{Pozzolana} + \text{Water} \rightarrow \text{C - S - H(gel)}$$

So, undoubtedly from the equation, we will construe that the surplus $Ca(OH)_2$ content made by hydration reaction of Cement is condensed by the utilization of silica content of fly ash. Due to the configuration of extra hydrated calcium silicate (C-S-H) binder, the strength of concrete will be increased.

- It produced reduced heat during hydration and that too at a low rate
- Densification of Microstructure in concrete due fineness of fly ash.

Table 1 Compound composition

Compound composition	Cement Per cent content [2]	Fly ash per cent content
$SiO_2 + Al_2O + Fe_2O_3$	–	89.91
CaO	60–67	–
SiO_2	17–25	58.34
Al_2O_3	3–8	–
Fe_2O_3	0.5–6	–
MgO	0.1–4	0.5
K_2O, Na_2O	0.4–1.3	1.05
SO_3	1.3–3	0.4

Table 2 Current position of fly ash production and consumption in India [3]

No. of thermal power station in India	Fly ash production (MT)	Total consumption (MT)
167	196.44	131.87
% Consumption		67.13

Enhanced impermeability of concrete mass will increases resistance against weathering actions leads to improved durability. Table 1 shows the compound properties of cement and fly ash.

1.2 *Current Position of Fly Ash Production in India*

See Table 2.

1.3 *Concrete Mix Design by Using Supplementary Cementious Material*

Using IS 10262:2009 [4], Concrete is aimed for following grades.

Design provisions	
Mix grade	M25, M30 and M40
Grade of cement	OPC 53
Degree of workability	Good

2 Experimental

2.1 *Material Properties [5]*

Fineness modulus of	
Fine aggregate	2.900
Coarse aggregate	7.130
Specific gravity of	
OPC 53	3.150
Fine aggregates	2.556
20 mm size coarse aggregates	2.920
10 mm size coarse aggregates	2.860
Combined coarse aggregate	2.890
Chemical admixture—superplasticizer	1.080
Fly ash	2.200

2.2 *Design Mix Fraction*

See Tables 3 and 4.

2.3 *Testing of Workability of Concrete Mix as Per EFNARC: 2005 [6]*

See Table 5.

2.4 *Effect of Supplementary Cementious Material on Strength as Per IS:516-1959 [7]*

- Early Age (3 Days) compressive strength of all GM found less when match with CM because pozzolonic reaction takes time.
- Strength gain with age is found more in GM at the age of 90 days; it is because of the use of low water to cement ratio and consumption of Pozzolonic stuff fly ash (Tables 6, 7 and 8).

Table 3 Mix fraction

Ingredients of concrete	Unit	CM25[a]	CM30[a]	GM25[a]	GM30[a]	GM40			
						M1	M2	M3	M4
Cementious material	kg/m^3	390.25	421	450	495	513.158	513.158	513.158	513.160
Cement	kg/m^3	390.25	421	270	300	410.526	307.395	256.579	205.260
Fly ash	kg/m^3	0	0	180	195	102.632	205.263	256.579	307.900
Water	kg/m^3	160	160	180	180	195	195	195	195
Fine aggregate	kg/m^3	647.5	638.3	920	904	836.84	836.34	836.84	836.84
Coarse aggregate	kg/m^3	1270.6	1253	862	846	823.02	781.94	761.4	740.85
20 mm aggregate	kg/m^3	1270.6	1253	432	423	0	0	0	0
10 mm aggregate	kg/m^3	–	–	430	423	823.02	781.94	761.4	740.85
Super plasticizer	kg/m^3	–	–	4.5	4.6	4.618	4.613	4.618	4.618
Viscosity modifying admixture	kg/m^3	–	–	0.9	0.95	0	0	0	0
Water to cement ratio		0.41	0.38	0.37	0.37	0.38	0.38	0.38	0.38
Total density of fresh concrete	kg/m^3	2468	2472	2465	2425	2373	2332	2311	2290
Total Powder (Cement + fly ash + fine aggregate)	kg/m^3	–	–	500	535	555	555	555	555

[a]Data from Ref. [5]

Table 4 Taxonomy of mix

Taxonomy	Designation
CM25, CM30	Conventional concrete mix without fly ash of grade 25 and 30
GM25, GM30, GM40	SCC Mix with Fly ash of grade 25 and 30
M1	Cement + 20% Fly ash
M2	Cement + 40% Fly ash
M3	Cement + 50% Fly ash
M4	Cement + 60% Fly ash

3 Conclusion

1. Experimental investigation shows that SCC possessed uniformity and self-compactability underneath its own mass, with none external tremor.
2. By using Fly ash and Superplasticizer the strength of SCC is enhanced considerably when match with conventional concrete mix of different grades.
3. It is seen that Fly ash consumption usually results in encouraging outcomes and is extremely suggested for all green SCC mixes. Fly ash that is finer than cement in SCC will increase filling, spreading and momentary quality of concrete. Consumption of those squander materials with cement will facilitate to preserve the precious resources.
4. pH of all concrete mix was observed to be more than 11, hence concrete will be durable.
5. The elimination of compacting instruments noise improves the environment protection close to erection sites wherever concrete is being poured.
6. In SCC, there's is no probabilities of honeycomb, therefore it offers better surface end.

Table 5 Workability test result

Test	Unit	Test Result								Permissible Values as per EFNARC-2005
		CM25[a]	CM30[a]	GM25[a]	GM30[a]	GM40				
						M1	M2	M3	M4	
Slump-cone test	mm	75	82	_b	_b	_b	_b	_b	_b	_b
Compaction factor		0.9	0.92	_b	_b	_b	_b	_b	_b	_b
Slump-flow by Abrams cone	mm	_b	_b	690	660	720	720	720	720	650–800
Slump-flow (T50 cm)	S	_b	_b			4.4	3.42	3.38	3.12	2–5
J-ring	Mm	_b	_b	8	10	8	10	10	10	0–10
V-funnel	S	_b	_b	11	13	10.9	9.6	8.6	8.2	8–12
V-funnel (T5 min)	S			14	15	12.8	10.2	9.63	8.5	0–3
L-Box	(h_2/h_1)	_b	_b	0.98	0.93	1	0.8	1	0.9	0.8–1.0
U-Box	(h_2-h_1) mm	_b	_b	22	18	25	15	25	22	0–30
Fill Box	%	_b	_b	100	100	100	100	100	100	90–100

[a]Data from Ref. [5]
—[b]Test not applicable

Table 6 Compressive strength result

Testing age ⇩	Mix ⇨	Average compressive strength (Mpa)							
		CM25[a]	GM25[a]	CM30[a]	GM30[a]	GM40			
						M1	M2	M3	M4
3 Days		12.93	11.94	16.6	15.55	29.95	25.8	22.08	19.76
7 Days		17.58	17.43	21.06	22.52	39.94	37.03	36.07	30.19
28 Days		29.2	30.80	31.98	35.47	50.04	49.18	47.55	42.04
90 Days		32.1	35.11	35.18	41.14	57.55	56.06	53.97	47.08

[a]Data from Ref. [5]

Table 7 Split tensile and flexural strength result

Testing age ⇩	Mix ⇨	Average split tensile strength (Mpa)							
		CM25	GM25	CM30	GM30	GM40			
						M1	M2	M3	M4
3 Days		2.31	2.2	2.71	2.59	2.54	1.95	1.78	1.4
7 Days		2.50	2.41	2.85	2.90	3.4	2.61	2.44	2.06
28 Days		2.69	2.80	3.40	3.64	3.98	3.24	3.04	2.98
90 Days		2.80	3.01	3.62	3.89	4.62	3.71	3.45	3.35
	Average Flexural Strength (MPa)								
3 Days	1.06	1.06	1.06	1.06	3.8	2.9	2.37	2.01	
7 Days	1.19	1.19	1.19	1.19	5.67	3.73	3.44	2.90	
28 Days	1.35	1.35	1.35	1.35	6.08	5.1	4.98	4.5	
90 Days	1.41	1.41	1.41	1.41	6.99	5.84	5.65	5.15	

Table 8 pH test result

Testing age ⇩	Mix ⇨	pH test							
		CM25	GM25	CM30	GM30	GM40			
						M1	M2	M3	M4
365 days		13	13	14.5	15	15.2	16	16	16

References

1. Okamura H (1997) Self-compacting high-performance concrete. In: Concrete International, pp 50–54
2. Sheety MS (2013) Concrete technology theory and practices, 7th edn. S. Chand & Company Pvt. Ltd., New Delhi
3. Report on fly ash generation at coal/lignite based thermal power stations and its utilization in the country for the year 2017–18 by Central Electricity Authority New Delhi, December 2018
4. Indian standard code IS 10262-2009 and IS 10262-1982 for concrete mix proportion-Guidelines, Bureau of Indian Standard, July 2009

5. Satone SR, Parbat DK, Singh DP (2015) Use of fly ash for manufacturing of self-compacting concrete. Int J Mod Trends Eng Res 295–299
6. The European guidelines for self-compacting concrete specification, production and use. May 2005
7. Indian standard code IS 516–1959 (reaffirmed 2004) for Method of test for strength of concrete. Bureau of Indian Standard, June 2006

Investigate the Temperature Effects on Curing of Reactive Powder Concrete Containing Silica

Hemantkumar G. Sonkusare and Prashant Y. Pawade

Abstract There is a burgeoning use of reactive powder concrete unsettled to the outstanding mechanical properties and resilience. Reactive powder concrete structural elements can resist chemical paroxysm, impact loading from vehicles and vessels, and sudden kinetic loading due to earthquakes. In addition, the lower maintenance requirements result in significant economic detriment. Moreover, in the composition of reactive powder concrete, a fractional substitution of cement by silica fume (which is a waste by-product of silicon alloy) results in less cement consumption (and hence less greenhouse gas emission). In this experimental investigation study, the material performance of reactive powder concrete (RPC) with two different curing techniques, normal water curing of 25 °C and accelerated steam curing at 60 °C, and 60% relative humidity, was studied experimentally. This paper also remits the investigation, application, and different preparation of silica fume in reactive powder concrete at transformed temperature. The silica fume was substituted with the cement and the dosage which bequeathed the maximum strength was used for further addition of silica sand and steel fibers. Steam curing is effective process to achieve high early strength, and in this narrative, the compressive strength at the age of 3, 7, and 28 days was determined for further compared with normal cured water. Also find split tensile test to see the performance of concrete cylinder under tension. To reduce the water content, super plasticizer may be used as an admixture. Thus, results are interpreted for different combination to achieve maximum compressive stress as well as tensile stress.

Keywords Reactive powder concrete (RPC) · Silica fumes (SF) · Admixtures · Accelerated temperatures · Compressive strength

---

H. G. Sonkusare (✉) · P. Y. Pawade
Department of Civil Engineering, G. H. Raisoni College of Engineering, Nagpur, India
e-mail: hemant.sonkusare@gmail.com

P. Y. Pawade
e-mail: prashant.pawade@raisoni.net

L. M. Gupta et al. (eds.), *Advances in Civil Engineering and Infrastructural Development*, Lecture Notes in Civil Engineering 87,
https://doi.org/10.1007/978-981-15-6463-5_71

1 Introduction

Reactive powder concrete (RPC), a bond-based complex material, is surely understood for its ultra-high quality, strength, and low porosity expressed in [1]. The upgraded mechanical and physical properties of RPC are gotten by improving the pressing thickness of concrete blend with the exact degree of all blend particles, and by utilizing exceedingly refined silica smoke to improve the microstructure of hydrated bond glues through the pozzolanic response [2]. In a common RPC blend proportioning, so as to upgrade the homogeneity and improve the thickness of concrete granular blend, the coarse totals of customary concrete are completely wiped out to such an extent that the measures of fine total, Portland bond, and silica smoke should be significantly expanded. The silica fumes (SF) is found from ferrosilicon combinations, and it very well may be either utilized in slurry or mixed structure. In any structure, the calculable measure of solidarity can be acquired. The silica raged concrete is the one which fewer occasions demonstrate the isolation and the issues like dying. Silica smoke does not by and the large reason any decrease or increment in setting time. Concretes relieved with quickened temperature have been found to have a customary structure in contrast with ordinary restored concrete. The silica-seethed concrete likewise results in a rich blend of the total bond proportion [3].

2 Experimental Investigation

The test examination in this arrangement was ordered into two sections. The initial segment was the varieties of material properties of reactive powder concrete (RPC) from two diverse relieving conditions, typical water restoring of 25 °C and quickened steam restoring at 60 °C and 60% relative moistness, and the second part was the assessment of different outcomes [4, 5].

2.1 Specimen Preparation

The mix proportion of the control mix (CM) with water-to-bond proportion of 0.4 is given in Table 2. Ordinary Portland pozzolana cement type I complying with IS1489 (Part 1): 1991 (Reaffirmed 2005) was used [6, 7]. Natural quarry crushed gravel (maximum size of 20 mm) and natural river sand (Specific gravity 2.74, fineness modulus 2.6 zone II) were used as coarse and fine aggregate, respectively. The CM was used to cast in a cube as well as in beam having dimensions 150 × 150 × 150 mm and 150 × 150 × 700 mm, respectively, as shown in Fig. 1. After 24 h demolding, the specimens were cured for 3, 7, and 28 days. In order to reduce the heat of hydration in

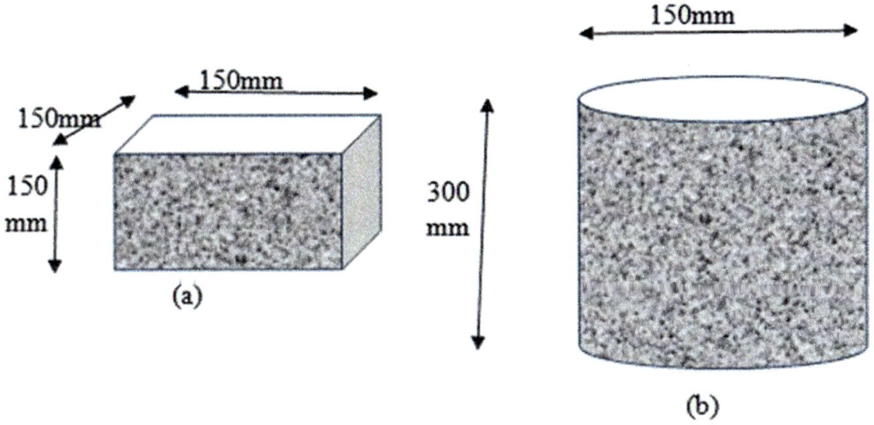

Fig. 1 Two types of mold cube and cylindrical shapes for control mix: **a** 150 × 150 mm **b** ø150 × 300 mm

cement and increase the capability of sulfate resistance, Portland pozzolana cement can be used [8]. Table 1 shows physical, chemical, and perfunctory properties of cement, silica fumes which can be used in an investigation [3, 4, 9–11].

To obtain good soundness of fresh RPC, proper order of mixing was done for all ingredients [14]. First, the fine granular components of RPC in a dry state [15], i.e., lightly grind cement and silica fumes to break up agglomerates [16, 17], add all dry powders and aggregate and start mixing for 2–5 min with low-speed gear to have a homogeneous mixture. The water and superplasticizers were mixed gradually to the dry granular mixture while it was still in a tray for another 3–5 min until the cement paste had a proper homogeneity. Finally, the steel fibers were spread and add (care should be taken that they should not mix with each other to form a complex structure) into the cement paste and continuously mixed for another 2–7 min with middle-speed gear to complete the mixing of fresh RPC [18, 19]. Stop mixing of all ingredients and prepare a test mold, the total elapsed time is nearly 25–30 min. Workability check is done in accordance with IS 1199:1959 (Reaffirmed 2004).

Square cube 150 × 150 mm form was utilized to cast the RPC examples for the trial of compressive quality. Cylindrical-shaped RPC example of 150 × 300 mm in a steel form was utilized to test the split rigidity. After these new RPC examples were relieved in a laboratory under encompassing temperature for 24 h, they were demolded. There were two sorts of restoring conditions utilized subsequently. The first was that the RPC examples were relieved in a quickened restoring tank at a temperature of 60 °C for 3, 7, and 28 days [20, 21]. The second RPC examples were ceaselessly relieved in typical water until the day of testing.

Table 1 Physical chemical and mechanical properties of cement, silica fumes

S. no.	Oxide composition	Chemical symbol	Cement (C)	Silica fume (SF)
Chemical composition (%)				
1	Calcium oxide	CaO	62.98	0.2
2	Silicon dioxide	SiO_2	20.24	93
3	Aluminum trioxide	Al_2O_3	4.61	2
4	Ferric oxide	Fe_2O_3	2.74	0.5
5	Potassium oxide	K_2O	0.27	1.0
6	Sodium oxide	Na_2O	0.13	0.4
7	Magnesium oxide	MgO	2.39	1.6
8	Sulfate	SO_3	2.24	0.5
9	Loss of ignition	L.O.I	3.29	2.6
10	Insoluble residue	Ins. Res.	0.86	–
11	Lime saturation factor	LSF	96.12	–
12	Free lime	FL	1.3	–
13	Silica ratio	SM	2.75	–
14	Chlorides	Cl	1.68	–
Physical composition [11]				
1	Fineness (by Residue on 45 μs)	F	396(m^2/kg)	7.5%
2	Specific gravity	Sp. Gr.	–	2.3
3	Specific surface	Sp. Sur	–	15 (m^2/gm)
4	Bulk density	γ	–	650 (kg/m^3)

Table 2 Composition of control mix or normal concrete (CM) [12, 13]

Mix	Cement (Kg/m^3)	% replacement (silica fume)	Silica fume (Kg/m^3)	Natural fine aggregates (Kg/m^3)	Coarse aggregates (Kg/m^3)	Super plasticizer (Kg/m^3)	Water (Kg/m^3)
CM	394	0	0	731	1137	1.57	158

2.2 *Investigational Variables for Performance Assessment*

One type of lists to examine the presentation of RPC at ages of 3, 7, and 28 days were directed: (1) Engineering properties including the compressive quality, part rigidity, and flexural quality. Strategies of IS 516:1959, IS1199:1959, and IS456:200 models were pursued to acquire the compressive qualities of RPC solid shape examples. The third point-loading technique was utilized to test the flexural quality with a progressive expanding pressure rate. The splitting tensile strength of RPC specimens was obtained according to IS 5816:1999 (reaffirmed 2004) standard.

3 Results and Discussion

3.1 Engineering Properties of RPC

The compressive strength of RPC has been carried out under two different conditions which are given in Tables 3 and 4. The results in Table 3 indicates the results of normal/conventional curing, whereas Table 4 indicates the steam curing results. It was discovered that RPC example relieved with the steam at three years old days compressive quality expanded by proportions 36–42%, 26–28% at seven years old days and 12–14% increments at 28 years old days. From previous investigation and

Table 3 Composition of reactive powder concrete (RPC)

Mix	Cement (Kg/m^3)	% replacement (SF)	Silica fume (Kg/m^3)	Natural fine aggregates (Kg/m^3)	Coarse aggregates (Kg/m^3)	Superplasticizer (Kg/m^3)	Water (Kg/m^3)
SF (10%)	353.6	10	39.4	731	1137	1.41	141.44
SF (12%)	346	12	47	731	1137	1.38	138.4
SF (14%)	338	14	55	731	1137	1.35	135.2
SF (16%)	330	16	63	731	1137	1.32	132
SF (18%)	323	18	70	731	1137	1.29	129.2
SF (20%)	315	20	78	731	1137	1.26	126

Table 4 Demonstration of % increase in compressive strength of steam curing with conventional curing

Mix	Increase in strength after 3 days (Mpa)	Increase in strength after 7 days (Mpa)	Increase in strength after 28 days (Mpa)
CM	41.86	32.09	9.95
SF(10%)	38.72	32.47	10.40
SF(12%)	40.60	30.29	10.72
SF(14%)	40.43	30.25	10.10
SF(16%)	42.61	29.85	8.62
SF(18%)	43.15	27.69	9.25
SF(20%)	45.09	28.31	8.61

other literature, it was found the percentage of silica fume. In this investigation, 2% silica fumes were added after 10% to achieve higher compressive strength. It is found that 14% silica fumes give an optimum compressive strength [22–24].

It was discovered that RPC example relieved with the steam at three years old days compressive quality expanded by proportions 36–42%, 26–28% at seven years old days, and 12–14% increments at 28 years old days

3.2 *Compressive Strength of Silica Concrete*

The compressive strength test was carried as per IS 516:1959 on a concrete cube of size 150 × 150 × 150 mm. For this purpose, compression testing machine was used of capacity 2000 KN. The steam curing was done at 60 °C. In Table 4, 14% silica fumes give optimum strength at 60 °C,and for each result, five molds have been prepared. The pozzolanic response coming about because of the elements of silica smolder, fly fiery debris, and impact heater slag in the RPC blend will be initiated enthusiastically by the high temperature and dampness of restoring steam. Such a pozzolanic response causes a denser microstructure of C-S-H bond hydrate and results in a quicker improvement of solidarity gain. The quality advancement for the RPC example relieved in the water under encompassing temperature is clearly much slower. It is noticed that, not at all like the typical weight concrete examples with the disappointment methods of either in squashed state or two isolated pieces, these fizzled RPC examples are still kept together by the steel strands.

From graphs, it tends to be seen that SF (14%) gives the maximum amount of strength in comparison with other mixes, so it is taken as an optimum dosage for further mixes. In further replacement, silica sand is added to 14% silica fume, and corresponding results are shown hereby (Figs. 2, 3, 4, 5, 6, and 7).

As the dosage of silica fume (14%) and silica sand (4%) gave the maximum strength, steel fibers are added in this mix in different proportion.

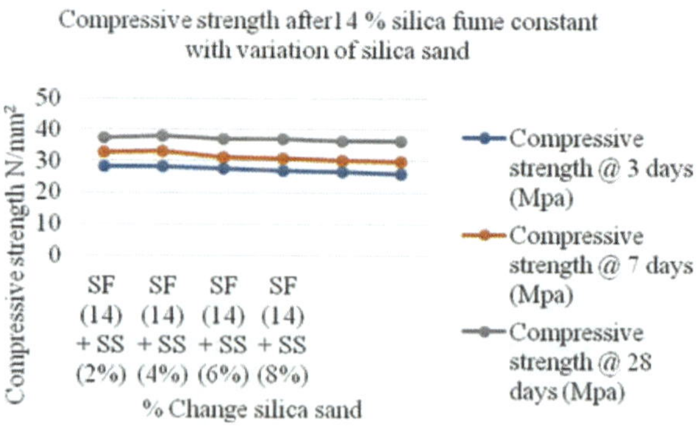

Fig. 2 Demonstrates the variety of compressive quality with addition of silica sand

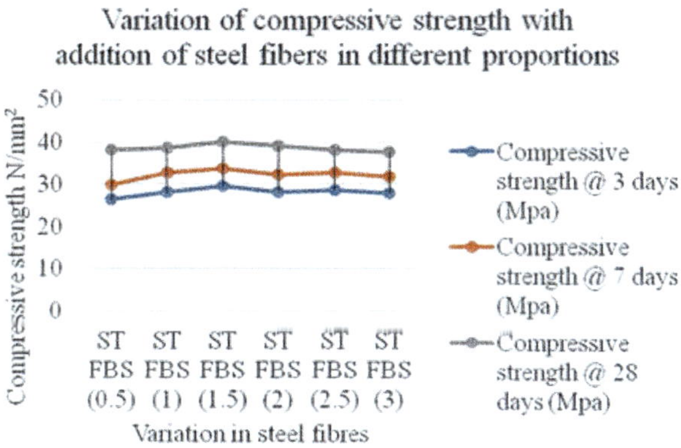

Fig. 3 Demonstrates the variety of compressive quality with expansion of fibers in various extents

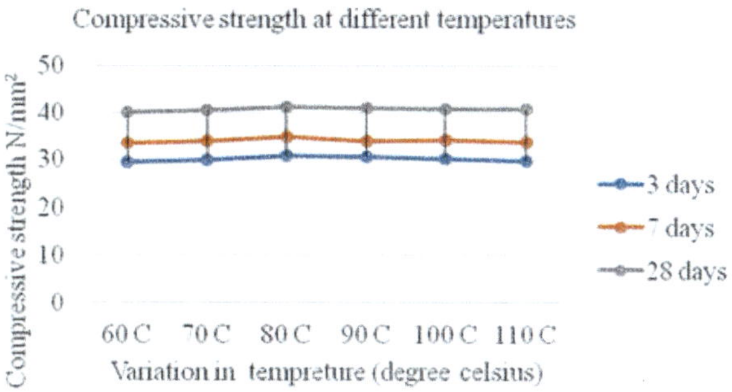

Fig. 4 Results of temperature curing at different temperatures

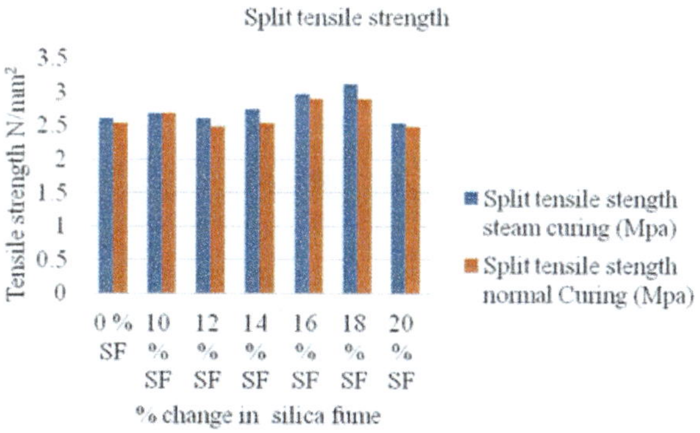

Fig. 5 Variation of split tensile strength with addition of silica fume

Reactive powder concrete gives optimum results containing silica fume (14%), silica sand (4%), and steel fibers (1.5%) which were further used, and the variation in temperature curing was done for different trial mixes.

In a variable curing, condition can be seen that up to certain temperature compressive strength increases (80 °C), there is a ceaseless increment in compressive quality, and further, there is ensuing lessening in compressive quality. It is further noticed that the top surface of the cube has convex shape due to silica combination not tested

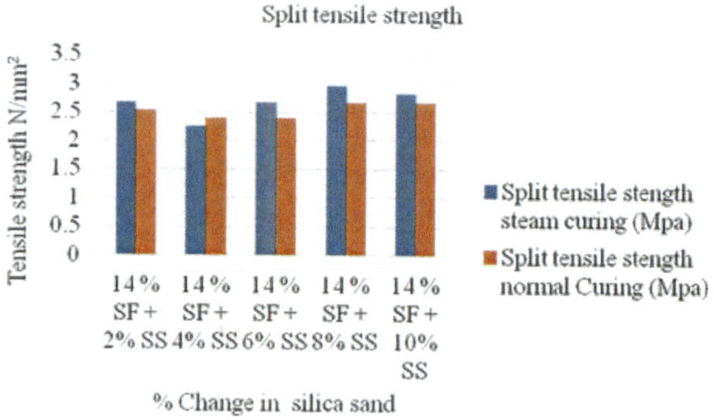

Fig. 6 Variation of split tensile strength with addition of silica fume and silica sand

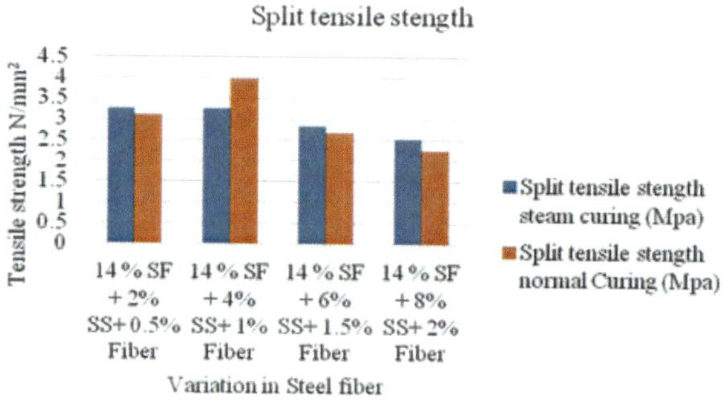

Fig. 7 Variation of split tensile strength with addition of steel fibers

chemically. Here, it is mentioned only for another significant research that what is the behavior of silica under different temperature condition chemically.

3.3 *Split Tensile Strength Results*

Split tensile strength was performed as per IS 5816:1999 (reaffirmed 2004). This test was done for all mixes those were done in compression test. The split tensile test was done at 28 days, and a comparison was made between the steam-cured results and conventional results.

Split tensile strength was executed according to IS 5816:1999 (reaffirmed 2004). This test was accomplished for all blends those were done in a pressure test. The split tensile test was done at 28 days, and correlation was made between the steam-cured outcomes and traditional outcomes.

4 Conclusion

The paper has investigated the optimal conditions for producing RPC using local materials by examining the effects of material composition, curing, and heating regimes.

A fuse of silica fume has expanded the water request radically or unfavorably influenced the consistency of the blend.

The blend structure of concrete appeared in written works was hard to accomplish because of the deviations in constituent materials, particularly from various geological status.

Major exploratory outcomes demonstrate that the compressive part tractable of RPC with steam is restoring increment generously.

A high-quality concrete achieving compressive quality 40 MPa at 28 days and high early quality advancement at the initial seven days of 29 MPa can be delivered utilizing the blend proportioning technique as depicted previously.

Influence of steel fibres content on RPC's mechanical properties increases drastically in tensile strength.

References

1. Roux N, Andrade C, Sanjuan MA (1996) Experimental study of durability of reactive powder concretes. J Mater Civil Eng (ASCE) 8:1–6
2. Chan Y-W, Chu S-H (2004) Effect of silica fume on steel fiber bond characteristics in reactive powder concrete. Cem Concr Res (Sci Direct Pergamon) 34:1167–1172
3. Chang TP, Chen BT, Wang JJ, Wu CS (2009) Performance of reactive powder concrete (RPC) with different curing conditions and its retrofitting effects on concrete member. In: Alexander et al (eds) Concrete repair, rehabilitation and retrofitting II. Taylor and Francis group, London. ISBN: 978-0-415-46850-3
4. Hoe KW, Ramli M (2010) Rational mix design approach for high strength concrete using sand with very high fineness modulus. Am J Appl Sci 7(12):1562–1568. ISSN: 1546-9239
5. Tai YS, Pan HH, Kung YN (2011) Mechanical properties of steel fiber reinforced reactive powder concrete following exposure to high temperature reaching 800 °C. Nucl Eng Des 241(7):2416–2424
6. IS: 2250 (1981) Code of practice for preparation and use of masonry mortars. Bureau of Indian Standards, New Delhi
7. IS: 8112 (1989) Ordinary Portland cement, 43 grade—specification. Bureau of Indian Standards, New Delhi
8. Khalil WI (2012) Some properties of modified reactive powder concrete. J. Eng. Dev 16(4):66–87. ISSN: 1813-7822
9. Yu R, Spiesz P, Brouwers HJH (2014) Mix design and properties assessment of ultra-high performance fibre reinforced concrete (UHPFRC)". Cem Concr Res 56:29–39
10. Tam CM, Tam VWY, Ng KM (2010) Optimal conditions for producing reactive powder concrete. Mag Concr Res 62:701–716
11. IS: 4031 (1988) (part 11) Methods of physical tests for hydraulic cement part ii determination of density. Bureau of Indian Standards, New Delhi
12. IS: 10262 (1999) Recommended guidelines for concrete mix design. Bureau of Indian Standards, New Delhi

13. IS: 383 (1970) Specification for coarse and fine aggregates from natural sources for concrete. Bureau of Indian Standards, New Delhi
14. Zdeb T (2013) Ultra-high performance concrete—properties and technology. Bull Pol Acad Sci Tech Sci 61(1):183–193
15. Kizilkanat AB, Oktay D, Kabay N, Tufekci MM (2016) Comparative experimental study of mortars incorporating pumice powder or fly ash. J Mater Civ Eng 28(2):1–7. (04015119)
16. IS: 3812 (1963) (part 1) Pulverized fuel ash specification part 1 for use as pozzolana in cement, cement mortar and concrete. Bureau of Indian Standards, New Delhi
17. IS: 3812 (2003) (part 2) Pulverized fuel ash—specification part 2 for use as admixture in cement mortar and concrete. Bureau of Indian Standards, New Delhi
18. Chung DDL (2005) Dispersion of short fibers in cement. J Mater Civ Eng 17(4):379–383
19. Carroll JC, Helminger N (2016) Fresh and hardened properties of fiber-reinforced rubber concrete. J Mater Civ Eng 28(7):1–9. (04016027)
20. Alkafaji MMK (2014) Performance of reactive powder concrete slabs with different curing conditions. J Eng Technol Res 6(6):81–93. ISSN: 2006-9790
21. Way RT, Wille K (2016) Effect of heat-induced chemical degradation on the residual mechanical properties of ultrahigh-performance fiber-reinforced concrete. J Mater Civ Eng 28(4):1–10. (04015164)
22. IS: 456 (2000) Plain and reinforced concrete-code of practice. Bureau of Indian Standards, New Delhi
23. IS: 516 (1959) Methods of tests for strength of concrete. Bureau of Indian Standards, New Delhi
24. IS: 2386 (1963) Indian standard code of practice for methods of test for aggregate for concrete. Indian Standard Institution, New Delhi

New Green Building Rating System for Residential Buildings in Tropical Countries

Taniya Thomas and Grace Mary Abraham

Abstract Green building is a sustainable system which promotes optimum use of natural resources and eco-friendly construction practices. Despite the benefits associated with green buildings, they are implemented only for 5% of the Indian construction industries. A number of researchers have investigated and found barriers which are hindering GBTs and practices in construction. Green building rating tools act as a yardstick to measure the greenness of buildings. Most of the GBRTs are country specific and energy specific. Even though many green building rating systems have been developed in past years by various countries, a system that is suitable for evaluating the residential buildings in the tropical countries has not been developed. So, it is essential to develop a rating system that is best for the context in the tropical countries. This paper reviews three prevailing GBRTSs across the world to study their differences and properties. The comparative study was conducted through case studies, and a new rating system was proposed.

Keywords Green building · Rating systems · Sustainability

1 Introduction

The population of the world is expected to increase from 7 to 9 billion by 2050. Undoubtedly, the growth in population will be associated with higher demand for water, energy, and other natural resources. The green building is a sustainable system that utilizes optimum use of the natural resources throughout its life cycle. A green building rating system is an evaluation tool that measures performance of a building based on environment-related aspects throughout its life cycle. During the past two

T. Thomas (✉) · G. M. Abraham
Amal Jyothi College of Engineering, Kanjirappally, Kerala, India
e-mail: taniyathomas@ce.ajce.in

G. M. Abraham
e-mail: gracemaryabraham@amaljyothi.ac.in

L. M. Gupta et al. (eds.), *Advances in Civil Engineering and Infrastructural Development*, Lecture Notes in Civil Engineering 87,
https://doi.org/10.1007/978-981-15-6463-5_72

decades, a large number of green building rating systems (GBRSs) have been developed by the countries around the world in order to rate and certify green buildings [1]. Some of them are designed for different types of projects and few for different phases of projects. They play a vital role in popularizing and implementing green building. Most of them are country specific or region specific and are designed to suit the climatic conditions and exposure conditions of a particular area. So, they are less effective when considered for other counties. The weather in India shows a dramatic variation. It ranges from tropical in south to temperate in north. The conditions in tropical areas are different, and a green building rating system to suit this region of the country has not been developed yet. The comparative analysis of three prevailing green building rating systems was conducted through case studies, and a new rating system to suit context in tropical region of the country was developed. It is designed in such a way that it suits other tropical countries too. Among the three rating systems considered for comparison, first was the oldest international rating system which is still accepted worldwide that is BREEAM, next is an energy-specific rating system LEED and then the national rating tool for India, GRIHA. The results from survey and experts opinion were combined in the process of green building rating system development.

1.1 Sustainable Development and Green Building

"Sustainability improves environmental, economic, and social conditions for present and future generations by enhancing quality of life and allows people to live in a healthy environment" (Ortiz et al. 2011). If a construction is environmentally, economically, and socially sustainable, only then it is considered as sustainable (Chethana et al 2016). Sustainable development in building sector can be achieved through adoption of green building. They help to increase building efficiency through maximum utilization and recycling of resources and thus reducing the negative impacts of construction industry on environment [2]. It is environmentally friendly, helps to eliminate and reduce the negative impacts from construction practices and material usage on environment (Peng et al. 2010).

2 Green Building Rating Tools

Green building rating tools are known as guidelines or standard that evaluates green buildings performance throughout its life cycle (Vivian et al. 2017). Parameters which are required to make a green building are included as main criteria of the rating tool (Vivian et al. 2017). There are many assessment tools and green building rating systems that are popular across the world (Gowri 2004). Buildings that are certified by GB rating systems are expected to consume lesser energy, utilize resources in the best manner, provide better living environment, and contribute to overall reputation of property (Dat et al. 2017). According to green building council (2015), there are 71 green building council members and 600 green building rating systems approximately (Vierra 2011). Among the existing GBRTs, Building Research Establishment's Environmental Assessment Method (BREEAM) was developed first, and it is the most widely used tool. Leadership in Energy and Environmental Design (LEED) was developed in 1998, and it is a widely accepted tool by more than 160 countries. GB rating tools are developed for different types of building such as residential buildings, educational buildings, office buildings, and industrial buildings. Some tools evaluate the building in design stage and some during the functional stage.

3 Comparison of GBRTs

The earliest green building rating system was launched in1990s and then in 2000s many more GBRTs were developed (Ming et al. 2018). Many versions of GBRTs were created for the assessment of sustainability of different types of buildings at different stages or phases of construction. GBRS are important in development of green buildings because they help building owners and engineers in following many aspects (Eisenstein et al. 2017). LEED, BREEAM, CASBEE are most frequently used and accepted rating systems (Ming et al. 2018). Sometimes GBRTs are formed for evaluating buildings in a particular location (country/region) only. If a rating tool is country specific, then their attributes and criteria will be best for that particular region and may not be the same for other regions. GRIHA is a national rating tool for India; it is a country-specific tool. And hence for the study BREEAM, LEED and GRIHA were chosen. Table 1 gives a comparison of the three rating tools based on their origin.

World's first sustainability rating system was BREEAM developed by UK in 1990. Then, in 1994, US Green Building Council released LEED [3]. LEED is the most widely used green building rating system in the world. It operates through US Green Building Council (USGBC) and takes triple bottom line approach. It not only considers energy but also people, planet, and profit and has seven major categories. It classifies buildings into certified, silver, gold, and platinum [4]. GRIHA rates building based on degree of its greenness. It is based on accepted energy and environmental principles and is revised every three years. It includes 31 criteria categorized into nine sections. It is a country-specific rating system. For the development of the green building rating system specifically for tropical countries, a worldwide accepted international tool, a national tool, and an energy-specific tool are compared. The comparative study was carried out through case studies. Three rating tools are assessing the rate of sustainability of a building by considering various aspects. The weightage provided for various criteria is different, and the categories are also different. The weightage varies depending on the regional priority for country-specific systems and building purpose for other systems. Country-specific tools mainly consider the

Table 1 Comparison of GBRTs

	Green building rating tools		
	BREEAM	LEED	GRIHA
Country	UK	USA	India
Organization	BRE	USGBC	Teri
Flexibility	77 Countries	160 Countries	1 Country
First version	1990	1998	2010
Latest version	2018	2018	2016
Rating approach	Pre-weighted categories	Additive credits	Additive credits
Rating level	5 levels	4 levels	5 levels
Criteria	Management Health and well-being Energy Transport Water Materials Waste Land use and ecology Pollution Innovation	Sustainable sites Water efficiency Energy and atmosphere Indoor environmental quality Innovation and design process Regional priority	Site planning Construction management Energy Occupant comfort and well-being Water Sustainable building materials Solid waste management Socioeconomic strategies Performance monitoring

climatic conditions and other features of the area, for example, green star for buildings in Australia, green mark for Singapore, etc. Criteria in each tool were grouped under same classification, and weightage allocated in each rating tool in percentage was calculated separately. Figure 1 gives a bar chart with criteria weightages of the three rating tools. Criteria weights were allocated by respondents based on their experience. The average of the weightage for each criterion is considered, and a new rating tool is developed. The main concern was to develop a tool that will suit the conditions of tropical countries and their construction practices. Sustainable architecture in tropical climate is still an unexplored field with many challenges.

In all the three systems, energy criteria score the highest followed by water in GRIHA, health and well-being, and site selection in LEED, and it is also health and well-being in BREEAM. So, the weightages for different categories vary for all rating tools. To study these differences and properties of rating tools, case studies on same buildings are conducted. Buildings were evaluated using all three tools and scores compared. Results confirmed that the differences in weightages contribute to changes in the rating level. Considering this as an issue, a new rating tool was developed after collecting suggestions and opinions from experts and experienced professionals in sustainable construction.

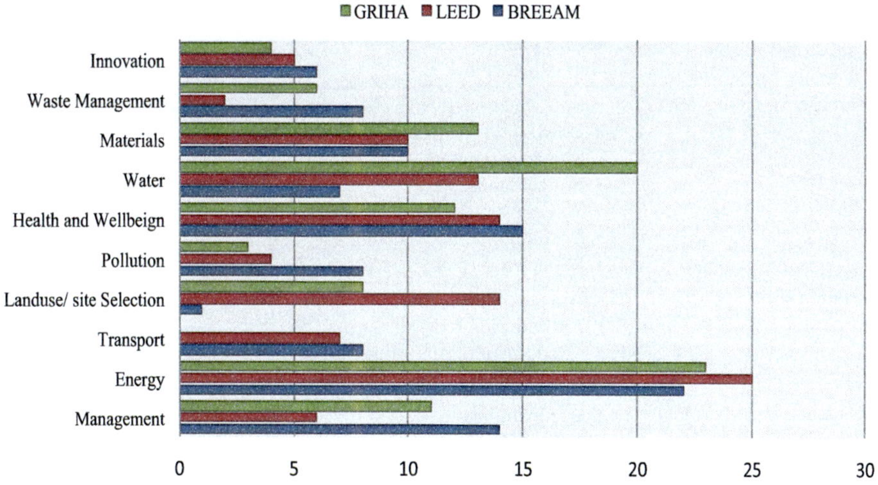

Fig. 1 GBRT weightage comparison

4 Proposed Green Building Rating System

Category	Sub-category	Credits
Management (6)	• Maintenance of facility • Life cycle cost and service planning • Green education for occupants	2 2 2
Energy (20)	• Motion sensor/occupancy sensor in common area • Energy-efficient fixtures • CFC free appliances • Utilization of renewable energy • Energy metering/monitoring • Solar water heating • Use low energy material at interiors • Energy-efficient cold storage • Energy-efficient transport • Measurement and verification	2 2 2 2 2 2 2 2 2 2
Transport (8)	• Public transport accessibility • Maximum car parking • Proximity to amenities • Design for differently abled • Alternative transport	2 1 2 1 2

(continued)

(continued)

Category	Sub-category	Credits
Land use/Site Selection (8)	• Site selection • Brownfield redevelopment • Site development • Protect landscape • Soil conservation • Design to include existing site features • Reduce hard paving	1 1 1 1 1 2 1
Pollution (8)	• Water pollution control • Air pollution reduction • • Reduction of night time pollution • Reduction of noise pollution	2 2 2 2
Health and Well-being (12)	• Occupant health • No smoking in common area • Minimize heat exposure • Indoor comfort requirement • Provision of low VOC paints/adhesives/sealants • Maintain good IAQ	2 2 2 3 2 1
Water (12)	• Rainwater harvesting • Water sub metering • Water-efficient fixtures • Automatic water level control • Wastewater treatment • Reduced landscape water demand • Water leak deflection	2 2 2 1 2 2 1
Waste Management (10)	• E-waste management • Waste segregation • Organic waste management • Storage and disposal of waste • Construction waste management	2 2 2 2 2
Materials (10)	• Utilization of waste material in building structure • Use of low environment impact material • Regionally available material • Material efficiency • Design for durability • Low emitting material	2 1 2 2 2 1
Innovation (6)	• Special credits for exceptional performance	6

5 Conclusion

Green building technologies (GBTs) are gaining an increasing interest in the construction industry globally because it is the way of enhancing the sustainability performance of buildings. Existing green building rating systems provide guidelines that promote sustainability throughout the life cycle of a construction project. They clearly define the goals necessary to minimize the environmental impact of a building. There are many GBRTs across the world, some of them are country-specific and few are building type specific. For example, BREEAM gives the highest weighting to the energy category, while LEED prioritizes the material and site. Reuse of the existing building is a strategy to minimize adverse impacts. In GRIHA, highest score is for energy-related aspects. And hence, we can conclude that different GBRTs focus on different aspects even though they serve the same purpose. And their weightage for same criteria also varies. So far, no rating systems have been developed to suit context in tropical regions. Since the climatic conditions and exposures are different, it is necessary to develop a separate system that specifically evaluates green buildings in the tropical region. For the development of a new a system, three of the prevailing green building rating systems were studied and compared. The comparative study was carried out through case studies, and the obtained results along with expert's opinion are combined to develop the new system. Criteria were chosen from the oldest, and most widely accepted tool and average weightages are calculated. By using this weightage, a new system with ten criteria is developed, and their subsection weightages are distributed with reference to the literature study and survey results.

All procedures performed in studies involving human participants were in accordance with the ethical standards of the institutional and/or national research committee and with the 1964 Helsinki declaration and its later amendments or comparable ethical standards. For this type of study, formal consent is not required.

References

1. Awadh O (2017) Sustainability and green building rating systems: LEED, BREEAM, GSAS and Estidama critical analysis. J Build Eng 11:25–29
2. Nguyen TH, Toroghi SH, Jacobs F (2016) Automated green building rating system for building designs. J Architect Eng 22(4):1–10
3. Zhang Y, Wang Y (2017) Comparison of evaluation standards for green building in China, Britain, United States. Renew Sustain Energy Rev 68:262–271
4. Ding Z, Fan Z, Tam VWY, Bian Y, Li S, Illankoon IMC, Moon S (2018) Green building evaluation system implementation. Build Environ 133:32–40

Utilization of Industrial By-Products in Geopolymer Prefabricated Blocks

Pallavi Sirohi, Abhishek Singh, and Chirag Varshney

Abstract This paper presents an experimental study on laboratory evaluation of industrial by-products that is copper slag and fly ash as a construction material for the production of prefabricated block used in construction. In order to determine the feasibility and optimum quantities of these industrial by-products for the formation of prefabricated blocks, concrete sample with different percentage of industrial by-products has been prepared. Different concentrations of NaOH and KOH also have been used in order to obtain the strength by inducing chemical reaction between the ingredients. All the specimens were cured for seven days at elevated temperature of 60 °C, and their unconfined compressive strength was determined. Test results indicated that mix with 30% fly ash and 70% copper slag with 8 M solution of alkali material is best suited for its commercial utilization and formation of prefabricated blocks. The utilization of copper slag and fly ash mix in construction will not only save the scarcity of conventional materials but also solve the problem related to its disposal.

Keywords Geopolymer · Copper slag · Fly ash · Prefabricated blocks

1 Introduction

1.1 General

Today in the world, the task of protecting the environment is of utmost importance. The exponential increase in the population and unsustainable utilization of natural resources has led to the tremendous degradation of the environment. Urbanization has further increased the demand from the construction industry, which results into

P. Sirohi · C. Varshney
Sardar Vallabhbhai National Institute of Technology, Surat, Gujarat, India

A. Singh (✉)
Inderprastha Engineering College, Ghaziabad, India
e-mail: abhi23193@gmail.com

L. M. Gupta et al. (eds.), *Advances in Civil Engineering and Infrastructural Development*, Lecture Notes in Civil Engineering 87,
https://doi.org/10.1007/978-981-15-6463-5_73

use of construction materials. Production of Portland cement leads to huge emissions of harmful gases such as CO_2 which is a major source of greenhouse effect.

Research studies [3] conducted in the past have led to the emergence of fly ash-based geopolymer as a promising new cement alternative as a construction material. Geopolymer can be produced with the basic raw materials containing silica and alumina-rich mineral composition. Several studies [5] have investigated the use of alkali activators, containing sodium hydroxide and sodium silicate or a potassium hydroxide and potassium silicate, in the geopolymer concrete. The results exhibit the formation of an excellent geopolymer with rapid setting properties. It should also be noted that the presence of calcium content in fly ash plays a significant by providing a faster reactivity and thus yielding early hardening and higher strength of geopolymer in shorter curing time.

In India, there is a great demand of infrastructures which can be erected within a small time period, thus leading to great opportunities for prefabricated structural elements. Also the growth in the industrial activities continue to produce a huge quantities of wastes and by-products such as fly ash, various slags, waste plastic, scrap tires, slate, and marble waste.

Using prefabricated blocks made from these industrial waste and by-products instead of the conventional burned clay bricks will not only prevent the emission of the harmful gasses but also gives sustainable solutions for the emerging demands. As per the classification, the compressive strength of first class brick should be 10.5 Mpa or more.

1.2 Literature Review

Davidovits et al. [1] explained that how an alkaline liquid reacts with the silicon (Si) and the aluminum (Al) in by-product materials like fly ash and rice husk ash to produce binding material as it leads to the process of polymerization.

Raijiwala et al. [4] investigated different mix proportions of fly ash and alkali activators, namely KOH and NaOH, and the samples were cured at elevated temperature of 60 °C. The study concluded that the compressive strength of geopolymer concrete (GPC) increases by 1.5 times as compared to conventional concrete.

Sanni et al. [3] examined the mechanical properties of the GPC with variation in the alkaline solution. The alkaline solution used for their study was the combination of sodium silicate and sodium hydroxide solution. For any grade of GPC, it was seen that as the ratio/concentration of alkaline solution increases, the compressive strength was also increased.

Deb et al. [2] investigated the effect of GGBFS with class F fly ash in presence of varying ratio of sodium silicate to sodium hydroxide on the mechanical properties of fresh and hardened state of geopolymer concrete. The study showed that workability of GPC decreased with the increase of GGBFS content together with fly ash in the binder when the other mixture variables remained the same.

Table 1 Physical and chemical composition of copper slag

Parameters	Test value
Physical properties	
Grain size analysis	Gravel—0%, sand—100%, silt + clay—0%
Granular shape	Angular and sharp
Specific gravity	3.61
Chemical properties	
FeO	42–48
SiO_2	26–30
Al_2O_3	1.0–3.0
CaO	1.0–2.0
MgO	0.8–1.5

Source Hindalco Industries Limited, Bharuch, Gujarat

2 Experimental Details

2.1 Materials

Copper slag. It is a by-product from copper extraction industries and is produced during matte smelting and converting steps of pyro-metallurgical production of copper. The chemical composition of the slag used in this experiment study was taken from Hindalco Industries Limited, Bharuch, Gujarat, as given in Table 1.

Fly ash. Fly ash is most commonly used by-product in which high amount of Al and Si compounds are present. For the experimental purpose, class *F* fly Ash was procured from Ashtech Industries Pvt Ltd., Ghaziabad. Its physical and chemical composition is mentioned in Table 2.

NaOH and KOH solutions. Alkali solution reacts with Si and Al compounds present in copper slag and fly ash, and polymerization takes place. Polymerization reaction is a substantially fast chemical reaction in the presence of alkaline condition on Si–Al minerals, which further results in a three-dimensional polymeric chain and ring-type structure which consists of Si–O–Al–O bond (hardening stage).

2.2 Experimental Setup

Step 1: Dry trial mixes were prepared of copper slag and fly ash at different proportioning as given in Table 3, and then, alkaline solution (NaOH or KOH) was added.

Table 2 Physical and chemical composition of fly ash

Parameters	Test value	Permissible limit (as per IS 1727:1967)
Physical properties		
Fineness	364	320 Min
Particle retained on 45 μ	23.5	34.0 Max
Chemical properties		
$SiO_2 + Al_2O_3 + Fe_2O_{3\%}$ by mass	90.3	70.0 Min
Silica %by mass	42.1	35.0 Min
MgO %by mass	2.8	5.0 Max
Total chlorides %by mass	0.017	0.05 Max

Source MCB Testing Labs, Ashtech Pvt Ltd

Table 3 Trial mixes along with its composition

Specimen	Copper slag (in %)	Fly ash (in %)
C80F20[a]	80	20
C70F30	70	30
C60F40	60	40
C50F50	50	50

[a]*where* C80F20 defines copper slag 80%, fly ash 20%

Step 2: Sixteen trial mixes were prepared for each alkaline solution, i.e., NaOH and KOH at different molarity of 4, 6, 8, and 10 M. This quantity will depend on OMC of the dry mix prepared.

Step 3: Six specimens were casted for each trial mixes having size 50 mm diameter and 100 mm height and cured for seven days at elevated temperature of 60 °C.

Step 4: These specimens were tested at the age of one and seven days for determining the unconfined compressive strength (UCS) at the point of failure.

3 Result and Discussion

3.1 Unconfined Compressive Strength

The average failure stress values for NaOH and KOH activated copper slag and fly ash mixes for one and seven days curing are presented in Tables 4 and 5.

Table 4 Average UCS failure stresses in MPa for NaOH activated copper slag and fly ash mixes for one and seven day curing

Specimen	One day curing (in MPa)				Seven day curing (in MPa)			
	4 M	6 M	8 M	10 M	4 M	6 M	8 M	10 M
C80F20	2.32	3.18	4.36	5.82	4.62	5.48	8.11	9.10
C70F30	**3.18**	**4.43**	**5.74**	**7.99**	**5.34**	**6.34**	**10.68**	**12.54**
C60F40	3.40	4.93	6.5	8.68	5.61	6.98	12.04	13.80
C50F50	3.48	5.21	6.8	8.81	5.80	7.24	12.22	14.08

Table 5 Average UCS failure stresses in MPa for KOH activated copper slag and fly ash mixes for one and seven day curing

Specimen	One day curing (in MPa)				Seven day curing (in MPa)			
	4 M	6 M	8 M	10 M	4 M	6 M	8 M	10 M
C80F20	3.8	4.2	5.32	6.37	5.44	7.15	8.99	9.49
C70F30	**4.46**	**5.42**	**6.71**	**8.56**	**6.11**	**8.43**	**11.45**	**12.66**
C60F40	4.77	6.10	7.16	9.19	6.79	9.22	13.11	14.09
C50F50	4.88	6.28	7.40	9.51	7.02	9.7	13.76	14.55

The tests were conducted at the age of one and seven day, in order to determine its early strength, which is a prime aspect of prefabricated construction. Mix C50F50 shows maximum gain in strength, whereas C80F20 shows the least

3.2 Discussion

Following are the graphs comparing the failure in UCS test values of NaOH and KOH activated copper slag and fly ash mixes for varied concentration of alkali solution. It can be seen from Tables 4 and 5 that on increasing the molarity of alkaline solution, the compressive strength for the same proportioning also increases.

Figure 1 shows that the compressive strength keeps on increasing as the fly ash percentage increases, but the rate of increase for one day strength becomes insignificant after 40% fly ash proportioning.

Figure 2 shows that compressive strength for seven day curing is also insignificant after 40% fly ash proportioning. The compressive strength increases with fly ash content because more Al and Si compounds will be available for reaction and more will be the formation of anhydrous alumino-silicate compounds that contributes in strengthening action.

It is observed from Figs. 3 and 4 that samples with KOH as activator show same trend, i.e., with NaOH as activator. Here also, the gain in strength after 40% fly ash proportion is insignificant. However, samples treated with KOH solution show better result for both one day and seven day compressive strength.

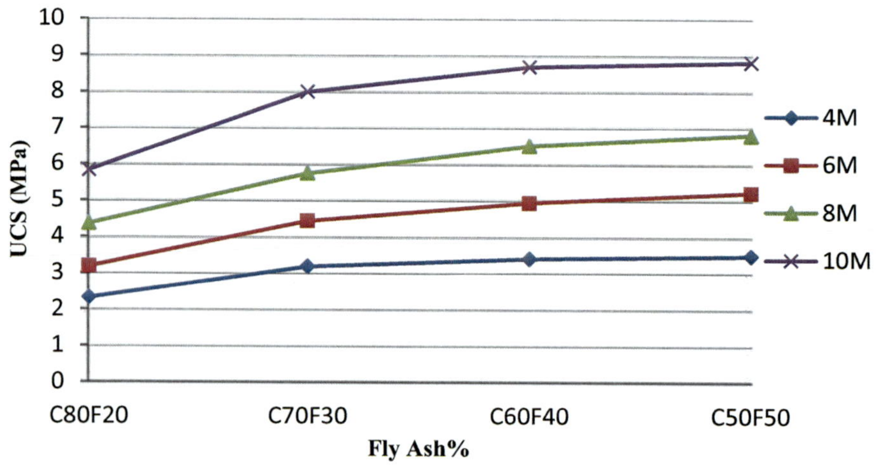

Fig. 1 Graphical representation b/w UCS (MPa) and fly ash content at different molarity of NaOH for one day curing

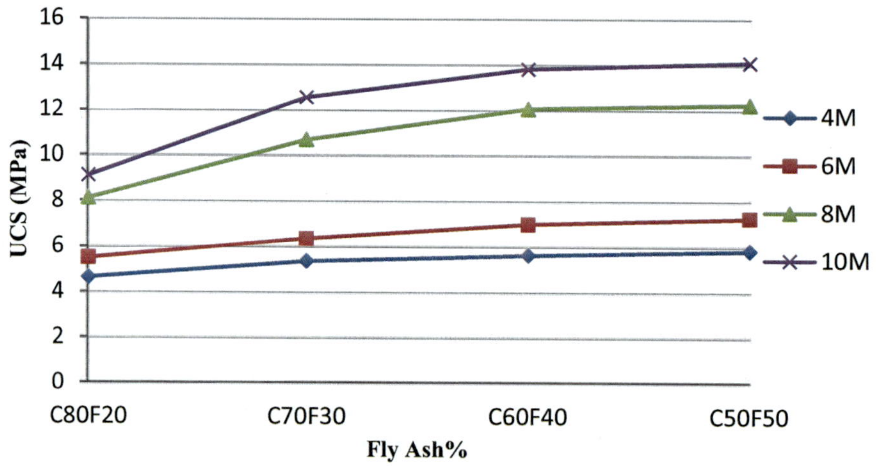

Fig. 2 Graphical representation b/w UCS (MPa) and fly ash content at different molarity of NaOH for seven day curing

Figure 5 shows that as molarity increase, the UCS value also increases. This implies that at high alkali medium, polymerization reaction will be more effective, and formation of amorphous alumina-silicate structure will be faster.

It also depicts that as molarity of alkaline solution increases, difference in strength gained by KOH and NaOH reduces.

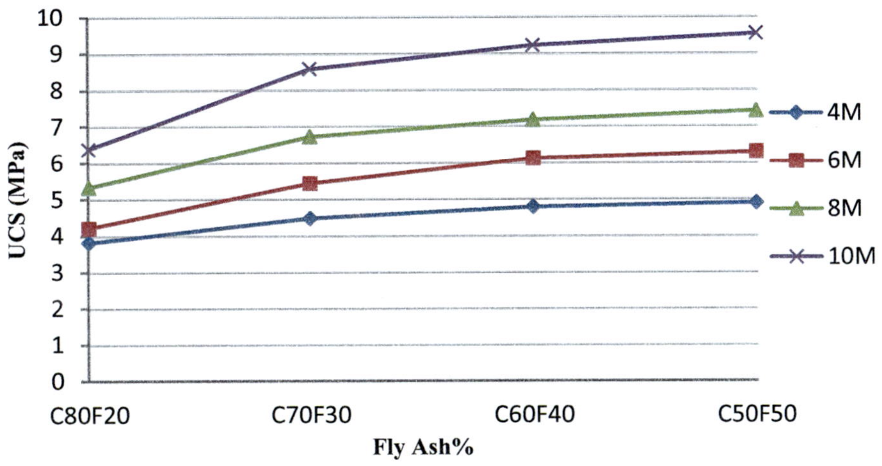

Fig. 3 Graphical representation b/w UCS (MPa) and fly ash content at different molarity of KOH for one day curing

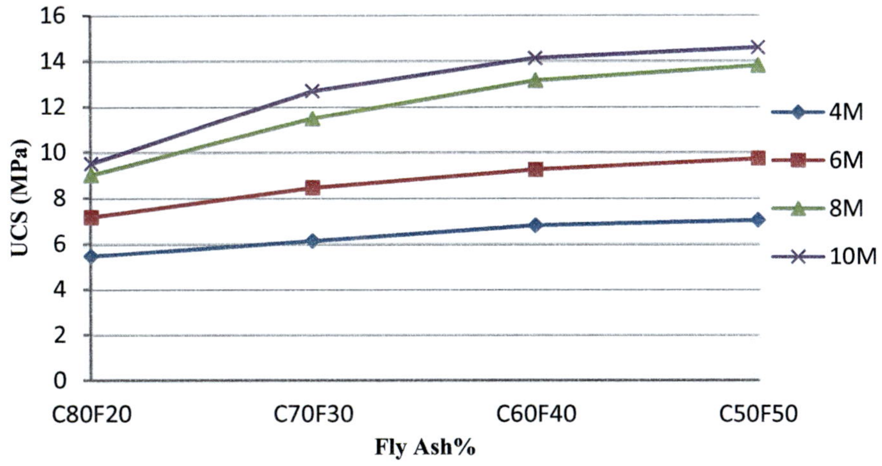

Fig. 4 Graphical representation b/w UCS (MPa) and fly ash content at different molarity of KOH for seven day curing

4 Conclusion

Experimental work has been performed to check the technical feasibility, practicability for use of prefabricated blocks for construction. To attain these results, various tests have been performed on industrial by-products:

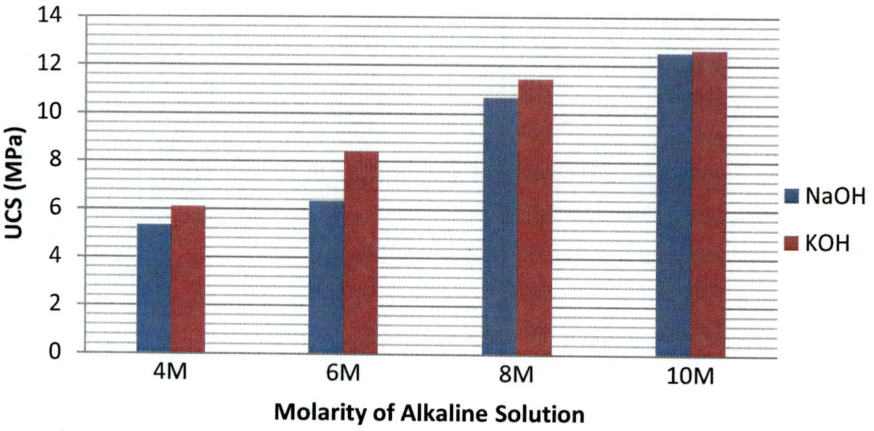

Fig. 5 Graphical representation b/w UCS and molarity of alkaline solution for C70F30 mix for seven day curing period

1. The maximum value of UCS for NaOH activated fly ash–copper slag mix is 8.81 MPa and 14.08 MPa, and that for KOH is 9.51 MPa and 14.55 MPa for one day and seven days curing period, respectively.
2. For better optimization of industrial by-products in the formation of precast bricks, mix of 30% fly ash and 70% copper slag with 8 M alkaline solution shows better results, that is, 12.54 Mpa at the age of seven day.
3. It is also suitable to use 8 M alkaline solution for conducting polymerization and poly-condensation reaction to achieve hardening state. Although, it has been observed that higher alkalinity will give better compressive strength.
4. The rate of increase of strength is insignificant after 40% fly ash proportion which may be due to the reason that fly ash content more than 40% is not properly utilized by the polymerization process.
5. This study compares the strength parameters of precast brick, made up of industrial waste, with conventional brick. Moreover, the procedure involved in the formation of precast brick is both economical and environment friendly.

References

1. Davidovits J, Davidovics M (1988) Geopolymer: room temperature ceramic matrix for composites. Ceram Eng Sci Proc 9:835–842
2. Sarthi DP, Pradip N, Kumar SP (2014) The effects of ground granulated blast-furnace slag blending with fly ash and activator content on the workability and strength properties of geopolymer concrete cured at ambient temperature. Mater Des 62:32–39
3. Mahendran K, Arunachelam N (2015) Study on utilization of copper slag as fine aggregate in geopolymer concrete. Int J Appl Eng Res 10(53):336–340

4. Raijiwala DB, Patil HS (2011) Geopolymer concrete: a concrete of next decade. J Eng Res Stud 2(1):19–25
5. Sanni SH, Khadiranaikar RB (2013) Performance of alkaline solutions on grades of geopolymer concrete. Int J Res Eng Technol 2(11):366–371